Permeability
And Other Film Properties
Of Plastics And Elastomers

Plastics Design Library

Plastics Design Library, 13 Eaton Avenue, NY 13815 Tel: 607/337-5080 Fax: 607/337-5090

Table of Contents

Introduction i

    Some Notes About The Information In This Book...ii

    How To Use This Book...ii

Barrier Materials - An Overview iii

Thermoplastics

Acetal Resin

Polyoxymethylene (Acetal) - Chapter 1

    Textual Information 1

    Tabular Information 1

Acrylonitrile

Polyacrylonitrile (PAN) - Chapter 2

    Tabular Information 5

Acrylic Resin

Acrylonitrile-Methyl Acrylate Copolymer (AMA) - Chapter 3

    Textual Information 7

    Tabular Information 8

    Graphical Information 13

Cellulosic Plastic

Cellophane Film - Chapter 4

    Tabular Information 15

Fluoroplastic

Fluoroplastic - Chapter 5

    Textual Information 17

Ethylene-Chlorotrifluoroethylene Copolymer (ECTFE) - Chapter 6

    Textual Information 19

    Tabular Information 19

    Graphical Information 21

Ethylene-Tetrafluoroethylene Copolymer (ETFE) - Chapter 7

    Tabular Information 25

Fluorinated Ethylene-Propylene Copolymer (FEP) - Chapter 8

    Tabular Information 27

    Graphical Information 29

Fluorinated Polyethylene (FPE) - Chapter 9

    Textual Information 33

    Tabular Information 33

Perfluoroalkoxy Resin (PFA) - Chapter 10
Textual Information 35
Tabular Information 35

Polychlorotrifluoroethylene (CTFE) - Chapter 11
Textual Information 37
Tabular Information 38
Graphical Information 44

Polytetrafluoroethylene (TFE) - Chapter 12
Tabular Information 45
Graphical Information 50

Polyvinyl Fluoride (PVF) - Chapter 13
Tabular Information 51

Polyvinylidene Fluoride (PVDF) - Chapter 14
Textual Information 53
Tabular Information 54
Graphical Information 59

Ionomer

Ionomer (EMA) - Chapter 15
Textual Information 63
Tabular Information 64

Parylene

Polyparaxylylene - Chapter 16
Textual Information 69
Tabular Information 70

Polyamide

Nylon - Chapter 17
Textual Information 73
Tabular Information 75
Graphical Information 78

Amorphous Nylon - Chapter 18
Textual Information 79
Tabular Information 81
Graphical Information 83

Nylon 6 - Chapter 19
Textual Information 85
Tabular Information 86

Nylon 66 - Chapter 20
Textual Information 95
Tabular Information 96

Nylon 6/66 - Chapter 21
Tabular Information 105

Nylon 66/610 - Chapter 22
Textual Information 109
Tabular Information 110

Nylon MXD6 - Chapter 23
Tabular Information 111
Graphical Information 112

Polycarbonate

Polycarbonate (PC) - Chapter 24
Textual Information 113
Tabular Information 113

Polyester

Polybutylene Terephthalate (PBT) - Chapter 25
Textual Information 117
Tabular Information 118

Polyethylene Naphthalate (PEN) - Chapter 26
Textual Information 121
Tabular Information 122

Polyethylene Terephthalate (PET) - Chapter 27
Textual Information 125
Tabular Information 127

Glycol Modified Polycyclohexylenedimethylene Terephthalate (PCTG) - Chapter 28
Tabular Information 137

Polycyclohexylenedimethylene Ethylene Terephthalate (PETG) - Chapter 29
Tabular Information 139

Liquid Crystal Polymer (LCP) - Chapter 30
Textual Information 143
Tabular Information 143

Polyimide

Polyimide (PI) - Chapter 31
Textual Information 145
Tabular Information 146

Polyolefin

Polyethylene (PE) - Chapter 32
Textual Information 149
Tabular Information 151
Graphical Information 152

Low Density Polyethylene (LDPE) - Chapter 33
Textual Information 157
Tabular Information 158
Graphical Information 174

Linear Low Density Polyethylene (LLDPE) - Chapter 34
Textual Information 175
Tabular Information 176

Ultra Low Density Ethylene-Octene Copolymer (ULDPE) - Chapter 35
Textual Information 187
Tabular Information 188

Medium Density Polyethylene (MDPE) - Chapter 36
Tabular Information 191

High Density Polyethylene (HDPE) - Chapter 37
Textual Information 195
Tabular Information 195
Graphical Information 209

Ethylene-Alpha Olefin Copolymer (POP) - Chapter 38
Textual Information 211
Tabular Information 213

Ethylene-Vinyl Acetate Copolymer (EVA) - Chapter 39
Textual Information 215
Tabular Information 216
Graphical Information 224

Ethylene-Vinyl Alcohol Copolymer (EVOH) - Chapter 40
Textual Information 225
Tabular Information 229
Graphical Information 252

Polyethylene-Acrylic Acid Copolymer (EAA) - Chapter 41
Textual Information 259
Tabular Information 260

Polyethylene-Ionomer Copolymer (PE-Ionomer) - Chapter 42
Textual Information 263
Tabular Information 263

Polypropylene (PP) - Chapter 43
Tabular Information 265
Graphical Information 271

Polypropylene Copolymer (PP Copolymer) - Chapter 44
Textual Information ....
Tabular Information ....
Graphical Information ....

Polybutylene - Chapter 45
Textual Information 273

Polymethylpentene (PMP) - Chapter 46
Tabular Information 277

Polyphenylene Sulfide

Polyphenylene Sulfide (PPS) - Chapter 47
Tabular Information 281

Polysulfone

Polysulfone (PSO) - Chapter 48
Tabular Information 283

Polyvinyl Alcohol

Polyvinyl Alcohol (PVA) - Chapter 49
Textual Information 287
Tabular Information 288

Contents

Styrenic Resin

Acrylonitrile-Butadiene-Styrene Copolymer (ABS) - Chapter 50
Textual Information 291
Tabular Information 291

Acrylonitrile-Styrene-Acrylate Copolymer (ASA) - Chapter 51
Textual Information 295
Tabular Information 295

Polystyrene (PS) - Chapter 52
Textual Information 299
Tabular Information 299
Graphical Information 302

General Purpose Polystyrene (GPPS) - Chapter 53
Tabular Information 303

Impact Resistant Polystyrene (IPS) - Chapter 54
Tabular Information 305

Styrene-Acrylonitrile Copolymer (SAN) - Chapter 55
Textual Information 307
Tabular Information 307
Graphical Information 309

Styrene-Butadiene Block Copolymer - Chapter 56
Textual Information 311
Tabular Information 311

Vinyl Resin

Polyvinyl Chloride (PVC) - Chapter 57
Tabular Information 313
Graphical Information 316

Polyvinyl Chloride-Polyvinylidene Chloride Copolymer (PVC-PVDC) - Chapter 58
Tabular Information 319

Polyvinylidene Chloride (PVDC) - Chapter 59
Textual Information 321
Tabular Information 322
Graphical Information 336

Thermoplastic Alloys

Plastic Alloy

Polyethylene/Polystyrene Alloy (PE/PS) - Chapter 60
Textual Information 339
Tabular Information 339

Multilayer Structures

Co-Continuous Lamellae Multilayer Structure

Co-Continuous Lamellar Injection Molded (LIM) Multilayer Structure - Chapter 61
Textual Information 341
Tabular Information 341

Laminar Multilayer Structure

Laminar Multilayer Structure - Chapter 62
Textual Information 343
Tabular Information 345
Graphical Information 351

Multilayer Films

Multilayer Films with Ethylene-Vinyl Alcohol Copolymer (EVOH) Barrier - Chapter 63
Tabular Information 353
Graphical Information 364
Multilayer Films with Polyvinylidene Chloride (PVDC) Barrier - Chapter 64
Tabular Information 367
Graphical Information 379
Multilayer Films - General - Chapter 65
Tabular Information 381

Thermosets

Epoxy Resin

Epoxy Resin (EP) - Chapter 66
Tabular Information 387

Polypyrrole

Polypyrrole - Chapter 67
Textual Information 389
Tabular Information 389

Thermoplastic Elastomers

Olefinic Thermoplastic Elastomer

Olefinic Thermoplastic Elastomer (TPO) - Chapter 68
Textual Information 391
Tabular Information 391

Polyamide Thermoplastic Elastomer

Polyamide Thermoplastic Elastomer (Polyamide TPE) - Chapter 69
Textual Information 395
Tabular Information 395

Polybutadiene Thermoplastic Elastomer

Syndiotactic 1,2-Polybutadiene Thermoplastic Elastomer (Polybutadiene TPE) - Chapter 70
Textual Information 401
Tabular Information 402
Graphical Information 405

Polyester Thermoplastic Elastomer

Polyester Thermoplastic Elastomer (Polyester TPE) - Chapter 71
Textual Information 407
Tabular Information 407

Polyurethane Thermoplastic Elastomer (TPUR)

Thermoplastic Polyester-Polyurethane Elastomer (TPAU) - Chapter 72
Tabular Information 411

Thermoplastic Polyether-Polyurethane Elastomer (TPEU) - Chapter 73
Tabular Information 415
Graphical Information 417

Styrenic Thermoplastic Elastomer

Styrenic Thermoplastic Elastomer (Styrenic TPE) - Chapter 74
Textual Information 419
Tabular Information 420

Vinyl Thermoplastic Elastomer

Polyvinyl Chloride Polyol (pPVC) - Chapter 75
Tabular Information 427

Rubbers

Acrylic Rubber

Ethylene-Acrylate Copolymer (EACM) - Chapter 76
Tabular Information 431

Butadiene Rubber

Polybutadiene - Chapter 77
Tabular Information 433
Graphical Information 435

Butyl Rubber

Isobutylene-Isoprene Copolymer (IIR) - Chapter 78
Textual Information 437
Tabular Information 438
Graphical Information 439

Bromobutyl Rubber

Bromoisobutylene-Isoprene Copolymer (BIIR) - Chapter 79
Textual Information 441
Graphical Information 442

Chlorobutyl Rubber

Chloroisobutylene-Isoprene Copolymer (CIIR) - Chapter 80
Textual Information 443
Tabular Information 444

Isobutylene Rubber

Isobutylene - Chapter 81
Tabular Information 445

Chlorosulfonated Polyethylene Rubber

Chlorosulfonated Polyethylene (CSM) - Chapter 82
Tabular Information 447

Epichlorohydrin Rubber

Polyepichlorohydrin (CO) - Chapter 83
Tabular Information 449

Polyepichlorohydrin Copolymer (CO Copolymer) - Chapter 84
Tabular Information 451

Ethylene-Propylene Rubber (EPR)

Ethylene-Propylene Copolymer (EPM) - Chapter 85
Tabular Information 453

Ethylene-Propylene-Diene Copolymer (EPDM) - Chapter 86
Tabular Information 455
Graphical Information 457

Fluoroelastomer

Vinylidene Fluoride-Hexafluoropropylene Copolymer (FKM) - Chapter 87
Textual Information 459
Tabular Information 459

Natural Rubber

Natural Rubber (NR) - Chapter 88
Textual Information 461
Tabular Information 462
Graphical Information 464

Neoprene Rubber

Polychloroprene (CR) - Chapter 89
Tabular Information 465

Nitrile Rubber

Acrylonitrile-Butadiene Copolymer (NBR) - Chapter 90
Textual Information 467
Tabular Information 468
Graphical Information 470

Polyisoprene Rubber

Polyisoprene Rubber (PI) - Chapter 91
Tabular Information 471

Polysulfide Rubber

Polysulfide Rubber (T) - Chapter 92
Tabular Information 473

Polyurethane

Polyester Urethane (AU) and Polyether Urethane (EU) - Chapter 93
Tabular Information 475

Propylene Oxide Rubber

Propylene Oxide (PO) - Chapter 94
Tabular Information 477

Silicone Rubber

Silicone (MQ) - Chapter 95
Textual Information 479
Tabular Information 480
Methylvinylfluorosilicone (FVMQ) - Chapter 96
Tabular Information 483

Styrene-Butadiene Rubber

Styrene-Butadiene Copolymer (SBR) - Chapter 97
Tabular Information 485
Graphical Information 487

Appendices

Penetrant Sort 491

Permeation Rates 615

Permeability of Rubber Glove Films 629

Permeability Units Conversion 655

Glossary of Terms 659

Indicies

Table and Graph Index 681

End Notes to Tables 697

Reference Index 701

Trade Name Index 705

Introduction

Plastics Design Library is pleased to introduce *Permeability And Other Film Properties Of Plastics And Elastomers*, a unique reference and data bank on the barrier and film properties of polymeric materials. The basic physical characteristics of polymers are generally well defined by manufacturers. However, data on the more capricious phenomenological issues such as permeability are difficult to find, especially in a comprehensive compilation. This volume serves to turn disparate information from wide ranging sources (i.e. conference proceedings, test laboratories, materials suppliers, monographs, trade and technical journals) into useful engineering knowledge.

The information provided ranges from a general overview of the barrier properties of plastics and elastomers to detailed discussions and test results. For users to whom the study and use of permeability and film property data are relatively new, the primer on barrier properties and detailed glossary of terms, including descriptions of test methods, will prove useful. For those who wish to delve beyond the data presented, source documentation is presented in detail.

Data presented in these pages detail differences in permeation between generic families of plastic and rubber materials. Also covered are differences within the same generic family due to environmental factors such as temperature and humidity or material characteristics such as sample preparation and material composition. This data serves as an indication of how one material is likely to behave relative to another material or relative to the same material exposed under different conditions.

The permeation of gases and vapors through thin films is dependent on the molecular size, shape, wettability and soundness of the fabricated membrane. Since permeation in well made items is a molecular transport phenomenon, it is affected by orientation, degree of crystallinity and temperature. Attempts have been made to relate permeation rates through thin films to absorption of thicker films, sheets, pipe, etc. This has been generally unsuccessful. Thicker films and sheets represent an average set of properties obtainable from many thin films produced under a variety of conditions. To produce a thin film representative of this average is not practical

In compiling data, the philosophy of Plastics Design Library is to provide as much information as is available. This means that complete information corresponding to each test result is provided. At the same time, an effort is made to provide information for as many tests, conditions, penetrants and materials combinations as possible. Therefore, even if detailed test metadata are not available, information is still provided. The belief is that some limited information serves as a reference point and is better than no information. In all cases, we undertake to provide information in as complete and detailed a form as it was presented in the source document. Flexibility and ease of use are also carefully considered in designing the layout of the book.

How a material performs in its end use environment is a critical consideration and the information here gives useful guidelines. However, this or any other information resource should not serve as a substitute for actual testing in determining the applicability of a particular part or material in a given end use environment.

We trust you will greet this reference publication with the same enthusiasm as other Plastics Design Library titles and that it will be a useful tool in your work. As always, your feedback on improving this volume or others in the PDL Handbook series is appreciated and encouraged.

Plastics Design Library
13 Eaton Avenue
Norwich, NY 13815
Tel: 607-337-5080   Fax: 607-337-5090

Some Notes About The Information In This Book

This publication contains data and information from many disparate sources. In order to make the product most useful to end users, Plastics Design Library normalizes presentation of the information. Permeability data, for example, are presented in many different units (greater than 60) throughout the literature. In this reference source, permeability data have been normalized into two units - $cm^3 \cdot mm/m^2 \cdot day \cdot atm$ for permeability coefficient and $g \cdot mm/m^2 \cdot day$ for vapor transmission rate.

Although substantial effort is exerted throughout the editorial process to maintain accuracy and consistency in unit conversion and presentation of information, possibility for error exists. Often these errors occur due to insufficient or inaccurate information in the source document. For this reason, values in the tables of permeability data are given in units as they appeared in the source document as well as in the converted (normalized) units. Appendix Four provides a conversion table detailing the conversion factors used and any assumptions which were made.

As with all PDL products, complete information as it was presented in the source document is provided. This includes details of test methods, test conditions, penetrant, sample size, material composition and other factors which may affect the resulting value. As a result, the user has all available information on which to make a judgment or comparison.

How To Use This Book

This publication is divided into 97 chapters and four appendices. Chapters are divided by generic families of plastics or elastomers. Within each chapter, information is presented as combinations of text, tables and graphs. Included are permeability data and data on the film properties (i.e. mechanical, optical, thermal, etc.) of the subject material. Information in the 97 chapters is in the most complete and detailed form.

Appendix I presents permeability data sorted by penetrant with a secondary sort on generic family. Appendix I is a resort and abridged presentation of information contained in the tables of the 97 chapters. (Data from Appendix III are also included for convenience.) If more information about a specific piece of information in Appendix I is needed, the user is directed to the chapter containing the appropriate material generic family.

Appendix II is useful in comparing permeation rates of penetrants through different materials at various temperature ranges. It is sorted by penetrant with a secondary sort on temperature range and a final sort on either permeability coefficient ($cm^3 \cdot mm/m^2 \cdot day \cdot atm$) or vapor transmission rate ($g \cdot mm/m^2 \cdot day$). Appendix Two presents data in the most concise form. Only normalized values for permeability coefficient or vapor transmission rate appear. As a result, the user is provided a convenient ranking of the permeability of a penetrant through various plastics and/or elastomers. Supporting test information, except for temperature, is not included. For more detailed information the user is directed to the chapter containing the appropriate material generic family.

Appendix III provides permeation data for penetrants through glove films. Appendix IV gives the conversions used in converting to permeation coefficient or vapor transmission rate.

Barrier Materials - An Overview

In the glossary, barrier materials are defined as materials which have low permeability to gases, vapors and liquids. There are many types of barrier materials including plastic films and sheeting, wood laminates, particle board, paper, fabrics and metallic foils. Important plastic and rubber barrier materials include ethylene-vinyl alcohol copolymers, polyvinylidene chloride, high nitrile resins, nylons, butyl rubber and acrylic latexes.

The two most important characteristics of barrier materials are Gas Permeability (transmission rate) and Water Vapor Transmission Rate. Although barrier polymeric materials are used widely in such applications as water vapor insulation in construction and protective clothing, their major use is in the packaging industry, especially the food packaging industry.

Material	Oxygen Transmission Rate (23°C, 0% Relative Humidity)	
	Source Document Units ($cm^3 \cdot \mu m / m^2 \cdot day \cdot day$)	Normalized Units ($cm^2 \cdot mm / m^2 \cdot day \cdot atm$)
Polyvinyl alcohol	2.5	0.0025
Ethylene vinyl alcohol based materials (EVOH)	4 - 60	0.0041 - 0.0608
Polyvinylidene chloride based materials (PVDC)	15 -250	0.0152 - 0.2533
Nylon MXD6 (Oriented)	52	0.0527
Nylon MXD6	250	0.2533
High nitrile resins	300	0.304
Cellophane	440	0.4458
Polyethylene terephthalate (Oriented)	1600	1.621
Nylon 66	2000	2.027
Nylon 6	2000	2.027
Polyvinylchloride (PVC), rigid	3100	3.141
Polychlorotrifluoroethylene (PCTFE)	4400	4.458
Polypropylene (Oriented)	44000	44.58
Polypropylene	81000	82.07

Material	Water Vapor Transmission Rate (38°C, 90% Relative Humidity)	
	Source Document Units ($g \cdot \mu m / m^2 \cdot day$)	Normalized Units ($g \cdot mm / m^2 \cdot day$)
Polyvinylidene chloride based materials (PVDC)	10 - 80	0.01 - 0.08
Polychlorotrifluoroethylene (PCTFE)	15	0.015
Polypropylene (Oriented)	160	0.16
Polypropylene	260	0.26
Polyethylene terephthalate (Oriented)	800	0.8
Polyvinylchloride (PVC), rigid	880	0.88
Nylon MXD6 (Oriented)	1100	1.1
Ethylene vinyl alcohol based materials (EVOH)	1300 - 3400	1.3 - 3.4
Nylon 66	1500	1.5
High nitrile resins	1600	1.6
Nylon MXD6	2000	2.0
Nylon 6	4300	4.3
Cellophane	137000	137
Polyvinyl alcohol	750000	750

Reference: *Food Contact Polymeric Materials,* review report (ISSN: 0889-3144) - RAPRA Technology Ltd., 1992.

Barrier Materials for Food Packaging

Barrier materials for food packaging must meet a variety of requirements. Most important, they must be non-toxic and chemically resistant to foodstuffs to comply with FDA regulations. Additionally, food packaging materials should have low permeability (to gases, vapors and liquids), high puncture strength, adequate transparency and gloss, low cost and good processability. Depending on the end use, resistance to sterilization, good heat sealing and shrinking properties, biodegradability, printability, antiblocking properties and toughness might be required as well. However, no single commercial resin could meet all these requirements. Desired properties are attained therefore, by combining two or more resins; creating a multilayer system. This multilayer system is achieved via co-extrusion, co-molding and extrusion coating techniques. In addition, surface modification such as plasma treatment and the use of special additives are often employed.

The principal component of a multilayer packaging material is the barrier resin. Among barrier resins, polyvinylidene chloride and ethylene-vinyl alcohol copolymers are used most. Typical values of water vapor transmission rate and oxygen transmission rate reported in the literature in comparison with other polymers, ranked by transmission rate, appear in the tables on the previous page.

Polyvinylidene Chloride

Polyvinylidene chloride (PVDC) resins are copolymers of >50% vinylidene chloride and other monomers such as methyl methacrylate or vinyl chloride. They are conveniently available as latexes for use in waterborne coatings, as solvent-soluble resins (for use in solvent-based coatings) and as (co)extrusion powders.

The permeability of PVDC is quite low. For example, water vapor transmission rate of a PVDC film is 0.01 - 0.08 $g \cdot mm/m^2 \cdot day$ at 38°C and 90% RH. The permeability of PVDC decreases with increasing mole fraction of vinylidene chloride due to an increase in crystallinity. Conversely, its toughness, flexibility at low temperatures and heat sealing properties improve with decreasing mole fraction of vinylidene chloride. PVDC has good chemical resistance and can be sealed by heat or high-frequency current. The main limitations of PVDC include corrosiveness during processing and limited stability to heat and sterilization by radiation. Its narrow processing temperature range prevents its co-extrusion with nylons and polycarbonate that require high processing temperatures.

PVDC latexes are mainly used for coating plastic and cellulose films and extrudable PVDC is used in heat-shrinkable and co-extruded films. Co-extruded PVDC has a good adhesion to polyvinyl chloride, ethylene-vinyl alcohol copolymers and acrylics.

Ethylene Vinyl Alcohol Copolymers

Ethylene-vinyl alcohol copolymers (EVOH) are obtained by hydrolysis of ethylene-vinyl acetate copolymers and contain 32 or 44 mol% ethylene. EVOH containing >50 or <25 mol% ethylene do not have optimum barrier properties. EVOH are processed easily by (co)extrusion into films and by injection molding.

The water vapor transmission rate of EVOH is higher than that of PVDC ranging from 1.3 to 3.4 $g \cdot mm/m^2 \cdot day$ at 38°C and 90% RH, but the dry gas permeability of EVOH is better than that of PVDC. The gas permeability of EVOH increases with increasing relative humidity.

EVOH give transparent, printable and glossy films but have a number of limitations such as lack of stretchability at temperatures <100-150°C and poor adhesion to some desirable co-extrudants. For example, they can be co-extruded with nylons but co-extrusion with polyolefins, polyesters and polycarbonates requires the use of adhesives.

EVOH-based multilayer films are manufactured by both blown and cast methods and are used in a range of applications, including modified atmosphere packaging, form-fill-seal pouches for high-fat foods, bag-in-box for juices and lidding on thermoformed trays. The films co-extruded with linear low-density polyethylene are especially popular.

Other Resins

High nitrile resins, nylons and polyesters are also high-volume food packaging materials.

High Nitrile Resins

High nitrile resins (HNR) are acrylonitrile-based copolymers. Because of the possible migration of acrylonitrile into foodstuff and its teratogenic effect, the use of HNR is restricted to the Barex brand. Barex resins are FDA-approved for food contact and comprise acrylonitrile and methyl acrylate grafted nitrile rubber. They have low permeability that is independent of humidity, good impact strength and somewhat limited thermal stability which restricts their use in hot fill or retort applications. HNR are resistant to sterilization by irradiation and are processed easily by (co)extrusion. Their largest application is in rigid and semi-rigid packaging of meat products.

Nylons

Nylons or polyamides (PA) are used in food packaging because of their thermal stability, toughness and resistance to cracking and puncture. They are good barriers to oils and fats but are inferior gas barriers compared to PVDC and EVOH. Because of this and high moisture absorption and low heat sealability, PA are used mostly co-extruded with such resins as low-density polyethylene. Nylon 6 and 66 are used widely, especially in vacuum packaging of meat and cheese. Other PA types used in food packaging include amorphous nylons and aromatic nylon MXD6. The former have an improved stiffness and barrier properties compared to nylon 6. The latter have an improved thermal stability and processability compared to conventional barrier resins. They are used mainly in co-injection molding of polyethylene terephthalate bottles.

Polyesters

Polyesters are very important packaging materials but due to their relatively high gas permeability they are not normally used as barrier layers, except for liquid-crystal polymers (LCP) and polyethylene naphthalate (PEN). LCP are aromatic polyesters that have very high stiffness, strength, toughness, chemical resistance, very low permeability to gases and vapors and good thermal stability. They are processed by (co)extrusion and injection molding. PEN has excellent clarity, strength, heat resistance and oxygen barrier properties. It can be used for hot fill applications. The main disadvantage of both LCP and PEN is their high cost.

Oriented Materials

EVOH (Biaxially Oriented)

Products with superior mechanical and optical properties can be produced from films that combine the barrier properties of EVOH with biaxial orientation. These films can be metalized, used in laminations and as monolayer films. They exhibit good dimensional stability and reduced moisture sensitivity.

PET (Biaxially Oriented)

Due to increased heat stability, biaxially oriented polyester can be used in a wide range of applications, including microwavable packaging. If metalized or coated, barrier properties can be enhanced and clarity maintained.

Nylon (Biaxially Oriented)

Better strength, transparency, printability and machinability are the characteristics of biaxially oriented Nylon as compared to non-oriented films.

Polyolefins (Biaxially Oriented)

Biaxially oriented polyolefin multilayer film products with characteristics such as low temperature sealing and the ability to print on both sides are available.

Metalized Films

If optical transparency is not needed, then barrier properties can be improved with metalizing. Barrier properties can be improved even further by laminating two metalized films. Metal to metal significantly reduces the problem of gas permeation through pinholes or defects.

Transport of Gases and Vapors in Barrier Materials

The most important properties of barrier materials are their transport properties. These properties are reviewed below.

Permeation Coefficient and Vapor Transmission Rate

The transport of gases and vapors in barrier materials such as polymeric films involves dissolving of the penetrant in the material, diffusion of dissolved penetrant through the material as a result of the concentration gradient and evaporation of the penetrant from the opposite surface of the material.

The transport of a penetrant in a barrier material can be described by Fick's first law stating that the volume (V) of a penetrant that penetrates a barrier wall is directly proportional to the area (A) of the wall, partial pressure differential (p) of the penetrant and time (t); and inversely proportional to the wall thickness(s), if the wall is homogeneous in the direction of penetration. The coefficient P in the equation representing Fick's first law, $V = P \cdot (A \cdot p \cdot t)/s$, is the permeability coefficient.

Fick's first law applies only to permanent gases that obey Henry's law on proportionality of penetrant solubility in the barrier to the partial pressure of the penetrant. Therefore, the permeability coefficient can be measured under standard conditions only for permanent gases, i.e., gases that become liquid at pressures and temperatures far from normal (1 atm and 0°C, respectively). These gases include air, oxygen, argon and carbon dioxide.

The vapors of substances, such as water and acetone, that are liquid at pressures and temperatures close to normal do not obey Henry's law. Consequently, the permeability coefficient for the vapors is not proportional to the pressure differential in Fick's first law. To account for this fact, Fick's first law for vapors is expressed as $W = VTR \cdot (A \cdot t)/s$, where W is the weight of the penetrant and VTR is the vapor transmission rate. VTR is used to characterize the transport properties of barrier materials with respect to the vapors.

Units of Measurement

Convenient units of measurement for the permeability coefficient and the vapor transmission rate in the metric system are $(cm^3 \cdot mm)/(m^2 \cdot day \cdot atm)$ and $(g \cdot mm)/(m^2 \cdot day)$, respectively.

The table on the next page gives conversion factors for the common units of measurement of the permeability coefficient and the vapor transmission rate. To convert a value from a common unit to the convenient metric unit for the permeability coefficient, $(cm^3 \cdot mm)/(m^2 \cdot day \cdot atm)$, or to do that for the vapor transmission rate, $(g \cdot mm)/(m^2 \cdot day)$, multiply this value by a factor provided in the corresponding column.

The values of the permeability coefficient and the vapor transmission rate in most of the units may be in the range of several powers of magnitude. However, these values are usually given in an easy-to-read decimal format (practical units), with the magnitude factor stated in a table or graph title or in the notes. Care should be taken, when converting, to account for this factor.

Original Unit	Common Units	
	Permeability Coefficient Unit ($cm^3 \cdot mm/m^2 \cdot day \cdot atm$)	Vapor Transmission Rate Unit ($g \cdot mm/m^2 \cdot day$)
gm mil/ 100 in^2 · day	-	3.937008×10^{-1}
cc · mil/ 100 in^2 · atm · day	3.937008×10^{-1}	-
m^2/ s · Pa	8.754480×10^{18}	-
cm^3 · mils/ m^2 · days · atm	2.54×10^{-2}	-
cc · mm/ m^2 · sec · cmHg	6.566397×10^{6}	-
cc · mm/ m^2 · sec · atm	8.64×10^{4}	-
in^3 · mil/ 100 in^2 · 24hrs · atm	6.4516	-
cm^3 · mm/ m^2 · day · bar	1.01325	-
mm^3 · mm/ m^2 · 24hrs · Pa	1.01325×10^{2}	-
μm^3 · mm/ m^2 · s · Pa	8.75448×10^{-3}	-
cm^3 · mm/ m^2 · 24hrs · Pa	1.01325×10^{5}	-
cm^3 (@STP) · cm/ atm · sec · cm^2	8.64×10^{9}	-
cm^3 · mil/ cm^2 · sec · atm	2.19456×10^{7}	-
ft^3 · mil/ ft^2 · day · psi	1.137749×10^{5}	-
m · MPa · day	1.01325×10^{-1}	-

Standard Test Methods

Depending on the barrier material, penetrant tested, desirable precision and availability of testing equipment among other factors, the permeability coefficient and the vapor transmission rate can be measured by a variety of US (e.g., ASTM), international (ISO) and foreign (e.g., DIN) standard methods. Below are brief descriptions of the most commonly used ASTM methods.

ASTM D1434

ASTM D1434 standard test method is intended for determining gas transmission rate, permeance and permeability (for homogeneous materials) of plastic film, sheeting, laminates and plastic-coated papers or fabrics under steady-state conditions.

The sample is mounted in a gas transmission cell to form a barrier between 2 chambers. One chamber contains the test gas at a high pressure and the other chamber receives gas at a lower pressure. The transmission rate is monitored either by an increase in pressure in the receiving chamber (Method M) or by a change in volume of gas (Method V).

ASTM D3985

ASTM D3985 standard test method is intended for determining the steady-state transmission rate of oxygen gas through a plastic film, sheeting, laminates, co-extrusions, or plastic-coated paper or fabric. In addition it provides for determination of the permeance of the film to oxygen gas and oxygen permeability coefficient.

The specimen is placed as a sealed semi-barrier between two chambers at ambient atmospheric pressure. One chamber is slowly purged with a stream of nitrogen; the other chamber contains oxygen. As oxygen penetrates through the film it is transported by nitrogen to a coulonometric detector, where it produces an electrical current, the magnitude of which is proportional to the amount of oxygen flowing through the detector per unit time.

ASTM E96

ASTM E96 standard test method is intended for determining water vapor transmission of materials such as paper, plastic film and sheeting, fiberboard, wood products, etc., that are less than 31 mm in thickness. Two basic methods, the Desiccant Method and the Water Method are used.

The specimens have either one side wetted or one side exposed to high humidity and another to low humidity. In the Desiccant Method, the specimen is placed air-tight on a test dish with a desiccant that is weighed to determine the gain of weight due to water vapor transmission. In the Water Method, the water is placed in the dish that is weighed to determine the loss of water due to evaporation through the specimen.

ASTM F1249

ASTM F1249 standard test method is intended for determining water vapor transmission rate through plastic film and sheeting up to 3 mm in thickness using a pressure-modulated infrared sensor. In addition it provides for determination of the permeance of the film to water vapor and the water vapor permeability coefficient.

The specimen is placed as a sealed semi-barrier between two chambers at ambient atmospheric pressure. One chamber is wet and another is dry. As water vapor penetrates through the film from wet chamber into dry one, it is carried by air into the sensor. It measures the fraction of infrared energy absorbed by the vapor and produces an electric signal that is proportional to water vapor concentration.

Effects of Environment

Effect of Temperature

The permeability coefficient P can be defined as $P = D \cdot S$, where D is the diffusion coefficient and S is the solubility coefficient of a penetrant. Both, the diffusion coefficient and the solubility coefficient depend on the temperature. Consequently, permeability coefficient is also a function of the temperature and a measurement of the permeability coefficient or the vapor transmission rate is not valid without a reference to the test temperature.

Solubility coefficients of permanent gases such as oxygen and sparingly soluble gases and vapors increase with increasing temperature, resulting in increased permeability. In contrast, solubility coefficients of readily condensable gases and vapors, such as sulfur dioxide and ammonia, decrease with increasing temperature, resulting in decreased permeability.

Permeability to water vapor usually increases with increasing temperature, depending on the moisture content of the barrier material and its nature. Permeability to organic vapors generally increases with increasing temperature, but is complicated by the swelling of the barrier material.

Effects of Pressure and Relative Humidity

Permanent gases at pressures close to normal obey Henry's law and their solubility coefficient is proportional to the partial pressure of the gas. As a result, their permeation rates, reduced to unit pressure, are generally independent of the pressure.

The permeation rates of sparingly soluble vapors, such as water vapor in polyolefins, may be proportional to the vapor pressure differential across the barrier wall. The permeation rates of readily soluble penetrants that do not obey Henry's law have a complex relation to pressure. For these penetrants, test pressure must be reported for the permeability data to be valid.

Absorbed water has a plasticizing effect on some barrier materials and can lead to increased permeability. On the other hand, the diffusion of water in some materials is concentration dependent and the water vapor transmission rate is affected by the relative humidity differential. Therefore, the relative humidity of the test environment has to be known to make a correct interpretation of the permeability measurements.

Effects of the Nature of Penetrant and Barrier Polymeric Material

Effect of the Nature of Penetrant

Permanent gases are usually inert towards barrier polymeric materials and their permeation rates are inversely proportional to their molecular size. The permeation of other gases and vapors depends strongly on the ease of their condensation and on their affinity to the barrier material. A readily soluble penetrant will produce swelling of the polymer, resulting in an increased permeability coefficient.

Effect of the Nature of Barrier Polymeric Material

The chemical structure and morphology of barrier polymeric materials have a strong effect on their permeability. Physical interaction between penetrant and barrier material such as the formation of hydrogen bonds or the interaction between polar functional groups may slow down the permeation. As was mentioned above, increased solubility of the penetrant in the barrier material may lead to material swelling and increased permeation. In contrast, improved packing order and increased crystallinity of the barrier polymeric material increases its density and decreases its permeability coefficient. A similar effect is observed on radiation crosslinking of the material such as polyethylene.

Other factors affecting penetrant transport in barrier polymeric materials include the effect of plasticizers and fillers. Plasticizers tend to increase permeability by "loosening" the polymer structure. Some fillers may retain the penetrant by sorption or create a physical barrier to its diffusion. This effect becomes more pronounced with decreasing particle size and increasing content of the filler. However, at high content levels, the fillers tend to increase the permeability of the barrier material due to increased porosity.

The permeability of barrier polymeric materials is also dependent on their pore properties, e.g., pore volume and structure; thickness; and design (for multilayer systems). For example, the permeability normally decreases with increasing thickness of the barrier material.

Chapter 1

Acetal Resin

Permeability

DuPont: Delrin

Delrin acetal resin has good impermeability to many substances including aliphatic, aromatic and halogenated hydrocarbons, alcohol, and esters. Its permeability to some small polar molecules such as water, methyl alcohol and acetone is relatively high. Permeability characteristics and strength properties of Delrin acetal resin make it a suitable material for containers, particularly of the aerosol type.

Reference: *Delrin Design Handbook For Du Pont Engineering Plastics,* supplier design guide (E-62619) - Du Pont Company, 1987.

TABLE 01: Cologne, Shampoo and Hair Spray Permeability Through DuPont Delrin Acetal Resin.

Material Family	ACETAL RESIN					
Material Supplier/ Grade	DUPONT DELRIN					
Reference Number	201	201	201	201	201	201

TEST CONDITIONS

Penetrant	cologne		hair spray		shampoo	
Penetrant Note	various formulations	various formulations	various formulations	various formulations	various formulations	various formulations
Temperature (°C)	23	38	23	38	23	38
Relative Humidity (%)	50		50		50	

PERMEABILITY (source document units)

Vapor Transmission Rate (g · mil/100 in^2 · day)	0.6	4.5	0.8	6.0	2.4	8.5
Vapor Transmission Rate (g · mm/m^2 · day)	0.24	1.77	0.32	2.36	0.95	3.35

PERMEABILITY (normalized units)

Vapor Transmission Rate (g · mm/m^2 · day)	0.24	1.77	0.31	2.36	0.94	3.35

TABLE 02: Gasoline, Freon Propellant, Motor Oil, and Ethyl Alcohol Permeability Through DuPont Delrin Acetal Resin.

Material Family	ACETAL RESIN									
Material Supplier/ Grade	DUPONT DELRIN									
Reference Number	201	201	201	201	201	201	201	201	201	201

TEST CONDITIONS

Penetrant	ethyl alcohol			Freon 12				gasoline	motor oils	
Concentration (%)	90	70	70	30	30	20	20			
Penetrant Note	with 10% water	with 30% water	with 30% water	with 70% Freon 11; propellant	with 70% Freon 11; propellant	with 80% Freon 114; propellant	with 80% Freon 114; propellant			
Temperature (°C)	23	23	38	23	38	23	38	23	23	38
Relative Humidity (%)	50	50		50		50		50	50	

PERMEABILITY (source document units)

Vapor Transmission Rate (g · mil/100 in^2 · day)	0.25	1.5	7.8	0.2	0.54	0.2	0.42	0.1	0	0

PERMEABILITY (normalized units)

Vapor Transmission Rate (g · mm/m^2 · day)	0.1	0.59	3.07	0.08	0.21	0.08	0.17	0.04	0	0

TABLE 03: Methyl Salicylate, Nitrogen, Perchloroethylene, Trichloroethylene, Toluene, Carbon Dioxide and Oxygen Permeability Through DuPont Delrin Acetal Resin.

Material Family	ACETAL RESIN							
Material Supplier/ Grade	DUPONT DELRIN							
Reference Number	201	201	201	201	201	201	201	201

TEST CONDITIONS

Penetrant	methyl salicylate	nitrogen (@ 620 kPa)	perchloro-ethylene	trichloroethylene		toluene	carbon dioxide	oxygen
Temperature (°C)	23	23	23	23	38	23	23	23
Relative Humidity (%)	50	50	50	50		50	50	50

PERMEABILITY (source document units)

Vapor Transmission Rate (g · mil/100 in^2 · day)	0.3	0.05	0.2	25	56	0.6		
Gas Permeability (cm^3 · mil/100 in^2 · day)							37 - 50	12 - 17

PERMEABILITY (normalized units)

Permeability Coefficient (cm^3 · mm/m^2 · day · atm)							14.6 - 19.7	4.7 - 6.7
Vapor Transmission Rate (g · mm/m^2 · day)	0.12	0.02	0.08	9.84	22.05	0.24		

TABLE 04: Mineral Oils, Vegetable Oils, Tar Remover and Road Oil Remover Permeability Through DuPont Delrin Acetal Resin.

Material Family	ACETAL RESIN							
Material Supplier/ Grade	DUPONT DELRIN							
Reference Number	201	201	201	201	201	201	201	201

TEST CONDITIONS

Penetrant	mineral oils		vegetable oils		tar remover		road oil remover	
Temperature (°C)	23	38	23	38	23	38	23	38
Relative Humidity (%)	50		50		50		50	

PERMEABILITY (source document units)

Vapor Transmission Rate (g · mil/100 in^2 · day)	0	0	0	0	0.03	0.19	0.03	0.19
Vapor Transmission Rate (g · mm/m^2 · day)	0	0	0	0	0.01	0.07	0.01	0.07

PERMEABILITY (normalized units)

Vapor Transmission Rate (g · mm/m^2 · day)	0	0	0	0	0.01	0.07	0.01	0.07

TABLE 05: Air and Oxygen Permeability Through Hoechst Celanese Celcon Acetal Copolymer Film.

Material Family	ACETAL COPOLYMER					
Material Supplier	HOECHST CELANESE					
Grade	CELCON M90	CELCON M25	CELCON M270	CELCON M90	CELCON M25	CELCON M270
Product Form	FILM					
Features	general purpose grade	high molecular weight	high flow, low molecular weight	general purpose grade	high molecular weight	high flow, low molecular weight
Reference Number	210	210	210	210	210	210

MATERIAL CHARACTERISTICS

Melt Flow Index	9.0 g/10 min.	2.5 g/10 min.	27.0 g/10 min.	9.0 g/10 min.	2.5 g/10 min.	27.0 g/10 min.
Sample Thickness	0.15 mm	0.15 mm	0.15 mm	0.15 mm	0.15 mm	0.15 mm

TEST CONDITIONS

Penetrant	air			oxygen		

PERMEABILITY (source document units)

Gas Permeability (cm^3 · mil/100 in^2 · day)	2.2 - 3.2	2.2 - 3.2	2.2 - 3.2	5.0 - 7.4	5.0 - 7.4	5.0 - 7.4

PERMEABILITY (normalized units)

Permeability Coefficient (cm^3 · mm/m^2 · day · atm)	0.87 - 1.3	0.87 - 1.3	0.87 - 1.3	2.0 - 2.9	2.0 - 2.9	2.0 - 2.9

TABLE 06: Nitrogen and Carbon Dioxide Permeability Through Hoechst Celanese Celcon Acetal Copolymer Film.

Material Family	ACETAL COPOLYMER					
Material Supplier	HOECHST CELANESE					
Grade	CELCON M90	CELCON M25	CELCON M270	CELCON M90	CELCON M25	CELCON M270
Product Form	FILM					
Features	general purpose grade	high molecular weight	high flow, low molecular weight	general purpose grade	high molecular weight	high flow, low molecular weight
Reference Number	210	210	210	210	210	210

MATERIAL CHARACTERISTICS

Melt Flow Index	9.0 g/10 min.	2.5 g/10 min.	27.0 g/10 min.	9.0 g/10 min.	2.5 g/10 min.	27.0 g/10 min.
Sample Thickness	0.15 mm	0.15 mm	0.15 mm	0.15 mm	0.15 mm	0.15 mm

TEST CONDITIONS

Penetrant	carbon dioxide			nitrogen		

PERMEABILITY (source document units)

Gas Permeability ($cm^3 \cdot mil/100\ in^2 \cdot day$)	144 - 174	144 - 174	144 - 174	2.2 - 3.2	2.2 - 3.2	2.2 - 3.2

PERMEABILITY (normalized units)

Permeability Coefficient ($cm^3 \cdot mm/m^2 \cdot day \cdot atm$)	56.7 - 68.5	56.7 - 68.5	56.7 - 68.5	0.87 - 1.3	0.87 - 1.3	0.87 - 1.3

Acetal

Chapter 2

Acrylonitrile

TABLE 07: Gas Permeability and Water Vapor Transmission Through Polyacrylonitrile.

Material Family	ACRYLONITRILE		
Product Form	FILM		
Reference Number	250	250	250

TEST CONDITIONS

Penetrant	oxygen	carbon dioxide	water vapor
Temperature (°C)	24	24	24

PERMEABILITY (source document units)

Vapor Transmission Rate (g · mil/100 in² · day)			0.6
Gas Permeability (cm³ · mil/100 in² · day)	0.04	0.1 - 0.2	

PERMEABILITY (normalized units)

Permeability Coefficient (cm³ · mm/m² · day · atm)	0.02	0.04 - 0.08	
Vapor Transmission Rate (g · mm/m² · day)			0.24

Chapter 3

Acrylonitrile-Methyl Acrylate Copolymer

Permeability

Acrylonitrile Copolymer (product form: film)

The permeability of nitrile polymers is determined by the polarity of the functional group, and decreases regularly with increasing nitrile concentration.

Reference: *Permeability Of Polymers To Gases And Vapors,* supplier technical report (P302-335-79, D306-115-79) - Dow Chemical Company, 1979.

Permeability to Oxygen

BP Chemicals: Barex 210 (applications: packaging; features: impact modified, barrier properties); **Barex 218** (applications: packaging; features: impact modified, high impact, barrier properties)

Barex resins have the lowest oxygen permeability of any plastic material used for single layer packages. Barex packages frequently outperform multilayer structures containing EVOH or PVDC and do so at lower costs. Barrier performance can be further enhanced by producing oriented containers, such as in the extrusion stretch blow molding process for bottles.

Barex resins offer a high barrier to oxygen at all levels of relative humidity. Not all barrier materials exhibit this property. Humidity can have a profound and often deleterious effect on the oxygen permeability of some competing barrier resins. Barex resins ensure that a consistently high barrier to oxygen is maintained regardless of the humidity.

Barrier performance is negatively impacted by increasing temperature.

Reference: *Barex Barrier Resins - Barrier Properties,* supplier technical report (Bx-555) - BP Chemicals Inc., 1992.

Water Vapor Permeability

BP Chemicals: Barex 210 (applications: packaging; features: impact modified, barrier properties); **Barex 218** (applications: packaging; features: impact modified, high impact, barrier properties)

The water vapor barrier properties of Barex resins are comparable to other plastic packaging materials except for polyolefins. In applications where exclusion of moisture is critical, the water vapor barrier of Barex packages can be enhanced by orientation or lamination to a polyolefin, giving an excellent combination of gas and moisture barrier.

Reference: *Barex Barrier Resins - Barrier Properties,* supplier technical report (Bx-555) - BP Chemicals Inc., 1992.

Effect of Temperature on Water Vapor Transmission Rate

Acrylonitrile Copolymer (product form: film)

For high nitrile polymers, temperature dependence of water vapor permeability is influenced by the equilibrium moisture content of the film.

Reference: *Permeability Of Polymers To Gases And Vapors,* supplier technical report (P302-335-79, D306-115-79) - Dow Chemical Company, 1979.

Film Properties and Applications

BP Chemicals: Barex 210 (applications: packaging; features: impact modified, barrier properties); **Barex 218** (applications: packaging; features: impact modified, high impact, barrier properties)

Barex 210 and 218 are high barrier, impact modified, acrylonitrile methylacrylate copolymer resins. Barex 218 contains a higher proportion of impact modifier for enhanced flexibility and toughness.

In food packging applications, the high barier properties of Barex resins ensure product freshness, extended shelf life, and retention of natural aromas and flavors without scalping. Extending shelf life is usually accomplished by sealing in beneficial gases such as nitrogen and carbon dioxide while preventing oxygen from entering the package.

Barex resins are in compliance with the US FDA regulations for direct food contact with non-beverage products. They are especially suitable for the packaging of processed meats, fish, cheese, spices, sauces, extracts, and juice concentrates. Their exceptional barrier properties make Barex resins an excellent choice for controlled or modified atmosphere packaging.

Reference: *Barex Barrier Resins - Barrier Properties,* supplier technical report (Bx-555) - BP Chemicals Inc., 1992.

TABLE 08: Gas Permeability and Water Vapor Transmission Through Acrylonitrile Copolymer.

Material Family	ACRYLONITRILE-METHYL ACRYLATE COPOLYMER		
Product Form	FILM		
Reference Number	250	250	250

TEST CONDITIONS

Penetrant	oxygen	carbon dioxide	water vapor
Temperature (°C)	24	24	24

PERMEABILITY (source document units)

Vapor Transmission Rate $(g \cdot mil/100\ in^2 \cdot day)$			0.9
Gas Permeability $(cm^3 \cdot mil/100\ in^2 \cdot day)$	0.2	0.5	

PERMEABILITY (normalized units)

Permeability Coefficient $(cm^3 \cdot mm/m^2 \cdot day \cdot atm)$	0.08	0.2	
Vapor Transmission Rate $(g \cdot mm/m^2 \cdot day)$			0.35

TABLE 09: Cyclohexanone, Chlorobenzene, Hexane, Butyl Alcohol, Trichloroethene, Methyl Salicylate and Tetrahydrofuran Permeability Through Acrylonitrile Copolymer Bottles.

Material Family	ACRYLONITRILE-METHYL ACRYLATE COPOLYMER						
Product Form	BOTTLES						
Features	barrier properties	barrier properties	barrier properties	barrier properties	barrier properties	barrier properties	barrier properties
Reference Number	293	293	293	293	293	293	293

TEST CONDITIONS

Penetrant	cyclohexanone	chlorobenzene	hexane	butyl alcohol	trichloroethene	methyl salicylate	tetrahydrofuran
Temperature (°C)	50	50	50	50	50	50	23
Exposure Time (days)	28	28	28	28	28	28	180

PERMEABILITY (source document units)

Penetrant Weight Loss (%)	failed	failed	0.93	0.03	1.75	0.11	0.01 (crazing)

TABLE 10: Ethyl Acetate, Isopropyl Acetate, Acetone, Butyl Acetate, Toluene, Xylene, Methyl Isobutyl Ketone and Methyl Ethyl Ketone Permeability Through Acrylonitrile Copolymer Bottles.

Material Family	ACRYLONITRILE-METHYL ACRYLATE COPOLYMER							
Product Form	BOTTLES							
Features	barrier properties	barrier properties	barrier properties	barrier properties	barrier properties	barrier properties	barrier properties	barrier properties
Reference Number	293	293	293	293	293	293	293	293

TEST CONDITIONS

Penetrant	ethyl acetate	isopropyl acetate	acetone	butyl acetate	toluene	xylene	methyl isobutyl ketone	methyl ethyl ketone
Temperature (°C)	50	50	23	50	50	50	50	50
Exposure Time (days)	28	28	180	28	28	28	28	28

PERMEABILITY (source document units)

Penetrant Weight Loss (%)	5.85 (crazing)	0.19	failed	0.19	0.07	0.06	0.09	failed

TABLE 11: Oxygen Permeability at Different Temperatures and Water Vapor Transmission Through BP Chemicals Barex 210 Acrylonitrile Copolymer.

Material Family	ACRYLONITRILE-METHYL ACRYLATE COPOLYMER				
Material Supplier/ Grade	BP CHEMICALS BAREX 210				
Features	barrier properties				
Reference Number	264	264	264	264	264

TEST CONDITIONS

Penetrant	oxygen				water vapor
Temperature (°C)	5	23	35	50	40
Relative Humidity (%)	0	0	0	0	90

PERMEABILITY (source document units)

Gas Permeability (cm$^3 \cdot$ mil/100 in$^2 \cdot$ day)	0.15	0.8	2	6.129	
Gas Permeability (cm$^3 \cdot$ 25μ/m$^2 \cdot$ day $\cdot$ atm)	2.325	12.4	31	95	
Vapor Transmission Rate (g $\cdot$ mil/100 in$^2 \cdot$ day)					6.1
Vapor Transmission Rate (g $\cdot$ 25μ/m$^2 \cdot$ day)					94.6

PERMEABILITY (normalized units)

Permeability Coefficient (cm$^3 \cdot$ mm/m$^2 \cdot$ day $\cdot$ atm)	0.06	0.31	0.79	2.41	
Vapor Transmission Rate (g $\cdot$ mm/m$^2 \cdot$ day)					2.4

TABLE 12: Water Vapor Transmission and Oxygen Permeability Through BP Chemicals Barex Acrylonitrile-Methyl Acrylate Copolymer.

Material Family	ACRYLONITRILE-METHYL ACRYLATE COPOLYMER							
Material Supplier/ Trade Name	BP CHEMICALS BAREX							
Grade	210	218	210	218	210	218	210	218
Features	barrier properties, impact modified	barrier properties, high impact, impact modified	barrier properties, impact modified	barrier properties, high impact, impact modified	barrier properties, impact modified	barrier properties, high impact, impact modified	barrier properties, impact modified	barrier properties, high impact, impact modified
Applications	packaging	packaging	packaging	packaging	packaging	packaging	packaging	packaging
Reference Number	296	296	296	296	296	296	296	296

TEST CONDITIONS

Penetrant	oxygen		nitrogen		carbon dioxide		water vapor	
Temperature (°C)	22.8	22.8	22.8	22.8	22.8	22.8	22.8	22.8
Relative Humidity (%)	100	100	100	100	0	0	100	100
Test Method	ASTM D3985	ASTM D3985					ASTM F1249	ASTM F1249

PERMEABILITY (source document units)

Gas Permeability ($cm^3 \cdot mil/\ 100\ in^2 \cdot bar \cdot day$)	0.8	1.6	0.2	0.4	1.6	1.6		
Vapor Transmission Rate ($g \cdot mil/\ 100\ in^2 \cdot bar \cdot day$)							5.0	7.5

PERMEABILITY (normalized units)

Permeability Coefficient ($cm^3 \cdot mm/m^2 \cdot day \cdot atm$)	0.32	0.64	0.08	0.16	0.64	0.64		
Vapor Transmission Rate ($g \cdot mm/m^2 \cdot day$)							1.99	2.99

TABLE 13: Water Vapor Transmission and Oxygen Permeability vs. Humidity Through BP Chemicals Barex Acrylonitrile-Methyl Acrylate Copolymer.

Material Family	ACRYLONITRILE-METHYL ACRYLATE COPOLYMER				
Material Supplier/ Grade	BP CHEMICALS BAREX 210			BP CHEMICALS BAREX 218	
Features	barrier properties, impact modified			barrier properties, high impact, impact modified	
Applications	packaging			packaging	
Reference Number	296	296	296	296	296

TEST CONDITIONS

Penetrant	oxygen		water vapor	oxygen	water vapor
Temperature (°C)	22.8	22.8	37.8	22.8	37.8
Relative Humidity (%)	0	90	90	0	90
Test Method	ASTM D3895	ASTM D3895	ASTM F1249	ASTM D1434	ASTM F1249

PERMEABILITY (source document units)

Gas Permeability ($cm^3 \cdot mil/\ 100\ in^2 \cdot bar \cdot day$)	0.8	0.8		1.6	
Vapor Transmission Rate ($g \cdot mil/\ 100\ in^2 \cdot bar \cdot day$)			5.5		7.5

PERMEABILITY (normalized units)

Permeability Coefficient ($cm^3 \cdot mm/m^2 \cdot day \cdot atm$)	0.32	0.32		0.64	
Vapor Transmission Rate ($g \cdot mm/m^2 \cdot day$)			2.19		2.99

GRAPH 01: Carbon Dioxide Permeability vs. Acrylonitrile Content through Acrylonitrile-Methyl Acrylate Copolymer.

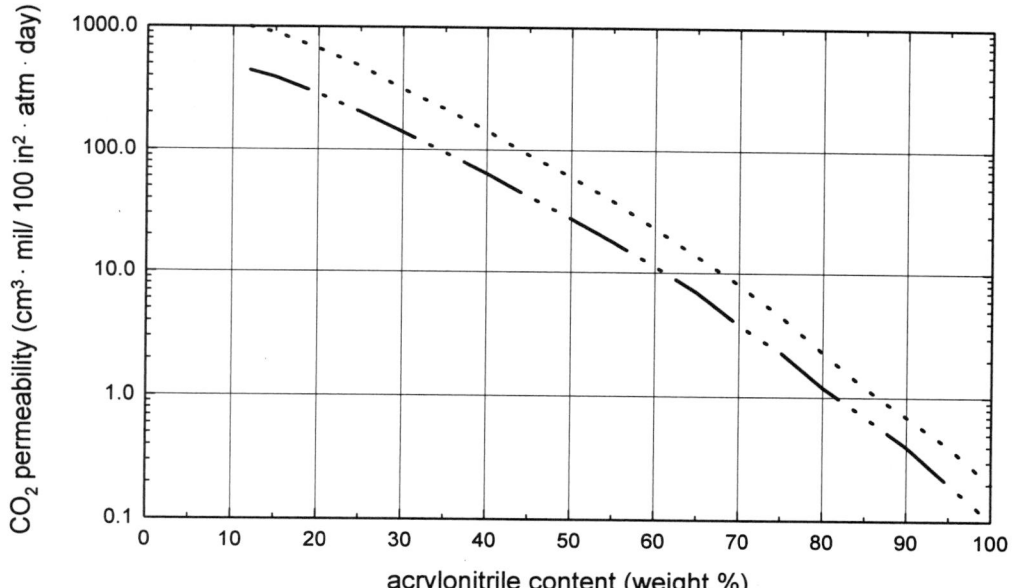

....	Acrylonitrile Copol. (film); penetrant: CO_2
— ·· — ·· —	Acrylonitrile Copol. (film); penetrant: CO_2
Reference No.	250

GRAPH 02: Carbon Dioxide and Oxygen Permeability vs. Relative Humidity through Acrylonitrile-Methyl Acrylate Copolymer.

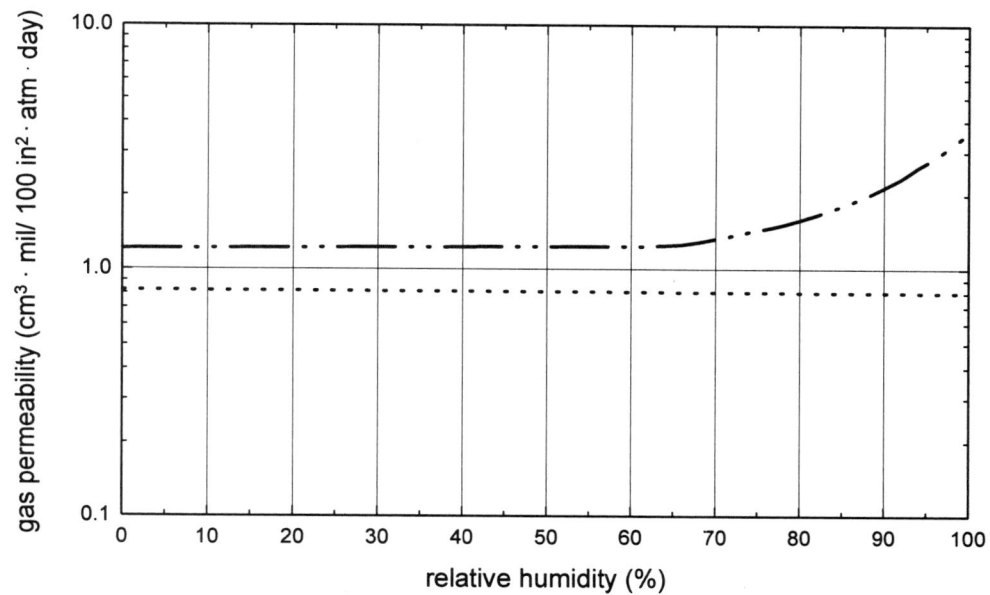

....	BP Chem. Barex 210 Acrylonitrile Copol. (barrier prop.); penetrant: O_2
— ·· — ··	BP Chem. Barex 210 Acrylonitrile Copol. (barrier prop.); penetrant: CO_2
Reference No.	264

GRAPH 03: Oxygen Permeability vs. Temperature through Acrylonitrile-Methyl Acrylate Copolymer.

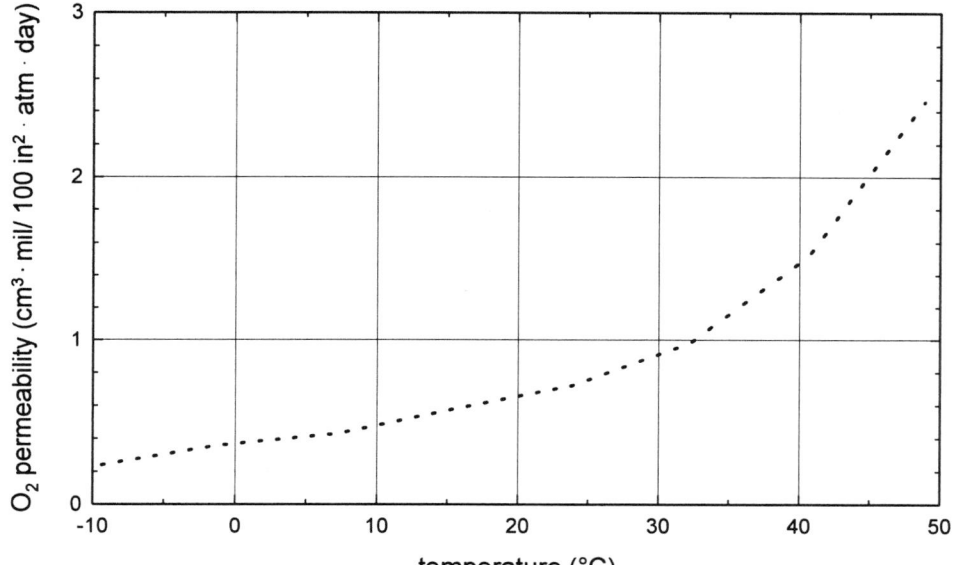

....	BP Chem. Barex 210 Acrylonitrile Copol. (packaging; impact modified, barrier prop.); penetrant: O₂
Reference No.	296

Chapter 4

Cellulosic Plastic

TABLE 14: Water Vapor Transmission and Oxygen Permeability Through Coated Cellophane Film.

Material Family	CELLULOSIC PLASTIC				
Product Form	FILM				
Reference Number	268	268	268	268	268

MATERIAL CHARACTERISTICS

Sample Thickness	0.023 mm	0.023 mm	0.023 mm	0.023 mm	0.023 mm

MATERIAL COMPOSITION

Note	PVDC coated				

TEST CONDITIONS

Penetrant	water vapor	oxygen			
Temperature (°C)	40	35	20	20	20
Relative Humidity (%)	90	0	65	85	100
Test Method	JIS Z0208	JIS Z1707	ASTM D3985	ASTM D3985	ASTM D3985

PERMEABILITY (source document units)

Vapor Transmission Rate (g · mil/100 in² · day)	1				
Gas Permeability (cm³ · mil/100 in² · day)		0.07	0.26	0.71	2.06

PERMEABILITY (normalized units)

Permeability Coefficient (cm³ · mm/m² · day · atm)		0.03	0.1	0.28	0.81
Vapor Transmission Rate (g · mm/m² · day)	0.39				

TABLE 15: Film Properties of Coated Cellophane Film.

Material Family	CELLULOSIC PLASTIC
Product Form	**FILM**
Reference Number	268

MATERIAL CHARACTERISTICS

Sample Thickness	0.023 mm
Material Composition Note	PVDC coated

TEST CONDITIONS

Temperature (°C)	20
Relative Humidity (%)	65

MECHANICAL PROPERTIES

Modulus Of Elasticity - MD (MPa)	1667 {r}
Modulus Of Elasticity - TD (MPa)	1371 {r}
Tensile Strength @ Break - MD (MPa)	117.9 {r}
Tensile Strength @ Break - TD (MPa)	88.3 {r}
Ultimate Elongation - MD (%)	30 {r}
Ultimate Elongation - TD (%)	60 {r}
Impact Strength (kg-cm)	3
Burst Strength (MPa)	0.20 {af}
Pinhole Strength (g)	400 {ag}
Elmendorf Tear Resistance - MD (g/mm)	200 {ad}
Elmendorf Tear Resistance - TD (g/mm)	300 {ad}
Tear Resistance - MD (g)	160
Tear Resistance - TD (g)	130

OTHER PROPERTIES

Water Absorption (%)	100 {ah}
Equilibrium Moist. Absorption (%)	13
Melting Point (°C)	150 (carbo.)
Haze (%)	3.0
Surface Resistivity (ohms)	1.4E+09 {aj}
Slip Factor (°)	34
Dimensional Stability - MD (%)	-3.2 {ai}
Dimensional Stability - TD (%)	-5.5 {ai}

Cellophane

Chapter 5

Fluoroplastic

Permeability

DuPont: Teflon (product form: pipe lining)

Permeation combines the properties of absorption and transport of diffusion. It is also a function of other physical effects, such as those caused by temperature. In over 20 years of experience with pipe lined with Teflon, the number of failures attributed to permeation of a corrosive vapor followed by corrosion of the support member have been remarkably few. The liner thicknesses of 1.27 to 6.35 mm (0.050 to 0.250 inches) necessary for physical strength at high temperatures, reduce permeation to the point that it is normally a minor consideration.

Quoting Harvey E. Atkinson*: "Because so many variables affect permeation, it is misleading to use laboratory permeability data obtained with thin polymer films as the basis for selection of specific fluoroplastic polymer linings. With few exceptions, differences in permeability among fluoroplastics have little bearing on performance of fabricated piping and equipment. Performance is controlled primarily by design, fabrication and quality control." Hence, the primary concern is usually with absorption, since this is the property most indicative of the serviceability of the fluorocarbon resins in a given chemical environment.

In unbonded linings, it is important that the space between liner and support member be vented to the atmosphere, not only to allow escape of minute quantities of permeant vapors but to prevent expansion of entrapped air from collapsing the liner. Also, these vents are used for quality control testing of lined pipe and as a safety device to indicate leakage in case of liner damage.

Liner collapse is often attributed to permeation when in fact the primary cause is the occurrence of vacuum in the process stream. Manufacturers of lined pipe publish the resistance to vacuum at rated temperature of their different sizes and liner thicknesses, but it is sometimes necessary to prevent excessive vacuum by design features and operating procedures.

*"Fluoroplastic Linings for Corrosive Service," by Harvey E. Atkinson, Principal Materials Consultant, Engineering Department, DuPont Company; Chemical Engineering, December 25, 1972.

Reference: *Teflon - A Performance Guide For The Chemical Processing Industry,* supplier technical report (E-21623-2) - DuPont Company.

Chapter 6

Ethylene-Chlorotrifluoroethylene Copolymer

Permeability to Gases and Water Vapor

Ausimont: Halar

HALAR fluoropolymer has low permeability to water vapor and various other gases. Water vapor permeability measured at 100°F (38°C) and 90% RH was found to be 0.15 g · mil/100 in^2 · 24 hrs.

At elevated surface temperature HALAR fluoropolymer has superior moisture vapor impermeability compared to certain other fluoropolymers at the same conditions.

Reference: *Chemical Resistance of Halar Fluoropolymer,* supplier technical report (AHH) - Ausimont.

<u>**TABLE 16:**</u> **Hydrogen Permeability vs. Temperature and Pressure Through Ausimont Halar Ethylene Chlorotrifluoroethylene Copolymer.**

Material Family	ETHYLENE-CHLOROTRIFLUOROETHYLENE COPOLYMER								
Material Supplier/ Grade	AUSIMONT HALAR								
Reference Number	306	306	306	306	306	306	306	306	306

MATERIAL CHARACTERISTICS

Sample Thickness	0.02 mm	0.02 mm	0.02 mm	0.02 mm	0.02 mm	0.02 mm	0.02 mm	0.02 mm	0.02 mm

TEST CONDITIONS

Penetrant	hydrogen								
Temperature (°C)	-22	25	66	-20	25	67	-21	25	68
Pressure Gradient (kPa)	1724	1724	1724	3447	3447	3447	6895	6895	6895
Test Method	mass spectrometry and calibrated standard gas leaks								
Test Note	developed by McDonnell Douglas Space Systems Company Chemistry Laboratory								

PERMEABILITY (source document units)

Gas Permeability (cm^3 · mm/ cm^2 · kPa · sec)	1.19×10^{-10}	1.21×10^{-9}	6.58×10^{-9}	1.18×10^{-10}	1.25×10^{-9}	6.65×10^{-9}	1.18×10^{-10}	1.23×10^{-9}	6.74×10^{-9}

PERMEABILITY (normalized units)

Permeability Coefficient (cm^3 · mm/m^2 · day · atm)	10.4	106	576	10.3	109	582	10.3	108	590

TABLE 17: Nitrogen Permeability vs. Temperature and Pressure Through Ausimont Halar Ethylene Chlorotrifluoroethylene Copolymer.

Material Family	ETHYLENE-CHLOROTRIFLUOROETHYLENE COPOLYMER								
Material Supplier/ Grade	AUSIMONT HALAR								
Reference Number	306	306	306	306	306	306	306	306	306

MATERIAL CHARACTERISTICS

Sample Thickness	0.02 mm	0.02 mm	0.02 mm	0.02 mm	0.02 mm	0.02 mm	0.02 mm	0.02 mm	0.02 mm

TEST CONDITIONS

Penetrant	nitrogen								
Temperature (°C)	11	25	71	10	25	72	10	25	68
Pressure Gradient (kPa)	1724	1724	1724	3447	3447	3447	6895	6895	6895
Test Method	mass spectrometry and calibrated standard gas leaks; developed by McDonnell Douglas Space Systems Company Chemistry Laboratory								

PERMEABILITY (source document units)

Gas Permeability ($cm^3 \cdot mm/ cm^2 \cdot kPa \cdot sec$)	5.53×10^{-12}	1.29×10^{-11}	2.43×10^{-10}	5.53×10^{-12}	1.49×10^{-11}	4.27×10^{-10}	6.09×10^{-12}	1.43×10^{-11}	2.48×10^{-10}

PERMEABILITY (normalized units)

Permeability Coefficient ($cm^3 \cdot mm/m^2 \cdot day \cdot atm$)	0.48	1.13	21.3	0.48	1.3	37.4	0.53	1.25	21.7

TABLE 18: Oxygen and Ammonia Permeability vs. Temperature and Pressure Through Ausimont Halar Ethylene-Chlorotrifluoroethylene Copolymer.

Material Family	ETHYLENE-CHLOROTRIFLUOROETHYLENE COPOLYMER								
Material Supplier/ Grade	AUSIMONT HALAR								
Reference Number	306	306	306	306	306	306	306	306	306

MATERIAL CHARACTERISTICS

Sample Thickness	0.02 mm	0.02 mm	0.02 mm	0.02 mm	0.02 mm	0.02 mm	0.02 mm	0.02 mm	0.02 mm

TEST CONDITIONS

Penetrant	ammonia			oxygen					
Temperature (°C)	-1	25	65	-18	25	55	-15	25	56
Pressure Gradient (kPa)	965	965	965	1724	1724	1724	3447	3447	3447
Test Method	mass spectrometry and calibrated standard gas leaks; developed by McDonnell Douglas Space Systems Company Chemistry Laboratory								

PERMEABILITY (source document units)

Gas Permeability ($cm^3 \cdot mm/ cm^2 \cdot kPa \cdot sec$)	3.73×10^{-10}	1.29×10^{-9}	7.05×10^{-9}	5.52×10^{-12}	1.16×10^{-10}	5.16×10^{-10}	5.73×10^{-12}	1.1×10^{-10}	5.26×10^{-10}

PERMEABILITY (normalized units)

Permeability Coefficient ($cm^3 \cdot mm/m^2 \cdot day \cdot atm$)	32.6	113	617	0.48	10.2	45.2	0.5	9.6	46.0

ECTFE

GRAPH 04: **Moisture Vapor Permeability Rate vs. Thickness through Ethylene-Chlorotrifluoroethylene Copolymer.**

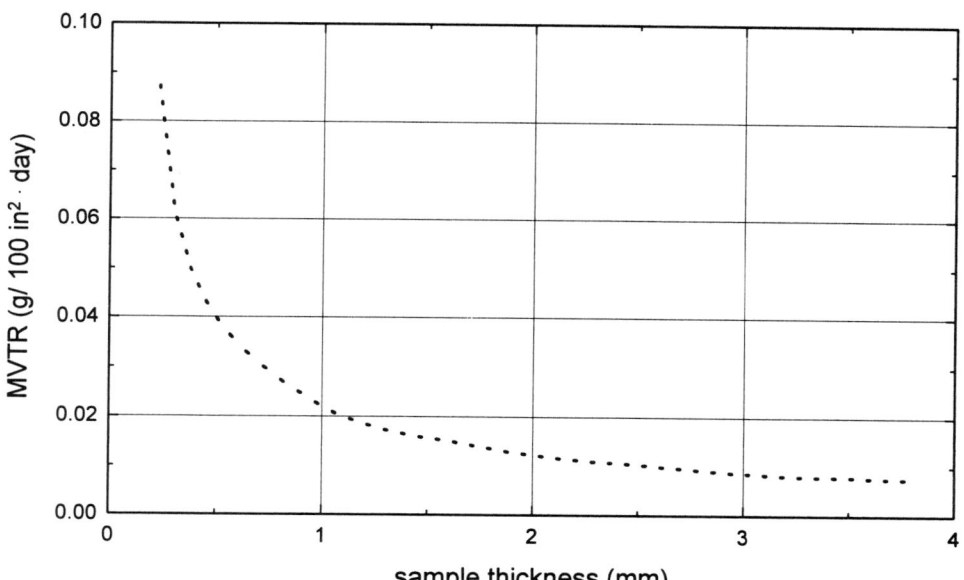

....	Ausimont Halar ECTFE; penetrant: moisture vapor; ΔP=134 mm Hg; 90% RH; 60°C
Reference No.	288

GRAPH 05: **Moisture Vapor Permeability Rate vs. Temperature through Ethylene-Chlorotrifluoroethylene Copolymer.**

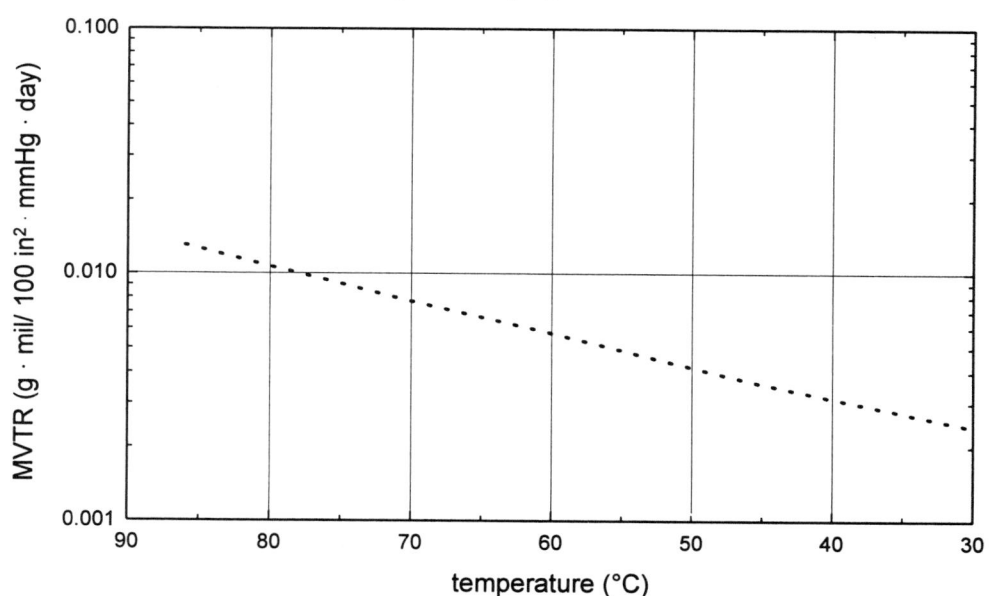

....	Ausimont Halar ECTFE; penetrant: moisture vapor
Reference No.	288

GRAPH 06: Carbon Dioxide and Oxygen Permeability vs. Temperature through Ethylene-Chlorotrifluoroethylene Copolymer.

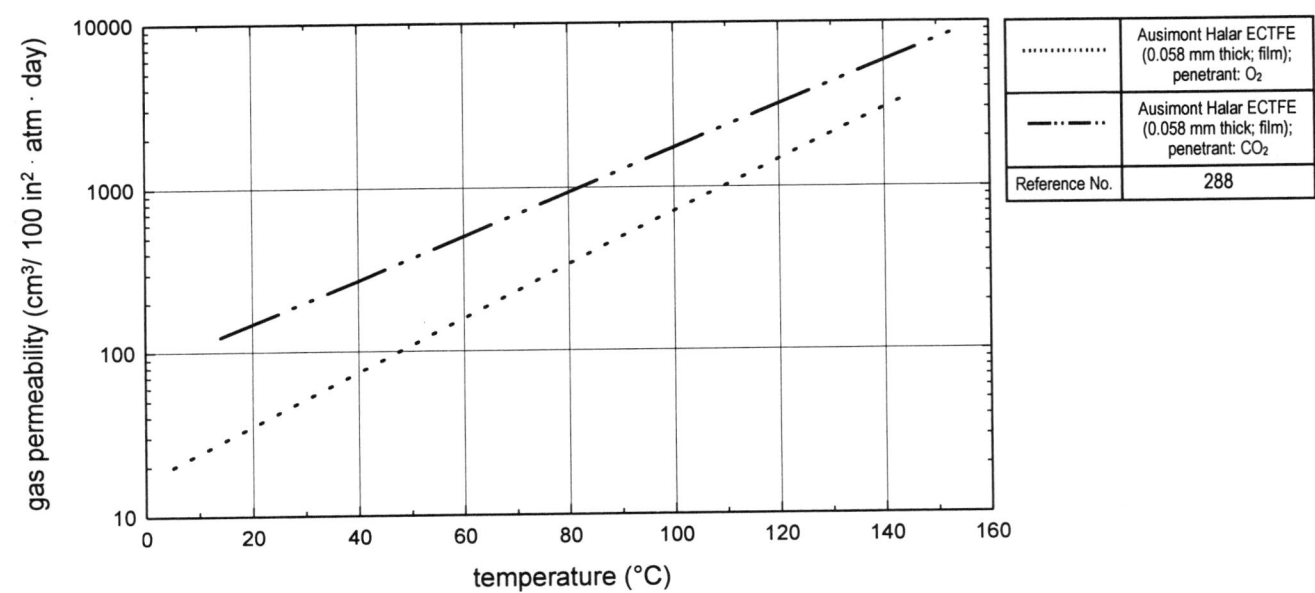

···············	Ausimont Halar ECTFE (0.058 mm thick; film); penetrant: O_2
—··—··—	Ausimont Halar ECTFE (0.058 mm thick; film); penetrant: CO_2
Reference No.	288

GRAPH 07: Nitrogen and Helium Permeability vs. Temperature through Ethylene-Chlorotrifluoroethylene Copolymer.

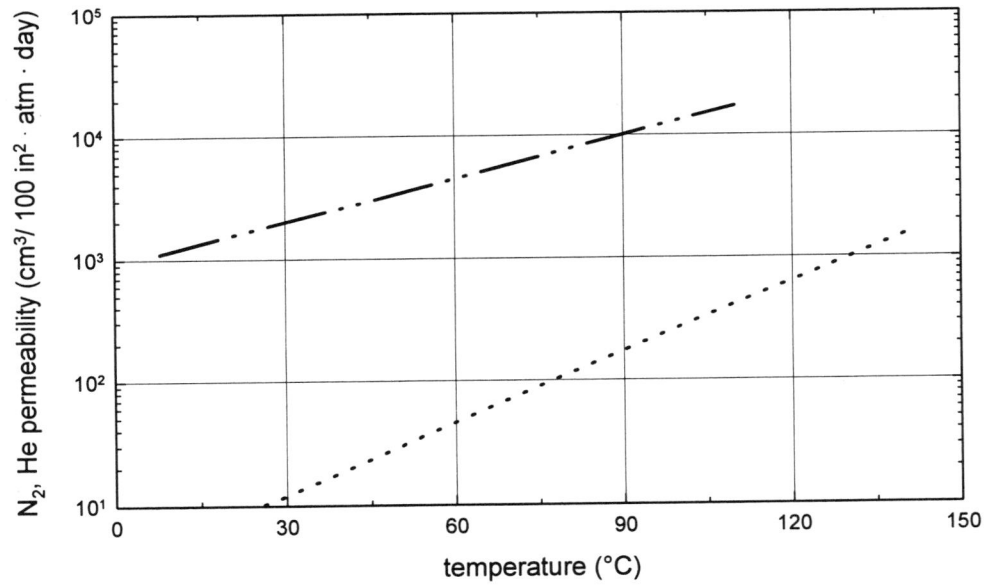

···············	Ausimont Halar ECTFE (0.058 mm thick; film); penetrant: N_2
—··—··—	Ausimont Halar ECTFE (0.058 mm thick; film); penetrant: He
Reference No.	288

ECTFE

<u>GRAPH 08:</u> Gas Permeability vs. Temperature through Ethylene-Chlorotrifluoroethylene Copolymer.

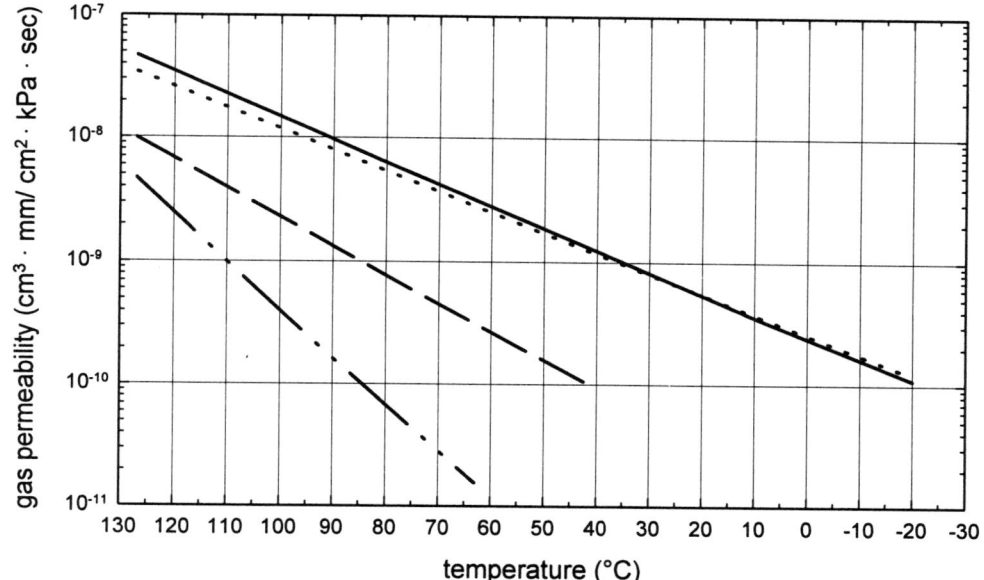

	Ausimont Halar ECTFE (0.02 mm thick); penetrant: H_2
	Ausimont Halar ECTFE (0.02 mm thick); penetrant: N_2
	Ausimont Halar ECTFE (0.02 mm thick); penetrant: O_2
	Ausimont Halar ECTFE (0.02 mm thick); penetrant: NH_3
Reference No.	306

Chapter 7

Ethylene-Tetrafluoroethylene Copolymer

TABLE 19: Carbon Dioxide, Nitrogen, Oxygen, Helium and Water Vapor Permeability Through DuPont

Material Family	ETHYLENE-TETRAFLUOROETHYLENE COPOLYMER				
Material Supplier/ Grade	DUPONT TEFZEL				
Product Form	FILM				
Features	developmental material				
Reference Number	205	205	205	205	205

MATERIAL CHARACTERISTICS

Sample Thickness	0.102 mm				

TEST CONDITIONS

Penetrant	carbon dioxide	nitrogen	oxygen	helium	water vapor
Temperature (°C)	25	25	25	25	25
Test Method	ASTM D1434	ASTM D1434	ASTM D1434	ASTM D1434	ASTM E96

PERMEABILITY (source document units)

	carbon dioxide	nitrogen	oxygen	helium	water vapor
Vapor Transmission Rate (g · mil/100 in^2 · day)					1.65
Gas Permeability (cm^3 · mil/100 in^2 · day)	250	30	100	900	

PERMEABILITY (normalized units)

	carbon dioxide	nitrogen	oxygen	helium	water vapor
Permeability Coefficient (cm^3 · mm/m^2 · day · atm)	98.4	11.8	39.4	354	
Vapor Transmission Rate (g · mm/m^2 · day)					0.65

TABLE 20: Oxygen, Nitrogen, Carbon Dioxide, Methane and Helium Permeability Through Ausimont Hyflon Ethylene Tetrafluoroehtylene Copolymer.

Material Family	ETHYLENE-TETRAFLUOROETHYLENE COPOLYMER									
Material Supplier/ Grade	AUSIMONT HYFLON 700					AUSIMONT HYFLON 800				
Features	high molecular weight					low molecular weight				
Reference Number	114	114	114	114	114	114	114	114	114	114

MATERIAL CHARACTERISTICS

Melt Flow Index	4 grams/10 min.					11 grams/10 min.				

TEST CONDITIONS

Penetrant	oxygen	nitrogen	carbon dioxide	methane	helium	oxygen	nitrogen	carbon dioxide	methane	helium
Temperature (°C)	23	23	23	23	23	23	23	23	23	23
Test Method	ASTM D1434	ASTM D1434	ASTM D1434	ASTM D1434	ASTM D1434	ASTM D1434	ASTM D1434	ASTM D1434	ASTM D1434	ASTM D1434
Test Note	activation energy = 6-8 kcal/mole									

PERMEABILITY (source document units)

Gas Permeability $(cm^3 \cdot mm/m^2 \cdot day \cdot atm)$	62.646	21.67	232.46	7.88	591	62.646	21.67	232.46	7.88	591

PERMEABILITY (normalized units)

Permeability Coefficient $(cm^3 \cdot mm/m^2 \cdot day \cdot atm)$	62.6	21.7	232	7.9	591	62.6	21.7	232	7.9	591

Fluorinated Ethylene-Propylene Copolymer

TABLE 21: Hydrogen Permeability vs. Temperature and Pressure Through DuPont Teflon Fluorinated Ethylene-Propylene Copolymer.

Material Family	FLUORINATED ETHYLENE-PROPYLENE COPOLYMER								
Material Supplier/ Grade	DUPONT TEFLON								
Reference Number	306	306	306	306	306	306	306	306	306

MATERIAL CHARACTERISTICS

Sample Thickness	0.05 mm	0.05 mm	0.05 mm	0.05 mm	0.05 mm	0.05 mm	0.05 mm	0.05 mm	0.05 mm

TEST CONDITIONS

Penetrant	hydrogen								
Temperature (°C)	-15	25	68	-13	25	67	-16	25	67
Pressure Gradient (kPa)	1724	1724	1724	3447	3447	3447	6895	6895	6895
Test Method	mass spectrometry and calibrated standard gas leaks								
Test Note	developed by McDonnell Douglas Space Systems Company Chemistry Laboratory								

PERMEABILITY (source document units)

Gas Permeability $(cm^3 \cdot mm/ cm^2 \cdot kPa \cdot sec)$	9.06×10^{-10}	4.41×10^{-9}	1.87×10^{-8}	9.64×10^{-10}	4.35×10^{-9}	1.77×10^{-8}	8.77×10^{-10}	4.4×10^{-9}	1.8×10^{-8}

PERMEABILITY (normalized units)

Permeability Coefficient $(cm^3 \cdot mm/m^2 \cdot day \cdot atm)$	79.3	386	1637	84.4	381	1550	76.8	385	1576

TABLE 22: Nitrogen Permeability vs. Temperature and Pressure Through DuPont Teflon Fluorinated Ethylene-Propylene Copolymer.

Material Family	FLUORINATED ETHYLENE-PROPYLENE COPOLYMER								
Material Supplier/ Grade	DUPONT TEFLON								
Reference Number	306	306	306	306	306	306	306	306	306

MATERIAL CHARACTERISTICS

Sample Thickness	0.05 mm	0.05 mm	0.05 mm	0.05 mm	0.05 mm	0.05 mm	0.05 mm	0.05 mm	0.05 mm

TEST CONDITIONS

Penetrant	nitrogen								
Temperature (°C)	-9	25	71	-7	25	66	-5	25	68
Pressure Gradient (kPa)	1724	1724	1724	3447	3447	3447	6895	6895	6895
Test Method	mass spectrometry and calibrated standard gas leaks; developed by McDonnell Douglas Space Systems Company Chemistry Laboratory								

PERMEABILITY (source document units)

Gas Permeability $(cm^3 \cdot mm/ cm^2 \cdot kPa \cdot sec)$	5.06×10^{-11}	3.8×10^{-10}	3.79×10^{-9}	5.64×10^{-11}	3.86×10^{-10}	3.85×10^{-9}	6.39×10^{-11}	3.85×10^{-10}	3.8×10^{-9}

PERMEABILITY (normalized units)

Permeability Coefficient $(cm^3 \cdot mm/m^2 \cdot day \cdot atm)$	4.4	33.3	332	4.9	33.8	337	5.6	33.7	333

TABLE 23: Oxygen and Ammonia Permeability vs. Temperature and Pressure Through DuPont Teflon Fluorinated Ethylene-Propylene Copolymer.

Material Family	FLUORINATED ETHYLENE-PROPYLENE COPOLYMER								
Material Supplier/ Grade	DUPONT TEFLON								
Reference Number	306	306	306	306	306	306	306	306	306

MATERIAL CHARACTERISTICS

Sample Thickness	0.05 mm	0.05 mm	0.05 mm	0.05 mm	0.05 mm	0.05 mm	0.05 mm	0.05 mm	0.05 mm

TEST CONDITIONS

Penetrant	ammonia			oxygen					
Temperature (°C)	0	25	66	-16	25	52	-16	25	53
Pressure Gradient (kPa)	965	965	965	1724	1724	1724	3447	3447	3447
Test Method	mass spectrometry and calibrated standard gas leaks; developed by McDonnell Douglas Space Systems Company Chemistry Laboratory								

PERMEABILITY (source document units)

Gas Permeability $(cm^3 \cdot mm/ cm^2 \cdot kPa \cdot sec)$	3.31×10^{-10}	1.15×10^{-9}	6.3×10^{-9}	1.04×10^{-10}	1.33×10^{-9}	5.16×10^{-9}	1.03×10^{-10}	1.15×10^{-9}	5.31×10^{-9}

PERMEABILITY (normalized units)

Permeability Coefficient $(cm^3 \cdot mm/m^2 \cdot day \cdot atm)$	29.0	101	552	9.1	116	452	9.0	101	465

FEP

TABLE 24: Water Vapor, Oxygen, Nitrogen and Carbon Dioxide Permeability Through Fluorinated Ethylene Propylene.

Material Family	FLUORINATED ETHYLENE-PROPYLENE COPOLYMER			
Reference Number	138	138	138	138

TEST CONDITIONS

Penetrant	water vapor	oxygen	nitrogen	carbon dioxide
Temperature (°C)	37.8	25	25	25
Relative Humidity (%)	90			
Test Note		STP conditions		

PERMEABILITY (source document units)

Gas Permeability $(cm^3 \cdot mil/100\ in^2 \cdot day)$		750	320	1670
Vapor Transmission Rate $(g \cdot mil/100\ in^2 \cdot day)$	0.4			

PERMEABILITY (normalized units)

Permeability Coefficient $(cm^3 \cdot mm/m^2 \cdot day \cdot atm)$		295	126	657
Vapor Transmission Rate $(g \cdot mm/m^2 \cdot day)$	0.16			

GRAPH 09: Moisture Vapor Permeability Rate vs. Thickness through Fluorinated Ethylene-Propylene Copolymer.

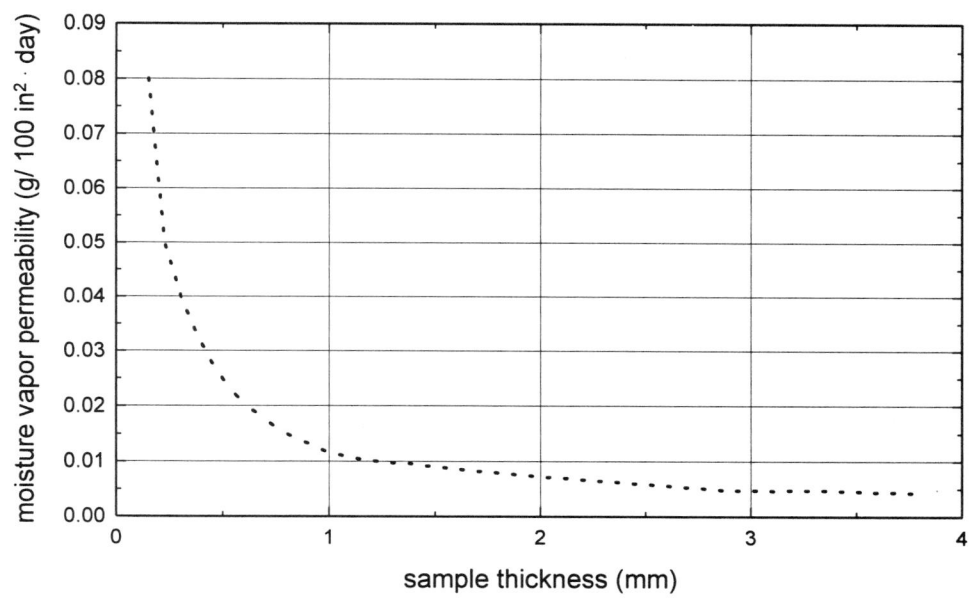

··············	FEP; penetrant: moisture vapor; ΔP=134 mm Hg; 90% RH; 60°C
Reference No.	288

GRAPH 10: **Moisture Vapor Permeability Rate vs. Temperature through Fluorinated Ethylene-Propylene Copolymer.**

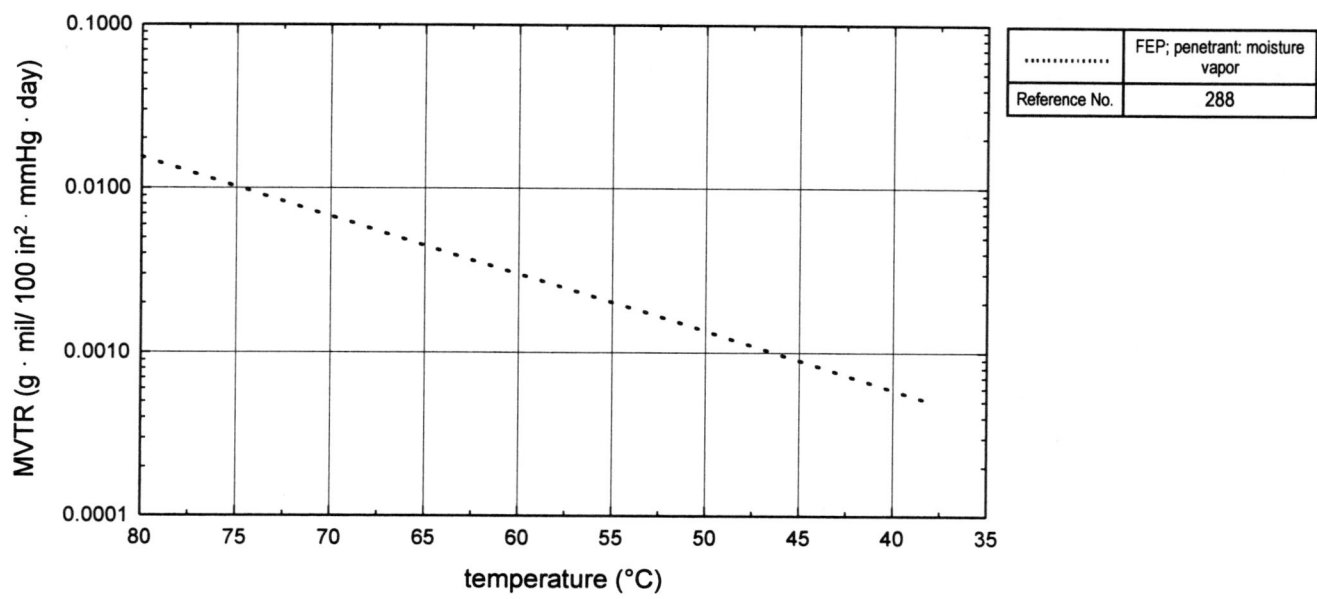

....	FEP; penetrant: moisture vapor
Reference No.	288

GRAPH 11: **Carbon Dioxide and Oxygen Permeability vs. Temperature through Fluorinated Ethylene-Propylene Copolymer.**

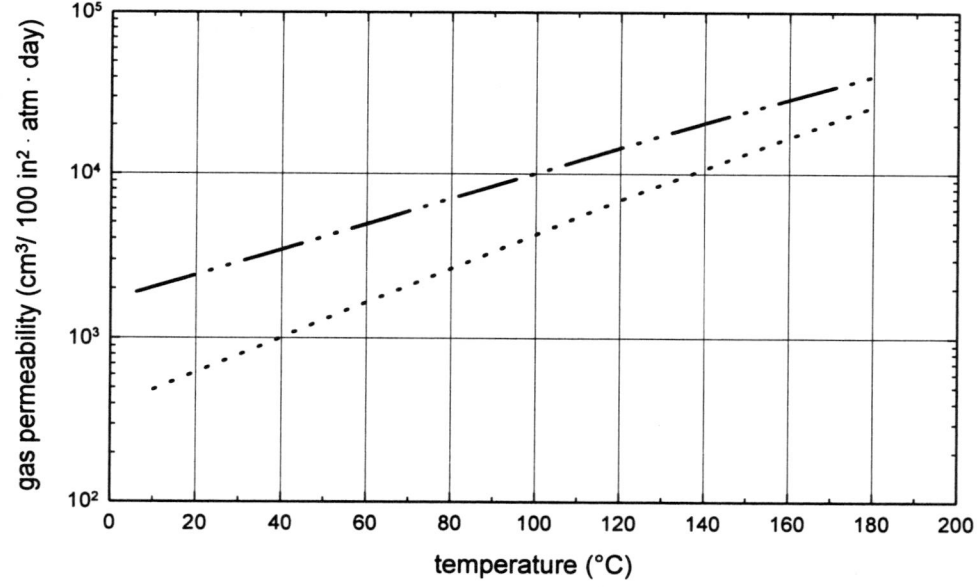

....	FEP (0.051 mm thick; film); penetrant: O_2
—··—··—	FEP (0.048 mm thick; film); penetrant: CO_2
Reference No.	288

GRAPH 12: Nitrogen and Helium Permeability vs. Time After Retort through Fluorinated Ethylene-Propylene Copolymer.

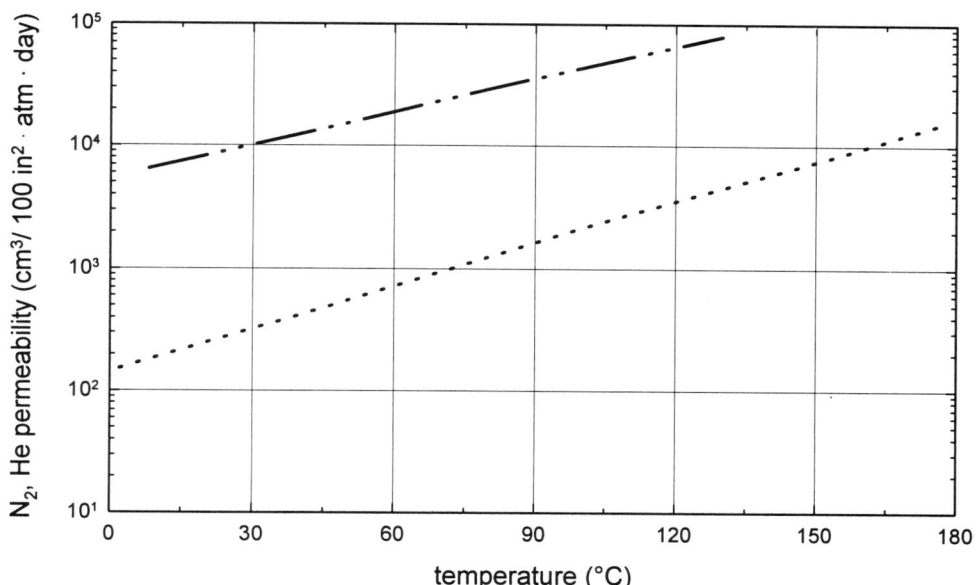

....	FEP (0.048 mm thick; film); penetrant: N₂
—··—··—	FEP (0.048 mm thick; film); penetrant: He
Reference No.	288

GRAPH 13: Gas Permeability vs. Temperature through Fluorinated Ethylene-Propylene Copolymer.

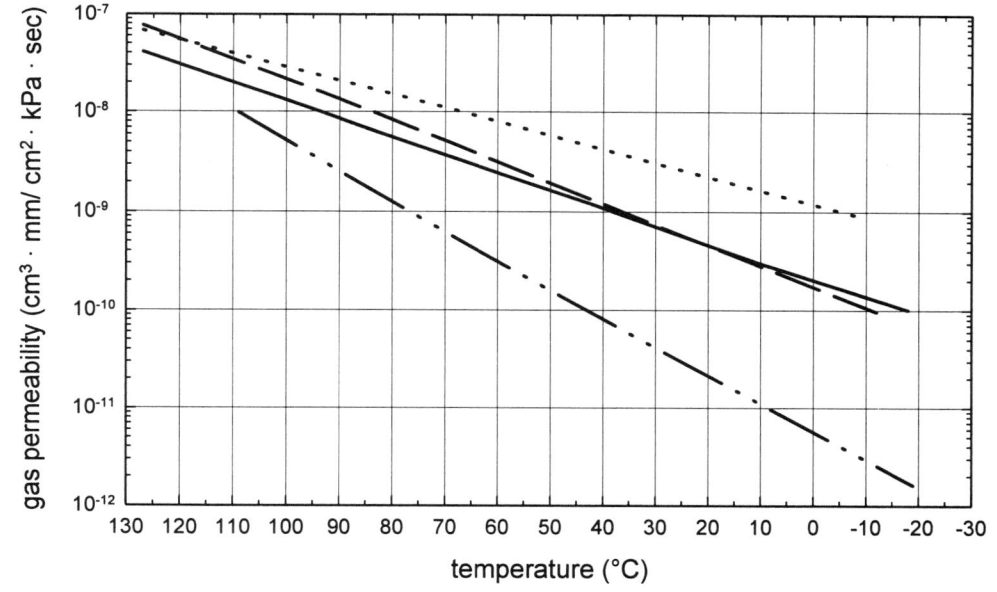

....	DuPont Teflon FEP (0.05 mm thick); penetrant: H₂
—··—··—	DuPont Teflon FEP (0.05 mm thick); penetrant: N₂
— — —	DuPont Teflon FEP (0.05 mm thick); penetrant: O₂
———	DuPont Teflon FEP (0.05 mm thick); penetrant: NH₃
Reference No.	306

Chapter 9

Fluorinated Polyethylene

Permeability to Oxygen

FPE (features: barrier properties; product form: bottles)

Fluorinated bottles do not provide any improvement in oxygen barrier over untreated straight HDPE bottles.

Reference: *Selar RB Barrier Resins - Resin Blend Technical Information,* supplier technical report (H-42016) - DuPont Company, 1992.

TABLE 25: Cyclohexanone, Chlorobenzene, Hexane, Butyl Alcohol, Trichloroethene, Methyl Salicylate and Tetrahydrofuran Permeability Through Fluorinated Polyethylene Bottles.

Material Family	FLUORINATED POLYETHYLENE						
Product Form	BOTTLES						
Features	barrier properties	barrier properties	barrier properties	barrier properties	barrier properties	barrier properties	barrier properties
Reference Number	293	293	293	293	293	293	293

TEST CONDITIONS

Penetrant	cyclohexanone	chlorobenzene	hexane	butyl alcohol	trichloroethene	methyl salicylate	tetrahydrofuran
Temperature (°C)	50	50	50	50	50	50	23
Exposure Time (days)	28	28	28	28	28	28	180

PERMEABILITY (source document units)

Penetrant Weight Loss (%)	0.17	0.65	0.42	0.1	0.58	0.03	8.89

TABLE 26: Ethyl Acetate, Isopropyl Acetate, Acetone, Butyl Acetate, Toluene, Xylene, Methyl Isobutyl Ketone and Methyl Ethyl Ketone Permeability Through Fluorinated Polyethylene Bottles.

Material Family	FLUORINATED POLYETHYLENE							
Product Form	BOTTLES							
Features	barrier properties	barrier properties	barrier properties	barrier properties	barrier properties	barrier properties	barrier properties	barrier properties
Reference Number	293	293	293	293	293	293	293	293

TEST CONDITIONS

Penetrant	ethyl acetate	isopropyl acetate	acetone	butyl acetate	toluene	xylene	methyl isobutyl ketone	methyl ethyl ketone
Temperature (°C)	50	50	23	50	50	50	50	50
Exposure Time (days)	28	28	180	28	28	28	28	28

PERMEABILITY (source document units)

Penetrant Weight Loss (%)	2.7	0.62	0.69	1.0	0.6	0.21	0.56	2.7

© Plastics Design Library

FPE

TABLE 27: Kerosine, d-Limonene, 2-Cycle Motor Oil, Pine Oil Cleaner, Diesel Fuel Conditioner and Brakleen Gas Additive Permeability Through Fluorinated Polyethylene Bottles.

Material Family	FLUORINATED POLYETHYLENE					
Product Form	BOTTLES					
Features	barrier properties	barrier properties	barrier properties	barrier properties	barrier properties	barrier properties
Reference Number	293	293	293	293	293	293

TEST CONDITIONS

Penetrant	kerosine	d-limonene	motor oils	pine oil	diesel fuel conditioner	gas additive
Penetrant Note			2 cycle	cleaner		Brakleen
Temperature (°C)	50	50	50	50	50	50
Exposure Time (days)	28	28	28	28	28	28

PERMEABILITY (source document units)

Penetrant Weight Loss (%)	0.04	0.11	0.07	0.10	0.08	0.06

TABLE 28: Mineral Spirits, Turpentine, STP Gas Treatment, Paint Thinner, Charcoal Starter and Naphtha Permeability Through Fluorinated Polyethylene Bottles.

Material Family	FLUORINATED POLYETHYLENE					
Product Form	BOTTLES					
Features	barrier properties	barrier properties	barrier properties	barrier properties	barrier properties	barrier properties
Reference Number	293	293	293	293	293	293

TEST CONDITIONS

Penetrant	mineral spirits	turpentine	STP gas treatment	paint thinner	charcoal starter	naphtha
Temperature (°C)	50	50	50	50	50	50
Exposure Time (days)	28	28	28	28	28	28

PERMEABILITY (source document units)

Penetrant Weight Loss (%)	0.02	0.06	0.12	0.08	0.03	0.06

Chapter 10

Perfluoroalkoxy Resin

Permeability to Gases

DuPont: Teflon PFA (product form: film)

The permeation of gases through thin film (0.08-0.13 mm) is dependant on the molecular size, shape, wettability and soundness of the fabricated membrane.

Attempts have been made to relate permeation rates through thin films to absorption of thicker films, sheets, tubes, pipe, etc. This has been generally unsuccessful. Thicker films and sheets represent an average set of properties obtainable from many thin films produced under a variety of conditions. To produce a thin film representative of this average is impossible from a practical viewpoint. Since permeation in well-fabricated articles is essentially a molecular transport phenomenon through fluorocarbon chains, it is affected by orientation, degree of crystallinity and temperature.

However, comparative data on identical tests can be used to predict performance in many thin film and coating applications.

Increased permeability with temperature parallels the decrease in specific gravity with increased temperature in the resin. This corresponds with increased spacing between molecules and increasing molecular activity which allows easier diffusion of the gas through the specimen.

Reference: *Handbook Of Properties For Teflon PFA*, supplier design guide (E-96679) - Du Pont Company, 1987.

TABLE 29: Gas Permeability of Oxygen, Carbon Dioxide and Nitrogen Through DuPont Company Teflon PFA Perfluoroalkoxy Film.

Material Family	PERFLUOROALKOXY RESIN		
Material Supplier/ Grade	DUPONT TEFLON PFA		
Product Form	FILM		
Reference Number	39	39	39

TEST CONDITIONS

Penetrant	carbon dioxide	nitrogen	oxygen
Temperature (°C)	25	25	25
Test Method	ASTM D1434	ASTM D1434	ASTM D1434

PERMEABILITY (source document units)

Gas Permeability (cm^3 · mil/100 in^2 · day)	2260	291	881
Gas Permeability (cm^3 · mm/m^2 · day · Pa)	0.00878	0.00113	0.00342

PERMEABILITY (normalized units)

Permeability Coefficient (cm^3 · mm/m^2 · day · atm)	890	115	347

Chapter 11

Polychlorotrifluoroethylene

Permeability

Allied Signal: Aclar (features: transparent; product form: film)

ACLAR has an outstanding ability to prevent the passage of water vapor and liquids. This means that ACLAR provides product protection and, because of its transparency, permits inspection viewing of the product at the same time. These combined properties have lead to new product designs for moisture-sensitive items.

Reference: *Aclar Performance Films,* supplier technical report (SFI-14 Rev. 9-89) - Allied-Signal Enineered Plastics, 1989.

Film Properties and Applications

Allied Signal: Aclar (features: transparent; product form: film)

ACLAR is a flexible thermoplastic film made from fluorinated-chlorinated resins. The film is produced in three basic types ACLAR 22, ACLAR 88 and ACLAR 33. These products are copolymers, and a terpolymer, respectively and consist primarily of CTFE. The presence of fluorine and chlorine in the structure leads to a clear, heat-sealable film with excellent thermal and chemical stability.

ACLAR can be heat-sealed, laminated, printed, thermoformed, metallized, and sterilized. The unsupported and laminated varieties can be handled and processed on most common converting and packaging machinery.

Reference: *Aclar Performance Films,* supplier technical report (SFI-14 Rev. 9-89) - Allied-Signal Enineered Plastics, 1989.

Allied Signal: Aclar 22A (features: transparent; product form: film)

ACLAR 22A is a copolymer film consisting primarily of CTFE. It is used primarily for pharmaceutical packaging applications. ACLAR 22A film thermoforms at a lower temperature than ACLAR 33C and may be formed on a vacuum forming machine. ACLAR 22A provides excellent moisture barier properties. Standard product thicknesses include 0.0381 mm and 0.127 mm.

Reference: *Aclar Performance Films,* supplier technical report (SFI-14 Rev. 9-89) - Allied-Signal Enineered Plastics, 1989.

Allied Signal: Aclar 22C (features: transparent; product form: film)

ACLAR 22C is a copolymer film consisting primarily of CTFE. It is used primarily as an encapsulating film for electroluminescent lamps and for clean room packaging. Standard product thicknesses include 0.051 mm.

Reference: *Aclar Performance Films,* supplier technical report (SFI-14 Rev. 9-89) - Allied-Signal Enineered Plastics, 1989.

Allied Signal: Aclar 33C (features: transparent; product form: film)

ACLAR 33C is a terpolymer film consisting primarily of CTFE. It is used for military and pharmaceutical packaging applications. It thermoforms satisfactorily on equipment having a pre-heat station and a plug assist system. ACLAR 33C provides superior moisture barrier properties. It is available in standard thicknesses of 0.019 mm, 0.025 mm, 0.051 mm, 0.076 mm, 0.127 mm and 0.19 mm.

Reference: *Aclar Performance Films,* supplier technical report (SFI-14 Rev. 9-89) - Allied-Signal Enineered Plastics, 1989.

Allied Signal: Aclar 88A (features: transparent; product form: film)

ACLAR 88A is a copolymer film consisting primarily of CTFE. It is used for pharmaceutical packaging applications. This product thermoforms at the same temperature as ACLAR 22A on a vacuum forming machine. It is available in a thickness of 0.019 mm.

Reference: *Aclar Performance Films,* supplier technical report (SFI-14 Rev. 9-89) - Allied-Signal Enineered Plastics, 1989.

TABLE 30: Carbon Dioxide, Hydrogen and Hydrogen Sulfide Permeability Through 3M Kel-F 81 Polychlorotrifluoroethylene Film.

Material Family	POLYCHLOROTRIFLUOROETHYLENE								
Material Supplier/ Grade	3M KEL-F 81								
Product Form	FILM								
Reference Number	96	96	96	96	96	96	96	96	96

MATERIAL COMPOSITION

Note	amorphous form of polymer

TEST CONDITIONS

Penetrant	carbon dioxide				hydrogen			hydrogen sulfide	
Temperature (°C)	0	25	50	75	0	25	50	50	75

PERMEABILITY (source document units)

Gas Permeability $(1 \times 10^{-10}\ cm^3 \cdot mm / cm^2 \cdot sec \cdot cm\ Hg)$	0.35	1.4	2.4	15	3.2	9.8	24	0.35	2.0

PERMEABILITY (normalized units)

Permeability Coefficient $(cm^3 \cdot mm/m^2 \cdot day \cdot atm)$	2.3	9.2	15.8	98.5	21.0	64.3	158	2.3	13.1

TABLE 31: Hydrogen Permeability vs. Temperature and Pressure Through 3M Kel-F
Polychlorotrifluoroethylene.

Material Family	POLYCHLOROTRIFLUOROETHYLENE								
Material Supplier/ Grade	3M KEL-F								
Reference Number	306	306	306	306	306	306	306	306	306

MATERIAL CHARACTERISTICS

Sample Thickness	0.01 mm	0.01 mm	0.01 mm	0.01 mm	0.01 mm	0.01 mm	0.01 mm	0.01 mm	0.01 mm

TEST CONDITIONS

Penetrant	hydrogen								
Temperature (°C)	-15	25	68	-12	25	67	-16	25	70
Pressure Gradient (kPa)	1724	1724	1724	3447	3447	3447	6895	6895	6895
Test Method	mass spectrometry and calibrated standard gas leaks; developed by McDonnell Douglas Space Systems Company Chemistry Laboratory								

PERMEABILITY (source document units)

Gas Permeability $(cm^3 \cdot mm/ cm^2 \cdot kPa \cdot sec)$	6.39×10^{-11}	4.07×10^{-10}	2.33×10^{-9}	6.69×10^{-11}	4.13×10^{-10}	2.25×10^{-9}	5.77×10^{-11}	4.14×10^{-10}	2.49×10^{-9}

PERMEABILITY (normalized units)

Permeability Coefficient $(cm^3 \cdot mm/m^2 \cdot day \cdot atm)$	5.6	35.6	204	5.9	36.2	197	5.1	36.2	218

TABLE 32: Nitrogen Permeability vs. Temperature and Pressure Through 3M Kel-F
Polychlorotrifluoroethylene.

Material Family	POLYCHLOROTRIFLUOROETHYLENE					
Material Supplier/ Grade	3M KEL-F					
Reference Number	306	306	306	306	306	306

MATERIAL CHARACTERISTICS

Sample Thickness	0.01 mm	0.01 mm	0.01 mm	0.01 mm	0.01 mm	0.01 mm

TEST CONDITIONS

Penetrant	nitrogen					
Temperature (°C)	25	68	25	69	25	70
Pressure Gradient (kPa)	1724	1724	3447	3447	6895	6895
Test Method	mass spectrometry and calibrated standard gas leaks; developed by McDonnell Douglas Space Systems Company Chemistry Laboratory					

PERMEABILITY (source document units)

Gas Permeability $(cm^3 \cdot mm/ cm^2 \cdot kPa \cdot sec)$	1.77×10^{-13}	4.15×10^{-11}	1.77×10^{-13}	4.36×10^{-11}	1.77×10^{-13}	4.45×10^{-11}

PERMEABILITY (normalized units)

Permeability Coefficient $(cm^3 \cdot mm/m^2 \cdot day \cdot atm)$	0.02	3.63	0.02	3.82	0.02	3.9

TABLE 33: Oxygen and Ammonia Permeability vs. Temperature and Pressure Through 3M Kel-F Polychlorotrifluoroethylene.

Material Family	POLYCHLOROTRIFLUOROETHYLENE					
Material Supplier/ Grade	3M KEL-F					
Reference Number	306	306	306	306	306	306

MATERIAL CHARACTERISTICS

Sample Thickness	0.01 mm	0.01 mm	0.01 mm	0.01 mm	0.01 mm	0.01 mm

TEST CONDITIONS

Penetrant	ammonia		oxygen			
Temperature (°C)	25	59	25	52	25	52
Pressure Gradient (kPa)	965	965	1724	1724	3447	3447
Test Method	mass spectrometry and calibrated standard gas leaks; developed by McDonnell Douglas Space Systems Company Chemistry Laboratory					

PERMEABILITY (source document units)

Gas Permeability $(cm^3 \cdot mm/ cm^2 \cdot kPa \cdot sec)$	1.2×10^{-11}	2.76×10^{-10}	2.95×10^{-12}	9.42×10^{-11}	2.84×10^{-12}	9.42×10^{-11}

PERMEABILITY (normalized units)

Permeability Coefficient $(cm^3 \cdot mm/m^2 \cdot day \cdot atm)$	1.05	24.2	0.26	8.2	0.25	8.25

TABLE 34: Oxygen, Nitrogen and Carbon Dioxide Permeability Through Allied Signal Aclar Polychlorotrifluoroethylene Film.

Material Family	POLYCHLOROTRIFLUOROETHYLENE							
Material Supplier/ Grade	ALLIED SIGNAL ACLAR							
Material Supplier/ Grade	33C		22C			22A		
Product Form	FILM							
Features	transparent	transparent	transparent	transparent	transparent	transparent	transparent	transparent
Reference Number	138	138	138	138	138	138	138	138

TEST CONDITIONS

Penetrant	oxygen	carbon dioxide	oxygen	nitrogen	carbon dioxide	oxygen	nitrogen	carbon dioxide
Temperature (°C)	25	25	25	25	25	25	25	25
Test Note	STP conditions							

PERMEABILITY (source document units)

Gas Permeability $(cm^3 \cdot mil/100 in^2 \cdot day)$	7	16	15	2.5	40	12	2.5	30

PERMEABILITY (normalized units)

Permeability Coefficient $(cm^3 \cdot mm/m^2 \cdot day \cdot atm)$	2.8	6.3	5.9	1.0	15.7	4.7	1.0	11.8

TABLE 35: Oxygen, Nitrogen and Helium Permeability Through 3M Kel-F 81 PCTFE Film.

Material Family	POLYCHLOROTRIFLUOROETHYLENE							
Material Supplier/ Grade	3M KEL-F 81							
Product Form	FILM							
Reference Number	96	96	96	96	96	96	96	96

MATERIAL COMPOSITION

Note	amorphous form of polymer

TEST CONDITIONS

Penetrant	nitrogen			helium	oxygen			
Temperature (°C)	25	50	75	25	0	25	50	75

PERMEABILITY (source document units)

Gas Permeability $(1 \times 10^{-10}$ cm$^3 \cdot$ mm / cm$^2 \cdot$ sec $\cdot$ cm Hg)	0.05	0.30	0.91	21.7	0.07	0.40	1.40	5.70

PERMEABILITY (normalized units)

Permeability Coefficient (cm$^3 \cdot$ mm/m$^2 \cdot$ day $\cdot$ atm)	0.33	1.97	5.98	142.5	0.46	2.63	9.19	37.43

TABLE 36: Water Vapor Permeability Through Allied Signal Aclar Polychlorotrifluoroethylene Film.

Material Family	POLYCHLOROTRIFLUOROETHYLENE						
Material Supplier/ Trade Name	ALLIED SIGNAL ACLAR						
Grade	33C		22C			22A	88A
Product Form	TRANSPARENT FILM						
Reference Number	138	138	138	138	138	138	138

MATERIAL CHARACTERISTICS

Sample Thickness	0.019 mm	0.051 mm	0.0254 mm	0.051 mm	0.19 mm	0.038 mm	0.019 mm

TEST CONDITIONS

Penetrant	water vapor						
Temperature (°C)	37.8	37.8	37.8	37.8	37.8	37.8	37.8
Relative Humidity (%)	90	90	90	90	90	90	90
Test Method	ASTM E96, method E; measured on sealed pouches						

PERMEABILITY (source document units)

Vapor Transmission Rate (g/m$^2 \cdot$ day)	0.43 - 0.59	0.15 - 0.31	0.47 - 0.93	0.24 - 0.62	0.09 - 0.13	0.32 - 0.62	0.70 - 0.86
Vapor Transmission Rate (g/day $\cdot$ 100 in^2)	0.028 - 0.038	0.010 - 0.020	0.030 - 0.060	0.016 - 0.040	0.006 - 0.007	0.020 - 0.040	0.045 - 0.055

PERMEABILITY (normalized units)

Vapor Transmission Rate (g $\cdot$ mm/m$^2 \cdot$ day)	0.008 - 0.011	0.0077 - 0.0158	0.0119 - 0.0236	0.0122 - 0.0316	0.0171 - 0.0247	0.0122 - 0.0236	0.0133 - 0.0163

TABLE 37: Water Vapor Transmission Through 3M Kel-F 81 Polychlorotrifluoroethylene Film.

Material Family	POLYCHLOROTRIFLUOROETHYLENE			
Material Supplier/ Grade	3M KEL-F 81			
Product Form	FILM			
Reference Number	96	96	96	96

MATERIAL COMPOSITION

Note	amorphous form of polymer			

TEST CONDITIONS

Penetrant	water vapor			
Temperature (°C)	25	50	75	100

PERMEABILITY (source document units)

Gas Permeability $(1 \times 10^{-10}\ cm^3 \cdot mm / cm^2 \cdot sec \cdot cm\ Hg)$	1	10	28	100
Vapor Transmission Rate $(g \cdot mil/ m^2 \cdot atm \cdot day)$	0.19	1.76	4.56	15.20

PERMEABILITY (normalized units)

Permeability Coefficient $(cm^3 \cdot mm/m^2 \cdot day \cdot atm)$	6.57	65.7	184	657
Vapor Transmission Rate $(g \cdot mm/ m^2 \cdot day)$	0.005	0.043	0.116	0.386

CTFE

TABLE 38: Film Properties of Allied Signal Aclar Polychlorotrifluoroethylene Film.

Material Family	POLYCHLOROTRIFLUOROETHYLENE			
Material Supplier/ Trade Name	ALLIED SIGNAL ACLAR			
Grade	88A	22A	22C	33C
Product Form	FILM			
Features	transparent	transparent	transparent	transparent
Reference Number	138	138	138	138

MATERIAL CHARACTERISTICS

Sample Thickness	0.019 mm	0.038 mm	0.19 mm	0.019 mm
Specific Gravity	2.10 {bs}	2.10 {bs}	2.11 {bs}	2.12 {bs}
Yield (in²/lb)	17,760	8880	1750	17,360

MECHANICAL PROPERTIES

Modulus Of Elasticity - MD (MPa)	970-1100 {r}	970-1100 {r}	900-1100 {r}	1310-1380 {r}
Modulus Of Elasticity - TD (MPa)	837-970 {r}	1040-1100 {r}	900-1100 {r}	1310-1380 {r}
Tensile Strength - MD (MPa)	48.3-69 {r}	51.7-75.9 {r}	27.6-41.4 {r}	65.5-79.3 {r}
Tensile Strength - TD (MPa)	27.6-41.4 {r}	37.9-55.2 {r}	27.6-41.4 {r}	37.9-55.2 {r}
Ultimate Elongation - MD (%)	150-250 {r}	115-225 {r}	200-300 {r}	50-150 {r}
Ultimate Elongation - TD (%)	200-300 {r}	200-300 {r}	200-300 {r}	50-150 {r}
Drop Dart Impact Strength (g)	93 {cc}	347 {cc}	1200 {dk}	<57 {cc}
Tear Strength, Propagated, Elmendorf - MD (g)	24 {k}	40 {k}	>1600 {k}	12 {k}
Tear Strength, Propagated, Elmendorf - transverse (g)	24 {k}	130 {k}	>1600 {k}	33 {k}
Tear Strength, Graves - MD (g)	240 {az}	380 {az}	465 {az}	400 {az}
Tear Strength, Graves - TD (g)	300 {az}	360 {az}	415 {az}	380 {az}
Abrasion Resistance - weight loss (mg)	14 {cd}	16 {cd}	7 {cd}	15 {ce}

THERMAL PROPERTIES

Melting Point (°C)	183-186	183-186	183-186	202-204
Thermal Conductivity (cal-cm/cm²-sec-°C)	5.3 x 10⁻⁴	5.3 x 10⁻⁴	5.3 x 10⁻⁴	4.7 x 10⁻⁴

OPTICAL PROPERTIES

Haze (%)	<1 {c}	<1 {c}	<1 {c}	<1 {c}

PERFORMANCE PROPERTIES

Unrestrained Shrink - 149°C Air, MD (%)	12-15 {bw}	12-15 {bw}	≤ 2 {bw}	≤ 2.5 {bw}
Unrestrained Shrink - 149°C Air, TD (%)	-12 to -15 {bw}	-12 to -15 {bw}	≤ 2 {bw}	≤ 2.5 {bw}

GRAPH 14: Gas Permeability vs. Temperature through Polychlorotrifluoroethylene.

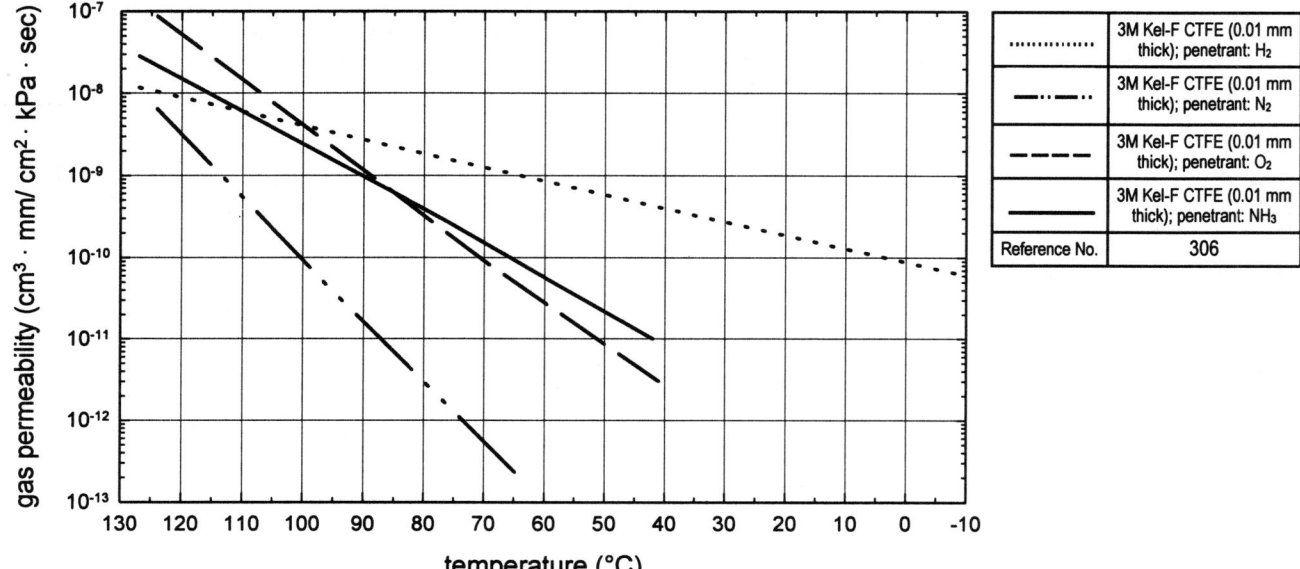

·················	3M Kel-F CTFE (0.01 mm thick); penetrant: H_2
— ·· — ··	3M Kel-F CTFE (0.01 mm thick); penetrant: N_2
— — —	3M Kel-F CTFE (0.01 mm thick); penetrant: O_2
———	3M Kel-F CTFE (0.01 mm thick); penetrant: NH_3
Reference No.	306

Chapter 12

Polytetrafluoroethylene

<u>TABLE 39:</u> **Hydrogen Permeability vs. Temperature and Pressure Through Carbon Filled DuPont Teflon Polytetrafluoroethylene.**

Material Family	POLYTETRAFLUOROETHYLENE								
Material Supplier/ Grade	DUPONT TEFLON								
Reference Number	306	306	306	306	306	306	306	306	306

MATERIAL CHARACTERISTICS

Sample Thickness	0.05 mm	0.05 mm	0.05 mm	0.05 mm	0.05 mm	0.05 mm	0.05 mm	0.05 mm	0.05 mm

MATERIAL COMPOSITION

Note	carbon filled	carbon filled	carbon filled	carbon filled	carbon filled	carbon filled	carbon filled	carbon filled	carbon filled

TEST CONDITIONS

Penetrant	hydrogen								
Temperature (°C)	-15	25	68	-11	25	67	-14	25	65
Pressure Gradient (kPa)	1724	1724	1724	3447	3447	3447	6895	6895	6895
Test Method	mass spectrometry and calibrated standard gas leaks; developed by McDonnell Douglas Space Systems Company Chemistry Laboratory								

PERMEABILITY (source document units)

Gas Permeability ($cm^3 \cdot mm/ cm^2 \cdot kPa \cdot sec$)	3.95×10^{-9}	1.34×10^{-8}	3.53×10^{-8}	4.51×10^{-9}	1.27×10^{-8}	3.42×10^{-8}	4.17×10^{-9}	1.23×10^{-8}	3.32×10^{-8}

PERMEABILITY (normalized units)

Permeability Coefficient ($cm^3 \cdot mm/m^2 \cdot day \cdot atm$)	346	1173	3090	395	1112	2994	365	1077	2906

TABLE 40: Hydrogen Permeability vs. Temperature and Pressure Through DuPont Teflon Polytetrafluoroethylene.

Material Family	POLYTETRAFLUOROETHYLENE								
Material Supplier/ Grade	DUPONT TEFLON								
Reference Number	306	306	306	306	306	306	306	306	306

MATERIAL CHARACTERISTICS

Sample Thickness	0.03 mm	0.03 mm	0.03 mm	0.03 mm	0.03 mm	0.03 mm	0.03 mm	0.03 mm	0.03 mm

TEST CONDITIONS

Penetrant	hydrogen								
Temperature (°C)	-16	25	68	-17	25	67	-18	25	63
Pressure Gradient (kPa)	1724	1724	1724	3447	3447	3447	6895	6895	6895
Test Method	mass spectrometry and calibrated standard gas leaks; developed by McDonnell Douglas Space Systems Company Chemistry Laboratory								

PERMEABILITY (source document units)

Gas Permeability $(cm^3 \cdot mm/ cm^2 \cdot kPa \cdot sec)$	1.7×10^{-9}	6.34×10^{-9}	1.88×10^{-8}	1.63×10^{-9}	5.9×10^{-9}	1.86×10^{-8}	1.59×10^{-9}	5.94×10^{-9}	1.64×10^{-8}

PERMEABILITY (normalized units)

Permeability Coefficient $(cm^3 \cdot mm/m^2 \cdot day \cdot atm)$	149	555	1646	143	516	1628	139	520	1436

TABLE 41: Nitrogen Permeability vs. Temperature and Pressure Through DuPont Teflon Polytetrafluoroethylene.

Material Family	POLYTETRAFLUOROETHYLENE								
Material Supplier/ Grade	DUPONT TEFLON								
Reference Number	306	306	306	306	306	306	306	306	306

MATERIAL CHARACTERISTICS

Sample Thickness	0.03 mm	0.03 mm	0.03 mm	0.03 mm	0.03 mm	0.03 mm	0.03 mm	0.03 mm	0.03 mm

TEST CONDITIONS

Penetrant	nitrogen								
Temperature (°C)	-23	25	71	-25	25	70	-23	25	68
Pressure Gradient (kPa)	1724	1724	1724	3447	3447	3447	6895	6895	6895
Test Method	mass spectrometry and calibrated standard gas leaks; developed by McDonnell Douglas Space Systems Company Chemistry Laboratory								

PERMEABILITY (source document units)

Gas Permeability $(cm^3 \cdot mm/ cm^2 \cdot kPa \cdot sec)$	9.46×10^{-11}	7.87×10^{-10}	2.9×10^{-9}	8.89×10^{-11}	7.88×10^{-10}	2.89×10^{-9}	9.47×10^{-11}	7.84×10^{-10}	2.87×10^{-9}

PERMEABILITY (normalized units)

Permeability Coefficient $(cm^3 \cdot mm/m^2 \cdot day \cdot atm)$	8.3	68.9	254	7.8	69	253	8.3	68.6	251

TABLE 42: Nitrogen Permeability vs. Temperature and Pressure Through Carbon Filled DuPont Teflon Polytetrafluoroethylene.

Material Family	POLYTETRAFLUOROETHYLENE								
Material Supplier/ Grade	DUPONT TEFLON								
Reference Number	306	306	306	306	306	306	306	306	306

MATERIAL CHARACTERISTICS

Sample Thickness	0.05 mm	0.05 mm	0.05 mm	0.05 mm	0.05 mm	0.05 mm	0.05 mm	0.05 mm	0.05 mm

MATERIAL COMPOSITION

Note	carbon filled	carbon filled	carbon filled	carbon filled	carbon filled	carbon filled	carbon filled	carbon filled	carbon filled

TEST CONDITIONS

Penetrant	nitrogen								
Temperature (°C)	-14	25	68	-17	25	71	-17	25	67
Pressure Gradient (kPa)	1724	1724	1724	3447	3447	3447	6895	6895	6895
Test Method	mass spectrometry and calibrated standard gas leaks; developed by McDonnell Douglas Space Systems Company Chemistry Laboratory								

PERMEABILITY (source document units)

Gas Permeability ($cm^3 \cdot mm/ cm^2 \cdot kPa \cdot sec$)	2.5×10^{-10}	1.46×10^{-9}	5.28×10^{-9}	2.34×10^{-10}	1.52×10^{-9}	5.32×10^{-9}	2.34×10^{-10}	1.42×10^{-9}	4.78×10^{-9}

PERMEABILITY (normalized units)

Permeability Coefficient ($cm^3 \cdot mm/m^2 \cdot day \cdot atm$)	21.9	128	462	20.5	133	466	20.5	124	418

TABLE 43: Oxygen and Ammonia Permeability vs. Temperature and Pressure Through DuPont Teflon Polytetrafluoroethylene.

Material Family	POLYTETRAFLUOROETHYLENE								
Material Supplier/ Grade	DUPONT TEFLON								
Reference Number	306	306	306	306	306	306	306	306	306

MATERIAL CHARACTERISTICS

Sample Thickness	0.03 mm	0.03 mm	0.03 mm	0.03 mm	0.03 mm	0.03 mm	0.03 mm	0.03 mm	0.03 mm

TEST CONDITIONS

Penetrant	ammonia			oxygen					
Temperature (°C)	-3	25	63	-17	25	51	-17	25	51
Pressure Gradient (kPa)	965	965	965	1724	1724	1724	3447	3447	3447
Test Method	mass spectrometry and calibrated standard gas leaks								
Test Note	developed by McDonnell Douglas Space Systems Company Chemistry Laboratory								

PERMEABILITY (source document units)

Gas Permeability ($cm^3 \cdot mm/ cm^2 \cdot kPa \cdot sec$)	4.71×10^{-10}	1.73×10^{-9}	8.62×10^{-9}	5.27×10^{-10}	2.55×10^{-9}	5.38×10^{-9}	4.55×10^{-10}	2.54×10^{-9}	5.46×10^{-9}

PERMEABILITY (normalized units)

Permeability Coefficient ($cm^3 \cdot mm/m^2 \cdot day \cdot atm$)	41.2	151	755	46.1	223	471	39.8	222	478

TFE

TABLE 44: Oxygen and Ammonia Permeability vs. Temperature and Pressure Through Carbon Filled DuPont Teflon Polytetrafluoroethylene.

Material Family	POLYTETRAFLUOROETHYLENE								
Material Supplier/ Grade	DUPONT TEFLON								
Reference Number	306	306	306	306	306	306	306	306	306

MATERIAL CHARACTERISTICS

Sample Thickness	0.05 mm	0.05 mm	0.05 mm	0.05 mm	0.05 mm	0.05 mm	0.05 mm	0.05 mm	0.05 mm

MATERIAL COMPOSITION

Note	carbon filled	carbon filled	carbon filled	carbon filled	carbon filled	carbon filled	carbon filled	carbon filled	carbon filled

TEST CONDITIONS

Penetrant	ammonia			oxygen					
Temperature (°C)	-2	25	62	-16	25	55	-15	25	53
Pressure Gradient (kPa)	965	965	965	1724	1724	1724	3447	3447	3447
Test Method	mass spectrometry and calibrated standard gas leaks								
Test Note	developed by McDonnell Douglas Space Systems Company Chemistry Laboratory								

PERMEABILITY (source document units)

Gas Permeability $(cm^3 \cdot mm/ cm^2 \cdot kPa \cdot sec)$	7.77×10^{-10}	2.75×10^{-9}	1.21×10^{-8}	9.28×10^{-10}	5.05×10^{-9}	1.16×10^{-8}	9.56×10^{-10}	5.15×10^{-9}	1.01×10^{-8}

PERMEABILITY (normalized units)

Permeability Coefficient $(cm^3 \cdot mm/m^2 \cdot day \cdot atm)$	68.0	241	1059	81.2	442	1015	83.7	451	884

GRAPH 15: Gas Permeability vs. Temperature through Polytetrafluoroethylene.

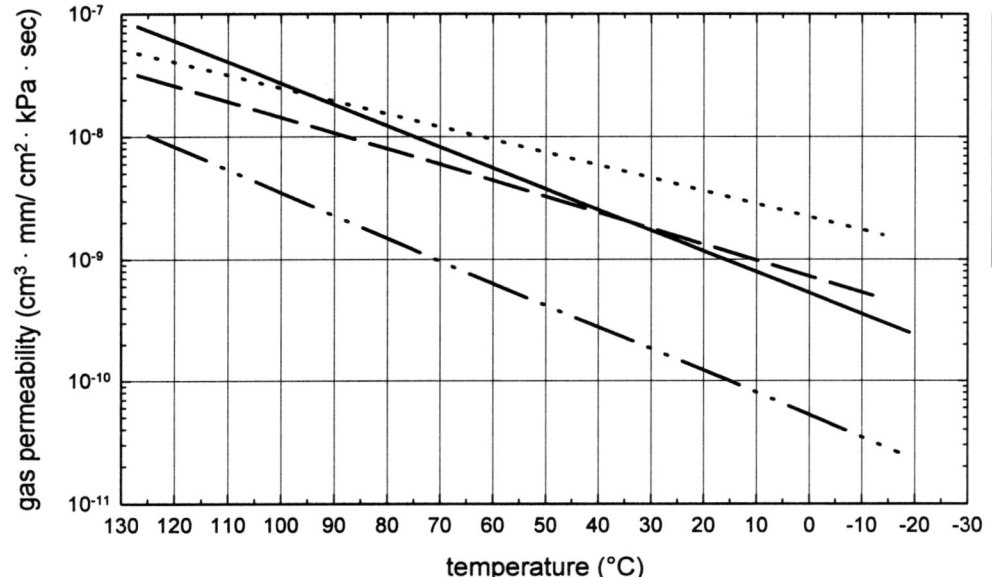

	DuPont Teflon TFE (0.03 mm thick); penetrant: H_2
	DuPont Teflon TFE (0.03 mm thick); penetrant: N_2
	DuPont Teflon TFE (0.03 mm thick); penetrant: O_2
	DuPont Teflon TFE (0.03 mm thick); penetrant: NH_3
Reference No.	306

GRAPH 16: Gas Permeability vs. Temperature through Carbon Filled Polytetrafluoroethylene.

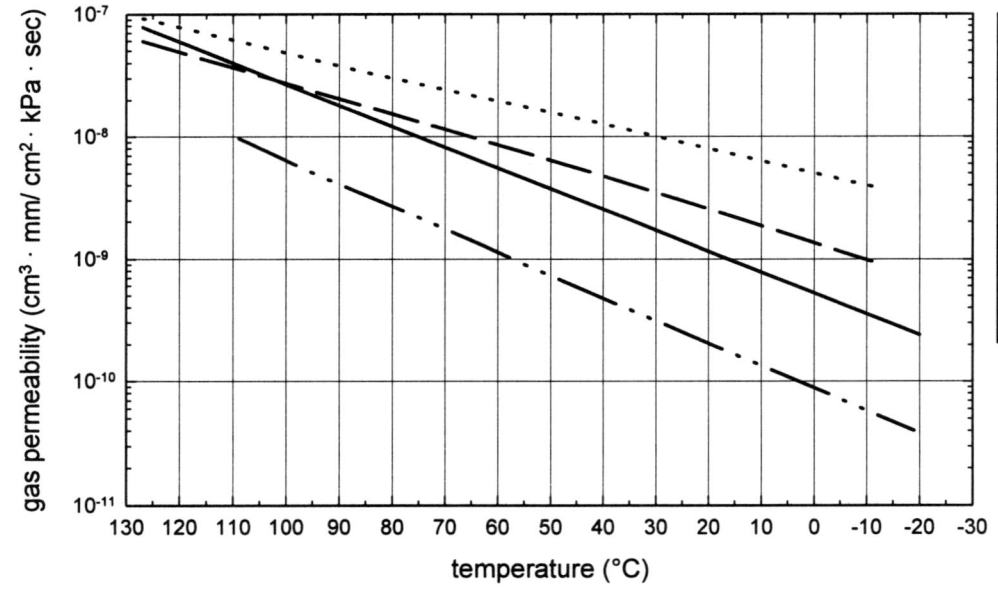

	DuPont Teflon TFE (0.05 mm thick; carbon filled); penetrant: H_2
	DuPont Teflon TFE (0.05 mm thick; carbon filled); penetrant: N_2
	DuPont Teflon TFE (0.05 mm thick; carbon filled); penetrant: O_2
	DuPont Teflon TFE (0.05 mm thick; carbon filled); penetrant: NH_3
Reference No.	306

TFE

Chapter 13

Polyvinyl Fluoride

TABLE 45: Water Vapor, Oxygen, Nitrogen and Carbon Dioxide Permeability Through Polyvinyl Fluoride.

Material Family	POLYVINYL FLUORIDE			
Reference Number	138	138	138	138

TEST CONDITIONS

Penetrant	water vapor	oxygen	nitrogen	carbon dioxide
Temperature (°C)	37.8	25	25	25
Relative Humidity (%)	90			
Test Note		STP conditions		

PERMEABILITY (source document units)

Gas Permeability (cm^3 · mil/100 in^2 · day)		3.0	0.25	11
Gas Permeability (cm^3 · mm/m^2 · day · atm)		1.2	0.10	4.3
Vapor Transmission Rate (g · mil/100 in^2 · day)	3.24			
Vapor Transmission Rate (g/day · 100 in^2)	1.3			

PERMEABILITY (normalized units)

Permeability Coefficient (cm^3 · mm/m^2 · day · atm)		1.2	0.1	4.3
Vapor Transmission Rate (g · mm/m^2 · day)	1.3			

Chapter 14

Polyvinylidene Fluoride

Permeability

Solvay: Solef

SOLEF PVDF has average permeability to small molecules such as carbon dioxide, nitrogen, oxygen, water and nitrous oxide.

Reference: *Solvay Polyvinylidene Fluoride,* supplier design guide (B-1292c-B-2.5-0390) - Solvay, 1992.

Permeability to Gases

Atochem: Foraflon

The permeability of a crystalline polymer to gases depends greatly on its crystallinity index and degree of order. As these depend on processing conditions, the preparation conditions and thermal history of the samples to be measured must be clearly stated.

Reference: *Foraflon PVDF,* supplier design guide (694.E/07.87/20) - Atochem S. A., 1987.

Permeability to Liquids

Atochem: Foraflon

The impermeability of solid FORAFLON has been verified with respect to several liquid compounds.

Mineral acids - A 20 mm diameter FORAFLON tube 2 mm thick has not allowed any trace of hydrochloric acid or sulfuric acid to pass through its walls during 6 months at 100°C. On the other hand, a certain permeability to hydrofluoric acid is noted above 70°C.

Liquid bromine - A 1.5 mm flask made by blow injection has not allowed any trace of bromine to pass through its walls during 2 years at ambient temperature.

Other liquids - FORAFLON is not at all or only very slightly permeable to aliphatic and aromatic hydrocarbons, perchloroethylene and trichloroethylene.

Reference: *Foraflon PVDF,* supplier design guide (694.E/07.87/20) - Atochem S. A., 1987.

TABLE 46: Ammonia, Helium, Chlorine and Hydrogen Permeability Through Solvay Solef Polyvinylidene Fluoride Film.

Material Family	POLYVINYLIDENE FLUORIDE			
Material Supplier/ Grade	SOLVAY SOLEF			
Product Form	FILM			
Manufacturing Method	cast film			
Reference Number	125	125	125	125

MATERIAL CHARACTERISTICS

Sample Thickness	0.1 mm	0.1 mm	0.1 mm	0.1 mm

TEST CONDITIONS

Penetrant	ammonia	helium	chlorine	hydrogen
Temperature (°C)	23	23	23	23
Test Method	ASTM D1434	ASTM D1434	ASTM D1434	ASTM D1434

PERMEABILITY (source document units)

Gas Permeability $(cm^3 \cdot N/m^2 \cdot bar \cdot day)$	65	850	12	210

PERMEABILITY (normalized units)

Permeability Coefficient $(cm^3 \cdot mm/m^2 \cdot day \cdot atm)$	6.6	86	1.2	21.3

PVDF

TABLE 47: Carbon Dioxide, Nitrogen, Oxygen and Water Vapor Permeability Through Solvay Solef 1008 Polyvinylidene Fluoride Film.

Material Family	POLYVINYLIDENE FLUORIDE			
Material Supplier/ Grade	SOLVAY SOLEF 1008			
Product Form	FILM			
Features	translucent	translucent	translucent	translucent
Reference Number	125	125	125	125

MATERIAL CHARACTERISTICS

Sample Thickness	0.1 mm	0.1 mm	0.1 mm	0.1 mm

TEST CONDITIONS

Penetrant	carbon dioxide	nitrogen	oxygen	water vapor
Temperature (°C)	23	23	23	38
Test Method	ASTM D1434	ASTM D1434	ASTM D1434	ASTM E96, proc. E

PERMEABILITY (source document units)

Vapor Transmission Rate ($g/m^2 \cdot day$)				7.5
Gas Permeability ($cm^3 \cdot N/ m^2 \cdot bar \cdot day$)	70	30	21	

PERMEABILITY (normalized units)

Permeability Coefficient ($cm^3 \cdot mm/m^2 \cdot day \cdot atm$)	7.09	3.04	2.13	
Vapor Transmission Rate ($g \cdot mm/m^2 \cdot day$)				0.75

TABLE 48: Freon, Nitrous Oxide, Hydrogen Sulfide, and Sulfur Dioxide Permeability Through Solvay Solef Polyvinylidene Fluoride Film.

Material Family	POLYVINYLIDENE FLUORIDE						
Material Supplier/ Grade	SOLVAY SOLEF						
Product Form	FILM						
Manufacturing Method	cast film						
Reference Number	125	125	125	125	125	125	125

MATERIAL CHARACTERISTICS

Sample Thickness	0.025 mm	0.025 mm	0.025 mm	0.025 mm	0.025 mm	0.025 mm	0.025 mm

TEST CONDITIONS

Penetrant	Freon 12	Freon 114	Freon 115	Freon 318	nitrous oxide	hydrogen sulfide	sulfur dioxide
Temperature (°C)	23	23	23	23	23	23	23
Test Method	ASTM D1434	ASTM D1434	ASTM D1434	ASTM D1434	ASTM D1434	ASTM D1434	ASTM D1434

PERMEABILITY (source document units)

Gas Permeability $(cm^3 \cdot N/m^2 \cdot bar \cdot day)$	6.3	10	4	7	900	60	60

PERMEABILITY (normalized units)

Permeability Coefficient $(cm^3 \cdot mm/m^2 \cdot day \cdot atm)$	0.16	0.25	0.1	0.18	22.8	1.52	1.52

PVDF

TABLE 49: Water Vapor, Oxygen, and Carbon Dioxide Permeability Through Atochem Foraflon Polyvinylidene Fluoride Film.

Material Family	POLYVINYLIDENE FLUORIDE				
Material Supplier/ Grade	ATOCHEM FORAFLON				
Product Form	EXTRUDED FILM				
Reference Number	89	89	89	89	89

MATERIAL CHARACTERISTICS

Sample Thickness	0.02 mm	0.028 mm	0.04 mm	0.037 mm	0.034 mm

TEST CONDITIONS

Penetrant	water vapor			oxygen	carbon dioxide
Temperature (°C)	38	38	38	30	30
Test Method	NFH 00044	NFH 00044	NFH 00044	ISO 2556	ISO 2556

PERMEABILITY (source document units)

Vapor Transmission Rate ($g/m^2 \cdot day$)	34	22	16		
Gas Permeability ($cm^3/m^2 \cdot day$)				140	890

PERMEABILITY (normalized units)

Permeability Coefficient ($cm^3 \cdot mm/m^2 \cdot day \cdot atm$)				5.18	30.26
Vapor Transmission Rate ($g \cdot mm/m^2 \cdot day$)	0.68	0.62	0.64		

TABLE 50: Water Vapor, Oxygen, Nitrogen and Carbon Dioxide Permeability Through Polyvinylidene Fluoride.

Material Family	POLYVINYLIDENE FLUORIDE			
Reference Number	138	138	138	138

TEST CONDITIONS

Penetrant	water vapor	oxygen	nitrogen	carbon dioxide
Temperature (°C)	23	25	25	25
Relative Humidity (%)	90			
Test Note		STP conditions		

PERMEABILITY (source document units)

Gas Permeability (cm^3 · mil/100 in^2 · day)		1.4	9	5.5
Gas Permeability (cm^3 · mm/m^2 · day · atm)		0.55	3.5	2.2
Vapor Transmission Rate (g · mil/100 in^2 · day)	2.6			
Vapor Transmission Rate (g/day · 100 in^2)	1.0			

PERMEABILITY (normalized units)

Permeability Coefficient (cm^3 · mm/m^2 · day · atm)		0.55	3.5	2.2
Vapor Transmission Rate (g · mm/m^2 · day)	1.0			

GRAPH 17: Moisture Vapor Permeability Rate vs. Thickness through Polyvinylidene Fluoride.

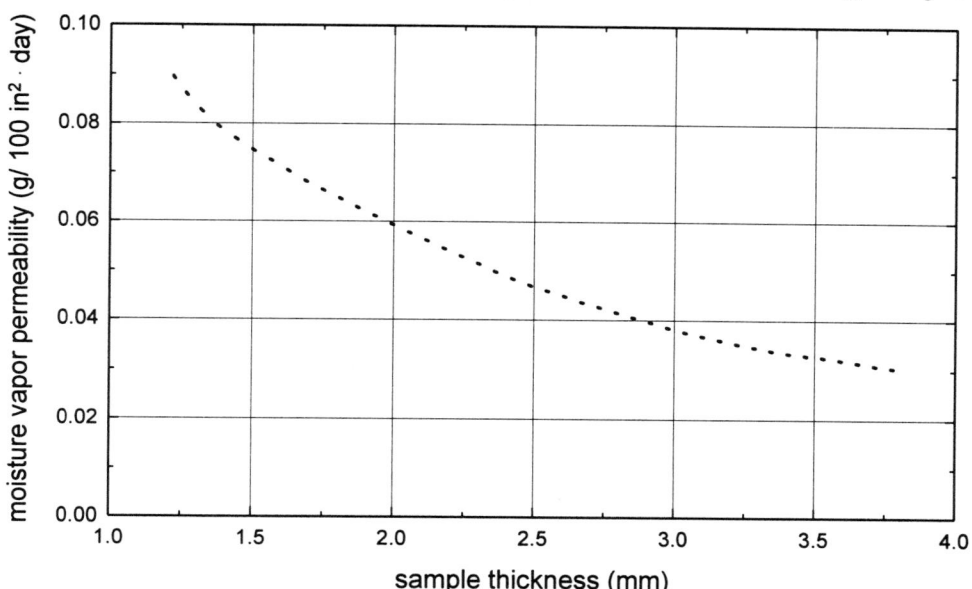

....	PVDF; penetrant: moisture vapor; ΔP=134 mm Hg; 90% RH; 60°C
Reference No.	288

GRAPH 18: Moisture Vapor Permeability Rate vs. Temperature through Polyvinylidene Fluoride.

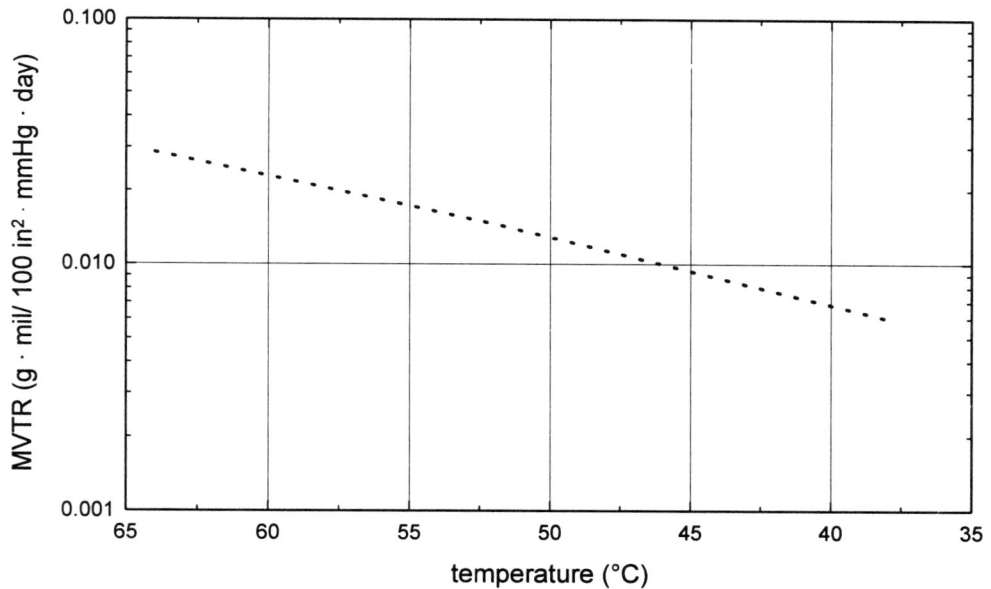

....	PVDF; penetrant: moisture vapor
Reference No.	288

GRAPH 19: Carbon Dioxide Permeability vs. Thickness through Polyvinylidene Fluoride.

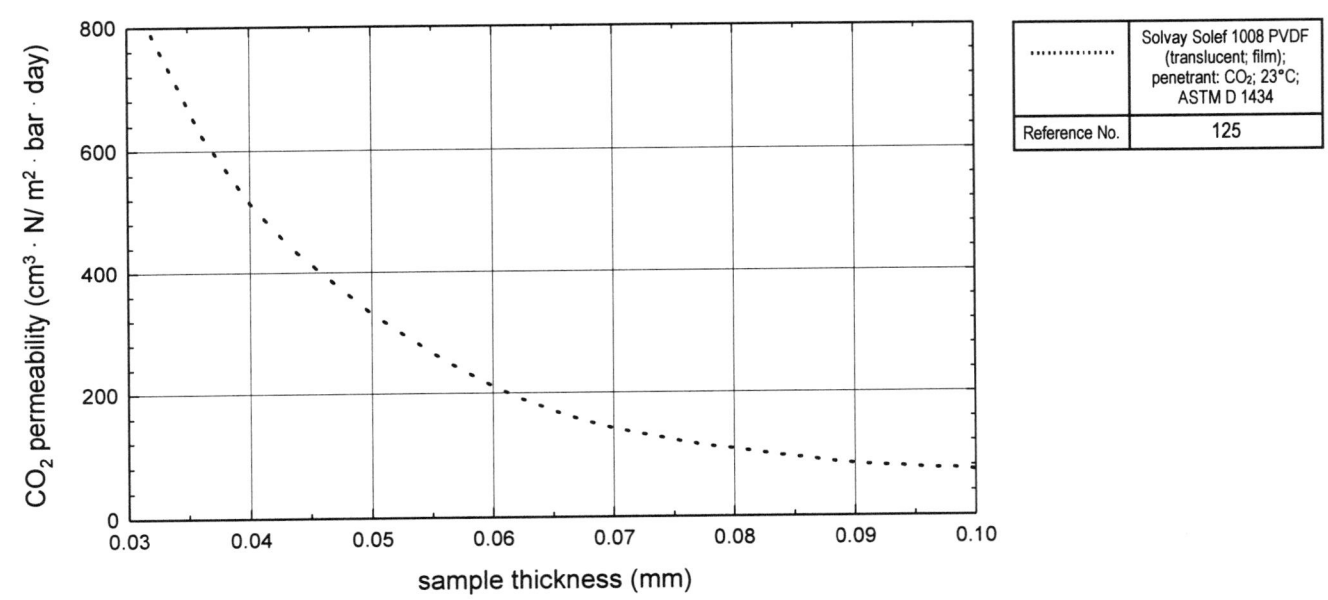

....	Solvay Solef 1008 PVDF (translucent; film); penetrant: CO_2; 23°C; ASTM D 1434
Reference No.	125

GRAPH 20: Water Vapor Permeability vs. Thickness through Polyvinylidene Fluoride.

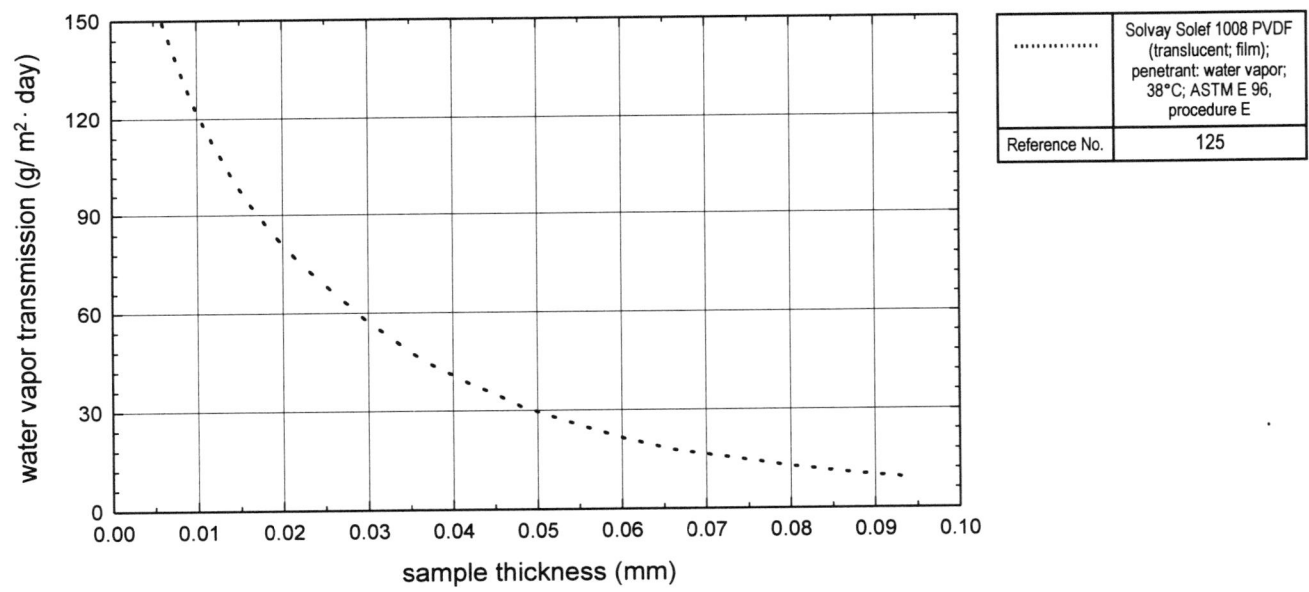

....	Solvay Solef 1008 PVDF (translucent; film); penetrant: water vapor; 38°C; ASTM E 96, procedure E
Reference No.	125

GRAPH 21: Water Vapor Permeability vs. Temperature through Polyvinylidene Fluoride.

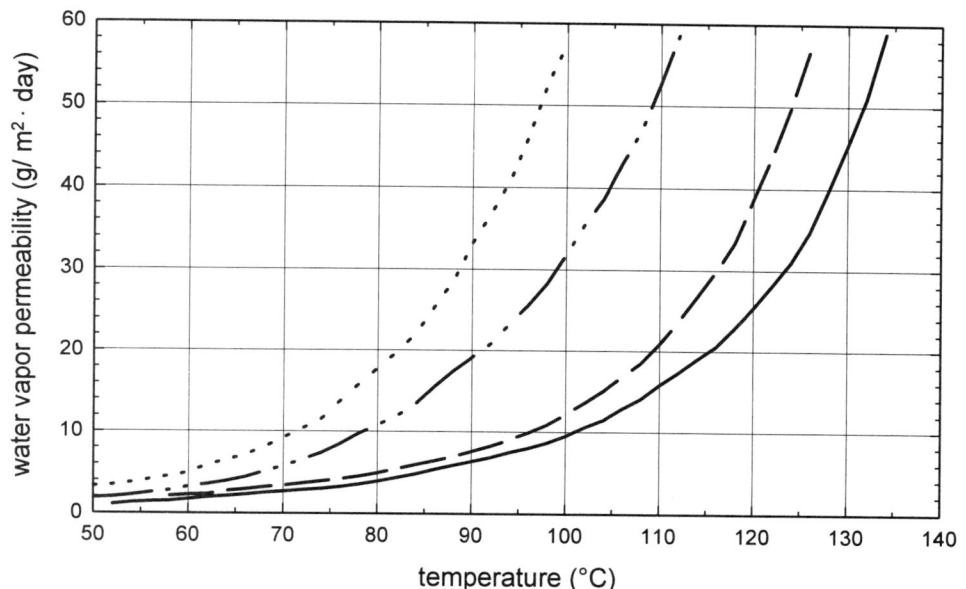

··············	Solvay Solef 1010 PVDF (0.5 mm thick, translucent; sheet); penetrant: water
—··—··—	Solvay Solef 1010 PVDF (translucent, 1 mm thick; sheet); penetrant: water
— — —	Solvay Solef 1010 PVDF (translucent, 2.0 mm thick; sheet); penetrant: water
———	Solvay Solef 1010 PVDF (translucent, 3.0 mm thick; sheet); penetrant: water
Reference No.	125

GRAPH 22: Nitrogen and Oxygen Permeability vs. Thickness through Polyvinylidene Fluoride.

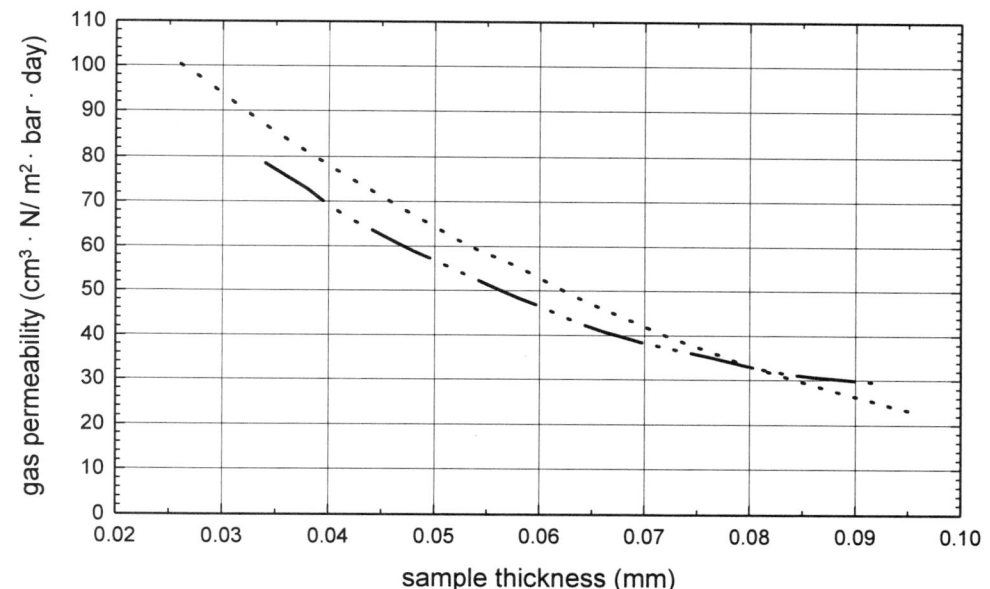

··············	Solvay Solef 1008 PVDF (translucent; film); penetrant: O_2; 23°C; ASTM D 1434
—··—··—	Solvay Solef 1008 PVDF (translucent; film); penetrant: N_2; 23°C; ASTM D 1434
Reference No.	125

GRAPH 23: Gas Permeability vs. Thickness through Polyvinylidene Fluoride.

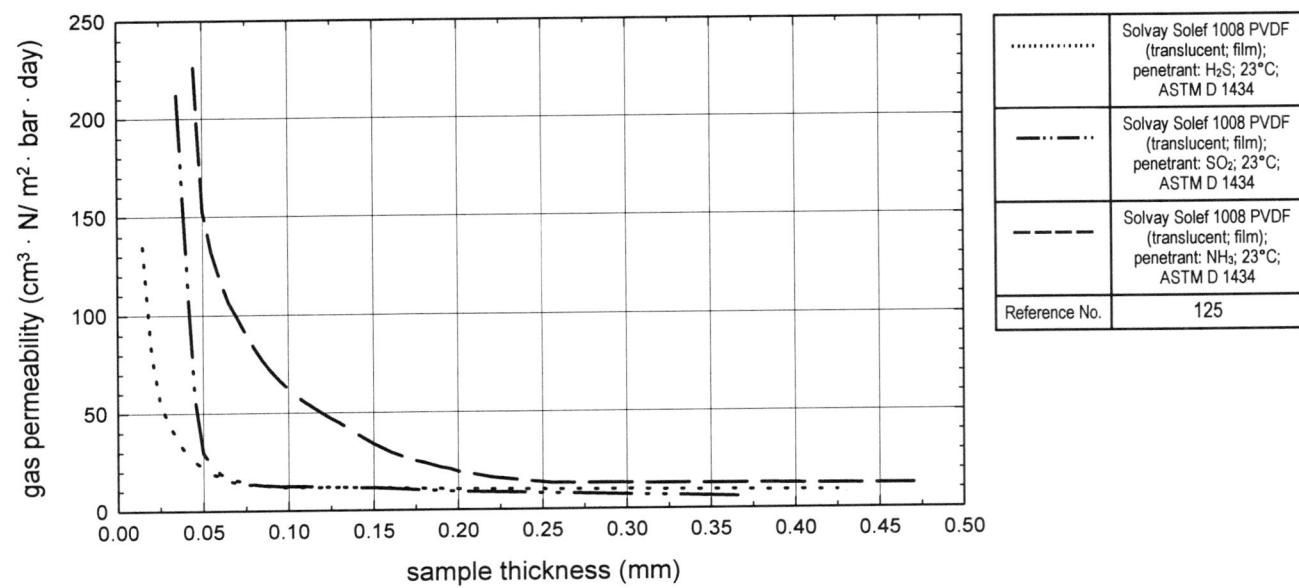

GRAPH 24: Helium and Hydrogen Permeability vs. Thickness through Polyvinylidene Fluoride.

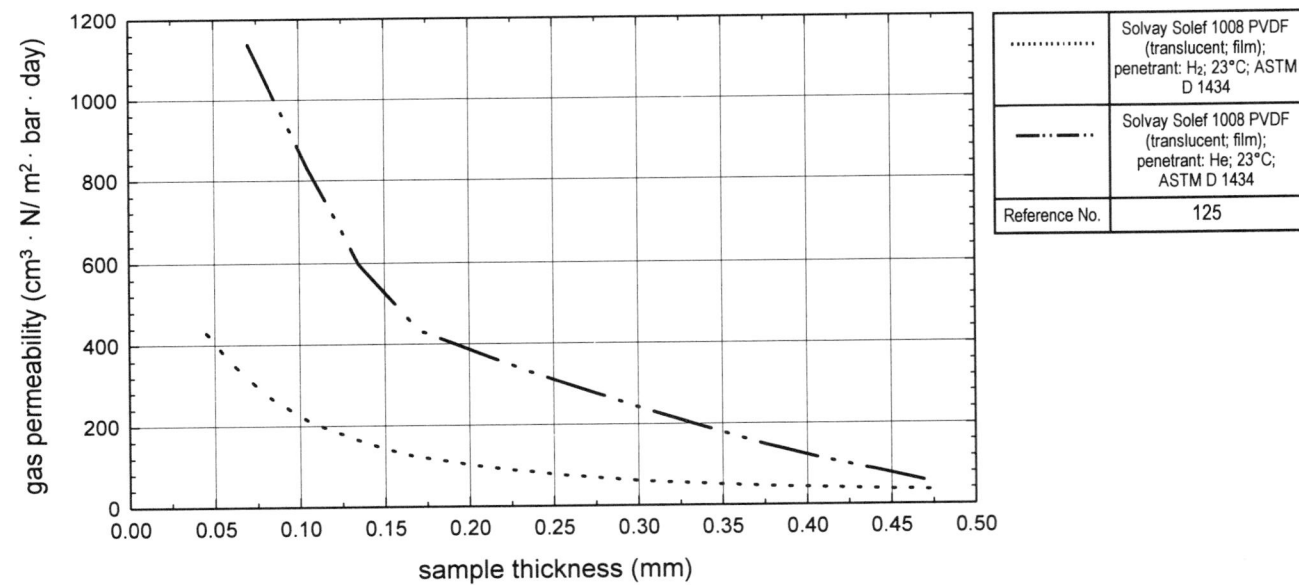

Chapter 15

Ionomer

Barrier Performance

DuPont: Surlyn

Although SURLYN resins do not possess high gas barrier properties, they can improve the barrier properties of structures containing foil or PVDC. In structures of paper/PVDC/SURLYN, the ionomer reduces the number of pinholes in the extremely thin foil used in flexible packaging. In the case of foil structures, SURLYN again reduces the number of pinholes which appear in the brittle PVDC layer when flexed.

SURLYN also improves the barrier of flexible structures against aggressive products and chemicals such as alcohols, sauces, toothpastes, grease and fruit juices. However, when this range of products is packaged in a foil composite where SURLYN 1652 is extrusion-coated onto unprimed foil, initial adhesion levels are in the order of 400 g/ 15mm and generally increase with age. Each aggressive product should be tested individually at normal exposure consitions. For example, a very aggressive chili pepper/oil mixture could not be packaged in a composite of foil/SURLYN, but was contained in a coextrusion of nylon/SURLYN.

SURLYN also improves the barrier performance of a companion thin PVDC layer by providing the same flex protection as with foil and by improving forming in vacuum packaging systems. But for processed meat and natural cheese, a forming web of nylon/SURLYN is generally sufficient and replaces nylon/PE. In the forming operation, the nylon thins out in the corners less with SURLYN than with PE. Since nylon acts as a barrier in this type of package, its increased thickness with SURLYN in the corners acts as a more effective gas barrier.

All grades of SURLYN ionomer resin have superior oil resistance in comparison to polyethylene and other common olefin copolymers.

Reference: *Surlyn Ionomer Resin Increases Packaging Efficiency and Package Performance,* supplier marketing literature (E-54995) - DuPont Company.

Film Properties and Applications

DuPont: Surlyn

Some of the main properties of SURLYN ionomer film are low temperature heat sealability, outstanding hot tack, formability, toughness, clarity, oil and solvent resistance and excellent adhesion to important packaging materials such as nylons and foils. Grades of SURLYN can be characterised by these physical properties as well as by melt index, ion type and application.

SURLYN is used today in film packaging where formability, toughness and visual appearance are important. It is also used extensively as a heat seal layer in composite films for processed meats (sausages, luncheon meat, hams, pates, etc.). When comparing SURLYN with the traditional polyethylene (PE) laminates, meat packers confirm that by using tough film made of SURLYN, lower leaker rates are experienced both off the packaging machine and during distribution. Lower leaker rates with SURLYN are due to improved seal performance, improved forming (less thinning of the formed web in deep draw applications) and toughness, particularly at refrigerated conditions.

SURLYN is also used in complex structures with polyester (PET) film, paper and foil for food (confectionery, powdered soups), for pharmaceuticals (powdered drugs, pills), for cosmetics (face creams, towlettes) and even for composite cans.

Additional properties of SURLYN ionomer film structures include:
- low temperature impact resistance
- puncture and abrasion resistance
- high melt strength
- easy deep-draw
- direct adhesion to foil and paper by extrusion coating, to nylon and other polymers by coextrusion
- low sealing temperature, high sealing strength, outstanding hot tack
- grease, oil and solvent resistance
- transparency with low haze

The ability of SURLYN to seal through contamination, at lower temperatures and at high packaging speeds, is particularly appreciated by users of these laminates.

Reference: *Surlyn Ionomer Resin Increases Packaging Efficiency and Package Performance*, supplier marketing literature (E-54995) - DuPont Company.

Heat Sealing

DuPont: Surlyn

Heat seals of SURLYN ionomer resin form tight, reliable closures while still molten - a key requirement for obtaining peak efficiency from high-speed, automated flexible packaging machines.

With high hot seal strength, often referred to as "hot tack", seal failures and leakers caused by stresses of high-speed machine vibration, heavy product weight, or rough off-machine handling on conveyor lines can be eliminated. Heat seal layers of SURLYN develop this "hot tack" at sealing bar temperatures 15°C to 25°C lower than those of PE, which pemits an increase in line speed at the same bar temperatures as used for PE - since SURLYN will seal faster with less dwell time.

Reference: *Surlyn Ionomer Resin Increases Packaging Efficiency and Package Performance*, supplier marketing literature (E-54995) - DuPont Company.

TABLE 51: Oxygen Gas Permeability Through DuPont Surlyn Zinc Ion Type Ionomoer Film.

Material Family	IONOMER						
Material Supplier/ Trade Name	DUPONT SURLYN						
Grade	**1650**	**1652**	**1702**	**1705**	**F1706**	**F1801**	**F1855**
Manufacturing Method	blown film	blown film	blown film	blown film	blown film	blown film	blown film
Reference Number	280	280	280	280	280	280	280

MATERIAL CHARACTERISTICS

Density	0.950 g/cm³	0.94 g/cm³	0.94 g/cm³	0.950 g/cm³	0.960 g/cm³	0.960 g/cm³	0.960 g/cm³
Melt Flow Index	1.6 g/10 min.	5.0 g/10 min.	14.0 g/10 min.	5.5 g/10 min.	0.7 g/10 min.	1.0 g/10 min.	1.0 g/10 min.
Sample Thickness	0.051 mm	0.051 mm	0.051 mm	0.051 mm	0.051 mm	0.051 mm	0.051 mm
Ion Type	zinc	zinc	zinc	zinc	zinc	zinc	zinc

TEST CONDITIONS

Penetrant	oxygen						

PERMEABILITY (source document units)

Gas Permeability (cm³/100 in² · day · atm)	220	180	175	170	185	215	295

PERMEABILITY (normalized units)

Permeability Coefficient (cm³ · mm/m² · day · atm)	174	142	138	134	146	170	233

Ionomer

TABLE 52: Water Vapor and Oxygen Gas Permeability Through DuPont Surlyn Sodium Ion Type Ionomoer Film.

Material Family	IONOMER									
Material Supplier/ Trade Name	DUPONT SURLYN									
Grade	**1601**	**1603**	**F1605**	**1707**	**F1856**	**1601**	**1603**	**F1605**	**1707**	**F1856**
Manufacturing Method	blown film	blown film	blown film	blown film	blown film	blown film	blown film	blown film	blown film	blown film
Reference Number	280	280	280	280	280	280	280	280	280	280

MATERIAL CHARACTERISTICS

Density	0.94 g/cm³	0.94 g/cm³	0.950 g/cm³	0.950 g/cm³	0.950 g/cm³	0.94 g/cm³	0.94 g/cm³	0.950 g/cm³	0.950 g/cm³	0.950 g/cm³
Melt Flow Index	1.3 g/10 min.	1.7 g/10 min.	2.8 g/10 min.	0.9 g/10 min.	1.0 g/10 min.	1.3 g/10 min.	1.7 g/10 min.	2.8 g/10 min.	0.9 g/10 min.	1.0 g/10 min.
Sample Thickness	0.051 mm	0.051 mm	0.051 mm	0.051 mm	0.051 mm	0.051 mm	0.051 mm	0.051 mm	0.051 mm	0.051 mm

MATERIAL COMPOSITION

Ion Type	sodium	sodium	sodium	sodium	sodium	sodium	sodium	sodium	sodium	sodium

TEST CONDITIONS

Penetrant	water vapor					oxygen				

PERMEABILITY (source document units)

	1601	1603	F1605	1707	F1856	1601	1603	F1605	1707	F1856
Vapor Transmission Rate (g/day · 100 in²)	0.8	0.65	0.8	0.8	1.2					
Gas Permeability (cm³/100 in² · day · atm)						265	190	200	165	290

PERMEABILITY (normalized units)

	1601	1603	F1605	1707	F1856	1601	1603	F1605	1707	F1856
Permeability Coefficient (cm³ · mm/m² · day · atm)						209	150	158	130	229
Vapor Transmission Rate (g · mm/m² · day)	0.63	0.51	0.63	0.63	0.95					

TABLE 53: Water Vapor Permeability Through DuPont Surlyn Zinc Ion Type Ionomoer Film.

Material Family	IONOMER						
Material Supplier/ Trade Name	DUPONT SURLYN						
Grade	**1650**	**1652**	**1702**	**1705**	**F1706**	**F1801**	**F1855**
Manufacturing Method	blown film	blown film	blown film	blown film	blown film	blown film	blown film
Reference Number	280	280	280	280	280	280	280

MATERIAL CHARACTERISTICS

Density	0.950 g/cm³	0.94 g/cm³	0.94 g/cm³	0.950 g/cm³	0.960 g/cm³	0.960 g/cm³	0.960 g/cm³
Melt Flow Index	1.6 g/10 min.	5.0 g/10 min.	14.0 g/10 min.	5.5 g/10 min.	0.7 g/10 min.	1.0 g/10 min.	1.0 g/10 min.
Sample Thickness	0.051 mm	0.051 mm	0.051 mm	0.051 mm	0.051 mm	0.051 mm	0.051 mm

MATERIAL COMPOSITION

Ion Type	zinc	zinc	zinc	zinc	zinc	zinc	zinc

TEST CONDITIONS

Penetrant	water vapor						

PERMEABILITY (source document units)

Vapor Transmission Rate (g/day · 100 in²)	0.75	0.6	0.7	0.7	0.7	0.7	1.0

PERMEABILITY (normalized units)

Vapor Transmission Rate (g · mm/m² · day)	0.59	0.47	0.55	0.55	0.55	0.55	0.79

Ionomer

TABLE 54: Film Properties of DuPont Surlyn Sodium Ion Type Ionomoer Film.

Material Family	IONOMER				
Material Supplier/ Trade Name	DUPONT SURLYN				
Material Supplier/ Grade	1601	1603	F1605	1707	F1856
Manufacturing Method	blown film	blown film	blown film	blown film	blown film
Reference Number	280	280	280	280	280

MATERIAL CHARACTERISTICS

Density	0.94 g/cm³	0.94 g/cm³	0.950 g/cm³	0.950 g/cm³	0.950 g/cm³
Melt Flow Index	1.3 g/10 min.	1.7 g/10 min.	2.8 g/10 min.	0.9 g/10 min.	1.0 g/10 min.
Sample Thickness	0.051 mm	0.051 mm	0.051 mm	0.051 mm	0.051 mm
Ion Type	sodium	sodium	sodium	sodium	sodium

MECHANICAL PROPERTIES

Secant Modulus - MD (MPa)	227 {r}	165 {r}	365 {r}	310 {r}	69 {r}
Secant Modulus - TD (MPa)	207 {r}	165 {r}	372 {r}	345 {r}	69 {r}
Tensile Strength - MD (MPa)	36.6 {r}	31.7 {r}	40.0 {r}	38.6 {r}	33.1 {r}
Tensile Strength - TD (MPa)	33.1 {r}	32.3 {r}	35.1 {r}	29.6 {r}	31.0 {r}
Tensile Strength @ Yield - MD (MPa)	17.2 {r}		19.3 {r}	18.6 {r}	
Tensile Strength @ Yield - TD (MPa)	13.8 {r}	11.0 {r}	17.9 {r}	17.9 {r}	7.6 {r}
Ultimate Elongation - MD (%)	350 {r}	350 {r}	350 {r}	300 {r}	350 {r}
Ultimate Elongation - TD (%)	450 {r}	500 {r}	350 {r}	400 {r}	400 {r}
Spencer Impact Strength (J/m)	0.16 - 0.19 {bp}	0.17 - 0.20 {bp}	0.12 - 0.15 {bp}	0.17 - 0.20 {bp}	0.19 - 0.22 {bp}
Elmendorf Tear Resistance - MD (g/mm)	709 - 866 {k}	945 - 1102 {k}	866 - 1024 {k}	787 - 945 {k}	2913 - 3071 {k}
Elmendorf Tear Resistance - TD (g/mm)	1102 - 1260 {k}	1181 - 1339 {k}	1181 - 1339 {k}	630 - 787 {k}	6299 - 6457 {k}
Tear Strength, Graves - MD (g)	400 {az}	560 {az}	600 {az}	700 {az}	530 {az}
Tear Strength, Graves - TD (g)	440 {az}	530 {az}	630 {az}	760 {az}	590 {az}

THERMAL PROPERTIES

DTA Melt Point (°C)	96 {bq}	96 {bq}	88 {bq}	88 {bq}	81 {bq}
DTA Freeze Point (°C)	74 {bq}	77 {bq}	70 {bq}	58 {bq}	68 {bq}

OPTICAL PROPERTIES

Haze (%)	1 - 3 {ac}	5 - 7 {ac}	1 - 3 {ac}	3 - 5 {ac}	4 - 6 {ac}
Gloss @ 20°	50 - 90 {br}	20 - 60 {br}	50 - 90 {br}	50 - 90 {br}	20 - 60 {br}
Gloss @ 45°	50 - 90 {br}	20 - 60 {br}	50 - 90 {br}	50 - 90 {br}	20 - 60 {br}
Clarity (%)	60 - 80 {s}	30 - 50 {s}	40 - 60 {s}	40 - 60 {s}	10 - 30 {s}

TABLE 55: Film Properties of DuPont Surlyn Zinc Ion Type Ionomoer Film.

Material Family	IONOMER						
Material Supplier/ Trade Name	DUPONT SURLYN						
Grade	**1650**	**1652**	**1702**	**1705**	**F1706**	**F1801**	**F1855**
Manufacturing Method	blown film	blown film	blown film	blown film	blown film	blown film	blown film
Reference Number	280	280	280	280	280	280	280

MATERIAL CHARACTERISTICS

Density	0.950 g/cm³	0.94 g/cm³	0.94 g/cm³	0.950 g/cm³	0.960 g/cm³	0.960 g/cm³	0.960 g/cm³
Melt Flow Index	1.6 g/10 min.	5.0 g/10 min.	14.0 g/10 min.	5.5 g/10 min.	0.7 g/10 min.	1.0 g/10 min.	1.0 g/10 min.
Sample Thickness	0.051 mm	0.051 mm	0.051 mm	0.051 mm	0.051 mm	0.051 mm	0.051 mm
Ion Type	zinc	zinc	zinc	zinc	zinc	zinc	zinc

MECHANICAL PROPERTIES

Secant Modulus - MD (MPa)	248 {r}	172 {r}	241 {r}	241 {r}	400 {r}	283 {r}	96 {r}
Secant Modulus - TD (MPa)	248 {r}	172 {r}	255 {r}	207 {r}	338 {r}	289 {r}	90 {r}
Tensile Strength - MD (MPa)	31.7 {r}	24.1 {r}	24.1 {r}	28.9 {r}	35.8 {r}	34.5 {r}	31.7 {r}
Tensile Strength - TD (MPa)	31.7 {r}	24.8 {r}	24.8 {r}	25.5 {r}	26.9 {r}	31.0 {r}	28.9 {r}
Tensile Strength @ Yield - MD (MPa)			15.9 {r}		25.5 {r}		
Tensile Strength @ Yield - TD (MPa)	13.8 {r}	10.3 {r}	15.9 {r}	14.5 {r}	22.1 {r}	14.5 {r}	9.0 {r}
Ultimate Elongation - MD (%)	400 {r}	400 {r}	300 {r}	300 {r}	200 {r}	350 {r}	350 {r}
Ultimate Elongation - TD (%)	450 {r}	450 {r}	300 {r}	350 {r}	250 {r}	400 {r}	400 {r}
Spencer Impact Strength (J/m)	0.16 - 0.19 {bp}	0.08 - 0.11 {bp}	0.08 - 0.11 {bp}	0.09 - 0.12 {bp}	0.21 - 0.24 {bp}	0.20 - 0.23 {bp}	0.24 - 0.27 {bp}
Elmendorf Tear Resistance - MD (g/mm)	1890 - 2047 {k}	3071 - 3228 {k}	1890 - 2047 {k}	1024 - 1181 {k}	315 - 472 {k}	945 - 1102 {k}	787 - 945 {k}
Elmendorf Tear Resistance - TD (g/mm)	2677 - 2835 {k}	4961 - 5118 {k}	3150 - 3307 {k}	2205 - 2362 {k}	551 - 709 {k}	1575 - 1732 {k}	1339 - 1496 {k}
Tear Strength, Graves - MD (g)	610 {az}	630 {az}	620 {az}	650 {az}	800 {az}	650 {az}	500 {az}
Tear Strength, Graves - TD (g)	700 {az}	610 {az}	630 {az}	680 {az}	760 {az}	660 {az}	550 {az}

THERMAL PROPERTIES

DTA Melt Point (°C)	94 {bq}	98 {bq}	86 {bq}	87 {bq}	81 {bq}	88 {bq}	85 {bq}
DTA Freeze Point (°C)	76 {bq}	86 {bq}	71 {bq}	73 {bq}	63 {bq}	74 {bq}	69 {bq}

OPTICAL PROPERTIES

Haze (%)	5 - 7 {ac}	4 - 6 {ac}	1 - 3 {ac}	5 - 7 {ac}	4 - 6 {ac}	7 - 9 {ac}	3 - 5 {ac}
Gloss @ 20°	20 - 60 {br}	20 - 60 {br}	50 - 90 {br}	20 - 60 {br}	20 - 60 {br}	10 - 60 {br}	20 - 60 {br}
Gloss @ 45°	50 - 90 {br}	50 - 90 {br}	50 - 90 {br}	20 - 60 {br}	50 - 70 {br}	20 - 60 {br}	50 - 90 {br}
Clarity (%)	40 - 60 {s}	20 - 40 {s}	50 - 70 {s}	25 - 45 {s}	45 - 65 {s}	20 - 40 {s}	30 - 50 {s}

Ionomer

Chapter 16

Parylene

Film Properties and Applications

Union Carbide Specialty Coating Systems: **Parylene C** (manufacturing method: vapor phase deposition process; product form: thin film); **Parylene D** (manufacturing method: vapor phase deposition process; product form: thin film); **Parylene N** (features: highly crystalline, high molecular weight, completely linear; manufacturing method: vapor phase deposition process; product form: thin film)

Parylene is the generic name for members of a unique polymer series developed by Union Carbide Corporation. The basic member of the series, called Parylene N, is poly-para-xylylene, a completely linear, highly crystalline material.

Parylene C, the second commercially available member of the series, is produced from the same monomer modified only by the substitution of a chlorine atom for one of the aromatic hydrogens. Parylene D, the third member of the series, is produced from the same monomer modified by the substitution of the chlorine atom for two of the aromatic hydrogens. Parylene D is similar in properties to Parylene C with the added ability to withstand higher use temperatures.

Parylene N is a primary dielectric, exhibiting a very low dissipation factor, high dielectric strength, and a dielectric constant invariant with frequency. It is currently used as a dielectric in extended foil capacitors marketed by Kemet Electronics. Parylene C has a useful combination of electrical and physical properties plus a very low permeability to moisture and other corrosive gases. Along with its ability to provide a true pinhole-free conformal insulation, Parylene C is the material of choice for coating critical electronic assemblies.

Due to the uniqueness of the vapor phase deposition, the Parylene polymers can be formed as structurally continuous films from as thin as a fraction of a micrometer to as thick as several mils.

Reference: *Parylene Conformal Coatings Specifications and Properties,* supplier technical report - Union Carbide Speciatly Coating Systems, 1992.

TABLE 56: Water Vapor, Oxygen, Nitrogen, Carbon Dioxide and Hydrogen Permeability Through Union Carbide Parylene N and Parylene C Parylene Film.

Material Family	PARYLENE									
Material Supplier	UNION CARBIDE SPECIALTY COATING SYSTEMS									
Trade Name/ Grade	PARYLENE N					PARYLENE C				
Product Form	THIN FILM									
Features	completely linear, high molecular weight, highly crystalline									
Manufacturing Method	vapor phase deposition process					vapor phase deposition process				
Reference Number	121	121	121	121	121	121	121	121	121	121

TEST CONDITIONS

Penetrant	nitrogen	oxygen	carbon dioxide	hydrogen	water vapor	nitrogen	oxygen	carbon dioxide	hydrogen	water vapor
Temperature (°C)	25	25	25	25	37	25	25	25	25	37
Relative Humidity (%)					90					90
Test Method	ASTM D1434-63T	ASTM D1434-63T	ASTM D1434-63T	ASTM D1434-63T	ASTM E96-63T	ASTM D1434-63T	ASTM D1434-63T	ASTM D1434-63T	ASTM D1434-63T	ASTM E96-63T

PERMEABILITY (source document units)

Vapor Transmission Rate (g · mil/100 in^2 · day)					1.5					0.21
Gas Permeability (cm^3 · mil/100 in^2 · day)	7.7	39	214	540		1.0	7.2	7.7	110	

PERMEABILITY (normalized units)

Permeability Coefficient (cm^3 · mm/m^2 · day · atm)	3.03	15.4	84.2	213		0.39	2.8	3.0	43.3	
Vapor Transmission Rate (g · mm/m^2 · day)					0.59					0.08

Parylene

TABLE 57: Water Vapor, Oxygen, Nitrogen, Carbon Dioxide and Hydrogen Permeability Through Union Carbide Parylene D Parylene Film.

Material Family	PARYLENE				
Material Supplier/ Grade	UNION CARBIDE SPECIALTY COATING SYSTEMS PARYLENE D				
Product Form	THIN FILM				
Manufacturing Method	vapor phase deposition process				
Reference Number	121	121	121	121	121

TEST CONDITIONS

Penetrant	nitrogen	oxygen	carbon dioxide	hydrogen	water vapor
Temperature (°C)	25	25	25	25	37
Relative Humidity (%)					90
Test Method	ASTM D1434-63T	ASTM D1434-63T	ASTM D1434-63T	ASTM D1434-63T	ASTM E96-63T

PERMEABILITY (source document units)

Vapor Transmission Rate (g · mil/100 in^2 · day)					0.25
Gas Permeability (cm^3 · mil/100 in^2 · day)	4.5	32	13	240	

PERMEABILITY (normalized units)

Permeability Coefficient (cm^3 · mm/m^2 · day · atm)	1.8	12.6	5.1	94.5	
Vapor Transmission Rate (g · mm/m^2 · day)					0.1

TABLE 58: Film Properties of Union Carbide Parylene Films.

Material Family	PARYLENE		
Material Supplier	UNION CARBIDE SPECIALTY COATING SYSTEMS		
Trade Name/ Grade	PARYLENE N	PARYLENE C	PARYLENE D
Product Form	THIN FILM		
Features	completely linear, high molecular weight, highly crystalline		
Manufacturing Method	vapor phase deposition process	vapor phase deposition process	vapor phase deposition process
Reference Number	121	121	121

MATERIAL CHARACTERISTICS

Sample Thickness	0.025 - 0.076 mm	0.025 - 0.076 mm	0.025 - 0.076 mm

TEST CONDITIONS

Test Condition Note	properties depend on deposition conditions		

PHYSICAL PROPERTIES

Density (g/cm³)	1.10 - 1.12 {dg}	1.289 {dg}	1.418 {dg}
Water Absorption (%)	<0.10 {dj}	<0.10 {dj}	<0.10 {dj}

MECHANICAL PROPERTIES

Secant Modulus @ 1% Elongation (MPa)	2412 {de}	2756 {de}	2618 {de}
Tensile Strength (MPa)	41 - 76 {df}	69 {df}	76 {df}
Tensile Strength @ Yield (MPa)	42 {df}	55 {df}	60 {df}
Ultimate Elongation (%)	20 - 250 {df}	200 {df}	10 {df}
Elongation @ Yield (%)	2.5 {df}	2.9 {df}	3.0 {df}
Coefficient Of Friction - Static	0.25 {di}	0.29 {di}	0.35 {di}
Coefficient Of Friction - Kinetic	0.25 {di}	0.29 {di}	0.31 {di}

OPTICAL PROPERTIES

Index Of Refraction		1.661 {dh}	1.639 {dh}

Chapter 17

Nylon

Product Summary

DuPont: Selar RB (applications: packaging; features: barrier properties, laminar technology)

SELAR RB resins are special polyamide or EVOH concentrates. They incorporate a highly reactive adhesive/compatibilizer for increasing the solvent and/or oxygen barrier of conventionally blow-molded HDPE containers using laminar technology.

Reference: *Selar Barrier Resin Selector Guide,* supplier marketing literature (H-38769-1) - DuPont Company, 1992.

DuPont: Selar RB 200 Series (applications: packaging; features: barrier properties, laminar technology)

SELAR RB 200 resins will increase the solvent barrier of HDPE by a factor of 40 to 100, or more, at concentrations of 4% to 10% in the HDPE matrix. Any shape and size achievable by extrusion blow-molding can easily be converted into a barrier container for solvent-based agrochemicals, paint chemicals, and industrial, household and automotive concentrates.

Reference: *Selar Barrier Resin Selector Guide,* supplier marketing literature (H-38769-1) - DuPont Company, 1992.

DuPont: Selar RB 300 Series (applications: packaging; features: barrier properties, laminar technology)

The nylon-based grades in the SELAR RB 300 series are designed for modest oxygen barrier including both food and non-food applications.

Reference: *Selar Barrier Resin Selector Guide,* supplier marketing literature (H-38769-1) - DuPont Company, 1992.

Film Properties and Applications

Allied Signal: Capron (product form: film)

Nylon films provide a barrier to oxygen, flavors and aromas, as well as offering mechanical properties including toughness, puncture and impact resistance and excellent thermoformability.

The term nylon represents a group of thermally processible polyamide polymers produced from a range of monomers which are the means used to identify the resulting nylon type. The different types produced include nylon 46; 6; 66; 612; 11 and 12. The two most common film types are 6 and 66. Type 6 nylon is produced by the polymerization of epsilon caprolactam and type 66 by combining hexamethylene diamine with adipic acid. The films produced from these two polymers represent nearly all of the nylon films commercially manufactured.

Nylon 6 is used where oxygen barrier, flexibility and thermoforming are required. Due to its lower melt temperature, nylon 6 is also favored in coextrusion. Nylon 66 is used where temperature resistance is needed. Nylon 6/66 is used where lower melt point and coextrusion compatibility are required.

Nylon finds its greatest use in flexible food packaging, primarily for processed meat and cheese products. Nylon is typically combined with other materials that enhance its barrier property performance providing the protection required to ensure preservation and freshness of package contents. Moreover, coated or laminated structures containing nylon can be heat sealed into pouches or thermoformed to provide cavities into which hot dogs, sliced processed meats, cheeses and similar food products can be positioned for aesthetic display and sales appeal in the supermarket.

Orientation improves the inherent barrier and mechanical properties of unoriented nylon film. After biaxial orientation, nylon film exhibits a significant improvement in tensile strength and modulus, in impact, puncture and flex-crack resistance and in oxygen and aroma barrier.

During processing, handling and shipping, packages are subject to forces which can cause flexible packaging materials to develop small flex cracks or pinholes. Any break in the packaging material obviously negates the barrier properties that have been designed into the package itself. Thus, a barrier film with excellent flex-crack resistance provides the greatest assurance of maintaining package integrity and the desired protection of the contents.

Nylon films, while providing barrier to many gases, aromas and flavors, are hygroscopic. Commonly, both unoriented and oriented nylon films are combined with moisture barrier materials to achieve optimum gas and water vapor protection. PVDC coating of the nylon films will also improve barrier properties for extended shelf life. Another technique used to increase the barrier performance of nylon films is the vacuum deposition of a thin metallic layer, usually aluminum. This process is commonly referred to as vacuum metallizing.

Packaging applications where oriented films perform best utilize either PVDC coatings, laminations to aluminum foil, polyethylene or ionomer film and/or metallized structures. Applications include portion pouch and vacuum brick coffee packages, soft cookies, bag in the box packages and snack food packages.

In addition to food packaging, oriented nylon is also used extensively in non-food packaging where migrating gases and odors are contained either within the package or prevented from entering from the adjacent products. Examples of non-food packaging include multi-wall bags for shipping materials impregnated with petroleum derivatives such as ready-to-light charcoal briquettes and agricultrual and industrial chemicals. Photographic film is also packaged in structures containing nylon to afford better protection for the contents.

Nylon films are resistant to grease and oils. Their ability to withstand temperatures up to 177°C (350°C) make them excellent for hot-fill packaging substances such as snack-food items and animal feeds which may contain oils or fatty substances.

Because nylons are crystalline polymers, control of the level of the crystallinity in nylon film can significantly improve the thermoforming characteristics which are so vital to many commercial packages. Most nylon film manufacturers now offer more amorphous films, especially useful when producing deep-drawn cavities and shapes. Another recent benefit to packages is the ability to form packages at temperatures lower than the forming temperatures of either nylon 6 or nylon 66. A new copolymer of nylon 6/66 has been developed which offers greater processing advantages in many application areas by forming at temperatures below those of either nylon 6 or 66.

Reference: *Capron Nylon Resins For Films - Operating Manual,* supplier technical report (SFF-08) - Allied Signal Inc., 1992.

BASF: Ultramid

Ultramid film is significant in the packaging sector because of its strength, the ease with which it can be thermoformed, its resistance to heat deformation including sterilizing, and good barrier properties towards gases, particularly to oxygen and aromatic gases.

The main application for Ultramid film is in composites with polyethylene (PE-LD, PE-MD and PE-VAC). Composite film of this nature is the ideal flexible material for the vacuum packaging of perishable foods, e.g. cold cuts, ham, cheese, fish, peanuts, etc.

The ease with which Ultramid film can be thermoformed ensures that the packaging can be well shaped, exert sales appeal, and is economic. Other interesting applications for Ultramid or Ultramid/polyethylene composite film are encountered in the medical and chemical engineering sectors; examples are bags for repacking or the primary packaging of sterile injectors.

Reference: *Ultramid Nylon Resins Product Line, Properties, Processing,* supplier design guide (B 568/1e/4.91) - BASF Corporation, 1991.

TABLE 59: Organic Solvents Permeability Through Oriented and PVDC Coated Nylon Film.

Material Family	NYLON							
Product Form	FILM							
Features	oriented							
Reference Number	266	266	266	266	266	266	266	266

MATERIAL CHARACTERISTICS

Sample Thickness	0.0254 mm	0.015 mm	0.0254 mm	0.015 mm	0.0254 mm	0.015 mm	0.0254 mm	0.015 mm

MATERIAL COMPOSITION

Note		PVDC coated		PVDC coated		PVDC coated		PVDC coated

TEST CONDITIONS

Penetrant	chloroform		xylene		methyl ethyl ketone		kerosine	
Temperature (°C)	20	20	20	20	20	20	20	20
Relative Humidity (%)	65	65	65	65	65	65	65	65

PERMEABILITY (source document units)

Vapor Transmission Rate (g/day · 100 in²)	0.87	0.56	0.06	0.05	0.17	0.10	0.02	<0.003

PERMEABILITY (normalized units)

Vapor Transmission Rate (g · mm/m² · day)	0.34	0.13	0.02	0.01	0.07	0.02	0.01	<0.0007

TABLE 60: Oxygen Permeability vs. Relative Humidity Through Oriented Nylon Film.

Material Family	NYLON							
Product Form	FILM							
Features	oriented							
Reference Number	265	265	265	265	265	265	265	265

MATERIAL COMPOSITION

Note		PVDC coated		PVDC coated		PVDC coated		PVDC coated

TEST CONDITIONS

Penetrant	oxygen							
Temperature (°C)	20	20	20	20	20	20	20	20
Relative Humidity (%)	65		85		100		0	

PERMEABILITY (source document units)

Gas Permeability (cm³ · mil/100 in² · day)	1.92	0.35	5.4	0.35	19.0	0.35	2.5	0.7

PERMEABILITY (normalized units)

Permeability Coefficient (cm³ · mm/m² · day · atm)	0.76	0.14	2.1	0.14	7.5	0.14	0.98	0.28

TABLE 61: Water Vapor Transmission and Oxygen Permeability Through Coated and Uncoated Oriented Nylon Film.

Material Family	NYLON									
Product Form	FILM									
Features	oriented									
Reference Number	268	268	268	268	268	268	268	268	268	268

MATERIAL CHARACTERISTICS

Sample Thickness	0.015 mm	0.017 mm	0.015 mm	0.017 mm	0.015 mm	0.017 mm	0.015 mm	0.017 mm	0.015 mm	0.017 mm

MATERIAL COMPOSITION

Note		PVDC coated		PVDC coated		PVDC coated		PVDC coated		PVDC coated

TEST CONDITIONS

Penetrant	water vapor		oxygen							
Temperature (°C)	40	40	35	35	20	20	20	20	20	20
Relative Humidity (%)	90	90	0	0	65	65	85	85	100	100
Test Method	JIS Z0208	JIS Z0208	JIS Z1707	JIS Z1707	ASTM D3985	ASTM D3985	ASTM D3985	ASTM D3985	ASTM D3985	ASTM D3985

PERMEABILITY (source document units)

Vapor Transmission Rate (g · mil/100 in^2 · day)	17	1								
Gas Permeability (cm^3 · mil/100 in^2 · day)			4.19	1.03	3.23	0.52	9.03	0.52	31.61	0.52

PERMEABILITY (normalized units)

Permeability Coefficient (cm^3 · mm/m^2 · day · atm)			1.65	0.41	1.27	0.2	3.56	0.2	12.44	0.2
Vapor Transmission Rate (g · mm/m^2 · day)	6.7	0.39								

TABLE 62: Film Properties of Coated and Uncoated Oriented Nylon Film.

Material Family	NYLON	
Product Form	FILM	
Features	oriented	oriented
Reference Number	268	268

MATERIAL CHARACTERISTICS

Sample Thickness	0.015 mm	0.017 mm
Note		PVDC coated

TEST CONDITIONS

Temperature (°C)	20	20
Relative Humidity (%)	65	65

MECHANICAL PROPERTIES

Modulus Of Elasticity - MD (MPa)	1667 {r}	1468 {r}
Modulus Of Elasticity - TD (MPa)	1468 {r}	1178 {r}
Tensile Str. @ Break - MD (MPa)	195.8 {r}	166.9 {r}
Tensile Str. @ Break - TD (MPa)	215.8 {r}	235.1 {r}
Ultimate Elongation - MD (%)	90 {r}	100 {r}
Ultimate Elongation - TD (%)	90 {r}	80 {r}
Impact Strength (kg-cm)	10	10
Burst Strength (MPa)	0.39 {af}	0.39 {af}
Pinhole Strength (g)	780 {ag}	860 {ag}
Elmendorf Tear Resistance - MD (g/mm)	500 {ad}	500 {ad}
Elmendorf Tear Resistance - TD (g/mm)	600 {ad}	400 {ad}
Tear Resistance - MD (g)	500	410
Tear Resistance - TD (g)	450	490

OTHER PROPERTIES

Water Absorption (%)	8.0 {ah}	7.7 {ah}
Equilibrium Moist. Absorption (%)	4.0	3.8
Melting Point (°C)	220	220
Haze (%)	2.0	3.3
Gloss	85	
Surface Resistivity (ohms)	4.6 x 10^{14} {aj}	9.3 x 10^{14} {aj}
Slip Factor (°)	44	26
Dimensional Stability - MD (%)	-1.5 {ai}	1.5 {ai}
Dimensional Stability - TD (%)	-0.9 {ai}	1.0 {ai}

GRAPH 25: Oxygen Permeability vs. Relative Humidity through Nylon.

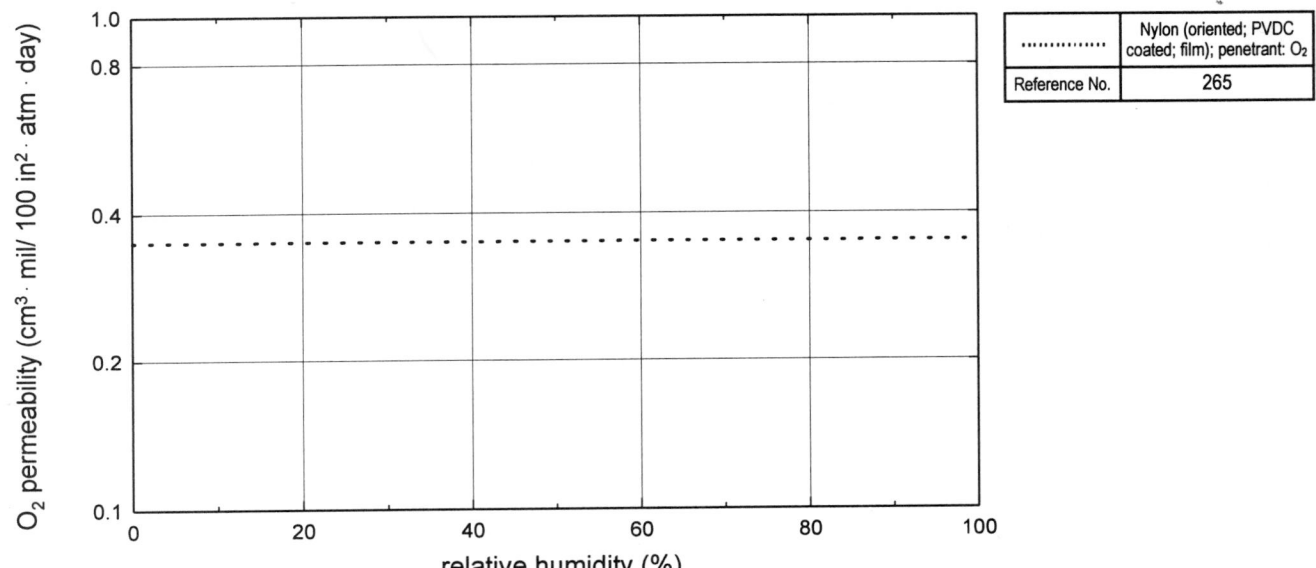

	Nylon (oriented; PVDC coated; film); penetrant: O_2
Reference No.	265

Chapter 18

Amorphous Nylon

Barrier Performance

DuPont: Selar PA   (features: barrier properties; product form: film)

In comparing typical barrier properties for films of nylon 6, amorphous nylon, and polycarbonate moisture permeability is significantly lower for the amorphous nylon than for either nylon 6 or polycarbonate. In addition to lower CO_2 and O_2 transmission rates, the barrier improves with increasing relative humidity. This is contrary to the behavior of other common polymers.

While there is no question that EVOH is the best barrier at low RH, most new packaging applications will require performance of the container at relative humidity ranges from 50% to 80%.

Both EVOH and PVDC will be used in multilayer structures and require adhesive tie layers and structural resin layers. The amorphous nylons, because of their combination of properties, can be used as monolayers as well as in coextruded structures. Thus, while the oxygen barrier of a container depends on the barrier polymer, it also depends on the thickness of that polymer layer.

At conditions of 30°C, 80% RH, the following container structures will all provide equivalent oxygen barrier: A 1 mil layer of high barrier PVDC or EVOH in a multilayer container, an 8 mil monolayer of amorphous nylon including recycled scrap generated in the fabrication process, or a 1.3 mil layer of amorphous nylon in a multilayer structure. The best experimental amorphous nylon is very close in barrier to both PVDC and EVOH at these conditions.

Reference: *High Barrier Amorphous Nylon Resins and Extensions of the Laminar Technology,* supplier technical report (E-73971) - DuPont Company, 1985.

Packaging Applications, Properties and Permeability

DuPont: Selar PA 3426   (density: 1.19 g/cm^3; features: transparent, barrier properties)

SELAR PA 3426 is an amorphous nylon (polyamide) resin which exhibits superior transparency, good barrier properties to gases, water, solvents and essential oils and high temperature structural properties which make is suitable for consideration in a number of packaging applications. It can be processed on conventional extrusion and injection equipment designed to process nylon or polyolefin resins.

SELAR PA 3426 barrier resin can be utilized in both flexible and rigid packaging structures and is characterized by very good gas (O_2, CO_2) and moisture barrier. Furthermore, SELAR PA is unique in that its gas barrier improves with increasing relative humidity. It has excellent physical properties, high temperature resistance and excellent optical properties desirable in a structural resin. Like other nylons, SELAR PA barrier resin is an excellent barrier for many types of solvent systems. In general it has good resistance to aliphatic hydrocarbons, aromatic hydrocarbons, dilute alkalis, higher molecular weight alcohols, and low concentrations of lower molecular weight alcohols. It is not recommended for acids and concentrated lower molecular weight alcohols. End use testing is recommended for each application. With its high degree of stiffness, very good gas and moisture barrier, solvent resistance and excellent gloss and clarity, SELAR PA used in monolayer provides a "glass like" container for rigid packaging.

Reference: *Selar PA 3426 Barrier Resin,* supplier technical report (E-73974) - DuPont Company, 1985.

EMS-American Grilon: Grivory G21 (features: transparent, barrier properties, 0.05 mm thick film)

Grivory G21 is an amorphous nylon copolymer intended for barrier films and barrier bottles. It complies with 21CFR 177.1500 requirements for a nylon 6I/6T in direct contact with all foods except those containing more than 8% alcohol. There are no restrictions on thickness or temperature of use.

Films of Grivory G21 have exceptional oxygen and carbon dioxide barrier properties, even under high humidity conditions. When Grivory G21 is mixed with other nylons, films can be produced with better transparency and gas barrier properties, resulting in long shelf-life for packaged foods. Mixtures of nylon 6 and 15-30% Grivory G21 yield films of good appearance (high gloss), better thermoforming, and higher shrinkage after stretching or thermoforming.

Grivory G21 can also be used to produce transparent bottles by blow molding. These bottles have good gas barrier properties and can be filled at higher temperatures than is possible with polyester (PET) bottles. Grivory G21 can also be used to produce multilayer bottles by multilayer blow molding with PET or polycarbonate, to improve the shelf-life of oxygen sensitive foods and drinks.

Reference: *Product Data Bulletin - Grivory G21,* supplier marketing literature (GV8-104) - EMS - American Grilon Inc..

Product Summary

DuPont: Selar PA (applications: packaging; features: barrier properties, flavor barrier, aroma barrier)

SELAR PA resins are amorphous and semicrystalline polyamides. The amorphous grades provide an improved gas barrier vs. semicrystalline nylons, and outstanding flavor and aroma barrier properties.

Reference: *Selar Barrier Resin Selector Guide,* supplier marketing literature (H-38769-1) - DuPont Company, 1992.

DuPont: Selar PA 3426 (applications: packaging)

SELAR PA 3426 can be a modifier for nylon 6 in high performance food packaging films. It enhances gloss, clarity, oxygen barrier, carbon dioxide barrier, ultraviolet light barrier and formability properties, as well as the processibility, of nylon 6.

Reference: *Selar Barrier Resin Selector Guide,* supplier marketing literature (H-38769-1) - DuPont Company, 1992.

DuPont: Selar PA 3508 (applications: packaging; features: high flow)

SELAR PA 3508 is a lower viscosity version of SELAR PA 3426.

Reference: *Selar Barrier Resin Selector Guide,* supplier marketing literature (H-38769-1) - DuPont Company, 1992.

DuPont: Selar PA 3901 (applications: packaging; features: high impact)

SELAR PA 3901, an impact-modified amorphous nylon, was developed for tough solvent and odor barrier agrochemical containers made by coextrusion blow-molding.

Reference: *Selar Barrier Resin Selector Guide,* supplier marketing literature (H-38769-1) - DuPont Company, 1992.

Film Properties and Applications

DuPont: Selar PA (features: barrier properties; product form: film)

The new amorphous nylons maintain all the advantages of nylon 6 resins, such as high temperature resistance, toughness and thermoformability. But they also exhibit a consistent high stiffness at all relative humidity levels. The barriers to oils and flavors are as good or better than nylon 6 and the barriers to gases and water are much higher. They are transparent and show a wide processing latitude in a variety of melt processes. In fact, these resins can be considered as similar to polycarbonates, but with vastly better barrier properties and closer to the polyolefins in ease of processing.

The flexural modulus of injection molded test bars actually increases slightly as the equilibrium moisture content goes up. This is in sharp contrast to the semicrystalline nylons which can lose three-fourths of their dry stiffness at high humidity. This factor of high and consistent stiffness combined with a high heat distortion temperature of 127°C (260°F) at 66 psi, allow us to speculate about the utility of these polymers in hot-fill and retort applications.

Reference: *High Barrier Amorphous Nylon Resins and Extensions of the Laminar Technology,* supplier technical report (E-73971) - DuPont Company, 1985.

TABLE 63: Oxygen, Carbon Dioxide, Nitrogen and Water Vapor Permeability Through EMS-American Grilon Grivory G21 Amorphous Nylon Copolymer Film.

Material Family	AMORPHOUS NYLON				
Material Supplier/ Grade	EMS-AMERICAN GRILON GRIVORY G21				
Product Form	FILM				
Features	barrier properties, transparent				
Reference Number	307	307	307	307	307

MATERIAL CHARACTERISTICS

Sample Thickness	0.05 mm	0.05 mm	0.05 mm	0.05 mm	0.05 mm

TEST CONDITIONS

Penetrant	oxygen		carbon dioxide	nitrogen	water vapor
Temperature (°C)	23	23	23	23	
Relative Humidity (%)	50	100	50	50	
Test Method	DIN 53380	DIN 53380	DIN 53380	DIN 53380	DIN 53122

PERMEABILITY (source document units)

Gas Permeability ($cm^3/m^2 \cdot day \cdot bar$)	30	8	75	10	
Vapor Transmission Rate ($g/m^2 \cdot day$)					7

PERMEABILITY (normalized units)

Permeability Coefficient ($cm^3 \cdot mm/m^2 \cdot day \cdot atm$)	1.5	0.41	3.8	0.51	
Vapor Transmission Rate ($g \cdot mm/m^2 \cdot day$)					0.35

TABLE 64: Water Vapor Transmission Through Du Pont Selar PA Amorphous Nylon Barrier Resin.

Material Family	AMORPHOUS NYLON
Material Supplier/ Grade	DUPONT SELAR PA
Features	barrier properties
Reference Number	264

TEST CONDITIONS

Penetrant	water vapor
Temperature (°C)	40
Relative Humidity (%)	90

PERMEABILITY (source document units)

Vapor Transmission Rate (g · mil/100 in^2 · day)	1.4
Vapor Transmission Rate (g · 25μ/m^2 · day)	21.7

PERMEABILITY (normalized units)

Vapor Transmission Rate (g · mm/m^2 · day)	0.55

TABLE 65: Water Vapor, Carbon Dioxide and Oxygen Permeability Through DuPont Selar PA Amorphous Nylon Film.

Material Family	AMORPHOUS NYLON				
Material Supplier/ Grade	DUPONT SELAR PA				
Product Form	FILM				
Features	barrier properties	barrier properties	barrier properties	barrier properties	barrier properties
Reference Number	294	294	294	294	294

TEST CONDITIONS

Penetrant	water vapor	carbon dioxide		oxygen	
Temperature (°C)	37.8	22.8	22.8	22.8	22.8
Relative Humidity (%)	90	0	80	0	80

PERMEABILITY (source document units)

Vapor Transmission Rate (g · mil/100 in^2 · day)	1.2				
Gas Permeability (cm^3 · mil/100 in^2 · day)		4.5	2.8	2.5	1.2

PERMEABILITY (normalized units)

Permeability Coefficient (cm^3 · mm/m^2 · day · atm)		1.8	1.1	0.98	0.47
Vapor Transmission Rate (g · mm/m^2 · day)	0.47				

GRAPH 26: Carbon Dioxide Permeability vs. Relative Humidity through Amorphous Nylon.

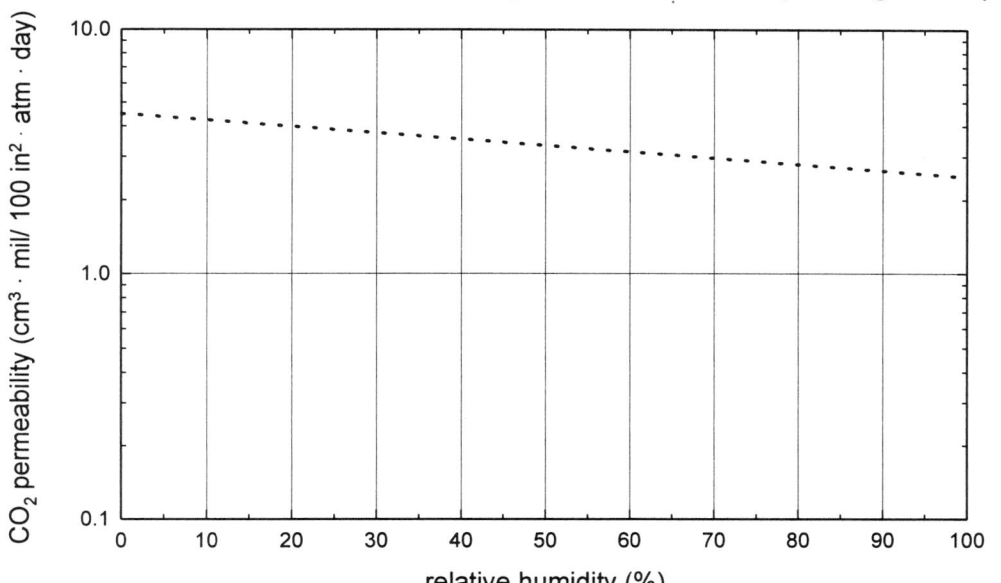

....	DuPont Selar PA Amorphous Nylon (barrier prop.); penetrant: CO_2; 20°C
Reference No.	264

GRAPH 27: Oxygen Permeability vs. Relative Humidity through Amorphous Nylon.

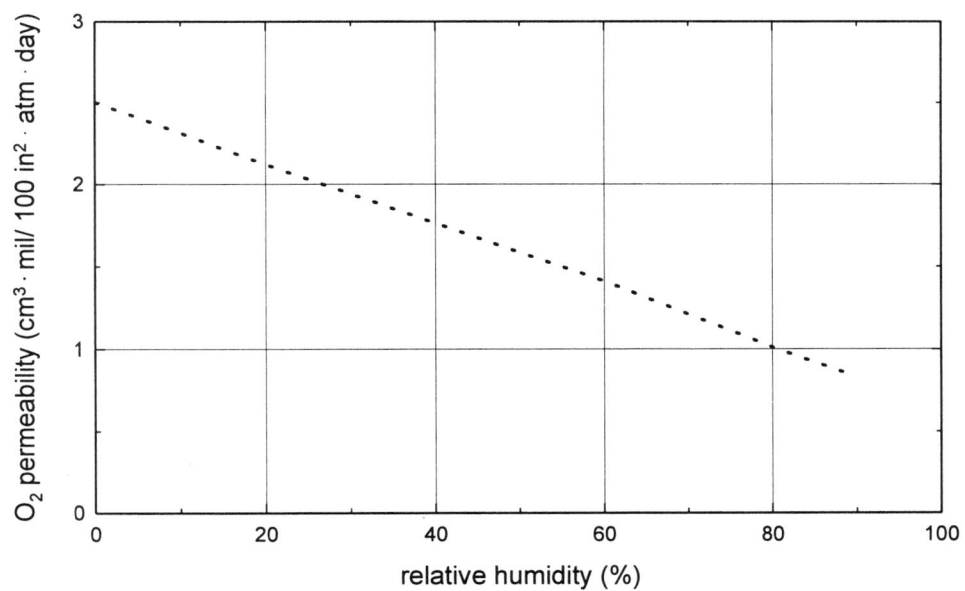

....	DuPont Selar PA 3426 Amorphous Nylon (1.19 g/cm³ density; transparent, barrier prop.); penetrant: O_2; 23°C
Reference No.	292

TABLE 82: Liquid Permeability Through DuPont Company Zytel 42 Nylon 66 Bottles.

Material Family	NYLON 66							
Material Supplier/ Grade	DUPONT ZYTEL 42							
Product Form	BOTTLES							
Features	low flow							
Reference Number	68	68	68	68	68	68	68	68

MATERIAL CHARACTERISTICS

Sample Thickness	2.54 mm							

TEST CONDITIONS

Penetrant	kerosine	methyl salicylate	motor oils	toluene	ASTM Fuel Oil B	water	carbon tetrachloride	naphtha
Concentration (%)								VMP naphtha
Penetrant Note			SAE 10		isooctane and toluene blend			

PERMEABILITY (source document units)

Vapor Transmission Rate (g · mm/m² · day)	0.08	0.08	0.08	0.08	0.2	1.2 - 2.4	2.0	2.4
Vapor Transmission Rate (g · mil/100 in² · day · atm)	0.2	0.2	0.2	0.2	0.5	3-6	5	6

PERMEABILITY (normalized units)

Vapor Transmission Rate (g · mm/m² · day)	0.08	0.08	0.08	0.08	0.2	1.2 - 2.4	2	2.4

TABLE 83: Mechanical Properties of BASF Ultramid B Nylon 66 Film.

Material Family	NYLON 66		
Material Supplier/ Grade	BASF ULTRAMID A5		
Features	low flow		
Manufacturing Method	flat film	tubular film	blown film
Reference Number	93	93	252

MATERIAL CHARACTERISTICS

Sample Thickness	0.02 - 0.1 mm	0.02 - 0.1 mm	0.05 mm

TEST CONDITIONS

Temperature (°C)	23	23	20

MECHANICAL PROPERTIES

Tensile Str. @ Break - MD (MPa)	75-90	70-90	80
Tensile Str. @ Break - TD (MPa)	70-80	50-75	60
Ultimate Elongation - MD (%)	350-400	250-400	300
Ultimate Elongation - TD (%)	350-400	200-350	250

GRAPH 28: Oxygen Permeability vs. Temperature through Amorphous Nylon.

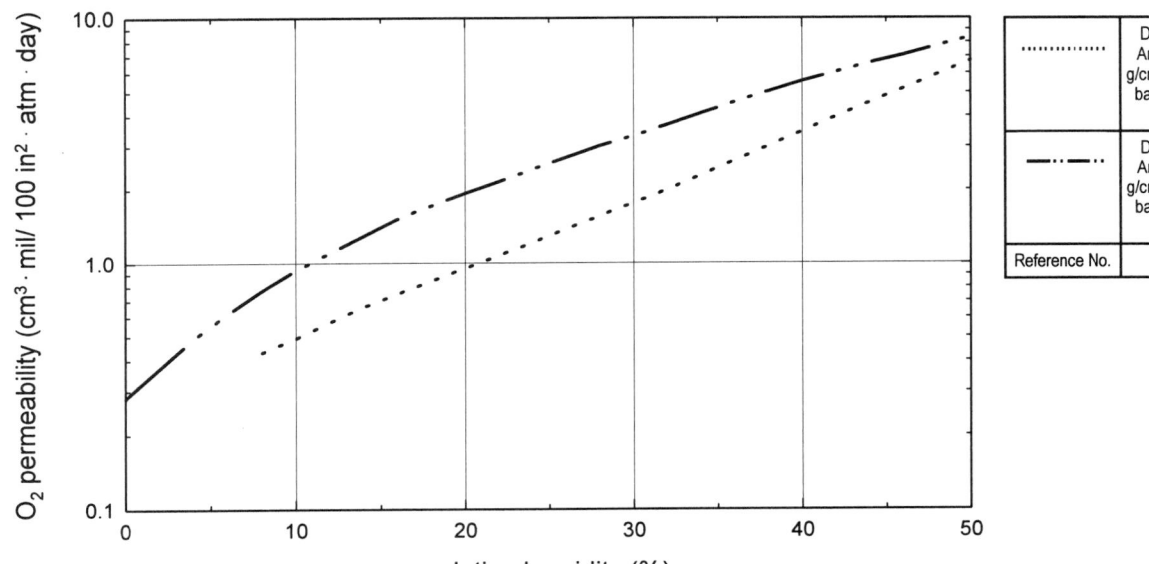

	DuPont Selar PA 3426 Amorphous Nylon (1.19 g/cm³ density; transparent, barrier prop.); penetrant: O₂; 80% RH
	DuPont Selar PA 3426 Amorphous Nylon (1.19 g/cm³ density; transparent, barrier prop.); penetrant: O₂; 0% RH
Reference No.	292

Amorphous Nylon

Chapter 19

Nylon 6

Permeability

Allied Signal: Capran   (product form: film)

CAPRAN has a low permeability to oxygen and nitrogen, a fact that suggests food packaging use, particularly where nitrogen flushing is common. Flavor and aroma are sealed in while the food is proected from contaminants.

Reference: *Capran Nylon Films,* supplier technical report - Allied Signal Inc..

Film Properties and Applications

Nylon 6   (features: barrier properties; product form: film)

The key advantages of nylon 6 film for packaging are its high temperature resistance, toughness, flex crack and puncture resistance, good thermoformability, stiffness, and dimensional stability. In addition, it provides excellent barriers to oils and fats and does not absorb or transmit most flavors. The gas barrier properties, while better than polyolefins, are only moderate when compared to saran (PVDC) and ethylene vinyl alcohol (EVOH). The main deficiencies of nylon are processing difficulties, poor water barrier, and tensile and stiffness variations with relative humidity (RH). These factors limit the use of nylon 6 in packaging.

Reference: *High Barrier Amorphous Nylon Resins and Extensions of the Laminar Technology,* supplier technical report (E-73971) - DuPont Company, 1985.

TABLE 66: Gas Permeability and Water Vapor Transmission Through BASF Ultramid B Nylon 6 Film.

Material Family	NYLON 6					
Material Supplier Trade Name	BASF ULTRAMID					
Grade	B4	B36F	B4	B36F	B4	B36F
Features	moderate flow	enhanced clarity, moderate flow	moderate flow	enhanced clarity, moderate flow	moderate flow	enhanced clarity, moderate flow
Manufacturing Method	flat film, tubular film	flat film, tubular film	flat film, tubular film	flat film, tubular film	flat film, tubular film	flat film, tubular film
Reference Number	93	93	93	93	93	93

MATERIAL CHARACTERISTICS

Sample Thickness	0.02 - 0.1 mm		0.02 - 0.1 mm		0.02 - 0.1 mm	

TEST CONDITIONS

Penetrant	oxygen		carbon dioxide		water vapor	
Temperature (°C)	23	23	23	23	23	23
Relative Humidity (%)	40	40	0	0	85%-0% gradient	85%-0% gradient
Test Method	DIN 53380	DIN 53380	DIN 53380	DIN 53380	DIN 53122	DIN 53122

PERMEABILITY (source document units)

Gas Permeability ($cm^3 \cdot 100 \, \mu m/m^2 \cdot day \cdot bar$)	6 - 7	6 - 7	40 - 45	40 - 45		
Vapor Transmission Rate ($g \cdot 100 \, \mu m/m^2 \cdot day$)					1.5 - 1.6	1.5 - 1.6

PERMEABILITY (normalized units)

Permeability Coefficient ($cm^3 \cdot mm/m^2 \cdot day \cdot atm$)	0.61 - 0.71	0.61 - 0.71	4.0 - 4.6	4.0 - 4.6		
Vapor Transmission Rate ($g \cdot mm/m^2 \cdot day$)					15 - 16	15 - 16

TABLE 67: Gas Permeability of Carbon Dioxide, Nitrogen, Helium and Water Vapor Transmission Through Oriented Nylon 6.

Material Family	NYLON 6			
Features	oriented			biaxially oriented
Reference Number	264	264	264	264

TEST CONDITIONS

Penetrant	carbon dioxide	nitrogen	helium	water vapor
Temperature (°C)	35	23	35	40
Relative Humidity (%)	0	0	0	90

PERMEABILITY (source document units)

Gas Permeability (cm^3 · mil/100 in^2 · day)	6.62	0.7	116	
Gas Permeability (cm^3 · 25μ/m^2 · day · atm)	102.6	10.8	1798	
Vapor Transmission Rate (g · mil/100 in^2 · day)				10.2
Vapor Transmission Rate (g · 25μ/m^2 · day)				158.1

PERMEABILITY (normalized units)

Permeability Coefficient (cm^3 · mm/m^2 · day · atm)	2.61	0.28	45.7	
Vapor Transmission Rate (g · mm/m^2 · day)				4.02

TABLE 68: Oxygen Permeability vs. Temperature Through Oriented and Non-Oriented Nylon 6.

Material Family	NYLON 6					
Features	oriented					
Reference Number	264	264	264	264	264	264

TEST CONDITIONS

Penetrant	oxygen					
Temperature (°C)	5	23	35	5	23	35
Relative Humidity (%)	0	0	0	0	0	0

PERMEABILITY (source document units)

Gas Permeability ($cm^3 \cdot mil/100\ in^2 \cdot day$)	0.49	1.78	3.3	1.439	5.08	10
Gas Permeability ($cm^3 \cdot 25\mu/m^2 \cdot day \cdot atm$)	7.59	25.59	51.15	22.3	78.74	154.9

PERMEABILITY (normalized units)

Permeability Coefficient ($cm^3 \cdot mm/m^2 \cdot day \cdot atm$)	0.19	0.7	1.3	0.57	2	3.9

TABLE 69: Oxygen Gas and Water Vapor Permeability Through BASF Ultramid B4 Nylon 6 Film.

Material Family	NYLON 6			
Material Supplier/ Grade	BASF ULTRAMID B4			
Features	unstretched	biaxially stretched	unstretched	biaxially stretched
Reference Number	252	252	252	252

MATERIAL CHARACTERISTICS

Sample Thickness	0.02 - 0.025 mm	0.02 mm	0.02 - 0.025 mm	0.02 mm

TEST CONDITIONS

Penetrant	water vapor		oxygen	
Temperature (°C)	20	20	20	20
Relative Humidity (%)	85-0% gradient	85-0% gradient	40	40
Test Method	DIN 53122	DIN 53122	DIN 53380	DIN 53380

PERMEABILITY (source document units)

Vapor Transmission Rate ($g/m^2 \cdot day$)	50 - 80	40 - 60		
Gas Permeability ($cm^3/m^2 \cdot day \cdot bar$)			25 - 35	12 - 15

PERMEABILITY (normalized units)

Permeability Coefficient ($cm^3 \cdot mm/m^2 \cdot day \cdot atm$)			0.57 - 0.8	0.24 - 0.3
Vapor Transmission Rate ($g \cdot mm/m^2 \cdot day$)	1.13 - 1.8	0.8 - 1.2		

TABLE 70: Oxygen, Nitrogen and Carbon Dioxide Permeability Through Allied Signal Capran 77C Nylon 6 Film.

Material Family	NYLON 6									
Material Supplier/ Grade	ALLIED SIGNAL CAPRAN 77C									
Product Form	FILM									
Reference Number	285	285	285	285	285	285	285	285	285	285

MATERIAL CHARACTERISTICS

Sample Thickness	0.0254 mm	0.0254 mm	0.019 mm	0.0254 mm	0.0254 mm	0.0254 mm	0.0254 mm	0.0254 mm	0.0254 mm	0.0254 mm

TEST CONDITIONS

Penetrant	oxygen				nitrogen			carbon dioxide		
Temperature (°C)	0	23	23	50	0	23	50	0	23	50
Relative Humidity (%)	0	0	0	0	0	0	0	0	0	0
Test Note	STP conditions									

PERMEABILITY (source document units)

Gas Permeability (cm^3/100 $in^2 \cdot day \cdot atm$)	0.5	2.6	3.2	14	0.2	0.9	12	0.6	4.7	44

PERMEABILITY (normalized units)

Permeability Coefficient ($cm^3 \cdot mm/m^2 \cdot day \cdot atm$)	0.2	1.02	0.94	5.5	0.08	0.35	4.7	0.24	1.8	17.3

TABLE 71: Water Vapor and Oxygen Permeability Through Nylon 6 Film.

Material Family	NYLON 6			
Material Supplier/ Trade Name	ALLIED SIGNAL CAPRAN			
Product Form	FILM			
Reference Number	284	284	296	296

MATERIAL CHARACTERISTICS

Sample Thickness	0.0254 mm	0.0254 mm		

TEST CONDITIONS

Penetrant	water vapor	oxygen	oxygen	water vapor
Temperature (°C)	37.8	23	22.8	37.8
Relative Humidity (%)	90	0	0	90
Test Method	pouch method	permeability cell	ASTM D1434	ASTM F1249
Test Note	STP conditions			

PERMEABILITY (source document units)

Vapor Transmission Rate ($g/m^2 \cdot day$)	295 - 310			
Vapor Transmission Rate ($g/day \cdot 100\ in^2$)	19 - 20			
Vapor Transmission Rate ($g \cdot mil/ 100\ in^2 \cdot bar \cdot day$)				23
Gas Permeability ($cm^3/m^2 \cdot day$)		40.3		
Gas Permeability ($cm^3 \cdot mil/ 100\ in^2 \cdot bar \cdot day$)			3	
Gas Permeability ($cm^3/100\ in^2 \cdot day \cdot atm$)		2.6		

PERMEABILITY (normalized units)

Permeability Coefficient ($cm^3 \cdot mm/m^2 \cdot day \cdot atm$)		1.02	1.2	
Vapor Transmission Rate ($g \cdot mm/m^2 \cdot day$)	7.5 - 7.9			9.2

TABLE 72: Water Vapor Permeability Through Allied Signal Capran 77C Nylon 6 Film.

Material Family	NYLON 6			
Material Supplier/ Grade	ALLIED SIGNAL CAPRAN 77C			
Product Form	FILM			
Reference Number	285	285	285	285

MATERIAL CHARACTERISTICS

Sample Thickness	0.019 mm	0.0254 mm	0.019 mm	0.019 mm

TEST CONDITIONS

Penetrant	water vapor			
Temperature (°C)	23	23	37.8	37.8
Relative Humidity (%)	50	50	90	90
Test Note	pouch method			

PERMEABILITY (source document units)

Vapor Transmission Rate (g/day · 100 in²)	0.8	0.6	24-26	19-20

PERMEABILITY (normalized units)

Vapor Transmission Rate (g · mm/m² · day)	0.24	0.24	7.1 - 7.7	5.6 - 5.9

TABLE 73: Water Vapor, Carbon Dioxide and Oxygen Permeability Through Nylon 6 Film.

Material Family	NYLON 6				
Product Form	FILM				
Features	barrier properties	barrier properties	barrier properties	barrier properties	barrier properties
Reference Number	294	294	294	294	294

TEST CONDITIONS

Penetrant	water vapor	carbon dioxide		oxygen	
Temperature (°C)	37.8	22.8	22.8	22.8	22.8
Relative Humidity (%)	90	0	80	0	80

PERMEABILITY (source document units)

Vapor Transmission Rate (g · mil/100 in² · day)	25				
Gas Permeability (cm³ · mil/100 in² · day)		4.7	8.0	3.6	7.0

PERMEABILITY (normalized units)

Permeability Coefficient (cm³ · mm/m² · day · atm)		1.8	3.2	1.4	2.8
Vapor Transmission Rate (g · mm/m² · day)	9.8				

TABLE 74: Water Vapor, Oxygen, Nitrogen and Carbon Dioxide Permeability Through Allied Signal Capran 77K PVDC Coated Nylon 6 Film.

Material Family	NYLON 6			
Material Supplier/ Grade	ALLIED SIGNAL CAPRAN 77K			
Product Form	FILM			
Reference Number	285	285	285	285

MATERIAL CHARACTERISTICS

Sample Thickness	0.0254 mm			

MATERIAL COMPOSITION

Note	PVDC coated			

TEST CONDITIONS

Penetrant	water vapor	oxygen	nitrogen	carbon dioxide
Temperature (°C)	37.8	23	23	23
Relative Humidity (%)	90	0	0	0
Test Note		STP conditions		

PERMEABILITY (source document units)

	water vapor	oxygen	nitrogen	carbon dioxide
Vapor Transmission Rate (g/day $\cdot$ 100 in^2)	0.2			
Gas Permeability (cm^3/100 in^2 $\cdot$ day $\cdot$ atm)		0.5	0.1	1.4

PERMEABILITY (normalized units)

	water vapor	oxygen	nitrogen	carbon dioxide
Permeability Coefficient (cm^3 $\cdot$ mm/m^2 $\cdot$ day $\cdot$ atm)		0.2	0.04	0.55
Vapor Transmission Rate (g $\cdot$ mm/m^2 $\cdot$ day)	0.08			

Nylon 6

TABLE 75: Mechanical Properties of BASF Ultramid B Nylon 6 Film.

Material Family	NYLON 6			
Material Supplier/ Grade	BASF ULTRAMID B4			BASF ULTRAMID B36F
Features	moderate flow	unstretched	biaxially stretched	enhanced clarity, moderate flow
Manufacturing Method	flat film, tubular film			flat film, tubular film
Reference Number	93	252	252	93

MATERIAL CHARACTERISTICS

Sample Thickness	0.02 - 0.1 mm	0.02 - 0.025 mm	0.02 mm	0.02 - 0.1 mm

TEST CONDITIONS

Temperature (°C)	23	20	20	23

MECHANICAL PROPERTIES

Tensile Strength @ Break (MPa)		75 - 80	240 - 250	
Tensile Strength @ Break - machine direction (MPa)	80 - 100			80 - 100
Tensile Strength @ Break - transverse direction (MPa)	70 - 90			70 - 90
Ultimate Elongation (%)		290 - 330	60 - 80	
Ultimate Elongation - Machine direction (%)	350 - 450			350 - 450
Ultimate Elongation - Transverse direction (%)	400 - 500			400 - 500

TABLE 76: Film Properties of Allied Signal Capran Nylon 6 Film.

Material Family	NYLON 6
Material Supplier/ Grade	ALLIED SIGNAL CAPRAN
Reference Number	284

MATERIAL CHARACTERISTICS

Sample Thickness	0.0254 mm
Specific Gravity	1.135 {bs}
Yield (in²/lb)	24,500

MECHANICAL PROPERTIES

Modulus Of Elasticity - MD (MPa)	621 - 759 {r}
Modulus Of Elasticity - TD (MPa)	725 - 863 {r}
Tensile Strength - MD (MPa)	69 - 110 {r}
Tensile Strength - TD (MPa)	69 - 110 {r}
Ultimate Elongation - MD (%)	375 - 500 {r}
Ultimate Elongation - TD (%)	375 - 500 {r}
Impact Strength (kg-cm)	4.4 {bu}
Burst Strength (MPa)	> 0.069 - 0.124 {bt}
Tear Strength, Propagated, Elmendorf - MD (g)	20 - 50 {k}
Tear Strength, Propagated, Elmendorf - TD (g)	20 - 50 {k}
Tear Strength, Graves - MD (g)	500 - 600 {az}
Tear Strength, Graves - TD (g)	470 - 520 {az}
Folding Endurance - MD (cycles)	> 250,000 {bx}
Folding Endurance - TD (cycles)	> 250,000 {bx}
Abrasion Resistance - weight loss (mg)	3 - 4 {bz}

THERMAL, OPTICAL AND PERFORMANCE PROPERTIES

Service Temperature (°C)	≤ 93 (continuous)
Service Temp. - Short Time (°C)	≤ 176 (15 min.)
Melting Point (°C)	220 {ca}
Brittleness Temperature (°C)	< -59 {cb}
Specific Heat (cal/g/°C)	0.4
Haze (%)	2.5 - 5.0 {c}
Gloss @ 20°	70 - 100 {m}
Shrink - 149°C Air, MD (%)	< 2 {bv}
Shrink - 149°C Air, TD (%)	< 2 {bv}

Chapter 20

Nylon 66

Permeability

DuPont: Zytel

Permeation rate is difficult to measure accurately. It will vary with pressure, temperature and even thickness of the container.

Zytel is a barrier to fuels and lubricants and to some gases, including most Freon gases.

Reference: *Design Handbook For Du Pont Engineering Plastics - Module II,* supplier design guide (E-42267) - Du Pont Engineering Polymers.

Film Properties and Applications

DuPont Canada: Dartek   (features: amber tint; product form: film)

Dartek is a strong transparent nylon film made from nylon 66 polymer - the condensation product of adipic acid and hexamethylinediamine. Type 66 nylon gives Dartek a broader heat seal range in laminate form than that of nylon 6 and greater resistance to thinning in thermo-forming operations. Dartek is durable and has a combination of properties, which make it suitable for packaging and industrial applications. Dartek exhibits excellent formability over a broad temperature range. It is a barrier to oils, greases and odors. Dartek has a low permeability to oxygen, nitrogen and carbon dioxide. It has good yield, elongation and fold endurance and it resists abrasion well. Dartek retains all these qualities, remaining flexible and tough from 26.7°C to 204.4°C (80°F to 400°F). Dartek can be printed, laminated or extrusion coated, and is available in a variety of thicknesses.

Reference: *Dartek Film Data Sheets,* supplier technical report (H-27768) - DuPont Canada, 1990.

DuPont Canada: Dartek B-601   (features: barrier properties; note: PVDC coated; product form: film)

Dartek B-601 is a strong transparent nylon type 66 film which has a PVDC coating applied to one side for enhanced barrier to oxygen, moisture, grease and odor. Because the base sheet is from the Dartek F series the coated film inherits many excellent properties such as broader heat resistance range, superior optics and good thermoforming characteristics.

Dartek B-601 can br printed, laminated, or extrusion coated. In converter combinations, it can be used for any packaging or industrial end use requiring high barrier properties. The Thermoforming ability of the film makes it applicable to assorted shapes and products such as meats and cheeses.

Reference: *Dartek Film Data Sheets,* supplier technical report (H-27768) - DuPont Canada, 1990.

DuPont Canada: Dartek B-602   (features: barrier properties; note: PVDC coated; product form: film)

Dartek B-602 is a strong transparent nylon 66 film which has a PVDC coating applied to one side for enhanced barrier to oxygen, moisture, grease and odor. B-602 is specially formulated for use in high humidity applications. Because the base sheet is from the Dartek F Series, the coated film inherits many excellent properties such as high heat resistance, superior optics and good thermoforming characteristics.

Dartek B-602 can be printed, laminated, or extrusion coated. In converter combinations, it can be used for any packaging or industrial end use requiring high barrier properties. B-602 can be easily thermoformed for assorted shapes and products such as meats and cheeses.

Reference: *Dartek Film Data Sheets,* supplier technical report (H-27768) - DuPont Canada, 1990.

<u>TABLE 77</u>: Oxygen, Carbon Dioxide and Water Vapor Permeability Through BASF Ultramid A5 Nylon 66 Film.

Material Family	NYLON 66					
Material Supplier/ Grade	BASF ULTRAMID A5					
Features	low flow	low flow	low flow	low flow	low flow	low flow
Manufacturing Method	flat film	tubular film	flat film	tubular film	flat film	tubular film
Reference Number	93	93	93	93	93	93

MATERIAL CHARACTERISTICS

Sample Thickness	0.02 - 0.1 mm	0.02 - 0.1 mm	0.02 - 0.1 mm	0.02 - 0.1 mm	0.02 - 0.1 mm	0.02 - 0.1 mm

TEST CONDITIONS

Penetrant	oxygen		carbon dioxide		water vapor	
Temperature (°C)	23	23	23	23	23	23
Relative Humidity (%)	40	40	0	0	85%-0% gradient	85%-0% gradient
Test Method	DIN 53380	DIN 53380	DIN 53380	DIN 53380	DIN 53122	DIN 53122

PERMEABILITY (source document units)

Gas Permeability ($cm^3 \cdot 100 \ \mu m/m^2 \cdot day \cdot bar$)	6 - 7	3 - 4	45	30		
Vapor Transmission Rate ($g \cdot 100 \ \mu m/m^2 \cdot day$)					11 - 12	8

PERMEABILITY (normalized units)

Permeability Coefficient ($cm^3 \cdot mm/m^2 \cdot day \cdot atm$)	0.61 - 0.71	0.3 - 0.41	4.6	3.0		
Vapor Transmission Rate ($g \cdot mm/m^2 \cdot day$)					1.1 - 1.2	0.8

Nylon 66

TABLE 78: Oxygen, Carbon Dioxide and Nitrogen Permeability Through DuPont Canada Dartek Nylon 66 Film.

Material Family	NYLON 66					
Material Supplier	DUPONT CANADA					
Trade Name/Grade	DARTEK		DARTEK B-601	DARTEK B-602	DARTEK	
Product Form	FILM					
Features	transparent	transparent	barrier properties	barrier properties	transparent	transparent
Reference Number	276	276	276	276	276	276

MATERIAL CHARACTERISTICS

Sample Thickness	0.0254 mm	0.0254 mm	0.0254 mm	0.038 mm	0.0254 mm	0.0254 mm

MATERIAL COMPOSITION

Note			PVDC coated	PVDC coated		

TEST CONDITIONS

Penetrant	oxygen				carbon dioxide	nitrogen
Temperature (°C)	23	23	23	23	23	23
Relative Humidity (%)	0	100	0	0	0	0
Test Method	ASTM D1434-66, method V	ASTM D1434-66, method V	ASTM D1434-66	ASTM D1434-66	ASTM D1434-66, method V	
Test Apparatus						isotactic gas permeability cell

PERMEABILITY (source document units)

Gas Permeability ($cm^3/m^2 \cdot day$)			7.7	7.7		
Gas Permeability ($cm^3/100\ in^2 \cdot day \cdot atm$)	3.5	16.0	0.5	0.5	16.0	0.7

PERMEABILITY (normalized units)

Permeability Coefficient ($cm^3 \cdot mm/m^2 \cdot day \cdot atm$)	1.4	6.3	0.2	0.29	6.3	0.28

TABLE 79: Oxygen, Carbon Dioxide, Nitrogen, Helium and Water Vapor Permeability Through DuPont Company Zytel 42 Nylon 66 Film.

Material Family	NYLON 66					
Material Supplier/ Grade	DUPONT ZYTEL 42					
Product Form	FILM					
Features	low flow					
Reference Number	68	68	68	68	68	68

TEST CONDITIONS

Penetrant	water vapor		oxygen	carbon dioxide	nitrogen	helium
Temperature (°C)	23	23	23	23	23	23
Relative Humidity (%)	50	100	50	50	50	50

PERMEABILITY (source document units)

Gas Permeability ($cm^3 \cdot mil/100\ in^2 \cdot day$)			2	9	0.7	150
Vapor Transmission Rate ($g \cdot mil/100\ in^2 \cdot day \cdot atm$)	1.0	20				

PERMEABILITY (normalized units)

Permeability Coefficient ($cm^3 \cdot mm/m^2 \cdot day \cdot atm$)			0.79	3.5	0.28	59.1
Vapor Transmission Rate ($g \cdot mm/m^2 \cdot day$)	0.39	7.9				

Nylon 66

TABLE 80: Oxygen Gas and Water Vapor Permeability Through BASF Ultramid A5 Nylon 66 Film.

Material Family	NYLON 66	
Material Supplier/ Grade	BASF ULTRAMID A5	
Manufacturing Method	blown film	blown film
Reference Number	252	252

MATERIAL CHARACTERISTICS

Sample Thickness	0.05 mm	0.05 mm

TEST CONDITIONS

Penetrant	water vapor	oxygen
Temperature (°C)	20	20
Relative Humidity (%)	85-0% gradient	40
Test Method	DIN 53122	DIN 53380

PERMEABILITY (source document units)

Vapor Transmission Rate (g/m² · day)	30	
Gas Permeability (cm³/m² · day · bar)		15

PERMEABILITY (normalized units)

Permeability Coefficient (cm³ · mm/m² · day · atm)		0.76
Vapor Transmission Rate (g · mm/m² · day)	1.5	

TABLE 81: Water Vapor Permeability Through DuPont Canada Dartek Nylon 66 Film.

Material Family	NYLON 66		
Material Supplier	DUPONT CANADA		
Trade Name/ Grade	DARTEK	DARTEK B-601	DARTEK B-602
Product Form	FILM		
Features	transparent	barrier properties	barrier properties
Reference Number	276	276	276

MATERIAL CHARACTERISTICS

Sample Thickness	0.0254 mm	0.0254 mm	0.038 mm

MATERIAL COMPOSITION

Note		PVDC coated	PVDC coated

TEST CONDITIONS

Penetrant	water vapor		
Temperature (°C)	23	38	38
Relative Humidity (%)	100	90	90
Test Method	ASTM E398-70	ASTM F372	ASTM F372
Test Apparatus	Honeywell MVTR tester		

PERMEABILITY (source document units)

Vapor Transmission Rate $(g/m^2 \cdot day)$		9.0	9.0
Vapor Transmission Rate $(g/day \cdot 100\ in^2)$	19.0	0.6	0.6

PERMEABILITY (normalized units)

Vapor Transmission Rate $(g \cdot mm/m^2 \cdot day)$	7.5	0.23	0.34

TABLE 84: Film Properties of DuPont Canada Dartek Nylon 66 Film.

Material Family	NYLON 66		
Material Supplier	DUPONT CANADA		
Trade Name/ Grade	DARTEK	DARTEK B-601	DARTEK B-602
Product Form	FILM		
Features	transparent	barrier properties	barrier properties
Reference Number	276	276	276

MATERIAL CHARACTERISTICS

Sample Thickness	0.0254 mm	0.0254 mm	
Specific Gravity	1.14 {aw}	1.15 {aw}	1.15 {aw}
Yield (in²/lb)	24,500	20,600	14,600
Note		PVDC coated	PVDC coated

MECHANICAL PROPERTIES

Modulus Of Elasticity (MPa)	689 {ay}		
Modulus Of Elasticity - MD (MPa)		689 {ay}	689 {ay}
Modulus Of Elasticity - TD (MPa)		689 {ay}	689 {ay}
Tensile Strength - MD (MPa)	82.7 {ay}	62 {ay}	68.9 {ay}
Tensile Strength - TD (MPa)	75.8 {ay}	62 {ay}	68.9 {ay}
Ultimate Elongation - MD (%)	350 {ay}	300 {ay}	300 {ay}
Ultimate Elongation - TD (%)	350 {ay}	300 {ay}	300 {ay}
Drop Dart Impact Strength (g)	600 {bc}	600 {bc}	600 {bc}
Burst Strength (MPa)	0.12 {bb}		
Tear Strength, Propagated, Elmendorf - machine (g/mm)	1378 {ba}	1378 {ba}	1378 {ba}
Tear Strength, Propagated, Elmendorf - transverse (g/mm)	1181 {ba}	1181 {ba}	1181 {ba}
Tear Strength, Initial, Graves - MD (g/mm)	23,622 {az}	23,622 {az}	23,622 {az}
Tear Strength, Initial, Graves - TD (g/mm)	23,622 {az}	23,622 {az}	23,622 {az}
Folding Endurance - MD (cycles)	1,000,000 {bd}		
Coefficient Of Friction - Static	0.6 {be}		
Coefficient Of Friction - Static	0.8 {bf}		
Coefficient Of Friction - Kinetic	0.45 {be}		
Coefficient Of Friction - Kinetic, film to coating		0.45 {be}	0.45 {be}

Nylon 66

<u>TABLE 84 (cont'd):</u> Film Properties of DuPont Canada Dartek Nylon 66 Film.

Material Family	NYLON 66		
Material Supplier	DUPONT CANADA		
Trade Name/ Grade	DARTEK	DARTEK B-601	DARTEK B-602
Product Form	FILM		
Features	transparent	barrier properties	barrier properties
Reference Number	276	276	276

THERMAL PROPERTIES

Service Temperature (°C)	-73.3 - 232.2		
Melting Point (°C)	265.6 {ax}		
Specific Heat (cal/g/°C)	0.4		
Coefficient Of Thermal Expansion (mm/mm/°C)	4.5E-5 {bg}		
Thermal Conductivity (cal-cm/cm²-sec-°C)	5.85 {bh}		

OPTICAL PROPERTIES

Haze (%)		1.5 {ac}	1.5 {ac}
Gloss @ 20°		150 {m}	150 {m}

PERFORMANCE CHARACTERISTICS

Dimensional Stability - MD (%)			1.5 {bi}
Dimensional Stability - TD (%)			0.5 {bi}

<u>TABLE 85</u>: **Gas Permeability and Water Vapor Transmission Through BASF Ultramid C Nylon 6/66 Film.**

Material Family	NYLON 6/66		
Material Supplier/ Grade	BASF ULTRAMID C35		
Features	moderate to high flow		
Manufacturing Method	flat film, tubular film		
Reference Number	93	93	93

MATERIAL CHARACTERISTICS

Sample Thickness	0.02 - 0.1 mm		

TEST CONDITIONS

Penetrant	oxygen	carbon dioxide	water vapor
Temperature (°C)	23	23	23
Relative Humidity (%)	40	0	85%-0% gradient
Test Method	DIN 53380	DIN 53380	DIN 53122

PERMEABILITY (source document units)

	oxygen	carbon dioxide	water vapor
Gas Permeability ($cm^3 \cdot 100\ \mu m/m^2 \cdot day \cdot bar$)	8 - 9	40 - 45	
Vapor Transmission Rate ($g \cdot 100\ \mu m/m^2 \cdot day$)			15 - 18

PERMEABILITY (normalized units)

	oxygen	carbon dioxide	water vapor
Permeability Coefficient ($cm^3 \cdot mm/m^2 \cdot day \cdot atm$)	0.81 - 0.91	4.0 - 4.6	
Vapor Transmission Rate ($g \cdot mm/m^2 \cdot day$)			1.5 - 1.8

TABLE 86: Water Vapor, Oxygen, Carbon Dioxide and Nitrogen Permeability Through Allied Signal Capran Nylon 6/66 Film.

Material Family	NYLON 6/66				
Material Supplier/ Trade Name	ALLIED SIGNAL CAPRAN				
Product Form	FILM				
Reference Number	284	284	284	284	284

MATERIAL CHARACTERISTICS

Sample Thickness	0.0254 mm	0.0254 mm	0.0254 mm	0.0254 mm	0.0254 mm

TEST CONDITIONS

Penetrant	water vapor	oxygen		carbon dioxide	nitrogen
Temperature (°C)	37.8		23	23	23
Relative Humidity (%)	90	0 (dry)	90 (wet)	dry	dry
Test Method	cup method	ASTM D3985	permeability cell	ASTM D1435; Dow Cell	ASTM D1435; Dow Cell

PERMEABILITY (source document units)

Vapor Transmission Rate (g/m² · day)	341				
Gas Permeability (cm³/m² · day)		37.2	232.5	113.2	7.75

PERMEABILITY (normalized units)

Permeability Coefficient (cm³ · mm/m² · day · atm)		0.94	5.91	2.88	0.2
Vapor Transmission Rate (g · mm/m² · day)	8.7				

TABLE 87: Air Conditioning Refrigerants Permeation Loss Through Nylon 6/66 Copolymer.

Material Family	NYLON 6/66		
Reference Number	275	275	275

MATERIAL CHARACTERISTICS

Sample Thickness	1 mm	1 mm	1 mm
Sample Length	305 mm	305 mm	305 mm
Sample Inside Diameter	15.9 mm	15.9 mm	15.9 mm

TEST CONDITIONS

Penetrant	Freon 12	HCFCX-134a	HCFC-22/ HCFC-124/ HFC-152a
Penetrant Note	air conditioning refrigerant;@ saturated vapor pressure		air conditioning refrigerant, ternary blend; @ saturated vapor pressure
Temperature (°C)	93	93	93
Test Note	calculated from permeation coefficient data		

PERMEABILITY (source document units)

Permeation Loss (lb/ft-yr)	0.067	0.077	0.178

TABLE 88: Film Properties of Nylon 66/6 Copolymer Film.

Material Family	NYLON 6/66	
Material Supplier/ Grade	ALLIED SIGNAL CAPRAN	BASF ULTRAMID C35
Product Form	FILM	
Features		moderate to high flow
Manufacturing Method		flat film, tubular film
Reference Number	284	93

MATERIAL CHARACTERISTICS

Sample Thickness	0.0254 mm	0.02 - 0.1 mm
Specific Gravity	1.1275 {bs}	
Yield (in²/lb)	24,570	

MECHANICAL PROPERTIES

Modulus Of Elasticity - MD (MPa)	345-483 {r}	
Modulus Of Elasticity - TD (MPa)	345-483 {r}	
Tensile Strength - MD (MPa)	96-138 {r}	80-90
Tensile Strength - TD (MPa)	83-124 {r}	70-90
Ultimate Elongation - MD (%)	350-450 {r}	500-600
Ultimate Elongation - TD (%)	350-450 {r}	500-600
Tear Strength, Propagated, Elmendorf -MD (g)	50-90 {k}	
Tear Strength, Propagated, Elmendorf - TD (g)	50-90 {k}	
Tear Strength, Graves - MD (g)	450-600 {az}	
Tear Strength, Graves - TD (g)	550-650 {az}	
Folding Endurance - MD (cycles)	>1000 {by}	
Folding Endurance - TD (cycles)	>1000 {by}	

OPTICAL PROPERTIES

Haze (%)	3.5-4.5 {c}	
Gloss @ 20°	95-100 {m}	

PERFORMANCE PROPERTIES

Unrestrained Shrink - 149°C Air, MD (%)	<2 {bw}	
Unrestrained Shrink - 149°C Air, TD (%)	<2 {bw}	

Chapter 22

Nylon 66/610

Film Properties and Applications

EMS-American Grilon: Grilon XE3303 (features: transparent, barrier properties; product form: film)

Grilon XE 3303 is a nylon 66/610 film grade copolymer with excellent processing and performance characteristics. It can be used as an inner, middle or outer layer in blown or cast, coextruded multi-layer film.

Films produced with Grilon XE 3303 exhibit excellent aroma barriers and resistance to the permeation of oxygen, carbon dioxide, nitrogen and water vapor. The oxygen barrier properties are not affected by steam sterilization or pasteurization. Grilon XE 3303 conforms to the requirements of 21 CFR 177.1500 for a nylon 66/610 in direct contact with food, with no temperature or thickness restrictions.

This product should be successful in cook-in film structures (i.e., for precooked ham, turkey, etc.). A variety of packaging applications are expected for Grilon XE 3303 due to its good barrier characteristics, clarity, high shrinkage, FDA status and suitability for use at cooking temperatures of 80°C to 100°C (180°F to 212°F).

Composite films that exhibit good resistance to steam sterilization and pasteurization can be produced including Grilon XE 3303 as part of the structure. Due to its unique properties, Grilon XE 3303 is particularly suited for use in blood, colostomy, and I.V. bags, as well as other medical packaging applications.

Reference: *Product Data Bulletin - Grivory G21,* supplier marketing literature (GV8-104) - EMS - American Grilon Inc.

TABLE 89: Oxygen, Carbon Dioxide, Nitrogen and Water Vapor Permeability Through EMS-American Grilon Grilon XE3303 Nylon 66/610 Copolymer Film.

Material Family	NYLON 66/610							
Material Supplier/ Grade	EMS-AMERICAN GRILON GRILON XE3303							
Product Form	FILM							
Features	barrier properties, transparent							
Reference Number	307	307	307	307	307	307	307	307

MATERIAL CHARACTERISTICS

Sample Thickness	0.05 mm	0.05 mm	0.05 mm	0.05 mm	0.05 mm	0.05 mm	0.05 mm	0.05 mm

TEST CONDITIONS

Penetrant	oxygen					carbon dioxide	nitrogen	water vapor
Temperature (°C)	23				23	23	23	23
Relative Humidity (%)	50	85	85	85	100	50	50	

PRE EXPOSURE CONDITIONING

Exposure Type			pasteurization	steam sterilization				

PERMEABILITY (source document units)

Gas Permeability ($cm^3/m^2 \cdot day \cdot bar$)	55	74	74	74	75	185	12	
Vapor Transmission Rate ($g/m^2 \cdot day$)								14

PERMEABILITY (normalized units)

Permeability Coefficient ($cm^3 \cdot mm/m^2 \cdot day \cdot atm$)	2.8	3.8	3.8	3.8	3.8	9.4	0.61	
Vapor Transmission Rate ($g \cdot mm/m^2 \cdot day$)								0.7

Chapter 23

Nylon MXD6

TABLE 90: Oxygen Permeability at Different Temperatures and Water Vapor Transmission Through Nylon MXD6.

Material Family	NYLON MXD6				
Features	barrier properties				
Reference Number	264	264	264	264	264

TEST CONDITIONS

Penetrant	oxygen				water vapor
Temperature (°C)	5	23	35	50	40
Relative Humidity (%)	0	0	0	0	90

PERMEABILITY (source document units)

Gas Permeability (cm$^3 \cdot$ mil/100 in$^2 \cdot$ day)	0.043	0.15	0.28	0.92	
Gas Permeability (cm$^3 \cdot$ 25μ/m$^2 \cdot$ day $\cdot$ atm)	0.67	2.325	4.43	14.26	
Vapor Transmission Rate (g $\cdot$ mil/100 in$^2 \cdot$ day)					3.2
Vapor Transmission Rate (g $\cdot$ 25μ/m$^2 \cdot$ day)					50

PERMEABILITY (normalized units)

Permeability Coefficient (cm$^3 \cdot$ mm/m$^2 \cdot$ day $\cdot$ atm)	0.02	0.06	0.11	0.36	
Vapor Transmission Rate (g $\cdot$ mm/m$^2 \cdot$ day)					1.26

TABLE 91: Oxygen Permeability vs. Relative Humidity Through Nylon MXD6.

Material Family	NYLON MXD6	
Reference Number	296	296

TEST CONDITIONS

Penetrant	oxygen	
Temperature (°C)	22.8	22.8
Relative Humidity (%)	0	90
Test Method	ASTM D3895	ASTM D3895

PERMEABILITY (source document units)

Gas Permeability ($cm^3 \cdot mil/ 100\ in^2 \cdot bar \cdot day$)	0.18	0.8

PERMEABILITY (normalized units)

Permeability Coefficient ($cm^3 \cdot mm/m^2 \cdot day \cdot atm$)	0.07	0.32

GRAPH 29: Oxygen Permeability vs. Relative Humidity through Nylon MXD6.

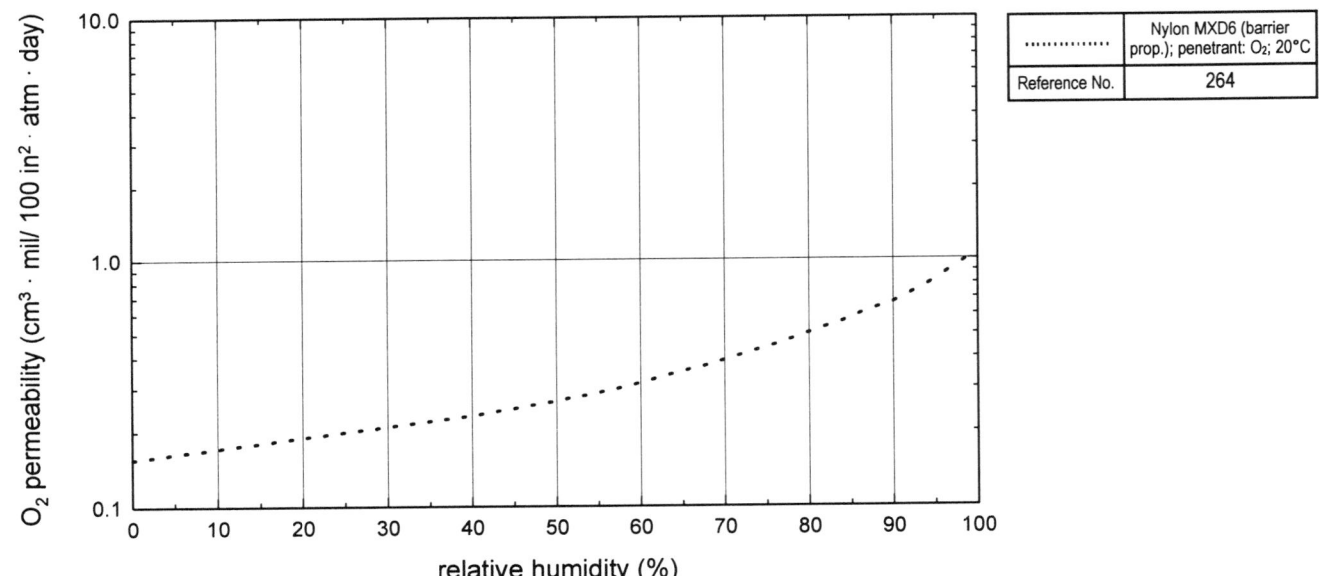

....	Nylon MXD6 (barrier prop.); penetrant: O₂; 20°C
Reference No.	264

Nylon MXD6

Chapter 24

Polycarbonate

Barrier Performance

Dow Chemical: Calibre (features: transparent)

Polycarbonate is not generally considered a good barrier material, although it is possible to use polycarbonate as the structural layer in a composite (coextruded) film for use in barrier applications. In such cases, polycarbonate contributes toughness and heat resistance to the final product while other components in the composite film may provide the barrier properties.

Reference: *Calibre Engineering Thermoplastics Basic Design Manual,* supplier design guide (301-1040-1288) - Dow Chemical Company, 1988.

<u>TABLE 92:</u> **Oxygen, Carbon Dioxide And Nitrogen Permeability Through Dow Chemical Calibre Polycarbonate.**

Material Family	POLYCARBONATE								
Material Supplier/ Trade Name	DOW CHEMICAL CALIBRE								
Grade	300-4	300-15	800-6	300-4	300-15	800-6	300-4	300-15	800-6
Features	general purpose grade, transparent	general purpose grade, transparent	flame retardant, transparent	general purpose grade, transparent	general purpose grade, transparent	flame retardant, transparent	general purpose grade, transparent	general purpose grade, transparent	flame retardant, transparent
Reference Number	78	78	78	78	78	78	78	78	78

MATERIAL CHARACTERISTICS

Melt Flow Index	4 g/10 min.	15 g/10 min.	6 g/10 min.	4 g/10 min.	15 g/10 min.	6 g/10 min.	4 g/10 min.	15 g/10 min.	6 g/10 min.

TEST CONDITIONS

Penetrant	nitrogen			oxygen			carbon dioxide		
Test Method	ASTM D2752								

PERMEABILITY (source document units)

Gas Permeability ($cm^3 \cdot mil/100\ in^2 \cdot day$)	31	27	57	260	230	314	1950	1720	2100

PERMEABILITY (normalized units)

Permeability Coefficient ($cm^3 \cdot mm/m^2 \cdot day \cdot atm$)	12.2	10.6	22.4	102	90.6	124	768	677	827

TABLE 93: Water Vapor, Carbon Dioxide and Oxygen Permeability Through Polycarbonate Film.

Material Family	POLYCARBONATE			
Product Form	FILM			
Reference Number	294	264	294	294

TEST CONDITIONS

Penetrant	water vapor	water vapor	carbon dioxide	oxygen
Temperature (°C)	37.8	40	22.8	22.8
Relative Humidity (%)	90	90	0	0

PERMEABILITY (source document units)

Vapor Transmission Rate ($g \cdot mil/100\ in^2 \cdot day$)	9.7	11		
Vapor Transmission Rate ($g \cdot 25\mu/m^2 \cdot day$)		170.5		
Gas Permeability ($cm^3 \cdot mil/100\ in^2 \cdot day$)			780	260

PERMEABILITY (normalized units)

Permeability Coefficient ($cm^3 \cdot mm/m^2 \cdot day \cdot atm$)			307	102
Vapor Transmission Rate ($g \cdot mm/m^2 \cdot day$)	3.82	4.33		

Polycarbonate

TABLE 94: Water Vapor, Oxygen, Nitrogen and Carbon Dioxide Permeability Through Bayer AG Makrolon Polycarbonate Film.

Material Family	POLYCARBONATE			
Material Supplier/ Grade	BAYER MAKROLON			
Product Form	FILM			
Reference Number	289	289	289	289

MATERIAL CHARACTERISTICS

Sample Thickness	0.1 mm	0.1 mm	0.1 mm	0.1 mm

TEST CONDITIONS

Penetrant	oxygen	nitrogen	carbon dioxide	water vapor
Temperature (°C)				23
Relative Humidity (%)				85
Test Method	DIN 53380, pt. 3	DIN 53380, pt. 3	DIN 53380, pt. 3	DIN 53122

PERMEABILITY (source document units)

Gas Permeability ($cm^3/m^2 \cdot day \cdot bar$)	670	110	4300	
Vapor Transmission Rate ($g/m^2 \cdot day$)				15 (approximate)

PERMEABILITY (normalized units)

Permeability Coefficient ($cm^3 \cdot mm/m^2 \cdot day \cdot atm$)	67.9	11.2	436	
Vapor Transmission Rate ($g \cdot mm/m^2 \cdot day$)				1.5 (approximate)

Chapter 25

Polybutylene Terephthalate

Permeability

BASF AG: Ultradur B4550 (features: 0.25 mm thick)

The barier properties of Ultradur B 4550 film can be greatly improved by vacuum metallizing with aluminum. The resulting bond is sufficiently strong to withstand the adhesive tape test.

Reference: *Ultradur Polybutylene Terephthalate (PBT) Product Line, Properties, Processing,* supplier design guide (B 575/1e - (819) 4.91) - BASF Aktiengesellschaft, 1991.

Heat Sealing

BASF AG: Ultradur B4550 (features: 0.25 mm thick)

The best means of heat-sealing Ultradur B 4550 monofilm is by ultrasonic techniques. Thermal impulse, slit-seal techniques are also feasible, but cause a white zone near the seam due to postcrystallization.

Reference: *Ultradur Polybutylene Terephthalate (PBT) Product Line, Properties, Processing,* supplier design guide (B 575/1e - (819) 4.91) - BASF Aktiengesellschaft, 1991.

TABLE 95: Gas Permeability and Water Vapor Transmission Through BASF AG Ultradur Polybutylene Terephthalate.

Material Family	POLYBUTYLENE TEREPHTHALATE			
Material Supplier/ Grade	BASF AG ULTRADUR B4550			
Reference Number	180	180	180	180

MATERIAL CHARACTERISTICS

Sample Thickness	0.25 mm	0.25 mm	0.25 mm	0.25 mm

TEST CONDITIONS

Penetrant	water vapor	nitrogen	oxygen	carbon dioxide
Temperature (°C)	23	23	23	23
Relative Humidity (%)	85%-0% gradient	50	50	50
Test Method	DIN 53122	DIN 53380	DIN 53380	DIN 53380
Test Condition Note	standard laboratory atmosphere	standard laboratory atmosphere	standard laboratory atmosphere	standard laboratory atmosphere

PERMEABILITY (source document units)

Gas Permeability ($cm^3/m^2 \cdot day \cdot bar$)		12	60	550
Vapor Transmission Rate ($g/m^2 \cdot day$)	10			

PERMEABILITY (normalized units)

Permeability Coefficient ($cm^3 \cdot mm/m^2 \cdot day \cdot atm$)		3.04	15.2	139
Vapor Transmission Rate ($g \cdot mm/m^2 \cdot day$)	2.5			

<u>TABLE 96:</u> **Mechanical and Optical Properties of BASF AG Ultradur Polybutylene Terephthalate Film.**

Material Family	POLYBUTYLENE TEREPHTHALATE
Material Supplier/ Grade	**BASF AG ULTRADUR B4550**
Reference Number	180

MATERIAL CHARACTERISTICS

Sample Thickness	0.25 mm

TEST CONDITIONS

Temperature (°C)	23
Relative Humidity	50

MECHANICAL PROPERTIES

Yield Strength - MD (MPa)	30-35 {a}
Yield Strength - TD (MPa)	30-35 {a}
Tensile Strength @ Break - MD (MPa)	75-80 {a}
Tensile Strength @ Break - TD (MPa)	75-80 {a}
Ultimate Elongation - MD (%)	450-500 {a}
Ultimate Elongation - TD (%)	450-500 {a}

OPTICAL PROPERTIES

Haze (%)	1 {b}

Chapter 26

Polyethylene Naphthalate

Product Summary

Eastman Chemical: PEN Homopolymer 14991 (features: transparent, barrier properties)

Eastman PEN Homopolymer is a high performance polymer which offers a high glass transition temperature, dimensional stability at elevated temperatures and excellent barrier to gases and moisture. Compared with PET homopolymer it also possesses greater solvent resistance, a higher melt point, greater modulus, higher tensile strength, improved hydrolytic and thermal stability and improved UV properties.

Potential Uses Include:

- Biaxially oriented films which offer improved physical properties compared with OPET and are UL listed for continuous use up to 155°C.

- Thermoformed and blow molded containers capable of undergoing pasteurization while maintaining their crystal clarity. Such containers would also offer improved gas and moisture barrier over containers made from PET homopolymer.

Reference: *Eastman Chemical Product Data Sheets,* supplier marketing literature - Eastman Chemical Company, 1994.

TABLE 97: Water Vapor and Oxygen Permeability Through Eastman Chemical Polyethylene Naphthalate (PEN) Homopolymer 14991.

Material Family	POLYETHYLENE NAPHTHALATE	
Material Supplier/ Grade	Eastman Chemical PEN Homopolymer 14991	
Product Form	FILM	
Features	transparent, barrier properties	
Reference Number	312	312

MATERIAL CHARACTERISTICS

Sample Thickness	0.25 mm	0.25 mm

TEST CONDITIONS

Penetrant	water vapor	oxygen
Test Method	ASTM F372	ASTM D3985

PERMEABILITY (source document units)

Gas Permeability ($cm^3 \cdot mm/m^2 \cdot day \cdot atm$)		1.5
Gas Permeability ($cm^3 \cdot mil/100\ in^2 \cdot day \cdot atm$)		3.8
Vapor Transmission Rate ($g/m^2 \cdot day$)	2.9	
Vapor Transmission Rate ($g/100\ in^2 \cdot day$)	0.2	

PERMEABILITY (normalized units)

Permeability Coefficient ($cm^3 \cdot mm/m^2 \cdot day \cdot atm$)		1.5
Vapor Transmission Rate ($g \cdot mm/m^2 \cdot day$)	0.73	

PEN

TABLE 98: Film Properties of Eastman Chemical PolyethyleneNaphthalate (PEN) Homopolymer 14991.

Material Family	Polyethylene Naphthalate
Material Supplier/ Grade	**Eastman Chemical PEN Homopolymer 14991**
Product Form	**FILM**
Features	transparent, barrier properties
Reference Number	312

MATERIAL CHARACTERISTICS

Sample Thickness	0.25 mm
Note	starting pellets I.V. (intrinsic viscosity) was 0.53 for the material used to extrude the film

MECHANICAL PROPERTIES

Modulus of Elasticity (MPa)	1900 {dl}
Tensile Strength @ Yield (MPa)	73 {r}
Tensile Strength @ Break (MPa)	56 {r}
Elongation @ Yield (%)	8.9 {r}
Elongation @ Break (%)	150 {r}

OPTICAL PROPERTIES

Haze (%)	2.5 {dm}

Chapter 27

Polyethylene Terephthalate

Product Summary

DuPont: Selar PT (applications: packaging; features: aroma barrier, flavor barrier)

An environmentally friendly, easy-to-recycle packaging material, SELAR PT polyester is an effective flavor and aroma barrier resin.

Reference: *Selar Barrier Resin Selector Guide,* supplier marketing literature (H-38769-1) - DuPont Company, 1992.

DuPont: Selar PT 4000 Series (applications: packaging; features: FDA grade, low flow)

The SELAR PT 4000 series features toughened PET polyester resins for monolayer and coextruded heat sealable film used for metallization and/or lamination. These toughened resins also can be used for ovenable trays. They are high melt viscosity (HMV) resins and comply with FDA regulations for direct food contact.

Reference: *Selar Barrier Resin Selector Guide,* supplier marketing literature (H-38769-1) - DuPont Company, 1992.

DuPont: Selar PT 5000 Series (applications: packaging; features: high impact)

SELAR PT 5000 impact-modified recycle-based PET resins are ideal for solvent barrier containers for non-food applications only.

Reference: *Selar Barrier Resin Selector Guide,* supplier marketing literature (H-38769-1) - DuPont Company, 1992.

DuPont: Selar PT 7000 Series (applications: packaging)

SELAR PT 7000 homopolymer PET resins include high melt viscosity (HMV) technology for sheeting, extrusion coating, and film applications. Intrinsic viscosities (IV) range from 0.68 to 1.02. Grades with high melt strength are suitable for CPET and extrusion blow-molded bottles up to 64 ounces. Standard rheology grades of 0.72 to 0.95 intrinsic viscosity are also available.

Reference: *Selar Barrier Resin Selector Guide,* supplier marketing literature (H-38769-1) - DuPont Company, 1992.

DuPont: Selar PT 8000 Series (applications: packaging; features: transparent, amorphous)

Copolymer grades of high melt viscosity (HMV) polyesters in the SELAR PT 8000 series are designed for amorphous applications such as thermoformed clear containers, blister packages, and coextruded sheet for heat seal use. A full range of high intrinsic viscosity (IV) grades are available for clear extrusion blow-molded bottles up to 16 ounces.

Reference: *Selar Barrier Resin Selector Guide,* supplier marketing literature (H-38769-1) - DuPont Company, 1992.

Shell Chemical: Cleartuf (applications: packaging; features: transparent, barrier properties)

Bottles made from CLEARTUF polyester resins, exhibit an overall balance of physical, mechanical, thermal and chemical properties. They provide important safety advantages such as superior impact and shatter resistance, making them ideal for packages used in bathrooms, kitchens and outdoors.

Bottles made from CLEARTUF resins offer a cost advantage due to their high strength and light weight. Total package costs for bottles made from CLEARTUF resins are less than what is possible with many competitive plastics and other conventional packaging material.

The wide range of properties offered by polyester resins make them versatile. Their low gas permeability, low weight, low cost and high strength and toughness make CLEARTUF resins the material of choice for many packaging applications. The popularity of packaging made from polyester resins began with soft drink bottling.

Low levels of extractables and minimal acetaldehyde generation also make containers made with CLEARTUF resins ideal packages for taste-sensitive foods and beverages. CLEARTUF resins provide the oil resistance needed for packaging products such as peanut butter and salad oils, and the flavor retention properties required for oral antiseptics.

Their ease of molding permits manufacturers to obtain the exotic shapes desirable for cosmetic products, air fresheners, and other products that rely on styling for added consumer appeal.

Reference: *Cleartuf PET Packaging Resins,* supplier technical report (SC: 1820-94) - Shell Chemical Company, 1994.

Packaging Applications, Properties and Permeability

DuPont: Selar PT (applications: food packaging; features: 0.23 mm thick, retort, hot fill, 60 mm diameter, 80 mm sample height; manufacturing method: thermoforming (DuPont Fortrex process); product form: cup)

The combination of package clarity, moderate oxygen barrier and thermal resistance for high temperature use has, until now, not been available for food packaging. With the development of a new patented process and modified polyester resin, now offered is a clear, heat-stable thermoformed polyester container for hot fill and retort applications. This new process, Du Pont's "Fortrex" process, will be licensed in conjunction with sale of proprietary SELAR PT PET resins to container manufacturers for food applications.

The Du Pont "Fortrex" process relies on a unique thermoforming process which must be combined with specially designed PET resins to give the combination of clarity and heat stability.

Food service opportunities, where product can be filled hot or reheated, are a logical choice for these clear polyester containers. Obviously the shelf life requirements of the food product must be considered, since the barrier of polyester is less than that of ethylene vinyl alcohol (EVOH), a resin of choice for high barrier plastic containers. The containers produced using the Du Pont "Fortrex" technology are particularly well suited for packages which currently have no barrier and require improved shelf life for distribution. The technology also addresses the need for convenience packages of products currently in glass or metal which require clarity, heat stability and moderate oxygen barrier. These containers have excellent toughness and resistance to deformation. These properties are retained at high temperatures. The containers' resistance to flexural deformation is also excellent up to 200°C. Thus containers will remain rigid and have good handling characteristics under hot fill, retort and reheating conditions.

The containers have moderate moisture barrier, oxygen barrier of about 5 cc · mil/100 in^2 · day · atm and this barrier performance is retained after retort. Sealing techniques for these containers should be similar to those used for other polyester thermoformed containers - including flexible lid stocks and double seamed metal ends.

Reference: *A New Clear Polyester Container For Hot fill And Retort Applications,* supplier technical report (H-44606) - DuPont Company, 1992.

Film Properties and Applications

DuPont: Mylar (product form: film)

Mylar is a strong durable film suited to many magnetic media, electrical, electronic, reprographic, industrial and packaging applications. It has excellent chemical, electrical and thermal properties and retains its physical properties over a wide temperature range typically 70°C to 150°C. The absence of plasticisers means that it does not become brittle with age.

Engineering applications include dielectric in capacitors; phase insulation for motor and generator field coils; a barrier and insulating tape in wire and cable; substrate for flexible printed circuits; membrane switches; hot stamping; thermal transfer materials; and separator sheet for reinforced plastics.

Reference: *Guide to Excellence- Research And Development,* supplier marketing literature - DuPont Company.

DuPont: **Mylar** (product form: film)

Food packaging is particularly suited to the strength, dimensional stability, barrier properties, heat-sealability and printability of Mylar. Mylar has good machinability and can be laminated or extrusion-coated to solve specialized packaging problems. Food packaging applications include "boil in bag" packages, lids for frozen food trays, microwave applications, oven wrap for meat and cheese and attractive bags for snack food.

Reference: *Mylar Polyester Film,* supplier technical report (E-99499) - DuPont Company, 1988.

TABLE 99: Carbon Dioxide, Oxygen and Water Vapor Permeability Through Shell Chemical Cleartuf Polyethylene Terephthalate Polyester (PET).

Material Family	POLYETHYLENE TEREPHTHALATE					
Material Supplier/ Grade	SHELL CHEMICAL CLEARTUF					
Features	barrier properties, transparent, unoriented	barrier properties, oriented, transparent	barrier properties, transparent, unoriented	barrier properties, oriented, transparent	barrier properties, transparent, unoriented	barrier properties, oriented, transparent
Applications	packaging	packaging	packaging	packaging	packaging	packaging
Reference Number	297	297	297	297	297	297

TEST CONDITIONS

Penetrant	carbon dioxide		oxygen		water vapor	
Temperature (°C)	25	25	25	25	37.8	37.8
Relative Humidity (%)	0	0	0	0	100	100
Test Method	ASTM D1434	ASTM D1434	ASTM D1434	ASTM D1434	ASTM E96	ASTM E96

PERMEABILITY (source document units)

Gas Permeability ($cm^3 \cdot mil/100\ in^2 \cdot day$)	20	12	10	5		
Vapor Transmission Rate ($g \cdot mil/100\ in^2 \cdot day$)					4	2

PERMEABILITY (normalized units)

Permeability Coefficient ($cm^3 \cdot mm/m^2 \cdot day \cdot atm$)	7.9	4.7	3.9	2.0		
Vapor Transmission Rate ($g \cdot mm/m^2 \cdot day$)					1.6	0.8

TABLE 100: Oxygen, Carbon Dioxide, Nitrogen and Hydrogen Permeability Through DuPont Company Mylar Polyester PET Film.

Material Family	POLYETHYLENE TEREPHTHALATE			
Material Supplier/ Grade	DUPONT MYLAR			
Product Form	FILM			
Reference Number	270	270	270	270

TEST CONDITIONS

Penetrant	carbon dioxide	hydrogen	nitrogen	oxygen
Temperature (°C)	25	25	25	25
Test Method	ASTM D1434-72	ASTM D1434-72	ASTM D1434-72	ASTM D1434-72

PERMEABILITY (source document units)

Gas Permeability ($cm^3 \cdot mil/100\ in^2 \cdot day$)	16	100	1	6
Gas Permeability ($\mu m^3 \cdot mm/m^2 \cdot sec \cdot Pa$)	1115	6970	70	418

PERMEABILITY (normalized units)

Permeability Coefficient ($cm^3 \cdot mm/m^2 \cdot day \cdot atm$)	6.3	39.4	0.39	2.4

TABLE 101: Organic Solvents Permeability Through Polyester PET Film.

Material Family	POLYETHYLENE TEREPHTHALATE			
Product Form	FILM			
Reference Number	266	266	266	266

MATERIAL CHARACTERISTICS

Sample Thickness	0.0254 mm	0.0254 mm	0.0254 mm	0.0254 mm

TEST CONDITIONS

Penetrant	chloroform	xylene	methyl ethyl ketone	kerosine
Temperature (°C)	20	20	20	20
Relative Humidity (%)	65	65	65	65

PERMEABILITY (source document units)

Vapor Transmission Rate ($g/day \cdot 100\ in^2$)	20.0	0.11	0.10	0.03

PERMEABILITY (normalized units)

Vapor Transmission Rate ($g \cdot mm/m^2 \cdot day$)	7.87	0.04	0.04	0.01

TABLE 102: Oxygen Permeability at Different Temperatures and Carbon Dioxide, Nitrogen and Helium Permeability Through Oriented Polyethylene Terephthalate Polyester.

Material Family	POLYETHYLENE TEREPHTHALATE						
Features	oriented						
Reference Number	264	264	264	264	264	264	264

TEST CONDITIONS

Penetrant	oxygen				carbon dioxide	nitrogen	helium
Temperature (°C)	5	23	35	50	35	23	35
Relative Humidity (%)	0	0	0	0	0	0	0

PERMEABILITY (source document units)

Gas Permeability ($cm^3 \cdot mil/100\ in^2 \cdot day$)	0.66	2.3	5.1	16.78	19.6	0.46	180
Gas Permeability ($cm^3 \cdot 25\mu/m^2 \cdot day \cdot atm$)	10.23	35.64	79.04	260	303.9	7.1	2790

PERMEABILITY (normalized units)

Permeability Coefficient ($cm^3 \cdot mm/m^2 \cdot day \cdot atm$)	0.26	0.91	2.01	6.61	7.72	0.18	70.9

TABLE 103: Oxygen Permeability vs. Relative Humidity Through Polyester PET Film.

Material Family	POLYETHYLENE TEREPHTHALATE			
Product Form	FILM			
Reference Number	265	265	265	265

TEST CONDITIONS

Penetrant	oxygen			
Temperature (°C)	20	20	20	20
Relative Humidity (%)	65	85	100	0

PERMEABILITY (source document units)

Gas Permeability ($cm^3 \cdot mil/100\ in^2 \cdot day$)	2.9	2.9	2.9	6.4

PERMEABILITY (normalized units)

Permeability Coefficient ($cm^3 \cdot mm/m^2 \cdot day \cdot atm$)	1.14	1.14	1.14	2.52

TABLE 104: Water Vapor Transmission and Oxygen Permeability Through Polyethylene Terephthalate Polyester (PET).

Material Family	POLYETHYLENE TEREPHTHALATE			
Features			biaxially oriented	
Reference Number	296	296	264	264

TEST CONDITIONS

Penetrant	oxygen	water vapor		
Temperature (°C)	22.8	37.8	40	40
Relative Humidity (%)	0	90	90	90
Test Method	ASTM D1434	ASTM F1249		

PERMEABILITY (source document units)

Gas Permeability (cm^3 · mil/ 100 in^2 · bar · day)	7			
Vapor Transmission Rate (g · mil/ 100 in^2 · bar · day)		4.25		
Vapor Transmission Rate (g · mil/100 in^2 · day)			1.2	1.3
Vapor Transmission Rate (g · 25µ/m^2 · day)			18.6	20.2

PERMEABILITY (normalized units)

Permeability Coefficient (cm^3 · mm/m^2 · day · atm)	2.79			
Vapor Transmission Rate (g · mm/m^2 · day)		1.7	0.47	0.51

TABLE 105: Water Vapor Transmission and Oxygen Permeability Through Coated and Uncoated Polyester PET Film.

Material Family	POLYETHYLENE TEREPHTHALATE									
Product Form	FILM									
Features		oriented		oriented		oriented		oriented		oriented
Reference Number	268	268	268	268	268	268	268	268	268	268

MATERIAL CHARACTERISTICS

Sample Thickness	0.012 mm	0.014 mm	0.012 mm	0.014 mm	0.012 mm	0.014 mm	0.012 mm	0.014 mm	0.012 mm	0.014 mm

MATERIAL COMPOSITION

Note		PVDC coated		PVDC coated		PVDC coated		PVDC coated		PVDC coated

TEST CONDITIONS

Penetrant	water vapor		oxygen							
Temperature (°C)	40	40	35	35	20	20	20	20	20	20
Relative Humidity (%)	90	90	0	0	65	65	85	85	100	100
Test Method	JIS Z0208	JIS Z0208	JIS Z1707	JIS Z1707	ASTM D3985	ASTM D3985	ASTM D3985	ASTM D3985	ASTM D3985	ASTM D3985

PERMEABILITY (source document units)

Vapor Transmission Rate (g · mil/100 in^2 · day)	3	1								
Gas Permeability (cm^3 · mil/100 in^2 · day)			12.9	1.1	5.81	0.52	5.81	0.52	5.81	0.52

PERMEABILITY (normalized units)

Permeability Coefficient (cm^3 · mm/m^2 · day · atm)			5.08	0.43	2.29	0.2	2.29	0.2	2.29	0.2
Vapor Transmission Rate (g · mm/m^2 · day)	1.2	0.39								

TABLE 106: Water Vapor, Acetone, Benzene, Carbon Tetrachloride, Ethyl Acetate and Hexane Permeability Through DuPont Company Mylar Polyester PET Film.

Material Family	POLYETHYLENE TEREPHTHALATE					
Material Supplier/ Grade	DUPONT MYLAR					
Product Form	FILM					
Reference Number	270	270	270	270	270	270

TEST CONDITIONS

Penetrant	acetone	benzene	carbon tetrachloride	ethyl acetate	hexane	water vapor
Temperature (°C)	40	25	40	40	40	37.8
Relative Humidity (%)						90
Test Method	ASTM E96-80	ASTM E96-80	ASTM E96-80	ASTM E96-80	ASTM E96-80	ASTM E96-80
Test Note	modified test, permeabilities determined at the partial pressure of the vapor at the test temperature					

PERMEABILITY (source document units)

	acetone	benzene	carbon tetrachloride	ethyl acetate	hexane	water vapor
Vapor Transmission Rate (g · mil/100 in^2 · day)	2.22	0.36	0.08	0.08	0.12	1.8
Vapor Transmission Rate (g · mm/m^2 · day)	0.87	0.14	0.03	0.03	0.05	0.7

PERMEABILITY (normalized units)

	acetone	benzene	carbon tetrachloride	ethyl acetate	hexane	water vapor
Vapor Transmission Rate (g · mm/m^2 · day)	0.87	0.14	0.03	0.03	0.05	0.71

PET

TABLE 107: Water Vapor, Oxygen, Nitrogen and Carbon Dioxide Permeability Through Oriented Polyethylene Terephthalate Polyester.

Material Family	POLYETHYLENE TEREPHTHALATE			
Features	oriented			
Reference Number	138	138	138	138

TEST CONDITIONS

Penetrant	water vapor	oxygen	nitrogen	carbon dioxide
Temperature (°C)	37.8	25	25	25
Relative Humidity (%)	90			
Test Note		STP conditions		

PERMEABILITY (source document units)

Vapor Transmission Rate (g · mil/100 in^2 · day)	1.0 - 1.3			
Gas Permeability (cm^3 · mil/100 in^2 · day)		3.0 - 6.0	0.7 - 1.0	15 - 25
Gas Permeability (cm^3 · mm/m^2 · day · atm)		1.2 - 2.4	0.28 - 0.39	5.9 - 9.8
Vapor Transmission Rate (g/day · 100 in^2)	0.39 - 0.51			

PERMEABILITY (normalized units)

Permeability Coefficient (cm^3 · mm/m^2 · day · atm)		1.2 - 2.4	0.28 - 0.39	5.9 - 9.8
Vapor Transmission Rate (g · mm/m^2 · day)	0.39 - 0.51			

TABLE 108: Water Vapor, Oxygen and Xylene Permeability Through DuPont Selar PT Polyethylene Terephthalate Polyester Containers.

Material Family	POLYETHYLENE TEREPHTHALATE			
Material Supplier/ Grade	DUPONT SELAR PT			
Product Form	CONTAINER		CUP	
Features	heat stabilized, hot fill, retort, transparent		hot fill, retort	
Applications	food packaging		food packaging	
Manufacturing Method	thermoforming (DuPont Fortrex process)		thermoforming (DuPont Fortrex process)	
Reference Number	290	290	290	290

MATERIAL CHARACTERISTICS

Sample Thickness			0.23 mm	0.23 mm

TEST CONDITIONS

Penetrant	water vapor	oxygen		
Temperature (°C)	25	25		
Relative Humidity (%)	50	50		
Test Note			container before retort	container after retort

PRE EXPOSURE CONDITIONING

Preconditioning Note				retort temperature: 121°C, retort time: 40 minutes

PERMEABILITY (source document units)

Permeation (cm³/pkg/day · atm)			0.204	C.228
Vapor Transmission Rate (g · mil/100 in² · day)	1.3			
Gas Permeability (cm³ · mil/100 in² · day)		5		

PERMEABILITY (normalized units)

Permeability Coefficient (cm³ · mm/m² · day · atm)		1.97		
Vapor Transmission Rate (g · mm/m² · day)	0.51			

TABLE 109: Film Properties of Coated and Uncoated Oriented Polyester Film.

Material Family	POLYETHYLENE TEREPHTHALATE	
Product Form	FILM	
Features		oriented; PVDC coated
Reference Number	268	268

MATERIAL CHARACTERISTICS

Sample Thickness	0.012 mm	0.014 mm

TEST CONDITIONS

Temperature (°C)	20	20
Relative Humidity (%)	65	65

MECHANICAL PROPERTIES

Modulus Of Elasticity - MD (MPa)	3431 {r}	2742 {r}
Modulus Of Elasticity - TD (MPa)	3920 {r}	2942 {r}
Tensile Strength - MD (MPa)	157.2 {r}	166.9 {r}
Tensile Strength - TD (MPa)	186.2 {r}	176.5 {r}
Ultimate Elongation - MD (%)	140 {r}	120 {r}
Ultimate Elongation - TD (%)	60 {r}	80 {r}
Impact Strength (kg-cm)	4	4
Burst Strength (MPa)	0.29 {af}	0.29 {af}
Pinhole Strength (g)	420 {ag}	520 {ag}
Elmendorf Tear Str. - TD (g/mm)	200 {ad}	400 {ad}
Elmendorf Tear Str. - TD (g/mm)	200 {ad}	500 {ad}
Tear Resistance - TD (g)	200	190
Tear Resistance - TD (g)	200	130

OTHER PROPERTIES

Water Absorption (%)	0.3 {ah}	0.3 {ah}
Equilibrium Moist. Absorption (%)	0.2	0.2
Melting Point (°C)	260	260
Haze (%)	2.5	3.9
Gloss	95	
Surface Resistivity (ohms)	>1.0E+16 {aj}	>1.0E+16 {aj}
Slip Factor (°)	30	24
Dimensional Stability - TD (%)	-1.6 {ai}	1.5 {ai}
Dimensional Stability - TD (%)	0.3 {ai}	0.3 {ai}

TABLE 110: Film Properties of DuPont Company Mylar Polyester PET Film.

Material Family	POLYETHYLENE TEREPHTHALATE		
Material Supplier/ Grade	**DUPONT MYLAR 92A**	**DUPONT MYLAR 57VB**	**DUPONT MYLAR**
Product Form	FILM		
Applications	industrial	video tape	
Reference Number	269	269	270

PHYSICAL PROPERTIES

Density (g/cm³)	1.392 {al}	1.391 {al}	1.39 {al}
Water Absorption (%)			<0.8 {av}

MECHANICAL PROPERTIES

Modulus Of Elasticity - MD (MPa)	3630 {r}	3473 {r}	3630 {ar}
Modulus Of Elasticity - TD (MPa)	3776 {r}	3562 {r}	
Tensile Strength - MD (MPa)	179 {r}	201 {r}	179 {ar}
Tensile Strength - TD (MPa)	214 {r}	232 {r}	
Strength @ 5% elongation - MD (MPa)	105 {r}	105 {r}	103 {ar}
Strength @ 5% elongation - TD (MPa)	101 {r}	95 {r}	
Ultimate Elongation - MD (%)	106 {r}	90 {r}	106 {ar}
Ultimate Elongation - TD (%)	84 {r}	76 {r}	
Elmendorf Tear Strength - MD (g/mm)			7.4 {av}
Tear Strength - Graves (N/mm)			294 {at}
Folding Endurance - MD (cycles)			100,000 {as}
Coefficient Of Friction - Kinetic			0.33 {ac}

THERMAL AND PERFORMANCE PROPERTIES

Melting Point (°C)	253 {am}	252 {am}	
Specific Heat (cal/g/°C)	0.28	0.28	
Thermal Expansion (mm/mm/°C)	1.7E-5 {ap}	1.7E-5 {ap}	
Thermal Conductivity (cal-cm/cm²-sec-°C)	3.7E-4 {aq}	3.7E-4 {aq}	
Surface Roughness Talysurf Ra (microns)	0.047 {ak}	0.031 {ak}	
Dimensional Stability - MD (%)	0.5 {an}	0.5 {an}	
Dimensional Stability - TD (%)	0.5 {an}	0.5 {an}	
Dimensional Stability - MD (%)	1.5 {ao}	1.8 {ao}	
Dimensional Stability - TD (%)	1.5 {ao}	1.8 {ao}	

PET

Chapter 28

Glycol Modified Polycyclohexylenedimethylene Terephthalate

TABLE 111: Water Vapor, Carbon Dioxide, Oxygen and Nitrogen Permeability Through Eastman Chemical Kodar PCTG 5445 Glycol Modified Polycyclohexylenedimethylene Terephthalate (PCTG) Film.

Material Family	GLYCOL MODIFIED POLYCYCLOHEXYLENEDIMETHYLENE TEREPHTHALATE			
Material Supplier/ Grade	EASTMAN KODAR PCTG 5445			
Product Form	FILM			
Features	transparent			
Reference Number	166	166	166	166

MATERIAL CHARACTERISTICS

Sample Thickness	0.25 mm	0.25 mm	0.25 mm	0.25 mm

TEST CONDITIONS

Penetrant	water vapor	carbon dioxide	oxygen	nitrogen
Temperature (°C)		23	23	23
Test Method	ASTM E96E	ASTM D1434	ASTM D1434	ASTM D1434

PERMEABILITY (source document units)

Vapor Transmission Rate (g/m$^2 \cdot$ day)	7			
Gas Permeability (cm$^3 \cdot$ mm/m$^2 \cdot$ day $\cdot$ atm)		50	10	3

PERMEABILITY (normalized units)

Permeability Coefficient (cm$^3 \cdot$ mm/m$^2 \cdot$ day $\cdot$ atm)		50	10	3
Vapor Transmission Rate (g $\cdot$ mm/m$^2 \cdot$ day)	1.75			

TABLE 112: Film Properties of Eastman Chemical Kodar PCTG 5445 Glycol Modified Polycyclohexylenedimethylene Terephthalate (PCTG) Film.

Material Family	GLYCOL MODIFIED POLYCYCLOHEXYLENEDIMETHYLENE TEREPHTHALATE
Material Supplier/ Grade	EASTMAN KODAR PCTG 5445
Product Form	FILM
Features	transparent
Reference Number	166

MATERIAL CHARACTERISTICS

Sample Thickness	0.25 mm
Density (g/cm³)	1.23

TEST CONDITIONS

Temperature (°C)	23
Relative Humidity (%)	50

MECHANICAL PROPERTIES

Modulus Of Elasticity (Mpa)	1520 {r}
Tensile Strength @ Yield - MD (MPa)	37 {r}
Tensile Strength @ Yield - TD (MPa)	34 {r}
Tensile Strength @ Break - MD (MPa)	52 {r}
Tensile Strength @ Break - TD (MPa)	39 {r}
Drop Dart Impact Strength (g)	480 {cz}
Drop Dart Impact Strength @ -18°C (g)	540 {cz}
Tear Strength, Propagated, Elmendorf - MD (g)	>3200 {k}
Tear Strength, Propagated, Elmendorf - transverse (g)	>3200 {k}
Coefficient Of Friction - Kinetic, film to film	0.3 {l}

OPTICAL PROPERTIES

Light Transmittance (%)	90 {c}
Haze (%)	0.2 {c}
Gloss @ 45°	106 {m}
Index Of Refraction	1.559 {cy}

Chapter 29

Polycyclohexylenedimethylene Ethylene Terephthalate

TABLE 113: Water Vapor Transmission and Oxygen Permeability Through Polycyclohexylenedimethylene Ethyelene Terephthalate Polyester (PETG).

Material Family	POLYCYCLOHEXYLENEDIMETHYLENE ETHYLENE TEREPHTHALATE	
Reference Number	296	296

TEST CONDITIONS

Penetrant	oxygen	water vapor
Temperature (°C)	22.8	37.8
Relative Humidity (%)	0	90
Test Method	ASTM D1434	ASTM F1249

PERMEABILITY (source document units)

Gas Permeability ($cm^3 \cdot$ mil/ 100 $in^2 \cdot$ bar $\cdot$ day)	25	
Vapor Transmission Rate (g $\cdot$ mil/ 100 $in^2 \cdot$ bar $\cdot$ day)		4

PERMEABILITY (normalized units)

Permeability Coefficient ($cm^3 \cdot mm/m^2 \cdot$ day $\cdot$ atm)	9.97	
Vapor Transmission Rate (g $\cdot mm/m^2 \cdot$ day)		1.6

TABLE 114: Water Vapor, Carbon Dioxide, Oxygen and Nitrogen Permeability Through Eastman Chemical Kodar PETG 6763 Polycyclohexylenedimethylene Ethylene Terephthalate (PETG) Film.

Material Family	POLYCYCLOHEXYLENEDIMETHYLENE ETHYLENE TEREPHTHALATE			
Material Supplier/ Grade	EASTMAN KODAR PETG 6763			
Product Form	FILM			
Features	amorphous, transparent			
Reference Number	165	165	165	165

MATERIAL CHARACTERISTICS

Sample Thickness	0.25 mm	0.25 mm	0.25 mm	0.25 mm

TEST CONDITIONS

Penetrant	water vapor	carbon dioxide	oxygen	nitrogen
Temperature (°C)		23	23	23
Test Method	ASTM E96E	ASTM D1434	ASTM D1434	ASTM D1434

PERMEABILITY (source document units)

Gas Permeability ($cm^3 \cdot mil/100 \ in^2 \cdot day$)		80	25	10
Gas Permeability ($cm^3 \cdot mm/m^2 \cdot day \cdot atm$)		30	10	5
Vapor Transmission Rate ($g/m^2 \cdot day$)	6			
Vapor Transmission Rate ($g/day \cdot 100 \ in^2$)	0.4			

PERMEABILITY (normalized units)

Permeability Coefficient ($cm^3 \cdot mm/m^2 \cdot day \cdot atm$)		31.5	9.84	3.94
Vapor Transmission Rate ($g \cdot mm/m^2 \cdot day$)	1.5			

TABLE 115: Film Properties of Eastman Chemical Kodar PETG 6763 Polycyclohexylenedimethylene Ethylene Terephthalate (PETG) Film.

Material Family	POLYCYCLOHEXYLENEDIMETHYLENE ETHYLENE TEREPHTHALATE
Material Supplier/ Grade	EASTMAN KODAR PETG 6763
Product Form	FILM
Features	amorphous, transparent
Reference Number	165

MATERIAL CHARACTERISTICS

Sample Thickness	0.25 mm

TEST CONDITIONS

Temperature (°C)	23
Relative Humidity (%)	50
Density (g/cm³)	1.27

MECHANICAL PROPERTIES

Modulus Of Elasticity (Mpa)	1720 {r}
Tensile Strength @ Yield - MD (MPa)	44.8 {r}
Tensile Strength @ Yield - TD (MPa)	43.4 {r}
Tensile Strength @ Break - MD (MPa)	57.2 {r}
Tensile Strength @ Break - TD (MPa)	56.5 {r}
Drop Dart Impact Strength (g)	425 {cz}
Drop Dart Impact Strength @ -18°C (g)	350 {cz}
Tear Strength, Propagated, Elmendorf - MD (g)	1600 {k}
Tear Strength, Propagated, Elmendorf - transverse (g)	1600 {k}
Coefficient Of Friction - Kinetic, film to film	>1.0 {l}

OPTICAL PROPERTIES

Light Transmittance (%)	90 {c}
Haze (%)	0.5 {c}
Gloss @ 45°	108 {m}
Index Of Refraction	1.567 {cy}

Chapter 30

Liquid Crystal Polymer

Permeability

Hoechst AG: Vectra A950 (features: 0.019 mm thick; product form: film)

Vectra has exceptional impermeability to gases and very good impermeability to water vapor.

Reference: *Vectra Polymer Materials,* supplier design guide (B 121 BR E 9102/014) - Hoechst AG, 1991.

<u>TABLE 116:</u> **Water Vapor Transmission and Oxygen Permeability Through Hoechst AG Liquid Crystal Polyester.**

Material Family	LIQUID CRYSTAL POLYMER					
Material Supplier/ Grade	HOECHST AG VECTRA A950					
Product Form	FILM					
Reference Number	70	70	70	70	70	70

MATERIAL CHARACTERISTICS

Sample Thickness	0.019 mm	0.019 mm	0.019 mm	0.019 mm	0.019 mm	0.019 mm

TEST CONDITIONS

Penetrant	oxygen				water vapor	
Temperature (°C)	23	23	38	38	23	38
Relative Humidity (%)	0	100	0	100	100	100
Test Note	test area: 5 cm²					

PERMEABILITY (source document units)

Gas Permeability (cm³ · µm/ m² · day · bar)	31.1	17.5	136.0	56.3		
Vapor Transmission Rate (g · µm/ m² · day)					27.6	51.2

PERMEABILITY (normalized units)

Permeability Coefficient (cm³ · mm/m² · day · atm)	0.03	0.02	0.14	0.06		
Vapor Transmission Rate (g · mm/m² · day)				.	0.03	0.05

Chapter 31

Polyimide

Film Properties and Applications

DuPont: **Kapton** (product form: film)

Kapton polymide film is a strong, tough, transparent amber-colored plastic film exhibiting good physical, chemical and electrical properties over a wide temperature range.

Kapton is produced in three forms, Type H, Type V, and Type F. Type H is the basic uncoated polymide film. Type V is similar to Type H but has superior dimensional stability. Type F is coated on one or both sides with Teflon FEP fluorocarbon resin which imparts heat sealability, provides a moisture barrier, and enhances chemical resistance.

Kapton is used as insulation for wire and cable, formed coils, magnet wire and transformers, and motor slot liners, among other uses. It also is used as a substrate for flexible printed circuits.

Reference: *Kapton Polyimide Film- Safe Handling,* supplier technical report (E-72084) - DuPont Company, 1988.

DuPont: **Kapton** (product form: film)

Kapton has good electrical and thermal properties. The film remains flexible from -200°C to +400°C, resists attack by all known organic solvents, does not melt when exposed to flame and is difficult to ignite. Kapton is specified for a wide range of applications, including phase insulation in high performance electric motors (as used in high speed trains) cable insulation for aircraft aerospace and military use; and flexible printed circuit boards for high technology applications.

Reference: *Guide to Excellence- Research And Development,* supplier marketing literature - DuPont Company.

DuPont: **Kapton Type-E** (product form: film); **Kapton Type-K** (product form: film)

Approaches to fabricate an advanced Tape Automated Bonding (TAB)/Flexible Printed Circuits (FPC) substrate have emphasized alteration of the polymide backbone structure. Out of these studies have come two production films, which are believed to have definite processing advantages, that are a direct consequence of the specific properties built into the films. These two films are designated as Kapton-K and Kapton-E, and are sold into both FPC and TAB end-uses. The Type-K is targeted at one and two sided circuitry via roll cladding. It is also intended for 3-layer TAB applications (use of adhesives) of medium complexity; e.g. 40-200 leads. The Type-E is targeted at multilayer and fine line circuitry where the condutor is bonded adhesivelessly. TAB applications are believed to lie in higher complexity systems with greater than 200 leads, as well as two conductor TAB.

Reference: Kreuz, J. A., Milligan, S. N., Sutton, R. F., *Kapton Polyimide Film- Advanced Flexible Dielectric Substrates For FPC/TAB Applications,* supplier technical report (H-24917) - DuPont Company, 1990.

TABLE 117: Water Vapor and Oxygen Permeability Through DuPont Company Kapton Polyimide Film.

Material Family	POLYIMIDE					
Material Supplier/ Trade Name	DUPONT KAPTON					
Grade	TYPE-V	TYPE-K	TYPE-E	TYPE-V	TYPE-K	TYPE-E
Product Form	FILM					
Reference Number	273	273	273	273	273	273

MATERIAL CHARACTERISTICS

Sample Thickness	0.076 mm	0.076 mm	0.076 mm	0.076 mm	0.076 mm	0.076 mm

TEST CONDITIONS

Penetrant	water vapor			oxygen		

PERMEABILITY (source document units)

Vapor Transmission Rate (g/m^2 · day)	22	28	4			
Gas Permeability (cm^3/m^2 · day)				114	105	4

PERMEABILITY (normalized units)

Permeability Coefficient (cm^3 · mm/m^2 · day · atm)				8.7	8.0	0.3
Vapor Transmission Rate (g · mm/m^2 · day)	1.7	2.1	0.3			

Polyimide

<u>TABLE 118:</u> **Water Vapor, Oxygen, Nitrogen, Carbon Dioxide and Helium Permeability Through Ube Upilex Polyimide Film.**

Material Family	POLYIMIDE							
Material Supplier/ Grade	UBE UPILEX R					UBE UPILEX S		
Product Form	FILM							
Reference Number	97	97	97	97	97	97	97	97

MATERIAL CHARACTERISTICS

Sample Thickness	0.025 mm					0.025 mm		

TEST CONDITIONS

Penetrant	water vapor	oxygen	nitrogen	carbon dioxide	helium	water vapor	oxygen	carbon dioxide
Temperature (°C)	38	30	30	30	30	38	30	30
Relative Humidity (%)	90					90		
Test Method	ASTM E96	ASTM D1434	ASTM D1434	ASTM D1434	ASTM D1434	ASTM E96	ASTM D1434	ASTM D1434

PERMEABILITY (source document units)

Gas Permeability ($cm^3 \cdot mil/m^2 \cdot day \cdot atm$)		100	30	115	2200		0.8	1.2
Vapor Transmission Rate ($g \cdot mil/m^2 \cdot atm \cdot day$)	22					1.7		

PERMEABILITY (normalized units)

Permeability Coefficient ($cm^3 \cdot mm/m^2 \cdot day \cdot atm$)		2.54	0.76	2.92	55.9		0.02	0.03
Vapor Transmission Rate ($g \cdot mm/m^2 \cdot day$)	0.56					0.04		

<u>TABLE 119:</u> **Film Properties of DuPont Company Kapton Polyimide Film.**

Material Family	POLYIMIDE		
Material Supplier/ Grade	DUPONT KAPTON TYPE-V	DUPONT KAPTON TYPE-K	DUPONT KAPTON TYPE-E
Product Form	FILM		
Reference Number	273	273	273

MATERIAL CHARACTERISTICS

Sample Thickness	0.076 mm	0.076 mm	0.076 mm

PHYSICAL PROPERTIES

Water Absorption (%)	3.0	3.7	2.4

MECHANICAL PROPERTIES

Modulus Of Elasticity (MPa)	2756	4341	5512
Ultimate Elongation (%)	80	80	40

TABLE 120: Film Properties of Ube Upilex Polyimide Film.

Material Family	POLYIMIDE	
Material Supplier/ Grade	UBE UPILEX R	UBE UPILEX S
Product Form	FILM	
Reference Number	97	97

MATERIAL CHARACTERISTICS

Sample Thickness	0.025 mm	0.025 mm
Density (g/cm³)	1.39 {bs}	1.47 {bs}

TEST CONDITIONS

Temperature (°C)	25	25

MECHANICAL PROPERTIES

Modulus Of Elasticity - MD (MPa)	3728 {r}	8190 {r}
Tensile Strength - MD (MPa)	245 {r}	392 {r}
Strength @ 5% Elongation - MD (MPa)	118 {r}	255 {r}
Ultimate Elongation - MD (%)	130 {r}	30 {r}
Tear Strength, propagated, Elmendorf - MD (g/mm)	750 {k}	330 {k}
Tear Strength, initial, Graves - MD (g/mm)	40,000 {az}	23,000 {az}
Folding Endurance (Cycles)	>100,000 {bx}	>100,000 {bx}
Coefficient Of Friction - Kinetic, film to film	0.4 {l}	0.4 {l}

THERMAL PROPERTIES

Melting Point (°C)	(none)	(none)
Glass Transition Temp. (°C)	285	>500
Specific Heat (cal/g/°C)	0.26 {am}	0.27 {am}

ELECTRICAL PROPERTIES

Surface Resistivity (ohms)	>10^{16} {cr}	>10^{16} {cr}
Volume Resistivity (ohm-cm)	10^{17} {cr}	10^{17} {cr}
Dielectric Constant	3.5 {ct}	3.5 {ct}

Polyimide

Chapter 32

Polyethylene

Permeability

BASF AG: Lupolen

Polyethylene is not entirely impermeable to gases and vapors, and the permeability for various gases depends on the crystallinity and density.

Reference: *Lupolen Polyethylene And Novolen Polypropylene Product Line, Properties, Processing,* supplier design guide (B 579 e / 4.92) - BASF Aktiengesellschaft, 1992.

Dow Chemical: PE (product form: film)

Radiation-induced crosslinking significantly reduces the permeability of polyethylene films.

Reference: *Permeability Of Polymers To Gases And Vapors,* supplier technical report (P302-335-79, D306-115-79) - Dow Chemical Company, 1979.

Permeability to Gases, Vapors, and Liquids

BASF AG: Lupolen (features: 0.1 mm thick)

Solubility and rate of diffusion through ethylene polymers determine their permeability to gases, vapors and liquids. The solubility depends on the chemical relationship between the diffusing molecule and the ethylene polymer; and the rate of diffusion in the amorphous zones of the partially crystalline polymer is governed by the size of the diffusing molecules. Since the molecules diffuse into the amorphous zones, a lower degree of crystallinity entails greater permeability. Other factors are the partial pressure difference, the temperature, and the area and thickness of the ethylene polymer.

Reference: *Lupolen, Lucalen Product Line, Properties, Processing,* supplier design guide (B 581 e/(8127) 10.91) - BASF Aktiengesellschaft, 1991.

Permeability to Gases

PE (product form: film)

In test method ASTM D1434 the polyethylene sample acts as a membrane between two pressure vessels. The vessel on one side of the sample is charged with the test gas and maintained at a specific temperature and at one atmosphere pressure. The vessel on the other side of the sample is evaluated and so maintained under vacuum. The volume of test gas which passes through the sample is recorded periodically until a steady rate is established. As density increases the polyethylene becomes more resistant to permeation by various gases and liquids.

Reference: *Engineering Properties Of Marlex Resins,* supplier design guide (TSM-243) - Phillips 66 Company, 1983.

Water Vapor Permeability

PE (product form: film)

ASTM D96 is the test most frequently used to measure the moisture barrier properties of polyethylene. The preferred procedure employs a desiccant, such as anhydrous calcium sulfate, sealed in an aluminum dish by a film sample with one surface exposed to the environment. The assembly is conditioned in a humidity cabinet which circulates air maintained at 37.8°C (100°F) having 90 % relative humidity. The weight gained over a period of time is attributed to water captured by the desiccant from which the water vapor transmission rate is determined. Increasing polymer density significantly reduces the rate at which moisture vapor permeates film.

Reference: *Engineering Properties Of Marlex Resins,* supplier design guide (TSM-243) - Phillips 66 Company, 1983.

Permeability to Gasoline

BASF AG: Lupolen

High-density polyethylene with a high molecular mass has been widely accepted as a material for fuel tanks. It permits substantial rationalization on automobile production lines because of the great scope that it allows in styling, the savings in weight that it achieves over its steel counterparts, and the ease with which it can be produced by extrusion blow molding and assembled in the vehicle. As compared to steel, polyethylene is not completely impermeable to gasoline, but it does not rust. The current European regulations on the permisible permeation rates at 23°C and 40°C through ready-for-assembly fuel tanks are laid down in the ECE (European Economic Council) Reglement R 34, Annex 5
Factors affecting the amount of fuel that permeates throughout the duration of the test are the specific surface, the weight of the fuel tank, and the uniformity of the wall thickness distribution. More severe test conditions can be anticipated in the future, and the permissible limits are likely to be lowered in order to meet growing demands for environmental protection.

The permeability to gasoline can be reduced by more than 90% by fluorinating or sulfonating the fuel tanks. Since the thickness of the impermeable fluorinated or sulfonated layer is of the order of only a few micrometers, the fuel tanks retain their high level of mechanical properties.

Other potential methods of reducing the permeability to engine fuels include special surface coatings, dispersions, films, modification of the material, and the coextrusion of composite fuel tanks.

Reference: *Lupolen Polyethylene And Novolen Polypropylene Product Line, Properties, Processing,* supplier design guide (B 579 e / 4.92) - BASF Aktiengesellschaft, 1992.

Permeation Reduction

BASF AG: Lupolen

Measures to reduce permeation are essential for special applications in the packaging sector, in which the contents of a polyethylene pack may suffer loss of aroma or undergo organoleptic changes on exposure to gases, particularly oxygen. Reductions in the permebility to gases of polyethylene bags, tubes, blow-molded containers, etc. can be achieved by applying coatings or laminating to thin metal foil, e.g. aluminum. Examples of polymers with good barrier effects that can be applied by coating techniques are polyamides, polyesters, poly(vinylidene chloride), ethylene/vinyl alcohol copolymers, and polymers containing acrylonitrile.

Another established method for solving the problem are coextruded composite film and multi-ply blow moldings in which the barrier layer is bonded to the polyethylene by means of suitable adhesion promotors, e.g. Lucalen.

Reference: *Lupolen, Lucalen Product Line, Properties, Processing,* supplier design guide (B 581 e/(8127) 10.91) - BASF Aktiengesellschaft, 1991.

Film Properties and Applications

Dow Chemical: Retain (features: recycle content; product form: film)

Formulated with 10 to 100% postconsumer recycle content, RETAIN postconsumer recycle content plastics are a family of resins engineered for high performance in a variety of polyethylene packaging.
By formulating the hightest quality postconsumer recycled materials with Dow virgin resins, Dow Plastics offers its converters postconsumer recycled content plastic resins with performance characteristics comparable to virgin resin. RETAIN plastics will be used for film applications such as grocery sacks, overwraps, and consumer liners.

Reference: *621 Ways To Succeed - 1993-1994 Materials Selection Guide,* supplier technical report (304-00286-1292X SMG) - Dow Chemical Company, 1992.

TABLE 121: Film Properties of Dow Chemical Retain Polyethylene Postconsumer Recycle Content Resins.

Material Family	POLYETHYLENE		
Material Supplier	DOW CHEMICAL RETAIN PE-		
Grade	1001	5000	5009
Product Form	FILM		
Features	natural resin, recycle content	pigmented, recycle content	natural resin, recycle content
Reference Number	254	254	254

MATERIAL CHARACTERISTICS

Density	0.960 g/cm³	0.921 g/cm³	0.934 g/cm³
Melt Flow Index	4.9 grams/10 min. (190/2.16)	0.7 grams/10 min. (190/2.16)	1.2 grams/10 min. (190/2.16)
Melt Flow Ratio		10.0	
Sample Thickness	0.0254 mm	0.0254 mm	0.0254 mm

MATERIAL COMPOSITION

Recycled Polyethylene	25%	25%	25%

MECHANICAL PROPERTIES

Secant Modulus @ 1% Elongation - MD (MPa)			279 {r}
Secant Modulus @ 1% Elongation - TD (MPa)			469 {r}
Secant Modulus @ 2% Elongation - machine dir. (MPa)	840 {r}	248 {r}	
Secant Modulus @ 2% Elongation - transverse dir. (MPa)		296 {r}	
Ultimate Tensile Strength - MD (MPa)	18 {r}	43.5 {r}	32 {r}
Ultimate Tensile Strength - TD (MPa)		39 {r}	21 {r}
Tensile Strength @ Yield - MD (MPa)	23 {r}	13 {r}	15 {r}
Tensile Strength @ Yield - TD (MPa)		15 {r}	17 {r}
Ultimate Elongation - Machine direction (%)	680 {r}	490 {r}	560 {r}
Ultimate Elongation - Transverse direction (%)		730 {r}	640 {r}
Drop Dart Impact Strength (G)		120 {f}	60 {f}

OPTICAL PROPERTIES

Gloss @ 45°			44 {m}

GRAPH 30: Gas Permeability vs. Density through Polyethylene.

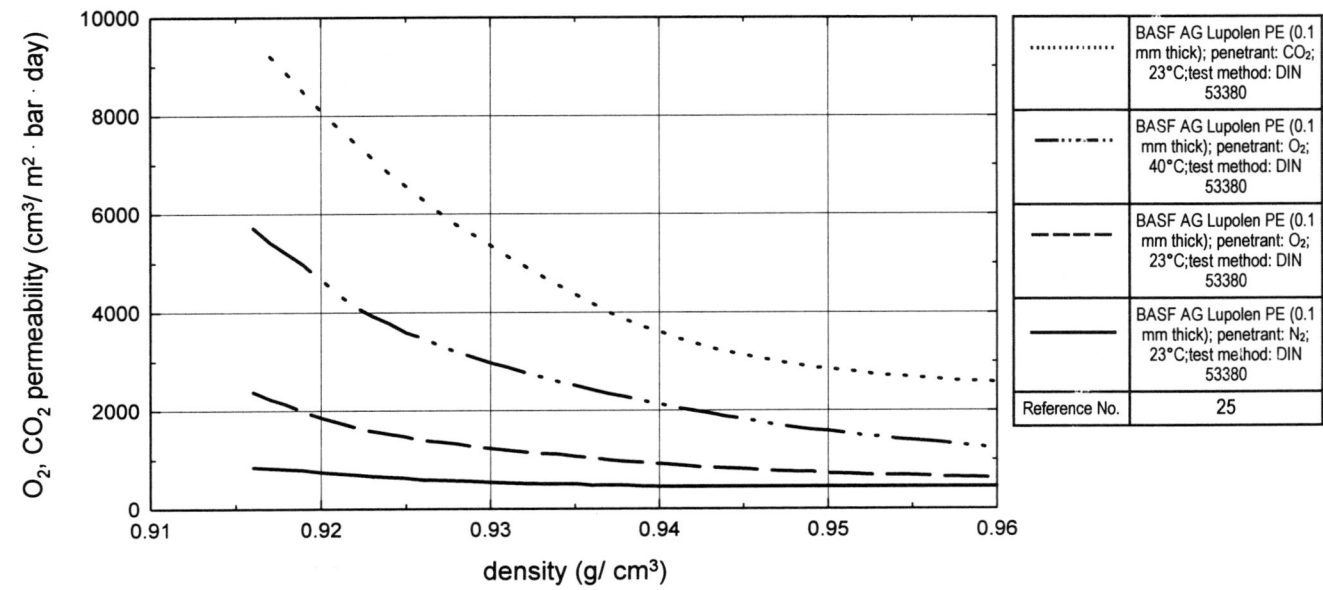

....	BASF AG Lupolen PE (0.1 mm thick); penetrant: CO_2; 23°C;test method: DIN 53380
─ ·· ─ ·· ─	BASF AG Lupolen PE (0.1 mm thick); penetrant: O_2; 40°C;test method: DIN 53380
─ ─ ─	BASF AG Lupolen PE (0.1 mm thick); penetrant: O_2; 23°C;test method: DIN 53380
─────	BASF AG Lupolen PE (0.1 mm thick); penetrant: N_2; 23°C;test method: DIN 53380
Reference No.	25

GRAPH 31: Oxygen Permeability vs. Density through Polyethylene.

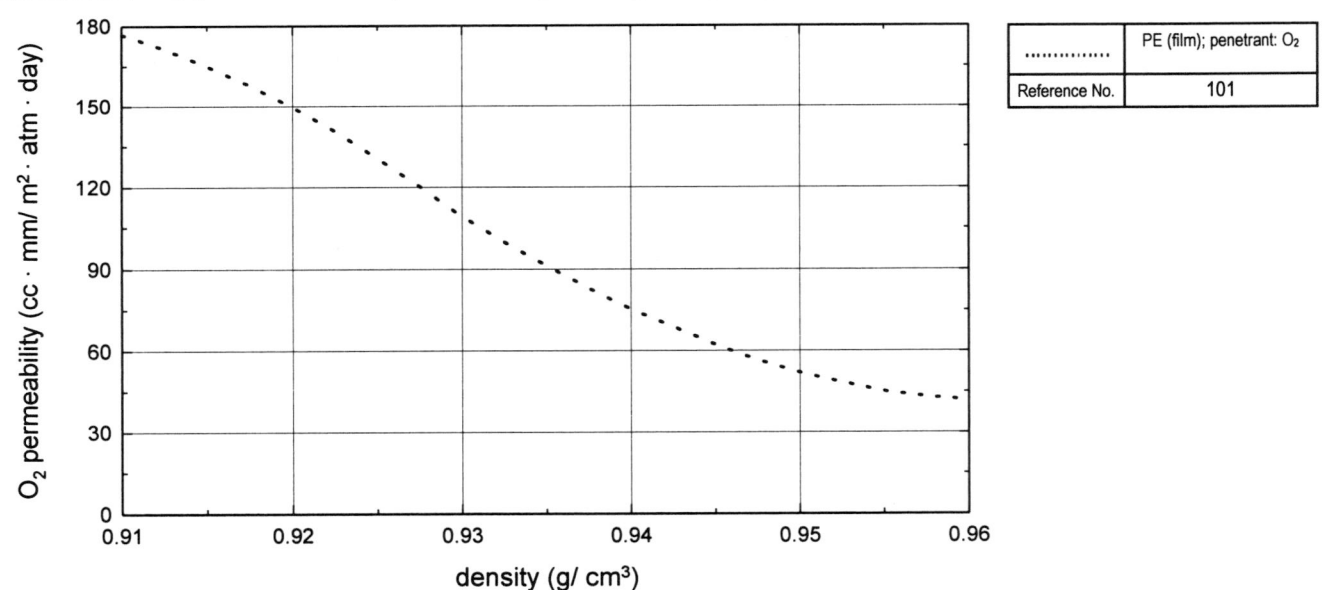

....	PE (film); penetrant: O_2
Reference No.	101

PE

GRAPH 32: Water Vapor Permeability vs. Density through Polyethylene.

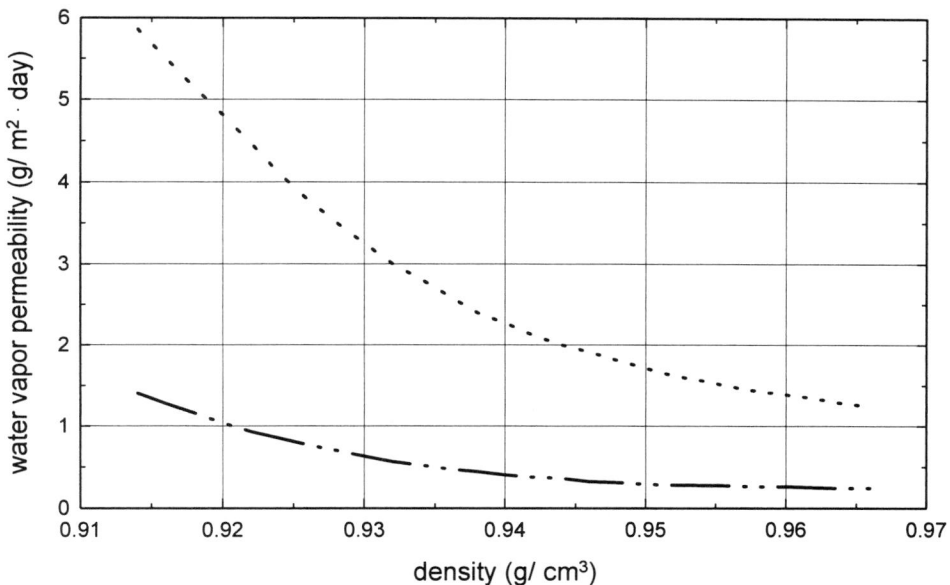

....	BASF AG Lupolen PE (0.1 mm thick); penetrant: water vapor; 40°C; test method DIN 53122
–··–··–	BASF AG Lupolen PE (0.1 mm thick); penetrant: water vapor; 23°C; 85% to 0% RH gradient; test method: DIN 53122
Reference No.	25

GRAPH 33: Moisture vapor transmission (MVT) vs. Density through Polyethylene.

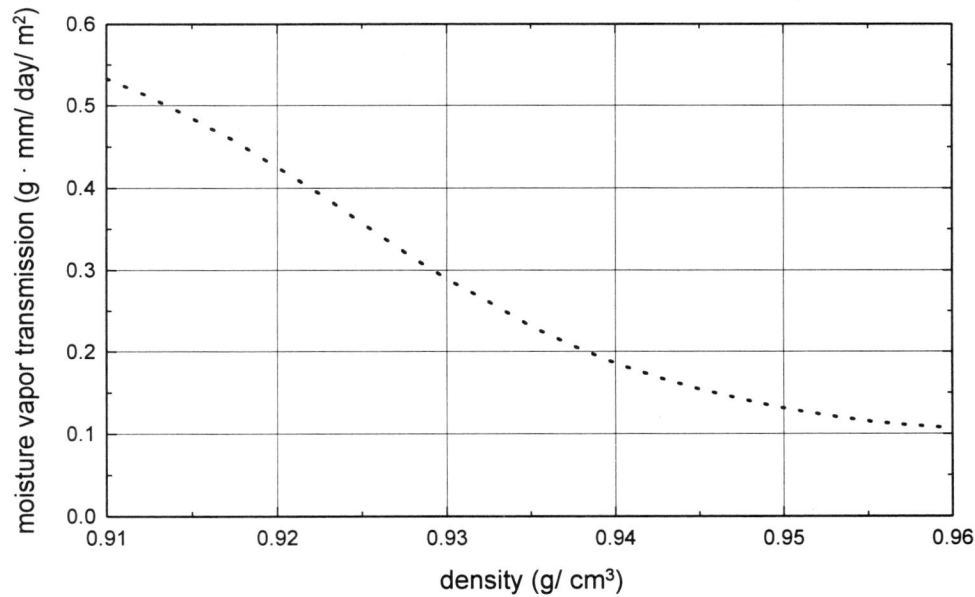

....	PE (film); penetrant: moisture vapor
Reference No.	101

GRAPH 34: Toluene and FAM Test Fluid Permeability vs. Density through Polyethylene.

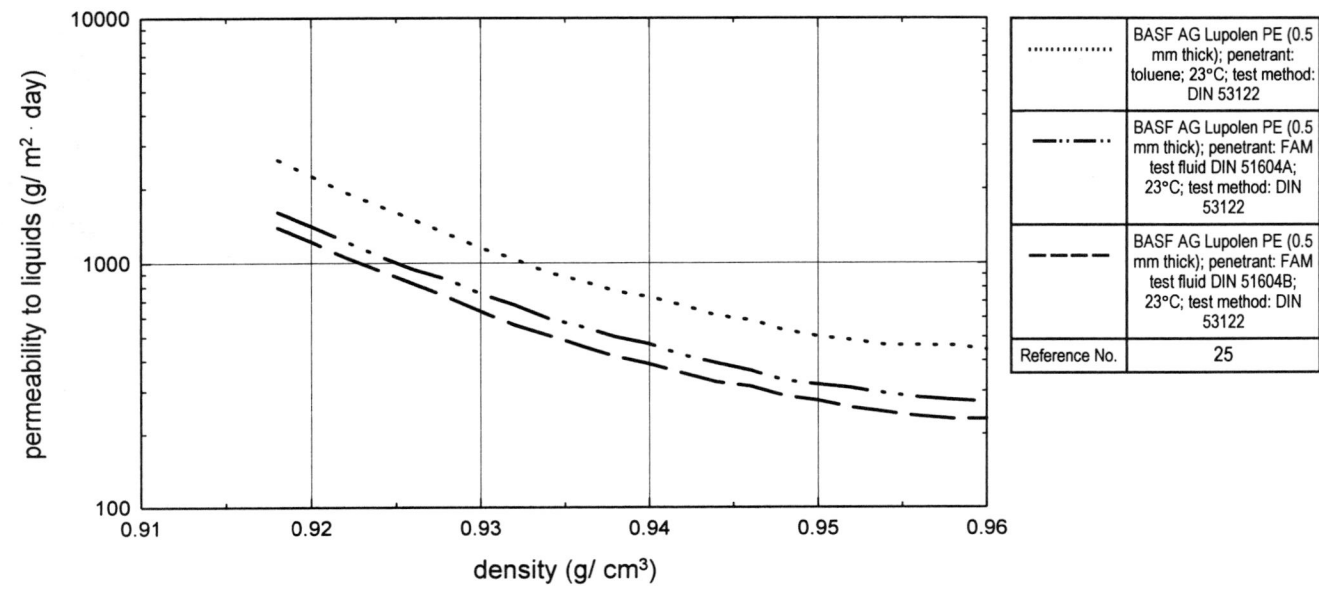

GRAPH 35: Methyl Alcohol, Ethyl Acetate, Fuel Oil and Chloroform Permeability vs. Density through Polyethylene.

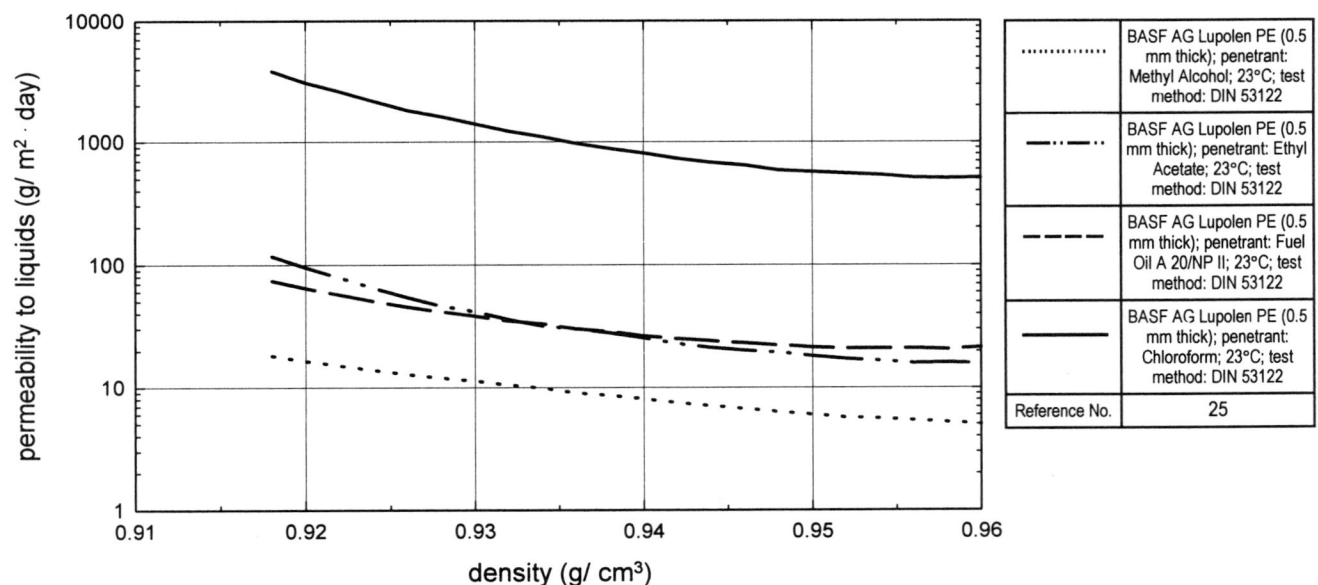

GRAPH 36: Oxygen Permeability vs. Temperature through Polyethylene.

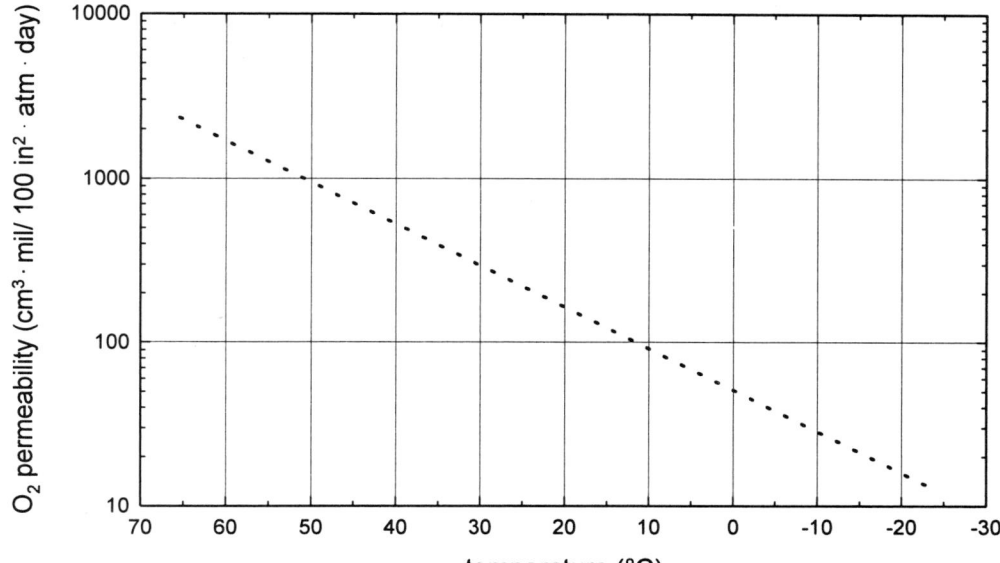

	Dow PE (film); penetrant: O_2
Reference No.	250

Chapter 33

Low Density Polyethylene

Film Properties and Applications

Dow Chemical: LDPE (features: fractional melt index; product form: film); (features: enhanced clarity; product form: liner film)

LDPE resins are the backbone polymers of the polyethylene film industry. To satisfy the needs of this industry, products must perform precisely, supply must be assured and cost-use economics must be competitive. Dow LDPE clarity, liner, fractional melt index (FMI) and heavy-duty film resins satisfy those basic needs and others as well. FMI resins give impact strength, puncture resistance and toughness. Medium density resins provide excellent clarity, high modulus and strength; draw-down resins not only permit film gauges down to 0.3 mil, but also maintain excellent film toughness. Using cast processing, the low-gel, flat film surfaces achieved with Dow resins make them ideal for laminating film applications.

Reference: *621 Ways To Succeed - 1993-1994 Materials Selection Guide,* supplier technical report (304-00286-1292X SMG) - Dow Chemical Company, 1992.

Extrusion Coating Properties and Applications

Dow Chemical: LDPE (manufacturing method: extrusion coating)

These resins provide optimal neck-in and draw-down performance, so you can achieve smoother processing and increased quality. Also, these LDPE resins provide a broad hot tack and heat seal window for consistent end-user performance. That means final packages can be formed, filled, and sealed for maximum integrity with minimum adjustments to the equipment. Dow LDPE extrusion coating resins are now being used for a variety of food, cosmetic, industrial, and medical packaging applications.

Reference: *621 Ways To Succeed - 1993-1994 Materials Selection Guide,* supplier technical report (304-00286-1292X SMG) - Dow Chemical Company, 1992.

TABLE 122: Carbon Dioxide, Oxygen, Ethylene Oxide and Water Vapor Permeability Through Low Density Polyethylene Film.

Material Family	LOW DENSITY POLYETHYLENE			
Product Form	FILM			
Features	2.5 blow up ratio			
Manufacturing Method	blown film			
Reference Number	216	216	216	216

MATERIAL CHARACTERISTICS

Density	0.920 g/cm³	0.920 g/cm³	0.920 g/cm³	0.920 g/cm³
Melt Flow Index	4 g/10 min.	4 g/10 min.	4 g/10 min.	4 g/10 min.
Sample Thickness	0.05 mm	0.05 mm	0.05 mm	0.05 mm

TEST CONDITIONS

Penetrant	water vapor	carbon dioxide	oxygen	ethylene oxide
Test Method	JIS Z0208	ASTM D1434	ASTM D1434	ASTM D1434

PERMEABILITY (source document units)

Vapor Transmission Rate (g · 100 µm/m² · day)	25			
Gas Permeability (cm³ · 100 µm/ m² · atm · day)		7900	1500	21,000

PERMEABILITY (normalized units)

Permeability Coefficient (cm³ · mm/m² · day · atm)		790	150	2100
Vapor Transmission Rate (g · mm/m² · day)	2.5			

TABLE 123: Oxygen, Nitrogen, Carbon Dioxide and Water Vapor Permeability Through Dow Chemical Low Density Polyethylene.

Material Family	LOW DENSITY POLYETHYLENE			
Material Supplier/ Grade	DOW CHEMICAL			
Product Form	FILM			
Reference Number	250	250	250	250

TEST CONDITIONS

Penetrant	oxygen	nitrogen	carbon dioxide	water vapor
Temperature (°C)	24	24	24	24

PERMEABILITY (source document units)

Vapor Transmission Rate (g · mil/100 in² · day)				1 - 1.5
Gas Permeability (cm³ · mil/100 in² · day)	250 - 350	100 - 200	1000 - 2000	

PERMEABILITY (normalized units)

Permeability Coefficient (cm³ · mm/m² · day · atm)	98 - 138	39 - 79	394 - 787	
Vapor Transmission Rate (g · mm/m² · day)				0.39 - 0.59

160

## TABLE 124: Oxygen Permeability at Different Temperatures and Water Vapor Transmission Through Low Density Polyethylene.

Material Family	LOW DENSITY POLYETHYLENE		
Reference Number	264	264	264

TEST CONDITIONS

Penetrant	oxygen		water vapor
Temperature (°C)	23	35	40
Relative Humidity (%)	0	0	90

PERMEABILITY (source document units)

Gas Permeability (cm$^3 \cdot$ mil/100 in$^2 \cdot$ day)	554	745	
Gas Permeability (cm$^3 \cdot$ 25μ/m$^2 \cdot$ day $\cdot$ atm)	8586	11,547	
Vapor Transmission Rate (g $\cdot$ mil/100 in$^2 \cdot$ day)			1.14
Vapor Transmission Rate (g $\cdot$ 25μ/m$^2 \cdot$ day)			17.7

PERMEABILITY (normalized units)

Permeability Coefficient (cm$^3 \cdot$ mm/m$^2 \cdot$ day $\cdot$ atm)	218.1	293.3	
Vapor Transmission Rate (g $\cdot$ mm/m$^2 \cdot$ day)			0.45

TABLE 125: Gas Permeability of Oxygen, Carbon Dioxide, Nitrogen and Helium Through Low Density Polyethylene Film.

Material Family	LOW DENSITY POLYETHYLENE			
Product Form	FILM			
Reference Number	63	63	63	63

TEST CONDITIONS

Penetrant	oxygen	nitrogen	carbon dioxide	helium
Temperature (°C)	25	25	25	25
Relative Humidity (%)	0	0	0	0

PERMEABILITY (source document units)

Gas Permeability (cm$^3 \cdot$ mil/100 in$^2 \cdot$ day)	696	180	2436	1624
Gas Permeability (cm$^3 \cdot$ 20μ/m$^2 \cdot$ day $\cdot$ atm)	12,000	3100	42,000	28,000

PERMEABILITY (normalized units)

Permeability Coefficient (cm$^3 \cdot$ mm/m$^2 \cdot$ day $\cdot$ atm)	274	71	959	639

trueLDPE

true© *Plastics Design Library*

TABLE 126: Vapor Transmission of Reagents Through Dow Chemical Low Density Polyethylene.

Material Family	LOW DENSITY POLYETHYLENE							
Material Supplier/ Grade	DOW CHEMICAL							
Product Form	FILM							
Reference Number	250	250	250	250	250	250	250	250

TEST CONDITIONS

Penetrant	methyl alcohol	ethyl alcohol	n-heptane	ethyl acetate	formaldehyde	tetrachloro-ethylene	acetone	benzene
Temperature (°C)	24	24	24	24	24	24	24	35

PERMEABILITY (source document units)

Vapor Transmission Rate (g · mil/100 in² · day)	6 - 8	2 - 4	300 - 500	30 - 300	2 - 5	500 - 750	10 - 40	600

PERMEABILITY (normalized units)

Vapor Transmission Rate (g · mm/m² · day)	2.4 - 3.1	0.8 - 1.6	118- 197	11.8 - 118	0.8 - 2.0	197 - 295	3.9 - 15.8	236

TABLE 127: Water Vapor Transmission Through Dow Chemical Low Density Polyethylene Extrusion Coating Resins.

Material Family	LOW DENSITY POLYETHYLENE			
Material Supplier	DOW CHEMICAL			
Grade	LDPE 722	LDPE 4005	LDPE 4012	LDPE 5004I
Manufacturing Method	extrusion coating	extrusion coating	extrusion coating	extrusion coating
Reference Number	254	254	254	254

MATERIAL CHARACTERISTICS

Density	0.916 g/cm³	0.916 g/cm³	0.916 g/cm³	0.923 g/cm³
Melt Flow Index	8 g/10 min. (190/2.16)	5.5 g/10 min. (190/2.16)	12 g/10 min. (190/2.16)	4 g/10 min. (190/2.16)
Sample Thickness	0.01 mm (minimum thickness)	0.015 mm (minimum thickness)	0.01 mm (minimum thickness)	0.01 mm (minimum thickness)

TEST CONDITIONS

Penetrant	water vapor			
Test Method	ASTM F1249	ASTM F1249	ASTM F1249	ASTM F1249

PERMEABILITY (source document units)

Vapor Transmission Rate (g/m² · day)	26.4	31	31	23.3
Vapor Transmission Rate (g/day · 100 in²)	1.7	2.0	1.8	1.5

PERMEABILITY (normalized units)

Vapor Transmission Rate (g · mm/m² · day)	0.26	0.46	0.31	0.23

TABLE 128: Water Vapor Transmission and Oxygen Permeability Through Low Density Polyethylene Film.

Material Family	LOW DENSITY POLYETHYLENE				
Product Form	FILM				
Reference Number	268	268	268	268	268

MATERIAL CHARACTERISTICS

Sample Thickness	0.05 mm	0.05 mm	0.05 mm	0.05 mm	0.05 mm

TEST CONDITIONS

Penetrant	water vapor	oxygen			
Temperature (°C)	40	35	20	20	20
Relative Humidity (%)	90	0	65	85	100
Test Method	JIS Z0208	JIS Z1707	ASTM D3985	ASTM D3985	ASTM D3985

PERMEABILITY (source document units)

Vapor Transmission Rate $(g \cdot mil/100\ in^2 \cdot day)$	1				
Gas Permeability $(cm^3 \cdot mil/100\ in^2 \cdot day)$		387	174	174	174

PERMEABILITY (normalized units)

Permeability Coefficient $(cm^3 \cdot mm/m^2 \cdot day \cdot atm)$		152	68.5	68.5	68.5
Vapor Transmission Rate $(g \cdot mm/m^2 \cdot day)$	0.39				

TABLE 129: Water Vapor, Oxygen, Nitrogen and Carbon Dioxide Permeability Through Low Density Polyethylene.

Material Family	LOW DENSITY POLYETHYLENE			
Reference Number	138	138	138	138

TEST CONDITIONS

Penetrant	water vapor	oxygen	nitrogen	carbon dioxide
Temperature (°C)	37.8	25	25	25
Relative Humidity (%)	90			
Test Note		STP conditions		

PERMEABILITY (source document units)

Gas Permeability (cm$^3 \cdot$ mil/100 in$^2 \cdot$ day)		500	180	2700
Gas Permeability (cm$^3 \cdot$ mm/m$^2 \cdot$ day $\cdot$ atm)		195	71	1060
Vapor Transmission Rate (g $\cdot$ mil/100 in$^2 \cdot$ day)	1.0 -1.5			
Vapor Transmission Rate (g/day $\cdot$ 100 in^2)	0.39 - 0.59			

PERMEABILITY (normalized units)

Permeability Coefficient (cm$^3 \cdot$ mm/m$^2 \cdot$ day $\cdot$ atm)		197	71	106.3
Vapor Transmission Rate (g $\cdot$ mm/m$^2 \cdot$ day)	0.39 - 0.59			

TABLE 130: Xylene and Oxygen Permeability Through Low Density Polyethylene.

Material Family	LOW DENSITY POLYETHYLENE	
Reference Number	293	293

TEST CONDITIONS

Penetrant	xylene	oxygen
Temperature (°C)	60	23
Exposure Time (days)	14	
Relative Humidity (%)		75

PERMEABILITY (source document units)

Vapor Transmission Rate (g · mil/100 in^2 · day)	3800	
Gas Permeability (cm^3 · mil/100 in^2 · day)		450

PERMEABILITY (normalized units)

Permeability Coefficient (cm^3 · mm/m^2 · day · atm)		177.2
Vapor Transmission Rate (g · mm/m^2 · day)	1496	

TABLE 131: Organic Solvents Permeability Through Low Density Polyethylene Film.

Material Family	LOW DENSITY POLYETHYLENE			
Product Form	FILM			
Reference Number	266	266	266	266

MATERIAL CHARACTERISTICS

Sample Thickness	0.051 mm	0.051 mm	0.051 mm	0.051 mm

TEST CONDITIONS

Penetrant	chloroform	xylene	methyl ethyl ketone	kerosine
Temperature (°C)	20	20	20	20
Relative Humidity (%)	65	65	65	65

PERMEABILITY (source document units)

Vapor Transmission Rate (g/day · 100 in^2)	178.1	20.97	4.77	4.9

PERMEABILITY (normalized units)

Vapor Transmission Rate (g · mm/m^2 · day)	141	16.6	3.8	3.9

TABLE 132: Mechanical and Optical Properties of BASF AG Lupolen Low Density Polyethylene Film.

Material Family	LOW DENSITY POLYETHYLENE										
Material Supplier	BASF AG LUPOLEN										
Grade	1810D	1840D	2441D	3020D	3040D	1810E	2410F	2414F	2420F	2424F	2426F
Product Form	FILM										
Reference Number	251	251	251	251	251	251	251	251	251	251	251

MATERIAL CHARACTERISTICS

Density	0.917 - 0.920 g/cm³	0.917 - 0.920 g/cm³	0.922 - 0.925 g/cm³	0.926 - 0.929 g/cm³	0.927 - 0.930 g/cm³	0.917 - 0.920 g/cm³	0.922 - 0.925 g/cm³	0.922 - 0.925 g/cm³	0.922 - 0.925 g/cm³	0.922 - 0.925 g/cm³	0.922 - 0.925 g/cm³
Melt Flow Index	0.15 - 0.35 g/10 min. (190/2.16)	0.15 - 0.35 g/10 min. (190/2.16)	0.15 - 0.35 g/10 min. (190/2.16)	0.15 - 0.35 g/10 min. (190/2.16)	0.15 - 0.35 g/10 min. (190/2.16)	0.3 - 0.6 g/10 min. (190/2.16)	0.6 - 0.9 g/10 min. (190/2.16)	0.6 - 0.9 g/10 min. (190/2.16)	0.6 - 0.9 g/10 min. (190/2.16)	0.6 - 0.9 g/10 min. (190/2.16)	0.6 - 0.9 g/10 min. (190/2.16)
Sample Thickness	0.07 mm	0.07 mm	0.07 mm	0.07 mm	0.07 mm	0.07 mm	0.05 mm	0.05 mm	0.05 mm	0.05 mm	0.05 mm

TEST CONDITIONS

Temperature (°C)	23	23	23	23	23	23	23	23	23	23	23

MECHANICAL PROPERTIES

Ultimate Tensile Strength - machine direction (MPa)	27 {a}	27 {a}	27 {a}	28 {a}	28 {a}	26 {a}	26 {a}	26 {a}	26 {a}	26 {a}	26 {a}
Ultimate Tensile Strength - transverse direction (MPa)	20 {a}	20 {a}	20 {a}	21 {a}	21 {a}	17 {a}	20 {a}	20 {a}	20 {a}	20 {a}	20 {a}
Ultimate Elongation - Machine direction (%)	200 {a}	200 {a}	200 {a}	250 {a}	250 {a}	200 {a}	250 {a}	250 {a}	300 {a}	300 {a}	300 {a}
Ultimate Elongation - Transverse direction (%)	600 {a}	600 {a}	600 {a}	600 {a}	600 {a}	600 {a}	600 {a}	600 {a}	600 {a}	600 {a}	600 {a}
Drop Dart Impact Strength (g)	250 {f}	250 {f}	250 {f}	180 {f}	180 {f}	200 {f}	130 {f}	130 {f}	130 {f}	130 {f}	130 {f}
Failure Energy (J/mm)	6.5 {g}	6.5 {g}	6.5 {g}	5 {g}	5 {g}	6 {g}	5.5 {g}	5.5 {g}	5.5 {g}	5.5 {g}	5.5 {g}
Coefficient Of Friction	75 {e}	75 {e}	65 {e}	55 {e}	55 {e}	75 {e}	65 {e}	40 {e}	85 {e}	25 {e}	25 {e}

OPTICAL PROPERTIES

Haze (%)	<13 {c}	<15 {c}	<11 {c}	<8 {c}	<8 {c}	<11 {c}	<8 {c}	<8 {c}	<7 {c}	<7 {c}	<7 {c}
Gloss @ 20°	>10 {d}	>10 {d}	>20 {d}	>35 {d}	>35 {d}	>10 {d}	>30 {d}	>30 {d}	>40 {d}	>40 {d}	>40 {d}

TABLE 133: Film Properties of Dow Chemical Low Density Polyethylene Clarity And Liner Film Resins.

Material Family	LOW DENSITY POLYETHYLENE						
Material Supplier	DOW CHEMICAL LDPE						
Grade	501I	503I	526I	527I	529I	530C	534C
Product Form	LINER FILM						
Features	enhanced clarity	enhanced clarity	enhanced clarity	enhanced clarity	enhanced clarity	enhanced clarity	enhanced clarity
Reference Number	254	254	254	254	254	254	254

MATERIAL CHARACTERISTICS

Density	0.922 g/cm³	0.923 g/cm³	0.922 g/cm³	0.923 g/cm³	0.920 g/cm³	0.922 g/cm³	0.922 g/cm³
Melt Flow Index	1.9 grams/10 min. (190/2.16)	1.9 grams/10 min. (190/2.16)	1.5 grams/10 min. (190/2.16)	1.0 grams/10 min.	2 grams/10 min. (190/2.16)	2 grams/10 min. (190/2.16)	2 grams/10 min. (190/2.16)
Sample Thickness	0.04 mm	0.04 mm	0.04 mm	0.04 mm	0.04 mm	0.04 mm	0.04 mm
Blow Up Ratio (BUR)	2.25	2.25	2.25	2.25	2.25	2.3	2.3

MECHANICAL PROPERTIES

Ultimate Tensile Strength - machine direction (MPa)	22.4 {p}	22.4 {p}	21.4 {p}	21.4 {p}	22.1 {p}	25.7 {p}	23.8 {p}
Ultimate Tensile Strength - transverse direction (MPa)	16.9 {p}	16.9 {p}	15.2 {p}	15.2 {p}	17.9 {p}	22.7 {p}	19.0 {p}
Tensile Strength @ Yield - machine direction (MPa)	10.7 {p}	10.7 {p}	10.0 {p}	10.0 {p}	10.3 {p}	11.8 {p}	11.6 {p}
Tensile Strength @ Yield - transverse direction (MPa)	10.3 {p}	10.3 {p}	9.7 {p}	9.7 {p}	10.3 {p}	11.6 {p}	11.4 {p}
Ultimate Elongation - machine direction (%)	400 {p}	400 {p}	370 {p}	370 {p}	275 {p}	300 {p}	230 {p}
Ultimate Elongation - transverse direction (%)	575 {p}	575 {p}	510 {p}	510 {p}	575 {p}	600 {p}	540 {p}
Drop Dart Impact Strength (G)	100 {j}	100 {j}	100 {j}	100 {j}	150 {j}	110 {j}	110 {j}
Elmendorf Tear Resistance - machine direction (g/mm)	14,760 {k}	14,760 {k}	16,930 {k}	16,930 {k}	12,795 {k}	11,800 {k}	10,200 {k}
Elmendorf Tear Resistance - transverse direction (g/mm)	11,810 {k}	11,810 {k}	11,020 {k}	11,020 {k}	5510 {k}	6690 {k}	5120 {k}
Coefficient Of Friction	0.35 {l}	0.1 {l}	0.5 {l}	0.5 {l}	<0.1 {l}	0.5 {l}	0.2 {l}

THERMAL PROPERTIES

Vicat Softening Temperature (°C)						93 {n}	93 {n}
Extrusion Temperature (°C)	177	177	204	204	177	188	188

OPTICAL PROPERTIES

Haze (%)	4 {c}	4.5 {c}	9 {c}	11 {c}	10 {c}	5.4 {c}	7.5 {c}
Gloss @ 45°	75 {m}	75 {m}	58 {m}	55 {m}	55 {m}	67 {m}	61 {m}

TABLE 134: Film Properties of Dow Chemical Low Density Polyethylene Clarity And Liner Film Resins.

Material Family	LOW DENSITY POLYETHYLENE					
Material Supplier/ Grade	DOW CHEMICAL LDPE					
Material Supplier/ Grade	535I	536I	607C	609C	611C	640I
Product Form	LINER FILM					
Features	enhanced clarity	enhanced clarity	enhanced clarity	enhanced clarity	enhanced clarity	enhanced clarity
Reference Number	254	254	254	254	254	254

MATERIAL CHARACTERISTICS

Density	0.926 g/cm³	0.922 g/cm³	0.924 g/cm³	0.924 g/cm³	0.924 g/cm³	0.922 g/cm³
Melt Flow Index	1.9 grams/10 min. (190/2.16)	2 grams/10 min. (190/2.16)	2 grams/10 min. (190/2.16)	0.9 grams/10 min.	0.9 grams/10 min.	2 grams/10 min. (190/2.16)
Sample Thickness	0.04 mm	0.04 mm	0.03 mm	0.03 mm	0.03 mm	0.04 mm
Blow Up Ratio (BUR)	2.25	2.25	2.3	2.3	2.3	2.25

MECHANICAL PROPERTIES

Ultimate Tensile Strength - machine direction (MPa)	26.7 {p}	22.1 {p}		24.1 {p}	24.1 {p}	29.6 {p}
Ultimate Tensile Strength - transverse direction (MPa)	19.5 {p}	17.9 {p}		19.8 {p}	19.8 {p}	17.9 {p}
Tensile Strength @ Yield - machine direction (MPa)	12.4 {p}	10.7 {p}	12.1 {p}	11.9 {p}	11.9 {p}	11.4 {p}
Tensile Strength @ Yield - transverse direction (MPa)	12.6 {p}	10.0 {p}	12.0 {p}	12.0 {p}	12.0 {p}	10.3 {p}
Ultimate Elongation - machine direction (%)	420 {p}	330 {p}		310 {p}	310 {p}	395 {p}
Ultimate Elongation - transverse direction (%)	585 {p}	600 {p}		590 {p}	590 {p}	530 {p}
Drop Dart Impact Strength (g)	65 {j}	120 {j}		80 {j}	80 {j}	95 {j}
Elmendorf Tear Resistance - machine direction (g/mm)	5710 {k}	15,750 {k}	11,800 {k}	14,600 {k}	14,600 {k}	16,140 {k}
Elmendorf Tear Resistance - transverse direction (g/mm)	8270 {k}	7870 {k}	9840 {k}	9450 {k}	11,800 {k}	9840 {k}
Coefficient Of Friction	1 {l}	0.6 {l}	0.2 {l}	0.2 {l}	0.5 {l}	0.7 {l}

THERMAL PROPERTIES

Extrusion Temperature (°C)	177	177	188	193	193	177

OPTICAL PROPERTIES

Haze (%)	6 {c}	6 {c}	3.8 {c}	4.4 {c}	4.4 {c}	7.5 {c}
Gloss @ 45°	50 {m}	65 {m}	77 {m}	75 {m}	75 {m}	63 {m}

TABLE 135: Film Properties of Dow Chemical Low Density Polyethylene Clarity And Liner Film Resins.

Material Family	LOW DENSITY POLYETHYLENE					
Material Supplier	DOW CHEMICAL LDPE					
Grade	641I	648I	681I	689I	748I	768I
Product Form	LINER FILM					
Features	enhanced clarity	enhanced clarity	enhanced clarity	enhanced clarity	enhanced clarity	enhanced clarity
Reference Number	254	254	254	254	254	254

MATERIAL CHARACTERISTICS

Density	0.922 g/cm³	0.922 g/cm³	0.922 g/cm³	0.923 g/cm³	0.919 g/cm³	0.930 g/cm³
Melt Flow Index	2 grams/10 min. (190/2.16)	2 grams/10 min. (190/2.16)	1.2 grams/10 min. (190/2.16)	0.7 grams/10 min. (190/2.16)	7 grams/10 min. (190/2.16)	2 grams/10 min. (190/2.16)
Sample Thickness	0.04 mm	0.04 mm	0.04 mm	0.04 mm	0.03 mm	0.04 mm
Blow Up Ratio (BUR)	2.25	2.25	2.25	2.25	2.25	2.25

MATERIAL COMPOSITION

Note	medium slip additive content	high slip additive content				

MECHANICAL PROPERTIES

Ultimate Tensile Strength - machine direction (MPa)	25.9 {p}	22.4 {p}	29.0 {p}	27.6 {p}	23.4 {p}	22.8 {p}
Ultimate Tensile Strength - transverse direction (MPa)	17.6 {p}	18.6 {p}	20.0 {p}	19.3 {p}	17.2 {p}	18.6 {p}
Tensile Strength @ Yield - machine direction (MPa)	11.4 {p}	11.0 {p}	11.7 {p}	10.3 {p}	10.3 {p}	15.2 {p}
Tensile Strength @ Yield - transverse direction (MPa)	10.3 {p}	10.7 {p}	10.7 {p}	10.3 {p}	9.7 {p}	15.2 {p}
Ultimate Elongation - machine direction (%)	290 {p}	325 {p}	425 {p}	360 {p}	260 {p}	425 {p}
Ultimate Elongation - transverse direction (%)	560 {p}	575 {p}	600 {p}	530 {p}	600 {p}	600 {p}
Drop Dart Impact Strength (g)	95 {j}	110 {j}	120 {j}	140 {j}	75 {j}	90 {j}
Elmendorf Tear Resistance - machine direction (g/mm)	20,470 {k}	15,750 {k}	6890 {k}	10,630 {k}	14,170 {k}	3350 {k}
Elmendorf Tear Resistance - transverse direction (g/mm)	9840 {k}	10,830 {k}	8860 {k}	11,420 {k}	7870 {k}	11,810 {k}
Coefficient Of Friction	0.1 {l}	<0.1 {l}	0.7 {l}	0.5 {l}	0.6 {l}	0.3 {l}

THERMAL PROPERTIES

Extrusion Temperature (°C)	191	177	177	218	149	177

OPTICAL PROPERTIES

Haze (%)	8.5 {c}	7 {c}	6 {c}	9.5 {c}	5 {c}	5 {c}
Gloss @ 45°	61 {m}	62 {m}	65 {m}	63 {m}	60 {m}	68 {m}

TABLE 136: Properties of Dow Chemical Low Density Polyethylene Extrusion Coating Resins.

Material Family	LOW DENSITY POLYETHYLENE			
Material Supplier	DOW CHEMICAL LDPE			
Grade	722	4005	4012	5004I
Manufacturing Method	extrusion coating	extrusion coating	extrusion coating	extrusion coating
Reference Number	254	254	254	254

MATERIAL CHARACTERISTICS

Density	0.916 g/cm³	0.916 g/cm³	0.916 g/cm³	0.923 g/cm³
Melt Flow Index	8 grams/10 min. (190/2.16)	5.5 grams/10 min. (190/2.16)	12 grams/10 min. (190/2.16)	4 grams/10 min. (190/2.16)
Sample Thickness	0.01 mm (minimum coating thickness)	0.015 mm (minimum coating thickness)	0.01 mm (minimum coating thickness)	0.01 mm (minimum coating thickness)

MECHANICAL PROPERTIES

Coefficient Of Friction	0.6 {l}	0.6 {l}	0.6 {l}	0.45 {l}

THERMAL PROPERTIES

Vicat Softening Temperature (°C)	85 {n}	88 {n}	84 {n}	95 {n}
Minimum Heat Seal Temp. (°C)	116 {q}	121 {q}	121 {q}	132 {q}

PERFORMANCE PROPERTIES

Minimum Coating Thick. (mm)	0.01	0.015	0.01	0.01
Neck-In @ 25 μm Thick. (mm)	40.6	35.6	45.7	51
Dynamic Slip (g)	1.1	1.1	1.1	0.8

TABLE 137: Film Properties of Dow Chemical Low Density Polyethylene Fractional Melt Index Film Resins.

Material Family	LOW DENSITY POLYETHYLENE									
Material Supplier/ Grade	DOW CHEMICAL LDPE									
	123C	124C	132I	133I	135I	493C	494C	6821	683I	685I
Product Form	FILM									
Features	fractional melt index									
Reference Number	254	254	254	254	254	254	254	254	254	254

MATERIAL CHARACTERISTICS

	123C	124C	132I	133I	135I	493C	494C	6821	683I	685I
Density	0.922 g/cm³	0.922 g/cm³	0.921 g/cm³	0.923 g/cm³	0.921 g/cm³	0.921 g/cm³	0.921 g/cm³	0.922 g/cm³	0.923 g/cm³	0.924 g/cm³
Melt Flow Index	0.3 g/10 min. (190/2.16)	0.3 g/10 min. (190/2.16)	0.22 g/10 min. (190/2.16)	0.22 g/10 min. (190/2.16)	0.22 g/10 min. (190/2.16)	0.8 g/10 min. (190/2.16)	0.8 g/10 min. (190/2.16)	0.7 g/10 min. (190/2.16)	0.7 g/10 min. (190/2.16)	0.7 g/10 min. (190/2.16)
Sample Thickness	0.05 mm	0.05 mm	0.05 mm	0.05 mm	0.05 mm	0.04 mm	0.04 mm	0.04 mm	0.04 mm	0.04 mm
Note				medium antiblock additive content	medium antiblock additive content	medium antiblock additive content	medium antiblock additive content		low antiblock additive content	medium antiblock additive content
Blow Up Ratio (BUR)	2.3	2.3	2.25	2.25	2.25	2.3	2.3	2.25	2.25	2.25

MECHANICAL PROPERTIES

	123C	124C	132I	133I	135I	493C	494C	6821	683I	685I
Ultimate Tensile Strength - MD (MPa)			22.1 {p}	23.4 {p}	22.1 {p}	24.2 {p}	27.0 {p}	30.3 {p}	30.3 {p}	30.3 {p}
Ultimate Tensile Strength - TD (MPa)			20.0 {p}	17.9 {p}	20.0 {p}	20.3 {p}	21.7 {p}	22.1 {p}	22.1 {p}	22.1 {p}
Tensile Strength @ Yield - MD (MPa)	11.7 {p}	11.7 {p}	10.3 {p}	10.3 {p}	10.3 {p}	11.4 {p}	12.0 {p}	11.4 {p}	11.4 {p}	11.4 {p}
Tensile Strength @ Yield - TD (MPa)	11.5 {p}	11.5 {p}	9.0 {p}	9.7 {p}	9.0 {p}	11.4 {p}	11.6 {p}	11.0 {p}	11.0 {p}	11.0 {p}
Ultimate Elongation - MD (%)			300 {p}	450 {p}	300 {p}	230 {p}	190 {p}	400 {p}	400 {p}	400 {p}
Ultimate Elongation - TD (%)			550 {p}	540 {p}	550 {p}	570 {p}	520 {p}	600 {p}	600 {p}	600 {p}
Drop Dart Impact Strength (g)	170 {j}	170 {j}	250 {j}	300 {j}	250 {j}	130 {j}	140 {j}	160 {j}	160 {j}	180 {j}
Elmendorf Tear Resistance - MD (g/mm)	14,600 {k}	14,600 {k}	5910 {k}	5710 {k}	5910 {k}	11,020 {k}	10,600 {k}	5910 {k}	5910 {k}	5910 {k}
Elmendorf Tear Resistance - TD (g/mm)	8270 {k}	8270 {k}	6890 {k}	7480 {k}	6890 {k}	4330 {k}	3940 {k}	6890 {k}	6890 {k}	6890 {k}
Coefficient Of Friction	0.5 {l}	0.3 {l}	0.7 {l}	0.5 {l}	0.2 {l}	0.3 {l}	0.5 {l}	0.7 {l}	0.6 {l}	<0.1 {l}

THERMAL PROPERTIES

	123C	124C	132I	133I	135I	493C	494C	6821	683I	685I
Vicat Softening Temperature (°C)	93 {n}	93 {n}				91 {n}	91 {n}			
Extrusion Temperature (°)	218	218	204	218	204	204	204	204	204	204

OPTICAL PROPERTIES

	123C	124C	132I	133I	135I	493C	494C	6821	683I	685I
Haze (%)			12 {c}	12 {c}	14 {c}			6 {c}	7.5 {c}	8 {c}
Gloss @45°			55 {m}	52 {m}	50 {m}			65 {m}	65 {m}	70 {m}

TABLE 138: Film Properties of Low Density Polyethylene Film.

Material Family	LOW DENSITY POLYETHYLENE	
Product Form	FILM	
Features		2.5 blow up ratio
Manufacturing Method		blown film
Reference Number	268	216

MATERIAL CHARACTERISTICS

Density		0.920 g/cm³
Melt Flow Index		4 g/10 min.
Sample Thickness	0.05 mm	0.05 mm

TEST CONDITIONS

Temperature (°C)	20	
Relative Humidity (%)	65	

PHYSICAL PROPERTIES

Water Absorption (%)	0.3 {ah}	
Equilibrium Moist. Absorption (%)	0.2	

MECHANICAL PROPERTIES

Modulus Of Elasticity - machine direction (MPa)	145 {r}	
Modulus Of Elasticity - transverse direction (MPa)	158 {r}	
Tensile Strength - machine direction (MPa)		16.7 {ci}
Tensile Strength - transverse direction (MPa)		13.7 {ci}
Tensile Strength @ Break - machine direction (MPa)	19.3 {r}	
Tensile Strength @ Break - transverse direction (MPa)	19.3 {r}	
Ultimate Elongation - machine direction (%)	340 {r}	290 {ci}
Ultimate Elongation - transverse direction (%)	600 {r}	410 {ci}
Impact Strength (kg-cm)	4	
Burst Strength (Mpa)	0.10 {af}	
Pinhole Strength (g)	150 {ag}	
Elmendorf Tear Resistance MD (g/mm)	14,000 {ad}	
Elmendorf Tear Resistance TD (g/mm)	6000 {ad}	

<u>TABLE 138:</u> **Film Properties of Low Density Polyethylene Film (cont'd).**

Material Family	LOW DENSITY POLYETHYLENE	
Product Form	FILM	
Features		2.5 blow up ratio
Manufacturing Method		blown film
Reference Number	268	216

MECHANICAL PROPERTIES

Tear Resistance - machine direction (g)	500	
Tear Resistance - transverse direction (g)	380	
Tear Strength - machine direction (g/mm)		1300 {cj}
Tear Strength - transverse direction (g/mm)		3300 {cj}

OTHER PROPERTIES

Light Transmittance (%)		80 {ck}
Haze (%)		14 {ck}
Gloss		15 {cl}
Melting Point (°C)	105-115	
Surface Resistivity (ohms)	3.2E+15 {aj}	
Slip Factor (°)	30	

GRAPH 37: Carbon Dioxide Permeability vs. Thickness through Low Density Polyethylene.

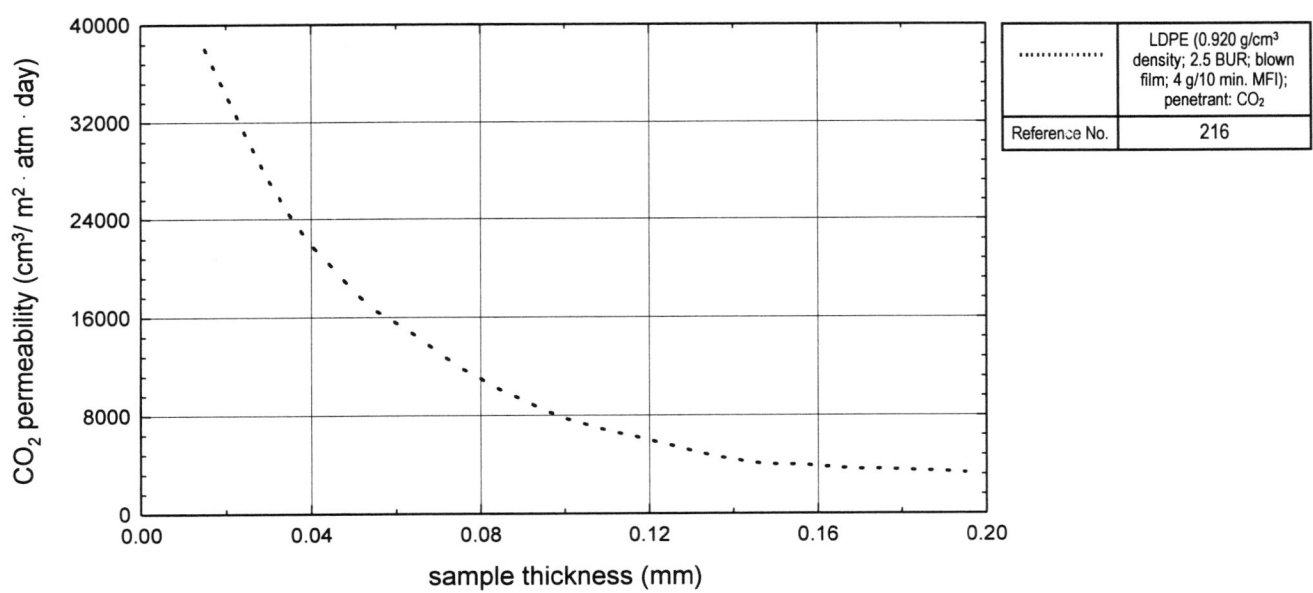

....	LDPE (0.920 g/cm³ density; 2.5 BUR; blown film; 4 g/10 min. MFI); penetrant: CO₂
Reference No.	216

GRAPH 38: Oxygen Permeability vs. Thickness through Low Density Polyethylene.

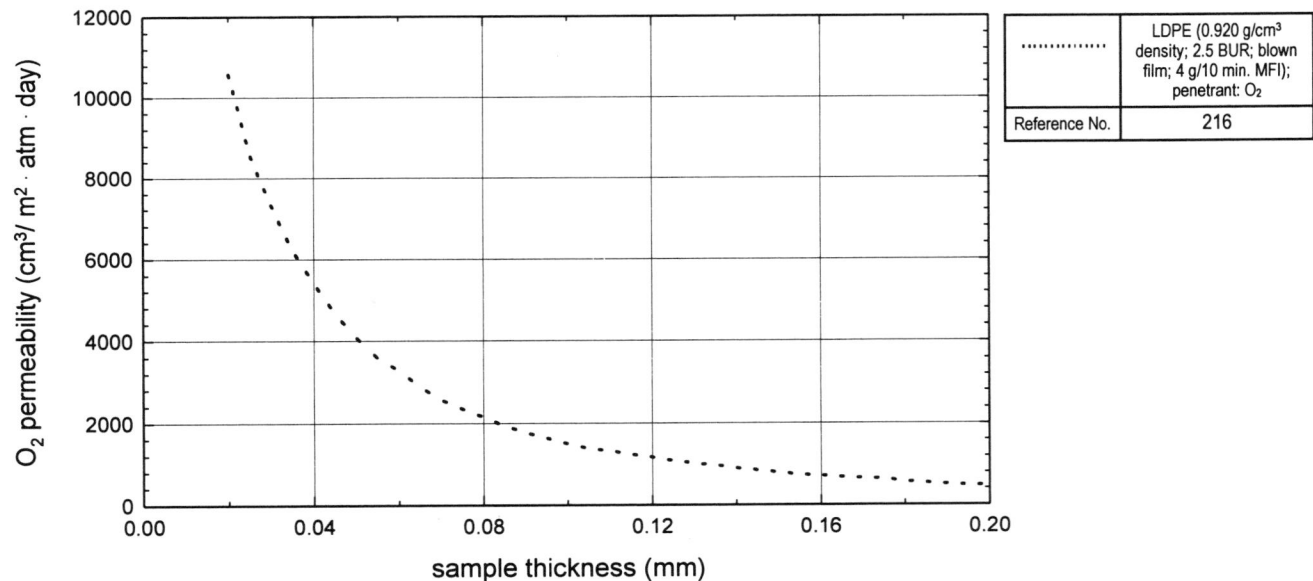

....	LDPE (0.920 g/cm³ density; 2.5 BUR; blown film; 4 g/10 min. MFI); penetrant: O₂
Reference No.	216

Chapter 34

Linear Low Density Polyethylene

Film Properties and Applications

Dow Chemical: Dowlex

In film applications, DOWLEX resins offer excellent processability, with enhanced output and lower energy demand. They offer proven performance: DOWLEX octene-based polyethylene resins have up to 126% better MD tear resistance than butene-based resins, up to 29% better CD tear resistance than butene-based resins, and up to 14% better impact resistance than hexene-based resins.

In the technically sophisticated and very competitive business of liner-grade polyethylenes, DOWLEX polyethylene resins designed for blown and cast film extrusion will deliver quality liner films and bags for many applications: consumer liner films, stretch films, shrink films, disposables, bag-in-box, delivery/doorknob bags, heavy-duty shipping sacks, grocery sacks, laundry and dry cleaning films, and more.

Reference: *621 Ways To Succeed - 1993-1994 Materials Selection Guide,* supplier technical report (304-00286-1292X SMG) - Dow Chemical Company, 1992.

Dow Chemical: Dowlex NG 2085 (density: 0.920 g/cm^3; features: thin gauge liner film, 0.03 mm thick; manufacturing method: blown film; melt flow index: 0.95 grams/10 min. (190/2.16); melt flow ratio: 8.0); **Dowlex NG 3347A** (density: 0.917 g/cm^3; features: stretch film, 0.02 mm thick; manufacturing method: cast film; melt flow index: 2.3 grams/10 min. (190/2.16); melt flow ratio: 7.35)

DOWLEX NG 3347A polyethylene resins provide the outstanding puncture resistance and load retention properties available with DOWLEX 2047A polyethylene resins, as well as improved extensibility and flaw resistance. With Extra Stretch Performance and this unique balance of properties, Next Generation DOWLEX resins are ideal for demanding stretch film applications.

DOWLEX NG 2085 polyethylene resins for thin film applications offer superior tensile strength, dart impact and the opportunity for converters to downgauge without sacrificing end-use properties. This Next Generation resin offers the key properties needed for high performance consumer trash liners.

Reference: *621 Ways To Succeed - 1993-1994 Materials Selection Guide,* supplier technical report (304-00286-1292X SMG) - Dow Chemical Company, 1992.

DuPont Canada: Sclair

SCLAIR (LLDPE) resins provide heat-seal characteristics, high tensile strength and tensile toughness and outstanding puncture resistance in finished products - all of which are attributable to the linear molecular structure and narrow molecular weight distribution. Slip and anti-block agents are also incorporated as appropriate.

Where the end use for film will include hot-tack and heat-seal applications, SCLAIR LLDPE provides resistance to the stresses tending to pull the film apart. Further, the resins provide a secure heat-seal even when traces of product being packaged are in the seal zone. This is true even for oils, powders and some greases.

As a 20 % to 30 % blend component with conventional polyethylene, SCLAIR LLDPE improves the physical properties of the finished film, particularly tensile strength and puncture resistance. Conversely, small amounts of conventional polyethylene added to SCLAIR LLDPE improves processibility and optical properties.

Reference: *Sclair Linear Polyethylene Resins For Film Packaging,* supplier marketing literature - DuPont Canada.

DuPont Canada: Sclairfilm SL1 (applications: laminations; density: 0.918 g/cm^3; product form: film);
Sclairfilm SL3 (applications: laminations; density: 0.918 g/cm^3; product form: film)

SCLAIR SL is a linear low density polyolefin film which offers a combination of toughness and durability plus clarity. In particular, types SL-1 and SL-3 have superior tensile strength, puncture strength, elongation and heat seal strength compared to conventional polyethylene laminating films. The "hot tack" strength of SL film permits seal areas to withstand stresses even while in the molten state. Its ability to seal under a wide variety of sealing conditions, including grease and oil contamination, make SL film ideal for use on vacuum packaging equipment.

SL will laminate to other substrates such as DARTEK nylon film or MYLAR polyester film for vacuum packaging applications. SL, in converter combinations, is suited for the vacuum packaging of processed meats.

SL has heat seal properties for vacuum packaging and other critical applications. The ability of SL to withstand pinholing because of stress flex also contributes to long lasting seals and lower leaker rates.

Reference: *Sclairfilm Polyolefin Film - SL-1 and SL-3 Laminating Film,* supplier technical report (H-27763) - DuPont Canada, 1990.

TABLE 139: Oxygen Gas and Water Vapor Permeability Through Dow Chemical Dowlex Linear Low Density Polyethylene.

Material Family	LINEAR LOW DENSITY POLYETHYLENE	
Material Supplier/ Grade	DOW CHEMICAL DOWLEX 2045	
Manufacturing Method	blown film	blown film
Reference Number	11	11

MATERIAL CHARACTERISTICS

Density	0.920 g/cm³	0.920 g/cm³
Melt Flow Index	1.0 grams/10 min.	1.0 grams/10 min.
Sample Thickness	0.0254 mm	0.0254 mm

TEST CONDITIONS

Penetrant	oxygen	water vapor
Relative Humidity (%)	<1% (dry test)	100
Test Method	ASTM D3985-81	Mocon Test Method
Test Apparatus		Mocon Permatron W-1

PERMEABILITY (source document units)

Gas Permeability (cm³ · mil/100 in² · day)	525	
Vapor Transmission Rate (g · mil/100 in² · day · atm)		0.7

PERMEABILITY (normalized units)

Permeability Coefficient (cm³ · mm/m² · day · atm)	207	
Vapor Transmission Rate (g · mm/m² · day)		0.28

TABLE 140: Oxygen Permeability Through DuPont Canada Sclair SL1 and Sclair SL3 Linear Low Density Polyethylene.

Material Family	LINEAR LOW DENSITY POLYETHYLENE					
Material Supplier/ Grade	DUPONT CANADA SCLAIRFILM SL1			DUPONT CANADA SCLAIRFILM SL3		
Product Form	FILM					
Applications	laminations					
Reference Number	278	278	278	278	278	278

MATERIAL CHARACTERISTICS

Density	0.918 g/cm³					
Sample Thickness	0.038 mm	0.051 mm	0.076 mm	0.038 mm	0.051 mm	0.076 mm

TEST CONDITIONS

Penetrant	oxygen					
Test Method	ASTM D3985	ASTM D3985	ASTM D3985	ASTM D3985	ASTM D3985	ASTM D3985

PERMEABILITY (source document units)

Gas Permeability (cm³/m² · day)	6200	3900	3100	6200	3900	3100
Gas Permeability (cm³/100 in² · day · atm)	400	250	200	400	250	200

PERMEABILITY (normalized units)

Permeability Coefficient (cm³ · mm/m² · day · atm)	236	199	236	236	199	236

TABLE 141: Water Vapor Transmission and Oxygen Permeability Through DuPont Canada Sclair Linear Low Density Polyethylene.

Material Family	LINEAR LOW DENSITY POLYETHYLENE			
Material Supplier/ Grade	**DUPONT CANADA SCLAIR 11F9**	**DUPONT CANADA SCLAIR 11H4**	**DUPONT CANADA SCLAIR 11R4**	**DUPONT CANADA SCLAIR 11F9**
Product Form	**BLOWN FILM**			
Applications	blending resin, multi-purpose bags	blending resin	blending resin, multi-purpose bags	blending resin, multi-purpose bags
Reference Number	277	277	277	277

MATERIAL CHARACTERISTICS

Density	0.921 g/cm³	0.921 g/cm³	0.921 g/cm³	0.921 g/cm³
Melt Flow Index	0.75 grams/10 min.	1.2 grams/10 min.	1.6 grams/10 min.	0.75 grams/10 min.
Sample Thickness	0.0254 mm	0.0254 mm	0.0254 mm	0.0254 mm

TEST CONDITIONS

Penetrant	water vapor			oxygen
Temperature (°C)	38	38	38	23
Relative Humidity (%)	90	90	90	
Test Method				ASTM D1434
Test Note	values are cooling rate dependent			
Test Apparatus	Honeywell model 825 apparatus	Honeywell model 825 apparatus	Honeywell model 825 apparatus	

PERMEABILITY (source document units)

Vapor Transmission Rate (g/m² · day)	15	18	20	
Gas Permeability (cm³/m² · day)				5200

PERMEABILITY (normalized units)

Permeability Coefficient (cm³ · mm/m² · day · atm)				132
Vapor Transmission Rate (g · mm/m² · day)	0.38	0.46	0.51	

TABLE 142: Water Vapor Transmission Through DuPont Canada Sclair SL1 and Sclair SL3 Linear Low Density Polyethylene.

Material Family	LINEAR LOW DENSITY POLYETHYLENE					
Material Supplier/ Grade	DUPONT CANADA SCLAIRFILM SL1			DUPONT CANADA SCLAIRFILM SL3		
Product Form	FILM					
Applications	laminations					
Reference Number	278	278	278	278	278	278

MATERIAL CHARACTERISTICS

Density	0.918 g/cm³					
Sample Thickness	0.038 mm	0.051 mm	0.076 mm	0.038 mm	0.051 mm	0.076 mm

TEST CONDITIONS

Penetrant	water vapor					
Test Method	ASTM F372	ASTM F372	ASTM F372	ASTM F372	ASTM F372	ASTM F372

PERMEABILITY (source document units)

Vapor Transmission Rate (g/m² · day)	12.4	9.3	4.7	12.4	9.3	4.7
Vapor Transmission Rate (g/day · 100 in²)	0.8	0.6	0.4	0.8	0.6	0.4

PERMEABILITY (normalized units)

Vapor Transmission Rate (g · mm/m² · day)	0.47	0.47	0.36	0.47	0.47	0.36

TABLE 143: Carbon Dioxide and Nitrogen Permeability Through DuPont Canada Sclair SL1 and Sclair SL3 Linear Low Density Polyethylene.

Material Family	LINEAR LOW DENSITY POLYETHYLENE			
Material Supplier/ Grade	DUPONT CANADA SCLAIRFILM SL1		DUPONT CANADA SCLAIRFILM SL3	
Product Form	FILM			
Applications	laminations			
Reference Number	278	278	278	278

MATERIAL CHARACTERISTICS

Density	0.918 g/cm³			
Sample Thickness	0.0254 mm			

TEST CONDITIONS

Penetrant	carbon dioxide	nitrogen	carbon dioxide	nitrogen
Test Method	ASTM D3985	ASTM D3985	ASTM D3985	ASTM D3985
Test Note	approximate values			

PERMEABILITY (source document units)

Gas Permeability (cm³/m² · day)	1400	150	1400	150

PERMEABILITY (normalized units)

Permeability Coefficient (cm³ · mm/m² · day · atm)	35.6	3.81	35.6	3.81

LLDPE

TABLE 144: Film Properties of Dow Chemical Dowlex Linear Low Density Polyethylene.

Material Family	LINEAR LOW DENSITY POLYETHYLENE					
Material Supplier	DOW CHEMICAL					
Grade	DOWLEX 2021	DOWLEX 2022	DOWLEX 2039	DOWLEX 2042	DOWLEX 2045	DOWLEX 2101
Manufacturing Method	blown film	blown film	blown film	blown film	blown film	blown film
Reference Number	254	254	254	254	254	254

MATERIAL CHARACTERISTICS

Density	0.920 g/cm³	0.926 g/cm³	0.938 g/cm³	0.930 g/cm³	0.920 g/cm³	0.924 g/cm³
Melt Flow Index	1.0 g/10 min.	1.0 g/10 min.	1.5 g/10 min. (190/2.16)	1.0 g/10 min.	1.0 g/10 min.	0.6 g/10 min. (190/2.16)
Sample Thickness	0.04 mm	0.04 mm	0.03 mm	0.04 mm	0.03 mm	0.03 mm
Blow Up Ratio (BUR)	2.5	2.5	2.3	2	2-3	2.5

MECHANICAL PROPERTIES

Secant Modulus @ 1% Elongation - MD (MPa)	210 {o}	275 {o}	440 {o}			
Secant Modulus @ 1% Elongation - TD (MPa)	248 {o}	315 {o}	530 {o}			
Ultimate Tensile Strength - MD (MPa)	43 {o}	52 {o}	40 {o}	43.4 {o}	57.9 {o}	34 {o}
Ultimate Tensile Strength - TD (MPa)	32 {o}	42 {o}	38 {o}	33.8 {o}	40 {o}	29 {o}
Tensile Strength @ Yield - MD (MPa)	11 {o}	13 {o}	18 {o}	14.5 {o}	11 {o}	11.7 {o}
Tensile Strength @ Yield - TD (MPa)	12 {o}	14 {o}	20 {o}	17.2 {o}	11 {o}	12.8 {o}
Ultimate Elongation - MD (%)	570 {o}	650 {o}	600 {o}	700 {o}	650 {o}	560 {o}
Ultimate Elongation - TD (%)	670 {o}	770 {o}	750 {o}	800 {o}	760 {o}	700 {o}
Drop Dart Impact Strength (d)	260 {j}	206 {j}	80 {j}	220 {j}	335 {j}	130 {j}
Puncture Resistance (J/cm³)	2.2	3.2				
Elmendorf Tear Resistance - MD (g/mm)	22,440 {k}	12,600 {k}	1970 {k}	7480 {k}	25,590 {k}	9060 {k}
Elmendorf Tear Resistance - TD (g/mm)	39370 {k}	37,400 {k}	7870 {k}	31,100 {k}	39,370 {k}	23,620 {k}
Coefficient Of Friction	0.6 {l}	0.5 {l}	0.35 {l}	0.5 {l}	0.6 {l}	0.21 {l}

THERMAL PROPERTIES

Extrusion Temperature (°)	190-220	190-220	190-220	221-243	204-246	191-227

OPTICAL PROPERTIES

Haze (%)	12 {c}	10 {c}	10 {c}	12 {c}	8 {c}	13 {c}
Gloss @ 45°	32 {m}	75 {m}	60 {m}	47 {m}	73 {m}	48 {m}

TABLE 145: Film Properties of Dow Chemical Dowlex Linear Low Density Polyethylene.

Material Family	LINEAR LOW DENSITY POLYETHYLENE					
Material Supplier	DOW CHEMICAL DOWLEX					
Grade	**2045A**	**2045.02**	**2056A**	**2070**	**2071**	**2073**
Features	heat stabilized		heat stabilized			
Manufacturing Method	blown film	blown film	blown film	blown film	blown film	blown film
Reference Number	254	254	254	254	254	254

MATERIAL CHARACTERISTICS

	2045A	2045.02	2056A	2070	2071	2073
Density	0.920 g/cm³	0.920 g/cm³	0.920 g/cm³	0.922 g/cm³	0.928 g/cm³	0.922 g/cm³
Melt Flow Index	1.0 grams/10 min.	1.0 grams/10 min.	1.0 grams/10 min.	1.0 grams/10 min.	0.85 grams/10 min. (190/2.16)	0.75 grams/10 min. (190/2.16)
Sample Thickness	0.03 mm	0.03 mm	0.03 mm	0.04 mm	0.04 mm	0.04 mm
Blow Up Ratio (BUR)	2	2	2	2.5	2.5	2.5
Material Composition		medium antiblock additive content, medium slip additive content				

MECHANICAL PROPERTIES

	2045A	2045.02	2056A	2070	2071	2073
Secant Modulus @ 1% Elongation - MD (MPa)				250 {o}	280 {o}	245 {o}
Secant Modulus @ 1% Elongation - TD (MPa)				260 {o}	340 {o}	266 {o}
Ultimate Tensile Strength - MD (MPa)	56.5 {o}	51 {o}	57.9 {o}	36 {o}	40 {o}	41 {o}
Ultimate Tensile Strength - TD (MPa)	37 {o}	37 {o}	40 {o}	35 {o}	31 {o}	35 {o}
Tensile Strength @ Yield - MD (MPa)	10.7 {o}	11 {o}	10.3 {o}	12 {o}	13 {o}	i2 {o}
Tensile Strength @ Yield - TD (MPa)	11 {o}	12 {o}	11 {o}	12 {o}	15 {o}	12 {o}
Ultimate Elongation - MD (%)	600 {o}	520 {o}	600 {o}	630 {o}	560 {o}	59U {o}
Ultimate Elongation - TD (%)	750 {o}	670 {o}	750 {o}	630 {o}	650 {o}	630 {o}
Drop Dart Impact Strength (g)	200 {j}	210 {j}	275 {j}	250 {j}	225 {j}	290 {j}
Puncture Resistance (J/cm³)				1.7	1.5	1.7
Elmendorf Tear Resistance - MD (g/mm)	13,780 {k}	14,170 {k}	17,760 {k}	22,050 {k}	13,780 {k}	20,870 {k}
Elmendorf Tear Resistance - TD (g/mm)	31,500 {k}	25,200 {k}	29,530 {k}	41,340 {k}	37,010 {k}	41,340 {k}
Coefficient Of Friction	0.7 {l}	0.2 {l}		0.17 {l}	0.2 {l}	0.16 {l}

THERMAL AND OPTICAL PROPERTIES

	2045A	2045.02	2056A	2070	2071	2073
Extrusion Temperature (°C)	216-260	204-246	216-249	204-246	204-246	204-246
Haze (%)	8 {c}	10 {c}	8 {c}	10 {c}	10 {c}	10 {c}
Gloss @45°	57 {m}	55 {m}	60 {m}	39 {m}	28 {m}	36 {m}

TABLE 146: Film Properties of Dow Chemical Dowlex And Next Generation Dowlex Linear Low Density Polyethylene.

Material Family	LINEAR LOW DENSITY POLYETHYLENE						
Material Supplier	DOW CHEMICAL DOWLEX					DOW CHEMICAL DOWLEX NG	
Grade	2035	2037	2047A	2247A	2032	2085	3347A
Features						thin gauge liner film	stretch film
Manufacturing Method	cast film	cast film	cast film	cast film	blown film, cast film	blown film	cast film
Reference Number	254	254	254	254	254	254	254

MATERIAL CHARACTERISTICS

Density	0.919 g/cm³	0.935 g/cm³	0.917 g/cm³	0.917 g/cm³	0.926 g/cm³	0.920 g/cm³	0.917 g/cm³
Melt Flow Index	6 g/10 min. (190/2.16)	2.5 g/10 min. (190/2.16)	2.3 g/10 min. (190/2.16)	2.3 g/10 min. (190/2.16)	2 g/10 min. (190/2.16)	0.95 g/10 min. (190/2.16)	2.3 g/10 min. (190/2.16)
Melt Flow Ratio						8.0	7.35
Sample Thickness	0.03 mm	0.03 mm	0.03 mm	0.03 mm	0.03 mm	0.03 mm	0.02 mm
Blow Up Ratio (BUR)					2.25	2-3	

MECHANICAL PROPERTIES

Ultimate Tensile Strength - MD (MPa)	46 {o}	35.2 {o}	58.6 {o}	62.1 {o}	62 {o}	58 {o}	60 {o}
Ultimate Tensile Strength - TD (MPa)	34 {o}	31 {o}	34.5 {o}	41.4 {o}	38 {o}	46 {o}	34.5 {o}
Tensile Strength @ Yield - MD (MPa)	8.9 {o}	15.2 {o}	10.3 {o}	9 {o}	11.7 {o}	9.9 {o}	9 {o}
Tensile Strength @ Yield - TD (MPa)	8.2 {o}	18.6 {o}	7.6 {o}	8.3 {o}	11.7 {o}	10.1 {o}	8.3 {o}
Ultimate Elongation - MD (%)	660 {o}	420 {o}	550 {o}	550 {o}	500 {o}	540 {o}	550 {o}
Ultimate Elongation - TD (%)	800 {o}	460 {o}	770 {o}	770 {o}	790 {o}	680 {o}	750 {o}
Drop Dart Impact Strength (g)	120 {j}	50 {j}	180 {j}	300 {j}	90 {j}		

MECHANICAL PROPERTIES

Elmendorf Tear Resistance - MD (g/mm)	11,810 {k}	2760 {k}	14,760 {k}	11,810 {k}	10,630 {k}		
Elmendorf Tear Resistance - TD (g/mm)	21,260 {k}	18,500 {k}	23,620 {k}	19,685 {k}	16,730 {k}		
Coefficient Of Friction					0.6 {l}		

THEMAL PROPERTIES

Extrusion Temperature (°C)	260-288	260-288	260-288	260-288	177-232	199-249	260-288

OPTICAL PROPERTIES

Haze (%)	1.3 {c}	3.5 {c}	1 {c}	0.8 {c}	1.2 {c}		0.8 {c}
Gloss @ 45°	95 {m}	85 {m}	90 {m}	95 {m}	90 {m}		92 {m}

TABLE 147: Film Properties of DuPont Canada Sclair Linear Low Density Polyethylene.

Material Family	LINEAR LOW DENSITY POLYETHYLENE							
Material Supplier/ Trade Name	DUPONT CANADA SCLAIR							
Grade	K 11C	11D6	11F9		11H4		11R4	
Product Form	BLOWN FILM							
Applications	stretch cling resin	stretch wrap resin	blending resin, multi-purpose bags	blending resin, multi-purpose bags	blending resin	blending resin	blending resin, multi-purpose bags	blending resin, multi-purpose bags
Reference Number	277	277	277	277	277	277	277	277

MATERIAL CHARACTERISTICS

	K 11C	11D6	11F9		11H4		11R4	
Density	0.920 g/cm³	0.920 g/cm³	0.921 g/cm³	0.921 g/cm³	0.921 g/cm³	0.921 g/cm³	0.921 g/cm³	0.921 g/cm³
Melt Flow Index	0.85 g/10 min.	0.6 g/10 min. (190/2.16)	0.75 g/10 min.	0.75 g/10 min.	1.2 g/10 min.	1.2 g/10 min.	1.6 g/10 min.	1.6 g/10 min.
Sample Thickness	0.0254 mm	0.0254 mm	0.0254 mm	0.125 mm	0.0254 mm	0.125 mm	0.0254 mm	0.125 mm

MECHANICAL PROPERTIES

	K 11C	11D6	11F9		11H4		11R4	
Modulus Of Elasticity - MD (MPa)		220 {r}	220 {r}		210 {r}		210 {r}	
Modulus Of Elasticity - TD (MPa)		275 {r}	240 {r}		240 {r}		240 {r}	
Ultimate Tensile Strength - MD (MPa)	41 {bk}	53.0 {bk}	41.5 {bk}	34.5 {bk}	39.5 {bk}	33.5 {bk}	38.0 {bk}	33.0 {bk}
Ultimate Tensile Strength - TD (MPa)	30 {bk}	34.5 {bk}	31.0 {bk}	31.0 {bk}	29.0 {bk}	30.0 {bk}	27.5 {bk}	29.5 {bk}
Tensile Strength @ Yield - MD (MPa)	10 {bk}	11.0 {bk}	11.5 {bk}	10.5 {bk}	11.0 {bk}	10.5 {bk}	10.5 {bk}	10.5 {bk}
Tensile Strength @ Yield - TD (MPa)	9 {bk}	11.0 {bk}	10.5 {bk}	10.5 {bk}	10.5 {bk}	10.5 {bk}	10.5 {bk}	10.5 {bk}
Ultimate Elongation - MD (%)	450 {bk}	350 {bk}	350 {bk}	800 {bk}	450 {bk}	800 {bk}	500 {bk}	750 {bk}
Ultimate Elongation - TD (%)	750 {bk}	850 {bk}	850 {bk}	900 {bk}	800 {bk}	850 {bk}	800 {bk}	850 {bk}
Impact Strength (g/mm)		3.2 {bl}	3.2 {bl}	3.4 {bl}	2.7 {bl}	3.1 {bl}	2.4 {bl}	2.9 {bl}
Puncture Strength (g-cm/mm)	490,000 {bm}	320,000 {bm}	270,000 {bm}	230,000 {bm}	230,000 {bm}	150,000 {bm}	200,000 {bm}	80,000 {bm}
Elmendorf Tear Resistance - MD (g/mm)	3300 {k}	1200 {k}	1800 {k}	4300 {k}	1900 {k}	3500 {k}	2000 {k}	3000 {k}
Elmendorf Tear Resistance - TD (g/mm)	21,000 {k}	26,000 {k}	22,000 {k}	9800 {k}	16,000 {k}	8600 {k}	11,000 {k}	7900 {k}

THERMAL PROPERTIES

	K 11C	11D6	11F9		11H4		11R4	
Vicat Softening Temperature (°C)		98 {n}	98 {n}	98 {n}	97 {n}	97 {n}	96 {n}	96 {n}
Brittleness Temperature (°C)		-68 {bj}	-68 {bj}	-68 {bj}	-68 {bj}	-68 {bj}	-68 {bj}	-68 {bj}

TABLE 148: Film Properties of DuPont Canada Sclair SL1 and Sclair SL3 Linear Low Density Polyethylene.

Material Family	LINEAR LOW DENSITY POLYETHYLENE					
Material Supplier/ Grade	DUPONT CANADA SCLAIRFILM SL1			DUPONT CANADA SCLAIRFILM SL3		
Product Form	FILM					
Applications	laminations	laminations	laminations	laminations	laminations	laminations
Reference Number	278	278	278	278	278	278

MATERIAL CHARACTERISTICS

Density	0.918 g/cm³	0.918 g/cm³	0.918 g/cm³	0.918 g/cm³	0.918 g/cm³	0.918 g/cm³
Sample Thickness	0.038 mm	0.051 mm	0.076 mm	0.038 mm	0.051 mm	0.076 mm

PHYSICAL PROPERTIES

Yield (in²/lb)	20,000	15,000	10,000	20,000	15,000	10,000

MECHANICAL PROPERTIES

Modulus Of Elasticity (MPa)	169 {r}	169 {r}	169 {r}	169 {r}	169 {r}	169 {r}
Ultimate Tensile Strength - MD (MPa)	34.5 {r}	34.5 {r}	34.5 {r}	34.5 {r}	34.5 {r}	34.5 {r}
Ultimate Elongation - MD (%)	600 {r}	600 {r}	600 {r}	600 {r}	600 {r}	600 {r}
Drop Dart Impact Strength (g)	100 {bn}	125 {bn}	180 {bn}	100 {bn}	125 {bn}	180 {bn}
Coefficient Of Friction - Kinetic	0.35 {l}	0.35 {l}	0.35 {l}	0.13 {l}	0.13 {l}	0.13 {l}

OPTICAL PROPERTIES

Haze (%)	8.0 {c}	9.0 {c}	11.0 {c}	8.0 {c}	9.0 {c}	11.0 {c}
Gloss @ 20°	100 {m}	100 {m}	100 {m}	100 {m}	100 {m}	100 {m}

PERFORMANCE CHARACTERISTICS

Heat Seal Strength (g/cm)	>710 {bo}	>790 {bo}	>950 {bo}	>710 {bo}	>790 {bo}	>950 {bo}

Chapter 35

Ultra Low Density Ethylene-Octene Copolymer

Permeability to Gases and Water Vapor

Dow Chemical: Attane   (product form: coating)

The ultra low density of ATTANE copolymers puts their permeability to oxygen on a par with or higher than that of EVAs. Further, ULDPE's are seen to offer improved (lower) moisture barrier compared with EVAs.

Reference: *Attane Ultra Low Density Ethylene-Octene Copolymers: Performance Plus Compared To LLDPE And EVA Resins In Flexible Packaging,* supplier marketing literature (305-1596-790) - Dow Chemical Company, 1989.

Film Properties and Applications

Dow Chemical: Attane   (product form: coating)

The performance margins enhanced by ATTANE polymers include a combination of strength, sealability, flexibility, and optical properties. Applications include bag-in-box liquids packaging, extruded tubing, profiles, films, film bags, laminations with various barrier layers for greater toughness, and coextruded structures with HDPE.

Reference: *621 Ways To Succeed - 1993-1994 Materials Selection Guide,* supplier technical report (304-00286-1292X SMG) - Dow Chemical Company, 1992.

Dow Chemical: Attane   (product form: coating)

ATTANE Ultra Low Density Ethylene-Octene Copolymers are linear polyethylenes having densities below 0.915 g/cm^3. Conventional density linear polyethylene resins have a density range of 0.916 - 0.965. ULDPEs are manufactured, as are LLDPEs, by copolymerizing ethylene and selected alphaolefins (commonly butene-1, octene-1, hexene-1, 4-methylpentene-1, or combinations of them) using a transition metal catalyst system. ATTANE Ultra Low Density copolymers, made by a proprietary Dow polymerization process, have octene-1 as the comonomer.

Comonomer content largely determines the polymer density. As comonomer content increases, density decreases, and property performance increases. Further, performance attributes also tend to increase as comonomer molecule length increases (i.e., octene vs butene). As density decreases, polymer softening point decreases, resulting in broader sealing range and lower effective sealing temperature for ULDPE vs LLDPE. These effects also result in ULDPE copolymers offering excellent hot tack strength and sealability through contamination, compared with LLDPEs and EVAs.

The non-polarity of ULDPEs provides better chemical resistance, and better thermal stability in processing, than is offered by polar copolymers such as EVAs. ULDPE adhesion to other polyethylenes, to acid copolymers (PRIMACOR resins, ionomers), and surprisingly even to polypropylene (PP) is excellent. However, adhesion to polar substrates such as PVDC (Saran) and EVOH barrier resins is predictably somewhat less than is attainable with polar resins.

The lower crystallinity of ULDPEs results in increased permeability to O$_2$ and other gases - permeability similar to that of EVAs. However, ULDPE water vapor permeability is much lower (selectively better) than that of EVAs. The characteristic narrow molecular weight distribution (MWD) of ATTANE ULDPE copolymers equates with improved properties, particularly in the strength and optics required in film and other critical applications.

The performance margins enhanced by the ultra low density of ATTANE copolymers include a unique and commercially useful combination of strength, sealability, flexibility, and optical properties. In puncture, tear, and impact resistance and in optics, ATTANE copolymers are superior to conventional-density LLDPEs; in physicals they are vastly superior to EVAs, and in optics they are equivalent or superior to EVAs. In their sealing and modulus properties, ATTANE copolymers are comparable to 7% to 18% EVAs.

Reference: *Attane Ultra Low Density Ethylene-Octene Copolymers: Performance Plus Compared To LLDPE And EVA Resins In Flexible Packaging,* supplier marketing literature (305-1596-790) - Dow Chemical Company, 1989.

TABLE 149: Oxygen Gas and Water Vapor Permeability Through Dow Chemical Attane Ultra Low Density Ethylene Octene Copolymers.

Material Family	ULTRA LOW DENSITY ETHYLENE-OCTENE COPOLYMER			
Material Supplier/ Grade	DOW CHEMICAL ATTANE 4003	DOW CHEMICAL ATTANE 4001	DOW CHEMICAL ATTANE 4003	DOW CHEMICAL ATTANE 4001
Manufacturing Method	blown film	blown film	blown film	blown film
Reference Number	11	11	11	11

MATERIAL CHARACTERISTICS

Density	0.912 g/cm³	0.905 g/cm³	0.912 g/cm³	0.905 g/cm³
Melt Flow Index	0.8 g/10 min. (190/2.16)	1.0 g/10 min.	0.8 g/10 min. (190/2.16)	1.0 g/10 min.
Sample Thickness	0.0254 mm	0.0254 mm	0.0254 mm	0.0254 mm

TEST CONDITIONS

Penetrant	oxygen		water vapor	
Relative Humidity (%)	<1% (dry test)	<1% (dry test)	100	100
Test Method	ASTM D3985-81	ASTM D3985-81	Mocon Test Method	Mocon Test Method
Test Apparatus			Mocon Permatron W-1	Mocon Permatron W-1

PERMEABILITY (source document units)

Gas Permeability (cm³ · mil/100 in² · day)	960	650		
Vapor Transmission Rate (g · mil/100 in² · day · atm)			1.2	0.8

PERMEABILITY (normalized units)

Permeability Coefficient (cm³ · mm/m² · day · atm)	378	256		
Vapor Transmission Rate (g · mm/m² · day)			0.47	0.31

TABLE 150: Film Properties of Dow Chemical Attane Ultra Low Density Ethylene Octene Copolymers.

Material Family	ULTRA LOW DENSITY ETHYLENE-OCTENE COPOLYMER				
Material Supplier/ Trade Name	DOW CHEMICAL ATTANE				
Grade	4601	4602	4603	4701	4402
Manufacturing Method	blown film	blown film	blown film	blown film	cast film
Reference Number	254	254	254	254	254

MATERIAL CHARACTERISTICS

Density	0.912 g/cm³	0.912 g/cm³	0.905 g/cm³	0.913 g/cm³	0.912 g/cm³
Melt Flow Index	1.0 grams/10 min.	3.3 grams/10 min. (190/2.16)	0.8 grams/10 min. (190/2.16)	1.0 grams/10 min.	3.3 grams/10 min. (190/2.16)
Melt Flow Ratio	8.7	7.5	8.9	8.7	7.5
Sample Thickness	0.0254 mm	0.0254 mm	0.0254 mm	0.0254 mm	0.0254 mm

MECHANICAL PROPERTIES

Secant Modulus @ 2% Elongation - MD (MPa)	122.7 {h}				
Secant Modulus @ 2% Elongation - TD (MPa)	135.1 {h}				
Ultimate Tensile Strength - MD (MPa)	57.9 {i}	56.5 {i}			56.5 {i}
Ultimate Tensile Strength - TD (MPa)	49.6 {i}	36.5 {i}			36.5 {i}
Tensile Strength @ Yield - MD (MPa)	9.2 {i}	6.9 {i}			6.9 {i}
Tensile Strength @ Yield - TD (MPa)	9.2 {i}	6.2 {i}			6.2 {i}
Ultimate Elongation - MD (%)	575 {i}	580 {i}			580 {i}
Ultimate Elongation - TD (%)	700 {i}	780 {i}			780 {i}
Drop Dart Impact Strength (g)	>885 {j}	460 {j}	>850 {j}	>855 {j}	460 {j}
Puncture Force (kJ/m³)	27,800 {j}	15,720 {j}	28,700 {j}	19,900 {j}	
Elmendorf Tear Resistance - MD (g/mm)	15,200 {k}	15,750 {k}	10,600 {k}	14,300 {k}	15,750 {k}
Elmendorf Tear Resistance - TD (g/mm)	19,600 {k}	23,230 {k}	22,240 {k}	21,400 {k}	23,230 {k}
Coefficient Of Friction	0.34 {l}		0.62 {l}	0.14 {l}	

OPTICAL AND PERFORMANCE PROPERTIES

Seal Initiation Temperature (°C)	102		96	99	
Vicat Softening Temperature (°C)	94	90	79	94	90
Haze (%)	6.0 {c}	0.8 {c}	4.6 {c}	7.3 {c}	0.8 {c}
Gloss @ 45°	60 {m}	95 {m}			95 {m}
Ultimate Hot Tack Strength (kg/m)	7.1		19.6	7.1	
Ultimate Seal Strength (kg/m)	141		163	120	

Chapter 36

Medium Density Polyethylene

TABLE 151: Water Vapor Transmission and Oxygen Permeability Through DuPont Canada Sclair 14D Medium Density Polyethylene.

Material Family	MEDIUM DENSITY POLYETHYLENE	
Material Supplier/ Grade	DUPONT CANADA SCLAIR 14D	
Product Form	BLOWN FILM	
Applications	merchandising bags	merchandising bags
Reference Number	277	277

MATERIAL CHARACTERISTICS

Density	0.935 g/cm³	0.935 g/cm³
Melt Flow Index	0.28 g/10 min.	0.28 g/10 min.
Shore D Hardness	59	59
Sample Thickness	0.0254 mm	0.0254 mm

TEST CONDITIONS

Penetrant	water vapor	oxygen
Temperature (°C)	38	23
Relative Humidity (%)	90	
Test Method		ASTM D1434
Test Note	values are cooling rate dependent	
Test Apparatus	Honeywell model 825 apparatus	

PERMEABILITY (source document units)

Vapor Transmission Rate (g/m² · day)	8.7	
Gas Permeability (cm³/m² · day)		3100

PERMEABILITY (normalized units)

Permeability Coefficient (cm³ · mm/m² · day · atm)		78.7
Vapor Transmission Rate (g · mm/m² · day)	0.22	

TABLE 152: Water Vapor, Oxygen, Nitrogen and Carbon Dioxide Permeability Through Medium Density Polyethylene.

Material Family	MEDIUM DENSITY POLYETHYLENE			
Reference Number	138	138	138	138

TEST CONDITIONS

Penetrant	water vapor	oxygen	nitrogen	carbon dioxide
Temperature (°C)	37.8	25	25	25
Relative Humidity (%)	90			
Test Note		STP conditions		

PERMEABILITY (source document units)

Gas Permeability ($cm^3 \cdot mil/100\ in^2 \cdot day$)		250 - 535	85 - 315	100 - 2500
Gas Permeability ($cm^3 \cdot mm/m^2 \cdot day \cdot atm$)		100 - 210	35 -125	40 - 985
Vapor Transmission Rate ($g \cdot mil/100\ in^2 \cdot day$)	0.7			
Vapor Transmission Rate ($g/day \cdot 100\ in^2$)	0.28			

PERMEABILITY (normalized units)

Permeability Coefficient ($cm^3 \cdot mm/m^2 \cdot day \cdot atm$)		98 - 211	33 - 124	39 - 984
Vapor Transmission Rate ($g \cdot mm/m^2 \cdot day$)	0.28			

MDPE

TABLE 153: Film Properties of DuPont Canada Sclair 14D Medium Density Polyethylene.

Material Family	MEDIUM DENSITY POLYETHYLENE
Material Supplier/ Grade	DUPONT CANADA SCLAIR 14D
Product Form	BLOWN FILM
Applications	merchandising bags
Reference Number	277

MATERIAL CHARACTERISTICS

Density	0.935 g/cm³
Melt Flow Index	0.28 grams/10 min.
Shore D Hardness	59
Sample Thickness	0.0254 mm

MECHANICAL PROPERTIES

Ultimate Tensile Strength - machine direction (MPa)	40 {bk}
Ultimate Tensile Strength - transverse direction (MPa)	37 {bk}
Tensile Strength @ Yield - machine direction (MPa)	15 {bk}
Tensile Strength @ Yield - transverse direction (MPa)	16 {bk}
Ultimate Elongation - machine direction (%)	600 {bk}
Ultimate Elongation - transverse direction (%)	800 {bk}
Elmendorf Tear Resistance - machine direction (g/mm)	1000 {k}
Elmendorf Tear Resistance - transverse direction (g/mm)	20,000 {k}

THERMAL PROPERTIES

Vicat Softening Temperature (°C)	112 {n}

Chapter 37

High Density Polyethylene

Film Properties and Applications

Dow Chemical: HDPE (product form: film)

Dow high density polyethylene resins for blown films are designed for thin gauge applications where product performance requirements can best be satisfied by high-technology resin design. These polymers incorporate state-of-the-art technology to provide high molecular weight; broad molecular weight distribution resins having an outstanding balance of superior toughness (tensile, tear, and impact); and excellent processability for high-quality, economical film products.

Reference: *621 Ways To Succeed - 1993-1994 Materials Selection Guide,* supplier technical report (304-00286-1292X SMG) - Dow Chemical Company, 1992.

TABLE 154: Water Vapor Transmission and Oxygen, Nitrogen and Carbon Dioxide Permeability Through Dow Chemical High Density Polyethylene.

Material Family	HIGH DENSITY POLYETHYLENE			
Material Supplier/ Grade	DOW CHEMICAL			
Product Form	FILM			
Reference Number	250	250	250	250

TEST CONDITIONS

Penetrant	oxygen	nitrogen	carbon dioxide	water vapor
Temperature (°C)	24	24	24	24

PERMEABILITY (source document units)

Vapor Transmission Rate (g · mil/100 in^2 · day)				0.4
Gas Permeability (cm^3 · mil/100 in^2 · day)	100 - 200	40 - 60	600 - 700	

PERMEABILITY (normalized units)

Permeability Coefficient (cm^3 · mm/m^2 · day · atm)	39.37 - 78.74	15.75 - 23.62	236.22 - 275.59	
Vapor Transmission Rate (g · mm/m^2 · day)				0.16

<u>TABLE 155:</u> **Oxygen Permeability at Different Temperatures and Water Vapor Transmission Through High Density Polyethylene.**

Material Family	HIGH DENSITY POLYETHYLENE		
Reference Number	264	264	264

TEST CONDITIONS

Penetrant	oxygen		water vapor
Temperature (°C)	23	35	40
Relative Humidity (%)	0	0	90

PERMEABILITY (source document units)

Gas Permeability (cm³ · mil/100 in² · day)	150	287	
Vapor Transmission Rate (g · mil/100 in² · day)			0.38
Gas Permeability (cm³ · 25μ/m² · day · atm)	2325	4448	
Vapor Transmission Rate (g · 25μ/m² · day)			5.9

PERMEABILITY (normalized units)

Permeability Coefficient (cm³ · mm/m² · day · atm)	59.1	113	
Vapor Transmission Rate (g · mm/m² · day)			0.15

<u>TABLE 156:</u> **Water Vapor Transmission Through Hoechst AG Hostalen Polyethylene.**

Material Family	HIGH DENSITY POLYETHYLENE				
Material Supplier/ Grade	HOECHST AG HOSTALEN				
Reference Number	94	94	94	94	94

TEST CONDITIONS

Penetrant	water vapor				
Temperature (°C)	20	25	30	40	50
Test Note	useable average for all Hostalen grades				

PERMEABILITY (source document units)

Vapor Transmission Rate (g · mm/m² · day)	0.034	0.043	0.068	0.14	0.324

PERMEABILITY (normalized units)

Vapor Transmission Rate (g · mm/m² · day)	0.03	0.04	0.07	0.14	0.32

HDPE

TABLE 157: Water Vapor Transmission and Oxygen Permeability Through DuPont Canada Sclair High Density Polyethylene.

Material Family	HIGH DENSITY POLYETHYLENE					
Material Supplier/ Trade Name	DUPONT CANADA SCLAIR					
Grade	15A	16A	19A	15A	16A	19A
Product Form	BLOWN FILM					
Applications	merchandising bags	merchandising bags	coextrusion, laminations	merchandising bags	merchandising bags	coextrusion, laminations
Reference Number	277	277	277	277	277	277

MATERIAL CHARACTERISTICS

Density	0.941 g/cm³	0.945 g/cm³	0.960 g/cm³	0.941 g/cm³	0.945 g/cm³	0.960 g/cm³
Melt Flow Index	0.35 grams/10 min.	0.28 grams/10 min.	0.75 grams/10 min.	0.35 grams/10 min.	0.28 grams/10 min.	0.75 grams/10 min.
Shore D Hardness	59	60	65	59	60	65
Sample Thickness	0.0254 mm	0.0254 mm	0.0254 mm	0.0254 mm	0.0254 mm	0.0254 mm

TEST CONDITIONS

Penetrant	water vapor			oxygen		
Temperature (°C)	38	38	38	23	23	23
Relative Humidity (%)	90	90	90			
Test Method				ASTM D1434	ASTM D1434	ASTM D1434
Test Note	values are cooling rate dependent					
Test Apparatus	Honeywell model 825 apparatus	Honeywell model 825 apparatus	Honeywell model 825 apparatus			

PERMEABILITY (source document units)

Vapor Transmission Rate (g/m² · day)	7.3	6.5	5.0			
Gas Permeability (cm³/m² · day)				2600	2200	1600

PERMEABILITY (normalized units)

Permeability Coefficient (cm³ · mm/m² · day · atm)				66.0	55.9	40.6
Vapor Transmission Rate (g · mm/m² · day)	0.19	0.17	0.13			

TABLE 158: Water Vapor Transmission and Oxygen Permeability Through High Density Polyethylene.

Material Family	HIGH DENSITY POLYETHYLENE	
Reference Number	296	296

TEST CONDITIONS

Penetrant	oxygen	water vapor
Temperature (°C)	22.8	37.8
Relative Humidity (%)	0	90
Test Method	ASTM D1434	ASTM F1249

PERMEABILITY (source document units)

Gas Permeability $(cm^3 \cdot mil/ 100\ in^2 \cdot bar \cdot day)$	>190	
Vapor Transmission Rate $(g \cdot mil/ 100\ in^2 \cdot bar \cdot day)$		0.25

PERMEABILITY (normalized units)

Permeability Coefficient $(cm^3 \cdot mm/m^2 \cdot day \cdot atm)$	>75.79	
Vapor Transmission Rate $(g \cdot mm/m^2 \cdot day)$		0.1

TABLE 159: Water Vapor, Oxygen, Nitrogen and Carbon Dioxide Permeability Through High Density Polyethylene.

Material Family	HIGH DENSITY POLYETHYLENE			
Reference Number	138	138	138	138

TEST CONDITIONS

Penetrant	water vapor	oxygen	nitrogen	carbon dioxide
Temperature (°C)	37.8	25	25	25
Relative Humidity (%)	90			
Test Note		STP conditions		

PERMEABILITY (source document units)

Gas Permeability $(cm^3 \cdot mil/100\ in^2 \cdot day)$		185	42	580
Vapor Transmission Rate $(g \cdot mil/100\ in^2 \cdot day)$	0.3			
Vapor Transmission Rate $(g/day \cdot 100\ in^2)$	0.12			

PERMEABILITY (normalized units)

Permeability Coefficient $(cm^3 \cdot mm/m^2 \cdot day \cdot atm)$		73	17	228
Vapor Transmission Rate $(g \cdot mm/m^2 \cdot day)$	0.12			

HDPE

TABLE 160: Oxygen and Carbon Dioxide Permeability at Various Temperatures Through Hoechst AG Hostalen Polyethylene.

Material Family	HIGH DENSITY POLYETHYLENE									
Material Supplier/ Grade	HOECHST AG HOSTALEN									
Reference Number	94	94	94	94	94	94	94	94	94	94

TEST CONDITIONS

Penetrant	oxygen					carbon dioxide				
Temperature (°C)	20	25	30	40	50	20	25	30	40	50
Test Condition Note	volume at standard temperature and pressure									
Test Note	useable average for all Hostalen grades									

PERMEABILITY (source document units)

Gas Permeability $(cm^3 \cdot mm/ m^2 \cdot bar \cdot day)$	72	76	92	140	230	280	290	340	520	800

PERMEABILITY (normalized units)

Permeability Coefficient $(cm^3 \cdot mm/m^2 \cdot day \cdot atm)$	72.9	77	93.2	141.9	233	283.7	293.8	344.5	526.9	810.6

TABLE 161: Air and Nitrogen Permeability at Various Temperatures Through Hoechst AG Hostalen Polyethylene.

Material Family	HIGH DENSITY POLYETHYLENE									
Material Supplier/ Grade	HOECHST AG HOSTALEN									
Reference Number	94	94	94	94	94	94	94	94	94	94

TEST CONDITIONS

Penetrant	air					nitrogen				
Temperature (°C)	20	25	30	40	50	20	25	30	40	50
Test Condition Note	volume at standard temperature and pressure									
Test Note	useable average for all Hostalen grades									

PERMEABILITY (source document units)

Gas Permeability $(cm^3 \cdot mm/ m^2 \cdot bar \cdot day)$	29	30	38	68	110	18	21	29	48	84

PERMEABILITY (normalized units)

Permeability Coefficient $(cm^3 \cdot mm/m^2 \cdot day \cdot atm)$	29.4	30.4	38.5	68.9	111.5	18.2	21.3	29.4	48.6	85.1

TABLE 162: Carbon Monoxide, Hydrogen and Helium Permeability at Various Temperatures Through Hoechst AG Hostalen Polyethylene.

Material Family	HIGH DENSITY POLYETHYLENE								
Material Supplier/ Grade	HOECHST AG HOSTALEN								
Reference Number	94	94	94	94	94	94	94	94	94

TEST CONDITIONS

Penetrant	carbon monoxide	hydrogen					helium		
Temperature (°C)	20	20	25	30	40	50	20	30	50
Test Condition Note	volume at standard temperature and pressure								
Test Note	useable average for all Hostalen grades								

PERMEABILITY (source document units)

Gas Permeability ($cm^3 \cdot mm/ m^2 \cdot bar \cdot day$)	36	220	240	290	440	670	150	210	460

PERMEABILITY (normalized units)

Permeability Coefficient ($cm^3 \cdot mm/m^2 \cdot day \cdot atm$)	36.5	222.9	243.2	293.8	445.8	678.9	152.0	212.8	466.1

TABLE 163: Hydrogen Permeability vs. Temperature and Pressure Through High Density Polyethylene.

Material Family	HIGH DENSITY POLYETHYLENE								
Reference Number	306	306	306	306	306	306	306	306	306

MATERIAL CHARACTERISTICS

Sample Thickness	0.03 mm	0.03 mm	0.03 mm	0.03 mm	0.03 mm	0.03 mm	0.03 mm	0.03 mm	0.03 mm

TEST CONDITIONS

Penetrant	hydrogen								
Temperature (°C)	-15	25	68	-16	25	67	-18	25	67
Pressure Gradient (kPa)	1724	1724	1724	3447	3447	3447	6895	6895	6895
Test Method	mass spectrometry and calibrated standard gas leaks								
Test Note	developed by McDonnell Douglas Space Systems Company Chemistry Laboratory								

PERMEABILITY (source document units)

Gas Permeability ($cm^3 \cdot mm/ cm^2 \cdot kPa \cdot sec$)	3.64×10^{-10}	1.78×10^{-9}	8.69×10^{-9}	3.49×10^{-10}	1.76×10^{-9}	8.54×10^{-9}	3.19×10^{-10}	1.84×10^{-9}	8.45×10^{-9}

PERMEABILITY (normalized units)

Permeability Coefficient ($cm^3 \cdot mm/m^2 \cdot day \cdot atm$)	31.9	156	761	30.6	154	748	27.9	161	740

TABLE 164: Nitrogen Permeability vs. Temperature and Pressure Through High Density Polyethylene.

Material Family	HIGH DENSITY POLYETHYLENE								
Reference Number	306	306	306	306	306	306	306	306	306

MATERIAL CHARACTERISTICS

Sample Thickness	0.03 mm	0.03 mm	0.03 mm	0.03 mm	0.03 mm	0.03 mm	0.03 mm	0.03 mm	0.03 mm

TEST CONDITIONS

Penetrant	nitrogen								
Temperature (°C)	-10	25	72	-19	25	69	-17	25	68
Pressure Gradient (kPa)	1724	1724	1724	3447	3447	3447	6895	6895	6895
Test Method	mass spectrometry and calibrated standard gas leaks								
Test Note	developed by McDonnell Douglas Space Systems Company Chemistry Laboratory								

PERMEABILITY (source document units)

Gas Permeability $(cm^3 \cdot mm/ cm^2 \cdot kPa \cdot sec)$	$1.81x10^{-11}$	$1.77x10^{-10}$	$1.98x10^{-9}$	$1.08x10^{-11}$	$1.6x10^{-10}$	$1.46x10^{-9}$	$1.13x10^{-11}$	$1.68x10^{-10}$	$1.71x10^{-9}$

PERMEABILITY (normalized units)

Permeability Coefficient $(cm^3 \cdot mm/m^2 \cdot day \cdot atm)$	1.6	15.5	173	0.95	14.0	127.8	0.99	14.7	150

TABLE 165: Oxygen and Ammonia Permeability vs. Temperature and Pressure Through High Density Polyethylene.

Material Family	HIGH DENSITY POLYETHYLENE								
Reference Number	306	306	306	306	306	306	306	306	306

MATERIAL CHARACTERISTICS

Sample Thickness	0.03 mm	0.03 mm	0.03 mm	0.03 mm	0.03 mm	0.03 mm	0.03 mm	0.03 mm	0.03 mm

TEST CONDITIONS

Penetrant	ammonia			oxygen					
Temperature (°C)	-3	25	61	-16	25	51	-15	25	52
Pressure Gradient (kPa)	965	965	965	1724	1724	1724	3447	3447	3447
Test Method	mass spectrometry and calibrated standard gas leaks								
Test Note	developed by McDonnell Douglas Space Systems Company Chemistry Laboratory								

PERMEABILITY (source document units)

Gas Permeability $(cm^3 \cdot mm/ cm^2 \cdot kPa \cdot sec)$	$3.71x10^{-10}$	$1.4x10^{-9}$	$7.12x10^{-9}$	$5.75x10^{-11}$	$5.75x10^{-10}$	$2.49x10^{-9}$	$5.91x10^{-11}$	$5.64x10^{-10}$	$2.03x10^{-9}$

PERMEABILITY (normalized units)

Permeability Coefficient $(cm^3 \cdot mm/m^2 \cdot day \cdot atm)$	32.5	122.6	623	5.0	50.3	218	5.2	49.4	178

<u>TABLE 166:</u> Xylene and Oxygen Permeability Through High Density Polyethylene.

Material Family	HIGH DENSITY POLYETHYLENE	
Reference Number	293	293

TEST CONDITIONS

Penetrant	xylene	oxygen
Temperature (°C)	60	23
Exposure Time (days)	14	
Relative Humidity (%)		75

PERMEABILITY (source document units)

Vapor Transmission Rate (g · mil/100 in^2 · day)	720	
Gas Permeability (cm^3 · mil/100 in^2 · day)		126

PERMEABILITY (normalized units)

Permeability Coefficient (cm^3 · mm/m^2 · day · atm)		49.6
Vapor Transmission Rate (g · mm/m^2 · day)	283	

TABLE 167: Water Vapor, Carbon Dioxide, Hydrogen, Oxygen, Helium, Ethane, Natural Gas, Freon 12 and Nitrogen Permeability Through Phillips Marlex High Density Polyethylene.

Material Family	HIGH DENSITY POLYETHYLENE								
Material Supplier/ Grade	PHILLIPS MARLEX								
Product Form	FILM								
Reference Number	101	101	101	101	101	101	101	101	101

TEST CONDITIONS

Penetrant	water vapor	carbon dioxide	hydrogen	oxygen	helium	ethane	natural gas	Freon 12	nitrogen
Temperature (°C)	37.8	23	23	23	23	23	23	23	23
Relative Humidity (%)	90								
Test Method	ASTM D96	ASTM D1434	ASTM D1434	ASTM D1434	ASTM D1434	ASTM D1434	ASTM D1434	ASTM D1434	ASTM D1434

PERMEABILITY (source document units)

	water vapor	carbon dioxide	hydrogen	oxygen	helium	ethane	natural gas	Freon 12	nitrogen
Gas Permeability ($cm^3 \cdot mil/100\ in^2 \cdot day$)		345	321	111	247	236	113	95	53
Gas Permeability ($cm^3 \cdot mm/m^2 \cdot day \cdot atm$)		136	126	44	97	93	44	37	21
Vapor Transmission Rate ($g \cdot mil/100\ in^2 \cdot day$)	0.3								
Vapor Transmission Rate ($g \cdot mm/ day/ m^2$)	0.12								

PERMEABILITY (normalized units)

	water vapor	carbon dioxide	hydrogen	oxygen	helium	ethane	natural gas	Freon 12	nitrogen
Permeability Coefficient ($cm^3 \cdot mm/m^2 \cdot day \cdot atm$)		136	126	44	97	93	44	37	21
Vapor Transmission Rate ($g \cdot mm/m^2 \cdot day$)	0.12								

TABLE 168: Argon, Methane, Ethane, Propane, Ethylene, Propylene and Sulfur Dioxide Permeability Through Hoechst AG Hostalen Polyethylene.

Material Family	HIGH DENSITY POLYETHYLENE								
Material Supplier/ Grade	HOECHST AG HOSTALEN								
Reference Number	94	94	94	94	94	94	94	94	94

TEST CONDITIONS

Penetrant	argon			methane	ethane	propane	ethylene	propylene	sulfur dioxide
Temperature (°C)	20	30	50	20	20	20	20	20	20
Test Condition Note	volume at standard temperature and pressure								
Test Note	useable average for all Hostalen grades								

PERMEABILITY (source document units)

Gas Permeability ($cm^3 \cdot mm/ m^2 \cdot bar \cdot day$)	66	89	230	56	89	35	110	76	430

PERMEABILITY (normalized units)

Permeability Coefficient ($cm^3 \cdot mm/m^2 \cdot day \cdot atm$)	66.9	90.2	233	56.7	90.2	35.5	112	77.0	436

TABLE 169: Cyclohexanone, Chlorobenzene, Hexane, Butyl Alcohol, Trichloroethene, Methyl Salicylate and Tetrahydrofuran Permeability Through High Density Polyethylene Bottles.

Material Family	HIGH DENSITY POLYETHYLENE						
Product Form	BOTTLES						
Reference Number	293	293	293	293	293	293	293

TEST CONDITIONS

Penetrant	cyclohexanone	chlorobenzene	hexane	butyl alcohol	trichloroethene	methyl salicylate	tetrahydrofuran
Temperature (°C)	50	50	50	50	50	50	23
Exposure Time (days)	28	28	28	28	28	28	180

PERMEABILITY (source document units)

Penetrant Weight Loss (%)	0.6	20.0	32.9	0.2	15.0	1.02	29.19

TABLE 170: Ethyl Acetate, Isopropyl Acetate, Acetone, Butyl Acetate, Toluene, Xylene, Methyl Isobutyl Ketone and Methyl Ethyl Ketone Permeability Through High Density Polyethylene Bottles.

Material Family	HIGH DENSITY POLYETHYLENE							
Product Form	BOTTLES							
Reference Number	293	293	293	293	293	293	293	293

TEST CONDITIONS

Penetrant	ethyl acetate	isopropyl acetate	acetone	butyl acetate	toluene	xylene	methyl isobutyl ketone	methyl ethyl ketone
Temperature (°C)	50	50	23	50	50	50	50	50
Exposure Time (days)	28	28	180	28	28	28	28	28

PERMEABILITY (source document units)

Penetrant Weight Loss (%)	4.0	2.4	0.91	3.7	45.1	38.1	1.8	2.8

TABLE 171: Kerosine, d-Limonene, 2-Cycle Motor Oil, Pine Oil Cleaner, Diesel Fuel Conditioner and Brakleen Gas Additive Permeability Through High Density Polyethylene Bottles.

Material Family	HIGH DENSITY POLYETHYLENE					
Product Form	BOTTLES					
Reference Number	293	293	293	293	293	293

TEST CONDITIONS

Penetrant	kerosine	d-limonene	motor oils	pine oil	diesel fuel conditioner	gas additive
Penetrant Note			2 cycle	cleaner		Brakleen
Temperature (°C)	50	50	50	50	50	50
Exposure Time (days)	28	28	28	28	28	28

PERMEABILITY (source document units)

Penetrant Weight Loss (%)	2.3	6.7	0.4	1.7 (oily surface)	5.5	10.6

<u>**TABLE 172:**</u> **Mineral Spirits, Turpentine, STP Gas Treatment, Paint Thinner, Charcoal Starter and Naphtha Permeability Through High Density Polyethylene Bottles.**

Material Family	HIGH DENSITY POLYETHYLENE					
Product Form	BOTTLES					
Reference Number	293	293	293	293	293	293

TEST CONDITIONS

Penetrant	mineral spirits	turpentine	STP gas treatment	paint thinner	charcoal starter	naphtha
Temperature (°C)	50	50	50	50	50	50
Exposure Time (days)	28	28	28	28	28	28

PERMEABILITY (source document units)

Penetrant Weight Loss (%)	0.8	2.4	16.4	10.3	14.8	8.8

<u>**TABLE 173:**</u> **Xylene, Propyl Alcohol, Methyl Alcohol, Xylene/ Propyl Alcohol and Xylene/ Methyl Alcohol Permeability Through High Density Polyethylene Bottles.**

Material Family	HIGH DENSITY POLYETHYLENE								
Material Supplier/ Grade	DUPONT								
Product Form	BOTTLE (1 LITER)								
Reference Number	293	293	293	293	293	293	293	293	293

TEST CONDITIONS

Penetrant	xylene			propyl alchohol		xylene			methyl alcohol
Penetrant Note		with 25% propyl alcohol	with 50% propyl alcohol	with 25% xylene		with 25% methyl alcohol	with 50% methyl alcohol	with 25% xylene	
Temperature (°C)	50	50	50	50	50	23	23	23	23
Exposure Time (days)	28	28	28	28	28	180	180	180	180

PERMEABILITY (source document units)

Penetrant Weight Loss (%)	28	23.45	16.27	4.71	0.15	20.30	14.99	4.90	0.29

TABLE 174: Gasoline Permeability Through High Density Polyethylene.

Material Family	HIGH DENSITY POLYETHYLENE
Reference Number	266

MATERIAL CHARACTERISTICS

Sample Thickness	1.27 mm

TEST CONDITIONS

Penetrant	gasoline

PERMEABILITY (source document units)

Vapor Transmission Rate (g/day · 100 in²)	1.29

PERMEABILITY (normalized units)

Vapor Transmission Rate (g · mm/m² · day)	25.39

TABLE 175: d-Limonene (flavor component) Permeability Through High Density Polyethylene.

Material Family	HIGH DENSITY POLYETHYLENE
Product Form	FILM
Reference Number	255

TEST CONDITIONS

Penetrant	d-limonene
Temperature (°C)	25
Relative Humidity (%)	dry

PERMEABILITY (source document units)

Vapor Transmission (10^{-20} kg · m/m² · sec · Pa)	1,700,000

PERMEABILITY (normalized units)

Vapor Transmission Rate (g · mm/m² · day)	149

TABLE 176: Film Properties of Dow Chemical High Density Polyethylene.

Material Family	HIGH DENSITY POLYETHYLENE			
Material Supplier/ Grade	DOW CHEMICAL HDPE 62020	DUPONT CANADA SCLAIR 15A	DUPONT CANADA SCLAIR 16A	DUPONT CANADA SCLAIR 19A
Product Form	FILM	BLOWN FILM		
Applications		merchandising bags	merchandising bags	coextrusion, laminations
Reference Number	254	277	277	277

MATERIAL CHARACTERISTICS

Density	0.949 g/cm^3	0.941 g/cm^3	0.945 g/cm^3	0.960 g/cm^3
Melt Flow Ratio	0.25			
Melt Flow Index		0.35 grams/10 min.	0.28 grams/10 min.	0.75 grams/10 min.
Shore D Hardness		59	60	65
Sample Thickness	0.02 mm	0.0254 mm	0.0254 mm	0.0254 mm

MECHANICAL PROPERTIES

Ultimate Tensile Strength - machine direction (MPa)	89.3 {p}	33 {bk}	40 {bk}	50 {bk}
Ultimate Tensile Strength - transverse direction (MPa)	45.9 {p}	37 {bk}	38 {bk}	36 {bk}
Tensile Strength @ Yield - machine direction (MPa)	39.3 {p}	19 {bk}	21 {bk}	28 {bk}
Tensile Strength @ Yield - transverse direction (MPa)	29 {p}	22 {bk}	23 {bk}	28 {bk}
Ultimate Elongation - machine direction (%)	330 {p}	550 {bk}	600 {bk}	700 {bk}
Ultimate Elongation - transverse direction (%)	415 {p}	750 {bk}	850 {bk}	400 {bk}
Drop Dart Impact Strength (G)	280 {j}			
Elmendorf Tear Resistance - machine direction (g/mm)	394 {k}	900 {k}	900 {k}	700 {k}
Elmendorf Tear Resistance - transverse direction (g/mm)	5905 {k}	23,000 {k}	23,000 {k}	14,000 {k}

THERMAL PROPERTIES

Vicat Softening Temperature (°C)		119 {n}	121 {n}	127 {n}

GRAPH 39: Gas Permeability vs. Temperature through High Density Polyethylene.

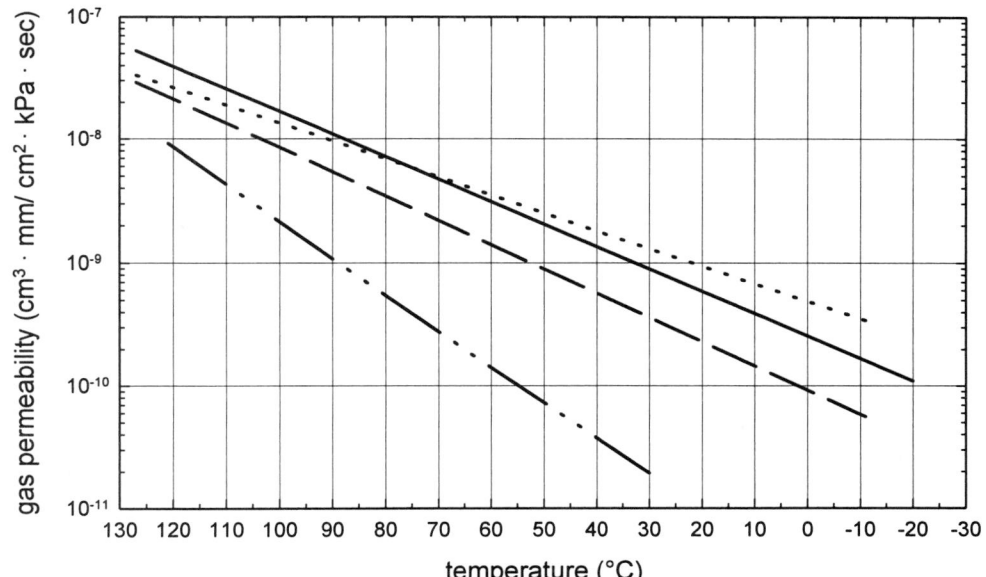

••••••••••••••	HDPE (0.03 mm thick); penetrant: H_2
—··—··—··	HDPE (0.03 mm thick); penetrant: N_2
— — —	HDPE (0.03 mm thick); penetrant: O_2
———	HDPE (0.03 mm thick); penetrant: NH_3
Reference No.	306

Chapter 38

Ethylene-Alpha Olefin Copolymer

Flavor and Aroma Transport

Dow Chemical: Affinity (applications: packaging; features: long chain branching, narrow molecular weight distribution, homogeneous)

Using sensory evaluation techniques, a 35 member test panel used an odor-intensity scale (for which 3 is most intense) to rate the odor characteristics of AFFINITY polymers versus EVAs and ionomers.

AFFINITY polymers received the most neutral response when compared with EVAs and ionomers, with confidence levels greater than 90 percent. In a separate test, confidence levels were at 100 percent when a 24 member test panel also selected an AFFINITY polymer over an EVA. In this test, an AFFINITY polymer containing 13 weight percent octene exhibited lower off-odor than an EVA containing 18 weight percent VA. The EVA was typically described as having an "acidic, sharp, or sour" odor.

Reference: *Affinity Polyolefin Plastomers,* supplier marketing literature (305-01953-893 SMG) - Dow Chemical Company, 1993.

Permeability to Oxygen

Dow Chemical: Affinity (applications: packaging; features: long chain branching, narrow molecular weight distribution, homogeneous)

Response has been very positive to higher oxygen transmission rates which permit more breathable films, and greater elasticity which enhances film recovery and flexibility.

Reference: *Affinity Polyolefin Plastomers,* supplier marketing literature (305-01953-893 SMG) - Dow Chemical Company, 1993.

Film Properties and Applications

Dow Chemical: Affinity (applications: packaging; features: long chain branching, narrow molecular weight distribution, homogeneous)

AFFINITY polyolefin plastomers (POPs) are a breed of homogeneous polymers that take a dramatic departure from conventional polyolefins, including LLDPE, ULDPE, EVA, and ionomer. Their novelty is due to INSITE Technology, a constrained geometry catalyst and process technology that markedly improves performance in ways that can enhance product protection, reduce package failures and allow downgauging.

As homogeneous resins, AFFINITY polymers feature narrower molecular weight distributions (MWDs), and comonomer distributions - both of which can significantly improve the toughness, clarity, sealability, taste/odor performance, and other critical properties of packaging products. This performance is enhanced by use of octene comonomer. The ability to incorporate up to 20 weight percent octene comonomer across a wide range of melt indexes allows AFFINITY polymers to offer a range of thermoplastic and elastic properties - further expanding performance and application envelopes.

This superior performance is achieved without sacrifices in processability, due to the unique ability to introduce long chain branching into a linear, shortchain branched polymer structure. This greatly improves the melt strength and flowability of AFFINITY polymers. This can result in enhanced shear flow, drawdown performance, and thermoformability, which can minimize the need for blending, equipment modifications, or costly processing aids. The benefits of narrower MWDs are not compromised.

AFINITY polymes provide three times the puncture resistance of EVA and comomer films, even surpassing ULDPE, which is well know for puncture resistance. This excellent puncture resistance, along with similar performance in impact strength, delivers toughness and package protection. Further, it provides the opportunity to downgauge products, reducing material costs while addressing environmental concerns for source reduction.

When comparing the percent haze of two AFFINITY polymers with traditional polyolefins, both AFFINITY polymers have less haze (improved optics) than the ULDPE, with one actually exceeding the clarity of EVA and ionomer films.

Reference: *Affinity Polyolefin Plastomers,* supplier marketing literature (305-01953-893 SMG) - Dow Chemical Company, 1993.

Heat Sealing

Dow Chemical: Affinity   (applications: packaging; features: long chain branching, narrow molecular weight distribution, homogeneous)

AFFINITY polymers' outstanding heat seal and hot tack strengths (even through contamination in the package seal), plus their lower sealing temperatures, are especially valuable in high-speed, form-fill-seal applications. These properties not only can reduce leakage, but also can improve line speeds, decrease blistering of the outer layer, and minimize product deterioration. An AFFINITY polymer at 12 weight percent octene seals at temperatures five to ten degrees below EVA and ionomer. Another AFFINITY polymer at 9 weight percent octene seals at temperatures lower than ULDPE and similar to the high-pressure copolymers. The same two AFFINITY polymers exhibit excellent hot tack strength over a wide range of temperatures.

Reference: *Affinity Polyolefin Plastomers,* supplier marketing literature (305-01953-893 SMG) - Dow Chemical Company, 1993.

TABLE 177: Film Properties of Dow Chemical Affinity Polyolefin Plastomer.

Material Family	ETHYLENE-ALPHA OLEFIN COPOLYMER				
Material Supplier	DOW CHEMICAL				
Grade	**AFFINITY HF 1030**	**AFFINITY PL 1880**	**AFFINITY PL 1845**	**AFFINITY PL 1840**	**AFFINITY FM 1570**
Features	drawdown, high modulus	enhanced clarity, hot tack strength, low heat seal	enhanced clarity, low heat seal	enhanced clarity, low heat seal, puncture resistance	enhanced clarity, hot tack strength, low heat seal, toughness
Applications	cast embossed films	form-fill-seal pouch, sealant layer	sealant layer	food pouch, sealant layer	laminations, sealant layer
Manufacturing Method	cast film	blown film	cast film	blown film	blown film
Reference Number	260	260	260	260	260

MATERIAL CHARACTERISTICS

Octene Content	2.0%	12.0%	9.5%	9.5%	7.5%
Density	0.935 g/cm³	0.902 g/cm³	0.910 g/cm³	0.908 g/cm³	0.915 g/cm³
Melt Flow Index	2.5 grams/10 min. (190/2.16)	1.0 grams/10 min.	3.5 grams/10 min.	1.0 grams/10 min.	1.0 grams/10 min.
Sample Thickness	0.0254 mm	0.051 mm	0.0254 mm	0.051 mm	0.051 mm

MECHANICAL PROPERTIES

Secant Modulus @ 2% Elongation - MD (MPa)	200 {r}				
Secant Modulus @ 2% Elongation - TD (MPa)	228 {r}				
Ultimate Tensile Str. - MD (MPa)	29.6 {r}	49.4 {r}	45.4 {r}	55 {r}	51.4 {r}
Ultimate Tensile Str. - TD (MPa)	22.8 {r}	26.2 {r}	33.4 {r}	52.7 {r}	43.9 {r}
Tensile Str. @ Yield - MD (MPa)	11.4 {r}				
Tensile Str. @ Yield - TD (MPa)	10.9 {r}				
Ultimate Elongation - MD (%)	530 {r}	570 {r}	527 {r}	660 {r}	640 {r}
Ultimate Elongation - TD (%)	580 {r}	560 {r}	664 {r}	730 {r}	670 {r}
Drop Dart Impact Strength (g)	55 {w}	>830 {x}	470 {x}	>830 {x}	550 {x}
Puncture Resistance (J/cm³)		26.0 {y}	18 {y}	27 {y}	26.5 {y}
Tear Resistance - MD (g)	42 {z}	355 {aa}	178 {aa}	480 {aa}	360 {aa}
Tear Resistance - TD (g)	118 {z}	500 {aa}	362 {aa}	650 {aa}	760 {aa}

THERMAL PROPERTIES

Seal Initiation Temperature (°C)		85 {ab}	99 {ab}	93 {ab}	100 {ab}

OPTICAL PROPERTIES

Haze (%)		1.1 {c}	0.7 {c}	3.0 {c}	4.9 {c}
Gloss @ 20°			145 {m}		
Gloss @ 45°		90 {m}		85 {m}	78 {m}
Clarity (%)		140 {s}	75 {s}	63 {s}	81 {s}

Chapter 39

Ethylene-Vinyl Acetate Copolymer

Film Properties and Applications

DuPont: Elvax

ELVAX is the DuPont registered trademark for its ethylene-vinyl acetate copolymer resins. The vinyl acetate units in the copolymer modify the basic polyethylene structure and properties. By varying the vinyl acetate content and the molecular weight (melt index), properties can be altered. ELVAX resins are utilized in four basic applications: 1) blown film, 2) cast film, 3) extrusion coating and 4) coextrusions. In addition, ELVAX resins are used in blends and other special applications.

The major effect of higher vinyl acetate content in a resin is to increase flexibility and reduce the softening point of the film extruded from it. Increased impact toughness, puncture resistance and improved flex-cracking resistance are important when packaging objects with sharp edges or corners that require a vacuum or moisture barrier or are subject to rough handling. For these reasons, ELVAX is used for pallet stretch wrapping, bundling, liquid packaging, and as a sealant in barrier bags for primal and subprimal cuts of meat. Low temperature sealability is a key property in its use as a sealant.

As the vinyl acetate content increases, optical properties improve. Film made from higher vinyl acetate content resins is more elastomeric and has a higher coefficient of friction. This "cling" or "tack" is desirable in pallet stretch wrap, bundling and poultry overwrap applications. For other uses, the level of slip may be adjusted through the use of slip and antiblock agents. The level of toughness can be increased by lowering the melt index. Higher melt index resins have the benefit of a lower heat sealing temperature.

Reference: *Elvax Ethylene Vinyl Acetate Copolymer Resins,* supplier technical report (E-45625) - DuPont Company, 1983.

TABLE 178: Carbon Dioxide, Oxygen and Water Vapor Permeability Through Ethylene-Vinyl Acetate Copolymer Film.

Material Family	ETHYLENE-VINYL ACETATE COPOLYMER		
Product Form	FILM		
Features	2.5 blow up ratio		
Manufacturing Method	blown film		
Reference Number	216	216	216

MATERIAL CHARACTERISTICS

Density	0.930 g/cm³	0.930 g/cm³	0.930 g/cm³
Sample Thickness	0.05 mm	0.05 mm	0.05 mm
Vinyl Acetate Content	12.0%	12.0%	12.0%

TEST CONDITIONS

Penetrant	water vapor	carbon dioxide	oxygen
Test Method	JIS Z0208	ASTM D1434	ASTM D1434

PERMEABILITY (source document units)

Vapor Transmission Rate (g · 100 µm/m² · day)	45		
Gas Permeability (cm³ · 100 µm/ m² · atm · day)		11,000	1800

PERMEABILITY (normalized units)

Permeability Coefficient (cm³ · mm/m² · day · atm)		1100	180
Vapor Transmission Rate (g · mm/m² · day)	4.5		

EVA

TABLE 179: Oxygen Gas Permeability Through DuPont Elvax Ethylene Vinyl Acetate Copolymer Film.

Material Family	ETHYLENE-VINYL ACETATE COPOLYMER							
Material Supplier/ Trade Nzame	DUPONT ELVAX							
Grade	3135	3135X	3135SB	3159	3165	3170	3170SB	3170SHB
Product Form	BLOWN FILM							
Reference Number	281	281	281	281	281	281	281	281

MATERIAL CHARACTERISTICS

Density	0.94 g/cm³	0.94 g/cm³	0.94 g/cm³	0.94 g/cm³	0.94 g/cm³	0.94 g/cm³	0.94 g/cm³	0.94 g/cm³
Melt Flow Index	0.25 - 0.35 g/10 min.	0.25 - 0.35 g/10 min.	0.25 - 0.35 g/10 min.	0.5 g/10 min.	0.7 g/10 min.	2.5 g/10 min.	2.5 g/10 min.	2.5 g/10 min.
Sample Thickness	0.051 mm	0.051 mm	0.051 mm	0.051 mm	0.051 mm	0.051 mm	0.051 mm	0.051 mm

MATERIAL COMPOSITION

Vinyl Acetate Content	12.0%	12.0%	12.0%	15.0%	18.0%	18.0%	18.0%	18.0%
Note			antiblock additive, slip additive				antiblock additive, slip additive	high antiblock additive, slip additive

TEST CONDITIONS

Penetrant	oxygen							
Test Method	ASTM D3985	ASTM D3985	ASTM D3985	ASTM D3985	ASTM D3985	ASTM D3985	ASTM D3985	ASTM D3985

PERMEABILITY (source document units)

Gas Permeability (cm³/100 in² · day · atm)	450	450	450	530	595	550	550	550

PERMEABILITY (normalized units)

Permeability Coefficient (cm³ · mm/m² · day · atm)	356	356	356	419	470	435	435	435

TABLE 180: Oxygen Gas Permeability Through DuPont Elvax Ethylene Vinyl Acetate Copolymer Film.

Material Family	ETHYLENE-VINYL ACETATE COPOLYMER						
Material Supplier/ Trade Name	DUPONT ELVAX						
Grade	3120	3121A	3128	3128SB	3130	3130SB	3137
Product Form	BLOWN FILM						
Reference Number	281	281	281	281	281	281	281

MATERIAL CHARACTERISTICS

Density	0.930 g/cm³	0.930 g/cm³	0.930 g/cm³	0.930 g/cm³	0.94 g/cm³	0.94 g/cm³	0.94 g/cm³
Melt Flow Index	1.2 g/10 min.	0.35 g/10 min.	2.0 g/10 min.	2.0 g/10 min.	2.5 g/10 min.	2.5 g/10 min.	0.3 g/10 min.
Sample Thickness	0.051 mm	0.051 mm	0.051 mm	0.051 mm	0.051 mm	0.051 mm	0.051 mm

MATERIAL COMPOSITION

Vinyl Acetate Content	7.5%	7.5%	8.9%	8.9%	12.0%	12.0%	12.0%
Note	antiblock additive, slip additive	antiblock additive, slip additive		antiblock additive, slip additive		antiblock additive, slip additive	

TEST CONDITIONS

Penetrant	oxygen						
Test Method	ASTM D3985	ASTM D3985	ASTM D3985	ASTM D3985	ASTM D3985	ASTM D3985	ASTM D3985

PERMEABILITY (source document units)

Gas Permeability (cm³/100 in² · day · atm)	300	325	380	380	450	450	550

PERMEABILITY (normalized units)

Permeability Coefficient (cm³ · mm/m² · day · atm)	237	257	300	300	356	356	435

TABLE 181: Water Vapor Permeability Through DuPont Elvax Ethylene Vinyl Acetate Copolymer Film.

Material Family	ETHYLENE-VINYL ACETATE COPOLYMER						
Material Supplier/ Trade Name	DUPONT ELVAX						
Grade	3120	3121A	3128	3128SB	3130	3130SB	3137
Product Form	BLOWN FILM						
Reference Number	281	281	281	281	281	281	281

MATERIAL CHARACTERISTICS

Density	0.930 g/cm³	0.930 g/cm³	0.930 g/cm³	0.930 g/cm³	0.94 g/cm³	0.94 g/cm³	0.94 g/cm³
Melt Flow Index	1.2 g/10 min.	0.35 g/10 min.	2.0 g/10 min.	2.0 g/10 min.	2.5 g/10 min.	2.5 g/10 min.	0.3 g/10 min.
Sample Thickness	0.051 mm	0.051 mm	0.051 mm	0.051 mm	0.051 mm	0.051 mm	0.051 mm

MATERIAL COMPOSITION

Vinyl Acetate	7.5%	7.5%	8.9%	8.9%	12.0%	12.0%	12.0%
Note	antiblock additive, slip additive	antiblock additive, slip additive		antiblock additive, slip additive		antiblock additive, slip additive	

TEST CONDITIONS

Penetrant	water vapor						
Test Method	ASTM E96-E	ASTM E96-E	ASTM E96-E	ASTM E96-E	ASTM E96-E	ASTM E96-E	ASTM E96-E

PERMEABILITY (source document units)

Vapor Transmission Rate (g/day · 100 in²)	1.8	1.8	1.8	1.8	2.0	2.0	2.0

PERMEABILITY (normalized units)

Vapor Transmission Rate (g · mm/m² · day)	1.4	1.4	1.4	1.4	1.6	1.6	1.6

TABLE 182: Water Vapor Permeability Through DuPont Elvax Ethylene Vinyl Acetate Copolymer Film.

Material Family	ETHYLENE-VINYL ACETATE COPOLYMER							
Material Supplier/ Trade Name	DUPONT ELVAX							
Grade	3135	3135X	3135SB	3159	3165	3170	3170SB	3170SHB
Product Form	BLOWN FILM							
Reference Number	281	281	281	281	281	281	281	281

MATERIAL CHARACTERISTICS

Density	0.94 g/cm³	0.94 g/cm³	0.94 g/cm³	0.94 g/cm³	0.94 g/cm³	0.94 g/cm³	0.94 g/cm³	0.94 g/cm³
Melt Flow Index	0.25 - 0.35 g/10 min.	0.25 - 0.35 g/10 min.	0.25 - 0.35 g/10 min.	0.5 g/10 min.	0.7 g/10 min.	2.5 g/10 min.	2.5 g/10 min.	2.5 g/10 min.
Sample Thickness	0.051 mm	0.051 mm	0.051 mm	0.051 mm	0.051 mm	0.051 mm	0.051 mm	0.051 mm

MATERIAL COMPOSITION

Vinyl Acetate	12.0%	12.0%	12.0%	15.0%	18.0%	18.0%	18.0%	18.0%
Note			antiblock additive, slip additive				antiblock additive, slip additive	high antiblock additive, slip additive

TEST CONDITIONS

Penetrant	water vapor							
Test Method	ASTM E96-E	ASTM E96-E	ASTM E96-E	ASTM E96-E	ASTM E96-E	ASTM E96-E	ASTM E96-E	ASTM E96-E

PERMEABILITY (source document units)

Vapor Transmission Rate (g/day · 100 in²)	2.2	2.2	2.2	4.8	6.0	6.0	6.0	6.0

PERMEABILITY (normalized units)

Vapor Transmission Rate (g · mm/m² · day)	1.7	1.7	1.7	3.8	4.7	4.7	4.7	4.7

TABLE 183: Film Properties of Ethylene-Vinyl Acetate Copolymer Film.

Material Family	ETHYLENE-VINYL ACETATE COPOLYMER	
Material Supplier/ Grade		BASF AG LUPOLEN V3910DX
Product Form	FILM	FILM
Features	2.5 blow up ratio	
Manufacturing Method	blown film	
Reference Number	216	251

MATERIAL CHARACTERISTICS

Density	0.930 g/cm³	0.938 - 0.943 g/cm³
Melt Flow Index		0.2 - 0.4 grams/10 min. (190/2.16)
Sample Thickness	0.05 mm	0.07 mm
Vinyl Acetate Content	12.0%	

TEST CONDITIONS

Temperature (°C)		23

MECHANICAL PROPERTIES

Tensile Strength - MD (MPa)	17.2 {ci}	24 {a}
Tensile Strength - TD (MPa)	17.7 {ci}	22 {a}
Ultimate Elongation - MD (%)	400 {ci}	300 {a}
Ultimate Elongation - TD (%)	560 {ci}	600 {a}
Tear Strength - MD (g/mm)	1500 {cj}	
Tear Strength - TD (g/mm)	1900 {cj}	
Drop Dart Impact Strength (g)		1000 {f}
Failure Energy (J/mm)		20 {g}

OPTICAL PROPERTIES

Light Transmittance (%)	88 {ck}	
Haze (%)	6 {ck}	
Gloss	70 {cl}	

TABLE 184: Film Properties of DuPont Elvax Ethylene Vinyl Acetate Copolymer Film.

Material Family	ETHYLENE-VINYL ACETATE COPOLYMER						
Material Supplier/ Trade Name	DUPONT ELVAX						
Grade	3120	3121A	3128	3128SB	3130	3130SB	3137
Product Form	BLOWN FILM						
Reference Number	281	281	281	281	281	281	281

MATERIAL CHARACTERISTICS

Density	0.930 g/cm³	0.930 g/cm³	0.930 g/cm³	0.930 g/cm³	0.94 g/cm³	0.94 g/cm³	0.94 g/cm³
Melt Flow Index	1.2 g/10 min.	0.35 g/10 min.	2.0 g/10 min.	2.0 g/10 min.	2.5 g/10 min.	2.5 g/10 min.	0.3 g/10 min.
Sample Thickness	0.051 mm	0.051 mm	0.051 mm	0.051 mm	0.051 mm	0.051 mm	0.051 mm

MATERIAL COMPOSITION

Vinyl Acetate Content	7.5%	7.5%	8.9%	8.9%	12.0%	12.0%	12.0%
Composition Note	antiblock additive, slip additive	antiblock additive, slip additive		antiblock additive, slip additive		antiblock additive, slip additive	

MECHANICAL PROPERTIES

Secant Modulus - MD (MPa)	97 {r}	124 {r}	65 {r}	65 {r}	76 {r}	76 {r}	48 {r}
Secant Modulus - TD (MPa)	117 {r}	152 {r}	72 {r}	72 {r}	76 {r}	76 {r}	50 {r}
Tensile Strength - MD (MPa)	20.0 {r}	22.7 {r}	17.9 {r}	17.9 {r}	22.7 {r}	22.7 {r}	25.5 {r}
Tensile Strength - TD (MPa)	17.9 {r}	19.3 {r}	19.3 {r}	19.3 {r}	24.1 {r}	24.1 {r}	30.3 {r}
Ultimate Elongation - MD (%)	410 {r}	365 {r}	450 {r}	450 {r}	470 {r}	470 {r}	450 {r}
Ultimate Elongation - TD (%)	580 {r}	540 {r}	530 {r}	530 {r}	550 {r}	550 {r}	520 {r}
Spencer Impact Strength (J/m)	0.09 {bp}	0.12 {bp}	0.10 {bp}	0.10 {bp}	0.14 {bp}	0.14 {bp}	0.31 {bp}
Elmendorf Tear Resistance - MD (g/mm)	2560 {k}	1654 {k}	3661 {k}	3661 {k}	3150 {k}	3150 {k}	3032 {k}
Elmendorf Tear Resistance - TD (g/mm)	3268 {k}	4488 {k}	5000 {k}	5000 {k}	4134 {k}	4134 {k}	3622 {k}
Tear Strength, Graves - MD (g)	400 {az}	470 {az}	336 {az}	336 {az}	1040 {az}	4040 {az}	360 {az}
Tear Strength, Graves - TD (g)	430 {az}	470 {az}	381 {az}	381 {az}	900 {az}	900 {az}	360 {az}

THERMAL PROPERTIES

DTA Melt Point (°C)	100	100	97	97	92	92	92
DTA Freeze Point (°C)	92	94	88	88	72	72	84

OPTICAL PROPERTIES

Haze (%)	10 {c}	14 {c}	3.5 {c}		10 {c}		1.6 {c}
Gloss @ 20°	18 {br}	21 {br}	72 {br}		23 {br}		94 {br}
Clarity (%)	8.4 {s}	4 {s}	55 {s}		3.9 {s}		55 {s}

EVA

TABLE 185: Film Properties of DuPont Elvax Ethylene Vinyl Acetate Copolymer Film.

Material Family	ETHYLENE-VINYL ACETATE COPOLYMER							
Material Supplier/ Trade Name	DUPONT ELVAX							
Material Supplier/ Grade	3135	3135X	3135SB	3159	3165	3170	3170SB	3170SHB
Product Form	BLOWN FILM							
Reference Number	281	281	281	281	281	281	281	281

MATERIAL CHARACTERISTICS

Density	0.94 g/cm³	0.94 g/cm³	0.94 g/cm³	0.94 g/cm³	0.94 g/cm³	0.94 g/cm³	0.94 g/cm³	0.94 g/cm³
Melt Flow Index	0.25 - 0.35 grams/10 min.	0.25 - 0.35 grams/10 min.	0.25 - 0.35 grams/10 min.	0.5 grams/10 min.	0.7 grams/10 min.	2.5 grams/10 min.	2.5 grams/10 min.	2.5 grams/10 min.
Sample Thickness	0.051 mm	0.051 mm	0.051 mm	0.051 mm	0.051 mm	0.051 mm	0.051 mm	0.051 mm

MATERIAL COMPOSITION

Vinyl Acetate Content	12.0%	12.0%	12.0%	15.0%	18.0%	18.0%	18.0%	18.0%
Note			antiblock additive, slip additive				antiblock additive, slip additive	high antiblock additive, slip additive

MECHANICAL PROPERTIES

Secant Modulus - MD (MPa)	62 {r}	62 {r}	62 {r}	55 {r}	29 {r}	27 {r}	27 {r}	27 {r}
Secant Modulus - TD (MPa)	62 {r}	62 {r}	62 {r}	55 {r}	29 {r}	28 {r}	28 {r}	28 {r}
Tensile Strength - MD (MPa)	28.9 {r}	28.9 {r}	28.9 {r}	29.6 {r}	31.7 {r}	20.7 {r}	20.7 {r}	20.7 {r}
Tensile Strength - TD (MPa)	29.6 {r}	29.6 {r}	29.6 {r}	29.6 {r}	31.0 {r}	21.4 {r}	21.4 {r}	21.4 {r}
Ultimate Elongation - MD (%)	600 {r}	600 {r}	600 {r}	550 {r}	520 {r}	490 {r}	490 {r}	490 {r}
Ultimate Elongation - TD (%)	550 {r}	550 {r}	550 {r}	590 {r}	540 {r}	510 {r}	510 {r}	510 {r}
Spencer Impact Strength (J/m)	0.28 {bp}	0.28 {bp}	0.28 {bp}	0.32 {bp}	0.35 {bp}	0.30 {bp}	0.30 {bp}	0.30 {bp}

MECHANICAL PROPERTIES

Elmendorf Tear Resistance - MD (g/mm)	5118 {k}	5118 {k}	5118 {k}	4173 {k}	3701 {k}	3622 {k}	3622 {k}	3622 {k}
Elmendorf Tear Resistance - TD (g/mm)	4291 {k}	4291 {k}	4291 {k}	5709 {k}	5236 {k}	4921 {k}	4921 {k}	4921 {k}
Tear Strength, Graves - MD (g)	590 {az}	590 {az}	590 {az}	420 {az}	360 {az}	320 {az}	320 {az}	320 {az}
Tear Strength, Graves - TD (g)	540 {az}	540 {az}	540 {az}	420 {az}	320 {az}	320 {az}	320 {az}	320 {az}

THERMAL PROPERTIES

DTA Melt Point (°C)	92	92	92	87	82	82	82	82
DTA Freeze Point (°C)	82	82	82	75	73	71	71	71

OPTICAL PROPERTIES

Haze (%)	2.9 {c}	2.9 {c}		1.7 {c}	3.1 {c}	2.5 {c}		
Gloss @ 20°	93 {br}	93 {br}		102 {br}	97 {br}	101 {br}		
Clarity (%)	73 {s}	73 {s}		66 {s}	68 {s}	72 {s}		

GRAPH 40: Oxygen Permeability vs. Vinyl Acetate Content through Ethylene-Vinyl Acetate Copolymer.

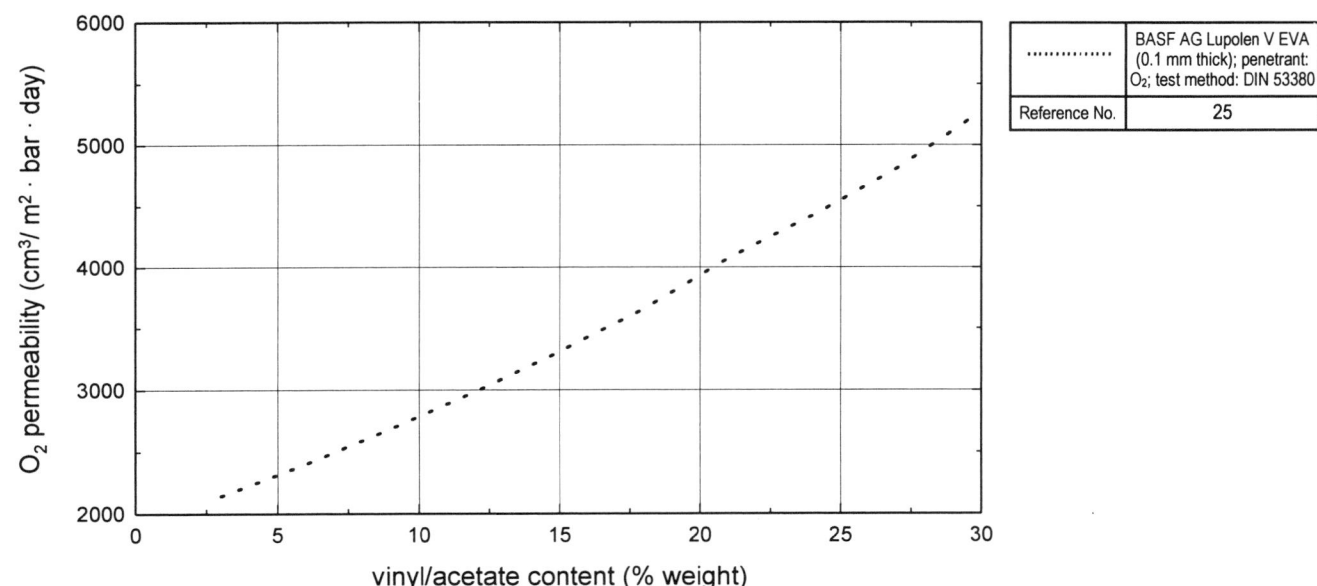

....	BASF AG Lupolen V EVA (0.1 mm thick); penetrant: O_2; test method: DIN 53380
Reference No.	25

GRAPH 41: Water Vapor Permeability vs. Vinyl Acetate Content through Ethylene-Vinyl Acetate Copolymer.

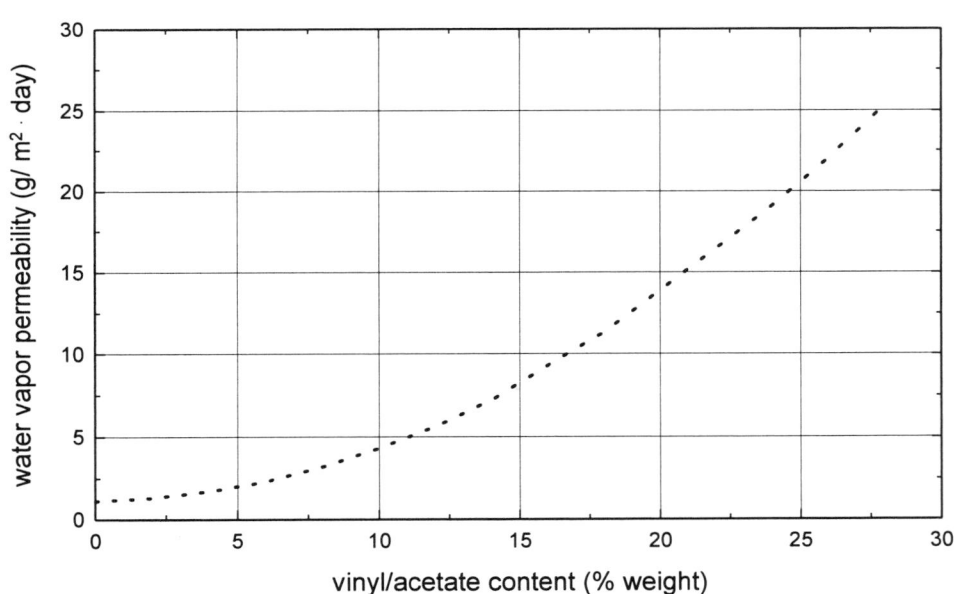

....	BASF AG Lupolen V EVA (0.1 mm thick); penetrant: water vapor; 23°C; 85% to 0% RH gradient; test method: DIN 53122
Reference No.	25

EVA

Chapter 40

Ethylene-Vinyl Alcohol Copolymer

Barrier Performance

Eval Company: Eval (features: barrier properties; product form: film)

The thickness of EVAL film is inversely proportional to its oxygen transmission rate. Since the polymer barrier properties vary in accordance with its thickness, a package can be designed to meet specific requirements by selecting the appropriate layer thickness. In addition to oxygen, EVAL resins offer good barrier against other gases such as carbon dioxide, nitrogen and helium.

The oxygen barrier properties of an ethylene vinyl alcohol copolymer will vary according to the ethylene content in the polymer. EVAL copolymers are produced at various levels of ethylene content to allow the selection of a grade that will fit barrier requirements and processing techniques.

Reference: *Kuraray Eval Resin,* supplier design guide (5-2,000-507) - Kuraray Co., Ltd..

Eval Company: Eval (features: barrier properties; product form: film)

EVAL resins are outstanding in their ability to provide a barrier to gases such as oxygen, carbon dioxide or nitrogen under a variety of environmental conditions. The use of EVAL resins in a packaging structure enhances flavor and quality retention by preventing oxygen from penetrating the package. In those applications where CAP (Controlled Atmosphere Packaging) or MAP (Modified Atmosphere Packaging) techniques are used, EVAL resins effectively retain the combination of gases, such as oxygen, nitrogen and carbon dioxide, used to blanket the product.

Reference: *Gas Barrier Properties Of Eval Resins - Technical Bulletin No. 110,* supplier technical report - Eval Company of America.

Flavor and Aroma Transport

Eval Company: Eval (features: barrier properties; product form: film)

EVAL films are used to preserve the natural aroma or fragrance of packaged products as well as preventing undesirable odors and flavors from reaching the product. The use of these films to package chemicals and highly aromatic materials effectively prevents the permeation of odors from reaching the environment.

Reference: *Eval Films The Ultimate Laminating Film For Barrier Packaging Applications - Technical Bulletin No. 160,* supplier technical report - Eval Company of America.

Effect of Humidity on Barrier Performance

Eval Company: Eval E (features: barrier properties; material compostion: 44 mol% ethylene; product form: film); **Eval** (features: barrier properties; product form: film)

EVAL resin, as indicated by the presence of hydroxy groups in its molecular structure, is hydrophilic and readily absorbs moisture. The oxygen barrier properties of the polymer are adversely affected by the amount of moisture absorbed. As the moisture absorption rate of the polymer increases, the oxygen transmission rate increases. Therefore, for applications involving nearly 100% relative humidity such as in a refrigerator, EVAL E grade is recommended.

Even at high humidities, however, EVAL resin still maintains its superior properties. Furthermore, by coextruding EVAL resin between layers of high moisture barrier resins like polyethylene or polypropylene, the loss of oxygen barrier properties is greatly diminished. Nevertheless, humidity should be considered in designing a high barrier structure.

Reference: *Kuraray Eval Resin,* supplier design guide (5-2,000-507) - Kuraray Co., Ltd..

Eval Company: Eval EF-XL (features: barrier properties, biaxially oriented; product form: film); **Eval** (features: barrier properties; product form: film)

Due to the hydrophilic nature of EVAL resins, EVAL films are moisture sensitive. As the moisture content of the film increases, the permeation rate increases. Significant improvement in gas barrier properties at high relative humidity are achieved with the biaxially oriented EVAL EF-XL.

Reference: *Eval Films The Ultimate Laminating Film For Barrier Packaging Applications - Technical Bulletin No. 160,* supplier technical report - Eval Company of America.

Effect of Temperature on Barrier Performance

Eval Company: Eval (features: barrier properties; product form: film)

The oxygen transmission rate of these copolymers increases with temperature. It increases by about 3.3 times its original value when the temperature rises from 20°C to 35°C. More specifically, it increases in accordance with the rise of both temperature and relative humidity. Thus, in designing a barrier structure both the temperature and the humidity of the environment must be taken into consideration.

Reference: *Kuraray Eval Resin,* supplier design guide (5-2,000-507) - Kuraray Co., Ltd..

Effect of Boiling on Barrier Performance

Eval Company: Eval (features: barrier properties; product form: film)

When Eval composite films are subjected to boiling sterilization, the gas barrier properties of the film are temporarily reduced. The films recover their original properties with time. The time to restore the original barrier properties depends on the polymers and thickness utilized as the outer layer. When polyamide is used for the outside layer the original gas barrier properties are completely restored in one day. It should be emphasized that in composite structures with Eval film employed as the intermediate layer, the Eval polymer never undergoes structural changes caused by the boiling process. The deterioration of the polymer's gas barrier properties is temporary and due to moisture absorption during boiling. As time passes, the moisture evaporates and the gas barrier properties are restored.

Reference: *Kuraray Eval Resin,* supplier design guide (5-2,000-507) - Kuraray Co., Ltd..

Permeation Reduction

Eval Company: Eval (features: barrier properties; product form: film)

In an effort to examine the performance of composite films based on EVAL resin, the following four cases were considered: 100% internal relative humidity (corresponding to high moisture food), 10% internal relative humidity (corresponding to dry food), 65% external relative humidity (corresponding to ordinary external atmosphere) and 85% relative humidity (corresponding to high external humidity). For all of these combinations, the corresponding percentage of RH of the intermediate EVAL resin layer was calculated and the oxygen transmission rate corresponding to that percentage of RH obtained. The results clearly indicate that when packaging high moisture products, the barrier properties of the EVAL resin layer will be optimized if a film of high moisture transmission rate such as polyamide is employed on the outside. In the case of packaging dry products, a film of low moisture transmission rate such as a polypropylene should be used on the outside.

Tests show that even in packaging wet foods, multi-layered structures containing an EVAL resin layer can be designed to give 10 times the oxygen barrier properties of PVDC.

Reference: *Kuraray Eval Resin,* supplier design guide (5-2,000-507) - Kuraray Co., Ltd..

Eval Company: Eval (features: barrier properties; product form: film)

EVAL resins are highly crystalline materials. It is this crystallinity that allows EVAL resins to offer superior barrier properties. Crystallinity may be affected by both heat treating and orientation (stretching). The following general improvements are seen when EVAL films are subjected to either heat treatment, orientation or a combination of both.

◆ Heat treatment alone can improve gas barrier properties, particularly those at high humidity conditions.

◆ A combination of heat treatment and orientation will further improve gas barrier properties at high humidity conditions.

◆ Improvement by orientation alone without heat treatment is marginal.

◆ Improvement in gas barrier properties by orientation and/or heat treatment is more pronounced with EVAL F rather than EVAL E.

Reference: *Gas Barrier Properties Of Eval Resins - Technical Bulletin No. 110,* supplier technical report - Eval Company of America.

Product Summary

DuPont: Selar OH (applications: packaging; features: barrier properties)

SELAR OH resins are designed for both rigid and flexible multilayer structures. They combine outstanding barrier properties with excellent chemical resistance. SELAR OH provides a barrier to permeation of flavors, aromas, and a variety of gases including oxygen, nitrogen, carbon dioxide and helium.

Reference: *Selar Barrier Resin Selector Guide,* supplier marketing literature (H-38769-1) - DuPont Company, 1992.

DuPont: Selar OH 3003 (applications: packaging; features: barrier properties; material compostion: 32 mol% ethylene; melt flow index: 3 g/10 min.); **Selar OH BX 220** (applications: packaging; features: barrier properties, heat stabilized; material compostion: 32 mol% ethylene; melt flow index: 3 g/10 min.)

SELAR OH 3003 and BX 220 provide exceptional barrier propeties for long shelf-life applications. They each have a low ethylene content and melt index which makes them ideal for coextruded sheet and blow-molded containers.

Reference: *Selar Barrier Resin Selector Guide,* supplier marketing literature (H-38769-1) - DuPont Company, 1992.

DuPont: Selar OH 4416 (applications: packaging; features: barrier properties; material compostion: 44 mol% ethylene; melt flow index: 12 g/10 min.); **Selar OH BX 228** (applications: packaging; features: barrier properties, heat stabilized; material compostion: 44 mol% ethylene; melt flow index: 11 g/10 min.)

SELAR OH 4416 and BX 228 are the resins of choice for flexible applications and deep-draw, solid-phase pressure forming (SPPF). BX 228 has excellent thermal stability and is an excellent choice for coextruded blown films.

Reference: *Selar Barrier Resin Selector Guide,* supplier marketing literature (H-38769-1) - DuPont Company, 1992.

DuPont: Selar OH BX 230 (applications: packaging; features: barrier properties, heat stabilized, formability; material compostion: 32 mol% ethylene; melt flow index: 1.5 g/10 min.)

SELAR OH BX 230 has better formability than standard EVOH resins, yet retains the same outstanding barrier properties of SELAR OH 3003 in both shallow and deep-draw applications. It is a good resin choice for retort as well as non-retort applications.

Reference: *Selar Barrier Resin Selector Guide,* supplier marketing literature (H-38769-1) - DuPont Company, 1992.

DuPont: Selar RB (applications: packaging; features: barrier properties, laminar technology)

SELAR RB resins are special polyamide or EVOH concentrates. They incorporate a highly reactive adhesive/compatibilizer for increasing the solvent and/or oxygen barrier of conventionally blow-molded HDPE containers using laminar technology.

Reference: *Selar Barrier Resin Selector Guide,* supplier marketing literature (H-38769-1) - DuPont Company, 1992.

DuPont: Selar RB 400 Series (applications: packaging; features: barrier properties, laminar technology)

The first generation of EVOH-based grades, which make up the SELAR RB 400 series, provide moderate oxygen barrier. These resins are alternatives to PVC and PET for both food and non-food applications, such as sauces, syrups and photochemicals.

Reference: *Selar Barrier Resin Selector Guide,* supplier marketing literature (H-38769-1) - DuPont Company, 1992.

Film Properties and Applications

Eval Company: Eval EF-E (features: barrier properties; material compostion: 44 mol% ethylene; product form: film); **Eval EF-F** (features: barrier properties; material compostion: 32 mol% ethylene; product form: film)

EVAL EF-F and EF-E films have good mechanical strength and high stiffness. Due to the stiffness, they offer excellent machineability. EF-F and EF-E films offer good dimensional stability and ease of printability. EF-E films can be used for deep draw thermoforming applications.

EVAL films are being used in many and varied applications. These include the following:

Product Classification	Product Type
Dried foods	Pet food, fish products, nuts, dehydrated food, cereals
Low Water Containing Foods	Bacon, salami, sausage, processed meats, fish paste, dry mixes
Flavorous or odorous products	Green tea, coffee, spices, toothpaste, pickles
Bag-in-box or pouches	Wine, tomato paste, sauces,
Oxidation sensitive products	Electronic parts, precision machinery parts

Reference: *Eval Films The Ultimate Laminating Film For Barrier Packaging Applications - Technical Bulletin No. 160,* supplier technical report - Eval Company of America.

Eval Company: Eval EF-XL (features: barrier properties, biaxially oriented; product form: film)

EVAL EF-XL combines the technology of biaxial orientation with the existing EVAL film manufacturing technology, producing a film with enhanced barrier and physical properties as well as decreased moisture sensitivity.

Through the technology of biaxial orientation, the mechanical properties of EVAL EF-XL films are greatly enhanced. Dimensional stability is greatly improved which makes possible high speed, high precision printing. The machineability allows high speed processing, such as bag fabrication and lamination. EF-XL films can be used as base films for monolayer packaging or coating applications. Shallow draw thermoforming is also possible with EF-XL films.

Reference: *Eval Films The Ultimate Laminating Film For Barrier Packaging Applications - Technical Bulletin No. 160,* supplier technical report - Eval Company of America.

Optical Properties

Eval Company: Eval (features: barrier properties; product form: film)

All EVAL films offer high gloss and sparkle together with a low haze. This together with the fact that they do not yellow, as do some PVDC coated products, guarantees a fresh appealing product image for a longer period of time.

Reference: *Eval Films The Ultimate Laminating Film For Barrier Packaging Applications - Technical Bulletin No. 160,* supplier technical report - Eval Company of America.

Heat Sealing

Eval Company: Eval   (features: barrier properties; product form: film)

EVAL films can be heat sealed to themselves. However, due to their high melting temperatures, higher heats and longer dwell times are needed. For best results, the use of a sealant layer is recommended.

Reference: *Eval Films The Ultimate Laminating Film For Barrier Packaging Applications - Technical Bulletin No. 160,* supplier technical report - Eval Company of America.

<u>TABLE 186:</u> **Gas Permeability of Carbon Dioxide, Nitrogen and Helium at Different Temperatures Through Kuraray Eval F Series Ethylene Vinyl Alcohol Copolymer.**

Material Family	ETHYLENE-VINYL ALCOHOL COPOLYMER							
Material Supplier/ Grade	EVAL COMPANY EVAL F							
Features	barrier properties							
Reference Number	264	264	264	264	264	264	264	264

MATERIAL COMPOSITION

Ethylene	32 mol%							

TEST CONDITIONS

Penetrant	carbon dioxide			nitrogen		helium		
Temperature (°C)	5	23	35	23	35	5	23	35
Relative Humidity (%)	0	0	0	0	0	0	0	0

PERMEABILITY (source document units)

Gas Permeability (cm^3 · mil/100 in^2 · day)	0.01	0.032	0.066	0.001	0.002	2.7	9.3	13.7
Gas Permeability (cm^3 · 25µ/m^2 · day · atm)	0.155	0.496	1.023	0.015	0.031	41.8	144.1	212.3

PERMEABILITY (normalized units)

Permeability Coefficient (cm^3 · mm/m^2 · day · atm)	0.0039	0.01	0.03	0.0004	0.0008	1.06	3.66	5.39

TABLE 187: Gas Permeability of Carbon Dioxide, Nitrogen and Helium at Different Temperatures Through Kuraray Eval H Series Ethylene Vinyl Alcohol Copolymer.

Material Family	ETHYLENE-VINYL ALCOHOL COPOLYMER							
Material Supplier/ Grade	EVAL COMPANY EVAL H							
Features	barrier properties							
Reference Number	264	264	264	264	264	264	264	264

MATERIAL COMPOSITION

Ethylene	38 mol%							

TEST CONDITIONS

Penetrant	carbon dioxide			nitrogen		helium		
Temperature (°C)	5	23	35	23	35	5	23	35
Relative Humidity (%)	0	0	0	0	0	0	0	0

PERMEABILITY (source document units)

Gas Permeability ($cm^3 \cdot mil/100\ in^2 \cdot day$)	0.017	0.067	0.214	0.004	0.008	4.6	16.6	23.8
Gas Permeability ($cm^3 \cdot 25\mu/m^2 \cdot day \cdot atm$)	0.263	1.04	3.32	0.062	0.124	71.3	257.3	381.3

PERMEABILITY (normalized units)

Permeability Coefficient ($cm^3 \cdot mm/m^2 \cdot day \cdot atm$)	0.01	0.03	0.08	0.0016	0.0031	1.81	6.54	9.37

EVOH

TABLE 188: Gas Permeability of Carbon Dioxide, Nitrogen and Helium at Different Temperatures Through Kuraray Eval E Series Ethylene Vinyl Alcohol Copolymer.

Material Family	ETHYLENE-VINYL ALCOHOL COPOLYMER							
Material Supplier/ Grade	EVAL COMPANY EVAL E							
Features	barrier properties							
Reference Number	264	264	264	264	264	264	264	264

MATERIAL COMPOSITION

Ethylene	44 mol%							

TEST CONDITIONS

Penetrant	carbon dioxide			nitrogen		helium		
Temperature (°C)	5	23	35	23	35	5	23	35
Relative Humidity (%)	0	0	0	0	0	0	0	0

PERMEABILITY (source document units)

Gas Permeability ($cm^3 \cdot mil/100\ in^2 \cdot day$)	0.056	0.214	0.498	0.008	0.015	6.6	23.8	35.6
Gas Permeability ($cm^3 \cdot 25\mu/m^2 \cdot day \cdot atm$)	0.87	3.32	7.72	0.124	0.232	102.3	368.9	551.8

PERMEABILITY (normalized units)

Permeability Coefficient ($cm^3 \cdot mm/m^2 \cdot day \cdot atm$)	0.02	0.08	0.2	0.0031	0.01	2.6	9.37	14.02

TABLE 189: Oxygen Permeability Through DuPont Selar OH Ethylene Vinyl Alcohol Copolymer.

Material Family	ETHYLENE-VINYL ALCOHOL COPOLYMER									
Material Supplier/ Grade	**DUPONT SELAR OH 3003**		**DUPONT SELAR OH 4416**		**DUPONT SELAR OH BX220**		**DUPONT SELAR OH BX228**		**DUPONT SELAR OH BX230**	
Features	barrier properties		barrier properties		barrier properties, heat stabilized		barrier properties, heat stabilized		barrier properties, formability, heat stabilized	
Applications	packaging		packaging		packaging		packaging		packaging	
Reference Number	295	295	295	295	295	295	295	295	295	295

MATERIAL CHARACTERISTICS

Melt Flow Index	3 g/10 min.	3 g/10 min.	12 g/10 min.	12 g/10 min.	3 g/10 min.	3 g/10 min.	11 g/10 min.	11 g/10 min.	1.5 g/10 min.	1.5 g/10 min.
Ethylene Content	32 mol%	32 mol%	44 mol%	44 mol%	32 mol%	32 mol%	44 mol%	44 mol%	32 mol%	32 mol%

TEST CONDITIONS

Penetrant	oxygen									
Temperature (°C)	30	30	30	30	30	30	30	30	30	30
Relative Humidity (%)	dry	80	dry	80	dry	80	dry	80	dry	80

PERMEABILITY (source document units)

Gas Permeability (cm^3 · mil/100 in^2 · day)	0.03	0.12	0.15	0.33	0.03	0.12	0.15	0.33	0.03	0.12

PERMEABILITY (normalized units)

Permeability Coefficient (cm^3 · mm/m^2 · day · atm)	0.01	0.05	0.06	0.13	0.01	0.05	0.06	0.13	0.01	0.05

EVOH

TABLE 190: Oxygen Permeability at Different Temperatures Through Kuraray Eval Ethylene Vinyl Alcohol Copolymer.

Material Family	ETHYLENE-VINYL ALCOHOL COPOLYMER							
Material Supplier/ Grade	EVAL COMPANY EVAL E				EVAL COMPANY EVAL G			
Features	barrier properties							
Reference Number	264	264	264	264	264	264	264	264

MATERIAL COMPOSITION

Ethylene Content	44 mol%							

TEST CONDITIONS

Penetrant	oxygen							
Temperature (°C)	5	23	35	50	5	23	35	50
Relative Humidity (%)	0	0	0	0	0	0	0	0

PERMEABILITY (source document units)

Gas Permeability ($cm^3 \cdot mil/100\ in^2 \cdot day$)	0.017	0.06	0.124	0.344	0.067	0.116	0.174	0.394
Gas Permeability ($cm^3 \cdot 25\mu/m^2 \cdot day \cdot atm$)	0.259	0.935	1.922	5.33	1.034	1.8	2.7	6.11

PERMEABILITY (normalized units)

Permeability Coefficient ($cm^3 \cdot mm/m^2 \cdot day \cdot atm$)	0.01	0.02	0.05	0.14	0.03	0.05	0.07	0.16

TABLE 191: Oxygen Permeability at Different Temperatures Through Kuraray Eval Ethylene Vinyl Alcohol Copolymer.

Material Family	ETHYLENE-VINYL ALCOHOL COPOLYMER							
Material Supplier/ Grade	EVAL COMPANY EVAL H				EVAL COMPANY EVAL K			
Features	barrier properties							
Reference Number	264	264	264	264	264	264	264	264

MATERIAL COMPOSITION

Ethylene Content	38 mol%							

TEST CONDITIONS

Penetrant	oxygen							
Temperature (°C)	5	23	35	50	5	23	35	50
Relative Humidity (%)	0	0	0	0	0	0	0	0

PERMEABILITY (source document units)

Gas Permeability ($cm^3 \cdot mil/100\ in^2 \cdot day$)	0.006	0.025	0.061	0.167	0.006	0.025	0.061	0.167
Gas Permeability ($cm^3 \cdot 25\mu/m^2 \cdot day \cdot atm$)	0.09	0.395	0.94	2.6	0.09	0.395	0.94	2.6

PERMEABILITY (normalized units)

Permeability Coefficient ($cm^3 \cdot mm/m^2 \cdot day \cdot atm$)	0.0024	0.01	0.02	0.07	0.0024	0.01	0.02	0.07

TABLE 192: Oxygen Permeability at Different Temperatures Through Kuraray Eval Ethylene Vinyl Alcohol Copolymer.

Material Family	ETHYLENE-VINYL ALCOHOL COPOLYMER							
Material Supplier/ Grade	EVAL COMPANY EVAL L				EVAL COMPANY EVAL F			
Features	barrier properties	barrier properties	barrier properties	barrier properties	barrier properties	barrier properties	barrier properties	barrier properties
Reference Number	264	264	264	264	264	264	264	264

MATERIAL COMPOSITION

Ethylene	27 mol%	27 mol%	27 mol%	27 mol%	32 mol%	32 mol%	32 mol%	32 mol%

TEST CONDITIONS

Penetrant	oxygen							
Temperature (°C)	5	23	35	50	5	23	35	50
Relative Humidity (%)	0	0	0	0	0	0	0	0

PERMEABILITY (source document units)

Gas Permeability ($cm^3 \cdot mil/100\ in^2 \cdot day$)	0.001	0.006	0.015	0.041	0.003	0.013	0.031	0.086
Gas Permeability ($cm^3 \cdot 25\mu/m^2 \cdot day \cdot atm$)	0.022	0.095	0.231	0.637	0.045	0.2	0.48	1.34

PERMEABILITY (normalized units)

Permeability Coefficient ($cm^3 \cdot mm/m^2 \cdot day \cdot atm$)	0.00039	0.0024	0.01	0.02	0.0012	0.01	0.01	0.03

TABLE 193: Oxygen Permeability vs. Relative Humidity Through Kuraray Eval EF-XL Ethylene Vinyl Alcohol Copolymer Biaxially Oriented Film.

Material Family	ETHYLENE-VINYL ALCOHOL COPOLYMER			
Material Supplier/ Grade	EVAL COMPANY EVAL EF-XL			
Product Form	FILM			
Features	biaxially oriented			
Reference Number	268	268	268	268

MATERIAL CHARACTERISTICS

Sample Thickness	0.015 mm	0.015 mm	0.015 mm	0.015 mm

TEST CONDITIONS

Penetrant	oxygen			
Temperature (°C)	35	20	20	20
Relative Humidity (%)	0	65	85	100
Test Method	JIS Z1707	ASTM D3985	ASTM D3985	ASTM D3985

PERMEABILITY (source document units)

Gas Permeability ($cm^3 \cdot mil/100\ in^2 \cdot day$)	0.03	0.02	0.07	0.39

PERMEABILITY (normalized units)

Permeability Coefficient ($cm^3 \cdot mm/m^2 \cdot day \cdot atm$)	0.01	0.01	0.03	0.15

TABLE 194: Oxygen Permeability vs. Relative Humidity Through Kuraray Eval Ethylene Vinyl Alcohol Copolymer Film.

Material Family	ETHYLENE-VINYL ALCOHOL COPOLYMER							
Material Supplier/ Trade Name	EVAL COMPANY EVAL							
Grade	EF-F	EF-E	EF-F	EF-E	EF-F	EF-E	EF-F	EF-E
Product Form	FILM							
Reference Number	268	268	268	268	268	268	268	268

MATERIAL CHARACTERISTICS

Sample Thickness	0.015 mm	0.02 mm	0.015 mm	0.02 mm	0.015 mm	0.02 mm	0.015 mm	0.02 mm

MATERIAL COMPOSITION

Ethylene Content	32 mol%	44 mol%	32 mol%	44 mol%	32 mol%	44 mol%	32 mol%	44 mol%

TEST CONDITIONS

Penetrant	oxygen							
Temperature (°C)	35	35	20	20	20	20	20	20
Relative Humidity (%)	0	0	65	65	85	85	100	100
Test Method	JIS Z1707	JIS Z1707	ASTM D3985	ASTM D3985	ASTM D3985	ASTM D3985	ASTM D3985	ASTM D3985

PERMEABILITY (source document units)

Gas Permeability $(cm^3 \cdot mil/100\ in^2 \cdot day)$	0.03	0.21	0.03	0.1	0.13	0.21	1.61	0.65

PERMEABILITY (normalized units)

Permeability Coefficient $(cm^3 \cdot mm/m^2 \cdot day \cdot atm)$	0.01	0.08	0.01	0.04	0.05	0.08	0.63	0.26

TABLE 195: Oxygen Permeability vs. Relative Humidity Through Kuraray Eval Ethylene Vinyl Alcohol Copolymer Film.

Material Family	ETHYLENE-VINYL ALCOHOL COPOLYMER											
Material Supplier/ Trade Name	EVAL COMPANY EVAL											
Grade	EF-XL	EF-F	EF-E	EF-XL	EF-F	EF-E	EF-XL	EF-F	EF-E	EF-XL	EF-F	EF-E
Product Form	FILM											
Features	barrier properties, biaxially oriented	barrier properties	barrier properties	barrier properties, biaxially oriented	barrier properties	barrier properties	barrier properties, biaxially oriented	barrier properties	barrier properties	barrier properties, biaxially oriented	barrier properties	barrier properties
Reference Number	265	265	265	265	265	265	265	265	265	265	265	265

MATERIAL COMPOSITION

Ethylene		32 mol%	44 mol%		32 mol%	44 mol%		32 mol%	44 mol%		32 mol%	44 mol%

TEST CONDITIONS

Penetrant	oxygen											
Temperature (°C)	20	20	20	20	20	20	20	20	20	20	20	20
Relative Humidity (%)	65			85			100			0		

PERMEABILITY (source document units)

Gas Permeability ($cm^3 \cdot mil/100\ in^2 \cdot day$)	0.01	0.02	0.08	0.04	0.08	0.17	0.23	1.0	0.52	0.02	0.02	0.16

PERMEABILITY (normalized units)

Permeability Coefficient ($cm^3 \cdot mm/m^2 \cdot day \cdot atm$)	0.004	0.01	0.03	0.02	0.03	0.07	0.09	0.39	0.2	0.01	0.01	0.06

TABLE 196: Oxygen Permeability at 0% Relative Humidity vs. Orientation and Heat Treatment Through Kuraray Eval F Ethylene Vinyl Alcohol Copolymer.

Material Family	ETHYLENE-VINYL ALCOHOL COPOLYMER						
Material Supplier/ Grade	EVAL COMPANY EVAL F						
Chill Roll Temperature	50°C	110°C	50°C	50°C	50°C	50°C	50°C
Heat Treatment	none	none	140°C	none	140°C	none	140°C
Orientation	none	none	none	uniaxially (3 times)	uniaxially (3 times)	biaxially (3x3)	biaxially (3x3)
Reference Number	264	264	264	264	264	264	264

MATERIAL COMPOSITION

Ethylene Content	32 mol%						

TEST CONDITIONS

Penetrant	oxygen						
Temperature (°C)	20	20	20	20	20	20	20
Relative Humidity (%)	0	0	0	0	0	0	0

PERMEABILITY (source document units)

Gas Permeability ($cm^3 \cdot mil/100\ in^2 \cdot day$)	0.008	0.0076	0.0066	0.0076	0.0006	0.0076	0.0061
Gas Permeability ($cm^3 \cdot 25\mu/m^2 \cdot day \cdot atm$)	0.126	0.118	0.102	0.118	0.094	0.118	0.094

PERMEABILITY (normalized units)

Permeability Coefficient ($cm^3 \cdot mm/m^2 \cdot day \cdot atm$)	0.0031	0.003	0.0026	0.003	0.0002	0.003	0.0024

TABLE 197: Oxygen Permeability at 0% Relative Humidity vs. Orientation and Heat Treatment Through Kuraray Eval E Ethylene Vinyl Alcohol Copolymer.

Material Family	ETHYLENE-VINYL ALCOHOL COPOLYMER						
Material Supplier/ Grade	EVAL COMPANY EVAL E						
Chill Roll Temperature	50°C	110°C	50°C	50°C	50°C	50°C	50°C
Heat Treatment	none	none	140°C	none	140°C	none	140°C
Orientation	none	none	none	uniaxially (3 times)	uniaxially (3 times)	biaxially (3x3)	biaxially (3x3)
Reference Number	264	264	264	264	264	264	264

MATERIAL COMPOSITION

Ethylene Content	44 mol%						

TEST CONDITIONS

Penetrant	oxygen						
Temperature (°C)	20	20	20	20	20	20	20
Relative Humidity (%)	0	0	0	0	0	0	0

PERMEABILITY (source document units)

Gas Permeability ($cm^3 \cdot mil/100\ in^2 \cdot day$)	0.076	0.066	0.061	0.071	0.061	0.071	0.061
Gas Permeability ($cm^3 \cdot 25\mu/m^2 \cdot day \cdot atm$)	1.18	1.02	0.94	1.02	0.94	1.02	0.94

PERMEABILITY (normalized units)

Permeability Coefficient ($cm^3 \cdot mm/m^2 \cdot day \cdot atm$)	0.03	0.026	0.024	0.028	0.024	0.028	0.024

TABLE 198: Oxygen Permeability at 100% Relative Humidity vs. Orientation and Heat Treatment Through Kuraray Eval F Ethylene Vinyl Alcohol Copolymer.

Material Family	ETHYLENE-VINYL ALCOHOL COPOLYMER						
Material Supplier/ Grade	EVAL COMPANY EVAL F						
Chill Roll Temperature	50°C	110°C	50°C	50°C	50°C	50°C	50°C
Heat Treatment	none	none	140°C	none	140°C	none	140°C
Orientation	none	none	none	uniaxially (3 times)	uniaxially (3 times)	biaxially (3x3)	biaxially (3x3)
Reference Number	264	264	264	264	264	264	264

MATERIAL COMPOSITION

Ethylene Content	32 mol%						

TEST CONDITIONS

Penetrant	oxygen						
Temperature (°C)	20	20	20	20	20	20	20
Relative Humidity (%)	100	100	100	100	100	100	100

PERMEABILITY (source document units)

Gas Permeability ($cm^3 \cdot mil/100\ in^2 \cdot day$)	2.6	2.2	0.71	2.1	0.25	2	0.15
Gas Permeability ($cm^3 \cdot 25\mu/m^2 \cdot day \cdot atm$)	40.9	33.8	11	32.3	3.9	31.5	2.3

PERMEABILITY (normalized units)

Permeability Coefficient ($cm^3 \cdot mm/m^2 \cdot day \cdot atm$)	1.02	0.87	0.28	0.83	0.1	0.79	0.06

TABLE 199: Oxygen Permeability at 100% Relative Humidity vs. Orientation and Heat Treatment Through Kuraray Eval E Ethylene Vinyl Alcohol Copolymer.

Material Family	ETHYLENE-VINYL ALCOHOL COPOLYMER						
Material Supplier/ Grade	EVAL COMPANY EVAL E						
Chill Roll Temperature	50°C	110°C	50°C	50°C	50°C	50°C	50°C
Heat Treatment	none	none	140°C	none	140°C	none	140°C
Orientation	none	none	none	uniaxially (3 times)	uniaxially (3 times)	biaxially (3x3)	biaxially (3x3)
Reference Number	264	264	264	264	264	264	264

MATERIAL COMPOSITION

Ethylene Content	44 mol%						

TEST CONDITIONS

Penetrant	oxygen						
Temperature (°C)	20	20	20	20	20	20	20
Relative Humidity (%)	100	100	100	100	100	100	100

PERMEABILITY (source document units)

Gas Permeability ($cm^3 \cdot mil/100\ in^2 \cdot day$)	0.76	0.61	0.41	0.71	0.2	0.71	0.15
Gas Permeability ($cm^3 \cdot 25\mu/m^2 \cdot day \cdot atm$)	11.8	9.4	6.3	10.2	3.1	10.2	2.4

PERMEABILITY (normalized units)

Permeability Coefficient ($cm^3 \cdot mm/m^2 \cdot day \cdot atm$)	0.299	0.24	0.16	0.28	0.079	0.28	0.06

EVOH

TABLE 200: Organic Solvents Permeability Through Kuraray Eval Ethylene Vinyl Alcohol Copolymer Film.

Material Family	ETHYLENE-VINYL ALCOHOL COPOLYMER								
Material Supplier/ Trade Name	EVAL COMPANY EVAL								
Grade	EF-F	EF-E	EF-XL	EF-F	EF-E	EF-XL	EF-F	EF-E	EF-XL
Product Form	FILM								
Features	barrier properties	barrier properties	barrier properties, biaxially oriented	barrier properties	barrier properties	barrier properties, biaxially oriented	barrier properties	barrier properties	barrier properties, biaxially oriented
Reference Number	265	265	265	265	265	265	265	265	265

MATERIAL COMPOSITION

Ethylene	32 mol%	44 mol%		32 mol%	44 mol%		32 mol%	44 mol%	

TEST CONDITIONS

Penetrant	chloroform			xylene			kerosine		
Temperature (°C)	20	20	20	20	20	20	20	20	20

PERMEABILITY (source document units)

Vapor Transmission Rate $(g \cdot mil/100\ in^2 \cdot day)$	0.1	0.16	0.006	0.054	0.074	0.016	>0.001	0.0025	0.001

PERMEABILITY (normalized units)

Vapor Transmission Rate $(g \cdot mm/m^2 \cdot day)$	0.04	0.06	0.0024	0.02	0.03	0.01	>0.0004	0.00098	0.0004

TABLE 201: Organic Solvents Permeability Through Kuraray Eval EF-XL Biaxially Oriented Ethylene Vinyl Alcohol Copolymer Film.

Material Family	ETHYLENE-VINYL ALCOHOL COPOLYMER			
Material Supplier/ Grade	EVAL COMPANY EVAL EF-XL			
Product Form	FILM			
Features	biaxially oriented			
Reference Number	266	266	266	266

MATERIAL CHARACTERISTICS

Sample Thickness	0.015 mm	0.015 mm	0.015 mm	0.015 mm

TEST CONDITIONS

Penetrant	chloroform	xylene	methyl ethyl ketone	kerosine
Temperature (°C)	20	20	20	20
Relative Humidity (%)	65	65	65	65

PERMEABILITY (source document units)

Vapor Transmission Rate (g/day · 100 in²)	0.01	0.03	0.02	<0.003

PERMEABILITY (normalized units)

Vapor Transmission Rate (g · mm/m² · day)	0.002	0.007	0.005	<0.0007

EVOH

TABLE 202: Organic Solvents Permeability Through Kuraray Eval F Ethylene Vinyl Alcohol Copolymer Film.

Material Family	ETHYLENE-VINYL ALCOHOL COPOLYMER							
Material Supplier/ Grade	EVAL COMPANY EVAL F							
Product Form	FILM							
Reference Number	266	266	266	266	266	266	266	266

MATERIAL CHARACTERISTICS

Sample Thickness	0.02 mm	0.032 mm	0.02 mm	0.032 mm	0.02 mm	0.032 mm	0.02 mm	0.032 mm

MATERIAL COMPOSITION

Ethylene Content	32 mol%							

TEST CONDITIONS

Penetrant	chloroform		xylene		methyl ethyl ketone		kerosine	
Temperature (°C)	20	20	20	20	20	20	20	20
Relative Humidity (%)	65	65	65	65	65	65	65	65

PERMEABILITY (source document units)

Vapor Transmission Rate (g/day · 100 in^2)	0.13	0.3	0.07	<0.003	0.25	0.02	<0.003	<0.003

PERMEABILITY (normalized units)

Vapor Transmission Rate (g · mm/m^2 · day)	0.04	0.2	0.02	<0.002	0.08	0.01	<0.001	<0.002

TABLE 203: Organic Solvents Permeability Through Kuraray Eval E Ethylene Vinyl Alcohol Copolymer Film.

Material Family	ETHYLENE-VINYL ALCOHOL COPOLYMER							
Material Supplier/ Grade	EVAL COMPANY EVAL E							
Product Form	FILM							
Reference Number	266	266	266	266	266	266	266	266

MATERIAL CHARACTERISTICS

Sample Thickness	0.02 mm	0.032 mm	0.02 mm	0.032 mm	0.02 mm	0.032 mm	0.02 mm	0.032 mm

MATERIAL COMPOSITION

Ethylene Content	44 mol%							

TEST CONDITIONS

Penetrant	chloroform		xylene		methyl ethyl ketone		kerosine	
Temperature (°C)	20	20	20	20	20	20	20	20
Relative Humidity (%)	65	65	65	65	65	65	65	65

PERMEABILITY (source document units)

Vapor Transmission Rate (g/day · 100 in^2)	0.2	0.06	0.09	0.04	0.31	0.03	<0.003	<0.003

PERMEABILITY (normalized units)

Vapor Transmission Rate (g · mm/m^2 · day)	0.06	0.03	0.03	0.02	0.12	0.01	<0.001	<0.002

EVOH

TABLE 204: Water Vapor Transmission Through Kuraray Eval Ethylene Vinyl Alcohol Copolymer Film.

Material Family	ETHYLENE-VINYL ALCOHOL COPOLYMER		
Material Supplier/ Grade	EVAL COMPANY EVAL EF-XL	EVAL COMPANY EVAL EF-F	EVAL COMPANY EVAL EF-E
Product Form	FILM		
Features	biaxially oriented		
Reference Number	268	268	268

MATERIAL CHARACTERISTICS

Sample Thickness	0.015 mm	0.015 mm	0.02 mm

MATERIAL COMPOSITION

Ethylene		32 mol%	44 mol%

TEST CONDITIONS

Penetrant	water vapor		
Temperature (°C)	40	40	40
Relative Humidity (%)	90	90	90
Test Method	JIS Z0208	JIS Z0208	JIS Z0208

PERMEABILITY (source document units)

Vapor Transmission Rate $(g \cdot mil/100 \ in^2 \cdot day)$	3	6	2

PERMEABILITY (normalized units)

Vapor Transmission Rate $(g \cdot mm/m^2 \cdot day)$	1.2	2.4	0.8

TABLE 205: Water Vapor Transmission Through Kuraray Eval Ethylene Vinyl Alcohol Copolymer.

Material Family	ETHYLENE-VINYL ALCOHOL COPOLYMER					
Material Supplier	EVAL COMPANY					
Grade	EVAL L	EVAL F	EVAL H	EVAL K	EVAL E	EVAL G
Features	barrier properties					
Reference Number	264	264	264	264	264	264

MATERIAL COMPOSITION

Ethylene	27 mol%	32 mol%	38 mol%	38 mol%	44 mol%	48 mol%

TEST CONDITIONS

Penetrant	water vapor					
Temperature (°C)	40	40	40	40	40	40
Relative Humidity (%)	90	90	90	90	90	90

PERMEABILITY (source document units)

Vapor Transmission Rate (g · mil/100 in^2 · day)	8	3.8	2.1	2.1	1.4	1.4
Vapor Transmission Rate (g · 25µ/m^2 · day)	124	58.9	32.6	32.6	21.7	21.7

PERMEABILITY (normalized units)

Vapor Transmission Rate (g · mm/m^2 · day)	3.2	1.5	0.8	0.8	0.6	0.6

TABLE 206: Water Vapor Transmission Through DuPont Selar OH Ethylene Vinyl Alcohol Copolymer.

Material Family	ETHYLENE-VINYL ALCOHOL COPOLYMER				
Material Supplier/ Trade Name	DUPONT SELAR OH				
Grade	3003	4416	BX220	BX228	BX230
Features	barrier properties	barrier properties	barrier properties, heat stabilized	barrier properties, heat stabilized	barrier properties, formability, heat stabilized
Applications	packaging	packaging	packaging	packaging	packaging
Reference Number	295	295	295	295	295

MATERIAL CHARACTERISTICS

Melt Flow Index	3 g/10 min. (210/2.16)	12 g/10 min. (210/2.16)	3 g/10 min. (210/2.16)	11 g/10 min. (210/2.16)	1.5 g/10 min. (210/2.16)

MATERIAL COMPOSITION

Ethylene Content	32 mol%	44 mol%	32 mol%	44 mol%	32 mol%

TEST CONDITIONS

Penetrant	water vapor				
Temperature (°C)	40	40	40	40	40
Relative Humidity (%)	80	80	80	80	80

PERMEABILITY (source document units)

Vapor Transmission Rate (g · mil/100 in^2 · day · atm)	3.1	1.4	3.1	1.4	1.4

PERMEABILITY (normalized units)

Vapor Transmission Rate (g · mm/m^2 · day)	1.2	0.6	1.2	0.6	0.6

TABLE 207: d-Limonene (flavor component) Permeability Through Eval Company Ethylene Vinyl Alcohol Barrier Resin.

Material Family	ETHYLENE-VINYL ALCOHOL COPOLYMER
Material Supplier/ Grade	EVAL COMPANY EVOH 44
Product Form	FILM
Features	barrier properties
Reference Number	255

MATERIAL CHARACTERISTICS

Sample Thickness	0.04 mm

TEST CONDITIONS

Penetrant	d-limonene
Temperature (°C)	25
Relative Humidity (%)	dry

PERMEABILITY (source document units)

Vapor Transmission (10^{-20} kg $\cdot$ m/m^2 $\cdot$ sec $\cdot$ Pa)	4700

PERMEABILITY (normalized units)

Vapor Transmission Rate (g $\cdot$ mm/m^2 $\cdot$ day)	0.41

TABLE 208: Film Properties of Kuraray Eval Ethylene Vinyl Alcohol Copolymer Film.

Material Family	ETHYLENE-VINYL ALCOHOL COPOLYMER		
Material Supplier/ Grade	EVAL COMPANY EVAL EF-XL	EVAL COMPANY EVAL EF-F	EVAL COMPANY EVAL EF-E
Product Form	FILM		
Features	biaxially oriented		
Reference Number	268	268	268

MATERIAL CHARACTERISTICS

Sample Thickness	0.015 mm	0.015 mm	0.02 mm
Yield (in²/lb)	39100	39106	30623
Ethylene Content		32 mol%	44 mol%

TEST CONDITIONS

Temperature (°C)	20	20	20
Relative Humidity (%)	65	65	65

MECHANICAL PROPERTIES

Modulus Of Elasticity - MD (MPa)	3528 {r}	2157 {r}	1764 {r}
Modulus Of Elasticity - TD (MPa)	3335 {r}	2157 {r}	1764 {r}
Tensile Strength @ Break - MD (MPa)	206.2 {r}	88.3 {r}	68.9 {r}
Tensile Strength @ Break - TD (MPa)	195.8 {r}	39.3 {r}	39.3 {r}
Ultimate Elongation - MD (%)	100 {r}	180 {r}	260 {r}
Ultimate Elongation - TD (%)	100 {r}	140 {r}	190 {r}
Impact Strength (kg-cm)	8	0.6	4.7
Burst Strength (Mpa)	0.39 {af}	0.13 {af}	0.13 {af}
Pinhole Strength (g)	800 {ag}	290 {ag}	310 {ag}
Elmendorf Tear Resistance - MD (g/mm)	900 {ad}	400 {ad}	500 {ad}
Elmendorf Tear Resistance - TD (g/mm)	1000 {ad}	800 {ad}	700 {ad}
Tear Resistance - MD (g)	260	380	460
Tear Resistance - TD (g)	330	300	440

TABLE 208: Film Properties of Kuraray Eval Ethylene Vinyl Alcohol Copolymer Film (cont'd).

Material Family	ETHYLENE-VINYL ALCOHOL COPOLYMER		
Material Supplier/ Grade	**EVAL COMPANY EVAL EF-XL**	**EVAL COMPANY EVAL EF-F**	**EVAL COMPANY EVAL EF-E**
Product Form	**FILM**		
Features	biaxially oriented		
Reference Number	268	268	268

OTHER PROPERTIES

Water Absorption (%)	5.9 {ah}	8.6 {ah}	6.7 {ah}
Equilibrium Moist. Absorption (%)	2.8	3.9	2.8
Melting Point (°C)	181	181	164
Haze (%)	0.5	1.5	1.7
Gloss	95	90	90
Surface Resistivity (ohms)	2.7×10^{15} {aj}	1.9×10^{15} {aj}	2.1×10^{15} {aj}
Slip Factor (°)	35	30	30
Dimensional Stability - MD (%)	-4.0 {ai}	-2.7 {ai}	-1.6 {ai}
Dimensional Stability - TD (%)	-0.5 {ai}	-0.9 {ai}	-1.2 {ai}

GRAPH 42: Oxygen Permeability vs. Relative Humidity through Ethylene-Vinyl Alcohol Copolymer.

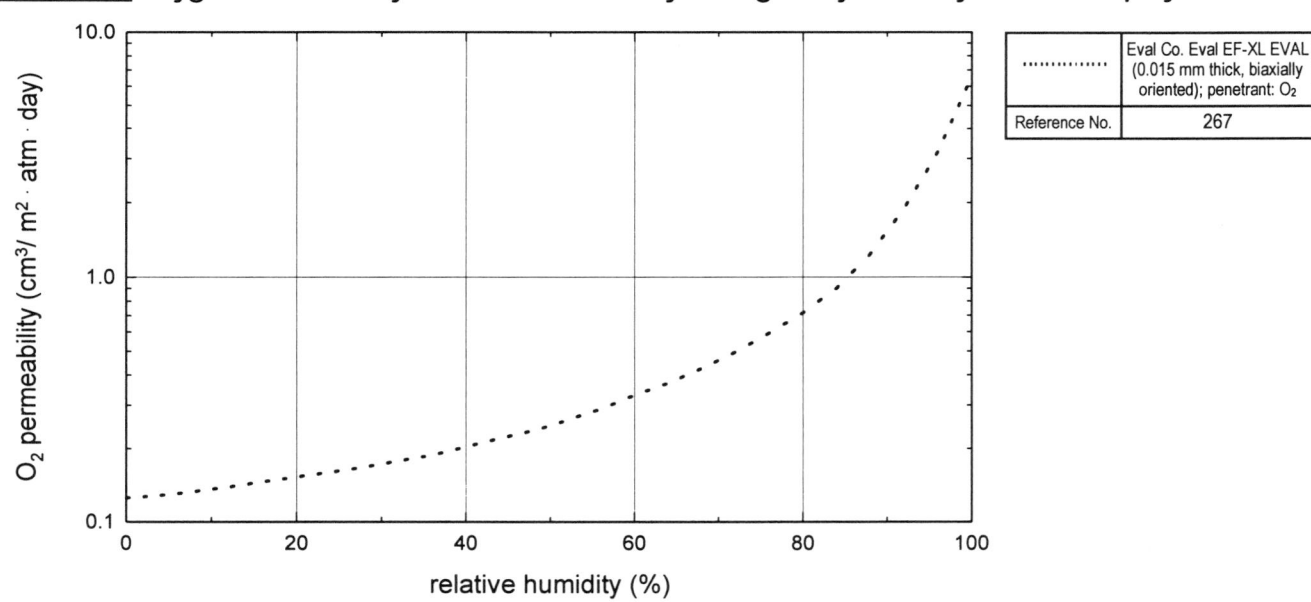

	Eval Co. Eval EF-XL EVAL (0.015 mm thick, biaxially oriented); penetrant: O_2
Reference No.	267

EVOH

GRAPH 43: Oxygen Permeability vs. Relative Humidity through Ethylene-Vinyl Alcohol Copolymer.

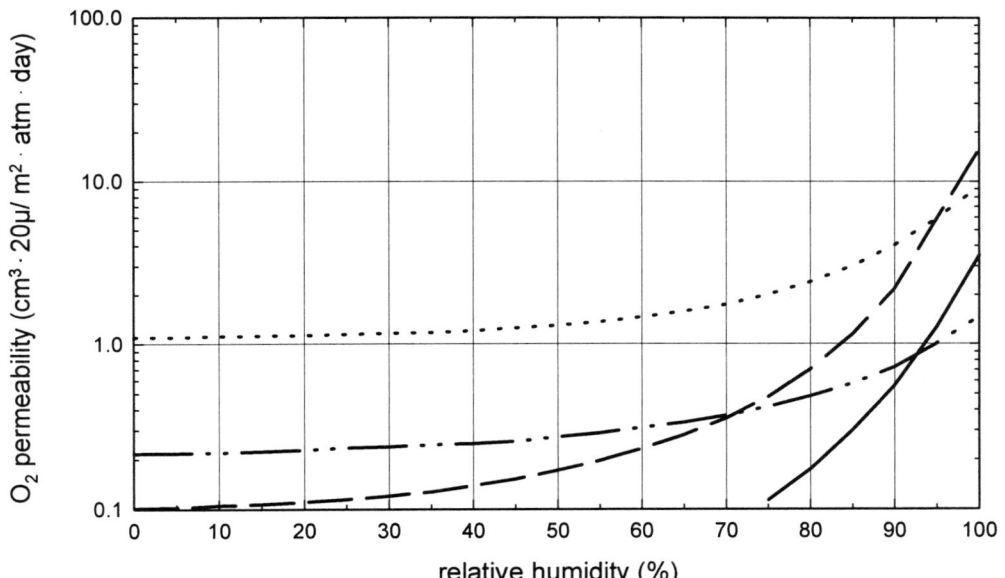

	Eval Co. Eval EF-E EVAL (film); penetrant: O₂; 20°C
	Eval Co. Eval EF-E EVAL (film); penetrant: O₂; 5°C
	Eval Co. Eval EF-F EVAL (film); penetrant: O₂; 20°C
	Eval Co. Eval EF-F EVAL (film); penetrant: O₂; 5°C
Reference No.	63

GRAPH 44: Oxygen Permeability vs. Relative Humidity through Ethylene-Vinyl Alcohol Copolymer.

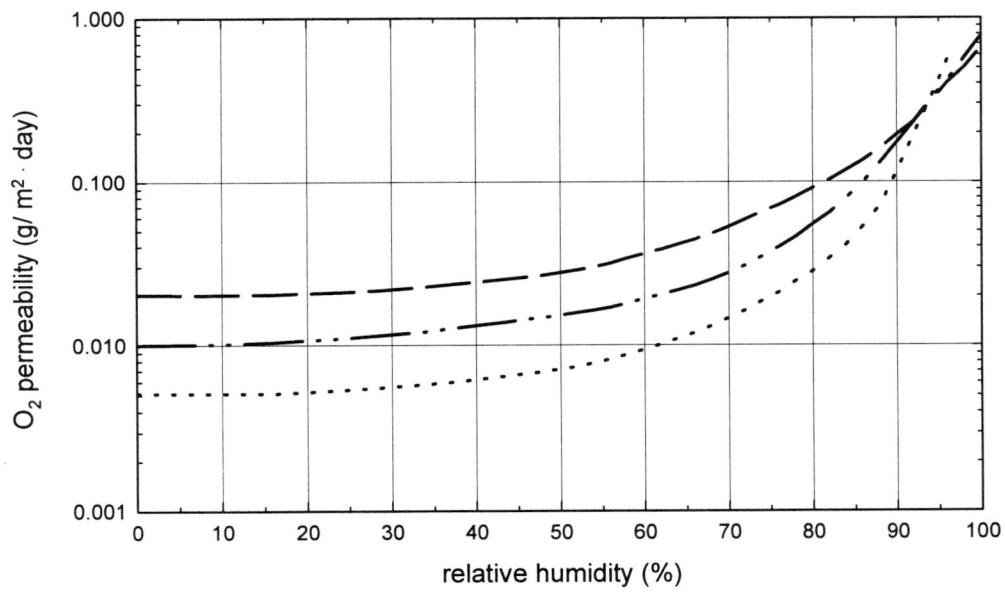

	Eval Co. Eval L EVAL (barrier prop.; 27 % ethylene); penetrant: O₂; 20°C
	Eval Co. Eval F EVAL (barrier prop.; 32% ethylene); penetrant: O₂; 20°C
	Eval Co. Eval H EVAL (barrier prop.; 38% ethylene); penetrant: O₂; 20°C
Reference No.	264

GRAPH 45: Oxygen Permeability vs. Relative Humidity through Ethylene-Vinyl Alcohol Copolymer.

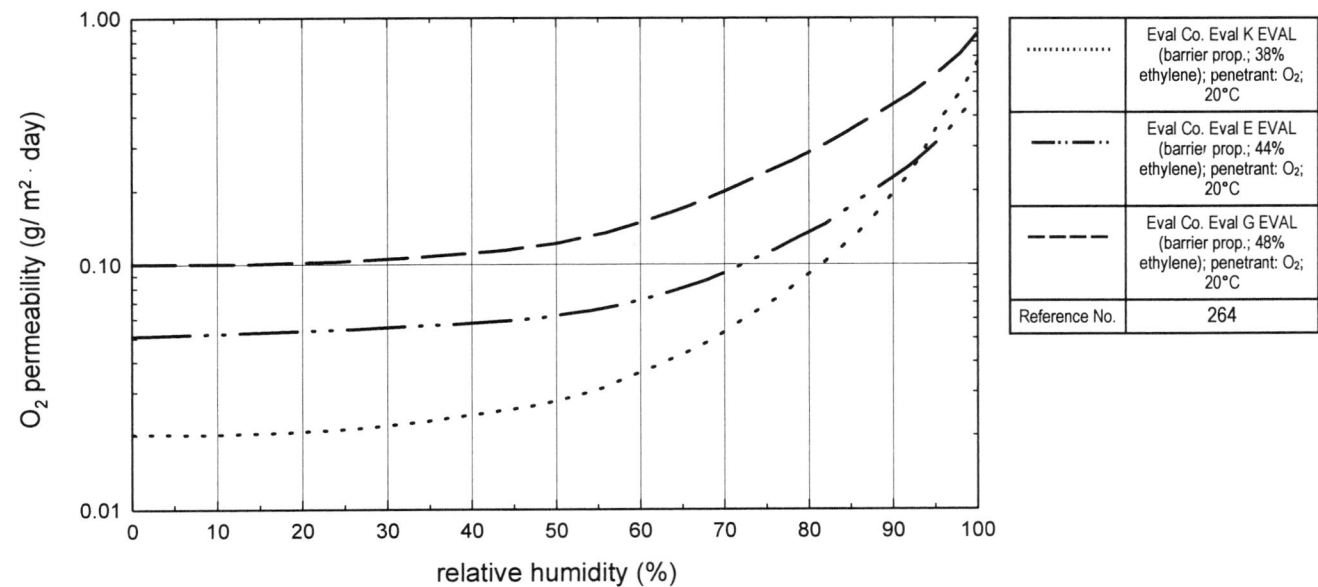

··············	Eval Co. Eval K EVAL (barrier prop.; 38% ethylene); penetrant: O$_2$; 20°C
— ·· — ·· —	Eval Co. Eval E EVAL (barrier prop.; 44% ethylene); penetrant: O$_2$; 20°C
— — —	Eval Co. Eval G EVAL (barrier prop.; 48% ethylene); penetrant: O$_2$; 20°C
Reference No.	264

GRAPH 46: Oxygen Permeability vs. Ethylene Content through Ethylene-Vinyl Alcohol Copolymer.

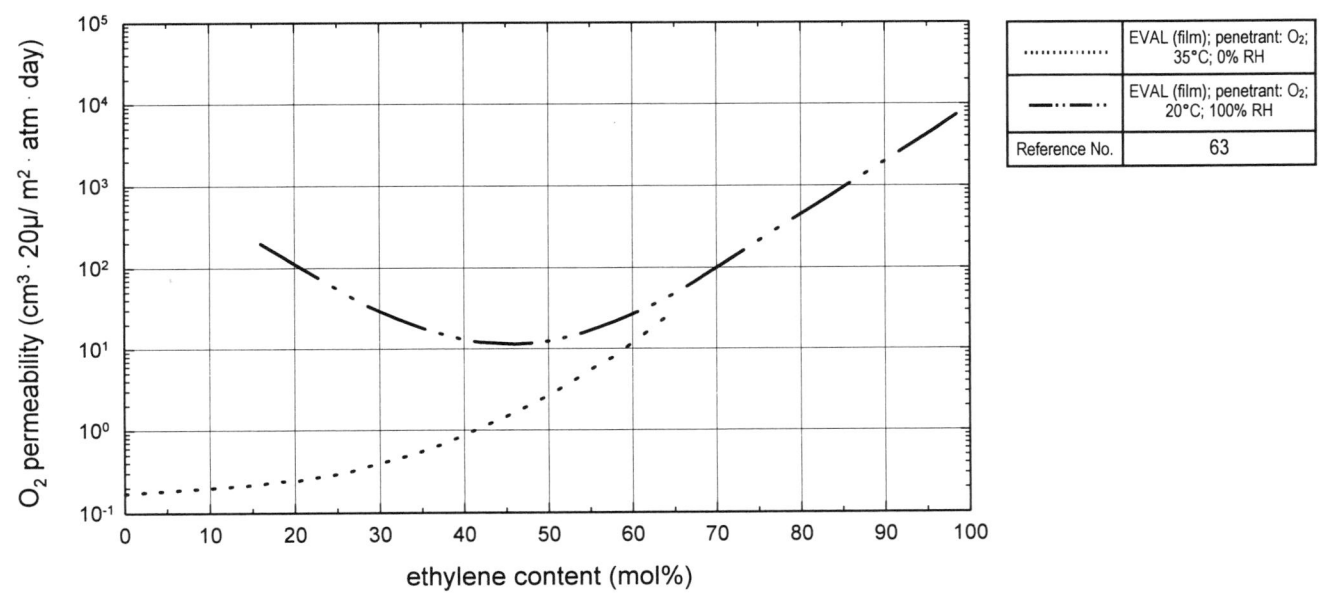

··············	EVAL (film); penetrant: O$_2$; 35°C; 0% RH
— ·· — ·· —	EVAL (film); penetrant: O$_2$; 20°C; 100% RH
Reference No.	63

EVOH

GRAPH 47: Oxygen Permeability vs. Moisture Absorption through Ethylene-Vinyl Alcohol Copolymer.

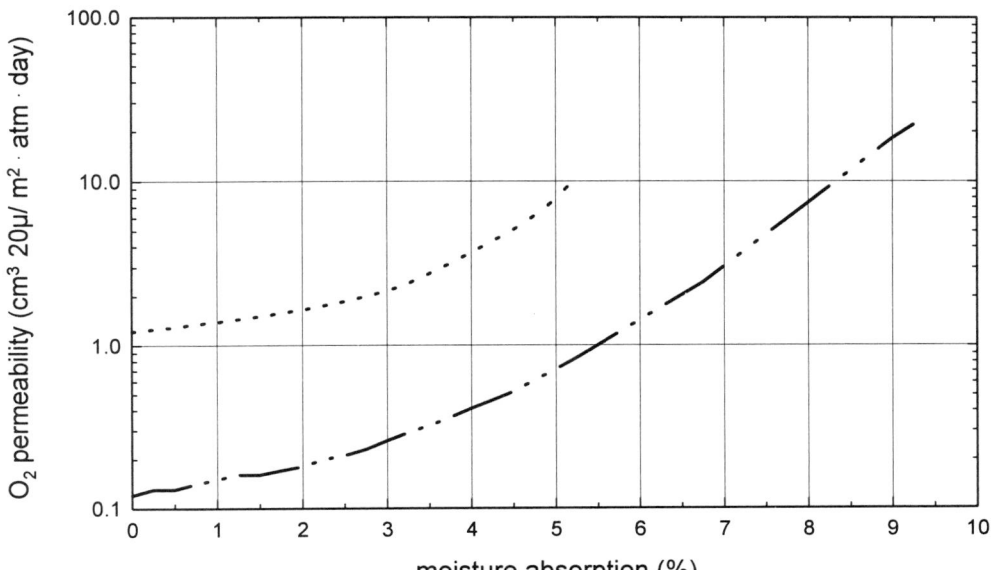

··············	Eval Co. Eval EF-E EVAL (film); penetrant: O₂; 20°C
—··—··—	Eval Co. Eval EF-F EVAL (film); penetrant: O₂; 20°C
Reference No.	63

GRAPH 48: Oxygen Permeability vs. Temperature through Ethylene-Vinyl Alcohol Copolymer.

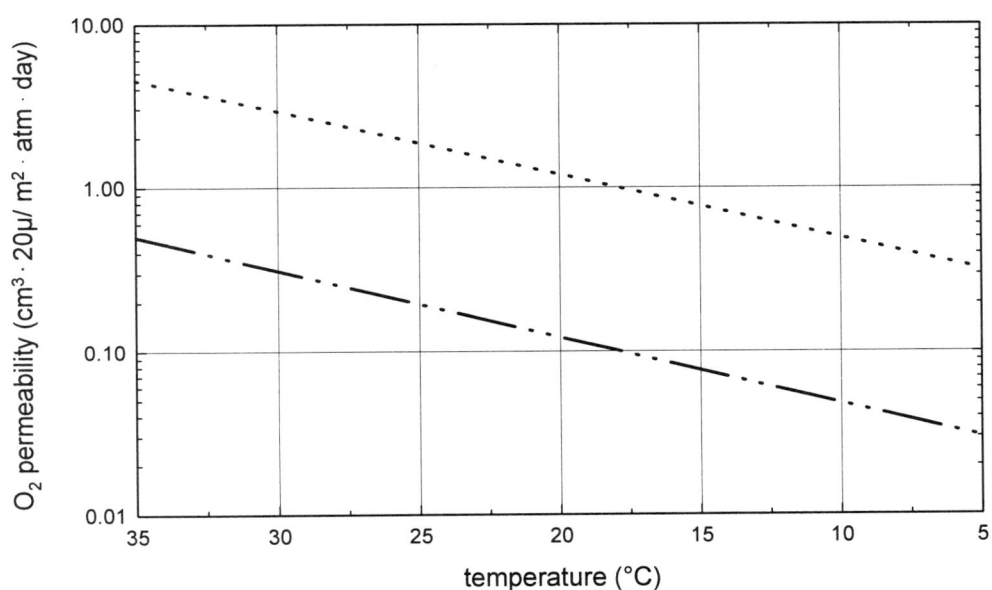

··············	Eval Co. Eval EF-E EVAL (film); penetrant: O₂; 0% RH
—··—··	Eval Co. Eval EF-F EVAL (film); penetrant: O₂; 0% RH
Reference No.	63

GRAPH 49: Oxygen Permeability vs. Temperature and Moisture Content through Ethylene-Vinyl Alcohol Copolymer.

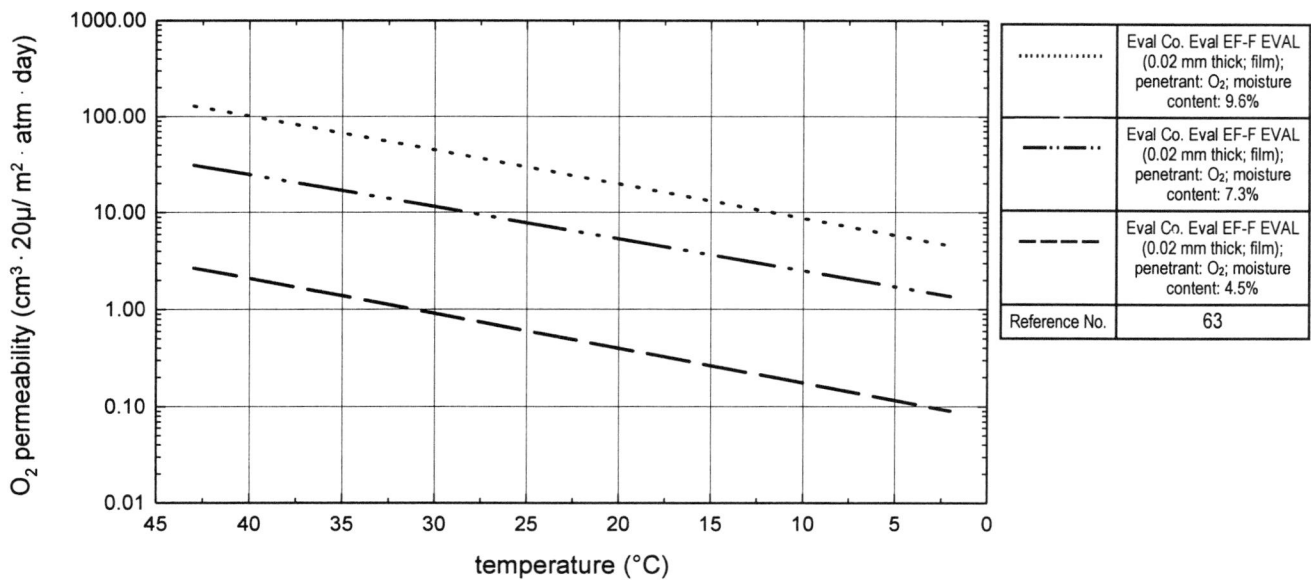

	Eval Co. Eval EF-F EVAL (0.02 mm thick; film); penetrant: O_2; moisture content: 9.6%
	Eval Co. Eval EF-F EVAL (0.02 mm thick; film); penetrant: O_2; moisture content: 7.3%
	Eval Co. Eval EF-F EVAL (0.02 mm thick; film); penetrant: O_2; moisture content: 4.5%
Reference No.	63

GRAPH 50: Equilibrium Moisture Abosrption vs. Relative Humidity of Ethylene-Vinyl Alcohol Copolymer.

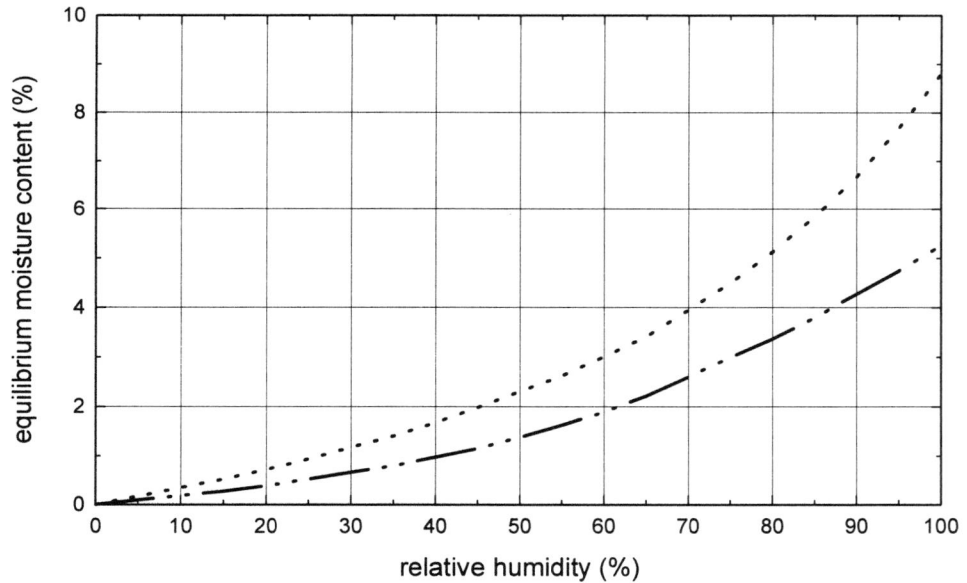

	Eval Co. Eval EF-F EVAL (barrier prop.; 32% ethylene; film); penetrant: water; 20°C
	Eval Co. Eval EF-E EVAL (barrier prop.; 44% ethylene; film); penetrant: water; 20°C
Reference No.	264

EVOH

GRAPH 51: Oxygen Transmission Rate vs. Time After Retort through Ethylene-Vinyl Alcohol Copolymer.

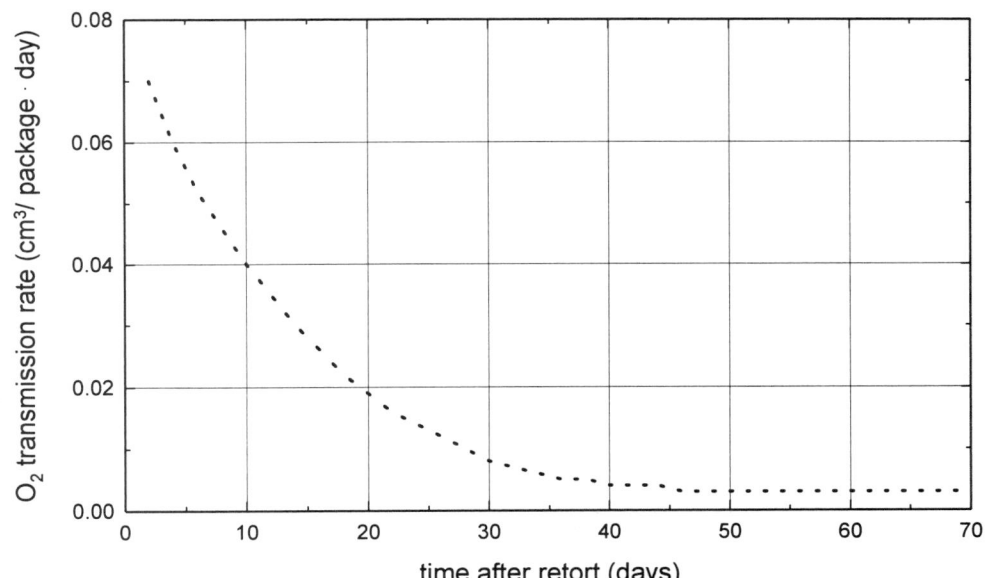

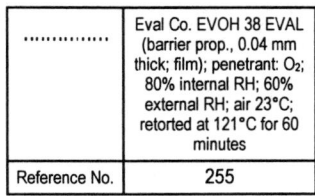

....	Eval Co. EVOH 38 EVAL (barrier prop., 0.04 mm thick; film); penetrant: O₂; 80% internal RH; 60% external RH; air 23°C; retorted at 121°C for 60 minutes
Reference No.	255

GRAPH 52: Oxygen Uptake vs. Time After Retort through Ethylene-Vinyl Alcohol Copolymer.

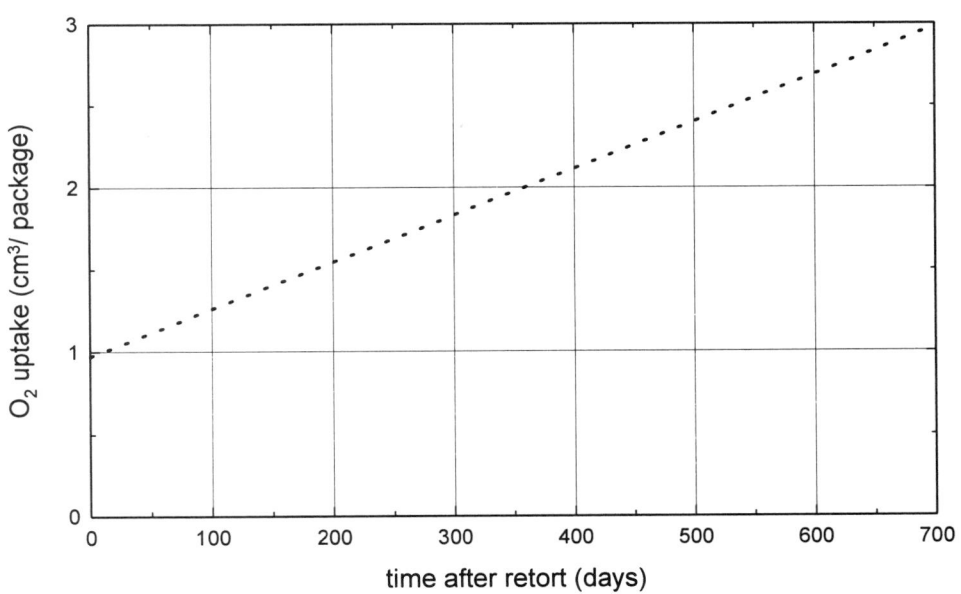

....	Eval Co. EVOH 38 EVAL (barrier prop., 0.04 mm thick; film); penetrant: O₂; 80% internal RH; 60% external RH; air 23°C; retorted at 121°C for 60 minutes
Reference No.	255

GRAPH 53: Carbon Dioxide Permeability vs. Relative Humidity through Ethylene-Vinyl Alcohol Copolymer.

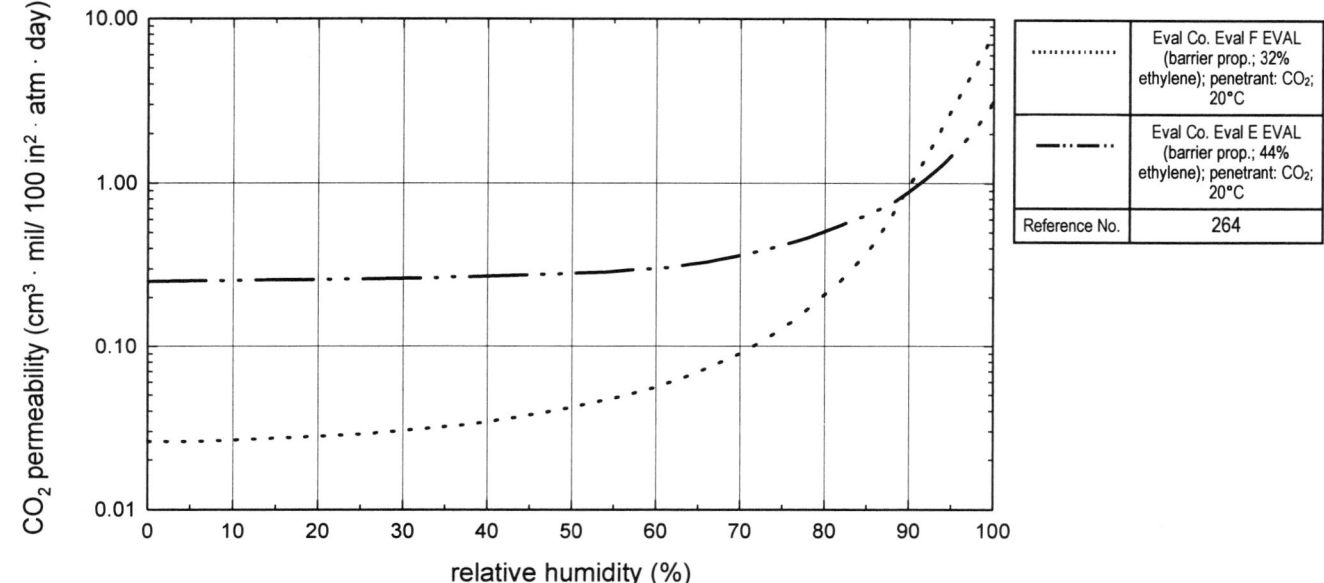

Chapter 41

Polyethylene-Acrylic Acid Copolymer

Permeability to Gases, Vapors, and Liquids

BASF AG: Lucalen A

A factor besides the temperature that influences a polymer's permeability to various gases is the amount of comonomer incorporated in the molecule. The higher the mass fraction of comonomer, the greater the permeability to gases.

Reference: *Lupolen, Lucalen Product Line, Properties, Processing,* supplier design guide (B 581 e/(8127) 10.91) - BASF Aktiengesellschaft, 1991.

Film Properties and Applications

Dow Chemical: Primacor   (manufacturing method: cast film, blown film)

PRIMACOR polymers are designed for the extrusion of blown and cast monolayer, coextruded and composite films, providing outstanding sealant and adhesive properties. PRIMACOR polymers for films combine sealability and adhesion with toughness, clarity, tear resistance, moisture insensitivity, and processability to provide excellent performance and value for many flexible packaging applications. And with recent development of low odor resins, PRIMACOR has become the best choice for demanding sealant applications.

PRIMACOR resins perform competitively with SURLYN ionomer resins in many applications, while offering the added benefit of bulk handling.

Reference: *621 Ways To Succeed - 1993-1994 Materials Selection Guide,* supplier technical report (304-00286-1292X SMG) - Dow Chemical Company, 1992.

TABLE 209: Oxygen Permeability and Water Vapor Transmission Through BASF AG Lucalen Polyethylene Acrylic Acid Copolymer.

Material Family	POLYETHYLENE-ACRYLIC ACID COPOLYMER					
Material Supplier Trade Name	BASF AG LUCALEN					
Grade	A2710H	A2910M	A3710MX	A2710H	A2910M	A3710MX
Reference Number	25	25	25	25	25	25

MATERIAL CHARACTERISTICS

Melt Flow Index	1.7 g/10 min. (190/2.16)	7 g/10 min. (190/2.16)	7 g/10 min. (190/2.16)	1.7 g/10 min. (190/2.16)	7 g/10 min. (190/2.16)	7 g/10 min. (190/2.16)
Sample Thickness	0.1 mm	0.1 mm	0.1 mm	0.1 mm	0.1 mm	0.1 mm

MATERIAL COMPOSITION

Acrylic Acid Content	17 wt.%	11 wt.%	8 wt.%	17 wt.%	11 wt.%	8 wt.%

TEST CONDITIONS

Penetrant	oxygen			water vapor		
Temperature (°C)	23	23	23	23	23	23
Relative Humidity (%)				85%-0% gradient	85%-0% gradient	85%-0% gradient
Test Method	DIN 53380	DIN 53380	DIN 53380	DIN 53122	DIN 53122	DIN 53122

PERMEABILITY (source document units)

Gas Permeability (cm^3/m^2 · day · bar)	5430	2400	1760			
Vapor Transmission Rate (g/m^2 · day)				6.8	2.3	0.8

PERMEABILITY (normalized units)

Permeability Coefficient (cm^3 · mm/m^2 · day · atm)	550	243	178			
Vapor Transmission Rate (g · mm/m^2 · day)				0.68	0.23	0.08

TABLE 210: Film Properties of Dow Chemical Primacor Ethylene Acrylic Acid Copolymer Extrusion Resins.

Material Family	POLYETHYLENE-ACRYLIC ACID COPOLYMER				
Material Supplier/ Trade Name	DOW CHEMICAL PRIMACOR				
Grade	1321	1410	1410-XT	1420	1430
Manufacturing Method	extrusion	extrusion	extrusion	extrusion	extrusion
Reference Number	254	254	254	254	254

MATERIAL CHARACTERISTICS

Density	0.935 g/cm³	0.938 g/cm³	0.938 g/cm³	0.938 g/cm³	0.938 g/cm³
Melt Flow Index	2.5 g/10 min. (190/2.16)	1.5 g/10 min. (190/2.16)	1.5 g/10 min. (190/2.16)	3 g/10 min. (190/2.16)	5 g/10 min. (190/2.16)
Sample Thickness	0.05 mm	0.05 mm	0.05 mm	0.05 mm	0.05 mm

MATERIAL COMPOSITION

Acrylic Acid	6.5%	9.5%	9.5%	9.5%	9.5%

PHYSICAL PROPERTIES

Blow Up Ratio (BUR)	2.5	2.5	2.5	2.5	2.5

MECHANICAL PROPERTIES

Ultimate Tensile Strength - MD (MPa)	32.4 {p}	42.7 {p}	42.7 {p}	37.9 {p}	35.9 {p}
Ultimate Tensile Strength - TD (MPa)	32.8 {p}	44.8 {p}	44.8 {p}	39.3 {p}	35.9 {p}
Tensile Strength @ Yield - MD (MPa)	9.4 {p}	8.6 {p}	8.6 {p}	8.6 {p}	8.1 {p}
Tensile Strength @ Yield - TD (MPa)	9.2 {p}	7.9 {p}	7.9 {p}	8.3 {p}	7.9 {p}
Ultimate Elongation - Machine direction (%)	470 {p}	470 {p}	470 {p}	515 {p}	525 {p}
Ultimate Elongation - Transverse direction (%)	570 {p}	510 {p}	510 {p}	545 {p}	535 {p}
Drop Dart Impact Strength (g)	410 {j}	610 {j}	610 {j}	425 {j}	400 {j}
Elmendorf Tear Resistance - MD (g/mm)	10,040 {k}	11,810 {k}	11,810 {k}	14,960 {k}	17,700 {k}
Elmendorf Tear Resistance - TD (g/mm)	12,010 {k}	15,750 {k}	15,750 {k}	19,685 {k}	20,100 {k}

THERMAL PROPERTIES

Extrusion Temperature (°C)	204	204	191	204	191

OPTICAL PROPERTIES

Haze (%)	3.7 {c}	5.5 {c}	9 {c}	4 {c}	3 {c}
Gloss @ 45°	75 {m}	65 {m}	50 {m}	65 {m}	70 {m}

Chapter 42

Polyethylene-Ionomer Copolymer

Permeability to Gases, Vapors, and Liquids

BASF AG: Lucalen I

A factor besides the temperature that influences a polymer's permeability to various gases is the amount of comonomer incorporated in the molecule. The higher the mass fraction of comonomer, the greater the permeability to gases.

Reference: *Lupolen, Lucalen Product Line, Properties, Processing,* supplier design guide (B 581 e/(8127) 10.91) - BASF Aktiengesellschaft, 1991.

TABLE 211: Oxygen Permeability and Water Vapor Transmission Through BASF AG Lucalen Polyethylene Ionomer Copolymer.

Material Family	POLYETHYLENE-IONOMER COPOLYMER	
Material Supplier/ Grade	BASF AG LUCALEN I4300MX	
Reference Number	25	25

MATERIAL CHARACTERISTICS

Melt Flow Index	7 g/10 min. (190/2.16)	7 g/10 min. (190/2.16)
Sample Thickness	0.1 mm	0.1 mm

MATERIAL COMPOSITION

Ionomer	8 wt.%	8 wt.%

TEST CONDITIONS

Penetrant	oxygen	water vapor
Temperature (°C)	23	23
Relative Humidity (%)		85%-0% gradient
Test Method	DIN 53380	DIN 53122

PERMEABILITY (source document units)

Gas Permeability ($cm^3/m^2 \cdot day \cdot bar$)	1570	
Vapor Transmission Rate ($g/m^2 \cdot day$)		0.8

PERMEABILITY (normalized units)

Permeability Coefficient ($cm^3 \cdot mm/m^2 \cdot day \cdot atm$)	159	
Vapor Transmission Rate ($g \cdot mm/m^2 \cdot day$)		0.08

Chapter 43

Polypropylene

TABLE 212: Gas Permeability of Oxygen, Carbon Dioxide, Nitrogen and Helium Through Oriented Polypropylene Film.

Material Family	POLYPROPYLENE		
Product Form	FILM		
Features	oriented		
Reference Number	63	63	63

TEST CONDITIONS

Penetrant	oxygen	nitrogen	carbon dioxide
Temperature (°C)	25	25	25
Relative Humidity (%)	0	0	0

PERMEABILITY (source document units)

Gas Permeability ($cm^3 \cdot mil/100\ in^2 \cdot day$)	174	42	528
Gas Permeability ($cm^3 \cdot 20\mu/m^2 \cdot day \cdot atm$)	3400	730	9100

PERMEABILITY (normalized units)

Permeability Coefficient ($cm^3 \cdot mm/m^2 \cdot day \cdot atm$)	68.5	16.5	208

TABLE 213: Oxygen Permeability at Different Temperatures and Water Vapor Transmission Through Oriented and Non-Oriented Polypropylene.

Material Family	POLYPROPYLENE			
Features	oriented	oriented	biaxially oriented	
Reference Number	264	264	264	264

TEST CONDITIONS

Penetrant	oxygen		water vapor	
Temperature (°C)	23	35	40	40
Relative Humidity (%)	0	0	90	90

PERMEABILITY (source document units)

Gas Permeability (cm^3 · mil/100 in^2 · day)	163	203		
Gas Permeability (cm^3 · 25μ/m^2 · day · atm)	2526	3146		
Vapor Transmission Rate (g · mil/100 in^2 · day)			0.38	0.69
Vapor Transmission Rate (g · 25μ/m^2 · day)			5.9	10.7

PERMEABILITY (normalized units)

Permeability Coefficient (cm^3 · mm/m^2 · day · atm)	64.2	80.0		
Vapor Transmission Rate (g · mm/m^2 · day)			0.15	0.27

TABLE 214: Oxygen Permeability vs. Relative Humidity Through Biaxially Oriented Polypropylene Film.

Material Family	POLYPROPYLENE			
Product Form	FILM			
Features	biaxially oriented; PVDC coated			
Reference Number	265	265	265	265

TEST CONDITIONS

Penetrant	oxygen			
Temperature (°C)	20	20	20	20
Relative Humidity (%)	65	85	100	0

PERMEABILITY (source document units)

Gas Permeability (cm^3 · mil/100 in^2 · day)	0.55	0.55	0.55	1.1

PERMEABILITY (normalized units)

Permeability Coefficient (cm^3 · mm/m^2 · day · atm)	0.22	0.22	0.22	0.43

TABLE 215: Water Vapor Transmission and Oxygen Permeability Through Polypropylene.

Material Family	POLYPROPYLENE			
Product Form	FILM			
Reference Number	101	101	296	296

TEST CONDITIONS

Penetrant	water vapor	oxygen	oxygen	water vapor
Temperature (°C)	37.8	23	22.8	37.8
Relative Humidity (%)	90		0	90
Test Method	ASTM D96	ASTM D1434	ASTM D1434	ASTM F1249

PERMEABILITY (source document units)

Gas Permeability (cm$^3 \cdot$ mil/100 in$^2 \cdot$ day)		272	>250	
Vapor Transmission Rate (g $\cdot$ mil/100 in$^2 \cdot$ day)	1.5			1
Vapor Transmission Rate (g $\cdot$ mm/ day/ m^2)	0.59			

PERMEABILITY (normalized units)

Permeability Coefficient (cm$^3 \cdot$ mm/m$^2 \cdot$ day $\cdot$ atm)		107	>99.7	
Vapor Transmission Rate (g $\cdot$ mm/m$^2 \cdot$ day)	0.59			0.4

TABLE 216: Xylene and Oxygen Permeability Through Polypropylene.

Material Family	POLYPROPYLENE	
Reference Number	293	293

TEST CONDITIONS

Penetrant	xylene	oxygen
Temperature (°C)	60	23
Exposure Time (days)	14	
Relative Humidity (%)		75

PERMEABILITY (source document units)

Vapor Transmission Rate (g $\cdot$ mil/100 in$^2 \cdot$ day)	2500	
Gas Permeability (cm$^3 \cdot$ mil/100 in$^2 \cdot$ day)		200

PERMEABILITY (normalized units)

Permeability Coefficient (cm$^3 \cdot$ mm/m$^2 \cdot$ day $\cdot$ atm)		78.7
Vapor Transmission Rate (g $\cdot$ mm/m$^2 \cdot$ day)	984	

TABLE 217: Water Vapor Transmission and Oxygen Permeability Through Coated and Uncoated Oriented Polypropylene Film.

Material Family	POLYPROPYLENE									
Product Form	FILM									
Features	oriented									
Reference Number	268	268	268	268	268	268	268	268	268	268

MATERIAL CHARACTERISTICS

Sample Thickness	0.02 mm	0.022 mm	0.02 mm	0.022 mm	0.02 mm	0.022 mm	0.02 mm	0.022 mm	0.02 mm	0.022 mm

MATERIAL COMPOSITION

Note		PVDC coated		PVDC coated		PVDC coated		PVDC coated		PVDC coated

TEST CONDITIONS

Penetrant	water vapor		oxygen							
Temperature (°C)	40	40	35	35	20	20	20	20	20	20
Relative Humidity (%)	90	90	0	0	65	65	85	85	100	100
Test Method	JIS Z0208	JIS Z0208	JIS Z1707	JIS Z1707	ASTM D3985	ASTM D3985	ASTM D3985	ASTM D3985	ASTM D3985	ASTM D3985

PERMEABILITY (source document units)

Vapor Transmission Rate (g · mil/100 in^2 · day)	<1	<1								
Gas Permeability (cm^3 · mil/100 in^2 · day)			226	1.23	135	0.65	135	0.65	135	0.65

PERMEABILITY (normalized units)

Permeability Coefficient (cm^3 · mm/m^2 · day · atm)			89.0	0.48	53.2	0.26	53.2	0.26	53.2	0.26
Vapor Transmission Rate (g · mm/m^2 · day)	<0.39	<0.39								

TABLE 218: Organic Solvents Permeability Through Oriented Polypropylene Film.

Material Family	POLYPROPYLENE			
Product Form	FILM			
Features	oriented	oriented	oriented	oriented
Reference Number	266	266	266	266

MATERIAL CHARACTERISTICS

Sample Thickness	0.02 mm	0.02 mm	0.02 mm	0.02 mm

TEST CONDITIONS

Penetrant	chloroform	xylene	methyl ethyl ketone	kerosine
Temperature (°C)	20	20	20	20
Relative Humidity (%)	65	65	65	65

PERMEABILITY (source document units)

Vapor Transmission Rate (g/day · 100 in^2)	241.3	22.58	0.77	3.42

PERMEABILITY (normalized units)

Vapor Transmission Rate (g · mm/m^2 · day)	74.8	7	0.24	1.06

TABLE 219: d-Limonene (flavor component) Permeability Through Polypropylene.

Material Family	POLYPROPYLENE
Product Form	FILM
Reference Number	255

TEST CONDITIONS

Penetrant	d-limonene
Temperature (°C)	25
Relative Humidity (%)	dry

PERMEABILITY (source document units)

Vapor Transmission (10^{-20} kg · m/m^2 · sec · Pa)	101,800

PERMEABILITY (normalized units)

Vapor Transmission Rate (g · mm/m^2 · day)	8.9

TABLE 220: Film Properties of Coated and Uncoated Oriented Polypropylene Film.

Material Family	POLYPROPYLENE	
Product Form	FILM	
Features	oriented	oriented; PVDC coated
Reference Number	268	268

MATERIAL CHARACTERISTICS

Sample Thickness	0.02 mm	0.022 mm

TEST CONDITIONS

Temperature (°C)	20	20
Relative Humidity (%)	65	65

PHYSICAL PROPERTIES

Water Absorption (%)	0.3 {ah}	0.3 {ah}
Equilibrium Moist. Absorption (%)	0.2	0.2

MECHANICAL PROPERTIES

Modulus Of Elasticity - MD (MPa)	1860 {r}	2157 {r}
Modulus Of Elasticity - TD (MPa)	3431 {r}	2549 {r}
Tensile Strength @ Break - MD (MPa)	127.6 {r}	137.2 {r}
Tensile Strength @ Break - TD (MPa)	245.5 {r}	215.8 {r}
Ultimate Elongation - MD (%)	140 {r}	140 {r}
Ultimate Elongation - TD (%)	50 {r}	60 {r}
Impact Strength (kg-cm)	9	9
Burst Strength (MPa)	0.39 {af}	0.39 {af}
Pinhole Strength (g)	690 {ag}	790 {ag}
Elmendorf Tear Resistance - MD (g/mm)	200 {ad}	300 {ad}
Elmendorf Tear Resistance - TD (g/mm)	100 {ad}	200 {ad}
Tear Resistance - MD (g)	260	300
Tear Resistance - TD (g)	220	200

OTHER PROPERTIES

Melting Point (°C)	165	165
Haze (%)	2.2	3.5
Gloss		90
Surface Resistivity (ohms)	3.0×10^{15} {aj}	4.2×10^{15} {aj}
Slip Factor (°)	31	25
Dimensional Stability - MD (%)	-13.0 {ai}	-10.4 {ai}
Dimensional Stability - TD (%)	-15.0 {ai}	-12.5 {ai}

GRAPH 54: Oxygen Permeability vs. Relative Humidity through Polypropylene.

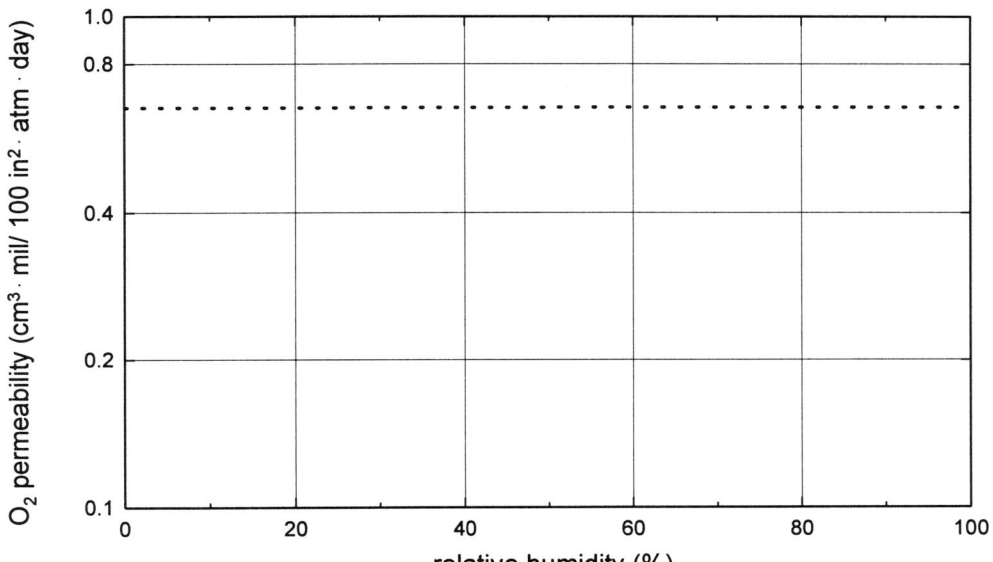

....	PP (biaxially oriented; PVDC coated; film); penetrant: O_2
Reference No.	265

Chapter 44

Polypropylene Copolymer

Film Properties and Applications

BASF AG: Novolen 3000 (chemical type: random copolymer; features: transparent)

Propylene random copolymers contain statistically distributed ethylene or butene comonomers. They are tougher and more flexible than the well-known highly isotactic polypropylene homopolymers.

Due to their reduced crystallinity and low melting points, random copolymers are easily processed and welded. However, the heat distortion temperature of articles made of Novolen 3000 with a medium comonomer content (32xx grades) is sufficient for hot filling and heat sterilization processes.

Random copolymers have increased in popularity due to their higher transparency. The transparency of nucleated Novolen 3000 grades, in thicknesses of up to 1 mm, is very close to that of glass, PVC and polystyrene.

Novolen random copolymers can be used to make strong, highly transparent films and sealing coats with outstanding weldability, e.g. for office and organizational applications, for food, textile and flower packaging as well as for the sanitary and medical sector. A wide range of products with various properties is available.

Reference: *Topics In Chemistry - BASF Plastics Research And Development,* supplier technical report - BASF Aktiengesellschaft, 1992.

Chapter 45

Polybutylene

TABLE 221: Water Vapor, Oxygen, Nitrogen and Carbon Dioxide Permeability Through Shell Chemical Duraflex Polybutylene Film.

Material Family	POLYBUTYLENE							
Material Supplier/ Grade	SHELL CHEMICAL DURAFLEX 1600				SHELL CHEMICAL DURAFLEX 1710			
Product Form	FILM							
Features	FDA grade, heat sealable				FDA grade, heat sealable			
Applications	peelable seals				peelable seals			
Manufacturing Method	blown film				blown film			
Reference Number	304	304	304	304	304	304	304	304

MATERIAL CHARACTERISTICS

Density	0.910 g/cm³				0.909 g/cm³			
Sample Thickness	0.051 mm				0.051 mm			

MATERIAL COMPOSITION

Zinc Oxide	5 phr	5 phr	5 phr	5 phr				
Note	slip and antiblock formulations				antiblock formulations, slip and antiblock formulations			

TEST CONDITIONS

Penetrant	oxygen	nitrogen	carbon dioxide	water vapor	oxygen	nitrogen	carbon dioxide	water vapor
Temperature (°C)	22.8	22.8	22.8	37.8	22.8	22.8	22.8	37.8
Relative Humidity (%)	50	50	50	90	50	50	50	90
Test Method	ASTM D1434, method M	ASTM D1434, method M	ASTM D1434, method M	ASTM D96, method E	ASTM D1434, method M	ASTM D1434, method M	ASTM D1434, method M	ASTM D96, method E

PERMEABILITY (source document units)

Gas Permeability ($cm^3 \cdot mil/100\ in^2 \cdot day$)	385	110	1425		400	110	1190	
Gas Permeability ($cm^3 \cdot \mu m/\ cm^2 \cdot atm \cdot day$)	15.1	4.3	55.8		16.0	4.3	46.9	
Vapor Transmission Rate ($g \cdot mil/100\ in^2 \cdot day$)				1.2				1.88
Vapor Transmission Rate ($g \cdot \mu m/\ cm^2 \cdot day$)				0.047				0.074

PERMEABILITY (normalized units)

Permeability Coefficient ($cm^3 \cdot mm/m^2 \cdot day \cdot atm$)	151.6	43.3	561		157.5	43.3	468	
Vapor Transmission Rate ($g \cdot mm/m^2 \cdot day$)				0.47				0.74

TABLE 222: Film Properties of Shell Chemical Duraflex Polybutylene Film.

Material Family	POLYBUTYLENE	
Material Supplier/ Grade	SHELL CHEMICAL DURAFLEX 1600	SHELL CHEMICAL DURAFLEX 1710
Product Form	FILM	
Features	FDA grade, heat sealable	FDA grade, heat sealable
Applications	peelable seals	peelable seals
Manufacturing Method	blown film	blown film
Reference Number	304	304

MATERIAL CHARACTERISTICS

Density	0.910 g/cm³	0.909 g/cm³
Sample Thickness	0.051 mm	0.051 mm

MATERIAL COMPOSITION

Zinc Oxide	5 phr	
Note	slip and antiblock formulations	antiblock formulations, slip and antiblock formulations

TEST CONDITIONS

Temperature (°C)	23	23

MECHANICAL PROPERTIES

Secant Modulus @ 2% Elongation - machine dir. (MPa)	234 {cx}	234 {cx}
Secant Modulus @ 2% Elongation - transverse dir. (MPa)	220 {cx}	220 {cx}
Tensile Strength @ Yield - machine direction (MPa)	13.8 {r}	15.2 {r}
Tensile Strength @ Yield - transverse direction (MPa)	13.1 {r}	15.2 {r}
Tensile Strength @ Break - machine direction (MPa)	37.9 {r}	44.8 {r}
Tensile Strength @ Break - transverse direction (MPa)	34.5 {r}	37.9 {r}
Ultimate Elongation - machine direction (%)	200 {r}	320 {r}
Ultimate Elongation - transverse direction (%)	220 {r}	300 {r}
Drop Dart Impact Strength (g)	381 {cw}	349 {cw}
Elmendorf Tear Resistance - machine direction (g/mm)	24,000 {k}	20,000 {k}
Elmendorf Tear Resistance - transverse direction (g/mm)	43,000 {k}	39,000 {k}

Chapter 46

Polymethylpentene

TABLE 223: Water Vapor Transmission and Oxygen Permeability Through Mitsui TPX Polymethylpentene Film.

Material Family	POLYMETHYLPENTENE					
Material Supplier/ Trade Name	MITSUI TPX					
Grade	X-22	X-44	X-88	X-22	X-44	X-88
Product Form	FILM					
Features	transparent	transparent	food grade, transparent	transparent	transparent	food grade, transparent
Reference Number	302	302	302	302	302	302

MATERIAL CHARACTERISTICS

Density	0.835 g/cm³	0.834 g/cm³	0.835 g/cm³	0.835 g/cm³	0.834 g/cm³	0.835 g/cm³
Sample Thickness	0.05 mm	0.05 mm	0.05 mm	0.05 mm	0.05 mm	0.05 mm

TEST CONDITIONS

Penetrant	water vapor			oxygen		
Temperature (°C)	23	23	23	23	23	23

PERMEABILITY (source document units)

Vapor Transmission Rate (g/m² · day)	65	60	50			
Gas Permeability (cm³/ m² · day)				40,000	38,000	32,000

PERMEABILITY (normalized units)

Permeability Coefficient (cm³ · mm/m² · day · atm)				2000	1900	1600
Vapor Transmission Rate (g · mm/m² · day)	3.25	3	2.5			

TABLE 224: Nitrogen and Carbon Dioxide Permeability Through Mitsui TPX Polymethylpentene Film.

Material Family	POLYMETHYLPENTENE					
Material Supplier/ Trade Name	MITSUI TPX					
Grade	X-22	X-44	X-88	X-22	X-44	X-88
Product Form	FILM					
Features	transparent	transparent	food grade, transparent	transparent	transparent	food grade, transparent
Reference Number	302	302	302	302	302	302

MATERIAL CHARACTERISTICS

Density	0.835 g/cm³	0.834 g/cm³	0.835 g/cm³	0.835 g/cm³	0.834 g/cm³	0.835 g/cm³
Sample Thickness	0.05 mm	0.05 mm	0.05 mm	0.05 mm	0.05 mm	0.05 mm

TEST CONDITIONS

Penetrant	nitrogen			carbon dioxide		
Temperature (°C)	23	23	23	23	23	23

PERMEABILITY (source document units)

Gas Permeability (cm³/ m² · day)	9500	9400	8000	120,000	110,000	110,000

PERMEABILITY (normalized units)

Permeability Coefficient (cm³ · mm/m² · day · atm)	475	470	400	6000	5500	5500

TABLE 225: Film Properties of Mitsui TPX Polymethylpentene Film.

Material Family	POLYMETHYLPENTENE				
Material Supplier/ Grade	**MITSUI TPX X-22**	**MITSUI TPX X-44**			**MITSUI TPX X-88**
Product Form	**FILM**				
Features	transparent	transparent			food grade, transparent
Reference Number	302	302	302	302	302

MATERIAL CHARACTERISTICS

Density	0.835 g/cm³	0.834 g/cm³			0.835 g/cm³
Sample Thickness	0.05 mm	0.05 mm	0.1 mm	0.2 mm	0.05 mm

TEST CONDITIONS

Temperature (°C)	23	23	23	23	23

MECHANICAL PROPERTIES

Modulus Of Elasticity - machine direction (MPa)	687 {r}	1079 {r}	1079 {r}	981 {r}	1472 {r}
Modulus Of Elasticity - transverse direction (MPa)	638 {r}	1079 {r}	1079 {r}	981 {r}	1570 {r}
Tensile Strength @ Yield - machine direction (MPa)	16.7 {r}	23.1 {r}	23.1 {r}	23.1 {r}	28.5 {r}
Tensile Strength @ Yield - transverse direction (MPa)	16.7 {r}	24.0 {r}	24.0 {r}	24.0 {r}	30.4 {r}
Tensile Strength @ Break - machine direction (MPa)	19.6 {r}	22.6 {r}	17.7 {r}	13.7 {r}	29.4 {r}
Tensile Strength @ Break - transverse direction (MPa)	18.6 {r}	17.7 {r}	15.7 {r}	13.7 {r}	
Ultimate Elongation - machine direction (%)	200 {r}	180 {r}	110 {r}	80 {r}	50 {r}
Ultimate Elongation - transverse direction (%)	300 {r}	100 {r}	90 {r}	80 {r}	5 {r}
Impact Strength (kg-cm/cm)	2300 {cn}	250 {cn}	400 {cn}	500 {cn}	200 {cn}
Elmendorf Tear Resistance - machine direction (g/mm)	3500 {k}	1000 {k}	2000 {k}	3500 {k}	500 {k}
Elmendorf Tear Resistance - transverse direction (g/mm)	8000 {k}	3500 {k}	4500 {k}	5500 {k}	1000 {k}

THERMAL PROPERTIES

Vicat Softening Temperature (°C)	145 {n}	160 {n}	160 {n}	160 {n}	185 {n}
Melting Point (°C)	235 {cn}	240 {cn}	240 {cn}	240 {cn}	235 {cn}
Specific Heat (cal/g/°C)	0.47 {cq}	0.47 {cq}	0.47 {cq}	0.47 {cq}	0.47 {cq}
Coefficient Of Thermal Expansion (mm/mm/°C)	1.17E-04 {co}	1.17E-04 {co}	1.17E-04 {co}	1.17E-04 {co}	1.17E-04 {co}
Thermal Conductivity (cal-cm/cm²-sec-°C)	4.0E-04 {cp}	4.0E-04 {cp}	4.0E-04 {cp}	4.0E-04 {cp}	4.0E-04 {cp}

<u>TABLE 225:</u> Film Properties of Mitsui TPX Polymethylpentene Film (cont'd).

Material Family	POLYMETHYLPENTENE				
Material Supplier/ Grade	MITSUI TPX X-22	MITSUI TPX X-44			MITSUI TPX X-88
Product Form	FILM				
Features	transparent	transparent			food grade, transparent
Reference Number	302	302	302	302	302

OPTICAL PROPERTIES

Light Transmittance (%)	94 {c}	94 {c}	92 {c}	90 {c}	93 {c}
Haze (%)	1.5 {c}	7 {c}	9 {c}	11 {c}	20 {c}
Gloss	120 {cg}	55 {cg}	45 {cg}	40 {cg}	5 {cg}

ELECTRICAL PROPERTIES

Volume Resistivity (ohm-cm)	>10^{16} {cr}	>10^{16} {cr}	>10^{16} {cr}	>10^{16} {cr}	>10^{16} {cr}
Dielectric Voltage Breakdown (kV/mm)	63 {cs}	65 {cs}	65 {cs}	65 {cs}	65 {cs}
Dielectric Constant	2.12 {ct}	2.12 {ct}	2.12 {ct}	2.12 {ct}	2.12 {ct}

PERFORMANCE PROPERTIES

Dimensional Stability - machine direction (%)	1.2 {cu}	1.1 {cu}	0.4 {cu}	0.3 {cu}	1.0 {cu}
Dimensional Stability - transverse direction (%)	-0.5 {cu}	-0.6 {cu}	-0.4 {cu}	-0.4 {cu}	-0.9 {cu}
Dimensional Stability - machine direction (%)	3.8 {cv}	2.8 {cv}	1.9 {cv}	1.1 {cv}	1.7 {cv}
Dimensional Stability - transverse direction (%)	-2.3 {cv}	-2.0 {cv}	-1.1 {cv}	-1.0 {cv}	-1.6 {cv}

Chapter 47

Polyphenylene Sulfide

TABLE 226: Oxygen, Carbon Dioxide, Hydrogen, Ammonia, Hydrogen Sulfide, Oxygen and Air Permeability Through Phillips Ryton Polyphenylene Sulfide Film.

Material Family	POLYPHENYLENE SULFIDE						
Material Supplier/ Grade	PHILLIPS RYTON						
Product Form	FILM						
Manufacturing Method	Formed as a baked coating by spraying unfilled PPS onto aluminum foil. Foil was then dissolved in a sodium hydroxide bath.						
Reference Number	102	102	102	102	102	102	102

MATERIAL CHARACTERISTICS

Sample Thickness	0.127 mm	0.127 mm	0.127 mm	0.127 mm	0.127 mm	0.875 mm	0.875 mm

TEST CONDITIONS

Penetrant	oxygen	carbon dioxide	hydrogen	ammonia	hydrogen sulfide	oxygen	air
Test Method	ASTM D1434, Method M	ASTM D1434, Method M	ASTM D1434, Method M	ASTM D1434, Method M	ASTM D1434, Method M	ASTM D1434, Method M	ASTM D1434, Method M

PERMEABILITY (source document units)

Gas Permeability ($cm^3 \cdot mil/100\ in^2 \cdot day$)	30	75	420	15	3	15 - 20	20 - 30
Gas Permeability ($cm^3 \cdot mm/m^2 \cdot day \cdot atm$)	11.8	29.6	165	5.9	1.2	5.9 - 7.9	7.9 - 11.8

PERMEABILITY (normalized units)

Permeability Coefficient ($cm^3 \cdot mm/m^2 \cdot day \cdot atm$)	11.8	29.5	165	5.9	1.2	5.9 - 7.9	7.9 - 11.8

TABLE 227: Water Vapor Transmission and Water, Hydrochloric Acid, Acetic Acid, Benzene and Methyl Alcohol Liquid Permeability Through Phillips Ryton Polyphenylene Sulfide Film.

Material Family	POLYPHENYLENE SULFIDE					
Material Supplier/ Grade	PHILLIPS RYTON					
Product Form	FILM					
Manufacturing Method	Formed as a baked coating by spraying unfilled PPS onto aluminum foil. Foil was then dissolved in a sodium hydroxide bath.					
Reference Number	102	102	102	102	102	102

MATERIAL CHARACTERISTICS

Sample Thickness	0.127 mm

TEST CONDITIONS

Penetrant	water	hydrochloric acid	acetic acid	benzene	methyl alcohol	water vapor
Concentration (%)		37				
Temperature (°C)	23	23	23	23	23	
Test Method	Die cut samples were fitted to tops of glass bottles by a rubber gasket and a lid, with a surface area of 96.8 cm².					ASTM E96, condition E
Test Note	Liquids were placed in bottles, gaskets and film put in place, and the lid screwed on. Apparatus was inverted to put liquid in direct contact with film. Weight loss measurements were made at 1 week intervals throughout 4 weeks of conditioning.					

PERMEABILITY (source document units)

Vapor Transmission Rate (g · mil/100 in² · day)	0.81	0.08	2.0	6.3	0.3	1.66
Vapor Transmission Rate (g · mm/m² · day)	0.35	0.03	0.79		0.12	0.66

PERMEABILITY (normalized units)

Vapor Transmission Rate (g · mm/m² · day)	0.32	0.03	0.79	2.48	0.12	0.65

Chapter 48

Polysulfone

TABLE 228: Ammonia, Carbon Dioxide, Helium, Hydrogen and Methane Permeability Through Amoco Performance Products Udel Polysulfone.

Material Family	POLYSULFONE				
Material Supplier/ Grade	AMOCO PERFORMANCE PRODUCTS UDEL				
Features	amber tint, transparent				
Reference Number	15	15	15	15	15

TEST CONDITIONS

Penetrant	ammonia	carbon dioxide	helium	hydrogen	methane
Temperature (°C)	23	23	23	23	23
Relative Humidity (%)	dry	dry	dry	dry	dry
Test Method	ASTM D1434	ASTM D1434	ASTM D1434	ASTM D1434	ASTM D1434

PERMEABILITY (source document units)

Gas Permeability ($cm^3 \cdot mil/100\ in^2 \cdot day$)	1070	950	1960	1800	37.5
Gas Permeability ($mm^3/m \cdot MPa \cdot day$)	4160	3690	7620	6990	146

PERMEABILITY (normalized units)

Permeability Coefficient ($cm^3 \cdot mm/m^2 \cdot day \cdot atm$)	421	374	772	709	14.8

TABLE 229: Nitrogen, Oxygen, Sulfur Hexafluoride, Dichlorodifluoromethane and Dichlorotetra-fluoroethane Permeability Through Amoco Performance Products Udel Polysulfone.

Material Family	POLYSULFONE				
Material Supplier/ Grade	AMOCO PERFORMANCE PRODUCTS UDEL				
Features	amber tint, transparent				
Reference Number	15	15	15	15	15

TEST CONDITIONS

Penetrant	nitrogen	oxygen	sulfur hexafluoride	dichlorodifluoromethane	dichlorotetrafluoroethane
Temperature (°C)	23	23	23	23	23
Relative Humidity (%)	dry	dry	dry	dry	dry
Test Method	ASTM D1434	ASTM D1434	ASTM D1434	ASTM D1434	ASTM D1434

PERMEABILITY (source document units)

Gas Permeability ($cm^3 \cdot mil/100\ in^2 \cdot day$)	40	230	1.8	0.59	0.25
Gas Permeability ($mm^3/m \cdot MPa \cdot day$)	155	894	6.99	2.29	0.97

PERMEABILITY (normalized units)

Permeability Coefficient ($cm^3 \cdot mm/m^2 \cdot day \cdot atm$)	15.7	90.5	0.71	0.23	0.098

TABLE 230: Water Vapor Permeability Through Amoco Performance Products Udel Polysulfone.

Material Family	POLYSULFONE	
Material Supplier/ Trade Name	AMOCO PERFORMANCE PRODUCTS UDEL	
Product Form	SLOT CAST THIN FILM	
Features	transparent	
Reference Number	15	15

TEST CONDITIONS

Penetrant	water vapor	
Temperature (°C)	38	71
Relative Humidity (%)	90	100
Test Method	ASTM E96	ASTM E96

PERMEABILITY (source document units)

Vapor Transmission Rate ($g \cdot mil/100\ in^2 \cdot day$)	18	69

PERMEABILITY (normalized units)

Vapor Transmission Rate ($g \cdot mm/m^2 \cdot day$)	7.1	27.2

Polysulfone

TABLE 231: Film Properties of Unoriented Amoco Performance Products Udel Polysulfone.

Material Family	POLYSULFONE
Material Supplier/ Grade	AMOCO PERFORMANCE PRODUCTS UDEL
Features	unoriented
Reference Number	15

MECHANICAL PROPERTIES

Secant Modulus @ 1% Elongation - machine dir. (MPa)	1972 {r}
Secant Modulus @ 1% Elongation - transverse dir.(MPa)	1903 {r}
Tensile Strength - machine direction (MPa)	73 {r}
Tensile Strength - transverse direction (MPa)	60 {r}
Tensile Strength @ Yield - machine direction (MPa)	61 {r}
Tensile Strength @ Yield - transverse direction (MPa)	57 {r}
Ultimate Elongation - machine direction (%)	110 {r}
Ultimate Elongation - transverse direction (%)	65 {r}
Burst Strength (MPa)	0.05 {ch}
Elmendorf Tear Resistance - machine direction (g/mm)	354 {cf}
Elmendorf Tear Resistance - transverse direction (g/mm)	433 {cf}
Tear Strength, Initial, Graves - machine direction (g/mm)	4290 {az}
Tear Strength, Initial, Graves - transverse direction (g/mm)	4290 {az}

OPTICAL PROPERTIES

Haze (%)	0.6 {c}
Gloss @ 45°	134 {cg}

Chapter 49

Polyvinyl Alcohol

Film Properties and Applications

Air Products and Chemicals: Vinex

Vinex resins have all the chemical, physical and mechanical properties of polyvinyl alcohol, including biodegradability, and the melt processability characteristics of a conventional thermoplastic. Vinex resins can be extruded or co-extruded into cast or tubular blown film with excellent tensile strength and dimensional stability. They are heat and impulse sealable.

Reference: *Vinex Thermoplastic Polyvinyl Alcohol Copolymer Resins Data Sheets,* supplier technical report (152-(9107, 9108, 9109)) - Air Products and Chemicals, 1991.

Air Products and Chemicals: Vinex 1000 Series

Vinex 1000 series resins are cold water insoluble and have good mechanical and chemical properties and water solubility above 37.8°C (100°F). They are structurally similar to fully hydrolyzed polyvinyl alcohol. Articles fabricated from these resins exhibit properties such as high tensile strength, good adhesion to cellulosics, heat sealability and resistance to solvents, grease and oils. Vinex 1000 resins can be converted into cast or blown films for use in numerous packaging applications. For example, hospital laundry bags made by heat sealing blown film can eliminate exposure to biologically contaminated clothing and bedding. Cold water resistance is achieved by proper downstream processing to impart high levels of crystallinity. Other Vinex grades do not crystallize as readily.

Reference: *Vinex Thermoplastic Polyvinyl Alcohol Copolymer Resins Data Sheets,* supplier technical report (152-(9107, 9108, 9109)) - Air Products and Chemicals, 1991.

Air Products and Chemicals: Vinex 2000 Series

Vinex 2000 series resins are cold water soluble and feature good mechanical and chemical properties. Vinex 2000 resins are structurally similar to partially hydrolyzed polyvinyl alcohol. Articles fabricated from these resins exhibit properties such as high tensile strength, good adhesion to cellulosics, heat sealability and resistance to solvents, grease and oils. They can be converted into cast or blown films or injection blow molded into containers for numerous packaging applications such as agricultural chemicals where solvent resistance and safe disposal of the container are required.

Reference: *Vinex Thermoplastic Polyvinyl Alcohol Copolymer Resins Data Sheets,* supplier technical report (152-(9107, 9108, 9109)) - Air Products and Chemicals, 1991.

Air Products and Chemicals: Vinex 5000 Series

Vinex 5000 series resins are impact modified and cold water soluble. Vinex 5000 resins are structurally similar to partially hydrolyzed polyvinyl alcohol. Articles fabricated from these resins exhibit properties such as high tensile strength, good adhesion to cellulosics, heat sealability and resistance to solvents, grease and oils. They can be converted into cast or blown films or injection blow molded into containers for packaging applications such as agricultural chemicals where solvent resistance and safe disposal of the container are required.

Reference: *Vinex Thermoplastic Polyvinyl Alcohol Copolymer Resins Data Sheets,* supplier technical report (152-(9107, 9108, 9109)) - Air Products and Chemicals, 1991.

TABLE 232: Oxygen and Carbon Dioxide Permeability and Water Vapor Transmission Through Polyvinyl Alcohol.

Material Family	POLYVINYL ALCOHOL		
Product Form	FILM		
Reference Number	250	250	250

TEST CONDITIONS

Penetrant	oxygen		carbon dioxide
Temperature (°C)	24	24	24
Relative Humidity (%)	0	75	0

PERMEABILITY (source document units)

Gas Permeability (cm$^3 \cdot$ mil/100 in$^2 \cdot$ day)	0.06	0.22	0.11

PERMEABILITY (normalized units)

Permeability Coefficient (cm$^3 \cdot$ mm/m$^2 \cdot$ day $\cdot$ atm)	0.02	0.09	0.04

TABLE 233: Oxygen Permeability Through Air Products and Chemicals Vinex Thermoplastic Polyvinyl Alcohol Copolymer.

Material Family	POLYVINYL ALCOHOL					
Material Supplier/ Trade Name	AIR PRODUCTS AND CHEMICALS VINEX					
Grade	1003	2144	5030	1003	2144	5030
Reference Number	283	283	283	283	283	283

MATERIAL CHARACTERISTICS

Specific Gravity	1.25	1.25	1.25	1.25	1.25	1.25
Melt Flow Index	5 - 7 g/10 min. (230/2.16)	6 - 8 g/10 min. (230/2.16)	9 - 11 g/10 min. (190/10.1)	5 - 7 g/10 min. (230/2.16)	6 - 8 g/10 min. (230/2.16)	9 - 11 g/10 min. (190/10.1)
Sample Thickness	0.03 mm	0.033 mm	0.038 mm	0.03 mm	0.033 mm	0.038 mm

TEST CONDITIONS

Penetrant	oxygen					
Relative Humidity (%)	0	0	0	50	50	50
Pressure Gradient (mmHg)	760	760	760	760	760	760

PERMEABILITY (source document units)

Gas Permeability (cm^3/100 in$^2 \cdot$ day $\cdot$ atm)	0.0350	0.0435	0.0327	1.2	1.45	1.5

PERMEABILITY (normalized units)

Permeability Coefficient (cm$^3 \cdot$ mm/m$^2 \cdot$ day $\cdot$ atm)	0.0163	0.0223	0.0193	0.56	0.74	0.88

PVA

TABLE 234: Water Vapor Transmission and Oxygen Permeability Through Polyvinyl Alcohol Film.

Material Family	POLYVINYL ALCOHOL				
Product Form	FILM				
Reference Number	268	268	268	268	268

MATERIAL CHARACTERISTICS

Sample Thickness	0.014 mm	0.014 mm	0.014 mm	0.014 mm	0.014 mm

TEST CONDITIONS

Penetrant	water vapor	oxygen			
Temperature (°C)	40	35	20	20	20
Relative Humidity (%)	90	0	65	85	100
Test Method	JIS Z0208	JIS Z1707	ASTM D3985	ASTM D3985	ASTM D3985

PERMEABILITY (source document units)

Vapor Transmission Rate (g · mil/100 in^2 · day)	71				
Gas Permeability (cm^3 · mil/100 in^2 · day)		0.02	0.16	0.9	38.71

PERMEABILITY (normalized units)

Permeability Coefficient (cm^3 · mm/m^2 · day · atm)		0.01	0.06	0.35	15.24
Vapor Transmission Rate (g · mm/m^2 · day)	28				

TABLE 235: Film Properties of Polyvinly Alcohol Film.

Material Family	POLYVINYL ALCOHOL
Product Form	**FILM**
Reference Number	268

MATERIAL CHARACTERISTICS

Sample Thickness	0.014 mm

TEST CONDITIONS

Temperature (°C)	20
Relative Humidity (%)	65

MECHANICAL PROPERTIES

Modulus Of Elasticity - MD (MPa)	4017 {r}
Modulus Of Elasticity - TD (MPa)	4217 {r}
Tensile Strength @ Break - MD (MPa)	166.9 {r}
Tensile Strength @ Break - TD (MPa)	186.2 {r}
Ultimate Elongation - MD (%)	70 {r}
Ultimate Elongation - TD (%)	60 {r}
Impact Strength (kg-cm)	9
Burst Strength (MPa)	0.39 {af}
Pinhole Strength (g)	680 {ag}
Elmendorf Tear Resistance - MD (g/mm)	1600 {ad}
Elmendorf Tear Resistance - TD (g/mm)	1400 {ad}
Tear Resistance - MD (g)	130
Tear Resistance - TD (g)	220

OTHER PROPERTIES

Water Absorption (%)	24 {ah}
Equilibrium Moist. Absorption (%)	5.0
Melting Point (°C)	225 (decomp.)
Haze (%)	1.5
Surface Resistivity (ohms)	9.9×10^{12} {aj}
Slip Factor (°)	30
Dimensional Stability - MD (%)	-2.0 {ai}
Dimensional Stability - TD (%)	-1.5 {ai}

PVA

Chapter 50

Acrylonitrile-Butadiene-Styrene Copolymer

Permeability to Gases and Water Vapor

BASF AG: Terluran

Terluran is impermeable to water, but gases and water vapor can diffuse through it. The rate of diffusion depends on the pressure gradient.

Reference: *Terluran Product Line, Properties, Processing,* supplier design guide (B 567e/ (8109) 9.90) - BASF Aktiengesellschaft, 1990.

TABLE 236: Oxygen, Nitrogen and Carbon Dioxide Permeability and Water Vapor Transmission Through Dow Chemical ABS.

Material Family	ACRYLONITRILE-BUTADIENE-STYRENE COPOLYMER						
Material Supplier/ Grade	DOW CHEMICAL						
Product Form	FILM						
Reference Number	250	250	250	250	250	250	250

MATERIAL COMPOSITION

Note	low acrylonitrile content	low acrylonitrile content	low acrylonitrile content	low acrylonitrile content	medium acrylonitrile content	medium acrylonitrile content	medium acrylonitrile content

TEST CONDITIONS

Penetrant	oxygen	nitrogen	carbon dioxide	water vapor	oxygen	nitrogen	carbon dioxide
Temperature (°C)	24	24	24	24-38	24	24	24

PERMEABILITY (source document units)

Vapor Transmission Rate (g · mil/100 in^2 · day)				5 - 16			
Gas Permeability (cm^3 · mil/100 in^2 · day)	200 - 260	25 - 35	900 - 1200		120 - 140	10 - 15	400 - 600

PERMEABILITY (normalized units)

Permeability Coefficient (cm^3 · mm/m^2 · day · atm)	79 - 102	9.8 - 13.8	354 - 472		47 - 55	3.9 - 5.9	157 - 236
Vapor Transmission Rate (g · mm/m^2 · day)				2.0 - 6.3			

TABLE 237: Oxygen, Nitrogen and Carbon Dioxide Permeability and Water Vapor Transmission Through BASF AG Terluran ABS Film.

Material Family	ACRYLONITRILE-BUTADIENE-STYRENE COPOLYMER							
Material Supplier/ Grade	BASF AG TERLURAN 967 K				BASF AG TERLURAN 877 M			
Product Form	FILM							
Features	moderate flow							
Reference Number	137	137	137	137	137	137	137	137

MATERIAL CHARACTERISTICS

Sample Thickness	0.1 mm	0.1 mm	0.1 mm	0.1 mm	0.1 mm	0.1 mm	0.1 mm	0.1 mm

MATERIAL COMPOSITION

Note	with butadiene acrylic rubber							

TEST CONDITIONS

Penetrant	water vapor	oxygen	nitrogen	carbon dioxide	water vapor	oxygen	nitrogen	carbon dioxide
Temperature (°C)	23	23	23	23	23	23	23	23
Relative Humidity (%)	85%-0% gradient				85%-0% gradient			
Test Method	DIN 53122	DIN 53380	DIN 53380	DIN 53380	DIN 53122	DIN 53380	DIN 53380	DIN 53380
Test Note	Values for permeability depend on the conditions under which the film was produced and may differ by as much as 50% from those given.							

PERMEABILITY (source document units)

Vapor Transmission Rate (g/m² · day)	27				31			
Gas Permeability (cm³ · 100 μm/m² · day · bar)		500	100	2000		450	100	2000

PERMEABILITY (normalized units)

Permeability Coefficient (cm³ · mm/m² · day · atm)		50.7	10.1	203		45.6	10	203
Vapor Transmission Rate (g · mm/m² · day)	2.7				3.1			

TABLE 238: Oxygen, Nitrogen and Carbon Dioxide Permeability and Water Vapor Transmission Through BASF AG Terluran ABS Film.

Material Family	ACRYLONITRILE-BUTADIENE-STYRENE COPOLYMER			
Material Supplier/ Grade	BASF AG TERLURAN 997 VE			
Product Form	FILM			
Features	low flow			
Reference Number	137	137	137	137

MATERIAL CHARACTERISTICS

Sample Thickness	0.1 mm	0.1 mm	0.1 mm	0.1 mm

MATERIAL COMPOSITION

Note	with butadiene acrylic rubber			

TEST CONDITIONS

Penetrant	water vapor	oxygen	nitrogen	carbon dioxide
Temperature (°C)	23	23	23	23
Relative Humidity (%)	85%-0% gradient			
Test Method	DIN 53122	DIN 53380	DIN 53380	DIN 53380
Test Note	Values for permeability depend on the conditions under which the film was produced and may differ by as much as 50% from those given.			

PERMEABILITY (source document units)

Vapor Transmission Rate (g/m² · day)	27			
Gas Permeability (cm³ · 100 mm/m² · day · bar)		800	200	3000

PERMEABILITY (normalized units)

Permeability Coefficient (cm³ · μm/m² · day · atm)		81	20.3	304
Vapor Transmission Rate (g · mm/m² · day)	2.7			

Chapter 51

Acrylonitrile-Styrene-Acrylate Copolymer

Permeability to Gases and Water Vapor

BASF AG: Luran S

Luran S moldings are impermeable to water. Depending on the pressure gradient gases and water vapor may diffuse through Luran S sheet.

The permeability to gases also depends on the processing conditions of the film or moldings.

Reference: *Luran S Acrylonitrile Styrene Acrylate Product Line, Properties, Processing,* supplier design guide (B 566 e / 11.90) - BASF Aktiengesellschaft, 1990.

<u>TABLE 239</u>: Nitrogen, Hydrogen and Methane Permeability Through BASF Luran S Acrylate Styrene Acrylonitrile Film.

Material Family	ACRYLONITRILE-STYRENE-ACRYLATE COPOLYMER							
Material Supplier/ Trade Name	BASF AG LURAN S							
Material Supplier/ Grade	776 S	757 R	776 S	757 R	776 S	797 S	776 S	757 R
Product Form	BLOWN FILM				FILM		BLOWN FILM	
Reference Number	143	143	143	143	142	142	143	143

MATERIAL CHARACTERISTICS

Sample Thickness	0.1 mm	0.1 mm	0.1 mm	0.1 mm	0.1 mm	0.1 mm	0.1 mm	0.1 mm

TEST CONDITIONS

Penetrant	hydrogen		methane		nitrogen			
Temperature (°C)	23	23	23	23	23	23	23	23
Test Method	DIN 53380	DIN 53380	DIN 53380	DIN 53380	DIN 53380 part 2 method M	DIN 53380 part 2 method M	DIN 53380	DIN 53380
Test Note					Values depend on conditions under which film was produced. Figures may differ by as much as 50%.			

PERMEABILITY (source document units)

Gas Permeability ($cm^3/m^2 \cdot day \cdot bar$)	5000	5000	110	100	100	75	70	60

PERMEABILITY (normalized units)

Permeability Coefficient ($cm^3 \cdot mm/m^2 \cdot day \cdot atm$)	507	507	11.1	10.1	10.1	7.6	7.1	6.1

TABLE 240: Oxygen and Carbon Dioxide Permeability Through BASF Luran S Acrylate Styrene Acrylonitrile Film.

Material Family	ACRYLONITRILE-STYRENE-ACRYLATE COPOLYMER							
Material Supplier/ Trade Name	BASF AG LURAN S							
Grade	776 S	797 S	776 S	757 R	776 S	797 S	776 S	757 R
Product Form	FILM		BLOWN FILM		FILM		BLOWN FILM	
Reference Number	142	142	143	143	142	142	143	143

MATERIAL CHARACTERISTICS

Sample Thickness	0.1 mm	0.1 mm	0.1 mm	0.1 mm	0.1 mm	0.1 mm	0.1 mm	0.1 mm

TEST CONDITIONS

Penetrant	oxygen				carbon dioxide			
Temperature (°C)	23	23	23	23	23	23	23	23
Test Method	DIN 53380 part 2 method M	DIN 53380 part 2 method M	DIN 53380	DIN 53380	DIN 53380 part 2 method M	DIN 53380 part 2 method M	DIN 53380	DIN 53380
Test Note	Values depend on conditions under which film was produced. Figures may differ by as much as 50%.				Values depend on conditions under which film was produced. Figures may differ by as much as 50%.			

PERMEABILITY (source document units)

Gas Permeability ($cm^3/m^2 \cdot day \cdot bar$)	550	500	180	150	2300	2000	1400	1000

PERMEABILITY (normalized units)

Permeability Coefficient ($cm^3 \cdot mm/m^2 \cdot day \cdot atm$)	55.7	50.7	18.2	15.2	233	203	142	101

TABLE 241: Water Vapor Transmission Through BASF Luran S Acrylate Styrene Acrylonitrile Film.

Material Family	ACRYLONITRILE-STYRENE-ACRYLATE COPOLYMER		
Material Supplier/ Grade	**BASF AG LURAN S 776 S**	**BASF AG LURAN S 797 S**	**BASF AG LURAN S 757 R**
Product Form	**FILM**		
Manufacturing Method			blown film
Reference Number	142	142	143

MATERIAL CHARACTERISTICS

Sample Thickness	0.1 mm	0.1 mm	0.1 mm

TEST CONDITIONS

Penetrant	water vapor		
Temperature (°C)	23	23	23
Relative Humidity (%)	85%-0% gradient	85%-0% gradient	85%-0% gradient
Pressure Gradient (Mbar)	23.87	23.87	19.86
Test Method	DIN 53122	DIN 53122	DIN 53122
Test Note	Values for permeability depend on the conditions under which the film was produced. Figures determined may differ by as much as 50% from those given.		

PERMEABILITY (source document units)

Vapor Transmission Rate (g/m^2 · day)	35	30	30

PERMEABILITY (normalized units)

Vapor Transmission Rate (g · mm/m^2 · day)	3.5	3	3

Chapter 52

Polystyrene

Permeability to Gases and Water Vapor

BASF AG: Polystyrol

In common with all other plastics, polystyrene has a certain permeability to gases and vapors, that increases with temperature.

Reference: *Polystyrol Product Line, Properties, Processing,* supplier design guide (B 564 e/2.93) - BASF Aktiengesellschaft, 1993.

Effect of Temperature on Water Vapor Transmission Rate

Dow Chemical: Styron   (product form: film)

For polystyrene, permeability to water vapor increases slightly with temperature.

Reference: *Permeability Of Polymers To Gases And Vapors,* supplier technical report (P302-335-79, D306-115-79) - Dow Chemical Company, 1979.

<u>TABLE 242:</u> **Gas Permeability and Water Vapor Transmission Through Dow Chemical Trycite Polystyrene Film.**

Material Family	POLYSTYRENE			
Material Supplier/ Grade	DOW CHEMICAL TRYCITE			
Product Form	FILM			
Reference Number	250	250	250	250

TEST CONDITIONS

Penetrant	oxygen	nitrogen	carbon dioxide	water vapor
Temperature (°C)	24	24	24	24

PERMEABILITY (source document units)

Vapor Transmission Rate (g $\cdot$ mil/100 in^2 $\cdot$ day)				9
Gas Permeability (cm^3 $\cdot$ mil/100 in^2 $\cdot$ day)	250 - 350	50 - 60	700 - 1100	

PERMEABILITY (normalized units)

Permeability Coefficient (cm^3 $\cdot$ mm/m^2 $\cdot$ day $\cdot$ atm)	98.4 - 138	19.7 - 23.6	276- 433	
Vapor Transmission Rate (g $\cdot$ mm/m^2 $\cdot$ day)				3.5

TABLE 243: Oxygen, Nitrogen, and Carbon Dioxide Permeability and Water Vapor Transmission Through Dow Chemical Styron Polystyrene.

Material Family	POLYSTYRENE			
Material Supplier/ Grade	DOW CHEMICAL STYRON			
Product Form	FILM			
Reference Number	250	250	250	250

TEST CONDITIONS

Penetrant	oxygen	nitrogen	carbon dioxide	water vapor
Temperature (°C)	24	24	24	24-38

PERMEABILITY (source document units)

Vapor Transmission Rate (g · mil/100 in² · day)				2 - 10
Gas Permeability (cm³ · mil/100 in² · day)	300 - 400	40 - 50	1000 - 1500	

PERMEABILITY (normalized units)

Permeability Coefficient (cm³ · mm/m² · day · atm)	118 - 157	15.8 - 19.7	394 - 590	
Vapor Transmission Rate (g · mm/m² · day)				0.79 - 3.9

TABLE 244: Oxygen Permeability and Water Vapor Transmission Through Polystyrene.

Material Family	POLYSTYRENE			
Reference Number	264	264	296	296

TEST CONDITIONS

Penetrant	oxygen	water vapor	oxygen	water vapor
Temperature (°C)	23	40	22.8	37.8
Relative Humidity (%)	0	90	0	90
Test Method			ASTM D1434	ASTM F1249

PERMEABILITY (source document units)

Gas Permeability (cm³ · mil/100 in² · day)	260		>350	
Gas Permeability (cm³ · 25µ/m² · day · atm)	4030			
Vapor Transmission Rate (g · mil/100 in² · day)		8.5		10
Vapor Transmission Rate (g · 25µ/m² · day)		131.8		

PERMEABILITY (normalized units)

Permeability Coefficient (cm³ · mm/m² · day · atm)	102.4		>140	
Vapor Transmission Rate (g · mm/m² · day)		3.4		4.0

TABLE 245: Vapor Transmission of Reagents Through Dow Chemical Styron Polystyrene Resin.

Material Family	POLYSTYRENE							
Material Supplier/ Grade	DOW CHEMICAL STYRON							
Product Form	FILM							
Reference Number	250	250	250	250	250	250	250	250

TEST CONDITIONS

Penetrant	methyl alcohol	ethyl alcohol	n-heptane	ethyl acetate	formaldehyde	tetrachloro ethylene	acetone	benzene
Temperature (°C)	24	24	24	24	24	24	24	35

PERMEABILITY (source document units)

Vapor Transmission Rate (g · mil/100 in² · day)	1 - 6	1 - sample failed during test	sample failed during test	sample failed during test	4 - 5	sample failed during test	sample failed during test	1200

PERMEABILITY (normalized units)

Vapor Transmission Rate (g · mm/m² · day)	0.39 - 2.4	0.39 - sample failed	sample failed	sample failed	1.6 - 2.0	sample failed	sample failed	472

TABLE 246: Film Properties of Dow Chemical Trycite Polystyrene Film.

Material Family	POLYSTYRENE		
Material Supplier	DOW CHEMICAL TRYCITE		
Grade	TRYCITE 8001	TRYCITE 8002	TRYCITE 8003
Product Form	FILM		
Features	high gloss, medium oriented, transparent	medium oriented, transluscent	high impact, medium oriented, transluscent
Applications	laminations	laminations	laminations
Reference Number	261	261	261

MECHANICAL PROPERTIES

Secant Modulus @ 1% Elongation - machine dir. (MPa)	3100 {r}	3030 {r}	2515 {r}
Secant Modulus @ 1% Elongation - transverse dir. (MPa)	3100 {r}	2960 {r}	2515 {r}
Tensile Strength @ Yield - MD (MPa)	75.8 {r}	72.3 {r}	55.1 {r}
Tensile Strength @ Yield - TD (MPa)	72.3 {r}	68.9 {r}	55.1 {r}
Ultimate Elongation - MD (%)	70 {r}	60 {r}	85 {r}
Ultimate Elongation - TD (%)	80 {r}	80 {r}	95 {r}

OPTICAL PROPERTIES

Haze (%)	<1.0 {ac}	6 {ac}	58 {ac}
Gloss @ 60°	165 (mach. dir., outside) {m}	125 (mach. dir.) {m}	30

GRAPH 55: Oxygen Permeability vs. Temperature through Polystyrene.

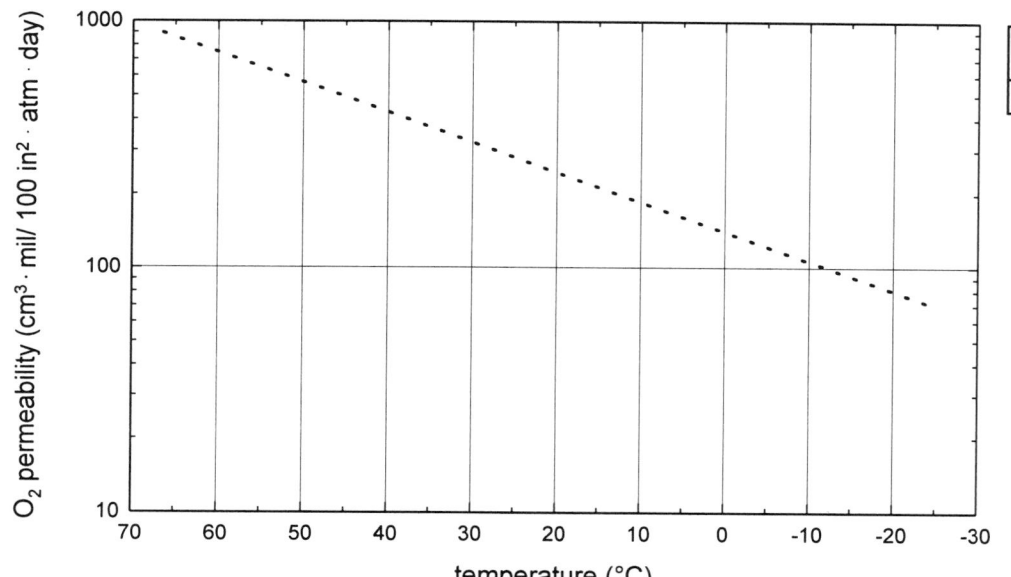

....	Dow Styron PS (film); penetrant: O₂
Reference No.	250

GRAPH 56: Water Vapor Permeability vs. Thickness through Polystyrene.

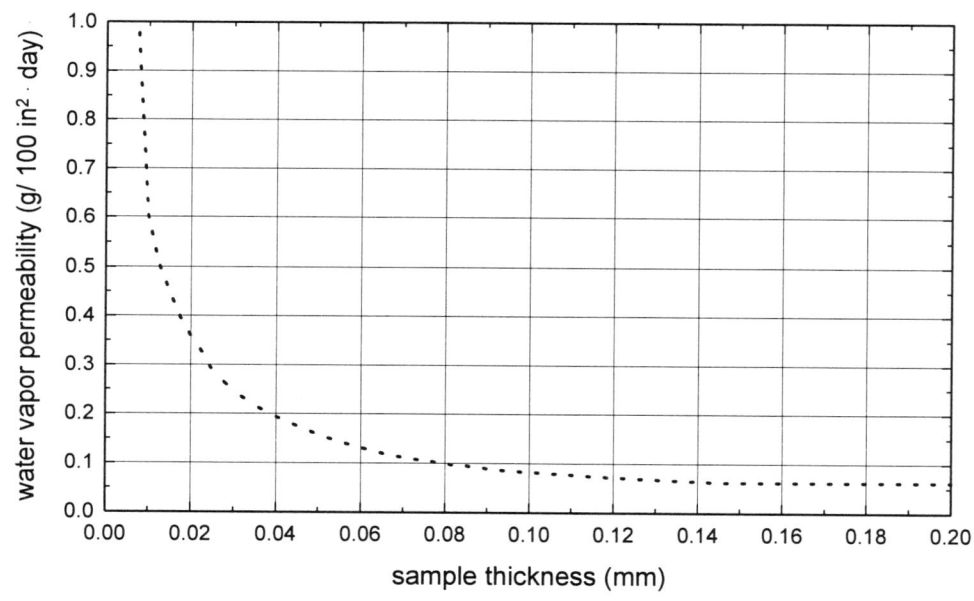

....	Dow Styron PS (film); penetrant: water vapor
Reference No.	250

Chapter 53

General Purpose Polystyrene

TABLE 247: Oxygen, Nitrogen and Carbon Dioxide Permeability and Water Vapor Transmission Through BASF AG Polystyrol General Purpose Polystyrene Film.

Material Family	GENERAL PURPOSE POLYSTYRENE			
Material Supplier/ Grade	BASF AG POLYSTYROL 168 N			
Product Form	FILM			
Features	transparent	transparent	transparent	transparent
Reference Number	26	26	26	26

MATERIAL CHARACTERISTICS

Sample Thickness	0.1 mm	0.1 mm	0.1 mm	0.1 mm

TEST CONDITIONS

Penetrant	water vapor	oxygen	nitrogen	carbon dioxide
Temperature (°C)	23	23	23	23
Relative Humidity (%)	85%-0% gradient			
Test Method	DIN 53122	DIN 53380	DIN 53380	DIN 53380

PERMEABILITY (source document units)

Vapor Transmission Rate ($g/m^2 \cdot day$)	12			
Gas Permeability ($cm^3/m^2 \cdot day \cdot bar$)		1000	250	5200

PERMEABILITY (normalized units)

Permeability Coefficient ($cm^3 \cdot mm/m^2 \cdot day \cdot atm$)		101	25.3	527
Vapor Transmission Rate ($g \cdot mm/m^2 \cdot day$)	1.2			

TABLE 248: Oxygen, Nitrogen and Carbon Dioxide Permeability and Water Vapor Transmission Through Dow Chemical Styron General Purpose Polystyrene.

Material Family	GENERAL PURPOSE POLYSTYRENE							
Material Supplier/ Trade Name	DOW CHEMICAL STYRON							
Product Form					SHEET			
Features					oriented			
Manufacturing Method	injection molding							
Reference Number	263	263	263	263	263	263	263	263

TEST CONDITIONS

Penetrant	oxygen	nitrogen	carbon dioxide	water vapor	oxygen	nitrogen	carbon dioxide	water vapor
Temperature (°C)	23	23	23	24-38	23	23	23	24-38
Test Method	ASTM D1434	ASTM D1434	ASTM D1434	ASTM E96	ASTM D1434	ASTM D1434	ASTM D1434	ASTM E96

PERMEABILITY (source document units)

	oxygen	nitrogen	carbon dioxide	water vapor	oxygen	nitrogen	carbon dioxide	water vapor
Vapor Transmission Rate (g · mil/100 in^2 · day)				2-10				9
Gas Permeability (cm^3 · mil/100 in^2 · day)	300-400	40-50	1000-1500		250-350	50-60	700-1100	

PERMEABILITY (normalized units)

	oxygen	nitrogen	carbon dioxide	water vapor	oxygen	nitrogen	carbon dioxide	water vapor
Permeability Coefficient (cm^3 · mm/m^2 · day · atm)	118 - 158	16 - 20	393 - 591		98- 138	20 - 24	276 - 433	
Vapor Transmission Rate (g · mm/m^2 · day)				0.79 - 3.9				3.5

Chapter 54

Impact Resistant Polystyrene

TABLE 249: Oxygen, Nitrogen and Carbon Dioxide Permeability and Water Vapor Transmission Through Dow Chemical Styron Impact Polystyrene.

Material Family	IMPACT RESISTANT POLYSTYRENE			
Material Supplier/ Trade Name	DOW CHEMICAL STYRON			
Manufacturing Method	injection molding	injection molding	injection molding	injection molding
Reference Number	262	262	262	262

TEST CONDITIONS

Penetrant	oxygen	nitrogen	carbon dioxide	water vapor
Temperature (°C)	23	23	23	24-38
Test Method	ASTM D1434	ASTM D1434	ASTM D1434	ASTM E96

PERMEABILITY (source document units)

Vapor Transmission Rate ($g \cdot mil/100 \ in^2 \cdot day$)				2-10
Gas Permeability ($cm^3 \cdot mil/100 \ in^2 \cdot day$)	300-400	40-50	1000-1500	

PERMEABILITY (normalized units)

Permeability Coefficient ($cm^3 \cdot mm/m^2 \cdot day \cdot atm$)	118 - 157	15.8 - 19.7	394 - 591	
Vapor Transmission Rate ($g \cdot mm/m^2 \cdot day$)				0.8 - 3.9

TABLE 250: Oxygen, Nitrogen and Carbon Dioxide Permeability and Water Vapor Transmission Through BASF AG Polystyrol Impact Polystyrene Film.

Material Family	IMPACT RESISTANT POLYSTYRENE			
Material Supplier/ Grade	BASF AG POLYSTYROL 476 L			
Product Form	FILM			
Reference Number	26	26	26	26

MATERIAL CHARACTERISTICS

Sample Thickness	0.1 mm	0.1 mm	0.1 mm	0.1 mm

TEST CONDITIONS

Penetrant	water vapor	oxygen	nitrogen	carbon dioxide
Temperature (°C)	23	23	23	23
Relative Humidity (%)	85%-0% gradient			
Test Method	DIN 53122	DIN 53380	DIN 53380	DIN 53380

PERMEABILITY (source document units)

Vapor Transmission Rate $(g/m^2 \cdot day)$	13			
Gas Permeability $(cm^3/m^2 \cdot day \cdot bar)$		1600	400	10,000

PERMEABILITY (normalized units)

Permeability Coefficient $(cm^3 \cdot mm/m^2 \cdot day \cdot atm)$		162	40.5	1013
Vapor Transmission Rate $(g \cdot mm/m^2 \cdot day)$	1.3			

Chapter 55

Styrene-Acrylonitrile Copolymer

Permeability to Gases and Water Vapor

BASF AG: Luran (features: transparent)

Luran is impermeable to water but allows water vapor and gases to permeate it in given amounts. The permeability to carbon dioxide is about five times higher, and the permeability to nitrogen about five times less, than that of oxygen.

The permeability to gases also depends on the conditions under which the film or moldings were produced.

Reference: *Luran Product Line, Properties, Processing,* supplier design guide (B 565 e/10.83) - BASF Aktiengesellschaft, 1983.

TABLE 251: Gas Permeability and Water Vapor Transmission Through Dow Chemical Tyril Styrene Acrylonitrile Copolymer.

Material Family	STYRENE-ACRYLONITRILE COPOLYMER				
Material Supplier/ Grade	DOW CHEMICAL TYRIL				
Product Form	FILM				
Reference Number	250	250	250	250	250

MATERIAL COMPOSITION

Note	low acrylonitrile content	medium acrylonitrile content	low acrylonitrile content	low acrylonitrile content	low acrylonitrile content

TEST CONDITIONS

Penetrant	oxygen		nitrogen	carbon dioxide	water vapor
Temperature (°C)	24	24	24	24	24-38

PERMEABILITY (source document units)

Vapor Transmission Rate (g · mil/100 in^2 · day)					5 - 14
Gas Permeability (cm^3 · mil/100 in^2 · day)	80 - 100	40 - 70	10	400	

PERMEABILITY (normalized units)

Permeability Coefficient (cm^3 · mm/m^2 · day · atm)	31.5 - 39.4	15.7 - 27.6	3.9	157	
Vapor Transmission Rate (g · mm/m^2 · day)					2.0 - 5.5

TABLE 252: Oxygen Gas and Water Vapor Permeability Through BASF Luran Styrene Acrylonitrile Copolymer Film.

Material Family	STYRENE-ACRYLONITRILE COPOLYMER							
Material Supplier/ Trade Name	BASF AG LURAN							
Grade	358 N	368 R	378 P	388 S	358 N	368 R	378 P	388 S
Product Form	FILM							
Features	high flow, transparent	moderate to low flow, transparent	moderate to high flow, transparent	low flow, transparent	high flow, transparent	moderate to low flow, transparent	moderate to high flow, transparent	low flow, transparent
Reference Number	30	30	30	30	30	30	30	30

MATERIAL CHARACTERISTICS

	358 N	368 R	378 P	388 S	358 N	368 R	378 P	388 S
Sample Thickness	0.1 mm	0.1 mm	0.1 mm	0.1 mm	0.1 mm	0.1 mm	0.1 mm	0.1 mm

TEST CONDITIONS

Penetrant	oxygen				water vapor			
Temperature (°C)	23	23	23	23	23	23	23	23
Relative Humidity (%)					85-0% gradient	85-0% gradient	85-0% gradient	85-0% gradient
Test Method	DIN 53380	DIN 53380	DIN 53380	DIN 53380	DIN 53122	DIN 53122	DIN 53122	DIN 53122

PERMEABILITY (source document units)

Gas Permeability ($cm^3/m^2 \cdot day \cdot bar$)	200 - 500	200 - 500	200 - 300	200 - 300				
Vapor Transmission Rate ($g/m^2 \cdot day$)					20 - 25	20 - 25	20-25	20 - 25

PERMEABILITY (normalized units)

Permeability Coefficient ($cm^3 \cdot mm/m^2 \cdot day \cdot atm$)	20.3 - 50.7	20.3 - 50.7	20.3 - 30.4	20.3 - 30.4				
Vapor Transmission Rate ($g \cdot mm/m^2 \cdot day$)					2 - 2.5	2 - 2.5	2 - 2.5	2 - 2.5

SAN

TABLE 253: Vapor Transmission of Reagents Through Dow Chemical Tyril Styrene Acrylonitrile.

Material Family	STYRENE-ACRYLONITRILE COPOLYMER						
Material Supplier/ Grade	DOW CHEMICAL TYRIL						
Product Form	FILM						
Reference Number	250	250	250	250	250	250	250

TEST CONDITIONS

Penetrant	methyl alcohol	ethyl alcohol	n-heptane	ethyl acetate	formaldehyde	tetrachloroethylene	acetone
Temperature (°C)	24	24	24	24	24	24	24

PERMEABILITY (source document units)

	methyl alcohol	ethyl alcohol	n-heptane	ethyl acetate	formaldehyde	tetrachloroethylene	acetone
Vapor Transmission Rate (g · mil/100 in² · day)	sample failed during test	sample failed during test	2 - 20	sample failed during test	5 - 10	sample failed during test	sample failed during test

PERMEABILITY (normalized units)

	methyl alcohol	ethyl alcohol	n-heptane	ethyl acetate	formaldehyde	tetrachloroethylene	acetone
Vapor Transmission Rate (g · mm/m² · day)	sample failed	sample failed	0.8 - 7.9	sample failed	2.0 - 3.9	sample failed	sample failed

GRAPH 57: Oxygen Permeability vs. Temperature through Styrene-Acrylonitrile Copolymer.

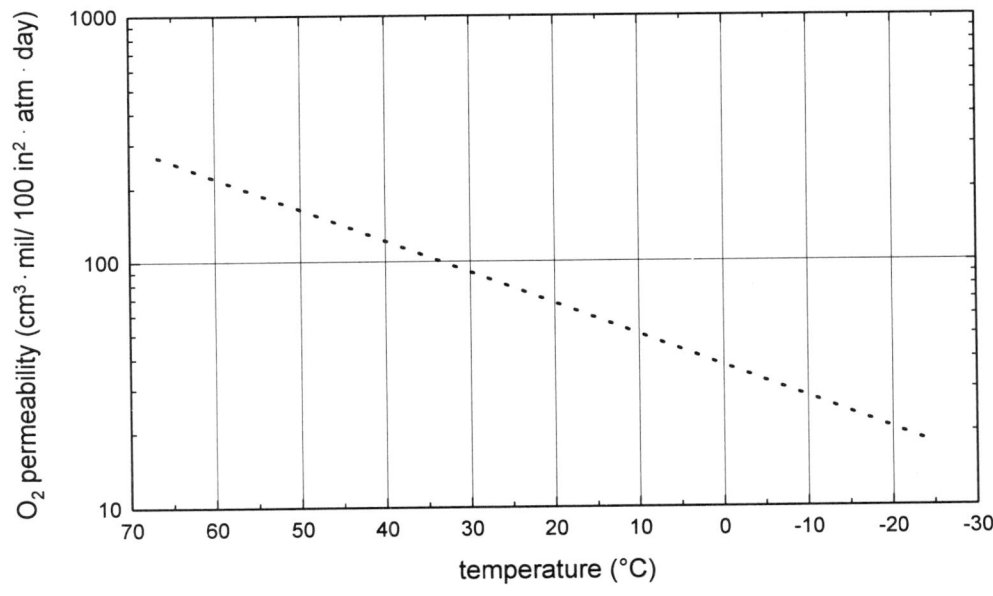

	Dow Tyril SAN (film); penetrant: O₂
Reference No.	250

GRAPH 58: Water Vapor Permeability vs. Thickness through Styrene-Acrylonitrile Copolymer.

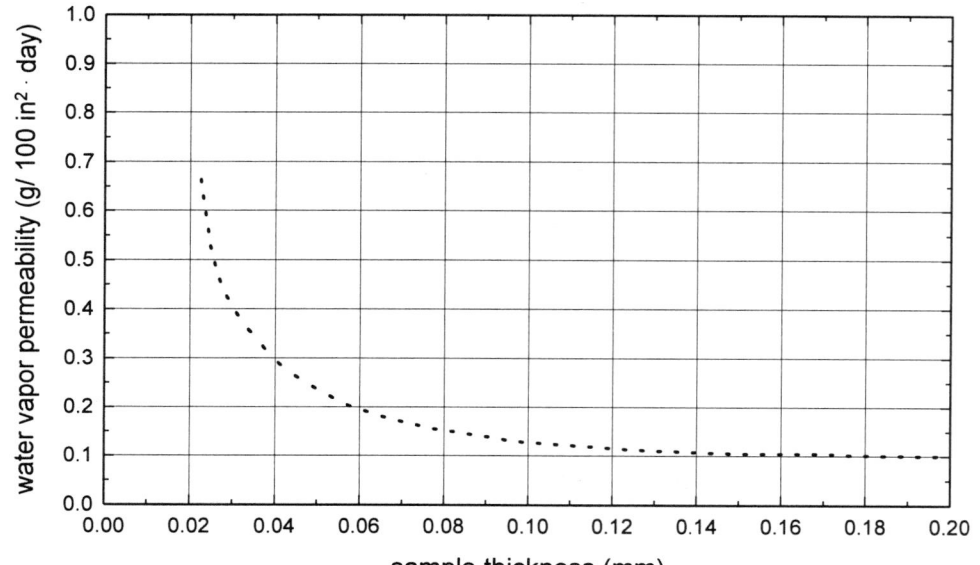

··············	Dow Tyril SAN (film); penetrant: water vapor
Reference No.	250

Chapter 56

Styrene-Butadiene Block Copolymer

Film Properties and Applications

BASF AG: Styrolux (features: transparent)

Styrolux is the trade name for a line of anionically produced styrene-butadiene block copolymers made by BASF. These products, possessing complex molecular structures, are characterized by their optical and mechanical properties.

Styrolux, introduced in 1983, has a combination of brilliant clarity and great toughness. The thermoformability of the material and its ability to blend with polystyrene establishes Styrolux as an important material in the packaging and medical sectors.

KR 2688 is an extrusion grade with optimized properties for use as a blend component with standard polystyrene. Films made from blends of KR2688 are tougher (better elongation at break) at all mixing ratios than equivalent films made from 684 D and standard polystyrene.

Because of its low gel particle content, KR 2688 is also suitable for blown film extrusion. Styrolux films with a thickness of 50 μm may be used for labeling and shrink wrap film.

Reference: *Topics In Chemistry - BASF Plastics Research And Development,* supplier technical report - BASF Aktiengesellschaft, 1992.

<u>**TABLE 254:**</u> **Gas Permeability Through BASF AG Styrolux Styrene Butadiene Block Copolymer.**

Material Family	STYRENE-BUTADIENE BLOCK COPOLYMER								
Material Supplier/ Grade	BASF AG STYROLUX 684 D			BASF AG STYROLUX 656 C			BASF AG STYROLUX 637 D		
Product Form	FILM								
Features				high flow	high flow	high flow			
Reference Number	29	29	29	29	29	29	29	29	29

MATERIAL CHARACTERISTICS

Sample Thickness	0.1 mm	0.1 mm	0.1 mm	0.1 mm	0.1 mm	0.1 mm	0.1 mm	0.1 mm	0.1 mm

TEST CONDITIONS

Penetrant	oxygen	nitrogen	carbon dioxide	oxygen	nitrogen	carbon dioxide	oxygen	nitrogen	carbon dioxide
Temperature (°C)	23	23	23	23	23	23	23	23	23

PERMEABILITY (source document units)

Gas Permeability ($cm^3/m^2 \cdot day \cdot bar$)	2600	700	15,000	1600	350	8000	1900	450	10,000

PERMEABILITY (normalized units)

Permeability Coefficient ($cm^3 \cdot mm/m^2 \cdot day \cdot atm$)	263	70.9	1520	162	35.5	811	192	45.6	1013

TABLE 255: Water Vapor Transmission Through BASF AG Styrolux Styrene Butadiene Block Copolymer.

Material Family	STYRENE-BUTADIENE BLOCK COPOLYMER		
Material Supplier/ Grade	BASF AG STYROLUX 684 D	BASF AG STYROLUX 656 C	BASF AG STYROLUX 637 D
Product Form	FILM		
Features		high flow	
Reference Number	29	29	29

MATERIAL CHARACTERISTICS

Sample Thickness	0.1 mm	0.1 mm	0.1 mm

TEST CONDITIONS

Penetrant	water vapor		
Temperature (°C)	23	23	23

PERMEABILITY (source document units)

Vapor Transmission Rate (g/m^2 · day)	13.8	11.3	12.7

PERMEABILITY (normalized units)

Vapor Transmission Rate (g · mm/m^2 · day)	1.38	1.13	1.27

TABLE 256: Mechanical Properties of BASF AG Styrolux Styrene Butadiene Block Copolymer Film.

Material Family	STYRENE-BUTADIENE BLOCK COPOLYMER	
Material Supplier/ Grade	BASF AG STYROLUX 684 D	BASF AG STYROLUX KR 2688
Product Form	FILM	
Features	transparent	transparent
Reference Number	182	182

MATERIAL COMPOSITION

Note	blended with 60% polystyrene 143E	blended with 60% polystyrene 143E

MECHANICAL PROPERTIES

Ultimate Elongation - transverse direction (%)	13	24.5

Chapter 57

Polyvinyl Chloride

TABLE 257: Gas Permeability and Water Vapor Transmission Through Polyvinyl Chloride.

Material Family	POLYVINYL CHLORIDE		
Product Form	FILM		
Reference Number	250	250	250

MATERIAL COMPOSITION

Note	unplasticized	unplasticized	unplasticized

TEST CONDITIONS

Penetrant	oxygen	carbon dioxide	water vapor
Temperature (°C)	24	24	38

PERMEABILITY (source document units)

Vapor Transmission Rate (g · mil/100 in² · day)			3
Gas Permeability (cm³ · mil/100 in² · day)	5 - 20	20 - 50	

PERMEABILITY (normalized units)

Permeability Coefficient (cm³ · mm/m² · day · atm)	2.0 - 7.9	7.9 - 19.7	
Vapor Transmission Rate (g · mm/m² · day)			1.2

TABLE 258: Oxygen Permeability vs. Temperature Through Rigid Polyvinyl Chloride Film.

Material Family	POLYVINYL CHLORIDE		
Product Form	FILM		
Reference Number	63	63	63

TEST CONDITIONS

Penetrant	oxygen		
Temperature (°C)	20	23	35
Relative Humidity (%)	dry	dry	dry

PERMEABILITY (source document units)

Gas Permeability ($cm^3 \cdot mil/100\ in^2 \cdot day$)	12.2	13.2	18.8
Gas Permeability ($cm^3 \cdot 20\mu/m^2 \cdot day \cdot atm$)	240	260	370

PERMEABILITY (normalized units)

Permeability Coefficient ($cm^3 \cdot mm/m^2 \cdot day \cdot atm$)	4.8	5.2	7.4

TABLE 259: Water Vapor Transmission Through Rigid Polyvinyl Chloride.

Material Family	POLYVINYL CHLORIDE		
Reference Number	264	296	296

TEST CONDITIONS

Penetrant	water vapor	oxygen	water vapor
Temperature (°C)	40	22.8	37.8
Relative Humidity (%)	90	0	90
Test Method		ASTM D1434	ASTM F1249

PERMEABILITY (source document units)

Gas Permeability ($cm^3 \cdot mil/100\ in^2 \cdot bar \cdot day$)		8	
Vapor Transmission Rate ($g \cdot mil/100\ in^2 \cdot day$)	3		4.25
Vapor Transmission Rate ($g \cdot 25\mu/m^2 \cdot day$)	46.5		

PERMEABILITY (normalized units)

Permeability Coefficient ($cm^3 \cdot mm/m^2 \cdot day \cdot atm$)		3.2	
Vapor Transmission Rate ($g \cdot mm/m^2 \cdot day$)	1.2		1.7

TABLE 260: Cyclohexanone, Chlorobenzene, Hexane, Butyl Alcohol, Trichloroethene, Methyl Salicylate and Tetrahydrofuran Permeability Through Rigid Polyvinyl Chloride Bottles.

Material Family	POLYVINYL CHLORIDE						
Product Form	BOTTLES						
Reference Number	293	293	293	293	293	293	293

TEST CONDITIONS

Penetrant	cyclohexanone	chlorobenzene	hexane	butyl alcohol	trichloroethene	methyl salicylate	tetrahydrofuran
Temperature (°C)	23	23	23	23	23	23	23
Exposure Time (days)	180	180	180	180	180	180	180

PERMEABILITY (source document units)

Penetrant Weight Loss (%)	failed	failed	6.06	0.18	failed	failed	failed

TABLE 261: Ethyl Acetate, Isopropyl Acetate, Acetone, Butyl Acetate, Toluene, Xylene, Methyl Isobutyl Ketone and Methyl Ethyl Ketone Permeability Through Rigid Polyvinyl Chloride Bottles.

Material Family	POLYVINYL CHLORIDE							
Product Form	BOTTLES							
Reference Number	293	293	293	293	293	293	293	293

TEST CONDITIONS

Penetrant	ethyl acetate	isopropyl acetate	acetone	butyl acetate	toluene	xylene	methyl isobutyl ketone	methyl ethyl ketone
Temperature (°C)	23	23	23	23	23	23	23	23
Exposure Time (days)	180	180	180	180	180	180	180	180

PERMEABILITY (source document units)

Penetrant Weight Loss (%)	failed	failed	failed	failed	failed	failed	failed	failed

GRAPH 59: Oxygen Permeability vs. Thickness through Polyvinyl Chloride.

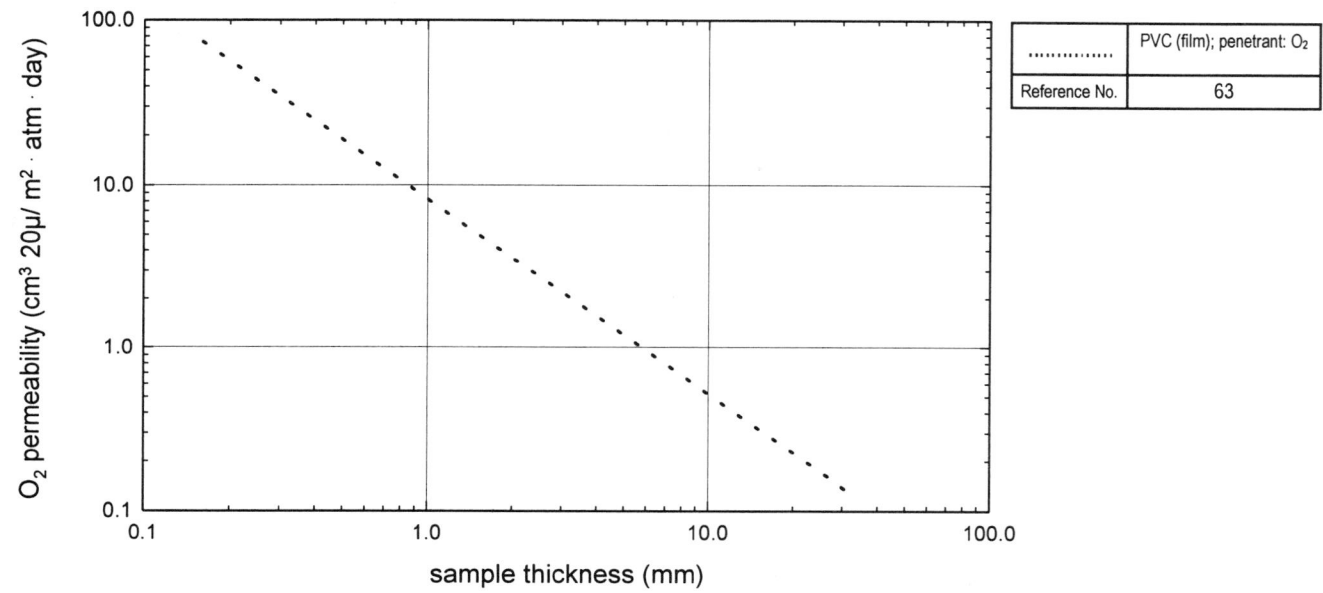

GRAPH 60: Water Vapor Permeability vs. Thickness through Polyvinyl Chloride.

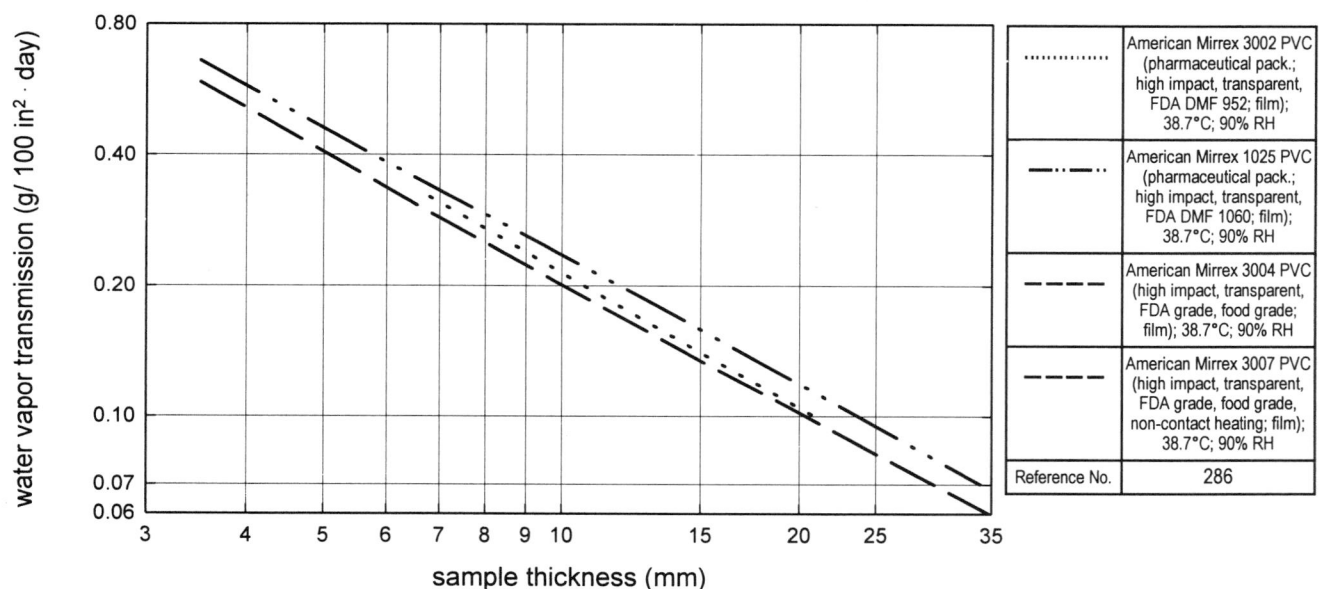

GRAPH 61: **Water Vapor Permeability vs. Thickness through Polyvinyl Chloride.**

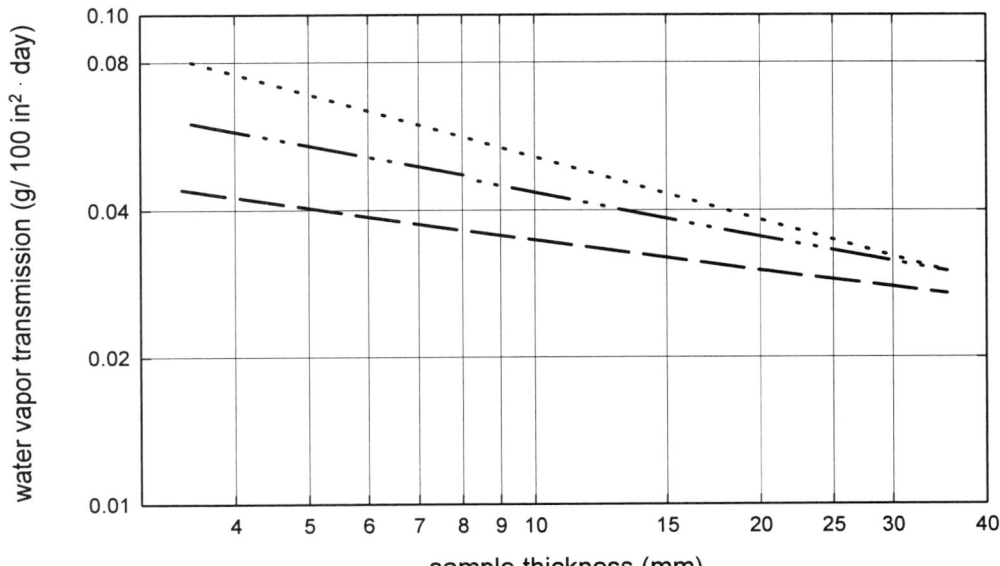

	American Mirrex EZ23 PVC (transparent; 40 g/m² PVDC coated; film); 38.7°C; 90% RH
	American Mirrex EZ23 PVC (transparent; 60 g/m² PVDC coated; film); 38.7°C; 90% RH
	American Mirrex EZ23 PVC (transparent; 80 g/m² PVDC coated; film); 38.7°C; 90% RH
Reference No.	286

Chapter 58

Polyvinyl Chloride-Polyvinylidene Chloride Copolymer

TABLE 262: Water Vapor, Oxygen, Nitrogen and Carbon Dioxide Permeability Through Polyvinyl Chloride Polyvinylidene Chloride Copolymer.

Material Family	POLYVINYL CHLORIDE-POLYVINYLIDENE CHLORIDE COPOLYMER			
Reference Number	138	138	138	138

TEST CONDITIONS

Penetrant	water vapor	oxygen	nitrogen	carbon dioxide
Temperature (°C)	37.8	25	25	25
Relative Humidity (%)	90			
Test Note		STP conditions		

PERMEABILITY (source document units)

Gas Permeability (cm^3 · mil/100 in^2 · day)		0.8-6.9	0.12-1.5	38-44
Gas Permeability (cm^3 · mm/m^2 · day · atm)		0.3-2.7	0.05-0.6	15-17
Vapor Transmission Rate (g · mil/100 in^2 · day)	0.20-0.6			
Vapor Transmission Rate (g · mm/m^2 · day)	0.08-0.24			

PERMEABILITY (normalized units)

Permeability Coefficient (cm^3 · mm/m^2 · day · atm)		0.3 - 2.7	0.05 - 0.6	15 - 17
Vapor Transmission Rate (g · mm/m^2 · day)	0.08 - 0.24			

Chapter 59

Polyvinylidene Chloride

Barrier Performance

Dow Chemical: Saran (chemical type: vinylidene chloride vinyl chloride copolymer; features: barrier properties; product form: film); **Saran MA** (chemical type: vinylidene chloride methyl acrylate copolymer; features: barrier properties; product form: film)

Saran MA Vinylidene Chloride Methyl Acrylate Copolymer provides twice the barrier properties, better color and color stability, and further improved flavor-aroma protection over conventional Saran vinylidene chloride vinyl chloride copolymers.

Saran resins, including the Saran MA form, can provide reliable barrier despite typical high humidity warehousing and storage conditions, and despite the moisture invariably encountered in retort and other packaging environments.

Reference: *Saran Barrier Polymers 1987 Update: Saran Barrier Polymer Dynamics,* supplier technical report (190-383-587) - Dow Chemical Company, 1987.

Flavor and Aroma Transport

Dow Chemical: Saran (chemical type: vinylidene chloride vinyl chloride copolymer; features: barrier properties; product form: film); **Saran MA** (chemical type: vinylidene chloride methyl acrylate copolymer; features: barrier properties; product form: film)

Saran polymers have an outstanding record, over their long commercial history, in terms of protecting the flavor and aroma of sensitive packaged products. The new higher barrier Saran MA resins have nevertheless significantly improved flavor and aroma barrier.

Reference: *Saran Barrier Polymers 1987 Update: Saran Barrier Polymer Dynamics,* supplier technical report (190-383-587) - Dow Chemical Company, 1987.

Film Properties and Applications

Dow Chemical: Saran (chemical type: vinylidene chloride vinyl chloride copolymer; features: barrier properties; manufacturing method: coating; product form: film); **Saran** (chemical type: vinylidene chloride vinyl chloride copolymer; features: barrier properties; manufacturing method: extrusion; product form: film)

Saran vinylidene chloride-based polymers exhibit excellent gas and moisture-barrier properties. These characteristics, combined with broad chemical and ignition-resistance properties, make Saran the preferred material in rigid and flexible packaging applications. Saran products are available in two separate forms. Saran F resins are solvent-soluble polymers used in barrier coatings. Saran extrusion resins are melt-processable resins used in rigid multilayer coextruded films, and monofilaments.

Saran products have proven commercial success in barrier packaging for a wide range of medicines, cosmetics, and foods. For example, Saran products are used in packages for red meat, processed meats like hot dogs and bacon, pudding, cookies, powdered detergents, and candy.

Reference: *621 Ways To Succeed - 1993-1994 Materials Selection Guide,* supplier technical report (304-00286-1292X SMG) - Dow Chemical Company, 1992.

TABLE 263: Air Permeability Through Dow Chemical Saran Wrap Polyvinylidene Chloride Barrier Film.

Material Family	POLYVINYLIDENE CHLORIDE				
Material Supplier/ Trade Name	DOW CHEMICAL SARAN WRAP				
Grade	18	18L	19	28	560
Product Form	MONOLAYER FILM				COEXTRUDED FILM
Features	barrier properties, biaxially oriented, transparent	barrier properties, biaxially oriented, preshrunk, transparent	barrier properties, biaxially oriented, transparent	barrier properties, biaxially oriented, transparent	barrier properties, biaxially oriented, transparent
Applications	chub packaging machines, laminations	laminations	chub packaging machines	chub packaging machines	unit packaging
Reference Number	256	256	256	256	256

MATERIAL CHARACTERISTICS

Sample Thickness	0.019 mm	0.019 mm	0.0254 mm	0.0254 mm	0.152 mm

TEST CONDITIONS

Penetrant	air				
Temperature (°C)	23	23	23	23	23
Relative Humidity (%)	10	10	10	10	10
Test Method	ASTM D1434	ASTM D1434	ASTM D1434	ASTM D1434	ASTM D1434

PERMEABILITY (source document units)

Gas Permeability ($cm^3 \cdot mil/100\ in^2 \cdot day$)	0.36	0.48	0.36	0.5	0.08

PERMEABILITY (normalized units)

Permeability Coefficient ($cm^3 \cdot mm/m^2 \cdot day \cdot atm$)	0.14	0.19	0.14	0.2	0.03

TABLE 264: Oxygen Permeability Through Dow Chemical Saran Polyvinylidene Chloride Barrier Resin.

Material Family	POLYVINYLIDENE CHLORIDE									
Material Supplier/ Trade Name	DOW CHEMICAL SARAN									
Grade	F-239	F-279	F-310	313	416	469	516	525	866	MA 119
Product Form	FILM									
Features	barrier properties									
Manufacturing Method	coating	coating	coating	extrusion	extrusion	extrusion	extrusion	extrusion	extrusion	extrusion
Reference Number	254	254	254	254	254	254	254	254	254	254

MATERIAL CHARACTERISTICS

Specific Gravity				1.69	1.73	1.76	1.73	1.73	1.70	1.78
Molecular Weight						43,000				90,000
Sample Thickness	0.003 mm	0.003 mm	0.003 mm							

MATERIAL COMPOSITION

Chemical Type	vinylidene chloride vinyl chloride copolymer	vinylidene chloride methyl acrylate copolymer

TEST CONDITIONS

Penetrant	oxygen									
Temperature (°C)	23	23	23	23	23	23	23	23	23	23
Relative Humidity (%)	75	75	75	75	75	75	75	75	75	75
Test Method	ASTM D3985	ASTM D3985	ASTM D3985	ASTM D3985	ASTM D3985	ASTM D3985	ASTM D3985	ASTM D3985	ASTM D3985	ASTM D3985

PERMEABILITY (source document units)

Gas Permeability ($cm^3 \cdot mil/100\ in^2 \cdot day$)	0.024	0.02	0.07	1.2	0.08	0.1	0.1	0.1	1.1	0.08
Gas Permeability ($cm^3 \cdot mm/m^2 \cdot day \cdot atm$)	0.009	0.008	0.03	0.47	0.03	0.04	0.04	0.04	0.43	0.03

PERMEABILITY (normalized units)

Permeability Coefficient ($cm^3 \cdot mm/m^2 \cdot day \cdot atm$)	0.01	0.01	0.03	0.47	0.03	0.04	0.04	0.04	0.43	0.03

PVDC

TABLE 265: Oxygen Permeability Through Dow Chemical Saran Wrap Polyvinylidene Chloride Barrier Film.

Material Family	POLYVINYLIDENE CHLORIDE				
Material Supplier/ Trade Name	DOW CHEMICAL SARAN WRAP				
Grade	18	18L	19	28	560
Product Form	MONOLAYER FILM				COEXTRUDED FILM
Features	barrier properties, biaxially oriented, transparent	barrier properties, biaxially oriented, preshrunk, transparent	barrier properties, biaxially oriented, transparent	barrier properties, biaxially oriented, transparent	barrier properties, biaxially oriented, transparent
Applications	chub packaging machines, laminations	laminations	chub packaging machines	chub packaging machines	unit packaging
Reference Number	256	256	256	256	256

MATERIAL CHARACTERISTICS

Sample Thickness	0.019 mm	0.019 mm	0.0254 mm	0.0254 mm	0.152 mm

TEST CONDITIONS

Penetrant	oxygen				
Temperature (°C)	23	23	23	23	23
Relative Humidity (%)	10	10	10	10	10
Test Method	ASTM D3985	ASTM D3985	ASTM D3985	ASTM D3985	ASTM D3985

PERMEABILITY (source document units)

Gas Permeability $(cm^3 \cdot mil/100\ in^2 \cdot day)$	1.2	1.6	1.2	1.8	0.25

PERMEABILITY (normalized units)

Permeability Coefficient $(cm^3 \cdot mm/m^2 \cdot day \cdot atm)$	0.47	0.63	0.47	0.71	0.1

TABLE 266: Oxygen Permeability at Different Temperatures Through Dow Chemical Saran Polyvinylidene Chloride Copolymer.

Material Family	POLYVINYLIDENE CHLORIDE							
Material Supplier/ Grade	DOW CHEMICAL SARAN VC				DOW CHEMICAL SARAN MA			
Features	barrier properties							
Reference Number	264	264	264	264	264	264	264	264

MATERIAL COMPOSITION

Chemical Type	vinylidene chloride vinyl chloride copolymer				vinyl chloride methyl acrylate copolymer			

TEST CONDITIONS

Penetrant	oxygen							
Temperature (°C)	5	23	35	50	5	23	35	50
Relative Humidity (%)	0	0	0	0	0	0	0	0

PERMEABILITY (source document units)

Gas Permeability (cm$^3 \cdot$ mil/100 in$^2 \cdot$ day)	0.012	0.15	0.429	1.9	0.006	0.08	0.288	1.28
Gas Permeability (cm$^3 \cdot$ 25μ/m$^2 \cdot$ day $\cdot$ atm)	0.186	2.325	6.65	29.4	0.093	1.24	4.464	19.8

PERMEABILITY (normalized units)

Permeability Coefficient (cm$^3 \cdot$ mm/m$^2 \cdot$ day $\cdot$ atm)	0.0047	0.06	0.17	0.75	0.0024	0.03	0.11	0.5

TABLE 267: Oxygen Permeability vs. Relative Humidity Through Polyvinylidene Chloride Film.

Material Family	POLYVINYLIDENE CHLORIDE				
Material Supplier/ Grade			DOW CHEMICAL SARAN HB		
Product Form			FILM		
Features			barrier properties	barrier properties	barrier properties
Reference Number	296	296	265	265	265

TEST CONDITIONS

Penetrant	oxygen				
Temperature (°C)	22.8	22.8	20	20	20
Relative Humidity (%)	0	90	65	85	100
Test Method	ASTM D3895	ASTM D3895			

PERMEABILITY (source document units)

Gas Permeability (cm^3 · mil/100 in^2 · day)			0.08	0.08	0.08
Gas Permeability (cm^3 · mil/ 100 in^2 · bar · day)	0.08	0.08			

PERMEABILITY (normalized units)

Permeability Coefficient (cm^3 · mm/m^2 · day · atm)	0.03	0.03	0.03	0.03	0.03

TABLE 268: Oxygen Permeability vs. Relative Humidity Through Dow Chemical Saran Polyvinylidene Chloride Barrier Resin.

Material Family	POLYVINYLIDENE CHLORIDE			
Material Supplier	DOW CHEMICAL			
Trade Name	SARAN MA	SARAN	SARAN MA	SARAN
Product Form	FILM			
Features	barrier properties			
Reference Number	255	255	255	255

MATERIAL COMPOSITION

Chemical Type	vinylidene chloride methyl acrylate copolymer	vinylidene chloride vinyl chloride copolymer	vinylidene chloride methyl acrylate copolymer	vinylidene chloride vinyl chloride copolymer

TEST CONDITIONS

Penetrant	oxygen			
Temperature (°C)	23	23	23	23
Relative Humidity (%)	75	75	100	100

PERMEABILITY (source document units)

Gas Permeability ($cm^3 \cdot mil/100\ in^2 \cdot day$)	0.07	0.15	0.07	0.15
Gas Permeability ($cm^3 \cdot 20\mu/m^2 \cdot day \cdot atm$)	1.2	3	1.2	3

PERMEABILITY (normalized units)

Permeability Coefficient ($cm^3 \cdot mm/m^2 \cdot day \cdot atm$)	0.03	0.06	0.03	0.06

TABLE 269: Gas Permeability and Water Vapor Transmission Through Dow Chemical Saran Polyvinylidene Chloride.

Material Family	POLYVINYLIDENE CHLORIDE			
Material Supplier/ Grade	DOW CHEMICAL SARAN			
Product Form	FILM			
Reference Number	250	250	250	250

TEST CONDITIONS

Penetrant	oxygen	nitrogen	carbon dioxide	water vapor
Temperature (°C)	24	24	24	24

PERMEABILITY (source document units)

Vapor Transmission Rate (g · mil/100 in^2 · day)				0.25 - 0.3
Gas Permeability (cm^3 · mil/100 in^2 · day)	0.8 - 1.1	0.1 - 0.2	4 - 6	

PERMEABILITY (normalized units)

Permeability Coefficient (cm^3 · mm/m^2 · day · atm)	0.31 - 0.43	0.04 - 0.08	1.6 - 2.4	
Vapor Transmission Rate (g · mm/m^2 · day)				0.1 - 0.12

TABLE 270: Gas Permeability of Carbon Dioxide, Nitrogen and Helium and Water Vapor Transmission Through Dow Chemical Saran Polyvinylidene Chloride Copolymer.

Material Family	POLYVINYLIDENE CHLORIDE				
Material Supplier/ Trade Name	DOW CHEMICAL SARAN				
Grade	5253				XU-32024
Features	barrier properties	barrier properties	barrier properties	barrier properties	
Reference Number	264	264	264	264	264

TEST CONDITIONS

Penetrant	carbon dioxide	nitrogen	helium	water vapor	
Temperature (°C)	35	23	35	40	40
Relative Humidity (%)	0	0	0	90	90

PERMEABILITY (source document units)

Gas Permeability ($cm^3 \cdot mil/100\ in^2 \cdot day$)	1.116	0.012	27.3		
Gas Permeability ($cm^3 \cdot 25\mu/m^2 \cdot day \cdot atm$)	17.3	0.186	423.2		
Vapor Transmission Rate ($g \cdot mil/100\ in^2 \cdot day$)				0.22	0.06
Vapor Transmission Rate ($g \cdot 25\mu/m^2 \cdot day$)				3.4	0.93

PERMEABILITY (normalized units)

Permeability Coefficient ($cm^3 \cdot mm/m^2 \cdot day \cdot atm$)	0.44	0.0047	10.8		
Vapor Transmission Rate ($g \cdot mm/m^2 \cdot day$)				0.09	0.02

TABLE 271: Carbon Dioxide Permeability Through Dow Chemical Saran Wrap Polyvinylidene Chloride Barrier Film.

Material Family	POLYVINYLIDENE CHLORIDE				
Material Supplier/ Trade Name	DOW CHEMICAL SARAN WRAP				
Grade	18	18L	19	28	560
Product Form	MONOLAYER FILM				COEXTRUDED FILM
Features	barrier properties, biaxially oriented, transparent	barrier properties, biaxially oriented, preshrunk, transparent	barrier properties, biaxially oriented, transparent	barrier properties, biaxially oriented, transparent	barrier properties, biaxially oriented, transparent
Applications	chub packaging machines, laminations	laminations	chub packaging machines	chub packaging machines	unit packaging
Reference Number	256	256	256	256	256

MATERIAL CHARACTERISTICS

Sample Thickness	0.019 mm	0.019 mm	0.0254 mm	0.0254 mm	0.152 mm

TEST CONDITIONS

Penetrant	carbon dioxide				
Temperature (°C)	23	23	23	23	23
Relative Humidity (%)	10	10	10	10	10
Test Method	ASTM D1434	ASTM D1434	ASTM D1434	ASTM D1434	ASTM D1434

PERMEABILITY (source document units)

Gas Permeability ($cm^3 \cdot mil/100\ in^2 \cdot day$)	5.4	7.2	5.4	8.0	1.2

PERMEABILITY (normalized units)

Permeability Coefficient ($cm^3 \cdot mm/m^2 \cdot day \cdot atm$)	2.1	2.8	2.1	3.2	0.47

TABLE 272: Nitrogen Permeability Through Dow Chemical Saran Wrap Polyvinylidene Chloride Barrier Film.

Material Family	POLYVINYLIDENE CHLORIDE				
Material Supplier/ Trade Name	DOW CHEMICAL SARAN WRAP				
Grade	18	18L	19	28	560
Product Form	MONOLAYER FILM				COEXTRUDED FILM
Features	barrier properties, biaxially oriented, transparent	barrier properties, biaxially oriented, preshrunk, transparent	barrier properties, biaxially oriented, transparent	barrier properties, biaxially oriented, transparent	barrier properties, biaxially oriented, transparent
Applications	chub packaging machines, laminations	laminations	chub packaging machines	chub packaging machines	unit packaging
Reference Number	256	256	256	256	256

MATERIAL CHARACTERISTICS

Sample Thickness	0.019 mm	0.019 mm	0.0254 mm	0.0254 mm	0.152 mm

TEST CONDITIONS

Penetrant	nitrogen				
Temperature (°C)	23	23	23	23	23
Relative Humidity (%)	10	10	10	10	10
Test Method	ASTM D1434	ASTM D1434	ASTM D1434	ASTM D1434	ASTM D1434

PERMEABILITY (source document units)

Gas Permeability ($cm^3 \cdot mil/100\ in^2 \cdot day$)	0.18	0.24	0.18	0.3	0.04

PERMEABILITY (normalized units)

Permeability Coefficient ($cm^3 \cdot mm/m^2 \cdot day \cdot atm$)	0.07	0.09	0.07	0.12	0.02

TABLE 273: Water Vapor Transmission Through Dow Chemical Saran Wrap Polyvinylidene Chloride Barrier Film.

Material Family	POLYVINYLIDENE CHLORIDE				
Material Supplier/ Trade Name	DOW CHEMICAL SARAN WRAP				
Grade	18	18L	19	28	560
Product Form	MONOLAYER FILM				COEXTRUDED FILM
Features	barrier properties, biaxially oriented, transparent	barrier properties, biaxially oriented, preshrunk, transparent	barrier properties, biaxially oriented, transparent	barrier properties, biaxially oriented, transparent	barrier properties, biaxially oriented, transparent
Applications	chub packaging machines, laminations	laminations	chub packaging machines	chub packaging machines	unit packaging
Reference Number	256	256	256	256	256

MATERIAL CHARACTERISTICS

Sample Thickness	0.019 mm	0.019 mm	0.0254 mm	0.0254 mm	0.152 mm

TEST CONDITIONS

Penetrant	water vapor				
Temperature (°C)	38	38	38	38	38
Relative Humidity (%)	90	90	90	90	90
Test Method	Permatran W	Permatran W	Permatran W	Permatran W	Permatran W

PERMEABILITY (source document units)

Vapor Transmission Rate $(g \cdot mil/100 \; in^2 \cdot day)$	0.27	0.3	0.25	0.4	0.04

PERMEABILITY (normalized units)

Vapor Transmission Rate $(g \cdot mm/m^2 \cdot day)$	0.11	0.12	0.1	0.16	0.02

PVDC

TABLE 274: Water Vapor Transmission Through Dow Chemical Saran Polyvinylidene Chloride Barrier Resin.

Material Family	POLYVINYLIDENE CHLORIDE									
Material Supplier/ Trade Name	DOW CHEMICAL SARAN									
Grade	F-239	F-279	F-310	313	416	469	516	525	866	MA 119
Product Form	FILM									
Features	barrier properties									
Manufacturing Method	coating	coating	coating	extrusion	extrusion	extrusion	extrusion	extrusion	extrusion	extrusion
Reference Number	254	254	254	254	254	254	254	254	254	254

MATERIAL CHARACTERISTICS

Specific Gravity				1.69	1.73	1.76	1.73	1.73	1.70	1.78
Molecular Weight						43,000				90,000
Sample Thickness	0.003 mm	0.003 mm	0.003 mm							

MATERIAL COMPOSITION

Chemical Type	vinylidene chloride vinyl chloride copolymer									vinylidene chloride methyl acrylate copolymer

TEST CONDITIONS

Penetrant	water vapor									
Temperature (°C)	30	30	30	30	30	30	30	30	30	30
Test Method	ASTM E96	ASTM E96	ASTM E96	ASTM E96	ASTM E96	ASTM E96	ASTM E96	ASTM E96	ASTM E96	ASTM E96

PERMEABILITY (source document units)

Vapor Transmission Rate ($g/m^2 \cdot day$)	4.65	3.1	13.95	4.19	1.55	2.02	2.02	2.02	3.1	0.78
Vapor Transmission Rate ($g/day \cdot 100\ in^2$)	0.3	0.2	0.9	0.27	0.1	0.13	0.13	0.13	0.2	0.05

PERMEABILITY (normalized units)

Vapor Transmission Rate ($g \cdot mm/m^2 \cdot day$)	0.01	0.01	0.04	not applicable without thickness						

TABLE 275: d-Limonene (flavor component) Permeability Through Dow Chemical Saran Polyvinylidene Chloride Barrier Resin.

Material Family	POLYVINYLIDENE CHLORIDE	
Material Supplier/ Grade	**DOW CHEMICAL SARAN MA**	**DOW CHEMICAL SARAN XU-32009**
Product Form	**FILM**	
Features	barrier properties	barrier properties, developmental material
Reference Number	255	255

MATERIAL COMPOSITION

Chemical Type	vinylidene chloride methyl acrylate copolymer	vinylidene chloride vinyl chloride copolymer

TEST CONDITIONS

Penetrant	d-limonene	
Temperature (°C)	25	25
Relative Humidity (%)	dry	dry

PERMEABILITY (source document units)

Vapor Transmission (10^{-20} kg $\cdot$ m/m^2 $\cdot$ sec $\cdot$ Pa)	100	180

PERMEABILITY (normalized units)

Vapor Transmission Rate (g $\cdot$ mm/m^2 $\cdot$ day)	0.0088	0.016

TABLE 276: Film Properties of Dow Chemical Saran Polyvinylidene Chloride Barrier Resin.

Material Family	POLYVINYLIDENE CHLORIDE		
Material Supplier/ Grade	**DOW CHEMICAL SARAN F-239**	**DOW CHEMICAL SARAN F-279**	**DOW CHEMICAL SARAN F-310**
Product Form	**FILM**		
Features	barrier properties	barrier properties	barrier properties
Manufacturing Method	coating	coating	coating
Reference Number	254	254	254

MATERIAL CHARACTERISTICS

Sample Thickness	0.003 mm	0.003 mm	0.003 mm

MATERIAL COMPOSITION

Chemical Type	vinylidene chloride vinyl chloride copolymer		

THERMAL PROPERTIES

Minimum Heat Seal Temp. (°C)	100	105	130

TABLE 277: Film Properties of Dow Chemical Saran Wrap Polyvinylidene Chloride Barrier Film.

Material Family	POLYVINYLIDENE CHLORIDE				
Material Supplier	DOW CHEMICAL				
Grade	SARAN WRAP 18	SARAN WRAP 18L	SARAN WRAP 19	SARAN WRAP 28	SARAN WRAP 560
Product Form	MONOLAYER FILM				COEXTRUDED FILM
Features	barrier properties, biaxially oriented, transparent	barrier properties, biaxially oriented, preshrunk, transparent	barrier properties, biaxially oriented, transparent	barrier properties, biaxially oriented, transparent	barrier properties, biaxially oriented, transparent
Applications	chub packaging machines, laminations	laminations	chub packaging machines	chub packaging machines	unit packaging
Reference Number	256	256	256	256	256

MATERIAL CHARACTERISTICS

Sample Thickness	0.019 mm	0.019 mm	0.0254 mm	0.0254 mm	0.152 mm

PHYSICAL PROPERTIES

Yield (in^2/lb)	21,700	21,700	16,200	16,600	2780

MECHANICAL PROPERTIES

Secant Modulus @ 2% Elongation - machine dir. (MPa)	583 {r}	469 {r}	597 {r}	427 {r}	470 {r}
Secant Modulus @ 2% Elongation - transverse dir. (MPa)	496 {r}	372 {r}	540 {r}	359 {r}	463 {r}
Ultimate Tensile Strength - machine direction (MPa)	69 {r}	97 {r}	97 {r}	96 {r}	97 {r}
Ultimate Tensile Strength - transverse direction (MPa)	110 {r}	124 {r}	124 {r}	116 {r}	97 {r}
Ultimate Elongation - machine direction (%)	70 {r}	80 {r}	90 {r}	80 {r}	95 {r}
Ultimate Elongation - transverse direction (%)	50 {r}	60 {r}	60 {r}	80 {r}	60 {r}
Coefficient Of Friction - Kinetic, film to metal	0.27 {l}	0.3 {l}	0.24 {l}	0.28 {l}	0.24 {l}

OPTICAL PROPERTIES

Haze (%)	7 {c}	5 {c}	6 {c}	7 {c}	2 {c}
Gloss @ 45°	99 {m}	100 {m}	96 {m}	92 {m}	103 {m}
Clarity (%)	56 {s}	63 {s}	60 {s}	52 {s}	75 {s}

PERFORMANCE PROPERTIES

Unrestrained Shrink - 100°C Air, machine direction (%)	16	12	16	20	8
Unrestrained Shrink - 100°C Air, transverse direction (%)	9	4	9	10	6

GRAPH 62: Oxygen Permeability vs. Temperature through Polyvinylidene Chloride.

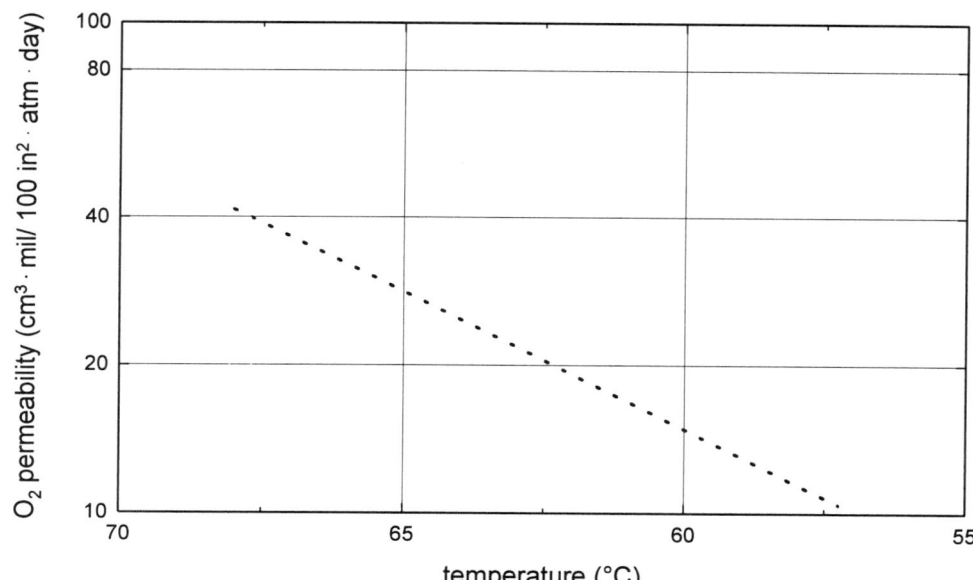

....	Dow Saran PVDC (film); penetrant: O_2
Reference No.	250

GRAPH 63: Oxygen Permeability vs. Relative Humidity through Polyvinylidene Chloride.

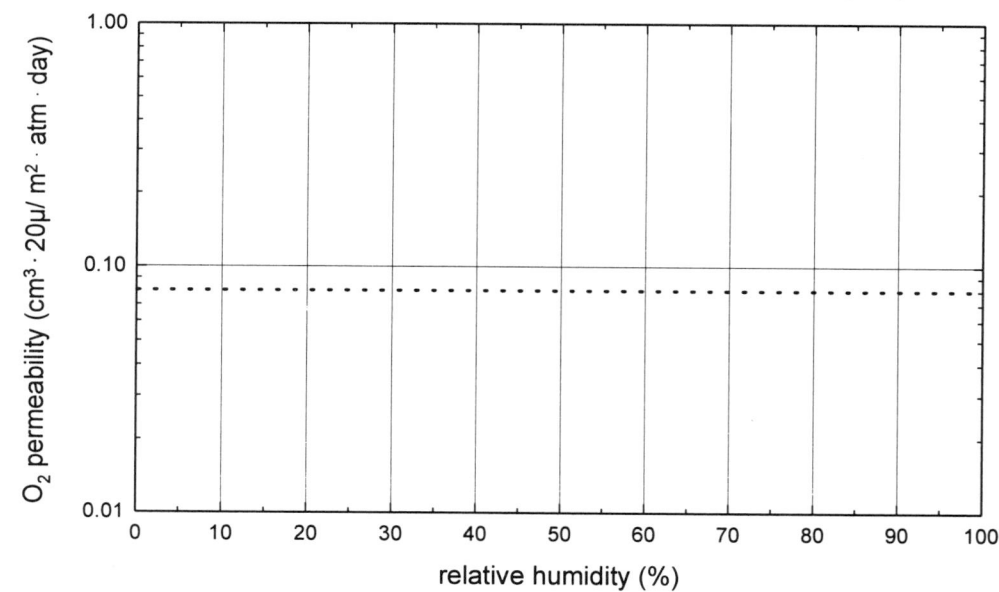

....	Dow Saran HB PVDC (barrier prop.; film); penetrant: O_2
Reference No.	265

PVDC

GRAPH 64: Carbon Dioxide and Oxygen Permeability vs. Relative Humidity through Polyvinylidene Chloride.

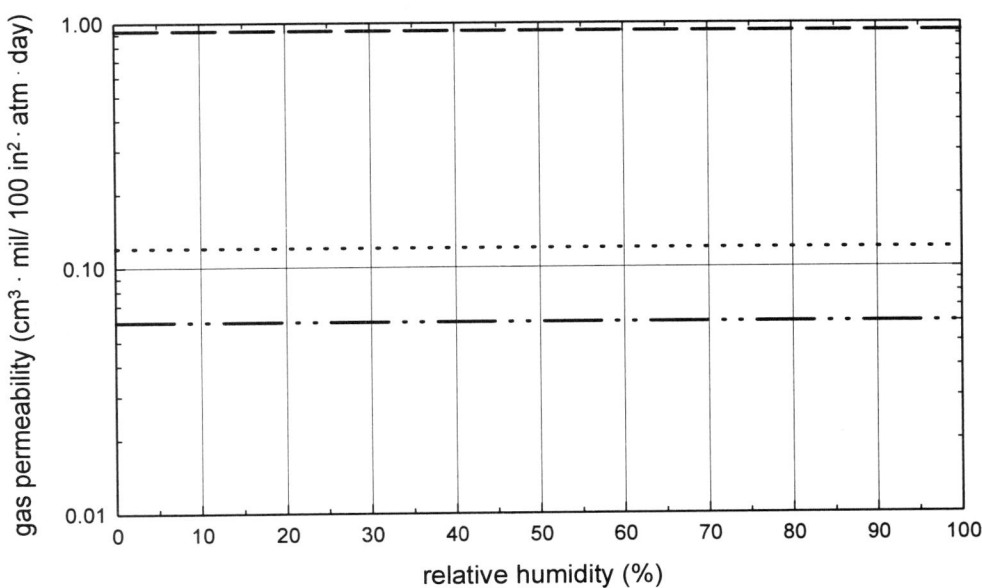

....	Dow Saran VC PVDC (VDC vinyl chloride; barrier prop.); penetrant: O_2
— ·· — ·· —	Dow Saran MA PVDC (VDC methyl acrylate; barrier prop.); penetrant: O_2
— — —	Dow Saran VC PVDC (VDC vinyl chloride; barrier prop.); penetrant: CO_2
Reference No.	264

GRAPH 65: Oxygen Transmission Rate vs. Time After Retort through Polyvinylidene Chloride.

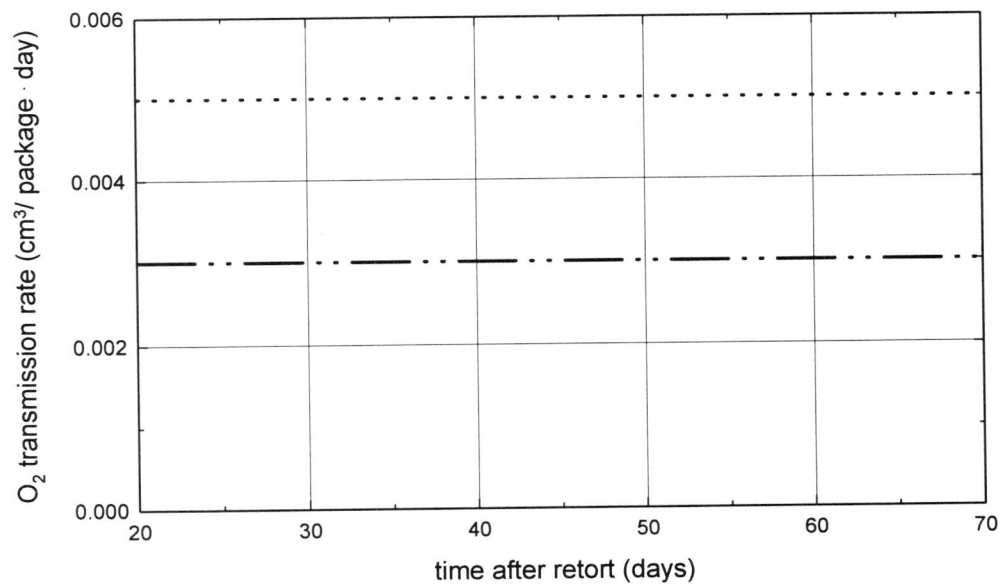

....	Dow Saran PVDC (VDC vinyl chloride; barrier prop., 0.04 mm thick; film); penetrant: O_2; 80% internal RH; 60% external RH; air 23°C; retorted at 121°C for 60 minutes
— ·· — ·· —	Dow Saran MA PVDC (VDC methyl acrylate; barrier prop., 0.04 mm thick; film); penetrant: O_2; 80% internal RH; 60% external RH; air 23°C; retorted at 121°C for 60 minutes
Reference No.	255

GRAPH 66: Oxygen Uptake vs. Time After Retort through Polyvinylidene Chloride.

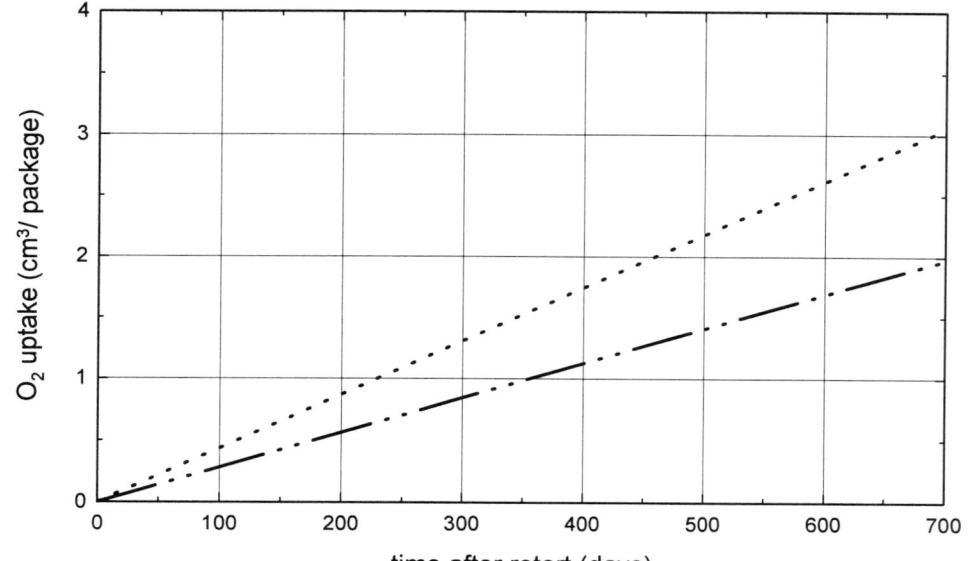

	Dow Saran PVDC (VDC vinyl chloride; barrier prop., 0.04 mm thick; film); penetrant: O_2; 80% internal RH; 60% external RH; air 23°C; retorted at 121°C for 60 minutes
	Dow Saran MA PVDC (VDC methyl acrylate; barrier prop., 0.04 mm thick; film); penetrant: O_2; 80% internal RH; 60% external RH; air 23°C; retorted at 121°C for 60 minutes
Reference No.	255

Chapter 60

Polyethylene/Polystyrene Alloy

Film Properties and Applications

BASF AG: Styroblend WS

The Styroblend WS range consists of polystyrene-polyethylene blends which combine the characteristic properties of the two individual constituents.

Styroblend has the following properties:
- great rigidity (due to PS)
- excellent toughness (due to PE)
- very good stress cracking resistance (due to PE)
- low water vapor permeability (due to PE)
- easy processing (due to PS)

Styroblend WS can be coextruded with polystyrene and polyethylene as well as with many other plastics including PETG, without additional coupling agents The resultant multilayer films combine the excellent properties of both component materials.

Reference: *Topics In Chemistry - BASF Plastics Research And Development,* supplier technical report - BASF Aktiengesellschaft, 1992.

<u>TABLE 278:</u> Water Vapor Transmission Through BASF AG Styroblend WS Polyethylene Polystyrene Blend.

Material Family	POLYETHYLENE/POLYSTYRENE ALLOY				
Material Supplier/ Trade Name	BASF AG STYROBLEND WS				
Grade	**KR 2773**	**KR 2774**	**KR 2775**	**KR 2776**	**KR 2777**
Reference Number	182	182	182	182	182

TEST CONDITIONS

Penetrant	water vapor

PERMEABILITY (source document units)

Vapor Transmission Rate (g/m^2 · day)	3	3	3	3	5

PERMEABILITY (normalized units)

Vapor Transmission Rate (g · mm/m^2 · day)	not applicable without thickness

Chapter 61

Co-Continuous Lamellae Multilayer Structure

Barrier Performance

LIM Structure   (features: barrier properties; manufacturing method: lamellar injection molding (LIM))

Lamellar Injection Molding (LIM) technology enables low levels (2 to 20 percent) of barrier polymer to be introduced and maintained as co-continuous lamellae in molded articles. A 300 fold reduction in oxygen permeability compared to conventional blends with minor amounts of barrier polymer (10 percent) has been achieved. Depending on the polymer used with LIM, improvements of between 10 and 100 times the oxygen permeability of monolayer PET have been achieved.

Reference: *Introducing Lamellar Injection Molding Technology - "The LIM Advantage" (Licensing Bulletin),* supplier marketing literature (304-00383-493 SMG) - Dow Chemical Company, 1993.

__TABLE 279:__ **Oxygen Permeability Through Multilayer Constructions Made With Lamellar Injection Molding.**

Material Family	CO-CONTINUOUS LAMELLAE MULTILAYER STRUCTURE					
Features	barrier properties, high impact, hot fill	barrier properties, high impact, retort	barrier properties	barrier properties, enhanced clarity, hot fill	barrier properties, compatibilizer not required, high impact, hot fill, transparent	barrier properties, compatibilizer not required, enhanced clarity, high impact
Manufacturing Method	lamellar injection molding (LIM)					
Reference Number	253	253	253	253	253	253

MATERIAL COMPOSITION

Host Polymer	high density polyehtylene	polypropylene	polyethylene terephthalate	polystyrene	impact polystyrene	thermoplastic polyurethane
Barrier Polymer	10% ethylene vinyl alcohol copolymer	10% ethylene vinyl alcohol copolymer	10% ethylene vinyl alcohol copolymer	10% nylon MXD6 (aromatic)	10% nylon MXD6 (aromatic)	10% nylon MXD6 (aromatic)
Compatibilizer	PE-g-maleic anhydride	PP-g-maleic anhydride	ethylene vinyl acetate	styrene maleic anhydride copolymer		

TEST CONDITIONS

Penetrant	oxygen					

PERMEABILITY (source document units)

Gas Permeability (cm^3 · mil/100 in^2 · day)	0.6	0.6	0.6	1.0	1.0	0.4

PERMEABILITY (normalized units)

Permeability Coefficient (cm^3 · mm/m^2 · day · atm)	0.24	0.24	0.24	0.39	0.39	0.16

TABLE 280: Oxygen Permeability Through Aromatic Nylon Multilayer Constructions Made With Lamellar Injection Molding.

Material Family	CO-CONTINUOUS LAMELLAE MULTILAYER STRUCTURE			
Features	barrier properties, enhanced clarity, hot fill			
Manufacturing Method	lamellar injection molding (LIM)			
Reference Number	253	253	253	253

MATERIAL COMPOSITION

Host Polymer	styrene-acrylonitrile copolymer			
Barrier Polymer	20% nylon MXD6 (aromatic)	10% nylon MXD6 (aromatic)	5% nylon MXD6 (aromatic)	2.5% nylon MXD6 (aromatic)
Compatibilizer	styrene maleic anhydride copolymer			

TEST CONDITIONS

Penetrant	oxygen			

PERMEABILITY (source document units)

Gas Permeability ($cm^3 \cdot mil/100\ in^2 \cdot day$)	non-detectable	1.0	2.3	4.7

PERMEABILITY (normalized units)

Permeability Coefficient ($cm^3 \cdot mm/m^2 \cdot day \cdot atm$)	non-detectable	0.39	0.91	1.85

Chapter 62

Laminar Multilayer Structure

Barrier Performance

DuPont: Selar RB/ HDPE (applications: solvent packaging; barrier polymer: 18% Selar RB; features: barrier properties; manufacturing method: extrusion blow molding (laminar technology); product form: container (one pint bottle))

Barrier to permeation is most effective with halogenated, aromatic and aliphatic hydrocarbons. The platelets of nylon copolymer (Selar RB) create a barrier to these chemicals up to 140 times more effective than that of a container made from HDPE alone, when Selar RB resin is processed under controlled mixing conditions. In addition, a tenfold reduction in migration is also attainable, particularly with products containing free disolved oxygen or oxygenated solvents such as ketones, esters and ethers.

Although improvements over conventional HDPE for oxygen-containing solvents are not of the same magnitude as those above, the weight loss may be acceptable in many packaging applications. Also, it's important to note that Selar RB provides a better barrier to halogenated and oxygen-containing solvents than surface-treated HDPE. With small, highly polar solvent molecules, such as water and some alcohols, neither Selar RB resin nor surface-treated polyethylene offer significant improvement over an all-polyolefin container.

Barrier properties of containers can be functionally and economically adjusted to suit individual barrier requirements by varying the concentration of Selar RB from about 4 to 18 percent.

Reference: Fetell, Arthur I., *Polyolefin + Nylon Yields Barrier Container*, Food & Drug Packaging, trade journal - Edgell Communications, 1986.

Packaging Applications and Properties

DuPont: Selar RB/ Polyolefin (applications: solvent packaging; features: barrier properties; manufacturing method: extrusion blow molding (laminar technology); product form: container, bottles)

SELAR RB barrier resins are specially modified resin blends available exclusively from Du Pont. The resins can be dry blended with a polyolefin resin (HDPE, LDPE, PP, etc.) and then extrusion blow molded on conventional single extruder extrusion blow molding machines by using Du Pont's patented Laminar Technology. In this unique one-step process, controlled mixing and shear allow the SELAR RB barrier resin to form dozens of overlapping, discontinuous, and elongated platelets. Packages made with this technology are capable of holding most common solvents (hydrocarbons, halogenated, and oxygenated solvents). The newly developed EVOH-based SELAR RB grades provide barrier not only to solvents but also to oxygen, aromas, and flavors. Depending on the concentration of SELAR RB resin added to the blend, the solvent barrier of laminar containers can be up to 300 times that of HDPE containers and the oxygen barrier can be up to 100 times that of HDPE containers.

Reference: *Selar RB Barrier Resins - Resin Blend Technical Information*, supplier technical report (H-42016) - DuPont Company, 1992.

Product Summary

DuPont: Selar RB/ Polyolefin Laminar Structure (applications: solvent packaging; features: barrier properties; manufacturing method: extrusion blow molding (laminar technology); product form: container, bottles)

Laminar bottles can be used to package solvent-based products such as household chemicals, industrial chemicals, cleaning solvents, adhesives, wood preservatives, automotive additives, photochemicals, and agricultural chemicals. Laminar bottles are an attractive alternative to glass, metal, fluorinated containers, multilayer bottles, and other single layer barrier resin containers such as Barex (polyacrylonitrile), polyester, and PVC. Glass and metal have excellent barrier properties, but have an unfortunate tendency to break, leak, or rust. The heavy weight of glass containers drives up the shipping costs and good quality metal containers are usually expensive. Both materials are not well suited to forming complicated shapes with handles and spouts. SELAR RB laminar containers are lightweight, impact resistant, and can be formed into a wide variety of complex shapes.

Laminar bottles have a number of advantages over fluorinated containers. Unlike fluorinated containers, the barrier level of laminar containers can be adjusted to the exact level needed for each application, making the package more cost effective. The barrier provided by laminar containers is permanent, i.e., it cannot be scratched off or damaged in any way. SELAR RB laminar containers have better barrier consistency than fluorinated containers. Laminar containers can be translucent or pigmented to any color, whereas fluorination is incompatible with some colors.

SELAR RB laminar containers have better impact resistance than polyester, Barex, and PVC, and are cheaper than Barex. SELAR RB laminar containers also provide a superior barrier to a wider range of solvents than polyester, Barex, PVC, and fluorinated containers. Containers made from SELAR RB barrier resins are most effective with products containing nonpolar hydrocarbons, halogenated solvents and oxygenated solvents such as ketones, esters, and ethers. SELAR RB laminar containers also provide more flexibility in shapes and sizes of containers that can be produced. In addition, the recently developed EVOH-based SELAR RB grades offer good flavor and aroma barrier and better oxygen barrier than PVC and polyester.

Reference: *Selar RB Barrier Resins - Resin Blend Technical Information,* supplier technical report (H-42016) - DuPont Company, 1992.

DuPont: Selar RB/ Polyolefin Nylon Copolymer/Polyolefin Laminar Stucture (applications: solvent packaging; features: barrier properties; manufacturing method: extrusion blow molding (laminar technology) product form: container)

Selar RB is a special modified nylon copolymer that can be dry-blended with a polyolefin resin such as high-density polyethylene (HDPE) and then extrusion-blow molded on conventional single-extruder machines.

In this unique one-step process, controlled mixing allows the nylon copolymer pellets to flatten and elongate into many discontinuous, but overlapping platelets that can attain sizes as large as 9 square inches.

A wide variety of matrix polyolefins can be used in this process. The packing of the nylon platelets and thus the barrier properties of the container can be varied by changing the proportion of nylon in the blend. The resultant containers, which can be molded with handles and other special features, can also be pigmented and decorated. In addition, the scrap can be ground and re-molded without loss of barrier effectiveness.

Reference: Fetell, Arthur I., *Polyolefin + Nylon Yields Barrier Container,* Food & Drug Packaging, trade journal - Edgell Communications, 1986.

TABLE 281: Xylene and Oxygen Permeability Through Nylon/ HDPE Laminar Structure.

Material Family	NYLON/HDPE LAMINAR STRUCTURE			
Material Supplier/ Grade	DUPONT SELAR RB 215/ HDPE		DUPONT SELAR RB 300/ HDPE	
Features	barrier properties		barrier properties	
Manufacturing Method	extrusion blow molding (laminar technology)		extrusion blow molding (laminar technology)	
Reference Number	293	293	293	293

MATERIAL COMPOSITION

Barrier Polymer	10% Selar RB 215		10% Selar RB 300	

TEST CONDITIONS

Penetrant	xylene	oxygen	xylene	oxygen
Temperature (°C)	60	23	60	23
Exposure Time (days)	14		14	
Relative Humidity (%)		75		75

PERMEABILITY (source document units)

	xylene	oxygen	xylene	oxygen
Vapor Transmission Rate (g · mil/100 in^2 · day)	1		9	
Gas Permeability (cm^3 · mil/100 in^2 · day)		38.2		34.0

PERMEABILITY (normalized units)

	xylene	oxygen	xylene	oxygen
Permeability Coefficient (cm^3 · mm/m^2 · day · atm)		15.0		13.4
Vapor Transmission Rate (g · mm/m^2 · day)	0.39		3.5	

TABLE 282: Xylene and Oxygen Permeability Through EVOH/ HDPE, Nylon/ LDPE and Nylon/ PP Laminar Structures.

Material Family	EVAL/HDPE LAMINAR STRUCTURE		NYLON/LDPE LAMINAR STRUCTURE		NYLON/PP LAMINAR STRUCTURE	
Material Supplier/ Grade	DUPONT SELAR RB 421/ HDPE		DUPONT SELAR RB 215/ LDPE		DUPONT SELAR RB 421/ PP	
Features	barrier properties		barrier properties		barrier properties	
Manufacturing Method	extrusion blow molding (laminar technology)		extrusion blow molding (laminar technology)		extrusion blow molding (laminar technology)	
Reference Number	293	293	293	293	293	293

MATERIAL COMPOSITION

Barrier Polymer	15% Selar RB 421		15% Selar RB 215		10% Selar RB 240	

TEST CONDITIONS

Penetrant	xylene	oxygen	xylene	oxygen	xylene	oxygen
Temperature (°C)	60	23	60	23	60	23
Exposure Time (days)	14		14		14	
Relative Humidity (%)		75		75		75

PERMEABILITY (source document units)

Vapor Transmission Rate (g · mil/100 in² · day)	2		12		18	
Gas Permeability (cm³ · mil/100 in² · day)		1.07		90		36

PERMEABILITY (normalized units)

Permeability Coefficient (cm³ · mm/m² · day · atm)		0.42		35.4		14.2
Vapor Transmission Rate (g · mm/m² · day)	0.79		4.7		7.1	

TABLE 283: Ethyl Acetate, Isopropyl Acetate, Acetone, Butyl Acetate, Toluene, Xylene, Methyl Isobutyl Ketone and Methyl Ethyl Ketone Permeability Through Nylon/ Polyolefin Laminar Structure.

Material Family	NYLON/ POLYOLEFIN LAMINAR STRUCTURE							
Material Supplier/ Grade	DUPONT SELAR RB/ POLYOLEFIN							
Product Form	BOTTLES							
Features	barrier properties							
Manufacturing Method	extrusion blow molding (laminar technology)							
Reference Number	293	293	293	293	293	293	293	293

MATERIAL COMPOSITION

Barrier Polymer	8% Selar RB	8% Selar RB	8% Selar RB	8% Selar RB	8% Selar RB	8% Selar RB	8% Selar RB	8% Selar RB

TEST CONDITIONS

Penetrant	ethyl acetate	isopropyl acetate	acetone	butyl acetate	toluene	xylene	methyl isobutyl ketone	methyl ethyl ketone
Temperature (°C)	50	50	23	50	50	50	50	50
Exposure Time (days)	28	28	180	28	28	28	28	28

PERMEABILITY (source document units)

Penetrant Weight Loss (%)	0.42	0.03	0.48	0.08	0.3	0.12	0.04	0.97

TABLE 284: Cyclohexanone, Chlorobenzene, Hexane, Butyl Alcohol, Trichloroethane, Methyl Salicylate and Tetrahydrofuran Permeability Through Nylon/ Polyolefin Laminar Structure.

Material Family	NYLON/ POLYOLEFIN LAMINAR STRUCTURE						
Material Supplier/ Grade	DUPONT SELAR RB/ POLYOLEFIN						
Product Form	BOTTLES						
Features	barrier properties						
Manufacturing Method	extrusion blow molding (laminar technology)						
Reference Number	293	293	293	293	293	293	293

MATERIAL COMPOSITION

Barrier Polymer	8% Selar RB

TEST CONDITIONS

Penetrant	cyclohexanone	chlorobenzene	hexane	butyl alcohol	trichloroethene	methyl salicylate	tetrahydrofuran
Temperature (°C)	50	50	50	50	50	50	23
Exposure Time (days)	28	28	28	28	28	28	180

PERMEABILITY (source document units)

Penetrant Weight Loss (%)	0.02	0.77	0.17	0.06	0.36	0.01	0.44

TABLE 285: Mineral Spirits, Turpentine, STP Gas Treatment, Paint Thinner, Charcoal Starter and Naphtha Permeability Through Nylon/ Polyolefin Laminar Structure.

Material Family	NYLON/POLYOLEFIN LAMINAR STRUCTURE					
Material Supplier/ Grade	DUPONT SELAR RB/ POLYOLEFIN					
Product Form	BOTTLES					
Features	barrier properties					
Manufacturing Method	extrusion blow molding (laminar technology)					
Reference Number	293	293	293	293	293	293

MATERIAL COMPOSITION

Barrier Polymer	8% Selar RB					

TEST CONDITIONS

Penetrant	mineral spirits	turpentine	STP gas treatment	paint thinner	charcoal starter	naphtha
Temperature (°C)	50	50	50	50	50	50
Exposure Time (days)	28	28	28	28	28	28

PERMEABILITY (source document units)

Penetrant Weight Loss (%)	0.01	0.03	0.07	0.06	0.04	0.03

TABLE 286: Kerosine, d-Limonene, Motor Oil, Pine Oil, Diesel Fuel Conditioner and Gas Additive Permeability Through Nylon/ Polyolefin Laminar Structure.

Material Family	NYLON/POLYOLEFIN LAMINAR STRUCTURE					
Material Supplier/ Grade	DUPONT SELAR RB/ POLYOLEFIN					
Product Form	BOTTLES					
Features	barrier properties					
Manufacturing Method	extrusion blow molding (laminar technology)					
Reference Number	293	293	293	293	293	293

MATERIAL COMPOSITION

Barrier Polymer	8% Selar RB					

TEST CONDITIONS

Penetrant	kerosine	d-limonene	motor oils	pine oil	diesel fuel conditioner	gas additive
Penetrant Note			2 cycle	cleaner		Brakleen
Temperature (°C)	50	50	50	50	50	50
Exposure Time (days)	28	28	28	28	28	28

PERMEABILITY (source document units)

Penetrant Weight Loss (%)	0.01	0.05	0.02	0.27	0.03	0.03

TABLE 287: Xylene, Propyl Alcohol and Methyl Alcohol Permeability Through Nylon/ Polyolefin Laminar Structure.

Material Family	NYLON/POLYOLEFIN LAMINAR STRUCTURE								
Material Supplier/ Grade	DUPONT SELAR RB/ POLYOLEFIN								
Product Form	BOTTLE (1 LITER)								
Features	barrier properties, laminar technology								
Manufacturing Method	extrusion blow molding (laminar technology)								
Reference Number	293	293	293	293	293	293	293	293	293

MATERIAL COMPOSITION

Barrier Polymer	8% Selar RB

TEST CONDITIONS

Penetrant	xylene			propyl alchohol		xylene		methyl alcohol	
Penetrant Note		with 25% propyl alcohol	with 50% propyl alcohol	with 25% xylene		with 25% methyl alcohol	with 50% methyl alcohol	with 25% xylene	
Temperature (°C)	50	50	50	50	50	23	23	23	23
Exposure Time (days)	28	28	28	28	28	180	180	180	180

PERMEABILITY (source document units)

Penetrant Weight Loss (%)	0.12	2.84	2.61	1.44	0.14	14.10	10.60	3.56	0.28

TABLE 288: Xylene, o-Dichlorobenzene, Toluene, Methyl Alcohol and Water Permeability Through Nylon/ HDPE Laminar Structure.

Material Family	NYLON/HDPE LAMINAR STRUCTURE				
Material Supplier/ Grade	DUPONT SELAR RB/ HDPE				
Product Form	CONTAINER (ONE PINT BOTTLE)				
Features	barrier properties				
Applications	solvent packaging				
Manufacturing Method	extrusion blow molding (laminar technology)				
Reference Number	291	291	291	291	291

MATERIAL COMPOSITION

Barrier Polymer	18% Selar RB

TEST CONDITIONS

Penetrant	xylene	o-dichlorobenzene	toluene	methyl alcohol	water
Penetrant Note	aromatic	halogenated	aromatic	oxygen containing	
Temperature (°C)	48.9	48.9	48.9	48.9	48.9
Exposure Time (days)	28	28	28	28	28

PERMEABILITY (source document units)

Penetrant Weight Loss (%)	0.3 (80% relative improvement vs. HDPE)	0.3 (70% relative improvement vs. HDPE)	0.43 (60% relative improvement vs. HDPE)	0.75 (1% relative improvement vs. HDPE)	0.08 (1% relative improvement vs. HDPE)

Laminar Structure

TABLE 289: Naphtha, Trichloroethane, Heptane, Ethyl Acetate and Methyl Ethyl Ketone Permeability Through Nylon/ HDPE Laminar Structure.

Material Family	NYLON/HDPE LAMINAR STRUCTURE				
Material Supplier/ Grade	DUPONT SELAR RB/ HDPE				
Product Form	CONTAINER (ONE PINT BOTTLE)				
Features	barrier properties				
Applications	solvent packaging				
Manufacturing Method	extrusion blow molding (laminar technology)				
Reference Number	291	291	291	291	291

MATERIAL COMPOSITION

Barrier Polymer	18% Selar RB

TEST CONDITIONS

Penetrant	naphtha	trichloroethane	heptane	ethyl acetate	methyl ethyl ketone
Penetrant Note	aromatic	halogenated	aliphatic	oxygen containing	
Temperature (°C)	48.9	48.9	48.9	48.9	48.9
Exposure Time (days)	28	28	28	28	28

PERMEABILITY (source document units)

Penetrant Weight Loss (%)	0.15 (140% relative improvement vs. HDPE)	0.18 (130% relative improvement vs. HDPE)	0.25 (100% relative improvement vs. HDPE)	0.75 (11% relative improvement vs. HDPE)	0.65 (7% relative improvement vs. HDPE)

TABLE 290: Xylene, Propyl Alcohol and Methyl Alcohol Permeability Through EVOH/ Polyolefin Laminar Structures.

Material Family	EVAL/POLYOLEFIN LAMINAR STRUCTURE								
Material Supplier/ Grade	DUPONT SELAR RB/ POLYOLEFIN								
Product Form	BOTTLE (1 LITER)								
Features	barrier properties, laminar technology								
Manufacturing Method	extrusion blow molding (laminar technology)								
Reference Number	293	293	293	293	293	293	293	293	293

MATERIAL COMPOSITION

Barrier Polymer	15% Selar RB

TEST CONDITIONS

Penetrant	xylene			propyl alchohol		xylene		methyl alcohol	
Penetrant Note		with 25% propyl alcohol	with 50% propyl alcohol	with 25% xylene		with 25% methyl alcohol	with 50% methyl alcohol	with 25% xylene	
Temperature (°C)	50	50	50	50	50	23	23	23	23
Exposure Time (days)	28	28	28	28	28	180	180	180	180

PERMEABILITY (source document units)

Penetrant Weight Loss (%)	0.11	0.16	0.11	0.02	0.01	2.27	1.51	0.74	0.3

GRAPH 67: Xylene Permeability through Various Bottle Sizes vs. Percentage (%) Weight Concentration of Generic Laminar Multilayer Structure.

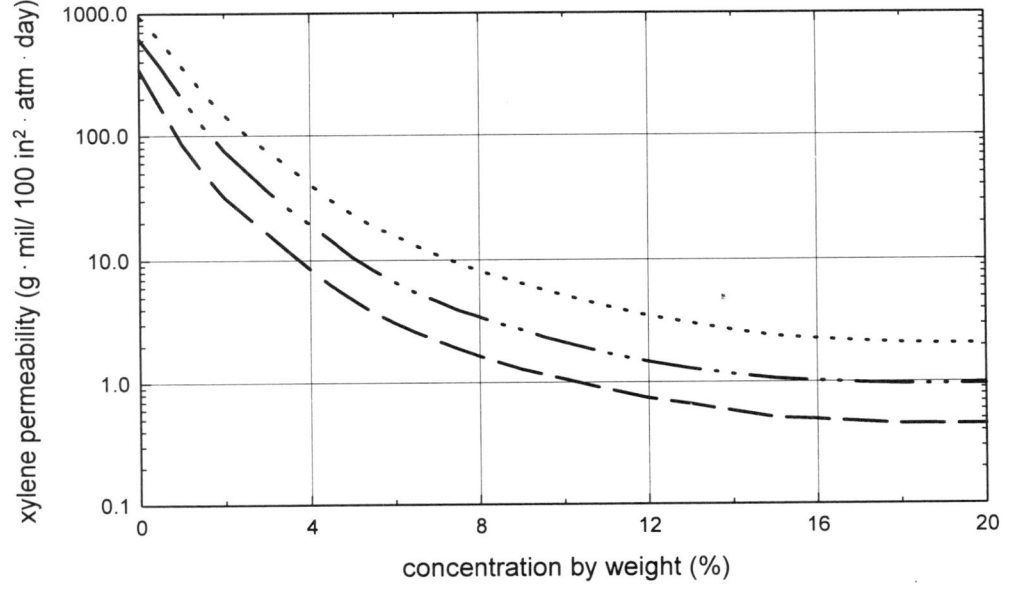

Laminar Structure

Chapter 63

Multilayer Films with Ethylene-Vinyl Alcohol Barrier

TABLE 291: Oxygen Permeability vs. Relative Humidity Through PP/ EVOH/ PP Multilayer Film.

Material Family	PP/ EVAL/ PP FILM									
Reference Number	264	264	264	264	264	264	265	265	265	265

MATERIAL CHARACTERISTICS

Sample Thickness	0.15 mm/ 0.025 mm/ 0.6 mm	0.15 mm/ 0.025 mm/ 0.6 mm	0.6 mm/ 0.025 mm/ 0.15 mm	0.15 mm/ 0.025 mm/ 0.6 mm	0.15 mm/ 0.025 mm/ 0.6 mm	0.6 mm/ 0.025 mm/ 0.15 mm	0.02 mm/ 0.015 mm/ 0.051 mm	0.02 mm/ 0.015 mm/ 0.051 mm	0.02 mm/ 0.015 mm/ 0.051 mm	0.02 mm/ 0.015 mm/ 0.051 mm

MATERIAL COMPOSITION

Barrier Layer	EVAL EP-F (EVOH)	EVAL EP-E (EVOH)	EVAL EP-F (EVOH)	EVAL EP-F (EVOH)	EVAL EP-E (EVOH)	EVAL EP-F (EVOH)	EVAL EF-F (EVOH)	EVAL EF-F (EVOH)	EVAL EF-F (EVOH)	EVAL EF-F (EVOH)
Inside Layer	polypropylene						coated polypropylene			
Outside Layer	polypropylene						oriented polypropylene			

TEST CONDITIONS

Penetrant	oxygen									
Temperature (°C)	20	20	20	20	20	20	20	20	20	20
Relative Humidity - Outside (%)	65	65	65	75	75	75	65	80	65	80
Relative Humidity - Inside (%)	100 (wet)	100 (wet)	10 (wet)	100 (wet)	100 (wet)	10 (wet)	100	100	10	10
Relative Humidity - Barrier, calculated (%)	72	72.2	24.2	80	80.1	25.9	80	41	89	49

PERMEABILITY (source document units)

Gas Permeability (cm^3/100 in^2 · day · atm)	0.028	0.097	0.011	0.048	0.13	0.011	0.08	0.19	0.02	0.02
Gas Permeability (cm^3/m^2 · day)							1.2	3	0.3	0.3

PERMEABILITY (normalized units)

Permeability Coefficient (cm^3 · mm/m^2 · day · atm)	0.34	1.17	0.16	0.58	1.56	0.16	0.11	0.25	0.03	0.03

TABLE 292: Oxygen Permeability vs. Relative Humidity Through PET/ EVOH/ PP and PC/ EVOH/ PP Multilayer Films.

Material Family	PET/ EVAL/ PP FILM		PC/ EVAL/ PP FILM			
Reference Number	264	264	264	264	264	264

MATERIAL CHARACTERISTICS

Sample Thickness	0.15 mm/ 0.025 mm/ 0.6 mm		0.15 mm/ 0.025 mm/ 0.6 mm			

MATERIAL COMPOSITION

Barrier Layer	EVAL EP-F (EVOH)	EVAL EP-F (EVOH)	EVAL EP-F (EVOH)	EVAL EP-E (EVOH)	EVAL EP-F (EVOH)	EVAL EP-E (EVOH)
Inside Layer	polypropylene		polypropylene			
Outside Layer	polyester PET		polycarbonate			

TEST CONDITIONS

Penetrant	oxygen					
Temperature (°C)	20	20	20	20	20	20
Relative Humidity - Outside (%)	65	75	65	65	75	75
Relative Humidity - Inside (%)	100 (wet)	100 (wet)	100 (wet)	100 (wet)	100 (wet)	100 (wet)
Relative Humidity - Barrier, calculated (%)	69.2	78	65.7	65.9	75.5	75.6

PERMEABILITY (source document units)

Gas Permeability (cm^3/100 $in^2 \cdot$ day $\cdot$ atm)	0.015	0.043	0.022	0.081	0.035	0.11

PERMEABILITY (normalized units)

Permeability Coefficient ($cm^3 \cdot mm/m^2 \cdot$ day $\cdot$ atm)	0.18	0.52	0.26	0.97	0.42	1.32

TABLE 293: Oxygen Permeability vs. Relative Humidity Through PS/ EVOH/ PP, HDPE/ EVOH/ PP and Nylon/ EVOH/ PP Multilayer Films.

Material Family	PS/ EVAL/ PP FILM		HDPE/ EVAL/ PP FILM		NYLON/ EVAL/ PP FILM	
Reference Number	264	264	264	264	264	264

MATERIAL CHARACTERISTICS

Sample Thickness	0.15 mm/ 0.025 mm/ 0.6 mm		0.15 mm/ 0.025 mm/ 0.6 mm		0.15 mm/ 0.025 mm/ 0.6 mm	

MATERIAL COMPOSITION

Barrier Layer	EVAL EP-F (EVOH)		EVAL EP-F (EVOH)		EVAL EP-E (EVOH)	
Inside Layer	polypropylene		polypropylene		polypropylene	
Outside Layer	polystyrene		high density polyethylene		nylon	

TEST CONDITIONS

Penetrant	oxygen					
Temperature (°C)	20	20	20	20	20	20
Relative Humidity - Outside (%)	65	75	65	75	65	75
Relative Humidity - Inside (%)	100 (wet)	100 (wet)	100 (wet)	100 (wet)	100 (wet)	100 (wet)
Relative Humidity - Barrier, calculated (%)	65.8	75.6	75.9	82.8	66.4	76

PERMEABILITY (source document units)

Gas Permeability $(cm^3/100\ in^2 \cdot day \cdot atm)$	0.022	0.035	0.035	0.059	0.083	0.11

PERMEABILITY (normalized units)

Permeability Coefficient $(cm^3 \cdot mm/m^2 \cdot day \cdot atm)$	0.26	0.42	0.42	0.71	1	1.32

TABLE 294: Oxygen Permeability vs. Relative Humidity Through LDPE/ EVOH/ PP, PP/ EVOH/ PET and PP/ EVOH/ PS Multilayer Films.

Material Family	LDPE/ EVAL/ PP FILM		PP/ EVAL/ PET FILM		PP/ EVAL/ PS FILM	
Reference Number	264	264	264	264	264	264

MATERIAL CHARACTERISTICS

Sample Thickness	0.15 mm/ 0.025 mm/ 0.6 mm		0.6 mm/ 0.025 mm/ 0.15 mm		0.6 mm/ 0.025 mm/ 0.15 mm	

MATERIAL COMPOSITION

Barrier Layer	EVAL EP-E (EVOH)		EVAL EP-F (EVOH)		EVAL EP-F (EVOH)	
Inside Layer	polypropylene		polyester PET		polystyrene	
Outside Layer	low density polyethylene		polypropylene		polypropylene	

TEST CONDITIONS

Penetrant	oxygen					
Temperature (°C)	20	20	20	20	20	20
Relative Humidity - Outside (%)	65	75	65	75	65	75
Relative Humidity - Inside (%)	100 (wet)	100 (wet)	10 (wet)	10 (wet)	10 (wet)	10 (wet)
Relative Humidity - Barrier, calculated (%)	69.8	78.4	16.6	17.8	16.9	17.1

PERMEABILITY (source document units)

Gas Permeability ($cm^3/100\ in^2 \cdot day \cdot atm$)	0.091	0.12	0.01	0.01	0.01	0.01

PERMEABILITY (normalized units)

Permeability Coefficient ($cm^3 \cdot mm/m^2 \cdot day \cdot atm$)	1.09	1.44	0.15	0.15	0.15	0.15

EVOH Multilayer Film

TABLE 295: Oxygen Permeability vs. Relative Humidity Through PP/ EVOH/ HDPE and PP/ EVOH/ LDPE Multilayer Films.

Material Family	PP/ EVAL/ HDPE FILM		PP/ EVAL/ LDPE FILM					
Reference Number	264	264	264	264	265	265	265	265

MATERIAL CHARACTERISTICS

Sample Thickness	0.6 mm/ 0.025 mm/ 0.15 mm		0.6 mm/ 0.025 mm/ 0.15 mm		0.02 mm/ 0.015 mm/ 0.051 mm			

MATERIAL COMPOSITION

Barrier Layer	EVAL EP-F (EVOH)		EVAL EP-F (EVOH)		EVAL EF-F (EVOH)			
Inside Layer	high density polyethylene		low density polyethylene		low density polyethylene			
Outside Layer	polypropylene		polypropylene		oriented polypropylene			

TEST CONDITIONS

Penetrant	oxygen							
Temperature (°C)	20	20	20	20	20	20	20	20
Relative Humidity - Outside (%)	65	75	65	75	65	80	65	80
Relative Humidity - Inside (%)	10 (wet)	10 (wet)	10 (wet)	10 (wet)	100	100	10	10
Relative Humidity - Barrier, calculated (%)	29.1	31.7	21.4	22.5	84	34	91	40

PERMEABILITY (source document units)

Gas Permeability ($cm^3/100\ in^2 \cdot day \cdot atm$)	0.011	0.011	0.011	0.01	0.11	0.24	0.01	0.02
Gas Permeability ($cm^3/m^2 \cdot day$)					1.7	3.8	0.2	0.3

PERMEABILITY (normalized units)

Permeability Coefficient ($cm^3 \cdot mm/m^2 \cdot day \cdot atm$)	0.16	0.16	0.16	0.15	0.15	0.32	0.01	0.03

EVOH Multilayer Film

TABLE 296: Oxygen Permeability vs. Relative Humidity Through PP/ EVOH/ PC and HDPE/ EVOH/ LDPE Multilayer Films.

Material Family	PP/ EVAL/ PC FILM		HDPE/ EVAL/ LDPE FILM	
Reference Number	264	264	264	264

MATERIAL CHARACTERISTICS

Sample Thickness	0.6 mm/ 0.025 mm/ 0.15 mm		0.6 mm/ 0.025 mm/ 0.15 mm	

MATERIAL COMPOSITION

Barrier Layer	EVAL EP-E (EVOH)		EVAL EP-E (EVOH)	
Inside Layer	polycarbonate		low density polyethylene	
Outside Layer	polypropylene		high density polyethylene	

TEST CONDITIONS

Penetrant	oxygen			
Temperature (°C)	20	20	20	20
Relative Humidity - Outside (%)	65	75	65	75
Relative Humidity - Inside (%)	10 (wet)	10 (wet)	10 (wet)	10 (wet)
Relative Humidity - Barrier, calculated (%)	21	21.3	14.5	15.3

PERMEABILITY (source document units)

Gas Permeability (cm^3/100 $in^2 \cdot$ day $\cdot$ atm)	0.041	0.042	0.041	0.041

PERMEABILITY (normalized units)

Permeability Coefficient ($cm^3 \cdot mm/m^2 \cdot$ day $\cdot$ atm)	0.6	0.62	0.6	0.6

TABLE 297: Oxygen Permeability vs. Relative Humidity Through Nylon/ EVOH/ LDPE Multilayer Film.

Material Family	NYLON/ EVAL/ LDPE FILM							
Reference Number	265	265	265	265	265	265	265	265

MATERIAL CHARACTERISTICS

Sample Thickness	0.015 mm/ 0.015 mm/ 0.051 mm	0.02 mm/ 0.02 mm/ 0.051 mm	0.015 mm/ 0.015 mm/ 0.051 mm	0.02 mm/ 0.02 mm/ 0.051 mm	0.015 mm/ 0.015 mm/ 0.051 mm	0.02 mm/ 0.02 mm/ 0.051 mm	0.015 mm/ 0.015 mm/ 0.051 mm	0.02 mm/ 0.02 mm/ 0.051 mm

MATERIAL COMPOSITION

Barrier Layer	EVAL EF-F (EVOH)	EVAL EF-E (EVOH)	EVAL EF-F (EVOH)	EVAL EF-E (EVOH)	EVAL EF-F (EVOH)	EVAL EF-E (EVOH)	EVAL EF-F (EVOH)	EVAL EF-E (EVOH)
Inside Layer	low density polyethylene							
Outside Layer	oriented nylon	coated nylon	oriented nylon	coated nylon	oriented nylon	coated nylon	oriented nylon	coated nylon

TEST CONDITIONS

Penetrant	oxygen							
Temperature (°C)	20	20	20	20	20	20	20	20
Relative Humidity - Outside (%)	65	65	80	80	65	65	80	80
Relative Humidity - Inside (%)	100	100	100	100	10	10	10	10
Relative Humidity - Barrier, calculated (%)	66	65	63	64	81	80	77	78

PERMEABILITY (source document units)

Gas Permeability (cm³/100 in² · day · atm)	0.04	0.1	0.08	0.16	0.03	0.1	0.06	0.15
Gas Permeability (cm³/m² · day)	0.6	1.5	1.3	2.5	0.5	1.5	1.0	2.3

PERMEABILITY (normalized units)

Permeability Coefficient (cm³ · mm/m² · day · atm)	0.05	0.14	0.1	0.2	0.04	0.14	0.08	0.21

TABLE 298: Oxygen Permeability vs. Relative Humidity Through EVOH/ LDPE Multilayer Film.

Material Family	EVAL/ LDPE FILM			
Reference Number	265	265	265	265

MATERIAL CHARACTERISTICS

Sample Thickness	0.015 mm/ 0.051 mm			

MATERIAL COMPOSITION

Barrier Layer	EVAL EF-XL (EVOH)			
Inside Layer	low density polyethylene			

TEST CONDITIONS

Penetrant	oxygen			
Temperature (°C)	20	20	20	20
Relative Humidity - Outside (%)	65	80	65	80
Relative Humidity - Inside (%)	100	100	10	10

PERMEABILITY (source document units)

Gas Permeability (cm^3/100 in^2 · day · atm)	0.02	0.04	0.02	0.04
Gas Permeability (cm^3/m^2 · day)	0.3	0.7	0.3	0.7

PERMEABILITY (normalized units)

Permeability Coefficient (cm^3 · mm/m^2 · day · atm)	0.02	0.04	0.02	0.04

TABLE 299: Oxygen Permeability vs. Relative Humidity, Vanilla, Peppermint, Piperonol and Camphor Permeability Through PET/ EVOH/ LDPE Multilayer Film.

Material Family	PET/ EVAL/ LDPE FILM							
Reference Number	265	265	265	265	265	265	265	265

MATERIAL CHARACTERISTICS

Sample Thickness	0.013 mm/ 0.015 mm/ 0.051 mm

MATERIAL COMPOSITION

Barrier Layer	EVAL EF-F (EVOH)
Inside Layer	low density polyethylene
Outside Layer	polyester PET

TEST CONDITIONS

Penetrant	oxygen				vanilla (vanillin)	peppermint (menthol)	piperonol (heliotropin)	camphor
Temperature (°C)	20	20	20	20				
Relative Humidity - Outside (%)	65	80	65	80				
Relative Humidity - Inside (%)	100	100	10	10				
Relative Humidity - Barrier, calculated (%)	70	57	83	69				

PERMEABILITY (source document units)

Days To Leakage					15	25	27	>30
Gas Permeability (cm³/100 in² · day · atm)	0.04	0.1	0.02	0.04				
Gas Permeability (cm³/m² · day)	0.7	1.6	0.4	0.7				

PERMEABILITY (normalized units)

Permeability Coefficient (cm³ · mm/m² · day · atm)	0.05	0.12	0.02	0.05				

TABLE 300: Vanilla, Peppermint, Piperonol and Camphor Permeability Through EVOH/ PE Multilayer Film.

Material Family	EVAL/ PE FILM			
Product Form	FILM			
Reference Number	265	265	265	265

MATERIAL CHARACTERISTICS

Sample Thickness	0.013 mm/ 0.051 mm

MATERIAL COMPOSITION

Inside Layer	polyethylene
Outside Layer	EVAL EF-XL (EVOH)

TEST CONDITIONS

Penetrant	vanilla (vanillin)	peppermint (menthol)	piperonol (heliotropin)	camphor

PERMEABILITY (source document units)

Days To Leakage	>30	>30	>30	>30

TABLE 301: Vanilla, Peppermint, Piperonol and Camphor Permeability Through PET/ EVOH and Nylon/ EVOH Multilayer Films.

Material Family	PET/ EVAL FILM				NYLON/ EVAL FILM			
Product Form	FILM							
Reference Number	265	265	265	265	265	265	265	265

MATERIAL CHARACTERISTICS

Sample Thickness	0.013 mm/ 0.015 mm				0.015 mm/ 0.015 mm			

MATERIAL COMPOSITION

Inside Layer	EVAL EF-F (EVOH)				EVAL EF-F (EVOH)			
Outside Layer	polyester PET				oriented nylon			

TEST CONDITIONS

Penetrant	vanilla (vanillin)	peppermint (menthol)	piperonol (heliotropin)	camphor	vanilla (vanillin)	peppermint (menthol)	piperonol (heliotropin)	camphor

PERMEABILITY (source document units)

Days To Leakage	>30	>30	30	>30	2	>30	27	30

EVOH Multilayer Film

TABLE 302: Gasoline Permeability Through HDPE/ EVOH Multilayer Film.

Material Family	HDPE/ EVAL FILM	
Reference Number	266	266

MATERIAL CHARACTERISTICS

Sample Thickness	1.27 mm/ 0.015 mm	1.27 mm/ 0.061 mm

MATERIAL COMPOSITION

Inside Layer	EVAL F (EVOH)	EVAL F (EVOH)
Outside Layer	high density polyethylene	high density polyethylene

TEST CONDITIONS

Penetrant	gasoline	

PERMEABILITY (source document units)

Vapor Transmission Rate (g/day · 100 in²)	0.29	0.07

PERMEABILITY (normalized units)

Vapor Transmission Rate (g · mm/m² · day)	6.4	1.4

TABLE 303: Chloroform and Xylene Permeability Through LDPE/ EVOH Multilayer Film.

Material Family	LDPE/ EVAL FILM					
Product Form	FILM					
Reference Number	266	266	266	266	266	266

MATERIAL CHARACTERISTICS

Sample Thickness	0.06 mm/ 0.025 mm	0.06 mm/ 0.015 mm	0.06 mm/ 0.015 mm	0.06 mm/ 0.025 mm	0.06 mm/ 0.015 mm	0.06 mm/ 0.015 mm

MATERIAL COMPOSITION

Inside Layer	EVAL EF-E (EVOH)	EVAL EF-F (EVOH)	EVAL EF-XL (EVOH)	EVAL EF-E (EVOH)	EVAL EF-F (EVOH)	EVAL EF-XL (EVOH)
Outside Layer	low density polyethylene					

TEST CONDITIONS

Penetrant	chloroform			xylene		
Temperature (°C)	20	20	20	20	20	20
Relative Humidity (%)	65	65	65	65	65	65

PERMEABILITY (source document units)

Vapor Transmission Rate (g/day · 100 in²)	0.01	0.02	<0.003	<0.003	<0.003	<0.003

PERMEABILITY (normalized units)

Vapor Transmission Rate (g · mm/m² · day)	0.013	0.023	<0.0035	<0.0035	<0.0035	<0.0035

EVOH Multilayer Film

TABLE 304: Methyl Ethyl Ketone and Kerosine Permeability Through LDPE/ EVOH Multilayer Film.

Material Family	LDPE/ EVAL FILM					
Product Form	FILM					
Reference Number	266	266	266	266	266	266

MATERIAL CHARACTERISTICS

Sample Thickness	0.06 mm/ 0.025 mm	0.06 mm/ 0.015 mm	0.06 mm/ 0.015 mm	0.06 mm/ 0.025 mm	0.06 mm/ 0.015 mm	0.06 mm/ 0.015 mm

MATERIAL COMPOSITION

Inside Layer	EVAL EF-E (EVOH)	EVAL EF-F (EVOH)	EVAL EF-XL (EVOH)	EVAL EF-E (EVOH)	EVAL EF-F (EVOH)	EVAL EF-XL (EVOH)
Outside Layer	low density polyethylene					

TEST CONDITIONS

Penetrant	methyl ethyl ketone			kerosine		
Temperature (°C)	20	20	20	20	20	20
Relative Humidity (%)	65	65	65	65	65	65

PERMEABILITY (source document units)

Vapor Transmission Rate (g/day · 100 in²)	0.01	0.01	0.003	<0.003	<0.003	<0.003

PERMEABILITY (normalized units)

Vapor Transmission Rate (g · mm/m² · day)	0.013	0.013	0.0035	<0.0035	<0.0035	<0.0035

GRAPH 68: Oxygen Permeability vs. Relative Humidity through Polyvinylidene Chloride coated Nylon/ Ethylene-Vinyl Alcohol Copolymer/ Polyethylene Multilayer Film.

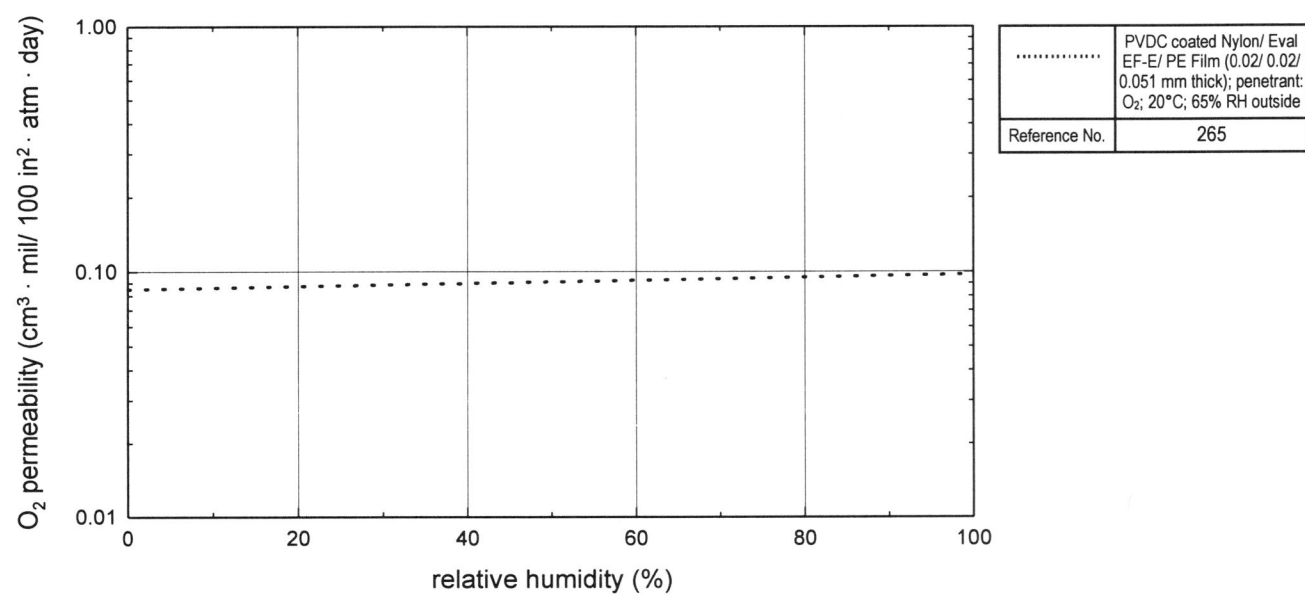

....	PVDC coated Nylon/ Eval EF-E/ PE Film (0.02/ 0.02/ 0.051 mm thick); penetrant: O₂; 20°C; 65% RH outside
Reference No.	265

EVOH Multilayer Film

GRAPH 69: Oxygen Permeability vs. Relative Humidity through Oriented Polypropylene/ Ethylene-Vinyl Alcohol Copolymer/ Polyethylene Multilayer Film.

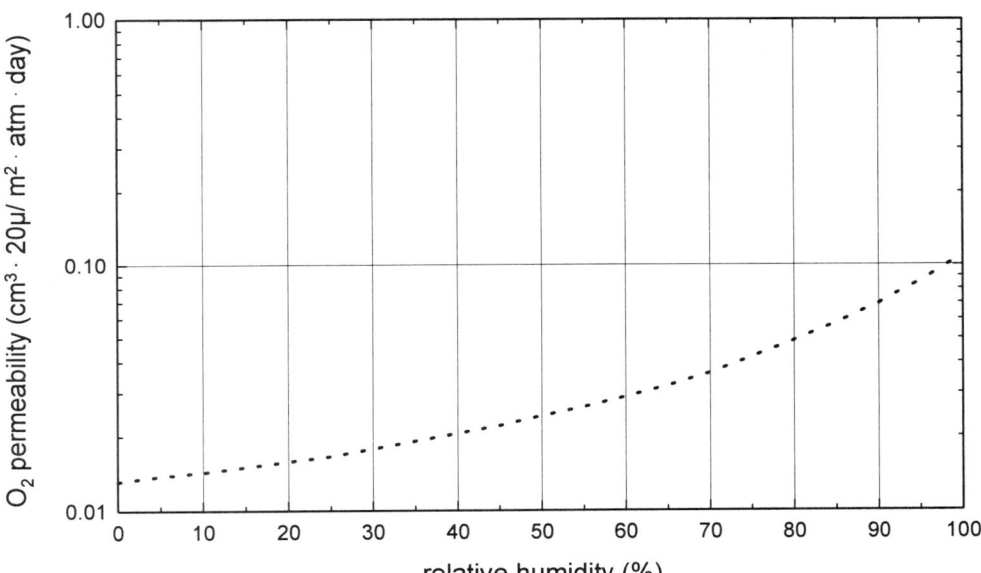

....	OPP/ Eval EF-F/ PE Film (0.02/ 0.015/ 0.051 mm thick); penetrant: O₂; 20°C; 65% RH outside
Reference No.	265

GRAPH 70: Oxygen Permeability vs. Relative Humidity through Ethylene-Vinyl Alcohol Copolymer/ Polyethylene Multilayer Film.

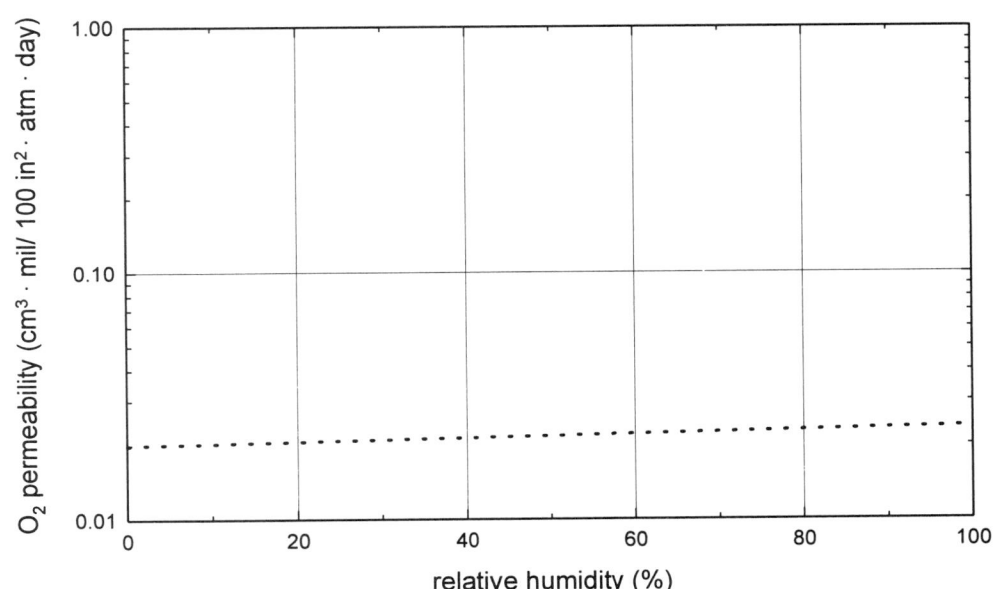

....	Eval EF-XL/ PE Film (0.015/ 0.051 mm thick); penetrant: O₂; 20°C; 65% RH outside
Reference No.	265

EVOH Multilayer Film

Chapter 64

Multilayer Films with Polyvinylidene Chloride Barrier

TABLE 305: Water Vapor and Oxygen Permeability Through Dow Chemical Saranex LDPE/ EVA/ PVDC/ EVA/ LDPE Multilayer Film.

Material Family	LDPE/ EVA/ PVDC/ EVA/ LDPE FILM							
Material Supplier/ Trade Name	DOW CHEMICAL SARANEX							
Grade	11	14	15	15	11	14	15	15
Product Form	COEXTRUDED FILM							
Features	barrier properties, neutral color							
Applications	laminations	form-fill-seal pouch	form-fill-seal pouch	form-fill-seal pouch	laminations	form-fill-seal pouch	form-fill-seal pouch	form-fill-seal pouch
Reference Number	257	257	257	257	257	257	257	257

MATERIAL CHARACTERISTICS

Sample Thickness	0.038 mm	0.051 mm	0.076 mm	0.102 mm	0.038 mm	0.051 mm	0.076 mm	0.102 mm

MATERIAL COMPOSITION

Note	medium slip additive content							

TEST CONDITIONS

Penetrant	water vapor				oxygen			
Temperature (°C)	38	38	38	38	23	23	23	23
Relative Humidity (%)	90	90	90	90	10	10	10	10
Test Method	Permatran W	Permatran W	Permatran W	Permatran W	ASTM D3985	ASTM D3985	ASTM D3985	ASTM D3985

PERMEABILITY (source document units)

Vapor Transmission Rate ($g \cdot mil/100\ in^2 \cdot day$)	0.4	0.2	0.15	0.11				
Gas Permeability ($cm^3/100\ in^2 \cdot day \cdot atm$)					1.5	0.5	0.5	0.5

PERMEABILITY (normalized units)

Permeability Coefficient ($cm^3 \cdot mm/m^2 \cdot day \cdot atm$)					0.88	0.4	0.59	0.79
Vapor Transmission Rate ($g \cdot mm/m^2 \cdot day$)	0.16	0.08	0.06	0.04				

TABLE 306: Carbon Dioxide and Nitrogen Permeability Through Dow Chemical Saranex LDPE/ EVA/ PVDC/ EVA/ LDPE Multilayer Film.

Material Family	LDPE/ EVA/ PVDC/ EVA/ LDPE FILM							
Material Supplier/ Trade Name	DOW CHEMICAL SARANEX							
Grade	11	14	15	15	11	14	15	15
Product Form	COEXTRUDED FILM							
Features	barrier properties, neutral color							
Applications	laminations	form-fill-seal pouch	form-fill-seal pouch	form-fill-seal pouch	laminations	form-fill-seal pouch	form-fill-seal pouch	form-fill-seal pouch
Reference Number	257	257	257	257	257	257	257	257

MATERIAL CHARACTERISTICS

Sample Thickness	0.038 mm	0.051 mm	0.076 mm	0.102 mm	0.038 mm	0.051 mm	0.076 mm	0.102 mm

MATERIAL COMPOSITION

Note	medium slip additive content							

TEST CONDITIONS

Penetrant	carbon dioxide				nitrogen			
Temperature (°C)	23	23	23	23	23	23	23	23
Relative Humidity (%)	10	10	10	10	10	10	10	10
Test Method	ASTM D1434	ASTM D1434	ASTM D1434	ASTM D1434	ASTM D1434	ASTM D1434	ASTM D1434	ASTM D1434

PERMEABILITY (source document units)

Gas Permeability (cm^3/100 $in^2 \cdot day \cdot atm$)	5.5	1.39	1.39	1.39	0.15	0.07	0.07	0.07

PERMEABILITY (normalized units)

Permeability Coefficient ($cm^3 \cdot mm/m^2 \cdot day \cdot atm$)	3.2	1.1	1.64	2.2	0.09	0.06	0.08	0.11

TABLE 307: Air and Oxygen Permeability Through Dow Chemical Saranex LDPE/ EVA/ PVDC/ EVA/ LDPE Multilayer Film.

Material Family	LDPE/ EVA/ PVDC/ EVA/ LDPE FILM									
Material Supplier/ Trade Name	DOW CHEMICAL SARANEX									
Grade	11	14	15	15	14	12	14	12	14	12
Product Form	COEXTRUDED FILM				FILM					
Features	barrier properties, neutral color				barrier properties					
Applications	laminations	form-fill-seal pouch	form-fill-seal pouch	form-fill-seal pouch						
Reference Number	257	257	257	257	265	265	265	265	265	265

MATERIAL CHARACTERISTICS

Sample Thickness	0.038 mm	0.051 mm	0.076 mm	0.102 mm	0.051 mm	0.051 mm	0.051 mm	0.051 mm	0.051 mm	0.051 mm

MATERIAL COMPOSITION

Note	medium slip additive content									

TEST CONDITIONS

Penetrant	air				oxygen					
Temperature (°C)	23	23	23	23	20	20	20	20	20	20
Relative Humidity (%)	10	10	10	10	65	65	85	85	100	100
Test Method	ASTM D1434	ASTM D1434	ASTM D1434	ASTM D1434						

PERMEABILITY (source document units)

Gas Permeability ($cm^3 \cdot mil/100\ in^2 \cdot day$)					0.45	1.0	0.45	1.0	0.45	1.0
Gas Permeability ($cm^3/100\ in^2 \cdot day \cdot atm$)	0.3	0.13	0.13	0.13						

PERMEABILITY (normalized units)

Permeability Coefficient ($cm^3 \cdot mm/m^2 \cdot day \cdot atm$)	0.18	0.1	0.15	0.21	0.18	0.39	0.18	0.39	0.18	0.39

TABLE 308: Water Vapor, Carbon Dioxide, Nitrogen, Air and Oxygen Permeability Through Dow Chemical Saranex 23 LDPE/ EVA/ PVDC/ EVA Multilayer Film.

Material Family	LDPE/ EVA/ PVDC/ EVA FILM				
Material Supplier/ Grade	DOW CHEMICAL SARANEX 23				
Product Form	COEXTRUDED FILM				
Features	barrier properties, neutral color				
Applications	laminations				
Reference Number	257	257	257	257	257

MATERIAL CHARACTERISTICS

Sample Thickness	0.051 mm	0.051 mm	0.051 mm	0.051 mm	0.051 mm

MATERIAL COMPOSITION

Note	medium slip additive content				

TEST CONDITIONS

Penetrant	water vapor	oxygen	carbon dioxide	nitrogen	air
Temperature (°C)	38	23	23	23	23
Relative Humidity (%)	90	10	10	10	10
Test Method	Permatran W	ASTM D3985	ASTM D1434	ASTM D1434	ASTM D1434

PERMEABILITY (source document units)

Vapor Transmission Rate (g · mil/100 in² · day)	0.25				
Gas Permeability (cm³/100 in² · day · atm)		0.5	1.39	0.15	0.3

PERMEABILITY (normalized units)

Permeability Coefficient (cm³ · mm/m² · day · atm)		0.4	1.1	0.12	0.24
Vapor Transmission Rate (g · mm/m² · day)	0.1				

TABLE 309: Water Vapor, Carbon Dioxide, Nitrogen, Air and Oxygen Permeability Through Dow Chemical Saranex 25 HDPE/ EVA/ PVDC/ EVA Multilayer Film.

Material Family	HDPE/ EVA/ PVDC/ EVA FILM				
Material Supplier/ Grade	DOW CHEMICAL SARANEX 25				
Product Form	COEXTRUDED FILM				
Features	barrier properties, neutral color				
Applications	bag-in-box, form-fill-seal pouch				
Reference Number	257	257	257	257	257

MATERIAL CHARACTERISTICS

Sample Thickness	0.051 mm

MATERIAL COMPOSITION

Note	high slip additive content

TEST CONDITIONS

Penetrant	water vapor	oxygen	carbon dioxide	nitrogen	air
Temperature (°C)	38	23	23	23	23
Relative Humidity (%)	90	10	10	10	10
Test Method	Permatran W	ASTM D3985	ASTM D1434	ASTM D1434	ASTM D1434

PERMEABILITY (source document units)

Vapor Transmission Rate (g · mil/100 in^2 · day)	0.16				
Gas Permeability (cm^3/100 in^2 · day · atm)		0.5	1.39	0.07	0.13

PERMEABILITY (normalized units)

Permeability Coefficient (cm^3 · mm/m^2 · day · atm)		0.4	1.1	0.06	0.1
Vapor Transmission Rate (g · mm/m^2 · day)	0.06				

TABLE 310: Oxygen Permeability vs. Relative Humidity Through PP/ PVDC/ PP Multilayer Film.

Material Family	PP/ PVDC/ PP FILM			
Reference Number	264	264	264	264

MATERIAL CHARACTERISTICS

Sample Thickness	0.15 mm/ 0.025 mm/ 0.6 mm	0.6 mm/ 0.025 mm/ 0.15 mm	0.15 mm/ 0.025 mm/ 0.6 mm	0.6 mm/ 0.025 mm/ 0.15 mm

MATERIAL COMPOSITION

Barrier Layer	Saran VC (PVDC)			
Inside Layer	polypropylene			
Outside Layer	polypropylene			

TEST CONDITIONS

Penetrant	oxygen			
Temperature (°C)	20	20	20	20
Relative Humidity - Outside (%)	65	65	75	75
Relative Humidity - Inside (%)	100 (wet)	10 (wet)	100 (wet)	10 (wet)
Relative Humidity - Barrier, calculated (%)	72	24.2	80	25.9

PERMEABILITY (source document units)

Gas Permeability ($cm^3/100$ $in^2 \cdot day \cdot atm$)	0.12	0.12	0.12	0.12

PERMEABILITY (normalized units)

Permeability Coefficient ($cm^3 \cdot mm/m^2 \cdot day \cdot atm$)	1.4	1.8	1.4	1.8

TABLE 311: Oxygen Permeability vs. Relative Humidity Through PET/ PVDC/ PP and PC/ PVDC/ PP Multilayer Films.

Material Family	PET/ PVDC/ PP FILM		PC/ PVDC/ PP FILM	
Reference Number	264	264	264	264

MATERIAL CHARACTERISTICS

Sample Thickness	0.15 mm/ 0.025 mm/ 0.6 mm		0.15 mm/ 0.025 mm/ 0.6 mm	

MATERIAL COMPOSITION

Barrier Layer	Saran VC (PVDC)		Saran VC (PVDC)	
Inside Layer	polypropylene		polypropylene	
Outside Layer	polyester PET		polycarbonate	

TEST CONDITIONS

Penetrant	oxygen			
Temperature (°C)	20	20	20	20
Relative Humidity - Outside (%)	65	75	65	75
Relative Humidity - Inside (%)	100 (wet)	100 (wet)	100 (wet)	100 (wet)
Relative Humidity - Barrier, calculated (%)	69.2	78	65.7	75.5

PERMEABILITY (source document units)

Gas Permeability (cm^3/100 $in^2 \cdot$ day $\cdot$ atm)	0.12	0.12	0.12	0.12

PERMEABILITY (normalized units)

Permeability Coefficient ($cm^3 \cdot$ mm/$m^2 \cdot$ day $\cdot$ atm)	1.4	1.4	1.4	1.4

TABLE 312: Oxygen Permeability vs. Relative Humidity Through PS/ PVDC/ PP, HDPE/ PVDC/ PP and Nylon/ PVDC/ PP Multilayer Films.

Material Family	PS/ PVDC/ PP FILM		HDPE/ PVDC/ PP FILM		NYLON/ PVDC/ PP FILM	
Reference Number	264	264	264	264	264	264

MATERIAL CHARACTERISTICS

Sample Thickness	0.15 mm/ 0.025 mm/ 0.6 mm		0.15 mm/ 0.025 mm/ 0.6 mm		0.15 mm/ 0.025 mm/ 0.6 mm	

MATERIAL COMPOSITION

Barrier Layer	Saran VC (PVDC)		Saran VC (PVDC)		Saran VC (PVDC)	
Inside Layer	polypropylene		polypropylene		polypropylene	
Outside Layer	polystyrene		high density polyethylene		nylon	

TEST CONDITIONS

Penetrant	oxygen					
Temperature (°C)	20	20	20	20	20	20
Relative Humidity - Outside (%)	65	75	65	75	65	75
Relative Humidity - Inside (%)	100 (wet)	100 (wet)	100 (wet)	100 (wet)	100 (wet)	100 (wet)
Relative Humidity - Barrier, calculated (%)	65.8	75.6	75.9	82.8	66.4	76

PERMEABILITY (source document units)

Gas Permeability ($cm^3/100\ in^2 \cdot day \cdot atm$)	0.12	0.12	0.12	0.12	0.12	0.12

PERMEABILITY (normalized units)

Permeability Coefficient ($cm^3 \cdot mm/m^2 \cdot day \cdot atm$)	1.4	1.4	1.4	1.4	1.4	1.4

PVDC Multilayer Film

TABLE 313: Oxygen Permeability vs. Relative Humidity Through LDPE/ PVDC/ PP, PP/ PVDC/ PET and PP/ PVDC/ PS Multilayer Films.

Material Family	LDPE/ PVDC/ PP FILM		PP/ PVDC/ PET FILM		PP/ PVDC/ PS FILM	
Reference Number	264	264	264	264	264	264

MATERIAL CHARACTERISTICS

Sample Thickness	0.15 mm/ 0.025 mm/ 0.6 mm		0.6 mm/ 0.025 mm/ 0.15 mm		0.6 mm/ 0.025 mm/ 0.15 mm	

MATERIAL COMPOSITION

Barrier Layer	Saran VC (PVDC)		Saran VC (PVDC)		Saran VC (PVDC)	
Inside Layer	polypropylene		polyester PET		polystyrene	
Outside Layer	low density polyethylene		polypropylene		polypropylene	

TEST CONDITIONS

Penetrant	oxygen					
Temperature (°C)	20	20	20	20	20	20
Relative Humidity - Outside (%)	65	75	65	75	65	75
Relative Humidity - Inside (%)	100 (wet)	100 (wet)	10 (wet)	10 (wet)	10 (wet)	10 (wet)
Relative Humidity - Barrier, calculated (%)	69.8	78.4	16.6	17.8	16.9	17.1

PERMEABILITY (source document units)

Gas Permeability (cm^3/100 in^2 · day · atm)	0.12	0.12	0.12	0.12	0.12	0.12

PERMEABILITY (normalized units)

Permeability Coefficient (cm^3 · mm/m^2 · day · atm)	1.4	1.4	1.8	1.8	1.8	1.8

TABLE 314: Oxygen Permeability vs. Relative Humidity Through PP/ PVDC/ HDPE and PP/ PVDC/ LDPE Multilayer Films.

Material Family	PP/ PVDC/ HDPE FILM		PP/ PVDC/ LDPE FILM	
Reference Number	264	264	264	264

MATERIAL CHARACTERISTICS

Sample Thickness	0.6 mm/ 0.025 mm/ 0.15 mm		0.6 mm/ 0.025 mm/ 0.15 mm	

MATERIAL COMPOSITION

Barrier Layer	Saran VC (PVDC)		Saran VC (PVDC)	
Inside Layer	high density polyethylene		low density polyethylene	
Outside Layer	polypropylene		polypropylene	

TEST CONDITIONS

Penetrant	oxygen			
Temperature (°C)	20	20	20	20
Relative Humidity - Outside (%)	65	75	65	75
Relative Humidity - Inside (%)	10 (wet)	10 (wet)	10 (wet)	10 (wet)
Relative Humidity - Barrier, calculated (%)	29.1	31.7	21.4	22.5

PERMEABILITY (source document units)

Gas Permeability (cm^3/100 in^2 · day · atm)	0.12	0.12	0.12	0.12

PERMEABILITY (normalized units)

Permeability Coefficient (cm^3 · mm/m^2 · day · atm)	1.8	1.8	1.8	1.8

TABLE 315: Oxygen Permeability vs. Relative Humidity Through PP/ PVDC/ PC and HDPE/ PVDC/ LDPE Multilayer Films.

Material Family	PP/ PVDC/ PC FILM		HDPE/ PVDC/ LDPE FILM	
Reference Number	264	264	264	264

MATERIAL CHARACTERISTICS

Sample Thickness	0.6 mm/ 0.025 mm/ 0.15 mm		0.6 mm/ 0.025 mm/ 0.15 mm	

MATERIAL COMPOSITION

Barrier Layer	Saran VC (PVDC)		Saran VC (PVDC)	
Inside Layer	polycarbonate		low density polyethylene	
Outside Layer	polypropylene		high density polyethylene	

TEST CONDITIONS

Penetrant	oxygen			
Temperature (°C)	20	20	20	20
Relative Humidity - Outside (%)	65	75	65	75
Relative Humidity - Inside (%)	10 (wet)	10 (wet)	10 (wet)	10 (wet)
Relative Humidity - Barrier, calculated (%)	21	21.3	14.5	15.3

PERMEABILITY (source document units)

Gas Permeability (cm^3/100 $in^2 \cdot$ day $\cdot$ atm)	0.12	0.12	0.12	0.12

PERMEABILITY (normalized units)

Permeability Coefficient ($cm^3 \cdot mm/m^2 \cdot$ day $\cdot$ atm)	1.8	1.8	1.8	1.8

TABLE 316: Film Properties of Dow Chemical Saranex Multilayer Film.

Material Family	LDPE/ EVA/ PVDC/ EVA/ LDPE FILM				LDPE/ EVA/ PVDC/ EVA FILM	HDPE/ EVA/ PVDC/ EVA FILM
Material Supplier/ Trade Name	DOW CHEMICAL SARANEX					
Grade	11	14	15	15	23	25
Product Form	COEXTRUDED FILM					
Features	barrier properties, neutral color					
Applications	laminations	form-fill-seal pouch	form-fill-seal pouch	form-fill-seal pouch	laminations	bag-in-box, form-fill-seal pouch
Reference Number	257	257	257	257	257	257

MATERIAL CHARACTERISTICS

Material Composition Note	medium slip additive content	medium slip additive content	medium slip additive content	medium slip additive content	medium slip additive content	high slip additive content
Sample Thickness	0.038 mm	0.051 mm	0.076 mm	0.102 mm	0.051 mm	0.051 mm
Yield (in²/lb)	18,350	13,500	9300	7100	13,250	13,250

MECHANICAL PROPERTIES

Secant Modulus @ 2% Elongation - MD (MPa)	200 {r}	221 {r}	186 {r}	193 {r}	179 {r}	358 {r}
Secant Modulus @ 2% Elongation - TD (MPa)	193 {r}	214 {r}	186 {r}	186 {r}	186 {r}	441 {r}
Ultimate Tensile Strength - MD (MPa)	25.5 {r}	24.1 {r}	24.1 {r}	23.4 {r}	20.7 {r}	33.1 {r}
Ultimate Tensile Strength - TD (MPa)	18.6 {r}	18.6 {r}	18.6 {r}	17.2 {r}	15.9 {r}	15.2 {r}
Tensile Strength @ Yield - MD (MPa)	11.7 {r}	13.1 {r}	11 {r}	11 {r}	11 {r}	18.6 {r}
Tensile Strength @ Yield - TD (MPa)	11.7 {r}	11.7 {r}	10.3 {r}	10.3 {r}	10.3 {r}	19.3 {r}
Ultimate Elongation - MD (%)	400 {r}	400 {r}	500 {r}	500 {r}	350 {r}	450 {r}
Ultimate Elongation - TD (%)	500 {r}	550 {r}	550 {r}	550 {r}	530 {r}	650 {r}
Drop Dart Impact Strength (g)	110 {t}	130 {t}	220 {t}	300 {t}	130 {t}	50 {t}
Tear Resistance -MD (g)	400 {k}	700 {k}	1000 {k}	1100 {k}	700 {k}	40 {k}
Tear Resistance -TD (g)	250 {k}	460 {k}	600 {k}	1000 {k}	550 {k}	600 {k}
Coefficient Of Friction - Kinetic, film to film	0.25 {l}	0.25 {l}	0.25 {l}	0.25 {l}	0.25 (PE/PE) {l}	0.35 (PE/PE) {l}
Coefficient Of Friction - Kinetic, film to film					>0.5 (EVA/EVA) {l}	0.2 (EVA/EVA) {l}

OPTICAL PROPERTIES

Haze (%)	15 {c}	26 {c}	30 {c}	40 {c}	12 {c}	20 {c}
Gloss @ 45°	62 {m}	55 {m}	47 {m}	48 {m}	65 {m}	56 {m}
Clarity (%)	21 {s}	17 {s}	30 {s}	5 {s}	14 {s}	

TABLE 316: Film Properties of Dow Chemical Saranex Multilayer Film (cont'd).

Material Family	LDPE/ EVA/ PVDC/ EVA/ LDPE FILM				LDPE/ EVA/ PVDC/ EVA FILM	HDPE/ EVA/ PVDC/ EVA FILM
Material Supplier/ Trade Name	DOW CHEMICAL SARANEX					
Grade	11	14	15	15	23	25
Product Form	COEXTRUDED FILM					
Features	barrier properties, neutral color					
Applications	laminations	form-fill-seal pouch	form-fill-seal pouch	form-fill-seal pouch	laminations	bag-in-box, form-fill-seal pouch
Reference Number	257	257	257	257	257	257

PERFORMANCE PROPERTIES

Seal Range (°C)	143-187 {u}	143-221 {u}	143-252 {u}	143-252 {u}	116-221 {u}	93-238 {u}
Ultimate Seal Strength (kg/m)	35.7 (minimum) {v}	58.9 {v}	89.3 {v}	98.2 {v}		
Unrestrained Shrink - boiling H_2O, MD (%)	0.5	0	0	0	0.5	0
Unrestrained Shrink - boiling H_2O, TD (%)	0	0	0	0	0	0
Unrestrained Shrink - 121°C oil, MD (%)	9	11	9	9	11	1.5
Unrestrained Shrink - 121°C oil, TD (%)	0	0	0	0	2	0

GRAPH 71: Oxygen Permeability vs. Relative Humidity through Low Density Polyethylene/ Ethylene-Vinyl Acetate Copolymer/ Polyvinylidene Chloride/ Ethylene-Vinyl Acetate Copolymer/ Low Density Polyethylene Multilayer Film.

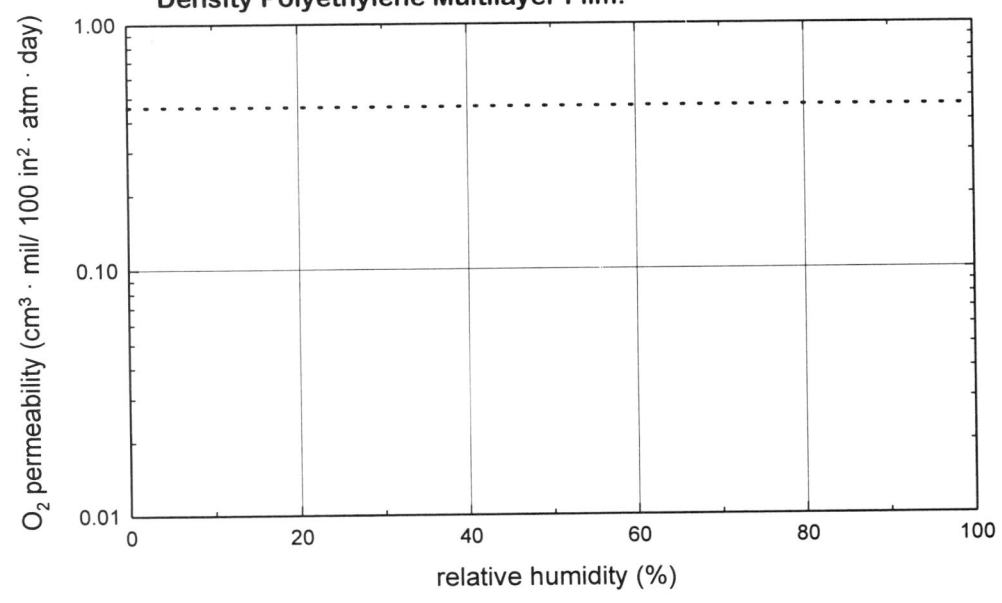

	Dow Saranex 14 LDPE/ EVA/ PVDC/ EVA/ LDPE Film (barrier prop., 0.051 mm thick; film); penetrant: O_2
Reference No.	265

PVDC Multilayer Film

Chapter 65

Multilayer Films - General

TABLE 317: Oxygen, Nitrogen, Carbon Dioxide and Water Vapor Permeability Through Dow Chemical Saranex PE/ PVC-PVDC Copolymer Multilayer Film.

Material Family	PE/ PVC-PVDC COPOLYMER MULTILAYER FILM			
Material Supplier/ Grade	DOW CHEMICAL SARANEX			
Product Form	MULTILAYER FILM			
Reference Number	250	250	250	250

TEST CONDITIONS

Penetrant	oxygen	nitrogen	carbon dioxide	water vapor
Temperature (°C)	24	24	24	24

PERMEABILITY (source document units)

Vapor Transmission Rate (g · mil/100 in^2 · day)				0.15 - 0.4
Gas Permeability (cm^3 · mil/100 in^2 · day)	0.5 - 2.0	0.1 - 0.5	1 - 8	

PERMEABILITY (normalized units)

Permeability Coefficient (cm^3 · mm/m^2 · day · atm)	0.2 - 0.79	0.04 - 0.2	0.39 - 3.15	
Vapor Transmission Rate (g · mm/m^2 · day)				0.06 - 0.16

TABLE 318: Water Vapor and Oxygen Permeability Through Nylon 6/ LDPE and HDPE/ EAA/ Nylon/ EAA Multilayer Films.

Material Family	NYLON 6/LDPE MULTILAYER FILM		HDPE/ EAA/ NYLON/ EAA MULTILAYER FILM	
Material Supplier/ Grade	BASF ULTRAMID B4/ BASF LUPOLEN 3020 D		DOW CHEMICAL NYLOPAK 570	
Product Form			COEXTRUDED FILM	
Features			barrier properties, neutral color	
Applications			form-fill-seal pouch	
Reference Number	252	252	258	258

MATERIAL CHARACTERISTICS

Sample Thickness	0.03 mm/ 0.05 mm	0.051 mm

TEST CONDITIONS

Penetrant	water vapor	oxygen	water vapor	oxygen
Temperature (°C)	20	20	38	23
Relative Humidity (%)	85-0% gradient	40	90	10
Test Method	DIN 53122	DIN 53380	Permatran W	ASTM D3985

PERMEABILITY (source document units)

Vapor Transmission Rate (g/m² · day)	1.6			
Vapor Transmission Rate (g/day · 100 in²)			0.3	
Gas Permeability (cm³/m² · day · bar)		20 - 25		
Gas Permeability (cm³/100 in² · day · atm)				12

PERMEABILITY (normalized units)

Permeability Coefficient (cm³ · mm/m² · day · atm)		1.6 - 2.0		9.5
Vapor Transmission Rate (g · mm/m² · day)	0.21		0.24	

Multilayer Film

TABLE 319: Water Vapor Permeability Through CTFE/ PE/ PVC and CTFE/ PVC Multilayer Films.

Material Family	CTFE/ PE/ PVC FILM		CTFE/ PVC FILM
Material Supplier/ Grade	ALLIED ACLAR 22C/ PE/ PVC	ALLIED ACLAR 88A/ PE/ PVC	ALLIED ACLAR 33C/ PVC
Product Form	LAMINATION		
Reference Number	138	138	138

MATERIAL CHARACTERISTICS

Sample Thickness	0.038 mm/ 0.051 mm/ 0.19 mm	0.019 mm/ 0.051 mm/ 0.19 mm	0.019 mm/ 0.254 mm

TEST CONDITIONS

Penetrant	water vapor		
Temperature (°C)	37.8	37.8	37.8
Relative Humidity (%)	90	90	90
Test Method	ASTM F372-78	ASTM F372-78	ASTM F372-78
Test Apparatus	Mocon Permatran	Mocon Permatran	Mocon Permatran

PERMEABILITY (source document units)

Vapor Transmission Rate (g/m^2 · day)	0.34	0.48	0.28
Vapor Transmission Rate (g/day · 100 in^2)	0.022	0.031	0.018

PERMEABILITY (normalized units)

Vapor Transmission Rate (g · mm/m^2 · day)	0.09	0.12	0.08

<u>TABLE 320:</u> **Vanilla, Peppermint, Piperonol and Camphor Permeability Through PET/ PE and Nylon/ PE Multilayer Films.**

Material Family	PET/ PE FILM				NYLON/ PE FILM			
Product Form	FILM							
Reference Number	265	265	265	265	265	265	265	265

MATERIAL CHARACTERISTICS

Sample Thickness	0.013 mm/ 0.051 mm				0.015 mm/ 0.051 mm			

MATERIAL COMPOSITION

Inside Layer	polyethylene				polyethylene			
Outside Layer	polyester PET				oriented nylon			

TEST CONDITIONS

Penetrant	vanilla (vanillin)	peppermint (menthol)	piperonol (heliotropin)	camphor	vanilla (vanillin)	peppermint (menthol)	piperonol (heliotropin)	camphor

PERMEABILITY (source document units)

Days To Leakage	2	16	5	>30	2	20	5	28

<u>TABLE 321:</u> **Vanilla, Peppermint, Piperonol and Camphor Permeability Through PP/ PE Multilayer Film.**

Material Family	PP/ PE FILM			
Product Form	FILM			
Reference Number	265	265	265	265

MATERIAL CHARACTERISTICS

Sample Thickness	0.018 mm/ 0.051 mm			

MATERIAL COMPOSITION

Inside Layer	polyethylene			
Outside Layer	PVDC coated biaxially oriented polypropylene			

TEST CONDITIONS

Penetrant	vanilla (vanillin)	peppermint (menthol)	piperonol (heliotropin)	camphor

PERMEABILITY (source document units)

Days To Leakage	6	2	1	13

Multilayer Film

TABLE 322: Film Properties of Dow Chemical Nylopak Multilayer Film.

Material Family	HDPE/ EAA/ NYLON/ EAA COPOLYMER MULTILAYER FILM
Material Supplier/ Grade	DOW CHEMICAL NYLOPAK 570
Product Form	COEXTRUDED FILM
Features	barrier properties, neutral color
Applications	form-fill-seal pouch
Reference Number	258

MATERIAL CHARACTERISTICS

Sample Thickness	0.051 mm
Yield (in²/lb)	14,200

MECHANICAL PROPERTIES

Secant Modulus @ 2% Elongation - MD (MPa)	324 {r}
Secant Modulus @ 2% Elongation - TD (MPa)	331 {r}
Ultimate Tensile Strength - MD (MPa)	30.3 {r}
Ultimate Tensile Strength - TD (MPa)	30.3 {r}
Tensile Strength @ Yield - MD (MPa)	17.2 {r}
Tensile Strength @ Yield - TD (MPa)	16.5 {r}
Ultimate Elongation - MD (%)	390 {r}
Ultimate Elongation - TD (%)	380 {r}
Drop Dart Impact Strength (g)	300 {t}
Tear Resistance - MD (g)	130 {k}
Tear Resistance - TD (g)	280 {k}
Coefficient Of Friction - Kinetic, film to film	0.35 (PE/PE) {l}
Coefficient Of Friction - Kinetic, film to film	0.13 (EAA/EAA) {l}

OPTICAL PROPERTIES

Haze (%)	19 {c}
Gloss @ 45°	44 {m}
Clarity (%)	3 {s}

TABLE 322: Film Properties of Dow Chemical Nylopak Multilayer Film (cont'd).

Material Family	HDPE/ EAA/ NYLON/ EAA COPOLYMER MULTILAYER FILM
Material Supplier/ Grade	DOW CHEMICAL NYLOPAK 570
Product Form	COEXTRUDED FILM
Features	barrier properties, neutral color
Applications	form-fill-seal pouch
Reference Number	258

PERFORMANCE PROPERTIES

Seal Range (°C)	107-218 {u}
Ultimate Seal Strength (kg/m)	71.4 {v}
Unrestrained Shrink - Boiling H$_2$O, machine dir. (%)	2
Unrestrained Shrink - Boiling H$_2$O, machine dir. (%)	1
Unrestrained Shrink - 121°C Oil, MD (%)	3
Unrestrained Shrink - 121°C Oil, TD (%)	2

Chapter 66

Epoxy Resin

TABLE 323: Water Vapor, Oxygen, Nitrogen, Carbon Dioxide and Hydrogen Permeability Through Epoxy Resin.

Material Family	EPOXY RESIN				
Reference Number	121	121	121	121	121

TEST CONDITIONS

Penetrant	nitrogen	oxygen	carbon dioxide	hydrogen	water vapor
Temperature (°C)	25	25	25	25	37
Relative Humidity (%)					90
Test Method	ASTM D1434-63T	ASTM D1434-63T	ASTM D1434-63T	ASTM D1434-63T	ASTM E96-63T

PERMEABILITY (source document units)

Vapor Transmission Rate (g · mil/100 in² · day)					1.79 - 2.38 {dc}
Gas Permeability (cm³ · mil/100 in² · day)	4	5 - 10	8	110	

PERMEABILITY (normalized units)

Permeability Coefficient (cm³ · mm/m² · day · atm)	1.6	2.0 - 3.9	3.2	43.3	
Vapor Transmission Rate (g · mm/m² · day)					0.7 - 0.94

Chapter 67

Polypyrrole

Film Properties and Applications

BASF AG: Lutamer ES 9567 (features: 0.031 - 0.037 mm thick, intrinsically conductive, high flexibility; note: with benzenesulfonate anions; product form: film)

Lutamer ES 9567 is a highly flexible polypyrrole film containing benzenesulphonate anions, giving it more elasticity and folding endurance. The black film is manufactured in the form of an endless sheet through the continuous anodic oxidation of Basotronic PYR. Lutamer ES 9567 is recommended for shielding against electromagnetic fields and waves under far field conditions and in the microwave range. These products may also be used in the fields of electrochemical charge accumulation and storage as well as in gas separation.

Reference: *Topics In Chemistry - BASF Plastics Research And Development,* supplier technical report - BASF Aktiengesellschaft, 1992.

TABLE 324: Water Vapor, Nitrogen, Oxygen, Carbon Dioxide, Methane, and Hydrogen Permeability Through BASF AG Lutamer Polypyrrole Film.

Material Family	POLYPYRROLE					
Material Supplier/ Grade	BASF AG LUTAMER ES 9567					
Product Form	FILM					
Features	high flexibility, intrinsically conductive					
Reference Number	182	182	182	182	182	182

MATERIAL CHARACTERISTICS

Sample Thickness	0.031 - 0.037 mm					

MATERIAL COMPOSITION

Note	with benzenesulfonate anions					

TEST CONDITIONS

Penetrant	water vapor	nitrogen	oxygen	carbon dioxide	methane	hydrogen

PERMEABILITY (source document units)

	water vapor	nitrogen	oxygen	carbon dioxide	methane	hydrogen
Gas Permeability ($cm^3 \cdot 100\ \mu m/m^2 \cdot day \cdot atm$)		4	23	220	4	690
Vapor Transmission Rate ($g \cdot 100\ \mu m/m^2 \cdot day$)	54					

PERMEABILITY (normalized units)

	water vapor	nitrogen	oxygen	carbon dioxide	methane	hydrogen
Permeability Coefficient ($cm^3 \cdot mm/m^2 \cdot day \cdot atm$)		0.41	2.33	22.3	0.41	69.9
Vapor Transmission Rate ($g \cdot mm/m^2 \cdot day$)	5.4					

TABLE 325: Physical, Mechanical, Electrical, and Ignition Resistance Properties of BASF AG Lutamer Polypyrrole Film.

Material Family	POLYPYRROLE
Material Supplier/ Grade	**BASF AG LUTAMER ES 9567**
Product Form	**FILM**
Features	high flexibility, intrinsically conductive
Reference Number	182

MATERIAL CHARACTERISTICS

Sample Thickness	0.031 - 0.037 mm

MATERIAL COMPOSITION

Note	with benzenesulfonate anions

PHYSICAL PROPERTIES

Weight Per Unit Area (g/m^2)	50
Density (g/ml)	1.4

MECHANICAL PROPERTIES

Modulus Of Elasticity (MPa)	1500
Tensile Strength (MPa)	55-75
Ultimate Elongation (%)	40
Tear Propagation Resistance (N/mm)	90

ELECTRICAL PROPERTIES

Specific Conductivity (S/cm)	120-180
Electromagnetic Shielding - 1 - 30 MHz, magnetic near field (dB)	<0.5
Electromagnetic Shielding - 5 - 1360 MHz, far field (dB)	39
Electromagnetic Shielding - 2 - 18 GHz (dB)	40
Potentiostatic Chargeagility (Ah/kg)	80

IGNITION CHARACTERISTICS

Flammability (UL 94)	V-0
Flammability (ISO 9773)	VF-0

Chapter 68

Olefinic Thermoplastic Elastomer

Permeability to Gases and Water Vapor

Advanced Elastomer Systems: Santoprene

The permeability of SANTOPRENE rubber to common gases is essentially the same as that of conventional EPDM thermosets. Vapor permeability values are in the range normally found for conventional thermoset rubbers. Thus, SANTOPRENE rubber merits consideration for applications employing conventional thermosets (such as EPDM) where vapor permeability is of concern.

SANTOPRENE rubber has been found to exhibit low water permeation rates. This permeability to water is comparable (and in many cases superior) to that of most common rubbers and plastics. Liquid contact data suggest that SANTOPRENE rubber may be considered for film material for lining pits and ponds for the retention of water.

Reference: *Santoprene Thermoplastic Elastomer Physical Properties,* supplier technical report (AES-1015) - Advanced Elastomer Systems, 1990.

<u>TABLE 326</u>: Air Permeability Through Advanced Elastomer Systems Trefsin 3281 Olefinic Thermoplastic Elastomer.

Material Family	OLEFINIC THERMOPLASTIC ELASTOMER	
Material Supplier/ Trade Name	ADVANCED ELASTOMER SYSTEMS TREFSIN	
Grade	**3281-50**	**3281-60**
Features	barrier properties, medical grade	barrier properties, medical grade
Reference Number	12	12

MATERIAL CHARACTERISTICS

Density	0.94 g/cm³	0.97 g/cm³
Shore A Hardness	50	60

TEST CONDITIONS

Penetrant	air	

PERMEABILITY (source document units)

Gas Permeability ($ft^3 \cdot mil/ft^2 \cdot day \cdot psi$)	0.0014	0.0019

PERMEABILITY (normalized units)

Permeability Coefficient ($cm^3 \cdot mm/m^2 \cdot day \cdot atm$)	159	216

TABLE 327: Air, Nitrogen, and Oxygen Permeability Through Advanced Elastomer Systems Santoprene Olefinic Thermoplastic Elastomer.

Material Family	OLEFINIC THERMOPLASTIC ELASTOMER								
Material Supplier/ Trade Name	ADVANCED ELASTOMER SYSTEMS SANTOPRENE								
Grade	201-73	201-87	203-50	201-73	201-87	203-50	201-73	201-87	203-50
Reference Number	282	282	282	282	282	282	282	282	282

MATERIAL CHARACTERISTICS

Shore A Hardness	73	87		73	87		73	87	
Shore D Hardness			50			50			50
Sample Thickness	0.5 mm	0.5 mm	0.5 mm	0.5 mm	0.5 mm	0.5 mm	0.5 mm	0.5 mm	0.5 mm

TEST CONDITIONS

Penetrant	air			nitrogen			oxygen		
Penetrant Note	gas at 0°C and 0.11 MPa								
Temperature (°C)	23	23	23	23	23	23	23	23	23
Pressure Gradient (kPa)	110	110	110	110	110	110	110	110	110
Test Method	ASTM D1434	ASTM D1434	ASTM D1434	ASTM D1434	ASTM D1434	ASTM D1434	ASTM D1434	ASTM D1434	ASTM D1434

PERMEABILITY (source document units)

Gas Permeability (cm^3 · 0.5 mm/ 100 in^2 · day)	31	39	18	25	34	12	65	76	36

PERMEABILITY (normalized units)

Permeability Coefficient (cm^3 · mm/m^2 · day · atm)	240	302	139	194	264	93	504	589	279

TABLE 328: Gas Permeability of Carbon Dioxide, Argon and Propane Through Advanced Elastomer Systems Santoprene Olefinic Thermoplastic Elastomer.

Material Family	OLEFINIC THERMOPLASTIC ELASTOMER								
Material Supplier/ Trade Name	ADVANCED ELASTOMER SYSTEMS SANTOPRENE								
Grade	201-73	201-87	203-50	201-73	201-87	203-50	201-73	201-87	203-50
Reference Number	282	282	282	282	282	282	282	282	282

MATERIAL CHARACTERISTICS

Shore A Hardness	73	87		73	87		73	87	
Shore D Hardness			50			50			50
Sample Thickness	0.5 mm	0.5 mm	0.5 mm	0.5 mm	0.5 mm	0.5 mm	0.5 mm	0.5 mm	0.5 mm

TEST CONDITIONS

Penetrant	carbon dioxide			argon			propane		
Penetrant Note	gas at 0°C and 0.11 MPa								
Temperature (°C)	23	23	23	23	23	23	23	23	23
Pressure Gradient (kPa)	110	110	110	110	110	110	110	110	110
Test Method	ASTM D1434	ASTM D1434	ASTM D1434	ASTM D1434	ASTM D1434	ASTM D1434	ASTM D1434	ASTM D1434	ASTM D1434

PERMEABILITY (source document units)

Gas Permeability ($cm^3 \cdot 0.5$ mm/ 100 $in^2 \cdot$ day)	390	260	170	67	77	51	150	430	250

PERMEABILITY (normalized units)

Permeability Coefficient ($cm^3 \cdot$ mm/$m^2 \cdot$ day $\cdot$ atm)	3022	2015	1318	519	597	395	1162	3332	1938

TABLE 329: Water Vapor Permeability Through Advanced Elastomer Systems Santoprene Olefinic Thermoplastic Elastomer.

Material Family	OLEFINIC THERMOPLASTIC ELASTOMER					
Material Supplier/ Trade Name	ADVANCED ELASTOMER SYSTEMS SANTOPRENE					
Grade	201-73	201-87	203-50	201-73	201-87	203-50
Reference Number	282	282	282	282	282	282

MATERIAL CHARACTERISTICS

Shore A Hardness	73	87		73	87	
Shore D Hardness			50			50
Sample Thickness	0.5 mm	0.5 mm	0.5 mm	0.5 mm	0.5 mm	0.5 mm

TEST CONDITIONS

Penetrant	water vapor					
Temperature (°C)	25	25	25	25	25	25
Test Method	ASTM E96, proc. A	ASTM E96, proc. A	ASTM E96, proc. A	ASTM E96, proc. BW	ASTM E96, proc. BW	ASTM E96, proc. BW
Test Note	saturated vapor over liquid water contacts the rubber, with 25% RH on the opposite side			liquid water contacts the rubber, with 75% RH on the opposite side		

PERMEABILITY (source document units)

Vapor Transmission Rate (g ·0.5 mm/ m² · day)	0.97	0.32	0.45	0.45	0.45	1.61

PERMEABILITY (normalized units)

Vapor Transmission Rate (g · mm/m² · day)	0.49	0.16	0.23	0.23	0.23	0.81

Chapter 69

Polyamide Thermoplastic Elastomer

Permeability to Refrigerants

DuPont: Zytel FN

Test data show low refrigerant permeation losses for Zytel FN when compared to conventional hose and tubing materials.

Reference: *Zytel FN Flexible Nylon Alloy Products and Properties Guide,* supplier technical report (H-14079-1) - DuPont Company, 1990.

<u>TABLE 330:</u> Oxygen, Carbon Dioxide, Nitrogen and Helium Permeability Through Shore D Hardness 25 and 35 Atochem Pebax Polyether Block Amide Thermoplastic Elastomer Film.

Material Family	POLYAMIDE THERMOPLASTIC ELASTOMER							
Material Supplier/ Grade	ATOCHEM PEBAX 3533				ATOCHEM PEBAX 2533			
Product Form	FILM							
Reference Number	287	287	287	287	287	287	287	287

MATERIAL CHARACTERISTICS

Shore D Hardness	35				25			
Sample Thickness	0.12 mm				0.12 mm			

TEST CONDITIONS

Penetrant	oxygen	carbon dioxide	nitrogen	helium	oxygen	carbon dioxide	nitrogen	helium
Temperature (°C)	23	23	23	23	23	23	23	23
Relative Humidity (%)	dry	dry	dry	dry	dry	dry	dry	dry

PERMEABILITY (source document units)

Gas Permeability ($cm^3 \cdot mm/cm^2 \cdot sec \cdot cm\,Hg$)	131×10^{-10}	1790×10^{-10}	100×10^{-10}	174×10^{-10}	150×10^{-10}	2600×10^{-10}	170×10^{-10}	235×10^{-10}

PERMEABILITY (normalized units)

Permeability Coefficient ($cm^3 \cdot mm/m^2 \cdot day \cdot atm$)	860	11,753	657	1142	985	17,073	1116	1543

TABLE 331: Oxygen, Carbon Dioxide, Nitrogen, Helium and Propane Permeability Through Shore D Hardness 40 and 55 Atochem Pebax Polyether Block Amide Thermoplastic Elastomer Film.

Material Family	POLYAMIDE THERMOPLASTIC ELASTOMER								
Material Supplier/ Grade	ATOCHEM PEBAX 5533					ATOCHEM PEBAX 4033			
Product Form	FILM								
Reference Number	287	287	287	287	287	287	287	287	287

MATERIAL CHARACTERISTICS

Shore D Hardness	55					40			
Sample Thickness	0.12 mm					0.12 mm			

TEST CONDITIONS

Penetrant	oxygen	carbon dioxide	nitrogen	helium	propane	oxygen	carbon dioxide	nitrogen	helium
Temperature (°C)	23	23	23	23	23	23	23	23	23
Relative Humidity (%)	dry	dry	dry	dry	dry	dry	dry	dry	dry

PERMEABILITY (source document units)

Gas Permeability ($cm^3 \cdot mm/cm^2 \cdot sec \cdot cmHg$)	35×10^{-10}	500×10^{-10}	11×10^{-10}	70×10^{-10}	120×10^{-10}	59×10^{-10}	780×10^{-10}	39×10^{-10}	147×10^{-10}

PERMEABILITY (normalized units)

Permeability Coefficient ($cm^3 \cdot mm/m^2 \cdot day \cdot atm$)	230	3283	72	460	789	387	5122	256	965

TABLE 332: Oxygen, Carbon Dioxide, Nitrogen, Helium and Propane Permeability Through Shore D Hardness 63 Atochem Pebax Polyether Block Amide Thermoplastic Elastomer Film.

Material Family	POLYAMIDE THERMOPLASTIC ELASTOMER				
Material Supplier/ Grade	ATOCHEM PEBAX 6333				
Product Form	FILM				
Reference Number	287	287	287	287	287

MATERIAL CHARACTERISTICS

Shore D Hardness	63
Sample Thickness	0.12 mm

TEST CONDITIONS

Penetrant	oxygen	carbon dioxide	nitrogen	helium	propane
Temperature (°C)	23	23	23	23	23
Relative Humidity (%)	dry	dry	dry	dry	dry

PERMEABILITY (source document units)

Gas Permeability ($cm^3 \cdot mm/cm^2 \cdot sec \cdot cm\ Hg$)	31×10^{-10}	420×10^{-10}	5×10^{-10}	46×10^{-10}	36×10^{-10}

PERMEABILITY (normalized units)

Permeability Coefficient ($cm^3 \cdot mm/m^2 \cdot day \cdot atm$)	204	2758	33	302	236

Polyamide TPE

TABLE 333: Water Vapor Permeability Through Atochem Pebax Polyether Block Amide Thermoplastic Elastomer Film.

Material Family	POLYAMIDE THERMOPLASTIC ELASTOMER				
Material Supplier/ Trade Name	ATOCHEM PEBAX				
Grade	6333	5533	4033	3533	2533
Product Form	FILM				
Reference Number	287	287	287	287	287

MATERIAL CHARACTERISTICS

Shore D Hardness	63	55	40	35	25
Sample Thickness	0.12 mm	0.12 mm	0.12 mm	0.12 mm	0.12 mm

TEST CONDITIONS

Penetrant	water vapor				
Temperature (°C)	38	38	38	38	38
Relative Humidity (%)	100	100	100	100	100

PERMEABILITY (source document units)

Vapor Transmission Rate (g · mm/m^2 · day)	31	34	38	67	89

PERMEABILITY (normalized units)

Vapor Transmission Rate (g · mm/m^2 · day)	31	34	38	67	89

Polyamide TPE

TABLE 334: Air Conditioning Refrigerants Permeation Loss Through DuPont Company Zytel FN 726 Polyamide Thermoplastic Elastomer.

Material Family	POLYAMIDE THERMOPLASTIC ELASTOMER		
Material Supplier/ Grade	DUPONT ZYTEL FN 726		
Reference Number	275	275	275

MATERIAL CHARACTERISTICS

Shore D Hardness	58	58	58
Sample Thickness	1 mm	1 mm	1 mm
Sample Length	305 mm	305 mm	305 mm
Sample Inside Diameter	15.9 mm	15.9 mm	15.9 mm

TEST CONDITIONS

Penetrant	Freon 12	HCFCX-134a	HCFC-22/ HCFC-124/ HFC-152a
Penetrant Note	air conditioning refrigerant		air conditioning refrigerant, ternary blend
Temperature (°C)	93	93	93
Test Condition Note	refrigerant at saturated vapor pressure	refrigerant at saturated vapor pressure	refrigerant at saturated vapor pressure
Test Note	calculated from permeation coefficient data		

PERMEABILITY (source document units)

Permeation Loss (lb/ft-yr)	0.012	0.015	0.086

Chapter 70

Polybutadiene Thermoplastic Elastomer

Permeability

Japan Synthetic Rubber Company: JSR RB820 (chemical type: syndiotactic 1,2-polybutadiene; features: approximately 15% crystallinity; product form: film)

Carbon dioxide gas and oxygen gas are permeable through JSR RB820 film, depending on film thickness. Thin JSR RB820 film of less than 50 μm has a permeability to these gases 3 to 5 times greater than low-density polyethylene film, but the permeability of JSR RB820 film about 1mm thick is about the same as that of low-density polyethylene film. Ethylene oxide gas permeates more easily through JSR RB820 film than carbon dioxide gas.

Reference: *Japan Synthetic Rubber JSR RB,* supplier design guide - Japan Synthetic Rubber Company.

Film Properties and Applications

Japan Synthetic Rubber Company: JSR RB820 (chemical type: syndiotactic 1,2-polybutadiene; features: biaxially oriented, shrinkable film, approximately 15% crystallinity; product form: film)

Biaxally oriented JSR RB820 film has the remarkable characteristic of low-temperature shrinkability. JSR RB820 film is improved in strength by biaxial orientation. Such advantages as pliability, gas permeability and heat-sealability do not change after bi-orientation. Owing to the nature of JSR RB820 which causes biaxially oriented film to shrink quickly at low temperature, it can be used for packaging materials like foods which cannot be packaged by conventional shrinkable film because they need high temperature to shrink. The shrinking force is subtle, such that it can be applied to packaging soft substances which tend to be deformed by the shrinking force when a conventional shrinkable film is used.

Reference: *Japan Synthetic Rubber JSR RB,* supplier design guide - Japan Synthetic Rubber Company.

Japan Synthetic Rubber Company: JSR RB820 (chemical type: syndiotactic 1,2-polybutadiene; features: approximately 15% crystallinity; product form: film)

JSR RB820 is suitable for food packaging. It is pliable and characterized by transparency, gloss, gas permeability, Elmendorf tear strength and density. Both JSR RB's transparency and gloss are as high in quality as those of soft polyvinylchloride which is regarded as having the best transparency and gloss among commercially available plastic films. Having excellent Elmendorf tear strength and elongation, it follows that JSR RB820 film has excellent puncture strength. Because of the pliability of JSR RB, it is not necessary to add plasticizer.

The density of JSR RB is 0.91. As compared with soft polyvinylchloride (density: 1.28 in compounds), JSR RB saves about 30 to 40% by weight per unit volume of the amount of resin. JSR RB820 film can be heat-sealed at a low temperature (75°C to 80°C) and its heat-sealed parts show good strength. The film also shows good seal strength when sealed with an ultrasonic welder.

Reference: *Japan Synthetic Rubber JSR RB,* supplier design guide - Japan Synthetic Rubber Company.

Japan Synthetic Rubber Company: JSR RB830 (density: 0.910 g/cm^3; features: stretch film, approximately 29% crystallinity; product form: film)

Stretch film is one of the applications which take advantage of such characteristics of JSR RB as safety for food hygiene, transparency, gloss, elongation, tackiness, and puncture strength. The 18 μm thick JSR RB830 film made by a blown film process is superior to a commercial stretch film of polyvinylchloride about 18 μm thick as it is lighter in weight and has greater elongation and tear resistance. Although the modulus is slightly lower, no trouble occurs in practical use.

When using JSR RB830 as stretch film, it is necessary to improve the surface properties of the film for workability of packaging operations as a commercial film. For this purpose, lubricants, anti-fogging agents and other surface improvers which are safe for food hygiene are added to JSR RB830. The results of wrapping tests with hand wrappers and various kinds of automatic wrapping machines show that the surface-improved stretch film of JSR RB830 does not raise any problem with the physical properties. JSR RB830 stretch film is suitable for packaging meats, vegetables, fruits, fish, shellfish, dried fish, and other foods.

Reference: *Japan Synthetic Rubber JSR RB,* supplier design guide - Japan Synthetic Rubber Company.

TABLE 335: Carbon Dioxide, Oxygen and Water Vapor Permeability Through Japan Synthetic Rubber JSR RB Polybutadiene Thermoplastic Elastomer Stretch Film.

Material Family	POLYBUTADIENE THERMOPLASTIC ELASTOMER		
Material Supplier/ Grade	JAPAN SYNTHETIC RUBBER COMPANY JSR RB830		
Product Form	FILM		
Features	4.8 blow up ratio, approximately 29% crystallinity, stretch film		
Manufacturing Method	blown film		
Reference Number	216	216	216

MATERIAL CHARACTERISTICS

Density	0.910 g/cm³	0.910 g/cm³	0.910 g/cm³
Sample Thickness	0.018 mm	0.018 mm	0.018 mm

TEST CONDITIONS

Penetrant	water vapor	carbon dioxide	oxygen
Test Method	JIS Z0208	ASTM D1434	ASTM D1434

PERMEABILITY (source document units)

Vapor Transmission Rate (g · 100 μm/m² · day)	70		
Gas Permeability (cm³ · 100 μm/ m² · atm · day)		29,000	5500

PERMEABILITY (normalized units)

Permeability Coefficient (cm³ · mm/m² · day · atm)		2900	550
Vapor Transmission Rate (g · mm/m² · day)	7		

TABLE 336: Carbon Dioxide, Oxygen, Ethylene Oxide and Water Vapor Permeability Through Japan Synthetic Rubber JSR RB Polybutadiene Thermoplastic Elastomer Film.

Material Family	POLYBUTADIENE THERMOPLASTIC ELASTOMER							
Material Supplier/ Grade	JAPAN SYNTHETIC RUBBER COMPANY JSR RB820				JAPAN SYNTHETIC RUBBER COMPANY JSR RB830			
Product Form	FILM							
Features	2.5 blow up ratio, approximately 15% crystallinity, transparent				2.5 blow up ratio, approximately 25% crystallinity, transparent			
Manufacturing Method	blown film	blown film	blown film	blown film	blown film	blown film	blown film	blown film
Reference Number	216	216	216	216	216	216	216	216

MATERIAL CHARACTERISTICS

Density	0.910 g/cm³	0.910 g/cm³	0.910 g/cm³	0.910 g/cm³	0.910 g/cm³	0.910 g/cm³	0.910 g/cm³	0.910 g/cm³
Sample Thickness	0.05 mm	0.05 mm	0.05 mm	0.05 mm	0.05 mm	0.05 mm	0.05 mm	0.05 mm
Chemical Type	syndiotactic 1,2-polybutadiene				syndiotactic 1,2-polybutadiene			

TEST CONDITIONS

Penetrant	water vapor	carbon dioxide	oxygen	ethylene oxide	water vapor	carbon dioxide	oxygen	ethylene oxide
Test Method	JIS Z0208	ASTM D1434	ASTM D1434	ASTM D1434	JIS Z0208	ASTM D1434	ASTM D1434	ASTM D1434

PERMEABILITY (source document units)

Vapor Transmission Rate (g · 100 μm/m² · day)	98				70			
Gas Permeability (cm³ · 100 μm/ m² · atm · day)		28,000	5500	320,000		29,000	5500	250,000

PERMEABILITY (normalized units)

Permeability Coefficient (cm³ · mm/m² · day · atm)		2800	550	32,000		2900	550	25,000
Vapor Transmission Rate (g · mm/m² · day)	9.8				7			

TABLE 337: Film Properties of Japan Synthetic Rubber JSR RB Polybutadiene Thermoplastic Elastomer Film and Stretch Film.

Material Family	POLYBUTADIENE THERMOPLASTIC ELASTOMER		
Material Supplier/ Grade	JAPAN SYNTHETIC RUBBER COMPANY		
Material Supplier/ Grade	JSR RB820	JSR RB830	
Product Form	FILM		
Features	2.5 blow up ratio, approximately 15% crystallinity, transparent	2.5 blow up ratio, approximately 25% crystallinity, transparent	4.8 blow up ratio, approximately 29% crystallinity, stretch film
Manufacturing Method	blown film	blown film	blown film
Reference Number	216	216	216

MATERIAL CHARACTERISTICS

Density	0.910 g/cm³	0.910 g/cm³	0.910 g/cm³
Sample Thickness	0.05 mm	0.05 mm	0.018 mm

MATERIAL COMPOSITION

Chemical Type	syndiotactic 1,2-polybutadiene	syndiotactic 1,2-polybutadiene	

MECHANICAL PROPERTIES

100% Modulus - MD (MPa)			10.3 {ci}
100% Modulus - TD (MPa)			7.4 {ci}
150% Modulus - MD (MPa)			11.8 {ci}
150% Modulus - TD (MPa)			8.8 {ci}
Tensile Strength - MD (MPa)	19.6 {ci}	24.0 {ci}	23.5 {ci}
Tensile Strength - TD (MPa)	19.6 {ci}	22.6 {ci}	24.5 {ci}
Ultimate Elongation - MD (%)	500 {ci}	510 {ci}	370 {ci}
Ultimate Elongation - TD (%)	570 {ci}	600 {ci}	450 {ci}
Tear Strength - MD (g/mm)	7800 {cj}	15000 {cj}	6500 {cm}
Tear Strength - TD (g/mm)	7600 {cj}	16000 {cj}	6500 {cm}

OPTICAL PROPERTIES

Light Transmittance (%)	92 {ck}	92 {ck}	92 {ck}
Haze (%)	1 {ck}	1 {ck}	1.0 {ck}
Gloss	130 {cl}	130 {cl}	130 {cl}

GRAPH 72: **Carbon Dioxide Permeability vs. Thickness through Polybutadiene Thermoplastic Elastomer.**

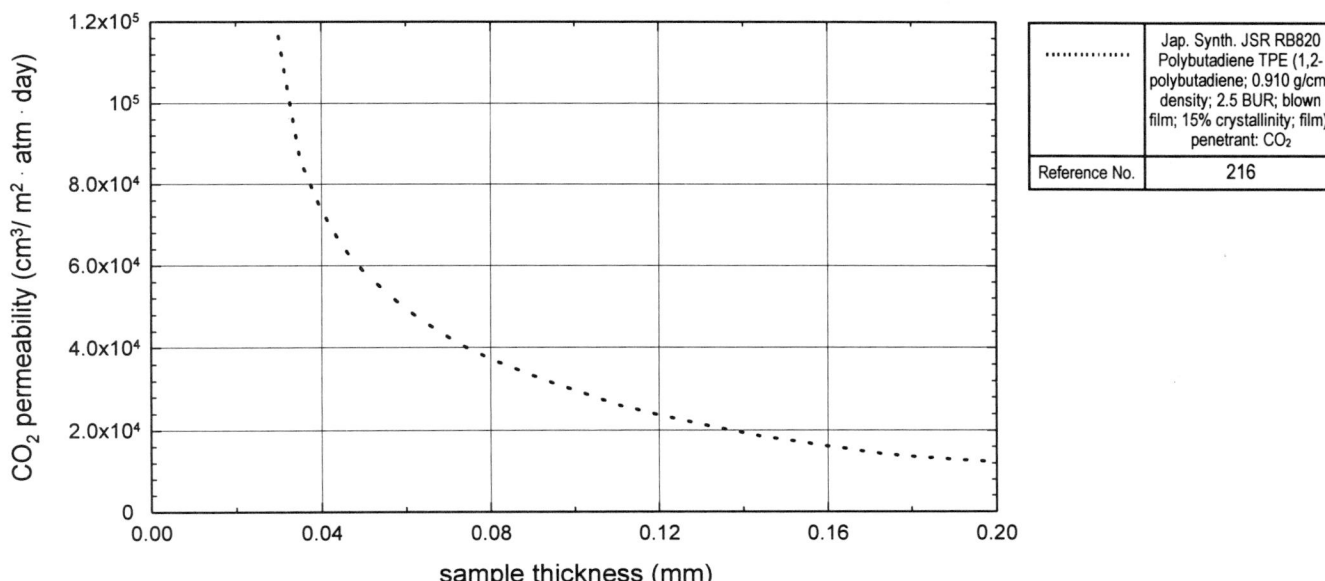

....	Jap. Synth. JSR RB820 Polybutadiene TPE (1,2-polybutadiene; 0.910 g/cm³ density; 2.5 BUR; blown film; 15% crystallinity; film); penetrant: CO_2
Reference No.	216

GRAPH 73: **Oxygen Permeability vs. Thickness through Polybutadiene Thermoplastic Elastomer.**

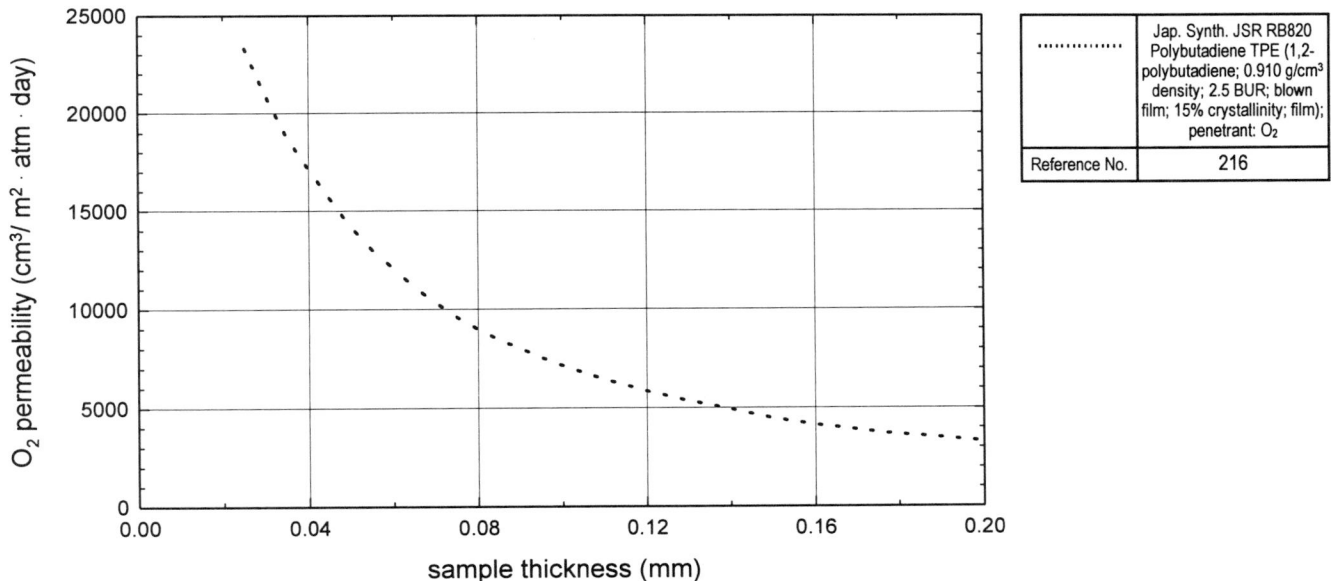

....	Jap. Synth. JSR RB820 Polybutadiene TPE (1,2-polybutadiene; 0.910 g/cm³ density; 2.5 BUR; blown film; 15% crystallinity; film); penetrant: O_2
Reference No.	216

Polybutadiene TPE

Chapter 71

Polyester Thermoplastic Elastomer

Permeability to Gases

DuPont: Hytrel

Hytrel polyester elastomers have a high degree of permeability to polar molecules, such as water, but are resistant to permeation by non-polar hydrocarbons and refrigerant gases.

In permeability to moisture, Hytrel is comparable to the polyether-based urethanes and, therefore, is useful as a fabric coating for apparel. Its low permeability to refrigerant gases and hydrocarbons such as propane makes Hytrel of interest for use in refrigerant hose or in flexible hose or tubing to transmit gas for heating and cooking.

Reference: *Hytrel Polyester Elastomer - Gas Permeability (HYT-506B),* supplier technical report (E-37763) - DuPont Company, 1984.

TABLE 338: **Gas Permeability of Various Gases Through DuPont Company Hytrel 4056 Polyester Thermoplastic Elastomer.**

Material Family	POLYESTER THERMOPLASTIC ELASTOMER								
Material Supplier/ Grade	DUPONT HYTREL 4056								
Reference Number	274	274	274	274	274	274	274	274	274

MATERIAL CHARACTERISTICS

Shore D Hardness	40								

TEST CONDITIONS

Penetrant	air	nitrogen	carbon dioxide	helium	propane	water	Freon 12	Freon 22	Freon 114
Temperature (°C)	21.5	21.5	21.5	21.5	21.5	25	21.5	21.5	21.5
Relative Humidity (%)						90			
Pressure Gradient (kPa)	34.5	34.5	34.5	34.5	34.5	34.5	34.5	34.5	34.5
Test Note						assuming that permeability laws hold for water			

PERMEABILITY (source document units)

Gas Permeability (cm^3 (STP) ·cm/ cm^2 · sec · atm)	2.4×10^{-8}	1.7×10^{-8}	3.5×10^{-7}	15.7×10^{-8}	$<0.2 \times 10^{-8}$	3.1×10^{-5}	1.4×10^{-8}	0.47×10^{-8}	41×10^{-8}

PERMEABILITY (normalized units)

Permeability Coefficient (cm^3 · mm/m^2 · day · atm)	207	147	3024	1356	<17	267,840	121	41	3542

TABLE 339: Gas Permeability of Various Gases Through DuPont Company Hytrel 5556 Polyester Thermoplastic Elastomer.

Material Family	POLYESTER THERMOPLASTIC ELASTOMER								
Material Supplier/ Grade	DUPONT HYTREL 5556								
Reference Number	274	274	274	274	274	274	274	274	274

MATERIAL CHARACTERISTICS

Shore D Hardness	55								

TEST CONDITIONS

Penetrant	air	nitrogen	carbon dioxide	helium	propane	water	Freon 12	Freon 22	Freon 114
Temperature (°C)	21.5	21.5	21.5	21.5	21.5	25	21.5	21.5	21.5
Relative Humidity (%)						90			
Pressure Gradient (kPa)	34.5	34.5	34.5	34.5	34.5	34.5	34.5	34.5	34.5
Test Note						assuming that permeability laws hold for water			

PERMEABILITY (source document units)

Gas Permeability $(cm^3 (STP) \cdot cm/ cm^2 \cdot sec \cdot atm)$	1.8×10^{-8}	1.4×10^{-8}	1.8×10^{-7}	9.9×10^{-8}	$<0.2 \times 10^{-8}$	2.4×10^{-5}	1.2×10^{-8}	0.59×10^{-8}	28×10^{-8}

PERMEABILITY (normalized units)

Permeability Coefficient $(cm^3 \cdot mm/m^2 \cdot day \cdot atm)$	156	121	1555	855	<17	207,360	104	51	2419

TABLE 340: Gas Permeability of Various Gases Through DuPont Company Hytrel 4056 and Hytrel 7246 Polyester Thermoplastic Elastomer.

Material Family	POLYESTER THERMOPLASTIC ELASTOMER						
Material Supplier/ Grade	DUPONT HYTREL 6346				DUPONT HYTREL 7246		
Reference Number	274	274	274	274	274	274	274

MATERIAL CHARACTERISTICS

Shore D Hardness	63	63	63	63	72	72	72

TEST CONDITIONS

Penetrant	propane	Freon 12	Freon 22	Freon 114	helium	Freon 12	Freon 114
Temperature (°C)	21.5	21.5	21.5	21.5	21.5	21.5	21.5
Pressure Gradient (kPa)	34.5	34.5	34.5	34.5	34.5	34.5	34.5

PERMEABILITY (source document units)

Gas Permeability $(cm^3 (STP) \cdot cm/ cm^2 \cdot sec \cdot atm)$	$<0.2 \times 10^{-8}$	1.2×10^{-8}	$<0.2 \times 10^{-8}$	4.6×10^{-8}	3.2×10^{-8}	0.82×10^{-8}	2.7×10^{-8}

PERMEABILITY (normalized units)

Permeability Coefficient $(cm^3 \cdot mm/m^2 \cdot day \cdot atm)$	<17	104	<17	397	276	71	233

TABLE 341: Water Vapor, Carbon Dioxide, Oxygen and Nitrogen Permeability Through Eastman Chemical Kodar Ecdel Copolyester Ether Thermoplastic Elastomer Film.

Material Family	POLYESTER THERMOPLASTIC ELASTOMER			
Material Supplier/ Grade	EASTMAN ECDEL			
Product Form	FILM			
Features	crystalline, transparent	crystalline, transparent	crystalline, transparent	crystalline, transparent
Reference Number	60	60	60	60

MATERIAL CHARACTERISTICS

Sample Thickness	0.11 - 0.14 mm	0.11 - 0.14 mm	0.11 - 0.14 mm	0.11 - 0.14 mm

MATERIAL COMPOSITION

Chemical Type	copolyester ether (COPE)	copolyester ether (COPE)	copolyester ether (COPE)	copolyester ether (COPE)

TEST CONDITIONS

Penetrant	water vapor	carbon dioxide	oxygen	nitrogen
Temperature (°C)	38	23	23	23
Relative Humidity (%)	90			
Test Method	ASTM F372	ASTM D1434	ASTM D1434	ASTM D1434
Test Note	Mocon value, confirmed by ASTM E96E			

PERMEABILITY (source document units)

Gas Permeability (cm^3 · mil/100 in^2 · day)		3225	323	65
Gas Permeability (cm^3 · mm/m^2 · day · atm)		1270	127	25
Vapor Transmission Rate (g/m^2 · day)	155			
Vapor Transmission Rate (g/day · 100 in^2)	10			

PERMEABILITY (normalized units)

Permeability Coefficient (cm^3 · mm/m^2 · day · atm)		1270	127.2	25.6
Vapor Transmission Rate (g · mm/m^2 · day)	19.4			

TABLE 342: Film Properties of Eastman Chemical Kodar Ecdel Copolyester Ether Thermoplastic Elastomer Film.

Material Family	POLYESTER THERMOPLASTIC ELASTOMER
Material Supplier/ Grade	EASTMAN ECDEL
Product Form	FILM
Features	crystalline, transparent
Reference Number	60

MATERIAL CHARACTERISTICS

Sample Thickness	0.11 - 0.14 mm
Density (g/cm³)	1.124 {da}

MATERIAL COMPOSITION

Chemical Type	copolyester ether (COPE)

MECHANICAL PROPERTIES

Modulus Of Elasticity (Mpa)	126 {db}
Tensile Strength @ Yield - machine direction (MPa)	12.4 {o}
Tensile Strength @ Yield - transverse direction (MPa)	11.7 {o}
Ultimate Elongation - machine direction (%)	700 {o}
Ultimate Elongation - transverse direction (%)	800 {o}
Tear Strength, Propagated, Elmendorf - machine direction (g)	1500 - 1700 {k}
Coefficient Of Friction - Kinetic, film to film	>1.0 {l}

OPTICAL PROPERTIES

Light Transmittance (%)	90 {c}
Haze (%)	1.0 {c}
Gloss @ 45°	85 {m}
Index Of Refraction	1.519 {cy}

Chapter 72

Thermoplastic Polyester-Polyurethane Elastomer

TABLE 343: Oxygen Gas and Water Vapor Permeability Through BASF Elastollan C Polyester Urethane Thermoplastic Elastomer.

Material Family	THERMOPLASTIC POLYESTER-POLYURETHANE ELASTOMER						
Material Supplier/ Trade Name	BASF ELASTOLLAN						
Grade	C80A	C85A	C90A	C95A	C80A	C95A	C64D
Reference Number	130	130	130	130	130	130	130

MATERIAL CHARACTERISTICS

Shore A Hardness	80	85	90	95	80	95	
Shore D Hardness							64

TEST CONDITIONS

Penetrant	oxygen				water vapor		
Temperature (°C)	20	20	20	20	23	23	23
Relative Humidity (%)					93% differential	93% differential	93% differential

PERMEABILITY (source document units)

Vapor Transmission Rate (g/m^2 · day)					14	6	2
Gas Permeability (1x10^{-18} m^2/sec · Pa)	14	10	7	4			

PERMEABILITY (normalized units)

Permeability Coefficient (cm^3 · mm/m^2 · day · atm)	123	87.5	61.3	35			
Vapor Transmission Rate (g · mm/m^2 · day)					not applicable without thickness	not applicable without thickness	not applicable without thickness

TABLE 344: Carbon Dioxide and Hydrogen Gas Permeability Through BASF Elastollan C Polyester Urethane Thermoplastic Elastomer.

Material Family	THERMOPLASTIC POLYESTER-POLYURETHANE ELASTOMER							
Material Supplier/ Trade Name	BASF ELASTOLLAN							
Grade	C80A	C85A	C90A	C95A	C80A	C85A	C90A	C95A
Reference Number	130	130	130	130	130	130	130	130
MATERIAL CHARACTERISTICS								
Shore A Hardness	80	85	90	95	80	85	90	95
TEST CONDITIONS								
Penetrant	carbon dioxide				hydrogen			
Temperature (°C)	20	20	20	20	20	20	20	20
PERMEABILITY (source document units)								
Gas Permeability (1×10^{-18} m^2/sec · Pa)	200	150	40	20	45	40	30	20
PERMEABILITY (normalized units)								
Permeability Coefficient (cm^3 · mm/m^2 · day · atm)	1751	1313	350	175	394	350	263	175

TABLE 345: Argon and Methane Gas Permeability Through BASF Elastollan C Polyester Urethane Thermoplastic Elastomer.

Material Family	THERMOPLASTIC POLYESTER-POLYURETHANE ELASTOMER							
Material Supplier/ Trade Name	BASF ELASTOLLAN							
Grade	C80A	C85A	C90A	C95A	C80A	C85A	C90A	C95A
Reference Number	130	130	130	130	130	130	130	130
MATERIAL CHARACTERISTICS								
Shore A Hardness	80	85	90	95	80	85	90	95
TEST CONDITIONS								
Penetrant	argon 3 X N₂				methane			
Temperature (°C)	20	20	20	20	20	20	20	20
PERMEABILITY (source document units)								
Gas Permeability (1×10^{-18} m^2/sec · Pa)	12	9	5	3	11	6	4	2
PERMEABILITY (normalized units)								
Permeability Coefficient (cm^3 · mm/m^2 · day · atm)	105	78.8	43.8	26.3	96.3	52.5	35	17.5

<u>TABLE 346:</u> Helium and Nitrogen Gas Permeability Through BASF Elastollan C Polyester Urethane Thermoplastic Elastomer.

Material Family	THERMOPLASTIC POLYESTER-POLYURETHANE ELASTOMER							
Material Supplier/ Trade Name	BASF ELASTOLLAN							
Grade	C80A	C85A	C90A	C95A	C80A	C85A	C90A	C95A
Reference Number	130	130	130	130	130	130	130	130

MATERIAL CHARACTERISTICS

Shore A Hardness	80	85	90	95	80	85	90	95

TEST CONDITIONS

Penetrant	helium				nitrogen			
Temperature (°C)	20	20	20	20	20	20	20	20

PERMEABILITY (source document units)

Gas Permeability (1×10^{-18} m^2/sec · Pa)	35	30	25	20	4	3	2	1

PERMEABILITY (normalized units)

Permeability Coefficient (cm^3 · mm/m^2 · day · atm)	306	263	219	175	35	26.3	17.5	8.8

Thermoplastic Polyether-Polyurethane Elastomer

TABLE 347: Oxygen Gas and Water Vapor Permeability Through BASF Elastollan 1100 Polyether Urethane Thermoplastic Elastomer.

Material Family	THERMOPLASTIC POLYETHER-POLYURETHANE ELASTOMER						
Material Supplier/ Trade Name	BASF ELASTOLLAN						
Grade	1180A	1185A	1190A	1195A	1180A	1195A	1164D
Reference Number	130	130	130	130	130	130	130

MATERIAL CHARACTERISTICS

Shore A Hardness	80	85	90	95	80	95	
Shore D Hardness							64

TEST CONDITIONS

Penetrant	oxygen				water vapor		
Temperature (°C)	20	20	20	20	23	23	23
Relative Humidity (%)					93% differential	93% differential	93% differential

PERMEABILITY (source document units)

Vapor Transmission Rate (g/m^2 · day)					18	9	3
Gas Permeability (1x10^{-18} m^2/sec · Pa)	21	16	12	8			

PERMEABILITY (normalized units)

Permeability Coefficient (cm^3 · mm/m^2 · day · atm)	184	140	105	70			
Vapor Transmission Rate (g · mm/m^2 · day)					not applicable without thickness	not applicable without thickness	not applicable without thickness

TABLE 348: Carbon Dioxide and Hydrogen Gas Permeability Through BASF Elastollan 1100 Polyether Urethane Thermoplastic Elastomer.

Material Family	THERMOPLASTIC POLYETHER-POLYURETHANE ELASTOMER							
Material Supplier/ Trade Name	BASF ELASTOLLAN							
Grade	**1180A**	**1185A**	**1190A**	**1195A**	**1180A**	**1185A**	**1190A**	**1195A**
Reference Number	130	130	130	130	130	130	130	130

MATERIAL CHARACTERISTICS

Shore A Hardness	80	85	90	95	80	85	90	95

TEST CONDITIONS

Penetrant	carbon dioxide				hydrogen			
Temperature (°C)	20	20	20	20	20	20	20	20

PERMEABILITY (source document units)

Gas Permeability (1×10^{-18} m^2/sec · Pa)	230	180	130	90	70	60	50	40

PERMEABILITY (normalized units)

Permeability Coefficient (cm^3 · mm/m^2 · day · atm)	2014	1576	1138	788	613	525	438	350

TABLE 349: Argon and Methane Gas Permeability Through BASF Elastollan 1100 Polyether Urethane Thermoplastic Elastomer.

Material Family	THERMOPLASTIC POLYETHER-POLYURETHANE ELASTOMER							
Material Supplier/ Trade Name	BASF ELASTOLLAN							
Grade	**1180A**	**1185A**	**1190A**	**1195A**	**1180A**	**1185A**	**1190A**	**1195A**
Reference Number	130	130	130	130	130	130	130	130

MATERIAL CHARACTERISTICS

Shore A Hardness	80	85	90	95	80	85	90	95

TEST CONDITIONS

Penetrant	argon				methane			
Temperature (°C)	20	20	20	20	20	20	20	20

PERMEABILITY (source document units)

Gas Permeability (1×10^{-18} m^2/sec · Pa)	14	9	7	6	18	14	9	5

PERMEABILITY (normalized units)

Permeability Coefficient (cm^3 · mm/m^2 · day · atm)	123	78.8	61.3	52.5	158	123	78.8	43.8

TABLE 350: Helium and Nitrogen Gas Permeability Through BASF Elastollan 1100 Polyether Urethane Thermoplastic Elastomer.

Material Family	THERMOPLASTIC POLYETHER-POLYURETHANE ELASTOMER							
Material Supplier/ Trade Name	BASF ELASTOLLAN							
Grade	**1180A**	**1185A**	**1190A**	**1195A**	**1180A**	**1185A**	**1190A**	**1195A**
Reference Number	130	130	130	130	130	130	130	130

MATERIAL CHARACTERISTICS

Shore A Hardness	80	85	90	95	80	85	90	95

TEST CONDITIONS

Penetrant	helium				nitrogen			
Temperature (°C)	20	20	20	20	20	20	20	20

PERMEABILITY (source document units)

Gas Permeability (1×10^{-18} m²/sec · Pa)	50	40	20	30	6	5	4	3

PERMEABILITY (normalized units)

Permeability Coefficient (cm³ · mm/m² · day · atm)	438	350	175	263	52.5	43.8	35.0	26.3

GRAPH 74: Nitrogen Permeability vs. Temperature through Thermoplastic Polyether-Polyurethane Elastomer.

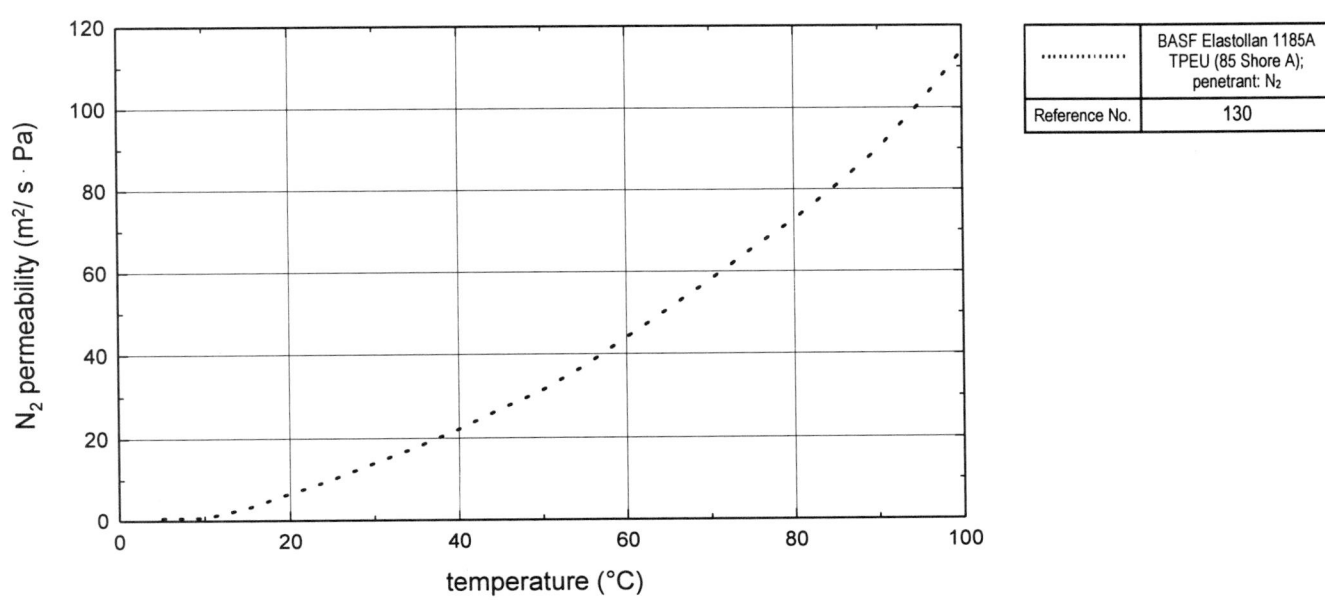

··············	BASF Elastollan 1185A TPEU (85 Shore A); penetrant: N₂
Reference No.	130

Chapter 74

Styrenic Thermoplastic Elastomer

Permeability

Shell Chemical: Kraton D (chemical type: styrene butadiene styrene block copolymer (SBS)); (chemical type: styrene isoprene styrene block copolymer (SIS)); **Kraton G** (chemical type: styrene ethylene butylene styrene block copolymer (SEBS))

The permeability of Kraton thermoplastic rubbers is similar to that of the type of rubber which forms the midblock. However, as the fraction of end block is increased, as for example when endblock compatible resins are added, permeability is somewhat reduced. Conversely, as the rubber phase fraction is increased, as when midblock compatible oils are added, permeability is increased. The permeability of Kraton D rubbers is comparable to that of SBR and natural rubber. The permeability of the Kraton G rubbers lies between that of butyl rubber and that of vinyl chloride polymers.

Reference: *Kraton Thermoplastic Rubber*, supplier design guide (SC:198-89) - Shell Chemical Company, 1989.

Styrenic TPE

TABLE 351: Oxygen Permeability Through Shell Chemical Kraton Neat Styrenic Thermoplastic Elastomer.

Material Family	STYRENIC THERMOPLASTIC ELASTOMER				
Material Supplier/ Trade Name	SHELL CHEMICAL KRATON				
Grade	D 1101	D 1107	G 1650	G 1651	G 1652
Features	FDA grade	FDA grade	FDA grade	FDA grade	FDA grade
Reference Number	303	303	303	303	303

MATERIAL CHARACTERISTICS

Shore A Hardness	71	37	75	75	75

MATERIAL COMPOSITION

Chemical Type	styrene-butadiene-styrene block copolymer (SBS)	styrene-isoprene-styrene block copolymer (SIS)	styrene-ethylene-butylene-styrene block copolymer (SEBS)	styrene-ethylene-butylene-styrene block copolymer (SEBS)	styrene-ethylene-butylene-styrene block copolymer (SEBS)
Note	31% styrene/ 69% rubber, neat rubber, unsaturated	14% styrene/ 86% rubber, neat rubber, unsaturated	29% styrene/ 71% rubber, neat rubber, saturated	29% styrene/ 71% rubber, neat rubber, saturated	29% styrene/ 71% rubber, neat rubber, saturated

TEST CONDITIONS

Penetrant	oxygen				
Temperature (°C)	23	23	23	23	23
Test Method/ Test Note	ASTM D1434-82; area: 50 cm²; gradient: 100% gas at 740 mm Hg				

PERMEABILITY (source document units)

Gas Permeability ($cm^3 \cdot mil/100\ in^2 \cdot day$)	4360	3170	2310	2160	2690
Gas Permeability ($1 \times 10^{-10}\ cm^3 \cdot cm/\ cm^2 \cdot s \cdot cmHg$)	26.8	19.5	14.2	13.3	16.6

PERMEABILITY (normalized units)

Permeability Coefficient ($cm^3 \cdot mm/m^2 \cdot day \cdot atm$)	1717	1248	909	850	1059

Styrenic TPE

TABLE 352: Oxygen Permeability Through Shell Chemical Kraton Styrenic Thermoplastic Elastomer Compounds.

Material Family	STYRENIC THERMOPLASTIC ELASTOMER					
Material Supplier/ Trade Name	SHELL CHEMICAL KRATON					
Grade	**D 2103**	**D 2104**	**D 2109**	**G 2701**	**G 2705**	**G 2706**
Features	FDA grade	FDA grade	FDA grade	FDA grade	FDA grade	FDA grade
Reference Number	303	303	303	303	303	303

MATERIAL CHARACTERISTICS

Shore A Hardness	70	27	44	67	55	28

MATERIAL COMPOSITION

Chemical Type	styrene-butadiene-styrene block copolymer (SBS)	styrene-butadiene-styrene block copolymer (SBS)	styrene-butadiene-styrene block copolymer (SBS)	styrene-ethylene-butylene-styrene block copolymer (SEBS)	styrene-ethylene-butylene-styrene block copolymer (SEBS)	styrene-ethylene-butylene-styrene block copolymer (SEBS)
Note	ready to use compound, unsaturated	ready to use compound, unsaturated	ready to use compound, unsaturated	ready to use compound, saturated	ready to use compound, saturated	ready to use compound, saturated

TEST CONDITIONS

Penetrant	oxygen					
Temperature (°C)	23	23	23	23	23	23
Test Method	ASTM D1434-82					
Test Condition Note	area: 50 cm^2; gradient: 100% gas at 740 mm Hg					

PERMEABILITY (source document units)

Gas Permeability (cm$^3 \cdot$ mil/100 in$^2 \cdot$ day)	4830	17,350	9000	2800	4180	5530
Gas Permeability (1x10^{-10} cm$^3 \cdot$ cm/ cm$^2 \cdot$ s $\cdot$ cmHg)	29.7	106	54.8	17.2	25.7	34.0

PERMEABILITY (normalized units)

Permeability Coefficient (cm$^3 \cdot$ mm/m$^2 \cdot$ day $\cdot$ atm)	1902	6831	3543	1102	1646	2177

TABLE 353: Carbon Dioxide Permeability Through Shell Chemical Kraton Styrenic Thermoplastic Elastomer Compounds.

Material Family	STYRENIC THERMOPLASTIC ELASTOMER					
Material Supplier/ Trade Name	SHELL CHEMICAL KRATON					
Grade	D 2103	D 2104	D 2109	G 2701	G 2705	G 2706
Features	FDA grade	FDA grade	FDA grade	FDA grade	FDA grade	FDA grade
Reference Number	303	303	303	303	303	303

MATERIAL CHARACTERISTICS

Shore A Hardness	70	27	44	67	55	28

MATERIAL COMPOSITION

Chemical Type	styrene-butadiene-styrene block copolymer (SBS)	styrene-butadiene-styrene block copolymer (SBS)	styrene-butadiene-styrene block copolymer (SBS)	styrene-ethylene-butylene-styrene block copolymer (SEBS)	styrene-ethylene-butylene-styrene block copolymer (SEBS)	styrene-ethylene-butylene-styrene block copolymer (SEBS)
Note	ready to use compound, unsaturated	ready to use compound, unsaturated	ready to use compound, unsaturated	ready to use compound, saturated	ready to use compound, saturated	ready to use compound, saturated

TEST CONDITIONS

Penetrant	carbon dioxide					
Temperature (°C)	23	23	23	23	23	23
Test Method/ Test Note	ASTM D1434-82; area: 50 cm²; gradient: 100% gas at 740 mm Hg					

PERMEABILITY (source document units)

Gas Permeability ($cm^3 \cdot mil/100\ in^2 \cdot day$)	13,410	62,630	17,830	6350	6450	18,420
Gas Permeability ($1 \times 10^{-10}\ cm^3 \cdot cm/cm^2 \cdot s \cdot cmHg$)	82.5	385	110	39.1	39.7	113

PERMEABILITY (normalized units)

Permeability Coefficient ($cm^3 \cdot mm/m^2 \cdot day \cdot atm$)	5280	24,657	7020	2500	2539	7252

Styrenic TPE

TABLE 354: Carbon Dioxide Permeability Through Shell Chemical Kraton Neat Styrenic Thermoplastic Elastomer.

Material Family	STYRENIC THERMOPLASTIC ELASTOMER				
Material Supplier/ Trade Name	SHELL CHEMICAL KRATON				
Grade	**D 1101**	**D 1107**	**G 1650**	**G 1651**	**G 1652**
Features	FDA grade	FDA grade	FDA grade	FDA grade	FDA grade
Reference Number	303	303	303	303	303

MATERIAL CHARACTERISTICS

Shore A Hardness	71	37	75	75	75

MATERIAL COMPOSITION

Chemical Type	styrene-butadiene-styrene block copolymer (SBS)	styrene-isoprene-styrene block copolymer (SIS)	styrene-ethylene-butylene-styrene block copolymer (SEBS)	styrene-ethylene-butylene-styrene block copolymer (SEBS)	styrene-ethylene-butylene-styrene block copolymer (SEBS)
Note	31% styrene/ 69% rubber, neat rubber, unsaturated	14% styrene/ 86% rubber, neat rubber, unsaturated	29% styrene/ 71% rubber, neat rubber, saturated	29% styrene/ 71% rubber, neat rubber, saturated	29% styrene/ 71% rubber, neat rubber, saturated

TEST CONDITIONS

Penetrant	carbon dioxide				
Temperature (°C)	23	23	23	23	23
Test Method	ASTM D1434-82	ASTM D1434-82	ASTM D1434-82	ASTM D1434-82	ASTM D1434-82
Test Condition Note	area: 50 cm^2; gradient: 100% gas at 740 mm Hg				

PERMEABILITY (source document units)

Gas Permeability (cm^3 · mil/100 in^2 · day)	17,040	19,300	5850	6100	8460
Gas Permeability (1x10^{-10} cm^3 · cm/ cm^2 · s · cmHg)	105	119	57.5	37.5	57.9

PERMEABILITY (normalized units)

Permeability Coefficient (cm^3 · mm/m^2 · day · atm)	6709	7598	2303	2402	3331

TABLE 355: Water Vapor Transmission Through Shell Chemical Kraton Styrenic Thermoplastic Elastomer Compounds.

Material Family	STYRENIC THERMOPLASTIC ELASTOMER					
Material Supplier/ Trade Name	SHELL CHEMICAL KRATON					
Grade	**D 2103**	**D 2104**	**D 2109**	**G 2701**	**G 2705**	**G 2706**
Features	FDA grade	FDA grade	FDA grade	FDA grade	FDA grade	FDA grade
Reference Number	303	303	303	303	303	303

MATERIAL CHARACTERISTICS

Shore A Hardness	70	27	44	67	55	28

MATERIAL COMPOSITION

Chemical Type	styrene-butadiene-styrene block copolymer (SBS)	styrene-butadiene-styrene block copolymer (SBS)	styrene-butadiene-styrene block copolymer (SBS)	styrene-ethylene-butylene-styrene block copolymer (SEBS)	styrene-ethylene-butylene-styrene block copolymer (SEBS)	styrene-ethylene-butylene-styrene block copolymer (SEBS)
Note	ready to use compound, unsaturated	ready to use compound, unsaturated	ready to use compound, unsaturated	ready to use compound, saturated	ready to use compound, saturated	ready to use compound, saturated

TEST CONDITIONS

Penetrant	water vapor					
Temperature (°C)	23	23	23	23	23	23
Relative Humidity (%)	90% gradient	90% gradient	90% gradient	90% gradient	90% gradient	90% gradient
Test Method	ASTM E96-80, procedure E					
Test Condition Note	area: 50 cm²					

PERMEABILITY (source document units)

Gas Permeability (1×10^{-10} cm³ · cm/ cm² · s · cmHg)	3810	7940	2040	650	900	1180
Vapor Transmission Rate (g · mil/ 100 in² · hr)	29.4	61.3	15.7	5.0	7.0	9.1

PERMEABILITY (normalized units)

Vapor Transmission Rate (g · mm/m² · day)	278	579	148	47.2	66.1	86.0

Styrenic TPE

TABLE 356: Water Vapor Transmission Through Shell Chemical Kraton Neat Styrenic Thermoplastic Elastomer.

Material Family	STYRENIC THERMOPLASTIC ELASTOMER				
Material Supplier/ Trade Name	SHELL CHEMICAL KRATON				
Grade	**D 1101**	**D 1107**	**G 1650**	**G 1651**	**G 1652**
Features	FDA grade	FDA grade	FDA grade	FDA grade	FDA grade
Reference Number	303	303	303	303	303

MATERIAL CHARACTERISTICS

Shore A Hardness	71	37	75	75	75

MATERIAL COMPOSITION

Chemical Type	styrene-butadiene-styrene block copolymer (SBS)	styrene-isoprene-styrene block copolymer (SIS)	styrene-ethylene-butylene-styrene block copolymer (SEBS)	styrene-ethylene-butylene-styrene block copolymer (SEBS)	styrene-ethylene-butylene-styrene block copolymer (SEBS)
Note	31% styrene/ 69% rubber, neat rubber, unsaturated	14% styrene/ 86% rubber, neat rubber, unsaturated	29% styrene/ 71% rubber, neat rubber, saturated	29% styrene/ 71% rubber, neat rubber, saturated	29% styrene/ 71% rubber, neat rubber, saturated

TEST CONDITIONS

Penetrant	water vapor				
Temperature (°C)	23	23	23	23	23
Relative Humidity (%)	90% gradient	90% gradient	90% gradient	90% gradient	90% gradient
Test Method	ASTM E96-80, procedure E				
Test Condition Note	area: 50 cm^2				

PERMEABILITY (source document units)

Gas Permeability (1×10^{-10} cm^3 · cm/ cm^2 · s · cmHg)	3520	2870	760	860	1140
Vapor Transmission Rate (g · mil/ 100 in^2 · hr)	27.2	22.1	5.8	6.6	8.8

PERMEABILITY (normalized units)

Vapor Transmission Rate (g · mm/m^2 · day)	257	208.8	54.8	62.4	83.1

Chapter 75

Vinyl Thermoplastic Elastomer

TABLE 357: Carbon Dioxide, Oxygen and Water Vapor Permeability Through Polyvinyl Chloride Film and Stretch Film.

Material Family	POLYVINYL CHLORIDE POLYOL					
Product Form	FILM					
Features	2.5 blow up ratio, transparent			4.8 blow up ratio, stretch film		
Manufacturing Method	blown film	blown film	blown film	blown film	blown film	blown film
Reference Number	216	216	216	216	216	216

MATERIAL CHARACTERISTICS

Density	1.26 g/cm³			1.23 - 1.31 g/cm³		
Sample Thickness	0.05 mm	0.05 mm	0.05 mm	0.018 mm	0.018 mm	0.018 mm

MATERIAL COMPOSITION

Note	50 phr plasticizer					

TEST CONDITIONS

Penetrant	water vapor	carbon dioxide	oxygen	water vapor	carbon dioxide	oxygen
Test Method	JIS Z0208	ASTM D1434	ASTM D1434	JIS Z0208	ASTM D1434	ASTM D1434

PERMEABILITY (source document units)

Vapor Transmission Rate (g · 100 μm/m² · day)	100			79 - 129		
Gas Permeability (cm³ · 100 μm/ m² · atm · day)		3000	930		14,000 - 27,000	1900 - 3600

PERMEABILITY (normalized units)

Permeability Coefficient (cm³ · mm/m² · day · atm)		300	93		1400 - 2700	190 - 360
Vapor Transmission Rate (g · mm/m² · day)	10			7.9 - 12.9		

TABLE 358: Oxygen and Carbon Dioxide Permeability Through Polyvinyl Chloride Polyol.

Material Family	POLYVINYL CHLORIDE POLYOL	
Product Form	FILM	
Reference Number	250	250

MATERIAL COMPOSITION

Note	plasticized	plasticized

TEST CONDITIONS

Penetrant	oxygen	carbon dioxide
Temperature (°C)	24	24

PERMEABILITY (source document units)

Gas Permeability ($cm^3 \cdot mil/100\ in^2 \cdot day$)	30 - 2000	100 - 3000

PERMEABILITY (normalized units)

Permeability Coefficient ($cm^3 \cdot mm/m^2 \cdot day \cdot atm$)	11.8 - 787	39.4 - 1181

TABLE 359: Film Properties of Polyvinyl Chloride Film and Stretch Film.

Material Family	POLYVINYL CHLORIDE POLYOL	
Product Form	FILM	
Features	2.5 blow up ratio, transparent	4.8 blow up ratio, stretch film
Manufacturing Method	blown film	blown film
Reference Number	216	216

MATERIAL CHARACTERISTICS

Density	1.26 g/cm³	1.23 - 1.31 g/cm³
Sample Thickness	0.05 mm	0.018 mm
Material Composition Note	50 phr plasticizer	

MECHANICAL PROPERTIES

100% Modulus - MD (MPa)		17.1 - 25.5 {ci}
100% Modulus - TD (MPa)		8.5 - 12.6 {ci}
150% Modulus - MD (MPa)		21.6 - 31.4 {ci}
150% Modulus - TD (MPa)		12.7 - 13.6 {ci}
Tensile Strength - MD (MPa)	24.5 {ci}	23.8 - 32.0 {ci}
Tensile Strength - TD (MPa)	24.5 {ci}	17.7 - 21.9 {ci}
Ultimate Elongation - MD (%)	240 {ci}	154 - 200 {ci}
Ultimate Elongation - TD (%)	240 {ci}	241 - 317 {ci}
Tear Strength - MD (g/mm)	5800 {cj}	4600 - 6600 {cm}
Tear Strength - TD (g/mm)	6700 {cj}	2900 - 3600 {cm}

OPTICAL PROPERTIES

Light Transmittance (%)	92 {ck}	89 - 91 {ck}
Haze (%)	1 {ck}	0.3 - 2.8 {ck}
Gloss	130 {cl}	118 - 150 {cl}

Chapter 76

Ethylene-Acrylate Copolymer

TABLE 360: Nitrogen Permeability Through Ethylene Acrylate Rubber.

Material Family	ETHYLENE-ACRYLATE COPOLYMER
Reference Number	309

TEST CONDITIONS

Penetrant	nitrogen
Temperature (°C)	21.1
Test Note	applies to general class of base polymer

PERMEABILITY (source document units)

Gas Permeability (1×10^{-8} cm^2/ s · atm)	0.88

PERMEABILITY (normalized units)

Permeability Coefficient (cm^3 · mm/m^2 · day · atm)	76

Chapter 77

Polybutadiene

TABLE 361: Air Permeability vs. Temperature Through Bayer Taktene 1220 Polybutadiene Rubber.

Material Family	POLYBUTADIENE		
Material Supplier/ Grade	BAYER TAKTENE 1220		
Cure	20 min. @ 145°C		
Reference Number	298	298	298

MATERIAL CHARACTERISTICS

Shore A Hardness	53 (30 second)		

MATERIAL COMPOSITION

Zinc Dimethyl Dithiocarbonate	1.5 phr		
Sulfur	1.4 phr		
Zinc Oxide	5 phr		
SRF Carbon Black	50 phr		
Chemical Type	cis polybutadiene		
Note	1.1 phr Santocure		

TEST CONDITIONS

Penetrant	air		
Temperature (°C)	40	60	80

PERMEABILITY (source document units)

Gas Permeability $(1 \times 10^{-8}$ cm$^3 \cdot$ cm/ cm$^2 \cdot$ sec $\cdot$ atm)	27.7	44.1	65.5

PERMEABILITY (normalized units)

Permeability Coefficient (cm$^3 \cdot$ mm/m$^2 \cdot$ day $\cdot$ atm)	2393	3810	5659

TABLE 362: Nitrogen, Carbon Dioxide and Water Vapor Permeability Through Polybutadiene Rubber.

Material Family	POLYBUTADIENE			
Product Form		FILM		
Reference Number	309	250	250	250

TEST CONDITIONS

Penetrant	nitrogen	nitrogen	carbon dioxide	water vapor
Temperature (°C)	21.1	24	24	39
Test Note	applies to general class of base polymer			

PERMEABILITY (source document units)

Gas Permeability (1×10^{-8} cm^2/ s · atm)	20.0			
Gas Permeability (cm^3 · mil/100 in^2 · day)		2000	20000	
Vapor Transmission Rate (g · mil/100 in^2 · day)				45

PERMEABILITY (normalized units)

Permeability Coefficient (cm^3 · mm/m^2 · day · atm)	1728	787	7874	
Vapor Transmission Rate (g · mm/m^2 · day)				17.7

Polybutadiene

TABLE 363: Water Vapor Transmission and Hydrogen, Oxygen, Nitrogen, Carbon Dioxide and Air Permeabiltiy Relative To Butyl Rubber Through Polybutadiene Rubber.

Material Family	POLYBUTADIENE					
Product Form	GUM VULCANIZATE					
Features	hot emulsion					
Reference Number	300	300	300	300	300	300

MATERIAL CHARACTERISTICS

Sample Thickness		0.51 mm

MATERIAL COMPOSITION

Paraffinic Oil		5 phr
SRF Carbon Black		50 phr

TEST CONDITIONS

Penetrant	hydrogen	oxygen	nitrogen	carbon dioxide	air	water vapor
Temperature (°C)	25	25	25	25	25	
Test Method						Tappi Standard T464 M-45

PERMEABILITY (source document units)

Permeability Relative to Butyl Rubber (%)	575	1430	1600	2620	1700	5125

GRAPH 75: Air Permeability vs. Temperature through Polybutadiene.

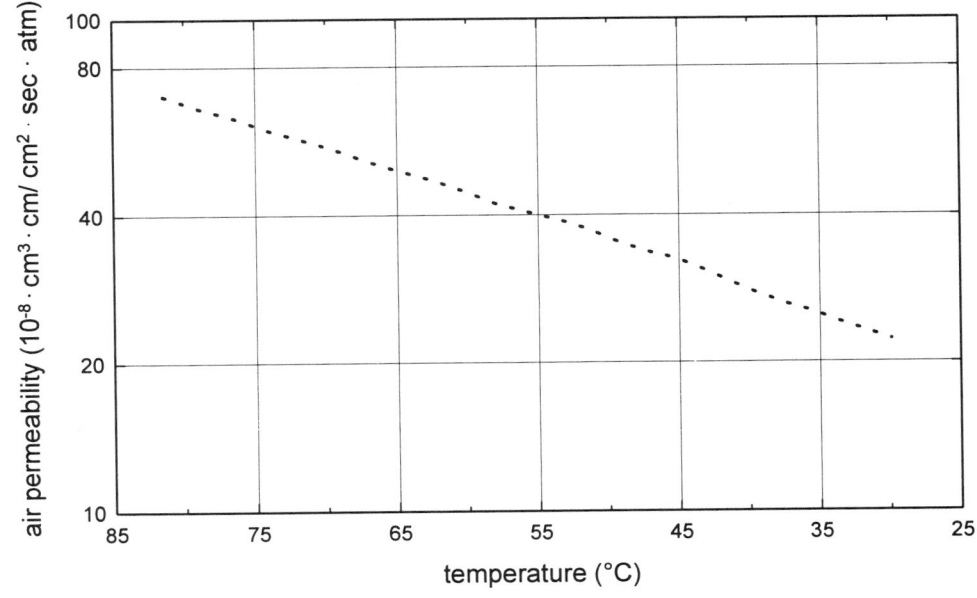

Bayer Taktene 1220 Polybutadiene (cis polybutadiene; 20 min. @ 145°C cure; 53 Shore A; 5 phr zinc oxide, 1.5 phr zinc dimethyl dithiocarbonate, 1.1 phr Santocure, 50 phr SRF black, 1.4 phr sulfur); penetrant: air; 60°C

Reference No.	298

Chapter 78

Butyl Rubber

Permeability

Bayer: Polysar Butyl

Among hydrocarbon elastomers, butyl rubber is outstanding in its low permeability to gases. Although compounding variables are secondary in this case, best results are obtained in unplasticized compounds containing a high volume loading of suitable fillers.

The solubility of gases in butyl rubber is similar to that in other hydrocarbon polymers, but the rate of diffusion through butyl rubber is exceptionally low compared to other rubbery materials. This phenomenon is one of the most important basic properties of butyl rubber, and is responsible for many of its major uses, for example, inner tubes.

The differences in permeability between butyl rubber and other common elastomers are large; butyl rubber is about one-eightieth as permeable as cis-polybutadiene. The relative differences diminish with increasing temperature, but remain large even at high temperatures.

In general, the effects of compounding variations are small compared to the inherent differences between polymeric types. The effect of state of cure is insignificant in a practical sense, but material changes do result from variation of oil and filler loadings. Air permeability is approximately tripled, at a constant black loading of 50 parts, by raising the plasticizer loading from zero to 30 parts. This result is typical of hydrocarbon plasticizers.

Increased black loading lowers the permeability, as the volume proportion of polymer through which the gas can permeate is reduced. This effect applies generally to fillers. Comparatively large reductions in permeability can be obtained with plate-like fillers such as finely divided mica, especially when the processing history (extrusion or calendering) is such that the platelets are aligned normal to the direction of permeation.

While essential for many of its applications, the low permeability of butyl rubber can be a source of difficulty during vulcanization. Pockets of entrapped gases will dissolve and diffuse into the surrounding rubber much less readily than is the case with other polymers, consequently porosity or pin-holing may occur. Attention should be given to ensuring a dense, gas-free condition in Polysar butyl stocks prior to vulcanization.

Reference: *Polysar Butyl Rubbers Handbook,* reference book - Miles Polysar.

Exxon: Exxon Butyl

Isobutylene-isoprene copolymers offer significant advantages over other commercial elastomers in low gas permeability. This is believed due to the steric hindrance afforded by bulky methyl groups in the isobutylene component. As a result butyl is widely used in innertubes. In tubed tires, Butyl innertubes provide outstanding air pressure rentention, thereby minimizing tire wear due to underinflation.

Reference: *Elastomers Technical Information - Elastomer Permeability,* supplier technical report (TI-20) - Exxon Chemical Company, 1974.

Butyl Rubber

The permeability of elastomeric films to the passage of gas is a function of the diffusion of gas molecules through the membrane and the solubility of the gas in the elastomer. The polyisobutylene portion of the butyl molecule provides a low degree of permeability to gases and is a familiar property, leading to an almost exclusive use in inner tubes. For example, the air permeability, at 65°C, of SBR is about 80% that of natural rubber, while butyl shows only 10% permeability on the same scale. The difference in air retention between a natural rubber and a butyl inner tube can be demonstrated by data from controlled road tests on cars driven 60 mph for 100 miles per day. Under these conditions, it was shown that butyl is at least 8 times better than natural rubber in air retention. Other gases such as helium, hydrogen, nitrogen, and carbon dioxide are also well retained by a butyl bladder membrane. While the significance of these properties in inner tubes is waning, it is of importance in air barriers for tubeless tires, air cushions, pneumatic springs, accumulator bags, air bellows, and the like.

Reference: Fusco, James V., Hous, Pierre, *Butyl and Halobutyl Rubbers,* Rubber Technology, Third Edition, reference book - Van Nostrand Reinhold Company, Inc., 1987.

TABLE 364: Air Permeability vs. Temperature Through Butyl Rubber.

Material Family	BUTYL RUBBER					
Material Supplier/ Grade	BAYER POLYSAR BUTYL 301			EXXON BUTYL 268		
Cure	20 min. @ 153°C					
Reference Number	298	298	298	299	299	299

MATERIAL CHARACTERISTICS

Shore A Hardness	50 (30 second)	

MATERIAL COMPOSITION

Stearic Acid		2 phr
Zinc Dimethyl Dithiocarbonate	1.5 phr	
Sulfur	2 phr	
Zinc Oxide	5 phr	5 phr
Channel Black		20 phr
MT Carbon Black		60 phr
SRF Carbon Black	50 phr	
Note	0.5 phr Altax, 1 phr Methyl Tuads	10 phr Flexon 845 oil (Exxon), 4 phr Amberol ST-149 (Rohm and Haas Co.), specific accelerator

TEST CONDITIONS

Penetrant	air					
Temperature (°C)	40	60	80	23.9	65.6	93.3
Test Apparatus				Aminco Permeability Apparatus		

PERMEABILITY (source document units)

Permeability Relative to Butyl Rubber (%)					100	
Gas Permeability ($ft^3 \cdot mil/ft^2 \cdot day \cdot psi$)				0.00032	0.0033	0.0105
Gas Permeability ($1 \times 10^{-8} cm^3 \cdot cm/cm^2 \cdot sec \cdot atm$)	0.6	1.8	46			

PERMEABILITY (normalized units)

Permeability Coefficient ($cm^3 \cdot mm/m^2 \cdot day \cdot atm$)	51.8	155	3974	36.4	375	1195

TABLE 365: Nitrogen Permeability Through Butyl Rubber.

Material Family	BUTYL RUBBER
Reference Number	309

TEST CONDITIONS

Penetrant	nitrogen
Temperature (°C)	21.1
Test Note	applies to general class of base polymer

PERMEABILITY (source document units)

Gas Permeability (1×10^{-8} cm^2/ s · atm)	0.25

PERMEABILITY (normalized units)

Permeability Coefficient (cm^3 · mm/m^2 · day · atm)	21.6

GRAPH 76: Air Permeability vs. Plasticizer (Polaris 45 Oil) Loading through Butyl Rubber.

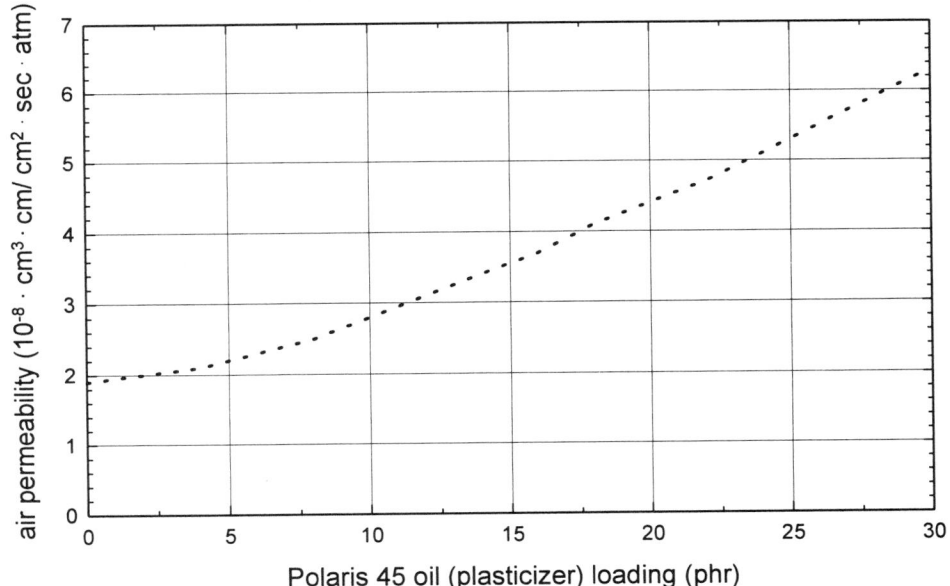

	Bayer Polysar Butyl 301 Butyl Rubber (40 min. @ 145°C cure; 5 phr zinc oxide, 2 phr sulfur, 1.5 phr zinc dimethyl dithiocarbonate, 50 phr SRF black, 0.5 phr Altax, 1 phr Methyl Tuads); penetrant: air; 60°C
Reference No.	298

440

GRAPH 77: Air Permeability vs. SRF Black Loading through Butyl Rubber.

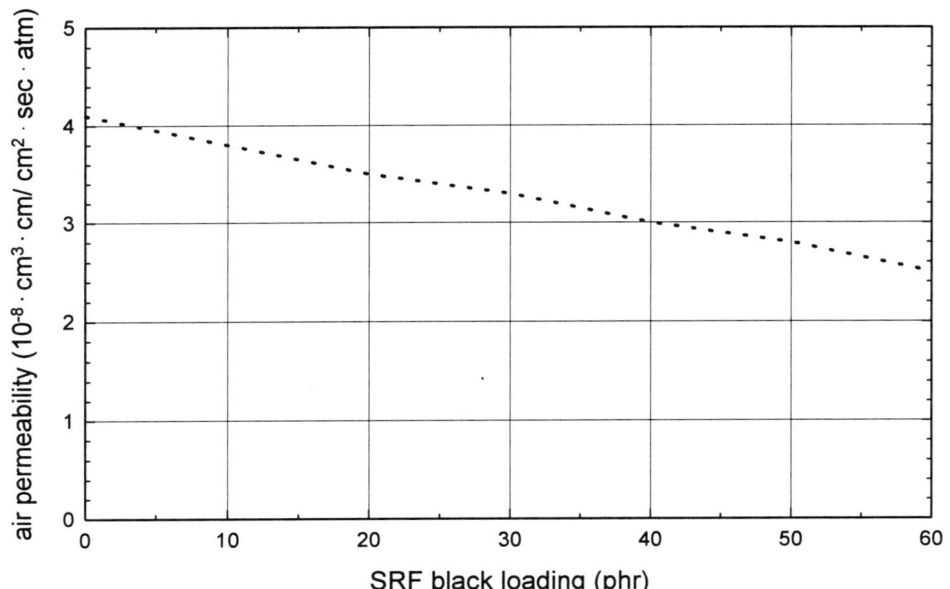

	Bayer Polysar Butyl 301 Butyl Rubber (40 min. @ 145°C cure; 5 phr zinc oxide, 2 phr sulfur, 1.5 phr zinc dimethyl dithiocarbonate, 10 phr Polar 45 oil, 1 phr Methyl Tuads, 0.5 phr Altax); penetrant: air
Reference No.	298

GRAPH 78: Air Permeability vs. Temperature through Butyl Rubber.

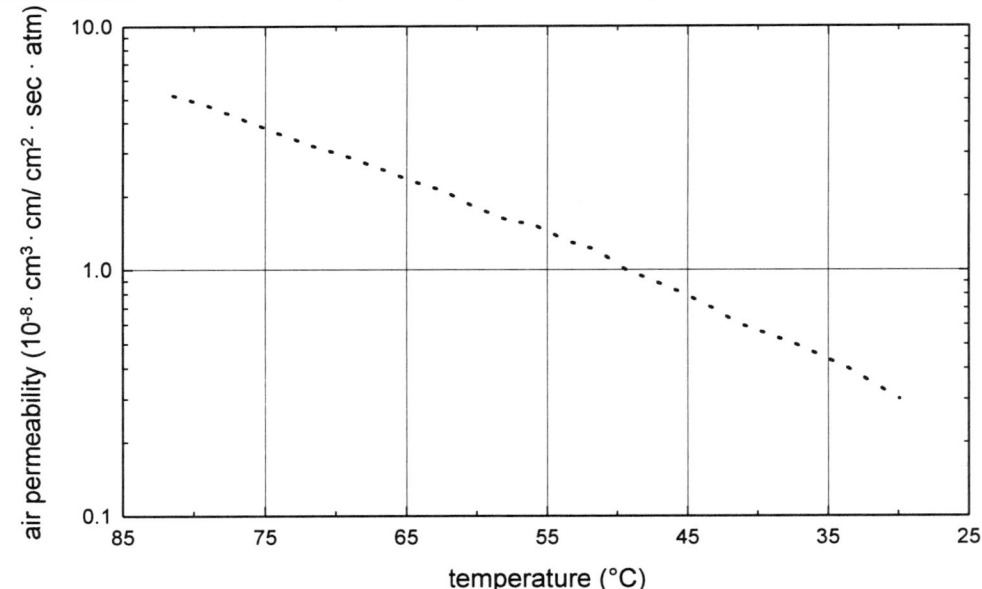

	Bayer Polysar Butyl 301 Butyl Rubber (20 min. @ 153°C cure; 50 Shore A; 5 phr zinc oxide, 2 phr sulfur, 1.5 phr zinc dimethyl dithiocarbonate, 1 phr Methyl Tuads, 0.5 phr Altax, 50 phr SRF black); penetrant: air; 60°C
Reference No.	298

Chapter 79

Bromobutyl Rubber

Permeability

BIIR Exxon:

The low permeability of bromobutyl and its ability to co-vulcanize with highly unsaturated rubbers makes it a polymer of choice for air and moisture barrier applications, like tire innerliners and pharmaceutical closures. At room temperature, for example, compounds based on all-bromobutyl are about ten fold less permeable than all-natural rubber compounds.

General guidelines for formulating low permeability bromobutyl compounds are as follows:

A.   Chlorobutyl, bromobutyl, and SB bromobutyl are equivalent in their effect on compound permeability.

B.   Compound permeability is predominantly a function of the bromobutyl/GPR polymer ratios, and secondarily affected by the remainder of the ingredients (fillers and plasticizers, primarily). All bromobutyl provides the lowest permeability to air and moisture. Blends of bromobutyl and BPR rubbers exhibit a near average permeability value based on the fraction of the polymers.

C.   Gas permeability can also be decreased by doing the following:

    1.   increase filler content up to about fifty percent by volume. Above this level, permeability increases sharply due to discontinuities in the rubber phase.

    2.   decrease process oil or other plasticizer levels.

    3.   use a filler with a larger particle size.

    4.   use platy fillers such as talc or ground oyster shells. These materials should be evaluated carefully for adverse effects on compound crack resistance.

All elastomers show an increase in permeability with temperature. Bromobutyl compounds exhibit air barrier advantages over natural rubber at temperatures in the typical operating range of 25°C to 100°C.

Reference: *Bromobutyl Rubber Optimizing Key Properties,* supplier marketing literature - Exxon Chemicals.

GRAPH 79: **Air Permeability vs. Bromobutyl Concentration through Bromoisobutylene-Isoprene Copolymer.**

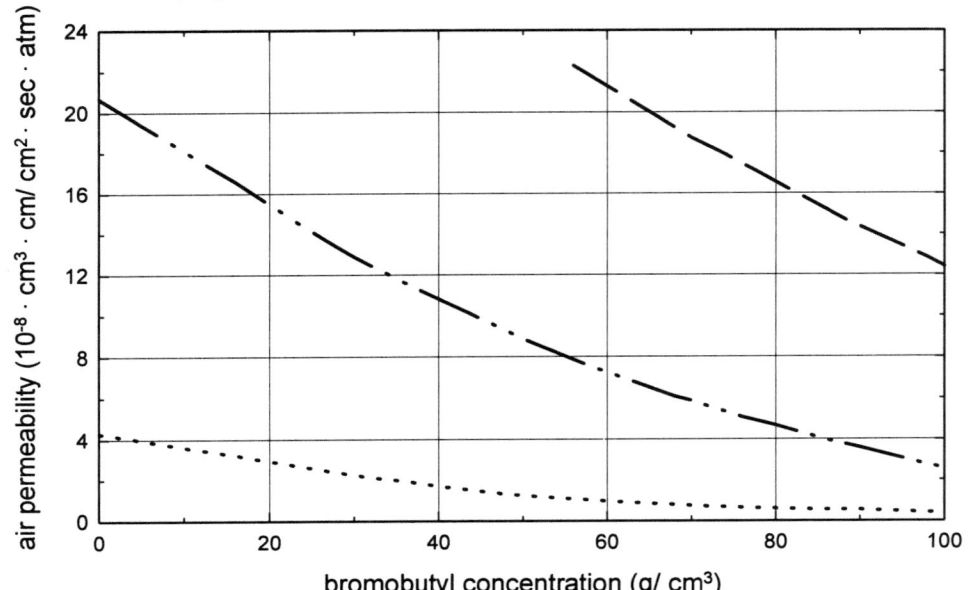

··············	Exxon BIIR; penetrant: air; 25°C
—··—··—	Exxon BIIR; penetrant: air; 65°C
— — —	Exxon BIIR; penetrant: air; 100°C
Reference No.	164

GRAPH 80: **Moisture Vapor Transmission vs. Bromobutyl Concentration through Bromoisobutylene-Isoprene Copolymer.**

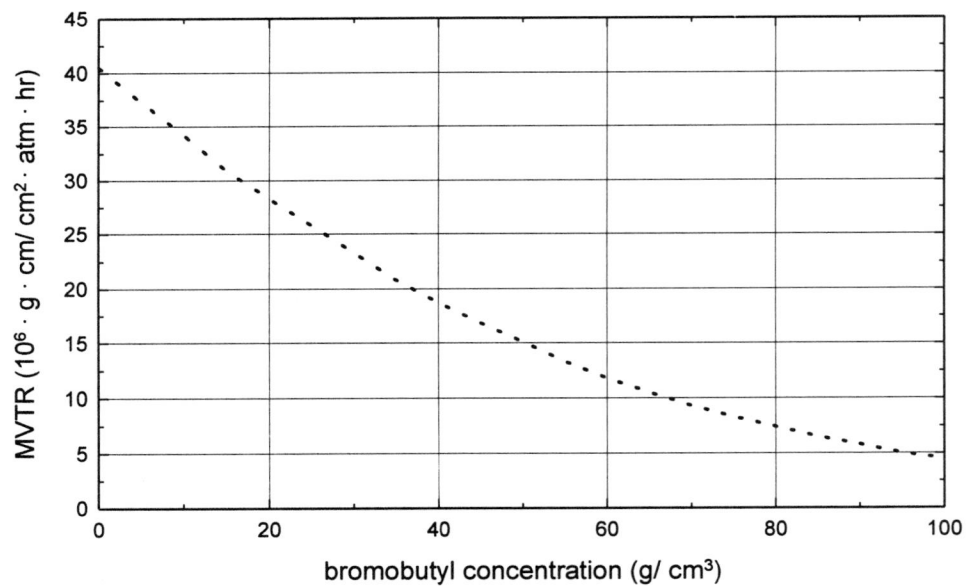

··············	Exxon BIIR; penetrant: moisture vapor
Reference No.	164

BIIR

Chlorobutyl Rubber

Permeability

Exxon: Exxon Chlorobutyl

Isobutylene-isoprene copolymers offer significant advantages over other commercial elastomers in low gas permeability. This is believed due to the steric hindrance afforded by bulky methyl groups in the isobutylene component. As a result, Chlorobutyl is widely used in innerliners for tubeless tires. In tubeless tires, the low permeability of the Chlorobutyl innerliner minimizes migration of air into the carcass fabric and improves tire durability.

Reference: *Elastomers Technical Information - Elastomer Permeability,* supplier technical report (TI-20) - Exxon Chemical Company, 1974.

Effect of Polymer Blends and Fillers on Permeability

Exxon Chlorobutyl   (product form: gum vulcanizate)

When elastomers are blended, the permeability of the resulting compound is intermediate between the permeabilities of the components. This is illustrated by the permeability values of Chlorobutyl and natural rubber. A typical tubeless tire innerliner formulation containing 65% (RHC) Chlorobutyl is one-half as permeable as a corresponding high styrene SBR formulation (SBR 1133). A large increase in the butyl reclaim content of the compound does not result in the same permeability as the Chlorobutyl based compound due to the poor state of cure achieved with the butyl reclaim stock.

The gas permeability of rubbers decreases gradually with increasing filler concentration of approximately 50% by volume or greater, the permeability increases sharply due to the development of discontinuities in the elastomer matrix. Considering various fillers, major effects on permeability are not apparent except between lamellar (platy) and acicular (more spherical) types. To a lesser extent, smaller particle size contributes to reduced permeability as shown by the comparison of thermal and furnace black.

Reference: *Elastomers Technical Information - Factors in the Gas Permeability of Elastomers,* supplier technical report (TI-28) - Exxon Chemical Company, 1974.

TABLE 366: Air Permeability vs. TemperatureThrough Chlorobutyl Rubber.

Material Family	CHLOROISOBUTYLENE-ISOPRENE COPOLYMER			
Material Supplier/ Grade		EXXON CHLOROBUTYL 1068		
Reference Number	12	299	299	299

MATERIAL CHARACTERISTICS

Density	1.13 g/cm³			
Shore A Hardness	55			

MATERIAL COMPOSITION

Stearic Acid		2 phr		
Zinc Oxide		5 phr		
Channel Black		20 phr		
MT Carbon Black		60 phr		
Note		10 phr Flexon 845 oil (Exxon), 4 phr Amberol ST-149 (Rohm and Haas Co.), specific accelerator		

TEST CONDITIONS

Penetrant	air			
Temperature (°C)		23.9	65.6	93.3
Test Apparatus		Aminco Permeability Apparatus		

PERMEABILITY (source document units)

Gas Permeability (ft³ · mil/ft² · day · psi)	0.00027	0.00034	0.0032	0.0104

PERMEABILITY (normalized units)

Permeability Coefficient (cm³ · mm/m² · day · atm)	30.7	38.7	364	1183

Chapter 81

Isobutylene Rubber

TABLE 367: Helium, Hydrogen, Oxygen, Nitrogen, Carbon Dioxide and Air Permeabiltiy Relative To Butyl Rubber Through Polyisobutylene Rubber.

Material Family	ISOBUTYLENE RUBBER					
Product Form	GUM VULCANIZATE					
Reference Number	300	300	300	300	300	300

TEST CONDITIONS

Penetrant	helium	hydrogen	oxygen	nitrogen	carbon dioxide	air
Temperature (°C)	25	25	25	25	25	25

PERMEABILITY (source document units)

Permeability Relative to Butyl Rubber (%)	86	87	91	72	93	88

Chapter 82

Chlorosulfonated Polyethylene Rubber

TABLE 368: Nitrogen Permeability Through Chlorosulfonated Polyethylene Rubber.

Material Family	CHLOROSULFONATED POLYETHYLENE RUBBER
Reference Number	309

TEST CONDITIONS

Penetrant	nitrogen
Temperature (°C)	21.1
Test Note	applies to general class of base polymer

PERMEABILITY (source document units)

Gas Permeability $(1 \times 10^{-8}\ cm^2/\ s \cdot atm)$	0.7 - 0.9

PERMEABILITY (normalized units)

Permeability Coefficient $(cm^3 \cdot mm/m^2 \cdot day \cdot atm)$	60.5 - 77.8

Chapter 83

Polyepichlorohydrin Rubber

TABLE 369: Nitrogen Permeability Through Epichlorohydrin Rubber.

Material Family	POLYEPICHLOROHYDRIN RUBBER
Reference Number	309

TEST CONDITIONS

Penetrant	nitrogen
Temperature (°C)	21.1
Test Note	applies to general class of base polymer

PERMEABILITY (source document units)

Gas Permeability (1×10^{-8} cm^2/ s · atm)	0.17

PERMEABILITY (normalized units)

Permeability Coefficient (cm^3 · mm/m^2 · day · atm)	14.7

Epichlorohydrin Copolymer Rubber

TABLE 370: Nitrogen Permeability Through Epichlorohydrin Copolymer Rubber.

Material Family	EPICHLOROHYDRIN COPOLYMER RUBBER
Reference Number	309

TEST CONDITIONS

Penetrant	nitrogen
Temperature (°C)	21.1
Test Note	applies to general class of base polymer

PERMEABILITY (source document units)

Gas Permeability (1×10^{-8} cm^2/ s · atm)	0.66

PERMEABILITY (normalized units)

Permeability Coefficient (cm^3 · mm/m^2 · day · atm)	57

Chapter 85

Ethylene-Propylene Copolymer

TABLE 371: Nitrogen Permeability Through Ethylene-Propylene Rubber.

Material Family	ETHYLENE-PROPYLENE COPOLYMER
Reference Number	309

TEST CONDITIONS

Penetrant	nitrogen
Temperature (°C)	21.1
Test Note	applies to general class of base polymer

PERMEABILITY (source document units)

Gas Permeability (1×10^{-8} cm^2/ s · atm)	6.4

PERMEABILITY (normalized units)

Permeability Coefficient (cm^3 · mm/m^2 · day · atm)	553

Chapter 86

Ethylene-Propylene-Diene Copolymer

TABLE 372: Air Permeability vs. Temperature Through Ethylene-Propylene-Diene Copolymer Rubber.

Material Family	ETHYLENE-PROPYLENE-DIENE COPOLYMER								
Material Supplier/ Grade				EXXON VISTALON 404			EXXON VISTALON 4608		
Cure	40 min. @ 153°C								
Reference Number	298	298	298	299	299	299	299	299	299

MATERIAL CHARACTERISTICS

Shore A Hardness	50 (30 second)		

MATERIAL COMPOSITION

Stearic Acid				2 phr			2 phr		
Zinc Dimethyl Dithiocarbonate	1.5 phr								
Sulfur	1.5 phr								
Zinc Oxide	5 phr			5 phr			5 phr		
Channel Black				20 phr			20 phr		
MT Carbon Black				60 phr			60 phr		
SRF Carbon Black	50 phr								
Note	0.5 phr Captax, 1.5 phr Santocure			10 phr Flexon 845 oil (Exxon), 4 phr Amberol ST-149 (Rohm and Haas Co.), specific accelerator			10 phr Flexon 845 oil (Exxon), 4 phr Amberol ST-149 (Rohm and Haas Co.), specific accelerator		

TEST CONDITIONS

Penetrant	air								
Temperature (°C)	40	60	80	23.9	65.6	93.3	23.9	65.6	93.3
Test Apparatus				Aminco Permeability Apparatus			Aminco Permeability Apparatus		

PERMEABILITY (source document units)

Permeability Relative to Butyl Rubber (%)					680			890	
Gas Permeability ($ft^3 \cdot mil/ft^2 \cdot day \cdot psi$)				0.00405	0.0225	0.0637	0.00587	0.029	0.0619
Gas Permeability ($1 \times 10^{-8} cm^3 \cdot cm/ cm^2 \cdot sec \cdot atm$)	7.9	17.1	33.0						

PERMEABILITY (normalized units)

Permeability Coefficient ($cm^3 \cdot mm/m^2 \cdot day \cdot atm$)	683	1477	2851	461	2560	7247	668	3299	7043

TABLE 373: Oxygen, Nitrogen, Carbon Dioxide, Air and Water Vapor Permeability Through DuPont Nordel Ethylene-Propylene-Diene Rubber.

Material Family	ETHYLENE-PROPYLENE-DIENE COPOLYMER				
Material Supplier/ Grade	DUPONT NORDEL 1040				
Reference Number	311	311	311	311	311

TEST CONDITIONS

Penetrant	oxygen	nitrogen	carbon dioxide	air	water vapor
Temperature (°C)	23	23	23	23	23
Test Condition Note	STP conditions	STP conditions	STP conditions	STP conditions	STP conditions

PERMEABILITY (source document units)

Vapor Transmission Rate (mg · mil/in^2 · day)					1.6
Gas Permeability (1x10^{-8} cm^2/ s · atm)	19 - 22	6.4 - 7.5	87 - 94	8.5 - 10.5	

PERMEABILITY (normalized units)

Permeability Coefficient (cm^3 · mm/m^2 · day · atm)	1641 - 1901	553 - 648	7516 - 8122	734 - 907	
Vapor Transmission Rate (g · mm/m^2 · day)					0.06

TABLE 374: Nitrogen Permeability Through Ethylene-Propylene-Diene Rubber.

Material Family	ETHYLENE-PROPYLENE-DIENE COPOLYMER
Reference Number	309

TEST CONDITIONS

Penetrant	nitrogen
Temperature (°C)	21.1
Test Note	applies to general class of base polymer

PERMEABILITY (source document units)

Gas Permeability (1x10^{-8} cm^2/ s · atm)	6.4

PERMEABILITY (normalized units)

Permeability Coefficient (cm^3 · mm/m^2 · day · atm)	553

TABLE 375: Oxygen, Nitrogen, Carbon Dioxide and Air Permeabiltiy Relative To Butyl Rubber Through Ethylene-Propylene-Diene Copolymer Rubber.

Material Family	ETHYLENE-PROPYLENE-DIENE COPOLYMER			
Product Form	GUM VULCANIZATE			
Reference Number	300	300	300	300

TEST CONDITIONS

Penetrant	oxygen	nitrogen	carbon dioxide	air
Temperature (°C)	25	25	25	25

PERMEABILITY (source document units)

Permeability Relative to Butyl Rubber (%)	1570	1600	1650	1750

GRAPH 81: Air Permeability vs. Temperature through Ethylene-Propylene-Diene Copolymer.

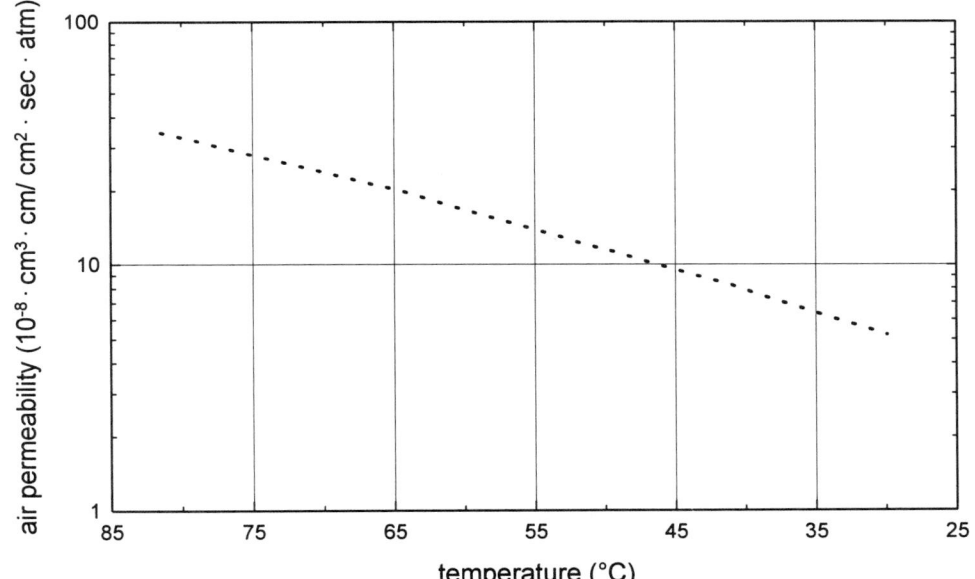

....	EPDM (40 min. @ 153°C cure; 50 Shore A; 5 phr zinc oxide, 1.5 phr sulfur, 1.5 phr zinc dimethyl dithiocarbonate, 100 phr EPDM, 0.5 phr Altax, 50 phr SRF black, 1.5 phr Santocure); penetrant: air; 60°C
Reference No.	298

Chapter 87

Vinylidene Fluoride-Hexafluoropropylene Copolymer

Permeability to Gases

DuPont: Viton   (note: compounded)

Viton is relatively impermeable to air and gases, ranking about midway between the best and the poorest elastomers in this respect. These comparative measurements were made using standard-sized specimens (1 cm^2 x 1 cm thick) of typical compounds, each exposed to a pressure differential of one atmosphere at 80°C (176°F).

The permeability of Viton can be modified considerably by the way it is compounded. But, in all cases, permeability increases rapidly with increasing temperature.

Reference: *The Engineering Properties Of Viton Fluoroelastomer,* supplier design guide (E-46315-1) - DuPont Company, 1987.

TABLE 376: Air, Carbon Dioxide, Helium, Nitrogen and Oxygen Permeability Through Compounded DuPont Viton Fluoroelastomer.

Material Family	VINYLIDENE FLUORIDE-HEXAFLUOROPROPYLENE COPOLYMER						
Material Supplier/ Grade	DUPONT VITON						
Reference Number	305	305	305	305	305	305	305

MATERIAL COMPOSITION

Note	compounded						

TEST CONDITIONS

Penetrant	air	carbon dioxide	helium			nitrogen	oxygen
Temperature (°C)	24	30	24	121	204	24	30
Test Condition Note	one atmosphere @ 80°C						
Test Note	specimen size: 1 cm² x 1 cm thick						

PERMEABILITY (source document units)

Gas Permeability (cm³ · cm/ cm² · sec · atm)	9.9×10^{-10}	5.9×10^{-8}	8.92×10^{-8}	1.74×10^{-6}	6.7×10^{-6}	5.4×10^{-10}	1.1×10^{-8}

PERMEABILITY (normalized units)

Permeability Coefficient (cm³ · mm/m² · day · atm)	8.55	508	771	15,034	57,888	4.67	95.0

TABLE 377: Nitrogen Permeability Through Fluoroelastomer.

Material Family	VINYLIDENE FLUORIDE-HEXAFLUOROPROPYLENE COPOLYMER
Reference Number	309

TEST CONDITIONS

Penetrant	nitrogen
Temperature (°C)	21.1
Test Note	applies to general class of base polymer

PERMEABILITY (source document units)

Gas Permeability (1×10^{-8} cm^2/ s · atm)	0.20

PERMEABILITY (normalized units)

Permeability Coefficient (cm^3 · mm/m^2 · day · atm)	17.3

Chapter 88

Natural Rubber

Effect of Polymeric Structure on Permeability

Natural Rubber   (product form: gum vulcanizate)

Generally, the presence of polar or methyl groups in the polymer molecule reduces the permeability of polymers while the presence of double bonds has the opposite effect. Cyclization, branching, and vinyl groups attached to the polymer chain all tend to reduce permeability. Natural rubber, which is more unsaturated and has fewer methyl groups than butyl rubber, is more than 20 times more permeable to air.

Reference: *Elastomers Technical Information - Factors in the Gas Permeability of Elastomers,* supplier technical report (TI-28) - Exxon Chemical Company, 1974.

TABLE 378: Air Permeability vs. Temperature Through Natural Rubber.

Material Family	NATURAL RUBBER					
Product Form	NO. 1 RIBBED SMOKE SHEETS					
Cure	20 min. @ 145°C					
Reference Number	298	298	298	299	299	299

MATERIAL CHARACTERISTICS

Shore A Hardness	51 (30 second)		

MATERIAL COMPOSITION

Stearic Acid		2 phr	
Zinc Dimethyl Dithiocarbonate	1.5 phr		
Sulfur	2.5 phr		
Zinc Oxide	5 phr	5 phr	
Channel Black		20 phr	
MT Carbon Black		60 phr	
SRF Carbon Black	50 phr		
Note	0.75 phr Altax	10 phr Flexon 845 oil (Exxon), 4 phr Amberol ST-149 (Rohm and Haas Co.), specific accelerator	

TEST CONDITIONS

Penetrant	air					
Temperature (°C)	40	60	80	23.9	65.6	93.3
Test Apparatus				Aminco Permeability Apparatus		

PERMEABILITY (source document units)

Permeability Relative to Butyl Rubber (%)					700	
Gas Permeability (ft^3 · mil/ft^2 · day · psi)				0.00436	0.0237	0.0402
Gas Permeability (1x10^{-8} cm^3 · cm/ cm^2 · sec · atm)	11.8	26.8	43.9			

PERMEABILITY (normalized units)

Permeability Coefficient (cm^3 · mm/m^2 · day · atm)	1020	2316	3793	496	2696	4574

TABLE 379: Nitrogen Permeability Through Natural Rubber.

Material Family	NATURAL RUBBER
Reference Number	309

TEST CONDITIONS

Penetrant	nitrogen
Temperature (°C)	21.1
Test Note	applies to general class of base polymer

PERMEABILITY (source document units)

Gas Permeability (1×10^{-8} cm²/ s · atm)	6.12

PERMEABILITY (normalized units)

Permeability Coefficient (cm³ · mm/m² · day · atm)	529

TABLE 380: Water Vapor Transmission and Helium, Hydrogen, Oxygen, Nitrogen, Carbon Dioxide and Air Permeabiltiy Relative To Butyl Rubber Through Natural Rubber.

Material Family	NATURAL RUBBER						
Product Form	GUM VULCANIZATE						
Reference Number	300	300	300	300	300	300	300

MATERIAL CHARACTERISTICS

Sample Thickness							0.51 mm

MATERIAL COMPOSITION

Paraffinic Oil							5 phr
SRF Carbon Black							50 phr

TEST CONDITIONS

Penetrant	helium	hydrogen	oxygen	nitrogen	carbon dioxide	air	water vapor
Temperature (°C)	25	25	25	25	25	25	
Test Method							Tappi Standard T464 M-45

PERMEABILITY (source document units)

Permeability Relative to Butyl Rubber (%)	357	667	1780	2000	2500	2100	2425

GRAPH 82: Air Permeability vs. Temperature through Natural Rubber.

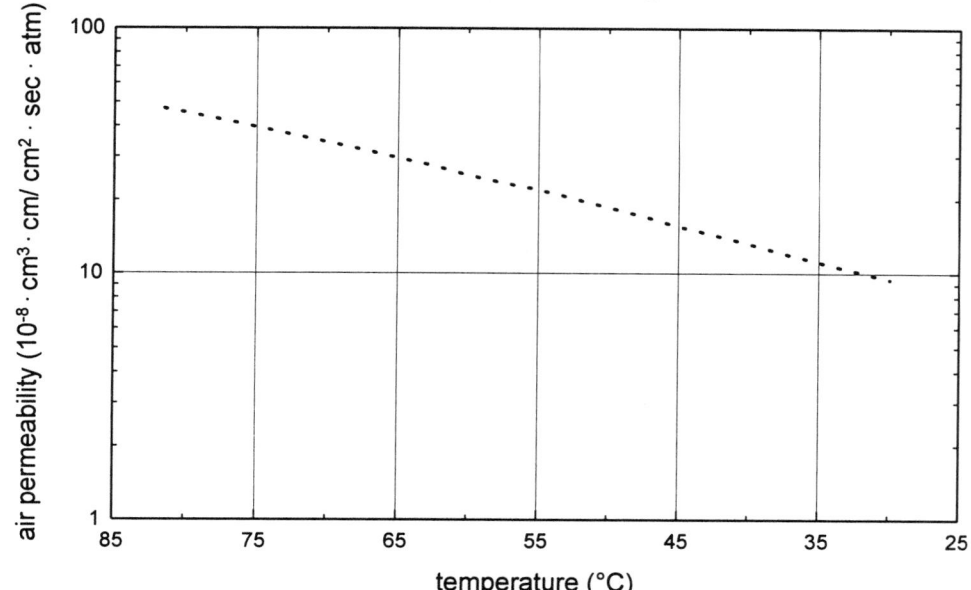

	Natural Rubber (20 min. @ 145°C cure; 51 Shore A; 5 phr zinc oxide, 1.5 phr zinc dimethyl dithiocarbonate, 2.5 phr sulfur, 50 phr SRF black, 0.75 phr Altax; ribbed smoked sheets); penetrant: air; 60°C
Reference No.	298

GRAPH 83: Gas Permeability vs. Mineral Filler Content through Natural Rubber.

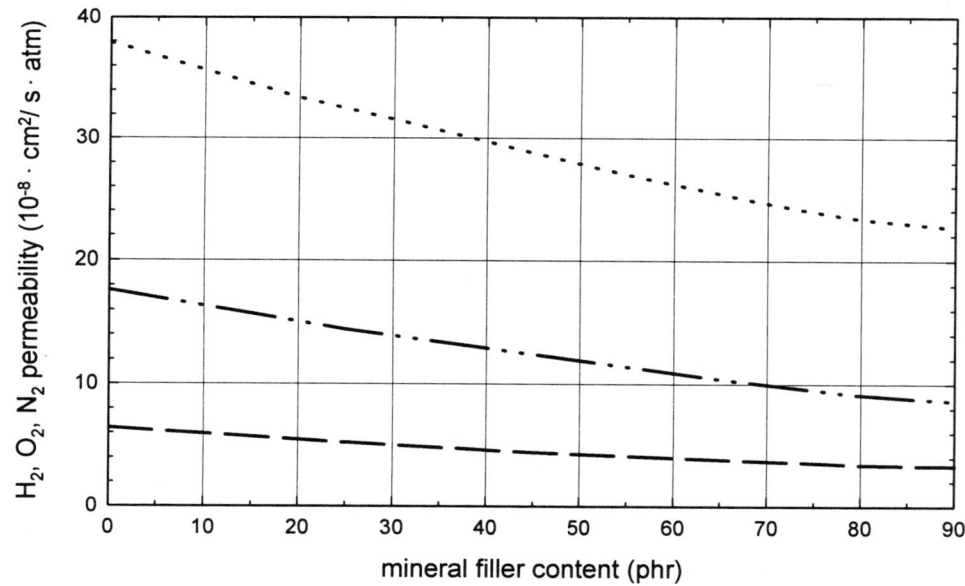

	Natural Rubber (gum vulcanizate); penetrant: H_2
	Natural Rubber (gum vulcanizate); penetrant: O_2
	Natural Rubber (gum vulcanizate); penetrant: N_2
Reference No.	300

Chapter 89

Polychloroprene Rubber

TABLE 381: Hydrogen, Oxygen, Nitrogen, Carbon Dioxide and Air Permeabiltiy Relative To Butyl Rubber Through Neoprene Rubber.

Material Family	POLYCHLOROPRENE RUBBER				
Product Form	GUM VULCANIZATE				
Reference Number	300	300	300	300	300

MATERIAL COMPOSITION

Note	Neoprene G type

TEST CONDITIONS

Penetrant	hydrogen	oxygen	nitrogen	carbon dioxide	air
Temperature (°C)	25	25	25	25	25

PERMEABILITY (source document units)

Permeability Relative to Butyl Rubber (%)	180	302	280	500	315

TABLE 382: Nitrogen Permeability Through Neoprene Rubber.

Material Family	POLYCHLOROPRENE RUBBER
Reference Number	309

TEST CONDITIONS

Penetrant	nitrogen
Temperature (°C)	21.1
Test Note	applies to general class of base polymer

PERMEABILITY (source document units)

Gas Permeability (1×10^{-8} cm^2/ s · atm)	0.89

PERMEABILITY (normalized units)

Permeability Coefficient (cm^3 · mm/m^2 · day · atm)	77

Chapter 90

Acrylonitrile-Butadiene Copolymer

Effect of Polymeric Structure on Permeability

Nitrile Rubber

Generally, the presence of polar or methyl groups in the polymer molecule reduces the permeability of polymers while the presence of double bonds has the opposite effect. Cyclization, branching, and vinyl groups attached to the polymer chain all tend to reduce permeability. In the case of the butadiene rubbers (NBR), as the acrylonitrile content is increased, the permeability is decreased due to the reduced number of double bonds and the increased number of polar groups.

Reference: *Elastomers Technical Information - Factors in the Gas Permeability of Elastomers,* supplier technical report (TI-28) - Exxon Chemical Company, 1974.

<u>TABLE 383:</u> Air Permeability vs. Temperature Through Bayer Krynac 800 Nitrile Rubber.

Material Family	ACRYLONITRILE-BUTADIENE COPOLYMER		
Material Supplier/ Grade	BAYER KRYNAC 800		
Cure	20 min. @ 145°C		
Reference Number	298	298	298

MATERIAL CHARACTERISTICS

Shore A Hardness	65 (30 second)		

MATERIAL COMPOSITION

Zinc Dimethyl Dithiocarbonate	1.5 phr		
Sulfur	1.5 phr		
Zinc Oxide	5 phr		
SRF Carbon Black	50 phr		
Note	0.5 phr Monex, 1.5 phr Altax		

TEST CONDITIONS

Penetrant	air		
Temperature (°C)	40	60	80

PERMEABILITY (source document units)

Gas Permeability $(1 \times 10^{-8}$ cm$^3 \cdot$ cm/ cm$^2 \cdot$ sec $\cdot$ atm)	1.1	4.1	9.9

PERMEABILITY (normalized units)

Permeability Coefficient (cm$^3 \cdot$ mm/m$^2 \cdot$ day $\cdot$ atm)	95	354	855

TABLE 384: Nitrogen Permeability Through Nitrile Rubber.

Material Family	ACRYLONITRILE-BUTADIENE COPOLYMER	
Reference Number	309	309

MATERIAL COMPOSITION

Note	38% acrylonitrile	34% acrylonitrile

TEST CONDITIONS

Penetrant	nitrogen	
Temperature (°C)	21.1	21.1
Test Note	applies to general class of base polymer	

PERMEABILITY (source document units)

Gas Permeability (1×10^{-8} cm^2/ s · atm)	0.18	0.46

PERMEABILITY (normalized units)

Permeability Coefficient (cm^3 · mm/m^2 · day · atm)	15.6	39.7

TABLE 385: Air Conditioning Refrigerants Permeation Loss Through Nitrile Rubber.

Material Family	ACRYLONITRILE-BUTADIENE COPOLYMER		
Reference Number	275	275	275

MATERIAL CHARACTERISTICS

Sample Thickness	1 mm	1 mm	1 mm
Sample Length	305 mm	305 mm	305 mm
Sample Inside Diameter	15.9 mm	15.9 mm	15.9 mm

TEST CONDITIONS

Penetrant	Freon 12	HCFCX-134a	HCFC-22/ HCFC-124/ HFC-152a
Penetrant Note	air conditioning refrigerant		air conditioning refrigerant, ternary blend
Temperature (°C)	93	93	93
Test Condition Note	refrigerant at saturated vapor pressure	refrigerant at saturated vapor pressure	refrigerant at saturated vapor pressure
Test Note	calculated from permeation coefficient data		

PERMEABILITY (source document units)

Permeation Loss (lb/ft-yr)	0.662	0.56	0.938

TABLE 386: Carbon Dioxide and Air Permeabiltiy Relative To Butyl Rubber Through Acrylonitrile Butadiene Copolymer Rubber.

Material Family	ACRYLONITRILE-BUTADIENE COPOLYMER							
Product Form	GUM VULCANIZATE							
Reference Number	300	300	300	300	300	300	300	300

MATERIAL COMPOSITION

Butadiene Content (%)	80	73	68	61	80	73	68	61
Acrylonitrile Content (%)	20	27	32	39	20	27	32	39

TEST CONDITIONS

Penetrant	air				carbon dioxide			
Temperature (°C)	25	25	25	25	25	25	25	25

PERMEABILITY (source document units)

Permeability Relative to Butyl Rubber (%)	693	315	179	72	1200	600	350	143

TABLE 387: Helium and Hydrogen Permeabiltiy Relative To Butyl Rubber Through Acrylonitrile Butadiene Copolymer Rubber.

Material Family	ACRYLONITRILE-BUTADIENE COPOLYMER							
Product Form	GUM VULCANIZATE							
Reference Number	300	300	300	300	300	300	300	300

MATERIAL COMPOSITION

Butadiene Content (%)	80	73	68	61	80	73	68	61
Acrylonitrile Content (%)	20	27	32	39	20	27	32	39

TEST CONDITIONS

Penetrant	helium				hydrogen			
Temperature (°C)	25	25	25	25	25	25	25	25

PERMEABILITY (source document units)

Permeability Relative to Butyl Rubber (%)	196	139	114	79	341	214	161	100

TABLE 388: Oxygen and Nitrogen Permeabiltiy Relative To Butyl Rubber Through Acrylonitrile Butadiene Copolymer Rubber.

Material Family	ACRYLONITRILE-BUTADIENE COPOLYMER							
Product Form	GUM VULCANIZATE							
Reference Number	300	300	300	300	300	300	300	300

MATERIAL COMPOSITION

Butadiene Content (%)	80	73	68	61	80	73	68	61
Acrylonitrile Content (%)	20	27	32	39	20	27	32	39

TEST CONDITIONS

Penetrant	nitrogen				oxygen			
Temperature (°C)	25	25	25	25	25	25	25	25

PERMEABILITY (source document units)

Permeability Relative to Butyl Rubber (%)	620	260	150	58	622	302	178	73

GRAPH 84: Air Permeability vs. Temperature through Acrylonitrile-Butadiene Copolymer.

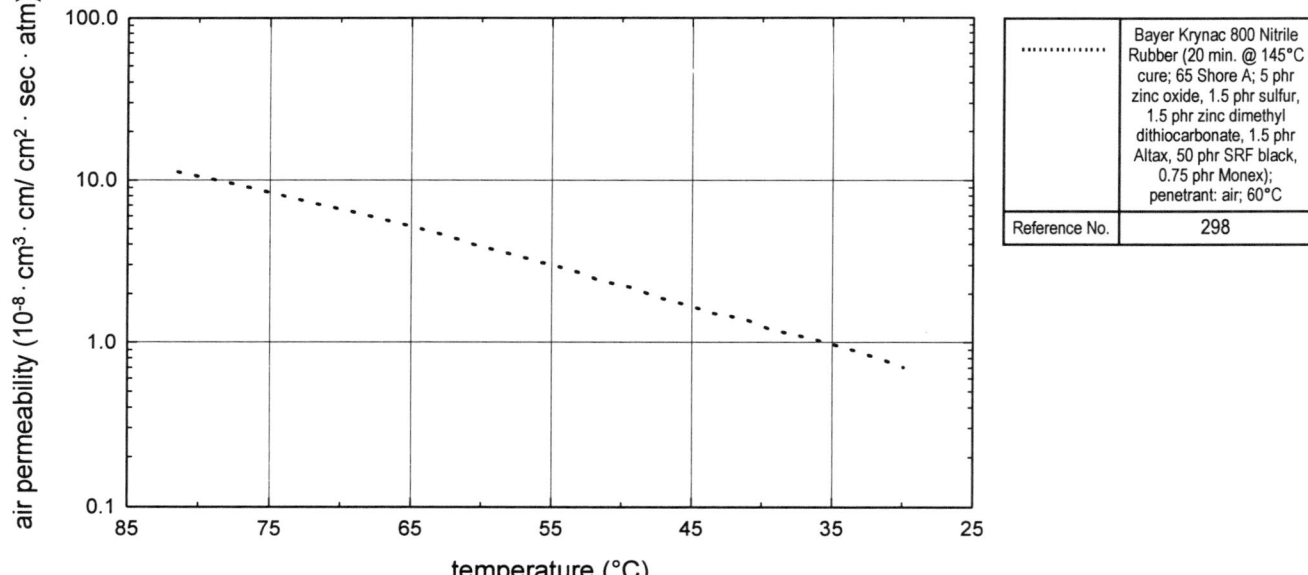

....	Bayer Krynac 800 Nitrile Rubber (20 min. @ 145°C cure; 65 Shore A; 5 phr zinc oxide, 1.5 phr sulfur, 1.5 phr zinc dimethyl dithiocarbonate, 1.5 phr Altax, 50 phr SRF black, 0.75 phr Monex); penetrant: air; 60°C
Reference No.	298

Chapter 91

Polyisoprene Rubber

TABLE 389: Nitrogen Permeability Through Polyisoprene Rubber.

Material Family	POLYISOPRENE RUBBER
Reference Number	309

TEST CONDITIONS

Penetrant	nitrogen
Temperature (°C)	21.1
Test Note	applies to general class of base polymer

PERMEABILITY (source document units)

Gas Permeability (1×10^{-8} cm^2/ s · atm)	6.12

PERMEABILITY (normalized units)

Permeability Coefficient (cm^3 · mm/m^2 · day · atm)	529

Chapter 92

Polysulfide Rubber

TABLE 390: Nitrogen Permeability Through Polysulfide Rubber.

Material Family	POLYSULFIDE RUBBER
Reference Number	309

TEST CONDITIONS

Penetrant	nitrogen
Temperature (°C)	21.1
Test Note	applies to general class of base polymer

PERMEABILITY (source document units)

Gas Permeability $(1\times10^{-8}$ cm^2/ s $\cdot$ atm)	0.66

PERMEABILITY (normalized units)

Permeability Coefficient (cm$^3 \cdot$ mm/m$^2 \cdot$ day $\cdot$ atm)	57

Chapter 93

Polyurethane

TABLE 391: Water Vapor, Oxygen, Nitrogen and Carbon Dioxide Permeability Through Polyurethanes.

Material Family	POLYURETHANE			
Reference Number	121	121	121	121

TEST CONDITIONS

Penetrant	nitrogen	oxygen	carbon dioxide	water vapor
Temperature (°C)	25	25	25	37
Relative Humidity (%)				90
Test Method	ASTM D1434-63T	ASTM D1434-63T	ASTM D1434-63T	ASTM E96-63T

PERMEABILITY (source document units)

Vapor Transmission Rate (g · mil/100 in^2 · day)				2.4 - 8.7 {dc}
Gas Permeability (cm^3 · mil/100 in^2 · day)	80	200	3000	

PERMEABILITY (normalized units)

Permeability Coefficient (cm^3 · mm/m^2 · day · atm)	31.5	78.7	1181	
Vapor Transmission Rate (g · mm/m^2 · day)				0.94 - 3.43

TABLE 392: Nitrogen Permeability Through Polyester and Polyether Urethane Rubber.

Material Family	POLYESTER URETHANE	POLYETHER URETHANE
Reference Number	309	309

TEST CONDITIONS

Penetrant	nitrogen	
Temperature (°C)	21.1	
Test Note	applies to general class of base polymer	

PERMEABILITY (source document units)

Gas Permeability (1x10^{-8} cm^2/ s · atm)	16	0.95

PERMEABILITY (normalized units)

Permeability Coefficient (cm^3 · mm/m^2 · day · atm)	1382	82.1

Chapter 94

Propylene Oxide Rubber

TABLE 393: Nitrogen Permeability Through Polypropylene Oxide Rubber.

Material Family	PROPYLENE OXIDE RUBBER
Reference Number	309

TEST CONDITIONS

Penetrant	nitrogen
Temperature (°C)	21.1
Test Note	applies to general class of base polymer

PERMEABILITY (source document units)

Gas Permeability (1×10^{-8} cm^2/ s · atm)	0.9

PERMEABILITY (normalized units)

Permeability Coefficient (cm^3 · mm/m^2 · day · atm)	77.8

Chapter 95

Silicone

Effect of Temperature on Barrier Performance

Silicone   (product form: gum vulcanizate)

Permeability is influenced not only by the changes in diffusive properties with temperature but by changes in solubility as well. Since the solubility of many gases decreases with increasing temperature, it sometimes occurs that this effect predominates over the increase in diffusivity with increasing temperature. Consequently, the permeability of some gases increases as the temperature is raised whereas for others, it decreases. Examples of this effect would be CO_2 in silicone rubber and C_4 and C_5 hydrocarbons in all rubbers which show decreases in permeability constants with increasing temperature. However, for most permanent gases (other than the illustrated silicone rubber CO_2 case), permeability will increase with temperature.

Reference: *Elastomers Technical Information - Factors in the Gas Permeability of Elastomers,* supplier technical report (TI-28) - Exxon Chemical Company, 1974.

Effect of Polymeric Structure on Permeability

Silicone   (product form: gum vulcanizate)

Generally, the presence of polar or methyl groups in the polymer molecule reduces the permeability of polymers while the presence of double bonds has the opposite effect. Cyclization, branching, and vinyl groups attached to the polymer chain all tend to reduce permeability. The seemingly anomalous behavior of silicone rubber, in view of the explanation above, is apparently connected with high internal mobility due to the presence of an -Si-O-Si- configuration in its molecular chain. This rubber has the highest known gas diffusivity.

Reference: *Elastomers Technical Information - Factors in the Gas Permeability of Elastomers,* supplier technical report (TI-28) - Exxon Chemical Company, 1974.

TABLE 394: Hydrogen, Oxygen, Nitrogen, Carbon Dioxide and Air Permeabiltiy Relative To Butyl Rubber Through Silicone Rubber.

Material Family	SILICONE				
Product Form	GUM VULCANIZATE				
Reference Number	300	300	300	300	300

TEST CONDITIONS

Penetrant	hydrogen	oxygen	nitrogen	carbon dioxide	air
Temperature (°C)	25	25	25	25	25

PERMEABILITY (source document units)

Permeability Relative to Butyl Rubber (%)	7150	3920	66,000	40,000	56,800

TABLE 395: Nitrogen Permeability Through Silicone Rubber.

Material Family	SILICONE
Reference Number	309

TEST CONDITIONS

Penetrant	nitrogen
Temperature (°C)	21.1
Test Note	applies to general class of base polymer

PERMEABILITY (source document units)

Gas Permeability $(1 \times 10^{-8}$ cm^2/ s $\cdot$ atm)	200

PERMEABILITY (normalized units)

Permeability Coefficient (cm^3 $\cdot$ mm/m^2 $\cdot$ day $\cdot$ atm)	17,280

<u>TABLE 396:</u> Water Vapor, Oxygen, Carbon Dioxide and Hydrogen Permeability Through Silicone Rubber.

Material Family	SILICONE			
Reference Number	121	121	121	121

TEST CONDITIONS

Penetrant	oxygen	carbon dioxide	hydrogen	water vapor
Temperature (°C)	25	25	25	37
Relative Humidity (%)				90
Test Method	ASTM D1434-63T	ASTM D1434-63T	ASTM D1434-63T	ASTM E96-63T

PERMEABILITY (source document units)

Vapor Transmission Rate (g · mil/100 in^2 · day)				4.4 - 7.9 {dc}
Gas Permeability (cm^3 · mil/100 in^2 · day)	50,000	300,000	45,000	

PERMEABILITY (normalized units)

Permeability Coefficient (cm^3 · mm/m^2 · day · atm)	19,685	118,110	17,716	
Vapor Transmission Rate (g · mm/m^2 · day)				1.73 - 3.11

Chapter 96

Methylvinylfluorosilicone

TABLE 397: Nitrogen Permeability Through Fluorosilicone.

Material Family	METHYLVINYLFLUOROSILICONE
Reference Number	309

TEST CONDITIONS

Penetrant	nitrogen
Temperature (°C)	21.1
Test Note	applies to general class of base polymer

PERMEABILITY (source document units)

Gas Permeability (1×10^{-8} cm²/ s · atm)	165

PERMEABILITY (normalized units)

Permeability Coefficient (cm³ · mm/m² · day · atm)	14,256

Chapter 97

Styrene-Butadiene Copolymer

TABLE 398: Air Permeability vs. Temperature Through Styrene-Butadiene Rubber.

Material Family	STYRENE-BUTADIENE COPOLYMER								
Material Supplier/ Grade	BAYER KRYLENE						EXXON SBR 1502		
Cure	40 min. @ 145°C								
Reference Number	298	298	298	299	299	299	299	299	299

MATERIAL CHARACTERISTICS

Shore A Hardness	53 (30 second)								

MATERIAL COMPOSITION

Styrene				43%					
Stearic Acid				2 phr			2 phr		
Zinc Dimethyl Dithiocarbonate	1.5 phr								
Sulfur	2.8 phr								
Zinc Oxide	5 phr			5 phr			5 phr		
Channel Black				20 phr			20 phr		
MT Carbon Black				60 phr			60 phr		
SRF Carbon Black	50 phr								
Note	1.2 phr Santocure			10 phr Flexon 845 oil (Exxon), 4 phr Amberol ST-149 (Rohm and Haas Co.), high styrene content, specific accelerator			10 phr Flexon 845 oil (Exxon), 4 phr Amberol ST-149 (Rohm and Haas Co.), specific accelerator		

TEST CONDITIONS

Penetrant	air								
Temperature (°C)	40	60	80	23.9	65.6	93.3	23.9	65.6	93.3
Test Apparatus				Aminco Permeability Apparatus			Aminco Permeability Apparatus		

PERMEABILITY (source document units)

Permeability Relative to Butyl Rubber (%)					300			550	
Gas Permeability ($ft^3 \cdot mil/ft^2 \cdot day \cdot psi$)				0.00121	0.0096	0.0239	0.00306	0.018	0.0382
Gas Permeability ($1 \times 10^{-8}\ cm^3 \cdot cm/\ cm^2 \cdot sec \cdot atm$)	4.6	12.5	24.2						

PERMEABILITY (normalized units)

Permeability Coefficient ($cm^3 \cdot mm/m^2 \cdot day \cdot atm$)	397	1080	2091	137.7	1092	2719	348	2048	4346

TABLE 399: Water Vapor Transmission and Helium, Hydrogen, Oxygen, Nitrogen, Carbon Dioxide and Air Permeabiltiy Relative To Butyl Rubber Through Styrene-Butadiene Rubber.

Material Family	STYRENE-BUTADIENE COPOLYMER						
Material Supplier/ Grade							**EXXON SBR 1500**
Product Form	GUM VULCANIZATE						
Features	hot						
Reference Number	300	300	300	300	300	300	300

MATERIAL CHARACTERISTICS

Sample Thickness		0.51 mm

MATERIAL COMPOSITION

Paraffinic Oil		5 phr
SRF Carbon Black		50 phr
Chemical Type		cis 1,4-polybutadiene

TEST CONDITIONS

Penetrant	helium	hydrogen	oxygen	nitrogen	carbon dioxide	air	water vapor
Temperature (°C)	25	25	25	25	25	25	
Test Method							Tappi Standard T464 M-45

PERMEABILITY (source document units)

Penetrant	helium	hydrogen	oxygen	nitrogen	carbon dioxide	air	water vapor
Permeability Relative to Butyl Rubber (%)	264	641	1290	1560	2360	1600	1875

TABLE 400: Nitrogen, Oxygen and Carbon Dioxide Permeability Through Styrene-Butadiene Rubber.

Material Family	STYRENE-BUTADIENE COPOLYMER		
		FILM	
Reference Number	309	250	250

TEST CONDITIONS

Penetrant	nitrogen	oxygen	carbon dioxide
Temperature (°C)	21.1	24	24
Test Note	applies to general class of base polymer		

PERMEABILITY (source document units)

Gas Permeability ($cm^3 \cdot mil/100\ in^2 \cdot day$)		3000	1000
Gas Permeability ($1 \times 10^{-8}\ cm^2/s \cdot atm$)	4.8		

PERMEABILITY (normalized units)

Permeability Coefficient ($cm^3 \cdot mm/in^2 \cdot day \cdot atm$)	415	1181	394

GRAPH 85: Air Permeability vs. Temperature through Styrene-Butadiene Copolymer.

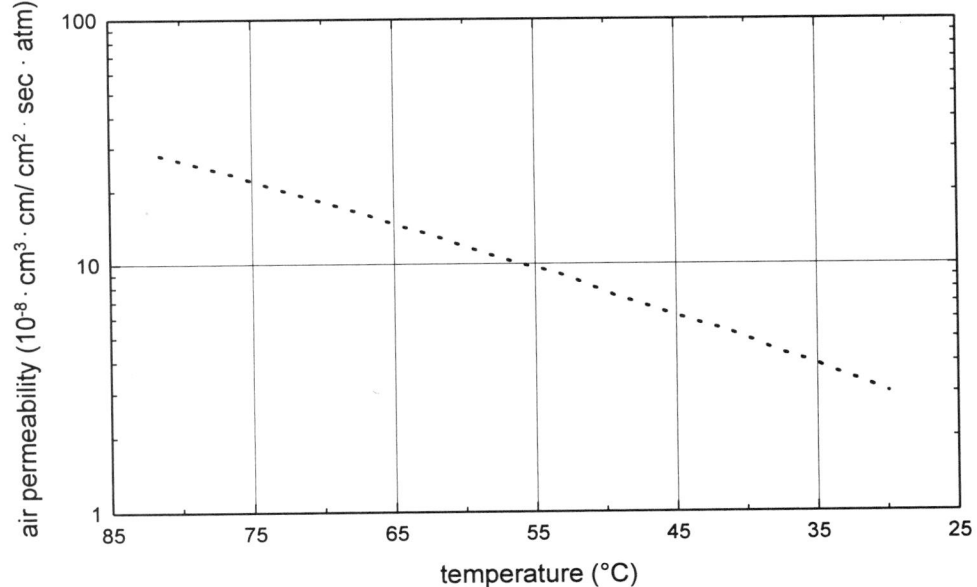

··············	Bayer Krylene SBR (40 min. @ 145°C cure; 53 Shore A; 5 phr zinc oxide, 1.5 phr zinc dimethyl dithiocarbonate, 50 phr SRF black, 2.8 phr sulfur, 1.2 phr Santocure); penetrant: air; 60°C
Reference No.	298

Appendix I

Tables Sorted By Penetrant

ACETALDEHYDE

Material Family / Material Note	Perm. Coefficient (cm³·mm/m²·day·atm)	VTR (g·mm/m²·day)	Non-normalized Data	Penetrant Note	Temp. (°C)	Time (days)	RH (%)	Pressure (kPa)	Test Note	Source
Rubber, Latex (NR)										
Ansell Edmont Canners and Handlers 392; unsupported glove film; 0.48 mm thick			fair permeation rate: 51 to 100 eyedropper size drops per hour; breakthrough time: 7 minutes; permeation rate: <900 µg/cm²/min		23				ASTM F739	313
Pioneer Industrial Products L-118; glove film			breakthrough time: 2 minutes; lower detection limit: 0.4 ppm; permeation rate: 78 µg/cm²/min						ASTM F739.85	312
Rubber, Neoprene (CR)										
Ansell Edmont Neoprene 29-840; unsupported glove film; 0.38 mm thick			poor permeation rate: 501 to 5000 eyedropper size drops per hour; breakthrough time: 10 minutes; permeation rate: <9000 µg/cm²/min		23				ASTM F739	313
Ansell Edmont Neox; supported (lined) glove film; specified by glove film weight			poor permeation rate: 501 to 5000 eyedropper size drops per hour; breakthrough time: 17 minutes; permeation rate: <9000 µg/cm²/min						ASTM F739	313
Pioneer Industrial Products Stanzoil N-44; glove film			breakthrough time: 21 minutes; lower detection limit: 0.1 ppm; permeation rate: 108 µg/cm²/min						ASTM F739.85	312
TPE, Vinyl										
Ansell Edmont Wet-Wear 600; clothing; nylon netting bonded between two layers of PVC			poor permeation rate: 501 to 5000 eyedropper size drops per hour; breakthrough time: 5 minutes; permeation rate: <9000 µg/cm²/min		23				ASTM F739	313
TPE, Vinyl coated fabric										
Ansell Edmont Wet-Wear 700; clothing; PVC coating with Nylon/Polyester lining			poor permeation rate: 501 to 5000 eyedropper size drops per hour; breakthrough time: 3 minutes; permeation rate: <9000 µg/cm²/min		23				ASTM F739	313

ACETIC ACID

Material Family / Material Note	Perm. Coefficient (cm³·mm/m²·day·atm)	VTR (g·mm/m²·day)	Non-normalized Data	Penetrant Note	Temp. (°C)	Time (days)	RH (%)	Pressure (kPa)	Test Note	Source
Polyphenylene Sulfide										
Phillips Ryton; 0.127 mm thick; baked coating; film		0.79			23				Die cut samples were fitted to tops of glass bottles by a rubber gasket and a lid, with a surface area of 96.8 cm²; Liquids were placed in bottles, gaskets and film put in place, and the lid screwed on. Apparatus was inverted to put liquid in direct contact with film. Weight loss measurements were made at 1 week intervals throughout 4 weeks of conditioning.	102
Rubber, Latex (NR)										
Ansell Edmont Canners and Handlers 392; unsupported glove film; 0.48 mm thick			breakthrough time: 110 minutes	glacial	23				ASTM F739	313
Pioneer Industrial Products L-118; glove film			breakthrough time: 21 minutes; lower detection limit: 0.1 ppm; permeation rate: 12 µg/cm²/min	glacial					ASTM F739.85	312
Pioneer Industrial Products L-118; glove film			breakthrough time: 31 minutes; lower detection limit: 0.02 ppm; permeation rate: 18 µg/cm²/min	50% conc.					ASTM F739.85	312
Rubber, Neoprene (CR)										
Ansell Edmont Neoprene 29-840; unsupported glove film; 0.38 mm thick			breakthrough time: 420 minutes	glacial	23				ASTM F739	313
Ansell Edmont Neox; supported (lined) glove film; specified by glove film weight			breakthrough time: >360 minutes	glacial	23				ASTM F739	313
Pioneer Industrial Products Stanzoil N-44; glove film			no permeation rate at steady state detected; breakthrough time: >480 minutes; lower detection limit: 0.1 ppm; permeation rate: 0 µg/cm²/min	50% conc.					ASTM F739.85	312
Rubber, Nitrile (NBR)										
Ansell Edmont Sol-Vex 37-165; unsupported glove film; 0.54 mm thick			breakthrough time: 270 minutes	glacial	23				ASTM F739	313
Pioneer Industrial Products Starsolv A-14; glove film			breakthrough time: 118 minutes; lower detection limit: 0.1 ppm; permeation rate: 1326 µg/cm²/min	glacial					ASTM F739.85	312
Pioneer Industrial Products Starsolv A-14; glove film			no permeation rate at steady state detected; breakthrough time: >480 minutes; lower detection limit: 0.02 ppm; permeation rate: 0 µg/cm²/min	50% conc.					ASTM F739.85	312
TPE, Vinyl										
Ansell Edmont Monkey Grip; supported (lined) glove film; specified by glove film weight			breakthrough time: 180 minutes	glacial	23				ASTM F739	313
Ansell Edmont Wet-Wear 550; clothing; PVC stretch outer layer bonded to a lightweight liner			breakthrough time: 30 minutes	glacial	23				ASTM F739	313

ACETIC ACID (continued)

Material Family	Material Note	Perm. Coefficient (cm³·mm/m²·day·atm)	VTR (g·mm/m²·day)	Non-normalized Data	Penetrant Note	Temp. (°C)	Time (days)	RH (%)	Pressure (kPa)	Test Note	Source
TPE, Vinyl	Ansell Edmont Wet-Wear 600; clothing; nylon netting bonded between two layers of PVC			breakthrough time: 6 minutes	glacial	23				ASTM F739	313
	Pioneer Industrial Products Pylox V-20; glove film			breakthrough time: 85 minutes; lower detection limit: 0.1 ppm; permeation rate: 1.8 µg/cm²/min	glacial					ASTM F739.85	312
	Pioneer Industrial Products Pylox V-20; glove film			breakthrough time: 47 minutes; lower detection limit: 0.02 ppm; permeation rate: 0.36 µg/cm²/min	50% conc.					ASTM F739.85	312
TPE, Vinyl coated fabric	Ansell Edmont Wet-Wear 700; clothing; PVC coating with Nylon/Polyester lining			breakthrough time: 16 minutes	glacial	23				ASTM F739	313

ACETONE

Material Family	Material Note	Perm. Coefficient (cm³·mm/m²·day·atm)	VTR (g·mm/m²·day)	Non-normalized Data	Penetrant Note	Temp. (°C)	Time (days)	RH (%)	Pressure (kPa)	Test Note	Source
Acrylonitrile Copolymer, AMA	barrier prop.; bottles			failed		23	180				293
Laminar, Nylon/Olefin	DuPont Selar RB/ Polyolefin; 8% Selar RB; barrier prop.; extrusion blow molded; bottles			0.48% penetrant weight loss		23	180				293
Polyester, PET	DuPont Mylar; film		0.87			40				ASTM E96-80; modified test, permeabilities determined at the partial pressure of the vapor at the test temperature	270
Polyethylene, Fluorinated	barrier prop.; bottles			0.69% penetrant weight loss		23	180				293
Polyethylene, HDPE	bottles			0.91% penetrant weight loss		23	180				293
Polyethylene, LDPE	Dow; film		3.9 - 15.8			24					250
Polystyrene	Dow Styron; film			sample failed		24					250
Polyvinyl Chloride	bottles			failed		23	180				293
Rubber, Latex (NR)	Ansell Edmont Canners and Handlers 392; unsupported glove film; 0.48 mm thick			fair permeation rate; 51 to 100 eyedropper size drops per hour; breakthrough time: 10 minutes; permeation rate: <900 µg/cm²/min		23				ASTM F739	313
	Pioneer Industrial Products L-118; glove film			breakthrough time: 7 minutes; lower detection limit: 0.05 ppm; permeation rate: 30 µg/cm²/min						ASTM F739.85	312
Rubber, Neoprene (CR)	Ansell Edmont Neoprene 29-840; unsupported glove film; 0.38 mm thick			fair permeation rate; 51 to 100 eyedropper size drops per hour; breakthrough time: 5 minutes; permeation rate: <900 µg/cm²/min		23				ASTM F739	313
	Ansell Edmont Neox; supported (lined) glove film; specified by glove film weight			fair permeation rate; 51 to 100 eyedropper size drops per hour; breakthrough time: 10 minutes; permeation rate: <900 µg/cm²/min		23				ASTM F739	313
	Pioneer Industrial Products Stanzoil N-44; glove film			breakthrough time: 12 minutes; permeation rate: 210 µg/cm²/min						ASTM F739.85	312
SAN	Dow Tyril; film			sample failed		24					250
TPE, Vinyl	Pioneer Industrial Products Pylox V-20; glove film			permeation rate too large to measure; breakthrough time: <1 minutes						ASTM F739.85	312

ACETONITRILE

Material Family	Material Note	Perm. Coefficient (cm³·mm/m²·day·atm)	VTR (g·mm/m²·day)	Non-normalized Data	Penetrant Note	Temp. (°C)	Time (days)	RH (%)	Pressure (kPa)	Test Note	Source
Polyvinyl Alcohol	Ansell Edmont PVA; supported (lined) glove film; specified by glove film weight			good permeation rate; 6 to 50 eyedropper size drops per hour; PVA coating is water soluble; breakthrough time: 150 minutes; permeation rate: <90 µg/cm²/min		23				ASTM F739	313
Rubber, Latex (NR)	Ansell Edmont Canners and Handlers 392; unsupported glove film; 0.48 mm thick			very good permeation rate; 1 to 5 eyedropper size drops per hour; breakthrough time: 4 minutes; permeation rate: <9 µg/cm²/min		23				ASTM F739	313
Rubber, Neoprene (CR)	Ansell Edmont Neoprene 29-840; unsupported glove film; 0.38 mm thick			very good permeation rate; 1 to 5 eyedropper size drops per hour; breakthrough time: 30 minutes; permeation rate: <9 µg/cm²/min		23				ASTM F739	313
	Ansell Edmont Neox; supported (lined) glove film; specified by glove film weight			low permeation rate; 0 to 1/2 eyedropper size drops per hour; breakthrough time: 90 minutes; permeation rate: <0.9 µg/cm²/min		23				ASTM F739	313

ACETONITRILE (continued)

Material Family	Material Note	Perm. Coefficient (cm³·mm/m²·day·atm)	VTR (g·mm/m²·day)	Non-normalized Data	Penetrant Note	Temp. (°C)	Time (days)	RH (%)	Pressure (kPa)	Test Note	Source
Rubber, Neoprene (CR)	Pioneer Industrial Products Stanzoil N-44; glove film			breakthrough time: 40 minutes; permeation rate: 42 µg/cm²/min						ASTM F739.85	312
Rubber, Nitrile (NBR)	Ansell Edmont Sol-Vex 37-165; unsupported glove film; 0.54 mm thick			fair permeation rate: 51 to 100 eyedropper size drops per hour; breakthrough time: 30 minutes; permeation rate: <900 µg/cm²/min		23				ASTM F739	313

ACRYLIC ACID

Material Family	Material Note	Perm. Coefficient (cm³·mm/m²·day·atm)	VTR (g·mm/m²·day)	Non-normalized Data	Penetrant Note	Temp. (°C)	Time (days)	RH (%)	Pressure (kPa)	Test Note	Source
Rubber, Latex (NR)	Ansell Edmont Canners and Handlers 392; unsupported glove film; 0.48 mm thick			breakthrough time: 80 minutes		23				ASTM F739	313
Rubber, Neoprene (CR)	Ansell Edmont Neoprene 29-840; unsupported glove film; 0.38 mm thick			breakthrough time: 70 minutes		23				ASTM F739	313
	Ansell Edmont Neox; supported (lined) glove film; specified by glove film weight			low permeation rate: 0 to 1/2 eyedropper size drops per hour; lower detection limit: 0 ppm; permeation rate: <0.9 µg/cm²/min		23	0.25			ASTM F739	313
Rubber, Nitrile (NBR)	Ansell Edmont Sol-Vex 37-165; unsupported glove film; 0.54 mm thick			breakthrough time: 120 minutes		23				ASTM F739	313

AIR

Material Family	Material Note	Perm. Coefficient (cm³·mm/m²·day·atm)	VTR (g·mm/m²·day)	Non-normalized Data	Penetrant Note	Temp. (°C)	Time (days)	RH (%)	Pressure (kPa)	Test Note	Source
Acetal Copolymer	Hoechst Cel. Celcon M25; high molecular weight, 0.15 mm thick; 2.5 g/10 min. MFI; film	0.87 - 1.3									210
	Hoechst Cel. Celcon M270; low molecular weight, high flow, 0.15 mm thick; 27.0 g/10 min. MFI; film	0.87 - 1.3									210
	Hoechst Cel. Celcon M90; gen. purp. grade, 0.15 mm thick; 9.0 g/10 min. MFI; film	0.87 - 1.3									210
Fluoroelastomer, FKM	DuPont Viton; compounded	8.6				24				one atmosphere @60°C; specimen size: 1 cm² x 1 cm thick	305
HDPE/ EVA/ PVDC/ EVA Film	Dow Saranex 25; form-fill-seal pouch, bag-in-box; neutral, barrier prop., 0.051 mm thick; high slip; coextruded film	0.1				23		10		ASTM D1434	257
LDPE/ EVA/ PVDC/ EVA Film	Dow Saranex 23; laminations; neutral, barrier prop., 0.051 mm thick; medium slip; coextruded film	0.24				23		10		ASTM D1434	257
LDPE/ EVA/ PVDC/ EVA/ LDPE Film	Dow Saranex 11; laminations; neutral, barrier prop., 0.038 mm thick; medium slip; coextruded film	0.18				23		10		ASTM D1434	257
	Dow Saranex 14; form-fill-seal pouch; neutral, barrier prop., 0.051 mm thick; medium slip; coextruded film	0.1				23		10		ASTM D1434	257
	Dow Saranex 15; form-fill-seal pouch; neutral, barrier prop., 0.076 mm thick; medium slip; coextruded film	0.15				23		10		ASTM D1434	257
	Dow Saranex 15; form-fill-seal pouch; neutral, barrier prop., 0.102 mm thick; medium slip; coextruded film	0.21				23		10		ASTM D1434	257
Polyethylene, HDPE	Hoechst AG Hostalen	29.4				20				volume at standard temperature and pressure; useable average for all Hostalen grades	94
	Hoechst AG Hostalen	30.4				25				"	94
	Hoechst AG Hostalen	38.5				30				"	94
	Hoechst AG Hostalen	68.9				40				"	94
	Hoechst AG Hostalen	111				50				"	94

AIR (continued)

Material Family	Material Note	Perm. Coefficient (cm³·mm/m²·day·atm)	VTR (g·mm/m²·day)	Non-normalized Data	Penetrant Note	Temp. (°C)	Time (days)	RH (%)	Pressure (kPa)	Test Note	Source
Polyphenylene Sulfide	Phillips Ryton; 0.875 mm thick; baked coating; film	7.9 - 11.8								ASTM D1434, Method M	102
Polyvinylidene Chloride	Dow Saran Wrap 18: chub packaging, laminations; transparent, barrier prop, 0.019 mm thick, biaxially oriented; monolayer film	0.14				23		10		ASTM D1434	256
	Dow Saran Wrap 18L; laminations; transparent, barrier prop, 0.019 mm thick, preshrunk, biaxially oriented; monolayer film	0.19				23		10		ASTM D1434	256
	Dow Saran Wrap 19: chub packaging; transparent, barrier prop, 0.0254 mm thick; biaxially oriented; monolayer film	0.14				23		10		ASTM D1434	256
	Dow Saran Wrap 28: chub packaging; transparent, barrier prop, 0.0254 mm thick; biaxially oriented; monolayer film	0.2				23		10		ASTM D1434	256
	Dow Saran Wrap 560: unit packaging; transparent, barrier prop, 0.152 mm thick, biaxially oriented; coextruded film	0.03				23		10		ASTM D1434	256
Rubber, Butyl (IIR)	Exxon Butyl 268; 5 phr zinc oxide, 2 phr stearic acid, 4 phr Amberol ST-149, accelerator, 10 phr Flexon 845 oil, 60 phr MT carb. bl., 20 phr channel black	36.4				23.9				Aminco Permeability Apparatus	299
	Bayer Polysar Butyl 301; 20 min. @ 153°C cure; 50 Shore A; 5 phr zinc oxide, 2 phr sulfur, 1.5 phr zinc dimethyl dithiocarbonate, 1 phr Methyl Tuads, 0.5 phr Altax, 50 phr SRF black	51.8				40					298
	Bayer Polysar Butyl 301; 20 min. @ 153°C cure; 50 Shore A; 5 phr zinc oxide, 2 phr sulfur, 1.5 phr zinc dimethyl dithiocarbonate, 1 phr Methyl Tuads, 0.5 phr Altax, 50 phr SRF black	156				60					298
	Exxon Butyl 268; 5 phr zinc oxide, 2 phr stearic acid, 4 phr Amberol ST-149, accelerator, 10 phr Flexon 845 oil, 60 phr MT carb. bl., 20 phr channel black	375		100% relative to Butyl Rubber		65.6				Aminco Permeability Apparatus	299
	Bayer Polysar Butyl 301; 20 min. @ 153°C cure; 50 Shore A; 5 phr zinc oxide, 2 phr sulfur, 1.5 phr zinc dimethyl dithiocarbonate, 1 phr Methyl Tuads, 0.5 phr Altax, 50 phr SRF black	3974				80					298
	Exxon Butyl 268; 5 phr zinc oxide, 2 phr stearic acid, 4 phr Amberol ST-149, accelerator, 10 phr Flexon 845 oil, 60 phr MT carb. bl., 20 phr channel black	1195				93.3				Aminco Permeability Apparatus	299
Rubber, Chlorobutyl (CIIR)	Exxon Chlorobutyl 1068; 5 phr zinc oxide, 2 phr stearic acid, 20 phr channel black, 60 phr MT carb. bl., 4 phr Amberol ST-149, 10 phr Flexon 845 oil, accelerator	38.7				23.9				Aminco Permeability Apparatus	299
	1.13 g/cm³ density; 55 Shore A	30.7									12
	Exxon Chlorobutyl 1068; 5 phr zinc oxide, 2 phr stearic acid, 20 phr channel black, 60 phr MT carb. bl., 4 phr Amberol ST-149, 10 phr Flexon 845 oil, accelerator	364		100% relative to Butyl Rubber		65.6				Aminco Permeability Apparatus	299
	Exxon Chlorobutyl 1068; 5 phr zinc oxide, 2 phr stearic acid, 20 phr channel black, 60 phr MT carb. bl., 4 phr Amberol ST-149, 10 phr Flexon 845 oil, accelerator	1183				93.3				Aminco Permeability Apparatus	299
Rubber, EPDM	DuPont Nordel 1040	734 - 907				23				STP conditions	311
	Exxon Vistalon 404; 5 phr zinc oxide, 2 phr stearic acid, accelerator, 10 phr Flexon 845 oil, 4 phr Amberol ST-149, 60 phr MT carb. bl., 20 phr channel black	461				23.9				Aminco Permeability Apparatus	299

AIR (continued)

Material Family	Material Note	Permeability Data			Test Conditions						Source
		Perm. Coefficient (cm³·mm/m²·day·atm)	VTR (g·mm/m²·day)	Non-normalized Data	Penetrant Note	Temp. (°C)	Time (days)	RH (%)	Pressure (kPa)	Test Note	
Rubber, EPDM	Exxon Vistalon 4608; 5 phr zinc oxide, 2 phr stearic acid, 20 phr channel black, 60 phr MT carb. bl, 4 phr Amberol ST-149, 10 phr Flexon 845 oil	668				23.9				Aminco Permeability Apparatus	299
	gum vulcanizate	683		1750% relative to Butyl Rubber		25					300
						40					298
	Exxon Vistalon 404; 5 phr zinc oxide, 2 phr stearic acid, 10 phr Flexon 845 oil, 4 phr Amberol ST-149, 60 phr MT carb. bl, 20 phr channel black	2560		680% relative to Butyl Rubber		65.6				Aminco Permeability Apparatus	299
	Exxon Vistalon 4608; 5 phr zinc oxide, 2 phr stearic acid, 20 phr channel black, 60 phr MT carb. bl, 4 phr Amberol ST-149, 10 phr Flexon 845 oil	3299		890% relative to Butyl Rubber		65.6				Aminco Permeability Apparatus	299
	40 min. @ 153°C cure; 50 Shore A; 5 phr zinc oxide, 1.5 phr sulfur, 1.5 phr zinc dimethyl dithiocarbonate, 100 phr EPDM, 0.5 phr Altax, 50 phr SRF black, 1.5 phr Santocure	1477				60					298
	Exxon Vistalon 404; 5 phr zinc oxide, 2 phr stearic acid, accelerator, 10 phr Flexon 845 oil, 4 phr Amberol ST-149, 60 phr MT carb. black	7247				93.3				Aminco Permeability Apparatus	299
	Exxon Vistalon 4608; 5 phr zinc oxide, 2 phr stearic acid, accelerator, 20 phr channel black, 60 phr MT carb. bl., 4 phr Amberol ST-149, 10 phr Flexon 845 oil	7043				93.3				Aminco Permeability Apparatus	299
	40 min. @ 153°C cure; 50 Shore A; 5 phr zinc oxide, 1.5 phr sulfur, 1.5 phr zinc dimethyl dithiocarbonate, 100 phr EPDM, 0.5 phr Altax, 50 phr SRF black, 1.5 phr Santocure	2851				80					298
Rubber, Isobutylene	gum vulcanizate			88% relative to Butyl Rubber		25					300
Rubber, Latex (NR)	5 phr zinc oxide, 2 phr stearic acid, 20 phr channel black, accelerator, 10 phr Flexon 845 oil, 4 phr Amberol ST-149, 60 phr MT carb. bl.	496				23.9				Aminco Permeability Apparatus	299
	gum vulcanizate	1020		2100% relative to Butyl Rubber		25					300
	20 min. @ 145°C cure; 51 Shore A; 5 phr zinc oxide, 1.5 phr zinc dimethyl dithiocarbonate, 2.5 phr sulfur, 50 phr SRF black, 0.75 phr Altax; ribbed smoked sheets	2315				40					298
	20 min. @ 145°C cure; 51 Shore A; 5 phr zinc oxide, 1.5 phr zinc dimethyl dithiocarbonate, 2.5 phr sulfur, 50 phr SRF black, 0.75 phr Altax; ribbed smoked sheets					60					298
	5 phr zinc oxide, 2 phr stearic acid, 20 phr channel black, accelerator, 10 phr Flexon 845 oil, 4 phr Amberol ST-149, 60 phr MT carb. bl.	2696		700% relative to Butyl Rubber		65.6				Aminco Permeability Apparatus	299
	20 min. @ 145°C cure; 51 Shore A; 5 phr zinc oxide, 1.5 phr zinc dimethyl dithiocarbonate, 2.5 phr sulfur, 50 phr SRF black, 0.75 phr Altax; ribbed smoked sheets	3793				80					298
	5 phr zinc oxide, 2 phr stearic acid, 20 phr channel black, accelerator, 10 phr Flexon 845 oil, 4 phr Amberol ST-149, 60 phr MT carb. bl.	4574				93.3				Aminco Permeability Apparatus	299
Rubber, Neoprene (CR)	Neoprene G type; gum vulcanizate			315% relative to Butyl Rubber		25					300
Rubber, Nitrile (NBR)	61% butadiene/ 39% acrylonitrile; gum vulcanizate			72% relative to Butyl Rubber		25					300
	68% butadiene/ 32% acrylonitrile; gum vulcanizate			179% relative to Butyl Rubber		25					300

AIR (continued)

Material Family	Material Note	Perm. Coefficient (cm³·mm/m²·day·atm)	VTR (g·mm/m²·day)	Non-normalized Data	Penetrant Note	Temp. (°C)	Time (days)	RH (%)	Pressure (kPa)	Test Note	Source
Rubber, Nitrile (NBR)	73% butadiene/ 27% acrylonitrile; gum vulcanizate			315% relative to Butyl Rubber		25					300
	80% butadiene/ 20% acrylonitrile; gum vulcanizate			693% relative to Butyl Rubber		25					300
	Bayer Krynac 800; 20 min. @ 145°C cure; 65 Shore A: 5 phr zinc oxide, 1.5 phr sulfur, 1.5 phr zinc dimethyl dithiocarbonate, 1.5 phr Altax, 50 phr SRF black, 0.75 phr Monex	95.0				40					298
	Bayer Krynac 800; 20 min. @ 145°C cure; 65 Shore A: 5 phr zinc oxide, 1.5 phr sulfur, 1.5 phr zinc dimethyl dithiocarbonate, 1.5 phr Altax, 50 phr SRF black, 0.75 phr Monex	354				60					298
	Bayer Krynac 800; 20 min. @ 145°C cure; 65 Shore A: 5 phr zinc oxide, 1.5 phr sulfur, 1.5 phr zinc dimethyl dithiocarbonate, 1.5 phr Altax, 50 phr SRF black, 0.75 phr Monex	855				80					298
Rubber, Polybutadiene	hot emulsion; gum vulcanizate			1700% relative to Butyl Rubber		25					300
	Bayer Taktene 1220; cis polybutadiene; 20 min. @ 145°C cure; 53 Shore A: 5 phr zinc oxide, 1.5 phr zinc dimethyl dithiocarbonate, 1.1 phr Santocure, 50 phr SRF black, 1.4 phr sulfur	2393				40					298
	Bayer Taktene 1220; cis polybutadiene; 20 min. @ 145°C cure; 53 Shore A: 5 phr zinc oxide, 1.5 phr zinc dimethyl dithiocarbonate, 1.1 phr Santocure, 50 phr SRF black, 1.4 phr sulfur	3810				60					298
	Bayer Taktene 1220; cis polybutadiene; 20 min. @ 145°C cure; 53 Shore A: 5 phr zinc oxide, 1.5 phr zinc dimethyl dithiocarbonate, 1.1 phr Santocure, 50 phr SRF black, 1.4 phr sulfur	5659				80					298
Rubber, Styrene Butadiene	Exxon SBR 1502; 5 phr zinc oxide, 2 phr stearic acid, accelerator, 20 phr channel black, 60 phr MT carb. bl., 4 phr Amberol ST-149, 10 phr Flexon 845 oil	348				23.9				Aminco Permeability Apparatus	299
	5 phr zinc oxide, 2 phr stearic acid, 20 phr channel black, high styrene, 60 phr MT carb. bl., 4 phr Amberol ST-149, accelerator, 43% styrene, 10 phr Flexon 845 oil	138				23.9				Aminco Permeability Apparatus	299
	hot; gum vulcanizate			1600% relative to Butyl Rubber		25					300
	Bayer Krylene; 40 min. @ 145°C cure; 53 Shore A: 5 phr zinc oxide, 1.5 phr zinc dimethyl dithiocarbonate, 50 phr SRF black, 2.8 phr sulfur, 1.2 phr Santocure	397.44				40					298
	Bayer Krylene; 40 min. @ 145°C cure; 53 Shore A: 5 phr zinc oxide, 1.5 phr zinc dimethyl dithiocarbonate, 50 phr SRF black, 2.8 phr sulfur, 1.2 phr Santocure	1080				60					298
	Exxon SBR 1502; 5 phr zinc oxide, 2 phr stearic acid, accelerator, 20 phr channel black, 60 phr MT carb. bl., 4 phr Amberol ST-149, 10 phr Flexon 845 oil	2048		550% relative to Butyl Rubber		65.6				Aminco Permeability Apparatus	299
	5 phr zinc oxide, 2 phr stearic acid, 20 phr channel black, high styrene, 60 phr MT carb. bl., 4 phr Amberol ST-149, accelerator, 43% styrene, 10 phr Flexon 845 oil	1092		300% relative to Butyl Rubber		65.6				Aminco Permeability Apparatus	299
	Bayer Krylene; 40 min. @ 145°C cure; 53 Shore A: 5 phr zinc oxide, 1.5 phr zinc dimethyl dithiocarbonate, 50 phr SRF black, 2.8 phr sulfur, 1.2 phr Santocure	2091				80					298

AIR (continued)

Material Family	Material Note	Perm. Coefficient (cm³·mm/m²·day·atm)	VTR (g·mm/m²·day)	Non-normalized Data	Penetrant Note	Temp. (°C)	Time (days)	RH (%)	Pressure (kPa)	Test Note	Source
Rubber, Styrene Butadiene	Exxon SBR 1502; 5 phr zinc oxide, 2 phr stearic acid, accelerator, 20 phr channel black, 60 phr MT carb. bl., 4 phr Amberol ST-149, 10 phr Flexon 845 oil	4346				93.3				Aminco Permeability Apparatus	299
	5 phr zinc oxide, 2 phr stearic acid, 20 phr channel black, high styrene, 60 phr MT carb. bl., 4 phr Amberol ST-149, accelerator, 43% styrene, 10 phr Flexon 845 oil	2719				93.3				Aminco Permeability Apparatus	299
Silicone	gum vulcanizate			56,800% relative to Butyl Rubber		25					300
TPE, Olefinic	Adv. Elast. Santoprene 201-73; 73 Shore A, 0.5 mm thick	240			gas at 0°C and 0.11 Mpa	23			110	ASTM D1434	282
	Adv. Elast. Santoprene 201-87; 87 Shore A, 0.5 mm thick	302			gas at 0°C and 0.11 Mpa	23			110	ASTM D1434	282
	Adv. Elast. Santoprene 203-50; 50 Shore D, 0.5 mm thick	140			gas at 0°C and 0.11 Mpa	23			110	ASTM D1434	282
	Adv. Elast. Trefsin 3281-50; 0.94 g/cm³ density; medical grade, barrier prop. 50 Shore A	159									12
	Adv. Elast. Trefsin 3281-60; 0.97 g/cm³ density; medical grade, barrier prop. 60 Shore A	216									12
TPE, Polyester	DuPont Hytrel 4056; 40 Shore D	207				21.5			34.5		274
	DuPont Hytrel 5556; 55 Shore D	156				21.5			34.5		274

AMMONIA

Material Family	Material Note	Perm. Coefficient (cm³·mm/m²·day·atm)	VTR (g·mm/m²·day)	Non-normalized Data	Penetrant Note	Temp. (°C)	Time (days)	RH (%)	Pressure (kPa)	Test Note	Source
Fluoropolymer, CTFE	3M Kel-F; 0.01 mm thick	1.05				25			965	mass spectrometry and calibrated standard gas leaks; developed by McDonnell Douglas Space Systems Caompany Chemistry Laboratory	306
Fluoropolymer, ECTFE	3M Kel-F; 0.01 mm thick	24.2				59			965	"	306
	Ausimont Halar; 0.02 mm thick	32.6				-1			965	"	306
	Ausimont Halar; 0.02 mm thick	113				25			965	"	306
	Ausimont Halar; 0.02 mm thick	617				65			965	"	306
Fluoropolymer, FEP	DuPont Teflon; 0.05 mm thick	29.0				0			965	"	306
	DuPont Teflon; 0.05 mm thick	101				25			965	"	306
	DuPont Teflon; 0.05 mm thick	551				66			965	"	306
Fluoropolymer, PVDF	Solvay Solef; 0.1 mm thick; cast film; film	6.6				23			965	ASTM D1434	125
Fluoropolymer, TFE	DuPont Teflon; 0.03 mm thick	41.2				-3			965	mass spectrometry and calibrated standard gas leaks; developed by McDonnell Douglas Space Systems Caompany Chemistry Laboratory	306
	DuPont Teflon; 0.05 mm thick; carbon filled	68.0				-2			965	"	306
	DuPont Teflon; 0.03 mm thick	151				25			965	"	306
	DuPont Teflon; 0.05 mm thick; carbon filled	241				25			965	"	306
	DuPont Teflon; 0.03 mm thick	755				63			965	"	306
	DuPont Teflon; 0.05 mm thick; carbon filled	1059				62			965	"	306

AMMONIA (continued)

Material Family	Material Note	Perm. Coefficient (cm³·mm/m²·day·atm)	VTR (g·mm/m²·day)	Non-normalized Data	Penetrant Note	Temp. (°C)	Time (days)	RH (%)	Pressure (kPa)	Test Note	Source
Polyethylene, HDPE	0.03 mm thick	32.5				-3			965	mass spectrometry and calibrated standard gas leaks; developed by McDonnell Douglas Space Systems Caompany Chemistry Laboratory	306
	0.03 mm thick	122.6				25			965	"	306
	0.03 mm thick	623				61			965	"	306
Polyphenylene Sulfide	Phillips Ryton; 0.127 mm thick; baked coating; film	5.9								ASTM D1434, Method M	102
Polysulfone	Amoco Udel; transparent, amber tint	421				23		dry		ASTM D1434	15

AMMONIUM FLUORIDE

Material Family	Material Note	Perm. Coefficient (cm³·mm/m²·day·atm)	VTR (g·mm/m²·day)	Non-normalized Data	Penetrant Note	Temp. (°C)	Time (days)	RH (%)	Pressure (kPa)	Test Note	Source
Rubber, Latex (NR)	Ansell Edmont Canners and Handlers 392; unsupported glove film; 0.48 mm thick			no permeation detected during a 6 hour test; lower detection limit: 0 ppm	40% conc.	23	0.25			ASTM F739	313
Rubber, Neoprene (CR)	Ansell Edmont Neoprene 29-840; unsupported glove film; 0.38 mm thick			no permeation detected during a 6 hour test; lower detection limit: 0 ppm	40% conc.	23	0.25			ASTM F739	313
	Ansell Edmont Neox; supported (lined) glove film; specified by glove film weight			no permeation detected during a 6 hour test; lower detection limit: 0 ppm	40% conc.	23	0.25			ASTM F739	313
Rubber, Nitrile (NBR)	Ansell Edmont Sol-Vex 37-165; unsupported glove film; 0.54 mm thick			no permeation detected during a 6 hour test; lower detection limit: 0 ppm	40% conc.	23	0.25			ASTM F739	313
TPE, Vinyl	Ansell Edmont Monkey Grip; supported (lined) glove film; specified by glove film weight			no permeation detected during a 6 hour test; lower detection limit: 0 ppm	40% conc.	23	0.25			ASTM F739	313

AMMONIUM HYDROXIDE

Material Family	Material Note	Perm. Coefficient (cm³·mm/m²·day·atm)	VTR (g·mm/m²·day)	Non-normalized Data	Penetrant Note	Temp. (°C)	Time (days)	RH (%)	Pressure (kPa)	Test Note	Source
Rubber, Latex (NR)	Ansell Edmont Canners and Handlers 392; unsupported glove film; 0.48 mm thick			breakthrough time: 90 minutes	concentrated	23				ASTM F739	313
	Pioneer Industrial Products L-118; glove film			breakthrough time: 58 minutes; lower detection limit: 1 ppm; permeation rate: 108 µg/cm²/min	29% conc.					ASTM F739.85	312
Rubber, Neoprene (CR)	Ansell Edmont Neoprene 29-840; unsupported glove film; 0.38 mm thick			breakthrough time: >360 minutes	concentrated	23				ASTM F739	313
	Ansell Edmont Neox; supported (lined) glove film; specified by glove film weight			breakthrough time: >360 minutes	concentrated	23				ASTM F739	313
	Pioneer Industrial Products Stanzoil N-44; glove film			no permeation rate at steady state detected; breakthrough time: >480 minutes; lower detection limit: 1 ppm; permeation rate: 0 µg/cm²/min	29% conc.					ASTM F739.85	312
Rubber, Nitrile (NBR)	Ansell Edmont Sol-Vex 37-165; unsupported glove film; 0.54 mm thick			no permeation detected during a 6 hour test; lower detection limit: 0 ppm	concentrated	23	0.25			ASTM F739	313
	Pioneer Industrial Products Stansolv A-14; glove film			no permeation rate at steady state detected; breakthrough time: >480 minutes; lower detection limit: 1 ppm; permeation rate: 0 µg/cm²/min	29% conc.					ASTM F739.85	312
TPE, Vinyl	Ansell Edmont Monkey Grip; supported (lined) glove film; specified by glove film weight			breakthrough time: 240 minutes	concentrated	23				ASTM F739	313
	Ansell Edmont Wet-Wear 550; clothing; PVC stretch outer layer bonded to a lightweight liner			no permeation detected during a 6 hour test; lower detection limit: 0 ppm	concentrated	23	0.25			ASTM F739	313
	Ansell Edmont Wet-Wear 600; clothing; nylon netting bonded between two layers of PVC			breakthrough time: 18 minutes	concentrated	23				ASTM F739	313
	Pioneer Industrial Products Pylox V-20; glove film			no permeation rate at steady state detected; breakthrough time: >480 minutes; lower detection limit: 1 ppm; permeation rate: 0 µg/cm²/min	29% conc.					ASTM F739.85	312
TPE, Vinyl coated fabric	Ansell Edmont Wet-Wear 700; clothing; PVC coating with Nylon/Polyester lining			breakthrough time: 11 minutes	concentrated	23				ASTM F739	313

Material Family	Material Note	Permeability Data			Test Conditions					Source
		Perm. Coefficient (cm³ · mm/m² · day · atm)	VTR (g · mm/m² · day)	Non-normalized Data	Temp. (°C)	Time (days)	RH (%)	Pressure (kPa)	Test Note	

AMYL ACETATE

Material Family	Material Note	Perm. Coefficient	VTR	Non-normalized Data	Temp. (°C)	Time (days)	RH (%)	Pressure (kPa)	Test Note	Source
Polyvinyl Alcohol	Ansell Edmont PVA; supported (lined) glove film; specified by glove film weight			low permeation rate; 0 to 1/2 eyedropper size drops per hour; PVA coating is water soluble; lower detection limit: 0 ppm; permeation rate: <0.9 µg/cm²/min	23	0.25			ASTM F739	313
Rubber, Nitrile (NBR)	Ansell Edmont Sol-Vex 37-165; unsupported glove film; 0.54 mm thick			good permeation rate: 6 to 50 eyedropper size drops per hour; breakthrough time: 60 minutes; permeation rate: <90 µg/cm²/min	23				ASTM F739	313

AMYL ALCOHOL

Material Family	Material Note	Perm. Coefficient	VTR	Non-normalized Data	Temp. (°C)	Time (days)	RH (%)	Pressure (kPa)	Test Note	Source
Polyvinyl Alcohol	Ansell Edmont PVA; supported (lined) glove film; specified by glove film weight			good permeation rate; 6 to 50 eyedropper size drops per hour; PVA coating is water soluble; breakthrough time: 180 minutes; permeation rate: <90 µg/cm²/min	23				ASTM F739	313
Rubber, Latex (NR)	Ansell Edmont Canners and Handlers 392; unsupported glove film; 0.48 mm thick			very good permeation rate: 1 to 5 eyedropper size drops per hour; breakthrough time: 25 minutes; permeation rate: <9 µg/cm²/min	23				ASTM F739	313
Rubber, Neoprene (CR)	Ansell Edmont Neoprene 29-840; unsupported glove film; 0.38 mm thick			low permeation rate; 0 to 1/2 eyedropper size drops per hour; breakthrough time: >360 minutes; permeation rate: <0.9 µg/cm²/min	23				ASTM F739	313
	Ansell Edmont Neox; supported (lined) glove film; specified by glove film weight			low permeation rate; 0 to 1/2 eyedropper size drops per hour; lower detection limit: 0 ppm; permeation rate: <0.9 µg/cm²/min	23	0.25			ASTM F739	313
Rubber, Nitrile (NBR)	Ansell Edmont Sol-Vex 37-165; unsupported glove film; 0.54 mm thick			low permeation rate; 0 to 1/2 eyedropper size drops per hour; breakthrough time: 30 minutes; permeation rate: <0.9 µg/cm²/min	23				ASTM F739	313
TPE, Vinyl	Ansell Edmont Monkey Grip; supported (lined) glove film; specified by glove film weight			low permeation rate; 0 to 1/2 eyedropper size drops per hour; breakthrough time: 12 minutes; permeation rate: <0.9 µg/cm²/min	23				ASTM F739	313

ANILINE

Material Family	Material Note	Perm. Coefficient	VTR	Non-normalized Data	Temp. (°C)	Time (days)	RH (%)	Pressure (kPa)	Test Note	Source
Polyvinyl Alcohol	Ansell Edmont PVA; supported (lined) glove film; specified by glove film weight			low permeation rate; 0 to 1/2 eyedropper size drops per hour; PVA coating is water soluble; lower detection limit: 0 ppm; permeation rate: <0.9 µg/cm²/min	23	0.25			ASTM F739	313
Rubber, Latex (NR)	Ansell Edmont Canners and Handlers 392; unsupported glove film; 0.48 mm thick			very good permeation rate; 1 to 5 eyedropper size drops per hour; breakthrough time: 25 minutes; permeation rate: <9 µg/cm²/min	23				ASTM F739	313
	Pioneer Industrial Products L-118; glove film			no permeation rate at steady state detected; breakthrough time: >480 minutes; lower detection limit: 0.008 ppm; permeation rate: 0 µg/cm²/min					ASTM F739,85	312
Rubber, Neoprene (CR)	Ansell Edmont Neoprene 29-840; unsupported glove film; 0.38 mm thick			very good permeation rate; 1 to 5 eyedropper size drops per hour; breakthrough time: 35 minutes; permeation rate: <9 µg/cm²/min	23				ASTM F739	313
	Ansell Edmont Neox; supported (lined) glove film; specified by glove film weight			very good permeation rate; 1 to 5 eyedropper size drops per hour; breakthrough time: 180 minutes; permeation rate: <9 µg/cm²/min	23				ASTM F739	313
	Pioneer Industrial Products Stanzoil N-44; glove film			no permeation rate at steady state detected; breakthrough time: >480 minutes; lower detection limit: 0.005 ppm; permeation rate: 0 µg/cm²/min					ASTM F739,85	312
Rubber, Nitrile (NBR)	Pioneer Industrial Products Stansolv A-14; glove film			breakthrough time: 72 minutes; lower detection limit: 0.001 ppm; permeation rate: 18 µg/cm²/min					ASTM F739,85	312
TPE, Vinyl	Ansell Edmont Monkey Grip; supported (lined) glove film; specified by glove film weight			very good permeation rate; 1 to 5 eyedropper size drops per hour; breakthrough time: 180 minutes; permeation rate: <9 µg/cm²/min	23				ASTM F739	313
	Ansell Edmont Wet-Wear 550; clothing; PVC stretch outer layer bonded to a lightweight liner			very good permeation rate; 1 to 5 eyedropper size drops per hour; breakthrough time: 12 minutes; permeation rate: <9 µg/cm²/min	23				ASTM F739	313
	Ansell Edmont Wet-Wear 600; clothing; nylon netting bonded between two layers of PVC			very good permeation rate; 1 to 5 eyedropper size drops per hour; breakthrough time: 30 minutes; permeation rate: <9 µg/cm²/min	23				ASTM F739	313

ANILINE (continued)

Material Family / Material Note	Perm. Coefficient (cm³·mm/m²·day·atm)	VTR (g·mm/m²·day)	Non-normalized Data	Penetrant Note	Temp. (°C)	Time (days)	RH (%)	Pressure (kPa)	Test Note	Source
TPE, Vinyl — Pioneer Industrial Products Pylox V-20; glove film			no permeation rate at steady state detected; breakthrough time: >480 minutes; lower detection limit: 0.009 ppm; permeation rate: 0 µg/cm²/min						ASTM F739.85	312
TPE, Vinyl coated fabric — Ansell Edmont Wet-Wear 700; clothing; PVC coating with Nylon/Polyester lining			very good permeation rate; 1 to 5 eyedropper size drops per hour; breakthrough time: 10 minutes; permeation rate: <9 µg/cm²/min		23				ASTM F739	313

AQUA REGIA

Material Family / Material Note	Perm. Coefficient (cm³·mm/m²·day·atm)	VTR (g·mm/m²·day)	Non-normalized Data	Penetrant Note	Temp. (°C)	Time (days)	RH (%)	Pressure (kPa)	Test Note	Source
Rubber, Neoprene (CR) — Ansell Edmont Neoprene 29-840; unsupported glove film; 0.38 mm thick			breakthrough time: 45 minutes		23				ASTM F739	313
Ansell Edmont Neox; supported (lined) glove film; specified by glove film weight			no permeation detected during a 6 hour test; lower detection limit: 0 ppm		23	0.25			ASTM F739	313
Rubber, Nitrile (NBR) — Ansell Edmont Sol-Vex 37-165; unsupported glove film; 0.54 mm thick			no permeation detected during a 6 hour test; lower detection limit: 0 ppm		23	0.25			ASTM F739	313
TPE, Vinyl — Ansell Edmont Monkey Grip; supported (lined) glove film; specified by glove film weight			breakthrough time: 120 minutes		23				ASTM F739	313

ARGON

Material Family / Material Note	Perm. Coefficient (cm³·mm/m²·day·atm)	VTR (g·mm/m²·day)	Non-normalized Data	Penetrant Note	Temp. (°C)	Time (days)	RH (%)	Pressure (kPa)	Test Note	Source
Polyethylene, HDPE — Hoechst AG Hostalen	66.9				20				volume at standard temperature and pressure; useable average for all Hostalen grades	94
Hoechst AG Hostalen	90.2				30				volume at standard temperature and pressure; useable average for all Hostalen grades	94
Hoechst AG Hostalen	233				50				volume at standard temperature and pressure; useable average for all Hostalen grades	94
TPE, Olefinic — Adv. Elast. Santoprene 201-73; 73 Shore A, 0.5 mm thick	519			gas at 0°C and 0.11 Mpa	23			110	ASTM D1434	282
Adv. Elast. Santoprene 201-87; 87 Shore A, 0.5 mm thick	597			gas at 0°C and 0.11 Mpa	23			110	ASTM D1434	282
Adv. Elast. Santoprene 203-50; 50 Shore D, 0.5 mm thick	395			gas at 0°C and 0.11 Mpa	23			110	ASTM D1434	282
TPE, Urethane (TPAU) — BASF Elastollan C80A; 80 Shore A	105				20					130
BASF Elastollan C85A; 85 Shore A	78.8				20					130
BASF Elastollan C90A; 90 Shore A	43.8				20					130
BASF Elastollan C95A; 95 Shore A	26.3				20					130
TPE, Urethane (TPEU) — BASF Elastollan 1180A; 80 Shore A	123				20					130
BASF Elastollan 1185A; 85 Shore A	78.8				20					130
BASF Elastollan 1190A; 90 Shore A	61.3				20					130
BASF Elastollan 1195A; 95 Shore A	52.5				20					130

AROCLOR 1254

Material Family / Material Note	Perm. Coefficient (cm³·mm/m²·day·atm)	VTR (g·mm/m²·day)	Non-normalized Data	Penetrant Note	Temp. (°C)	Time (days)	RH (%)	Pressure (kPa)	Test Note	Source
Rubber, Neoprene (CR) — Pioneer Industrial Products Stanzoil N-44; glove film			no permeation rate at steady state detected; breakthrough time: >480 minutes; lower detection limit: 1 ppm; permeation rate: 0 µg/cm²/min	50% TCB					ASTM F739.85	312
Rubber, Nitrile (NBR) — Pioneer Industrial Products Stansolv A-14; glove film			breakthrough time: 343 minutes; lower detection limit: 1 ppm; permeation rate: 216 µg/cm²/min	50% TCB					ASTM F739.85	312

Material Family / Material Note	Perm. Coefficient (cm³·mm/m²·day·atm)	VTR (g·mm/m²·day)	Non-normalized Data	Penetrant Note	Temp (°C)	Time (days)	RH (%)	Pressure (kPa)	Test Note	Source

ASTM FUEL OIL B

Material Family / Material Note	Perm. Coef.	VTR	Non-normalized Data	Penetrant Note	Temp	Time	RH	Pressure	Test Note	Source
Nylon 66 — DuPont Zytel 42; low flow, 2.54 mm thick; bottles		0.2		isooctane and toluene blend						68

BENZALDEHYDE

Material Family / Material Note	Perm. Coef.	VTR	Non-normalized Data	Penetrant Note	Temp	Time	RH	Pressure	Test Note	Source
Polyvinyl Alcohol — Ansell Edmont PVA; supported (lined) glove film; specified by glove film weight			low permeation rate: 0 to 1/2 eyedropper size drops per hour; PVA coating is water soluble; lower detection limit: 0 ppm; permeation rate: <0.9 µg/cm²/min		23	0.25			ASTM F739	313
Rubber, Latex (NR) — Ansell Edmont Canners and Handlers 392; unsupported glove film; 0.48 mm thick			very good permeation rate: 1 to 5 eyedropper size drops per hour; breakthrough time: 10 minutes; permeation rate: <9 µg/cm²/min		23				ASTM F739	313

BENZENE

Material Family / Material Note	Perm. Coef.	VTR	Non-normalized Data	Penetrant Note	Temp	Time	RH	Pressure	Test Note	Source
Polyester, PET — DuPont Mylar; film		0.14			25				ASTM E96-80; modified test, permeabilities determined at the partial pressure of the vapor at the test temperature	270
Polyethylene, LDPE — Dow, film		236			35					250
Polyphenylene Sulfide — Phillips Ryton; 0.127 mm thick; baked coating; film		2.5			23				Die cut samples were fitted to tops of glass bottles by a rubber gasket and a lid, with a surface area of 96.8 cm²; Liquids were placed in bottles, gaskets and film put in place, and the lid screwed on. Apparatus was inverted to put liquid in direct contact with film. Weight loss measurements were made at 1 week intervals throughout 4 weeks of conditioning.	102
Polystyrene — Dow Styron; film		472			35					250
Polyvinyl Alcohol — Ansell Edmont PVA; supported (lined) glove film; specified by glove film weight			low permeation rate: 0 to 1/2 eyedropper size drops per hour; PVA coating is water soluble; lower detection limit: 0 ppm; permeation rate: <0.9 µg/cm²/min	benzol	23	0.25			ASTM F739	313
Rubber, Neoprene (CR) — Pioneer Industrial Products Stanzoil N-44; glove film			breakthrough time: 16 minutes; permeation rate: 798 µg/cm²/min						ASTM F739.85	312
Rubber, Nitrile (NBR) — Pioneer Industrial Products Stansolv A-14; glove film			breakthrough time: 27 minutes; lower detection limit: 0.03 ppm; permeation rate: 582 µg/cm²/min						ASTM F739.85	312
TPE, Vinyl — Pioneer Industrial Products Pylox V-20; glove film			breakthrough time: 2 minutes; permeation rate: 1500 µg/cm²/min						ASTM F739.85	312

BENZENE CHLORIDE

Material Family / Material Note	Perm. Coef.	VTR	Non-normalized Data	Penetrant Note	Temp	Time	RH	Pressure	Test Note	Source
Rubber, Nitrile (NBR) — Pioneer Industrial Products Stansolv A-14; glove film			breakthrough time: 15 minutes; permeation rate: 960 µg/cm²/min						ASTM F739.85	312

BROMOPROPIONIC ACID

Material Family / Material Note	Perm. Coef.	VTR	Non-normalized Data	Penetrant Note	Temp	Time	RH	Pressure	Test Note	Source
Rubber, Latex (NR) — Ansell Edmont Canners and Handlers 392; unsupported glove film; 0.48 mm thick			breakthrough time: 190 minutes		23				ASTM F739	313
Rubber, Neoprene (CR) — Ansell Edmont Neoprene 29-840; unsupported glove film; 0.38 mm thick			breakthrough time: 180 minutes		23				ASTM F739	313
Rubber, Neoprene (CR) — Ansell Edmont Neox; supported (lined) glove film; specified by glove film weight			breakthrough time: 240 minutes		23				ASTM F739	313
Rubber, Nitrile (NBR) — Ansell Edmont Sol-Vex 37-165; unsupported glove film; 0.54 mm thick			breakthrough time: 120 minutes		23				ASTM F739	313

BROMOPROPIONIC ACID (continued)

Material Family	Material Note	Permeability Data — Pem. Coefficient (cm³·mm/m²·day·atm)	VTR (g·mm/m²·day)	Non-normalized Data	Penetrant Note	Temp. (°C)	Time (days)	RH (%)	Pressure (kPa)	Test Note	Source
TPE, Vinyl	Ansell Edmont Monkey Grip; supported (lined) glove film; specified by glove film weight			breakthrough time: 180 minutes		23				ASTM F739	313

BUTANONE (2-)

Material Family	Material Note			Non-normalized Data	Penetrant Note	Temp. (°C)	Time (days)	RH (%)	Pressure (kPa)	Test Note	Source
Rubber, Latex (NR)	Pioneer Industrial Products L-118; glove film			breakthrough time: 6 minutes; permeation rate: 522 µg/cm²/min						ASTM F739.85	312
Rubber, Neoprene (CR)	Pioneer Industrial Products Stanzoil N-44; glove film			breakthrough time: 22 minutes; permeation rate: 930 µg/cm²/min						ASTM F739.85	312
Rubber, Nitrile (NBR)	Pioneer Industrial Products Stansolv A-14; glove film			breakthrough time: 6 minutes; permeation rate: 522 µg/cm²/min						ASTM F739.85	312
TPE, Vinyl	Pioneer Industrial Products Pylox V-20; glove film			permeation rate too large to measure; breakthrough time: 1 minutes						ASTM F739.85	312

BUTOXYETHANOL (2-)

Material Family	Material Note			Non-normalized Data	Penetrant Note	Temp. (°C)	Time (days)	RH (%)	Pressure (kPa)	Test Note	Source
Rubber, Latex (NR)	Pioneer Industrial Products L-118; glove film			breakthrough time: 12 minutes; lower detection limit: 1 ppm; permeation rate: 162 µg/cm²/min						ASTM F739.85	312
Rubber, Neoprene (CR)	Pioneer Industrial Products Stanzoil N-44; glove film			breakthrough time: 147 minutes; lower detection limit: 1 ppm; permeation rate: 30 µg/cm²/min						ASTM F739.85	312
Rubber, Nitrile (NBR)	Pioneer Industrial Products Stansolv A-14; glove film			no permeation rate at steady state detected; breakthrough time: >480 minutes; lower detection limit: 0.5 ppm; permeation rate: 0 µg/cm²/min						ASTM F739.85	312

BUTYL ACETATE

Material Family	Material Note			Non-normalized Data	Penetrant Note	Temp. (°C)	Time (days)	RH (%)	Pressure (kPa)	Test Note	Source
Acrylonitrile Copolymer, AMA	barrier prop.; bottles			0.19% penetrant weight loss		50	28				293
Laminar, Nylon/Olefin	DuPont Selar RB/ Polyolefin; 8% Selar RB; barrier prop.; extrusion blow molded; bottles			0.08% penetrant weight loss		50	28				293
Polyethylene, Fluorinated	barrier prop.; bottles			1.0% penetrant weight loss		50	28				293
Polyethylene, HDPE	bottles			3.7% penetrant weight loss		50	28				293
Polyvinyl Alcohol	Ansell Edmont PVA; supported (lined) glove film; specified by glove film weight			low permeation rate; 0 to 1/2 eyedropper size drops per hour; PVA coating is water soluble; lower detection limit: 0 ppm; permeation rate: <0.9 µg/cm²/min		23	0.25			ASTM F739	313
Polyvinyl Chloride	bottles			failed		23	180				293
Rubber, Neoprene (CR)	Pioneer Industrial Products Stanzoil N-44; glove film			breakthrough time: 52 minutes; permeation rate: 318 µg/cm²/min						ASTM F739.85	312
Rubber, Nitrile (NBR)	Ansell Edmont Sol-Vex 37-165; unsupported glove film; 0.54 mm thick			fair permeation rate; 51 to 100 eyedropper size drops per hour; breakthrough time: 75 minutes; permeation rate: <900 µg/cm²/min		23				ASTM F739	313
	Pioneer Industrial Products Stansolv A-14; glove film			breakthrough time: 101 minutes; lower detection limit: 0.1 ppm; permeation rate: 144 µg/cm²/min						ASTM F739.85	312

BUTYL ALCOHOL

Material Family	Material Note			Non-normalized Data	Penetrant Note	Temp. (°C)	Time (days)	RH (%)	Pressure (kPa)	Test Note	Source
Acrylonitrile Copolymer, AMA	barrier prop.; bottles			0.03% penetrant weight loss		50	28				293
Laminar, Nylon/Olefin	DuPont Selar RB/ Polyolefin; 8% Selar RB; barrier prop.; extrusion blow molded; bottles			0.06% penetrant weight loss		50	28				293
Polyethylene, Fluorinated	barrier prop.; bottles			0.1% penetrant weight loss		50	28				293
Polyethylene, HDPE	bottles			0.2% penetrant weight loss		50	28				293

BUTYL ALCOHOL (continued)

Material Family	Material Note	Perm. Coefficient (cm³ · mm/m² · day · atm)	VTR (g · mm/m² · day)	Non-normalized Data	Penetrant Note	Temp. (°C)	Time (days)	RH (%)	Pressure (kPa)	Test Note	Source
Polyvinyl Alcohol	Ansell Edmont PVA; supported (lined) glove film; specified by glove film weight			good permeation rate: 6 to 50 eyedropper size drops per hour; PVA coating is water soluble; breakthrough time: 75 minutes; permeation rate: <90 µg/cm²/min		23				ASTM F739	313
Polyvinyl Chloride	bottles			0.18% penetrant weight loss		23	180				293
Rubber, Latex (NR)	Ansell Edmont Canners and Handlers 392; unsupported glove film; 0.48 mm thick			very good permeation rate: 1 to 5 eyedropper size drops per hour; breakthrough time: 200 minutes; permeation rate: <9 µg/cm²/min		23				ASTM F739	313
Rubber, Neoprene (CR)	Ansell Edmont Neoprene 29-840; unsupported glove film; 0.38 mm thick			very good permeation rate: 1 to 5 eyedropper size drops per hour; breakthrough time: 240 minutes; permeation rate: <9 µg/cm²/min		23				ASTM F739	313
	Ansell Edmont Neox; supported (lined) glove film; specified by glove film weight			low permeation rate: 0 to 1/2 eyedropper size drops per hour; breakthrough time: >480 minutes; permeation rate: <0.9 µg/cm²/min		23				ASTM F739	313
Rubber, Nitrile (NBR)	Ansell Edmont Sol-Vex 37-165; unsupported glove film; 0.54 mm thick			low permeation rate: 0 to 1/2 eyedropper size drops per hour; lower detection limit: 0 ppm; permeation rate: <0.9 µg/cm²/min		23	0.25			ASTM F739	313
TPE, Vinyl	Ansell Edmont Monkey Grip; supported (lined) glove film; specified by glove film weight			very good permeation rate: 1 to 5 eyedropper size drops per hour; breakthrough time: 180 minutes; permeation rate: <9 µg/cm²/min		23				ASTM F739	313
TPE, Vinyl	Ansell Edmont Wet-Wear 550; clothing; PVC stretch outer layer bonded to a lightweight liner			good permeation rate: 6 to 50 eyedropper size drops per hour; breakthrough time: 8 minutes; permeation rate: <90 µg/cm²/min		23				ASTM F739	313
TPE, Vinyl	Ansell Edmont Wet-Wear 600; clothing; nylon netting bonded between two layers of PVC			good permeation rate: 6 to 50 eyedropper size drops per hour; breakthrough time: 40 minutes; permeation rate: <90 µg/cm²/min		23				ASTM F739	313
TPE, Vinyl coated fabric	Ansell Edmont Wet-Wear 700; clothing; PVC coating with Nylon/Polyester lining			very good permeation rate: 1 to 5 eyedropper size drops per hour; breakthrough time: 35 minutes; permeation rate: <9 µg/cm²/min		23				ASTM F739	313

BUTYL CELLOSOLVE

Material Family	Material Note	Perm. Coefficient (cm³ · mm/m² · day · atm)	VTR (g · mm/m² · day)	Non-normalized Data	Penetrant Note	Temp. (°C)	Time (days)	RH (%)	Pressure (kPa)	Test Note	Source
Polyvinyl Alcohol	Ansell Edmont PVA; supported (lined) glove film; specified by glove film weight			good permeation rate: 6 to 50 eyedropper size drops per hour; PVA coating is water soluble; breakthrough time: 120 minutes; permeation rate: <90 µg/cm²/min		23				ASTM F739	313
Rubber, Latex (NR)	Ansell Edmont Canners and Handlers 392; unsupported glove film; 0.48 mm thick			good permeation rate: 6 to 50 eyedropper size drops per hour; breakthrough time: 45 minutes; permeation rate: <90 µg/cm²/min		23				ASTM F739	313
	Pioneer Industrial Products L-118; glove film			breakthrough time: 12 minutes; lower detection limit: 1 ppm; permeation rate: 162 µg/cm²/min						ASTM F739.85	312
Rubber, Neoprene (CR)	Ansell Edmont Neoprene 29-840; unsupported glove film; 0.38 mm thick			very good permeation rate: 1 to 5 eyedropper size drops per hour; breakthrough time: 90 minutes; permeation rate: <9 µg/cm²/min		23				ASTM F739	313
	Ansell Edmont Neox; supported (lined) glove film; specified by glove film weight			low permeation rate: 0 to 1/2 eyedropper size drops per hour; lower detection limit: 0 ppm; permeation rate: <0.9 µg/cm²/min		23	0.25			ASTM F739	313
	Pioneer Industrial Products Stanzoil N-44; glove film			breakthrough time: 147 minutes; lower detection limit: 1 ppm; permeation rate: 30 µg/cm²/min						ASTM F739.85	312
Rubber, Nitrile (NBR)	Ansell Edmont Sol-Vex 37-165; unsupported glove film; 0.54 mm thick			very good permeation rate: 1 to 5 eyedropper size drops per hour; breakthrough time: 90 minutes; permeation rate: <9 µg/cm²/min		23				ASTM F739	313
	Pioneer Industrial Products Stansolv A-14; glove film			no permeation rate at steady state detected; breakthrough time: >480 minutes; lower detection limit: 0.5 ppm; permeation rate: 0 µg/cm²/min						ASTM F739.85	312

Material Note	Perm. Coefficient (cm³ · mm/m² · day · atm)	VTR (g · mm/m² · day)	Non-normalized Data	Penetrant Note	Temp. (°C)	Time (days)	RH (%)	Pressure (kPa)	Test Note	Source

BUTYROLACTONE (γ-)

Material Note	Perm. Coefficient	VTR	Non-normalized Data	Penetrant Note	Temp. (°C)	Time	RH	Pressure	Test Note	Source
Polyvinyl Alcohol			very good permeation rate: 1 to 5 eyedropper size drops per hour; PVA coating is water soluble; breakthrough time: 120 minutes; permeation rate: <9 µg/cm²/min		23				ASTM F739	313
Ansell Edmont PVA; supported (lined) glove film; specified by glove film weight										
Rubber, Latex (NR)			good permeation rate: 6 to 50 eyedropper size drops per hour; breakthrough time: 60 minutes; permeation rate: <90 µg/cm²/min		23				ASTM F739	313
Ansell Edmont Canners and Handlers 392; unsupported glove film; 0.48 mm thick										
Rubber, Neoprene (CR)			very good permeation rate: 1 to 5 eyedropper size drops per hour; breakthrough time: 10 minutes; permeation rate: <9 µg/cm²/min		23				ASTM F739	313
Ansell Edmont Neoprene 29-840; unsupported glove film; 0.38 mm thick										

CAMPHOR

Material Note	Perm. Coefficient	VTR	Non-normalized Data	Penetrant Note	Temp.	Time	RH	Pressure	Test Note	Source
EVAL/ PE Film 0.013/ 0.051 mm thick; PE inside; EVAL EF-XL outside; film			>30 days to leakage							265
Nylon/ EVAL Film 0.015/ 0.015 mm thick; EVAL EF-F inside; oriented nylon outside; film			30 days to leakage							265
Nylon/ PE Film 0.015/ 0.051 mm thick; PE inside; oriented nylon outside; film			28 days to leakage							265
PET/ EVAL Film 0.013/ 0.015 mm thick; EVAL EF-F inside; PET outside; film			>30 days to leakage							265
PET/ EVAL/ LDPE Film EVAL EF-F barrier; 0.013/ 0.015/ 0.051 mm thick; LDPE inside; PET outside			>30 days to leakage							265
PET/ PE Film 0.013/ 0.051 mm thick; PE inside; PET outside; film			>30 days to leakage							265
PP/ PE Film 0.018/ 0.051 mm thick; PE inside; PVDC coated BOPP outside; film			13 days to leakage							265

CARBOLIC ACID

Material Note	Perm. Coefficient	VTR	Non-normalized Data	Penetrant Note	Temp.	Time	RH	Pressure	Test Note	Source
Rubber, Latex (NR) Pioneer Industrial Products L-118; glove film			no permeation rate at steady state detected; breakthrough time: >480 minutes; lower detection limit: 0.06 ppm; permeation rate: 0 µg/cm²/min						ASTM F739.85	312
Rubber, Neoprene (CR) Pioneer Industrial Products Stanzoil N-44; glove film			no permeation rate at steady state detected; breakthrough time: >480 minutes; lower detection limit: 0.2 ppm; permeation rate: 0 µg/cm²/min						ASTM F739.85	312
Rubber, Nitrile (NBR) Pioneer Industrial Products Stansolv A-14; glove film			no permeation rate at steady state detected; breakthrough time: >480 minutes; permeation rate: 0 µg/cm²/min						ASTM F739.85	312
TPE, Vinyl Pioneer Industrial Products Pylox V-20; glove film			breakthrough time: 32 minutes; permeation rate: 78 µg/cm²/min						ASTM F739.85	312

CARBON BICHLORIDE

Material Note	Perm. Coefficient	VTR	Non-normalized Data	Penetrant Note	Temp.	Time	RH	Pressure	Test Note	Source
Rubber, Neoprene (CR) Pioneer Industrial Products Stanzoil N-44; glove film			breakthrough time: 28 minutes; lower detection limit: 0.0002 ppm; permeation rate: 453 µg/cm²/min	carbon dichloride					ASTM F739.85	312
Rubber, Nitrile (NBR) Pioneer Industrial Products Stansolv A-14; glove film			breakthrough time: 373 minutes; lower detection limit: 0.0002 ppm; permeation rate: 27 µg/cm²/min	carbon dichloride					ASTM F739.85	312

CARBON BISULFIDE

Material Note	Perm. Coefficient	VTR	Non-normalized Data	Penetrant Note	Temp.	Time	RH	Pressure	Test Note	Source
Rubber, Nitrile (NBR) Pioneer Industrial Products Stansolv A-14; glove film			breakthrough time: 20 minutes; lower detection limit: 0.2 ppm; permeation rate: 516 µg/cm²/min	carbon disulfide					ASTM F739.85	312

CARBON DICHLORIDE

Material Family	Material Note	Permeability Data			Test Conditions						Source
		Perm. Coefficient (cm³·mm/m²·day·atm)	VTR (g·mm/m²·day)	Non-normalized Data	Penetrant Note	Temp (°C)	Time (days)	RH (%)	Pressure (kPa)	Test Note	
Rubber, Neoprene (CR)	Pioneer Industrial Products Stanzoil N-44; glove film			breakthrough time: 28 minutes; lower detection limit: 0.0002 ppm; permeation rate: 453 µg/cm²/min	carbon bichloride					ASTM F739.85	312
Rubber, Nitrile (NBR)	Pioneer Industrial Products Stansolv A-14; glove film			breakthrough time: 373 minutes; lower detection limit: 0.0002 ppm; permeation rate: 27 µg/cm²/min	carbon bichloride					ASTM F739.85	312

CARBON DIOXIDE

Material Family	Material Note	Perm. Coefficient (cm³·mm/m²·day·atm)	VTR (g·mm/m²·day)	Non-normalized Data	Penetrant Note	Temp (°C)	Time (days)	RH (%)	Pressure (kPa)	Test Note	Source
ABS	BASF AG Terluran 877 M; moderate flow, 0.1 mm thick; film	202.6				23				DIN 53380; Values for permeability depend on the conditions under which the film was produced. Figures determined may differ by as much as 50% from those given.	137
	BASF AG Terluran 967 K; moderate flow, 0.1 mm thick; w/ butadiene acrylic rubber; film	202.6				23				DIN 53380; Values for permeability depend on the conditions under which the film was produced. Figures determined may differ by as much as 50% from those given.	137
	BASF AG Terluran 997 VE; 0.1 mm thick, low flow, w/ butadiene acrylic rubber; film	304				23				DIN 53380; Values for permeability depend on the conditions under which the film was produced. Figures determined may differ by as much as 50% from those given.	137
	Dow; low nitrile content; film	354 - 472				24					250
	Dow; medium nitrile content; film	157 - 236				24					250
Acetal	DuPont Delrin	14.6 - 19.7				23		50			201
Acetal Copolymer	Hoechst Cel. Celcon M25; high molecular weight, 0.15 mm thick; 2.5 g/10 min. MFI; film	56.7 - 68.5									210
	Hoechst Cel. Celcon M270; low molecular weight, high flow, 0.15 mm thick; 27.0 g/10 min. MFI; film	56.7 - 68.5									210
	Hoechst Cel. Celcon M90; gen. purp. grade, 0.15 mm thick; 9.0 g/10 min. MFI; film	56.7 - 68.5									210
Acrylonitrile	film	0.04 - 0.08				24					250
Acrylonitrile Copolymer, AMA	film	0.2				24					250
	BP Chem. Barex 210; packaging; impact modified, barrier prop.	0.64				22.8		0			296
	BP Chem. Barex 218; packaging; impact modified, high impact, barrier prop.	0.64				22.8		0			296
ASA	BASF AG Luran S 757 R; 0.1 mm thick; blown film	101				23				DIN 53380	143
	BASF AG Luran S 776 S; 0.1 mm thick; blown film	142				23				DIN 53380	143
	BASF AG Luran S 776 S; 0.1 mm thick; film	233				23				DIN 53380 part 2 method M; Values for permeability depend on the conditions under which the film was produced. Figures determined may differ by as much as 50% from those given.	142

CARBON DIOXIDE (continued)

Material Family	Material Note	Permeability Data			Penetrant Note	Test Conditions				Test Note	Source
		Perm. Coefficient (cm³ · mm/m² · day · atm)	VTR (g · mm/m² · day)	Non-normalized Data		Temp. (°C)	Time (days)	RH (%)	Pressure (kPa)		
ASA	BASF AG Luran S 797 S: 0.1 mm thick; film	203				23				DIN 53380 part 2 method M; Values for permeability depend on the conditions under which the film was produced. Figures determined may differ by as much as 50% from those given.	142
Epoxy		3.2				25				ASTM D1434-63T	121
EVOH	Eval Co. Eval E; barrier prop; 44% ethylene	0.02				5		0			264
	Eval Co. Eval F; barrier prop.; 32% ethylene	0.0039				5		0			264
	Eval Co. Eval H; barrier prop.; 38% ethylene	0.01				5		0			264
	Eval Co. Eval E; barrier prop.; 44% ethylene	0.08				23		0			264
	Eval Co. Eval F; barrier prop; 32% ethylene	0.01				23		0			264
	Eval Co. Eval H; barrier prop.; 38% ethylene	0.03				23		0			264
	Eval Co. Eval E; barrier prop; 44% ethylene	0.2				35		0			264
	Eval Co. Eval F; barrier prop; 32% ethylene	0.03				35		0			264
	Eval Co. Eval H; barrier prop.; 38% ethylene	0.08				35		0			264
Fluoroelastomer, FKM	DuPont Viton; compounded	510				30				one atmosphere @80°C; specimen size: 1 cm² x 1 cm thick	305
Fluoropolymer, CTFE	3M Kel-F 81; amorphous form; film	2.3				0					96
	3M Kel-F 81; amorphous form; film	9.2				25					96
	Allied Sig. Aclar 22A; transparent; film	11.8				25				STP conditions	138
	Allied Sig. Aclar 22C; transparent; film	15.8				25				STP conditions	138
	Allied Sig. Aclar 33C; transparent; film	6.3				25				STP conditions	138
	3M Kel-F 81; amorphous form; film	15.8				50					96
	3M Kel-F 81; amorphous form; film	98.5				75					96
Fluoropolymer, ETFE	Ausimont Hyflon 700; high molecular weight; 4 g/10 min. MFI	232				23				ASTM D1434; activation energy = 6-8 kcal/mole	114
	Ausimont Hyflon 800; low molecular weight; 11 g/10 min. MFI	232				23				ASTM D1434; activation energy = 6-8 kcal/mole	114
	DuPont Tefzel; developmental material; 0.102 mm thick; film	98.4				25				ASTM D1434	205
Fluoropolymer, FEP		657				25				STP conditions	138
Fluoropolymer, PFA	DuPont Teflon PFA; film	890				25				ASTM D1434	39
Fluoropolymer, PVDF	Solvay Solef 1008; 0.1 mm thick; translucent; film	7.1				23				ASTM D1434	125
		2.2				25				STP conditions	138
Fluoropolymer, PVF	Atochem Foraflon; 0.034 mm thick; extruded film	30.3				30				ISO 2556	89
		4.3				25				STP conditions	138
HDPE/ EVA/ PVDC/ EVA Film	Dow Saranex 25, form-fill-seal pouch, bag-in-box, neutral, barrier prop., 0.051 mm thick; high slip; coextruded film	1.1				23		10		ASTM D1434	257
LDPE/ EVA/ PVDC/ EVA Film	Dow Saranex 23; laminations; neutral, barrier prop. 0.051 mm thick; medium slip; coextruded film	1.1				23		10		ASTM D1434	257
LDPE/ EVA/ PVDC/ EVA/ LDPE Film	Dow Saranex 11; laminations; neutral, barrier prop., 0.038 mm thick; medium slip; coextruded film	3.24				23		10		ASTM D1434	257

CARBON DIOXIDE (continued)

Material Family	Material Note	Perm. Coefficient (cm³·mm/m²·day·atm)	VTR (g·mm/m²·day)	Non-normalized Data	Penetrant Note	Temp. (°C)	Time (days)	RH (%)	Pressure (kPa)	Test Note	Source
LDPE/EVA/PVDC/EVA/LDPE Film	Dow Saranex 14; form-fill-seal pouch; neutral, barrier prop., 0.051 mm thick; medium slip; coextruded film	1.1				23		10		ASTM D1434	257
	Dow Saranex 15; form-fill-seal pouch; neutral, barrier prop., 0.076 mm thick; medium slip; coextruded film	1.64				23		10		ASTM D1434	257
	Dow Saranex 15; form-fill-seal pouch; neutral, barrier prop., 0.102 mm thick; medium slip; coextruded film	2.2				23		10		ASTM D1434	257
Nylon 6	Allied Sig. Capran 77C; 0.0254 mm thick; film	0.24				0		0		STP conditions	285
	Allied Sig. Capran 77C; 0.0254 mm thick; film	1.8				23		0		STP conditions	285
	Allied Sig. Capran 77K; 0.0254 mm thick; PVDC coated; film	0.55				23		0		STP conditions	285
	BASF Ultramid B36F; moderate flow, clarity; flat film, tubular film	4.1 - 4.6				23		0		DIN 53380	93
	BASF Ultramid B4; moderate flow, 0.02-0.1 mm thick; flat film, tubular film	4.1 - 4.6				23		0		DIN 53380	93
	barrier prop.; film	1.8				22.8		0			294
	barrier prop.; film	3.2				22.8		80			294
	Allied Sig. Capran 77C; 0.0254 mm thick; film	17.3				50		0		STP conditions	285
	oriented	2.61				35		0			264
Nylon 66	BASF Ultramid A5; 0.02-0.1 mm thick, low flow; tubular film	3.1				23		0		DIN 53380	93
	BASF Ultramid A5; low flow, 0.02-0.1 mm thick; flat film	4.6				23		0		DIN 53380	93
	DuPont Zytel 42; low flow; film	3.5				23		50			68
	DuPont Can. Dartek; transparent, 0.0254 mm thick; film	6.3				23		0		ASTM D1434-66, method V	276
Nylon 66/610	Emser Grilon XE3303; transparent, barrier prop., 0.05 mm thick; film	9.4				23		50			307
Nylon 6/66	Allied Sig. Capran; 0.0254 mm thick; film	2.9				23				ASTM D1435; Dow Cell	284
	BASF Ultramid C35; mod.-high flow, 0.02-0.1 mm thick; flat film, tubular film	4.1 - 4.6				23		0		DIN 53380	93
Nylon, Amorphous	DuPont Selar PA; barrier prop.; film	1.8				22.8		0			294
	DuPont Selar PA; barrier prop.; film	1.1				22.8		80			294
	Emser Grivory G21; transparent, barrier prop., 0.05 mm thick; film	3.8				23		50		DIN 53380	307
Parylene	Union Carbide Parylene C; vapor phase deposition; thin film	3.0				25				ASTM D1434-63T	121
	Union Carbide Parylene D; vapor phase deposition; thin film	5.1				25				ASTM D1434-63T	121
	Union Carbide Parylene N; highly crystalline, high molecular weight, completely linear; vapor phase deposition; thin film	84.2				25				ASTM D1434-63T	121
PE/PVC-PVDC Copolymer Multilayer Film	Dow Saranex; multilayer film	0.39 - 3.2				24					250
Polybutylene	Shell Duraflex 1600; peelable seals; FDA grade, 0.051 mm thick, heat sealable; blown film; 5 phr zinc oxide, slip and antiblock formulations; film	561				22.8		50		ASTM D1434, method M	304
	Shell Duraflex 1710; peelable seals; 0.910 g/cm³ density, 0.909 g/cm³ density; FDA grade, heat sealable, 0.051 mm thick; blown film; slip and antiblock formulations, antiblock formulations; film	468				22.8		50		ASTM D1434, method M	304

CARBON DIOXIDE (continued)

Material Family	Material Note	Perm. Coefficient (cm³·mm/m²·day·atm)	VTR (g·mm/m²·day)	Non-normalized Data	Penetrant Note	Temp. (°C)	Time (days)	RH (%)	Pressure (kPa)	Test Note	Source
Polycarbonate	Bayer Makrolon: 0.1 mm thick; film	436								DIN 53380, pt. 3	289
	Dow Calibre 300-15; transparent, gen. purp. grade; 15 g/10 min. MFI	677								ASTM D2752	78
	Dow Calibre 300-4; transparent, gen. purp. grade; 4 g/10 min. MFI	768								ASTM D2752	78
	Dow Calibre 800-6; transparent, flame retardant; 6 g/10 min. MFI	827								ASTM D2752	78
	film	307				22.8		0			294
Polyester, PBT	BASF AG Ultradur B4550; 0.25 mm thick	139				23		50		DIN 53380; measured in standard laboratory atmosphere	180
Polyester, PCTG	Eastman Kodar PCTG 5445; transparent, 0.25 mm thick; film	50				23				ASTM D1434	166
Polyester, PET	DuPont Mylar; film	6.3				25				ASTM D1434-72	270
	Shell Cleartuf; packaging; transparent, barrier prop., oriented	4.7				25		0		ASTM D1434	297
	Shell Cleartuf; packaging; transparent, barrier prop., unoriented	7.9				25		0		ASTM D1434	297
	oriented	5.9 - 9.8				25				STP conditions	138
	oriented	7.7				35		0			264
Polyester, PETG	Eastman Kodar PETG 6763; transparent, amorphous, 0.25 mm thick; film	31.5				23				ASTM D1434	165
Polyethylene, HDPE	Dow; film	236 - 276				24					250
	Hoechst AG Hostalen	284				20				volume at standard temperature and pressure; useable average for all Hostalen grades	94
	Hoechst AG Hostalen	294				25				volume at standard temperature and pressure; useable average for all Hostalen grades	94
	Phillips Marlex; film	136				23				ASTM D1434	101
	Hoechst AG Hostalen	344				30				volume at standard temperature and pressure; useable average for all Hostalen grades	94
	Hoechst AG Hostalen	527				40				volume at standard temperature and pressure; useable average for all Hostalen grades	94
	Hoechst AG Hostalen	811				50				volume at standard temperature and pressure; useable average for all Hostalen grades	94
Polyethylene, LDPE	Dow; film	394 - 787				24					250
	0.920 g/cm³ density; 0.05 mm thick, 2.5 BUR; blown film; 4 g/10 min. MFI; film	790								ASTM D1434	216
		1063				25				STP conditions	138
Polyethylene, LLDPE	DuPont Can. Sclairfilm SL1; laminations; 0.918 g/cm³ density; 0.0254 mm thick; film	35.6								ASTM D3985; approximate values	278
	DuPont Can. Sclairfilm SL3; laminations; 0.918 g/cm³ density; 0.0254 mm thick; film	35.6								ASTM D3985; approximate values	278
Polyethylene, MDPE		39.4 - 984				25				STP conditions	138
Polyethylene, PE/EVA Cop	0.930 g/cm³ density; 0.05 mm thick, 2.5 BUR; blown film; 12.0% VA; film	1100								ASTM D1434	216

CARBON DIOXIDE (continued)

Material Family	Material Note	Perm. Coefficient (cm³·mm/m²·day·atm)	VTR (g·mm/m²·day)	Non-normalized Data	Penetrant Note	Temp. (°C)	Time (days)	RH (%)	Pressure (kPa)	Test Note	Source
Polyimide	Ube Upilex R; 0.025 mm thick; film	2.9				30				ASTM D1434	97
	Ube Upilex S; 0.025 mm thick; film	0.03				30				ASTM D1434	97
Polymethylpentene	Mitsui TPX X-22; 0.835 g/cm³ density; transparent, 0.05 mm thick; film	6000				23					302
	Mitsui TPX X-44; 0.834 g/cm³ density; transparent, 0.05 mm thick; film	5500				23					302
	Mitsui TPX X-88; 0.835 g/cm³ density; transparent, food grade, 0.05 mm thick; film	5500				23					302
Polyphenylene Sulfide	Phillips Ryton; 0.127 mm thick; baked coating; film	29.5								ASTM D1434, Method M	102
Polypropylene	oriented; film	208				25		0			63
Polypyrrole	BASF AG Lutamer ES 9567; 0.031-0.037 mm thick, intrinsically conductive, high flexibility; w/ benzenesulfonate anions; film	22.3									182
Polystyrene	Dow Styron; film	394 - 590				24					250
	Dow Trycite; film	276 - 433				24					250
Polystyrene, GP	BASF AG Polystyrol 168 N; transparent, 0.1 mm thick; film	527				23				DIN 53380	26
	Dow Styron; injection molding	394 - 590				23				ASTM D1434	263
	Dow Styron; oriented; sheet	276 - 433				23				ASTM D1434	263
Polystyrene, IPS	BASF AG Polystyrol 476 L; 0.1 mm thick; film	1013				23				DIN 53380	26
	Dow Styron; injection molding	394 - 590				23				ASTM D1434	262
Polysulfone	Amoco Udel; transparent, amber tint	374				23		dry		ASTM D1434	15
Polyurethane		1181				25				ASTM D1434-63T	121
Polyvinyl Alcohol	film	0.04				24		0			250
Polyvinyl Chloride	film	959				25		0			63
	unplasticized; film	7.9 - 19.7				24					250
Polyvinylidene Chloride	Dow Saran; film	1.6 - 2.4				24					250
	Dow Saran Wrap 18; chub packaging, laminations: transparent, barrier prop., 0.019 mm thick, biaxially oriented; monolayer film	2.13				23		10		ASTM D1434	256
	Dow Saran Wrap 18L; laminations: transparent, barrier prop., 0.019 mm thick, preshrunk, biaxially oriented; monolayer film	2.83				23		10		ASTM D1434	256
	Dow Saran Wrap 19; chub packaging; transparent, barrier prop., 0.0254 mm thick, biaxially oriented; monolayer film	2.13				23		10		ASTM D1434	256
	Dow Saran Wrap 28; chub packaging; transparent, barrier prop., 0.0254 mm thick, biaxially oriented; monolayer film	3.15				23		10		ASTM D1434	256
	Dow Saran Wrap 560; unit packaging; transparent, barrier prop., 0.152 mm thick, biaxially oriented; coextruded film	0.47				23		10		ASTM D1434	256
	Dow Saran 5253; barrier prop.	0.44				35		0			264
PVC-PVDC Copolymer		15.0 - 17.3				25				STP conditions	138
Rubber, EPDM	DuPont Nordel 1040	7516 - 8122				23				STP conditions	311
	gum vulcanizate			1650% relative to Butyl Rubber		25					300
Rubber, Isobutylene	gum vulcanizate			93% relative to Butyl Rubber		25					300
Rubber, Latex (NR)	gum vulcanizate			2500% relative to Butyl Rubber		25					300

CARBON DIOXIDE (continued)

Material Family	Material Note	Perm. Coefficient (cm³·mm/m²·day·atm)	VTR (g·mm/m²·day)	Non-normalized Data	Penetrant Note	Temp. (°C)	Time (days)	RH (%)	Pressure (kPa)	Test Note	Source
Rubber, Neoprene (CR)	Neoprene G type; gum vulcanizate			500% relative to Butyl Rubber		25					300
Rubber, Nitrile (NBR)	61% butadiene/ 39% acrylonitrile; gum vulcanizate			143% relative to Butyl Rubber		25					300
	68% butadiene/ 32% acrylonitrile; gum vulcanizate			350% relative to Butyl Rubber		25					300
	73% butadiene/ 27% acrylonitrile; gum vulcanizate			600% relative to Butyl Rubber		25					300
	80% butadiene/ 20% acrylonitrile; gum vulcanizate			1200% relative to Butyl Rubber		25					300
Rubber, Polybutadiene	film	7874				24					250
Rubber, Styrene Butadiene	hot emulsion; gum vulcanizate			2620% relative to Butyl Rubber		25					300
	film	394				24					250
SAN	hot; gum vulcanizate			2360% relative to Butyl Rubber		25					300
	Dow Tyril; low nitrile content; film	157				24					250
Silicone	gum vulcanizate			40,000% relative to Butyl Rubber		25					300
	gum vulcanizate	118,110				25				ASTM D1434-63T	121
Styrene-Butadiene Block Copol.	BASF AG Styrolux 637 D; 0.1 mm thick; film	1013				23					29
	BASF AG Styrolux 656 C; high flow, 0.1 mm thick; film	811				23					29
	BASF AG Styrolux 684 D; 0.1 mm thick; film	1520				23					29
TPE, Olefinic	Adv. Elast. Santoprene 201-73; 73 Shore A, 0.5 mm thick	3022			gas at 0°C and 0.11 Mpa	23			110	ASTM D1434	282
	Adv. Elast. Santoprene 201-87; 87 Shore A, 0.5 mm thick	2015			gas at 0°C and 0.11 Mpa	23			110	ASTM D1434	282
	Adv. Elast. Santoprene 203-50; 50 Shore D, 0.5 mm thick	1318			gas at 0°C and 0.11 Mpa	23			110	ASTM D1434	282
TPE, Polyamide	Atochem Pebax 2533; 25 Shore D, 0.12 mm thick; film	17,073				23		dry			287
	Atochem Pebax 3533; 35 Shore D, 0.12 mm thick; film	11,753				23		dry			287
	Atochem Pebax 4033; 40 Shore D, 0.12 mm thick; film	5122				23		dry			287
	Atochem Pebax 5533; 55 Shore D, 0.12 mm thick; film	3283				23		dry			287
	Atochem Pebax 6333; 63 Shore D, 0.12 mm thick; film	2758				23		dry			287
TPE, Polybutadiene	Jap. Synth. JSR RB820; 1,2-polybutadiene; 0.910 g/cm³ density; transparent, 2.5 BUR, 0.05 mm thick; blown film; 15% crystallinity; film	2800									216
	Jap. Synth. JSR RB830; 0.910 g/cm³ density; stretch film, 0.018 mm thick, 4.8 BUR; blown film; 29% crystallinity; film	2900								ASTM D1434	216
	Jap. Synth. JSR RB830; 1,2-polybutadiene; 0.910 g/cm³ density; transparent, 2.5 BUR, 0.05 mm thick; blown film; 25% crystallinity; film	2900								ASTM D1434	216
TPE, Polyester	DuPont Hytrel 4056; 40 Shore D	3024				21.5			34.5		274
	DuPont Hytrel 5556; 55 Shore D	1555				21.5			34.5		274
	Eastman Ecdel; copolyester ether; transparent, crystalline, 0.11 - 0.14 mm thick; film	1267				23				ASTM D1434	60

Material Family	Material Note	Permeability Data			Test Conditions						Source
		Perm. Coefficient (cm³·mm/m²·day·atm)	VTR (g·mm/m²·day)	Non-normalized Data	Penetrant Note	Temp. (°C)	Time (days)	RH (%)	Pressure (kPa)	Test Note	

CARBON DIOXIDE (continued)

Material Family	Material Note	Perm. Coefficient	VTR	Non-normalized Data	Penetrant Note	Temp. (°C)	Time (days)	RH (%)	Pressure (kPa)	Test Note	Source
TPE, Styrenic	Shell Kraton D 1101; SBS; FDA grade, 71 Shore A; unsaturated; 31% styrene/ 69% rubber; neat rubber	6709				23				ASTM D1434-82; area: 50 cm²; gradient: 100% gas at 740 mm Hg	303
	Shell Kraton D 1107; SIS; FDA grade, 37 Shore A; unsaturated; 14% styrene/ 86% rubber; neat rubber	7598				23				ASTM D1434-82; area: 50 cm²; gradient: 100% gas at 740 mm Hg	303
	Shell Kraton D 2103; SBS; FDA grade, 70 Shore A; unsaturated; ready to use compound	5280				23				ASTM D1434-82; area: 50 cm²; gradient: 100% gas at 740 mm Hg	303
	Shell Kraton D 2104; SBS; FDA grade, 27 Shore A; unsaturated; ready to use compound	24,657				23				ASTM D1434-82; area: 50 cm²; gradient: 100% gas at 740 mm Hg	303
	Shell Kraton D 2109; SBS; FDA grade, 44 Shore A; unsaturated; ready to use compound	7020				23				ASTM D1434-82; area: 50 cm²; gradient: 100% gas at 740 mm Hg	303
	Shell Kraton G 1650; SEBS; FDA grade, 75 Shore A; saturated; 29% styrene/ 71% rubber; neat rubber	2303				23				ASTM D1434-82; area: 50 cm²; gradient: 100% gas at 740 mm Hg	303
	Shell Kraton G 1651; SEBS; FDA grade, 75 Shore A; saturated; 29% styrene/ 71% rubber; neat rubber	2402				23				ASTM D1434-82; area: 50 cm²; gradient: 100% gas at 740 mm Hg	303
	Shell Kraton G 1652; SEBS; FDA grade, 75 Shore A; saturated; 29% styrene/ 71% rubber; neat rubber	3331				23				ASTM D1434-82; area: 50 cm²; gradient: 100% gas at 740 mm Hg	303
	Shell Kraton G 2701; SEBS; FDA grade, 67 Shore A; saturated; ready to use compound	2500				23				ASTM D1434-82; area: 50 cm²; gradient: 100% gas at 740 mm Hg	303
	Shell Kraton G 2706; SEBS; FDA grade, 55 Shore A; saturated; ready to use compound	2539				23				ASTM D1434-82; area: 50 cm²; gradient: 100% gas at 740 mm Hg	303
	Shell Kraton G 2706; SEBS; FDA grade, 28 Shore A; saturated; ready to use compound	7252				23				ASTM D1434-82; area: 50 cm²; gradient: 100% gas at 740 mm Hg	303
TPE, Urethane (TPAU)	BASF Elastollan C80A; 80 Shore A	1751				20					130
	BASF Elastollan C85A; 85 Shore A	1313				20					130
	BASF Elastollan C90A; 90 Shore A	350				20					130
	BASF Elastollan C95A; 95 Shore A	175				20					130
TPE, Urethane (TPEU)	BASF Elastollan 1180A; 80 Shore A	2014				20					130
	BASF Elastollan 1185A; 85 Shore A	1576				20					130
	BASF Elastollan 1190A; 90 Shore A	1138				20					130
	BASF Elastollan 1195A; 95 Shore A	788				20					130
TPE, Vinyl	1.23 - 1.31 g/cm³ density; stretch film; 0.018 mm thick; 4.8 BUR; blown film; film	1400 - 2700								ASTM D1434	216
	1.26 g/cm³ density; transparent; 2.5 BUR; 0.05 mm thick; blown film; 50 phr plasticizer; film	300								ASTM D1434	216
	plasticized; film	39.4 - 1181				24					250

CARBON DISULFIDE

Material Family	Material Note	Perm. Coefficient	VTR	Non-normalized Data	Penetrant Note	Temp. (°C)	Time (days)	RH (%)	Pressure (kPa)	Test Note	Source
Polyvinyl Alcohol	Ansell Edmont PVA; supported (lined) glove film; specified by glove film weight			low permeation rate; 0 to 1/2 eyedropper size drops per hour; PVA coating is water soluble; lower detection limit: 0 ppm; permeation rate: <0.9 µg/cm²/min		23	0.25			ASTM F739	313
Rubber, Nitrile (NBR)	Ansell Edmont Sol-Vex 37-165; unsupported glove film; 0.54 mm thick			fair permeation rate; 51 to 100 eyedropper size drops per hour; breakthrough time: 30 minutes; permeation rate: <900 µg/cm²/min		23				ASTM F739	313

CARBON DISULFIDE (continued)

Material Family	Material Note	Perm. Coefficient (cm³·mm/m²·day·atm)	VTR (g·mm/m²·day)	Non-normalized Data	Penetrant Note	Temp. (°C)	Time (days)	RH (%)	Pressure (kPa)	Test Note	Source
Rubber, Nitrile (NBR)	Pioneer Industrial Products Stansolv A-14; glove film			breakthrough time: 20 minutes; lower detection limit: 0.2 ppm; permeation rate: 516 µg/cm²/min	carbon bisulfide					ASTM F739.85	312

CARBON MONOXIDE

Material Family	Material Note	Perm. Coefficient	VTR	Non-normalized Data	Penetrant Note	Temp. (°C)	Time (days)	RH (%)	Pressure (kPa)	Test Note	Source
Polyethylene, HDPE	Hoechst AG Hostalen	36.5				20				volume at standard temperature and pressure; useable average for all Hostalen grades	94

CARBON TETRACHLORIDE

Material Family	Material Note	Perm. Coefficient	VTR	Non-normalized Data	Penetrant Note	Temp. (°C)	Time (days)	RH (%)	Pressure (kPa)	Test Note	Source
Nylon 66	DuPont Zytel 42; low flow, 2.54 mm thick; bottles		2								68
Polyester, PET	DuPont Mylar film		0.03			40				ASTM E96-80; modified test, permeabilities determined at the partial pressure of the vapor at the test temperature	270
Polyvinyl Alcohol	Ansell Edmont PVA; supported (lined) glove film; specified by glove film weight			low permeation rate: 0 to 1/2 eyedropper size drops per hour; PVA coating is water soluble; lower detection limit: 0 ppm; permeation rate: <0.9 µg/cm²/min		23	0.25			ASTM F739	313
Rubber, Neoprene (CR)	Pioneer Industrial Products Stanzoil N-44; glove film			breakthrough time: 31 minutes; permeation rate: 1512 µg/cm²/min						ASTM F739.85	312
Rubber, Nitrile (NBR)	Ansell Edmont Sol-Vex 37-165; unsupported glove film; 0.54 mm thick			good permeation rate; 6 to 50 eyedropper size drops per hour; breakthrough time: 150 minutes; permeation rate: <90 µg/cm²/min		23				ASTM F739	313
	Pioneer Industrial Products Stansolv A-14; glove film			breakthrough time: 341 minutes; lower detection limit: 1 ppm; permeation rate: 48 µg/cm²/min						ASTM F739.85	312
TPE, Vinyl	Ansell Edmont Monkey Grip; supported (lined) glove film; specified by glove film weight			fair permeation rate; 51 to 100 eyedropper size drops per hour; breakthrough time: 25 minutes; permeation rate: <900 µg/cm²/min		23				ASTM F739	313

CELLOSOLVE

Material Family	Material Note	Perm. Coefficient	VTR	Non-normalized Data	Penetrant Note	Temp. (°C)	Time (days)	RH (%)	Pressure (kPa)	Test Note	Source
Polyvinyl Alcohol	Ansell Edmont PVA; supported (lined) glove film; specified by glove film weight			good permeation rate; 6 to 50 eyedropper size drops per hour; PVA coating is water soluble; breakthrough time: 75 minutes; permeation rate: <90 µg/cm²/min	solvent	23				ASTM F739	313
Rubber, Latex (NR)	Ansell Edmont Canners and Handlers 392; unsupported glove film; 0.48 mm thick			very good permeation rate; 1 to 5 eyedropper size drops per hour; breakthrough time: 25 minutes; permeation rate: <9 µg/cm²/min	solvent	23				ASTM F739	313
	Pioneer Industrial Products L-118; glove film			breakthrough time: 10 minutes; lower detection limit: 0.03 ppm; permeation rate: 12 µg/cm²/min						ASTM F739.85	312
Rubber, Neoprene (CR)	Ansell Edmont Neoprene 29-840; unsupported glove film; 0.38 mm thick			low permeation rate: 0 to 1/2 eyedropper size drops per hour; breakthrough time: 45 minutes; permeation rate: <0.9 µg/cm²/min	solvent	23				ASTM F739	313
	Ansell Edmont Neox; supported (lined) glove film; specified by glove film weight			low permeation rate: 0 to 1/2 eyedropper size drops per hour; breakthrough time: 240 minutes; permeation rate: <0.9 µg/cm²/min	solvent	23				ASTM F739	313
	Pioneer Industrial Products Stanzoil N-44; glove film			breakthrough time: 352 minutes; lower detection limit: 0.06 ppm; permeation rate: 18 µg/cm²/min						ASTM F739.85	312
Rubber, Nitrile (NBR)	Ansell Edmont Sol-Vex 37-165; unsupported glove film; 0.54 mm thick			good permeation rate; 6 to 50 eyedropper size drops per hour; permeation rate: <90 µg/cm²/min	solvent	23				ASTM F739	313
	Pioneer Industrial Products Stansolv A-14; glove film			breakthrough time: 416 minutes; lower detection limit: 0.03 ppm; permeation rate: 24 µg/cm²/min						ASTM F739.85	312

Material Family	Material Note	Perm. Coefficient (cm³·mm/m²·day·atm)	VTR (g·mm/m²·day)	Non-normalized Data	Penetrant Note	Temp. (°C)	Time (days)	RH (%)	Pressure (kPa)	Test Note	Source

CELLOSOLVE ACETATE

Polyvinyl Alcohol	Ansell Edmont PVA; supported (lined) glove film; specified by glove film weight			low permeation rate: 0 to 1/2 eyedropper size drops per hour; PVA coating is water soluble; lower detection limit: 0 ppm; permeation rate: <0.9 µg/cm²/min		23	0.25			ASTM F739	313
Rubber, Latex (NR)	Ansell Edmont Canners and Handlers 392; unsupported glove film; 0.46 mm thick			good permeation rate: 6 to 50 eyedropper size drops per hour; breakthrough time: 10 minutes; permeation rate: <90 µg/cm²/min		23				ASTM F739	313
	Pioneer Industrial Products L-118; glove film			breakthrough time: 13 minutes						ASTM F739.85	312
Rubber, Neoprene (CR)	Ansell Edmont Neoprene 29-840; unsupported glove film; 0.38 mm thick			good permeation rate: 6 to 50 eyedropper size drops per hour; breakthrough time: 25 minutes; permeation rate: <90 µg/cm²/min		23				ASTM F739	313
	Ansell Edmont Neox; supported (lined) glove film; specified by glove film weight			very good permeation rate: 1 to 5 eyedropper size drops per hour; breakthrough time: 75 minutes; permeation rate: <9 µg/cm²/min		23				ASTM F739	313
	Pioneer Industrial Products Stanzoil N-44; glove film			breakthrough time: 76 minutes; lower detection limit: 0.07 ppm; permeation rate: 252 µg/cm²/min						ASTM F739.85	312
Rubber, Nitrile (NBR)	Ansell Sol-Vex 37-165; unsupported glove film; 0.54 mm thick			good permeation rate: 6 to 50 eyedropper size drops per hour; breakthrough time: 90 minutes; permeation rate: <90 µg/cm²/min		23				ASTM F739	313
	Pioneer Industrial Products Stansolv A-14; glove film			breakthrough time: 162 minutes; lower detection limit: 0.1 ppm; permeation rate: 72 µg/cm²/min						ASTM F739.85	312

CHARCOAL STARTER

Laminar, Nylon/Olefin	DuPont Selar RB/ Polyolefin; 8% Selar RB; barrier prop.; extrusion blow molded; bottles			0.04% penetrant weight loss		50	28				293
Polyethylene, Fluorinated	barrier prop.; bottles			0.03% penetrant weight loss		50	28				293
Polyethylene, HDPE	bottles			14.8% penetrant weight loss		50	28				293

CHLORINE

Fluoropolymer, PVDF	Solvay Solef; 0.1 mm thick; cast film; film	1.2				23				ASTM D1434	125

CHLOROBENZENE

Acrylonitrile Copolymer, AMA	barrier prop.; bottles			failed		50	28				293
Laminar, Nylon/Olefin	DuPont Selar RB/ Polyolefin; 8% Selar RB; barrier prop.; extrusion blow molded; bottles			0.77% penetrant weight loss		50	28				293
Polyethylene, Fluorinated	barrier prop.; bottles			0.65% penetrant weight loss		50	28				293
Polyethylene, HDPE	bottles			20.0% penetrant weight loss		50	28				293
Polyvinyl Alcohol	Ansell Edmont PVA; supported (lined) glove film; specified by glove film weight			low permeation rate: 0 to 1/2 eyedropper size drops per hour; PVA coating is water soluble; lower detection limit: 0 ppm; permeation rate: <0.9 µg/cm²/min		23	0.25			ASTM F739	313
Polyvinyl Chloride	bottles			failed		23	180				293
Rubber, Nitrile (NBR)	Pioneer Industrial Products Stansolv A-14; glove film			breakthrough time: 15 minutes; permeation rate: 960 µg/cm²/min		23				ASTM F739.85	312

CHLOROFORM

EVOH	Eval Co. Eval E; 0.02 mm thick; 44% ethylene; film		0.06			20		65			266
	Eval Co. Eval E; 0.032 mm thick; 44% ethylene; film		0.03			20		65			266

CHLOROFORM (continued)

Material Family	Material Note	Permeability Data			Test Conditions						Source
		Perm. Coefficient (cm³·mm/m²·day·atm)	VTR (g·mm/m²·day)	Non-normalized Data	Penetrant Note	Temp. (°C)	Time (days)	RH (%)	Pressure (kPa)	Test Note	
EVOH	Eval Co. Eval EF-E; barrier prop.: 44% ethylene; film		0.06			20					265
	Eval Co. Eval EF-F; barrier prop.: 32% ethylene; film		0.04			20					265
	Eval Co. Eval EF-XL; 0.015 mm thick, biaxially oriented; film		0.0023			20		65			266
	Eval Co. Eval EF-XL; barrier prop., biaxially oriented; film		0.0024			20					265
	Eval Co. Eval F; 0.02 mm thick; 32% ethylene; film		0.04			20		65			266
	Eval Co. Eval F; 0.032 mm thick; 32% ethylene; film		0.15			20		65			266
LDPE/ EVAL Film	0.06/ 0.015 mm thick; EVAL EF-F inside; LDPE outside; film		0.02			20		65			266
	0.06/ 0.015 mm thick; EVAL EF-XL inside; LDPE outside; film		<0.0035			20		65			266
	0.06/ 0.025 mm thick; EVAL EF-E inside; LDPE outside; film		0.01			20		65			266
Nylon	0.0254 mm thick, oriented; film		0.34			20		65			266
	oriented, 0.015 mm thick; PVDC coated; film		0.13			20		65			266
Polyester, PET	0.0254 mm thick; film		7.9			20		65			266
Polyethylene, LDPE	0.051 mm thick; film		140.79			20		65			266
	film		138			20					265
Polypropylene	0.02 mm thick, oriented; film		74.8			20		65			266
	biaxially oriented; film		74.8			20					265
Polyvinyl Alcohol	Ansell Edmont PVA; supported (lined) glove film; specified by glove film weight			low permeation rate; 0 to 1/2 eyedropper size drops per hour; PVA coating is water soluble; lower detection limit: 0 ppm; permeation rate: <0.9 µg/cm²/min		23	0.25			ASTM F739	313
Rubber, Neoprene (CR)	Pioneer Industrial Products Stanzoil N-44; glove film			breakthrough time: 12 minutes; lower detection limit: 0.2 ppm; permeation rate: 1368 µg/cm²/min						ASTM F739.85	312

CHLORONAPHTHALENE

Material Family	Material Note	Perm. Coefficient	VTR	Non-normalized Data	Penetrant Note	Temp.	Time	RH	Pressure	Test Note	Source
Polyvinyl Alcohol	Ansell Edmont PVA; supported (lined) glove film; specified by glove film weight			low permeation rate; 0 to 1/2 eyedropper size drops per hour; PVA coating is water soluble; lower detection limit: 0 ppm; permeation rate: <0.9 µg/cm²/min		23	0.25			ASTM F739	313

CHLOROTHENE

Material Family	Material Note	Perm. Coefficient	VTR	Non-normalized Data	Penetrant Note	Temp.	Time	RH	Pressure	Test Note	Source
Rubber, Neoprene (CR)	Pioneer Industrial Products Stanzoil N-44; glove film			breakthrough time: 27 minutes; permeation rate: 1182 µg/cm²/min						ASTM F739.85	312
Rubber, Nitrile (NBR)	Pioneer Industrial Products Stansolv A-14; glove film			breakthrough time: 131 minutes; lower detection limit: 0.05 ppm; permeation rate: 264 µg/cm²/min						ASTM F739.85	312

CHLOROTOLUENE (o-)

Material Family	Material Note	Perm. Coefficient	VTR	Non-normalized Data	Penetrant Note	Temp.	Time	RH	Pressure	Test Note	Source
Rubber, Nitrile (NBR)	Pioneer Industrial Products Stansolv A-14; glove film			breakthrough time: 52 minutes; permeation rate: 984 µg/cm²/min						ASTM F739.85	312

CHLOROTOLUENE (p-)

Material Family	Material Note	Perm. Coefficient	VTR	Non-normalized Data	Penetrant Note	Temp.	Time	RH	Pressure	Test Note	Source
Rubber, Nitrile (NBR)	Pioneer Industrial Products Stansolv A-14; glove film			breakthrough time: 25 minutes; permeation rate: 888 µg/cm²/min						ASTM F739.85	312

CHROMIC ACID

Material Family	Material Note	Perm. Coefficient (cm³·mm/m²·day·atm)	VTR (g·mm/m²·day)	Non-normalized Data	Penetrant Note	Temp (°C)	Time (days)	RH (%)	Pressure (kPa)	Test Note	Source
Rubber, Nitrile (NBR)	Ansell Edmont Sol-Vex 37-165; unsupported glove film; 0.54 mm thick			breakthrough time: 240 minutes	50% conc.	23				ASTM F739	313
	Pioneer Industrial Products Stansolv A-14; glove film			permeation rate too large to measure; breakthrough time: >175 minutes; lower detection limit: 0.1 ppm	50% conc.					ASTM F739.85	312
TPE, Vinyl	Ansell Edmont Monkey Grip; supported (lined) glove film; specified by glove film weight			no permeation detected during a 6 hour test; lower detection limit: 0 ppm	50% conc.	23	0.25			ASTM F739	313
	Ansell Edmont Wet-Wear 550; clothing; PVC stretch outer layer bonded to a lightweight liner			no permeation detected during a 6 hour test; lower detection limit: 0 ppm	50% conc.	23	0.25			ASTM F739	313
	Ansell Edmont Wet-Wear 600; clothing; nylon netting bonded between two layers of PVC			no permeation detected during a 6 hour test; lower detection limit: 0 ppm	50% conc.	23	0.25			ASTM F739	313
	Pioneer Industrial Products Pylox V-20; glove film			no permeation rate at steady state detected; breakthrough time: >480 minutes; lower detection limit: 0.1 ppm; permeation rate: 0 µg/cm²/min	50% conc.					ASTM F739.85	312
TPE, Vinyl coated fabric	Ansell Edmont Wet-Wear 700; clothing; PVC coating with Nylon/Polyester lining			no permeation detected during a 6 hour test; lower detection limit: 0 ppm	50% conc.	23	0.25			ASTM F739	313

CITRIC ACID

Material Family	Material Note	Perm. Coefficient (cm³·mm/m²·day·atm)	VTR (g·mm/m²·day)	Non-normalized Data	Penetrant Note	Temp (°C)	Time (days)	RH (%)	Pressure (kPa)	Test Note	Source
Polyvinyl Alcohol	Ansell Edmont PVA; supported (lined) glove film; specified by glove film weight			PVA coating is water soluble; breakthrough time: 50 minutes	10% conc.	23				ASTM F739	313
Rubber, Latex (NR)	Ansell Edmont Canners and Handlers 392; unsupported glove film; 0.48 mm thick			no permeation detected during a 6 hour test; lower detection limit: 0 ppm	10% conc.	23	0.25			ASTM F739	313
Rubber, Neoprene (CR)	Ansell Edmont Neoprene 29-840; unsupported glove film; 0.38 mm thick			no permeation detected during a 6 hour test; lower detection limit: 0 ppm	10% conc.	23	0.25			ASTM F739	313
	Ansell Edmont Neox; supported (lined) glove film; specified by glove film weight			no permeation detected during a 6 hour test; lower detection limit: 0 ppm	10% conc.	23	0.25			ASTM F739	313
Rubber, Nitrile (NBR)	Ansell Edmont Sol-Vex 37-165; unsupported glove film; 0.54 mm thick			no permeation detected during a 6 hour test; lower detection limit: 0 ppm	10% conc.	23	0.25			ASTM F739	313
TPE, Vinyl	Ansell Edmont Monkey Grip; supported (lined) glove film; specified by glove film weight			no permeation detected during a 6 hour test; lower detection limit: 0 ppm	10% conc.	23	0.25			ASTM F739	313

COLOGNE

Material Family	Material Note	Perm. Coefficient (cm³·mm/m²·day·atm)	VTR (g·mm/m²·day)	Non-normalized Data	Penetrant Note	Temp (°C)	Time (days)	RH (%)	Pressure (kPa)	Test Note	Source
Acetal	DuPont Delrin		0.24		various formulations	23		50			201
	DuPont Delrin		1.8		various formulations	38					201

CRESOL (3-)

Material Family	Material Note	Perm. Coefficient (cm³·mm/m²·day·atm)	VTR (g·mm/m²·day)	Non-normalized Data	Penetrant Note	Temp (°C)	Time (days)	RH (%)	Pressure (kPa)	Test Note	Source
Rubber, Latex (NR)	Pioneer Industrial Products L-118; glove film			breakthrough time: 150 minutes; lower detection limit: 5 ppm; permeation rate: 12 µg/cm²/min	m-Cresol					ASTM F739.85	312
Rubber, Neoprene (CR)	Pioneer Industrial Products Stanzoil N-44; glove film			no permeation rate at steady state detected; breakthrough time: >480 minutes; lower detection limit: 5 ppm; permeation rate: 0 µg/cm²/min	m-Cresol					ASTM F739.85	312
Rubber, Nitrile (NBR)	Pioneer Industrial Products Stansolv A-14; glove film			breakthrough time: 210 minutes; lower detection limit: 5 ppm; permeation rate: 126 µg/cm²/min	m-Cresol					ASTM F739.85	312
TPE, Vinyl	Pioneer Industrial Products Pylox V-20; glove film			breakthrough time: 150 minutes; lower detection limit: 5 ppm; permeation rate: 36 µg/cm²/min	m-Cresol					ASTM F739.85	312

Appendix I - Permeability Sort

Material Family / Material Note	Permeability Data — Non-normalized Data	Penetrant Note	Temp. (°C)	Time (days)	RH (%)	Pressure (kPa)	Test Note	Source
CRESOL (m-)								
Rubber, Latex (NR) — Pioneer Industrial Products L-118; glove film	breakthrough time: 150 minutes; lower detection limit: 5 ppm; permeation rate: 12 µg/cm²/min	3-Cresol					ASTM F739.85	312
Rubber, Neoprene (CR) — Pioneer Industrial Products Stanzoil N-44; glove film	no permeation rate at steady state detected; breakthrough time: >480 minutes; lower detection limit: 5 ppm; permeation rate: 0 µg/cm²/min	3-Cresol					ASTM F739.85	312
Rubber, Nitrile (NBR) — Pioneer Industrial Products Stansolv A-14; glove film	breakthrough time: 210 minutes; lower detection limit: 5 ppm; permeation rate: 126 µg/cm²/min	3-Cresol					ASTM F739.85	312
TPE, Vinyl — Pioneer Industrial Products Pylox V-20; glove film	breakthrough time: 150 minutes; lower detection limit: 5 ppm; permeation rate: 36 µg/cm²/min	3-Cresol					ASTM F739.85	312
CUMENE								
Rubber, Neoprene (CR) — Pioneer Industrial Products Stanzoil N-44; glove film	breakthrough time: 41 minutes; lower detection limit: 0.4 ppm; permeation rate: 216 µg/cm²/min						ASTM F739.85	312
Rubber, Nitrile (NBR) — Pioneer Industrial Products Stansolv A-14; glove film	breakthrough time: 271 minutes; lower detection limit: 0.03 ppm; permeation rate: 48 µg/cm²/min						ASTM F739.85	312
CYCLOHEXANE								
Rubber, Neoprene (CR) — Pioneer Industrial Products Stanzoil N-44; glove film	breakthrough time: 159 minutes; lower detection limit: 0.03 ppm; permeation rate: 42 µg/cm²/min						ASTM F739.85	312
Rubber, Nitrile (NBR) — Pioneer Industrial Products Stansolv A-14; glove film	no permeation rate at steady state detected; breakthrough time: >480 minutes; lower detection limit: 0.02 ppm; permeation rate: 0 µg/cm²/min						ASTM F739.85	312
TPE, Vinyl — Pioneer Industrial Products Pylox V-20; glove film	breakthrough time: 16 minutes; permeation rate: 102 µg/cm²/min						ASTM F739.85	312
CYCLOHEXANONE								
Acrylonitrile Copolymer, AMA — barrier prop.; bottles	failed		50	28				293
Laminar, Nylon/Olefin — DuPont Selar RB/ Polyolefin; 8% Selar RB; barrier prop.; extrusion blow molded; bottles	0.02% penetrant weight loss		50	28				293
Polyethylene, Fluorinated — barrier prop.; bottles	0.17% penetrant weight loss		50	28				293
Polyethylene, HDPE — bottles	0.6% penetrant weight loss		50	28				293
Polyvinyl Chloride — bottles	failed		23	180				293
CYCLOHEXYL ALCOHOL								
Polyvinyl Alcohol — Ansell Edmont PVA; supported (lined) glove film; specified by glove film weight	low permeation rate: 0 to 1/2 eyedropper size drops per hour; PVA coating is water soluble; lower detection limit: 0 ppm; permeation rate: <0.9 µg/cm²/min		23	0.25			ASTM F739	313
Rubber, Latex (NR) — Ansell Edmont Canners and Handlers 392; unsupported glove film; 0.48 mm thick	good permeation rate: 6 to 50 eyedropper size drops per hour; breakthrough time: 10 minutes; permeation rate: <90 µg/cm²/min		23				ASTM F739	313
Rubber, Neoprene (CR) — Ansell Edmont Neoprene 29-840; unsupported glove film; 0.38 mm thick	very good permeation rate: 1 to 5 eyedropper size drops per hour; breakthrough time: 150 minutes; permeation rate: <9 µg/cm²/min		23				ASTM F739	313
Rubber, Neoprene (CR) — Ansell Edmont Neox; supported (lined) glove film; specified by glove film weight	low permeation rate: 0 to 1/2 eyedropper size drops per hour; breakthrough time: 180 minutes; permeation rate: <0.9 µg/cm²/min		23				ASTM F739	313
Rubber, Nitrile (NBR) — Ansell Edmont Sol-Vex 37-165; unsupported glove film; 0.54 mm thick	low permeation rate: 0 to 1/2 eyedropper size drops per hour; lower detection limit: 0 ppm; permeation rate: <0.9 µg/cm²/min		23	0.25			ASTM F739	313

CYCLOHEXYL ALCOHOL (continued)

Material Family / Material Note	Perm. Coefficient (cm³·mm/m²·day·atm)	VTR (g·mm/m²·day)	Non-normalized Data	Penetrant Note	Temp (°C)	Time (days)	RH (%)	Pressure (kPa)	Test Note	Source
TPE, Vinyl										
Ansell Edmont Monkey Grip; supported (lined) glove film; specified by glove film weight			low permeation rate; 0 to 1/2 eyedropper size drops per hour; breakthrough time: 360 minutes; permeation rate: <0.9 µg/cm²/min		23				ASTM F739	313
Ansell Edmont Wet-Wear 550; clothing; PVC stretch outer layer bonded to a lightweight liner			low permeation rate; 0 to 1/2 eyedropper size drops per hour; breakthrough time: 200 minutes; permeation rate: <0.9 µg/cm²/min		23				ASTM F739	313
Ansell Edmont Wet-Wear 600; clothing; nylon netting bonded between two layers of PVC			low permeation rate; 0 to 1/2 eyedropper size drops per hour; lower detection limit: 0 ppm; permeation rate: <0.9 µg/cm²/min		23	0.25			ASTM F739	313
TPE, Vinyl coated fabric										
Ansell Edmont Wet-Wear 700; clothing; PVC coating with Nylon/Polyester lining			very good permeation rate; 1 to 5 eyedropper size drops per hour; breakthrough time: 15 minutes; permeation rate: <9 µg/cm²/min		23				ASTM F739	313

CHLOROTHENE VG

Material Family / Material Note	Perm. Coefficient (cm³·mm/m²·day·atm)	VTR (g·mm/m²·day)	Non-normalized Data	Penetrant Note	Temp (°C)	Time (days)	RH (%)	Pressure (kPa)	Test Note	Source
Polyvinyl Alcohol										
Ansell Edmont PVA; supported (lined) glove film; specified by glove film weight			low permeation rate; 0 to 1/2 eyedropper size drops per hour; PVA coating is water soluble; lower detection limit: 0 ppm; permeation rate: <0.9 µg/cm²/min		23	0.25			ASTM F739	313
Rubber, Nitrile (NBR)										
Ansell Edmont Sol-Vex 37-165; unsupported glove film; 0.54 mm thick			poor permeation rate; 501 to 5000 eyedropper size drops per hour; breakthrough time: 90 minutes; permeation rate: <9000 µg/cm²/min		23				ASTM F739	313

D-LIMONENE

Material Family / Material Note	Perm. Coefficient (cm³·mm/m²·day·atm)	VTR (g·mm/m²·day)	Non-normalized Data	Penetrant Note	Temp (°C)	Time (days)	RH (%)	Pressure (kPa)	Test Note	Source
EVOH										
Eval Co. EVOH 44; barrier prop.; 0.04 mm thick; film		0.41			25		dry			255
Laminar, Nylon/Olefin										
DuPont Selar RB/ Polyolefin; 8% Selar RB; barrier prop.; extrusion blow molded; bottles			0.05% penetrant weight loss		50	28				293
Polyethylene, Fluorinated										
barrier prop.; bottles			0.11% penetrant weight loss		50	28				293
Polyethylene, HDPE										
film		149			25		dry			255
bottles			6.7% penetrant weight loss		50	28				293
Polypropylene										
film		8.9			25		dry			255
Polyvinylidene Chloride										
Dow Saran MA; VDC methyl acrylate; barrier prop.; film		0.0088			25		dry			255
Dow Saran XU-32009; VDC vinyl chloride; barrier prop.; developmental material; film		0.016			25		dry			255

DIACETONE ALCOHOL

Material Family / Material Note	Perm. Coefficient (cm³·mm/m²·day·atm)	VTR (g·mm/m²·day)	Non-normalized Data	Penetrant Note	Temp (°C)	Time (days)	RH (%)	Pressure (kPa)	Test Note	Source
Polyvinyl Alcohol										
Ansell Edmont PVA; supported (lined) glove film; specified by glove film weight			good permeation rate; 6 to 50 eyedropper size drops per hour; PVA coating is water soluble; breakthrough time: 150 minutes; permeation rate: <90 µg/cm²/min		23				ASTM F739	313
Rubber, Latex (NR)										
Ansell Edmont Canners and Handlers 392; unsupported glove film; 0.48 mm thick			very good permeation rate; 1 to 5 eyedropper size drops per hour; breakthrough time: 15 minutes; permeation rate: <9 µg/cm²/min		23				ASTM F739	313
Rubber, Neoprene (CR)										
Ansell Edmont Neoprene 29-840; unsupported glove film; 0.38 mm thick			low permeation rate; 0 to 1/2 eyedropper size drops per hour; breakthrough time: 300 minutes; permeation rate: <0.9 µg/cm²/min		23				ASTM F739	313
Ansell Edmont Neox; supported (lined) glove film; specified by glove film weight			low permeation rate; 0 to 1/2 eyedropper size drops per hour; lower detection limit: 0 ppm; permeation rate: <0.9 µg/cm²/min		23	0.25			ASTM F739	313

Appendix I - Permeability Sort

Material Family	Material Note	Perm. Coefficient (cm³·mm/m²·day·atm)	VTR (g·mm/m²·day)	Non-normalized Data	Penetrant Note	Temp. (°C)	Time (days)	RH (%)	Pressure (kPa)	Test Note	Source
DIACETONE ALCOHOL (continued)											
Rubber, Nitrile (NBR)	Ansell Edmont Sol-Vex 37-165; unsupported glove film; 0.54 mm thick			low permeation rate; 0 to 1/2 eyedropper size drops per hour; breakthrough time: 240 minutes; permeation rate: <0.9 μg/cm²/min		23				ASTM F739	313
DIAMINE											
Rubber, Latex (NR)	Pioneer Industrial Products L-118; glove film			breakthrough time: 218 minutes; lower detection limit: 0.7 ppm; permeation rate: 12 μg/cm²/min						ASTM F739.85	312
Rubber, Neoprene (CR)	Pioneer Industrial Products Starzoil N-44; glove film			no permeation rate at steady state detected; breakthrough time: >480 minutes; lower detection limit: 0.7 ppm; permeation rate: 0 μg/cm²/min						ASTM F739.85	312
Rubber, Nitrile (NBR)	Pioneer Industrial Products Stansolv A-14; glove film			no permeation rate at steady state detected; breakthrough time: >480 minutes; lower detection limit: 0.7 ppm; permeation rate: 0 μg/cm²/min						ASTM F739.85	312
TPE, Vinyl	Pioneer Industrial Products Pylox V-20; glove film			no permeation rate at steady state detected; breakthrough time: >480 minutes; lower detection limit: 0.7 ppm; permeation rate: 0 μg/cm²/min						ASTM F739.85	312
DIBUTYL PHTHALATE											
Polyvinyl Alcohol	Ansell Edmont PVA; supported (lined) glove film; specified by glove film weight			low permeation rate; 0 to 1/2 eyedropper size drops per hour; PVA coating is water soluble; lower detection limit: 0 ppm; permeation rate: <0.9 μg/cm²/min		23	0.25			ASTM F739	313
Rubber, Latex (NR)	Ansell Edmont Canners and Handlers 392; unsupported glove film; 0.48 mm thick			breakthrough time: 200 minutes		23				ASTM F739	313
Rubber, Neoprene (CR)	Ansell Edmont Neoprene 29-840; unsupported glove film; 0.38 mm thick			low permeation rate; 0 to 1/2 eyedropper size drops per hour; breakthrough time: 120 minutes; permeation rate: <0.9 μg/cm²/min		23				ASTM F739	313
	Ansell Edmont Neox; supported (lined) glove film; specified by glove film weight			very good permeation rate; 1 to 5 eyedropper size drops per hour; breakthrough time: 300 minutes; permeation rate: <9 μg/cm²/min		23				ASTM F739	313
Rubber, Nitrile (NBR)	Ansell Edmont Sol-Vex 37-165; unsupported glove film; 0.54 mm thick			low permeation rate; 0 to 1/2 eyedropper size drops per hour; lower detection limit: 0 ppm; permeation rate: <0.9 μg/cm²/min		23	0.25			ASTM F739	313
TPE, Vinyl	Ansell Edmont Wet-Wear 600; clothing; nylon netting bonded between two layers of PVC			no permeation detected during a 6 hour test; lower detection limit: 0 ppm		23	0.25			ASTM F739	313
TPE, Vinyl coated fabric	Ansell Edmont Wet-Wear 700; clothing; PVC coating with Nylon/Polyester lining			no permeation detected during a 6 hour test; lower detection limit: 0 ppm		23	0.25			ASTM F739	313
DICHLOROBENZENE (1,2-)											
Rubber, Nitrile (NBR)	Pioneer Industrial Products Stansolv A-14; glove film			breakthrough time: 37 minutes; permeation rate: 1140 μg/cm²/min	o-dichlorobenzene					ASTM F739.85	312
DICHLOROBENZENE (1,3-)											
Rubber, Nitrile (NBR)	Pioneer Industrial Products Stansolv A-14; glove film			breakthrough time: 73 minutes; lower detection limit: 0.3 ppm; permeation rate: 174 μg/cm²/min						ASTM F739.85	312
DICHLOROBENZENE (o-)											
Laminar, Nylon/HDPE	DuPont Selar RB/HDPE; solvent packaging; 18% Selar RB; barrier prop.; extrusion blow molded one pint bottle			0.3% penetrant weight loss (70% relative improvement vs. HDPE)	halogenated	48.9	28				291
Rubber, Nitrile (NBR)	Pioneer Industrial Products Stansolv A-14; glove film			breakthrough time: 37 minutes; permeation rate: 1140 μg/cm²/min	1,2-dichlorobenzene					ASTM F739.85	312

Material Family / Material Note	Perm. Coefficient (cm³·mm/m²·day·atm)	VTR (g·mm/m²·day)	Non-normalized Data	Penetrant Note	Temp (°C)	Time (days)	RH (%)	Pressure (kPa)	Test Note	Source

DICHLORODIFLUOROMETHANE

Polysulfone										
Amoco Udel; transparent, amber tint	0.23				23		dry		ASTM D1434	15

DICHLOROETHANE (1,2-)

Rubber, Neoprene (CR)										
Pioneer Industrial Products Stanzoil N-44; glove film			breakthrough time: 33 minutes; lower detection limit: 0.09 ppm; permeation rate: 1482 µg/cm²/min						ASTM F739.85	312
Rubber, Nitrile (NBR)										
Pioneer Industrial Products Stansolv A-14; glove film			breakthrough time: 16 minutes; lower detection limit: 0.06 ppm; permeation rate: 1752 µg/cm²/min						ASTM F739.85	312

DICHLOROMETHANE

Rubber, Neoprene (CR)										
Pioneer Industrial Products Stanzoil N-44; glove film			breakthrough time: 6 minutes; permeation rate: 1434 µg/cm²/min						ASTM F739.85	312

DICHLOROTETRAFLUOROETHANE

Polysulfone										
Amoco Udel; transparent, amber tint	0.1				23		dry		ASTM D1434	15

DIESEL FUEL CONDITIONER

Laminar, Nylon/Olefin										
DuPont Selar RB/ Polyolefin; 8% Selar RB; barrier prop.; extrusion blow molded; bottles			0.03% penetrant weight loss		50	28				293
Polyethylene, Fluorinated										
barrier prop.; bottles			0.08% penetrant weight loss		50	28				293
Polyethylene, HDPE										
bottles			5.5% penetrant weight loss		50	28				293

DIETHANOLAMINE

Rubber, Latex (NR)										
Pioneer Industrial Products L-118; glove film			no permeation rate at steady state detected; breakthrough time: >480 minutes; lower detection limit: 1.1 ppm; permeation rate: 0 µg/cm²/min						ASTM F739.85	312
Rubber, Neoprene (CR)										
Pioneer Industrial Products Stanzoil N-44; glove film			no permeation rate at steady state detected; breakthrough time: >480 minutes; lower detection limit: 1.1 ppm; permeation rate: 0 µg/cm²/min						ASTM F739.85	312
Rubber, Nitrile (NBR)										
Pioneer Industrial Products Stansolv A-14; glove film			no permeation rate at steady state detected; breakthrough time: >480 minutes; lower detection limit: 1.1 ppm; permeation rate: 0 µg/cm²/min						ASTM F739.85	312
TPE, Vinyl										
Pioneer Industrial Products Pylox V-20; glove film			no permeation rate at steady state detected; breakthrough time: >480 minutes; lower detection limit: 1.1 ppm; permeation rate: 0 µg/cm²/min						ASTM F739.85	312

DIETHYL ETHER

Rubber, Neoprene (CR)										
Pioneer Industrial Products Stanzoil N-44; glove film			breakthrough time: 18 minutes; lower detection limit: 0.1 ppm; permeation rate: 496 µg/cm²/min						ASTM F739.85	312
Rubber, Nitrile (NBR)										
Pioneer Industrial Products Stansolv A-14; glove film			breakthrough time: 64 minutes; lower detection limit: 0.1 ppm; permeation rate: 78 µg/cm²/min						ASTM F739.85	312

DIETHYLAMINE

Rubber, Nitrile (NBR)										
Ansell Edmont Sol-Vex 37-165; unsupported glove film; 0.54 mm thick			fair permeation rate: 51 to 100 eyedropper size drops per hour; breakthrough time: 45 minutes; permeation rate: <900 µg/cm²/min		23				ASTM F739	313

DIETHYLENE DIOXIDE (1,4-)

Material Family / Material Note	Perm. Coefficient (cm³·mm/m²·day·atm)	VTR (g·mm/m²·day)	Non-normalized Data	Penetrant Note	Temp. (°C)	Time (days)	RH (%)	Pressure (kPa)	Test Note	Source
Rubber, Neoprene (CR)										
Pioneer Industrial Products Stanzoil N-44; glove film			breakthrough time: 28 minutes; permeation rate: 372 μg/cm²/min						ASTM F739.85	312
TPE, Vinyl										
Pioneer Industrial Products Pylox V-20; glove film			breakthrough time: 6 minutes; permeation rate: 1500 μg/cm²/min						ASTM F739.85	312

DIISOBUTYL KETONE

Material Family / Material Note	Perm. Coefficient (cm³·mm/m²·day·atm)	VTR (g·mm/m²·day)	Non-normalized Data	Penetrant Note	Temp. (°C)	Time (days)	RH (%)	Pressure (kPa)	Test Note	Source
Polyvinyl Alcohol										
Ansell Edmont PVA; supported (lined) glove film; specified by glove film weight			low permeation rate; 0 to 1/2 eyedropper size drops per hour; PVA coating is water soluble; lower detection limit: 0 ppm; permeation rate: <0.9 μg/cm²/min	DIBK	23	0.25			ASTM F739	313
Rubber, Nitrile (NBR)										
Ansell Edmont Sol-Vex 37-165; unsupported glove film; 0.54 mm thick			fair permeation rate; 51 to 100 eyedropper size drops per hour; breakthrough time: 120 minutes; permeation rate: <900 μg/cm²/min	DIBK	23				ASTM F739	313

DIMETHYL SULFOXIDE

Material Family / Material Note	Perm. Coefficient (cm³·mm/m²·day·atm)	VTR (g·mm/m²·day)	Non-normalized Data	Penetrant Note	Temp. (°C)	Time (days)	RH (%)	Pressure (kPa)	Test Note	Source
Rubber, Latex (NR)										
Ansell Edmont Canners and Handlers 392; unsupported glove film; 0.48 mm thick			low permeation rate; 0 to 1/2 eyedropper size drops per hour; breakthrough time: 180 minutes; permeation rate: <0.9 μg/cm²/min	DMSO	23				ASTM F739	313
Pioneer Industrial Products L-118; glove film			no permeation rate at steady state detected; breakthrough time: 240 minutes; lower detection limit: 0.004 ppm; permeation rate: 0 μg/cm²/min	DMSO					ASTM F739.85	312
Rubber, Neoprene (CR)										
Ansell Edmont Neoprene 29-840; unsupported glove film; 0.38 mm thick			low permeation rate; 0 to 1/2 eyedropper size drops per hour; breakthrough time: 0 ppm; permeation rate: <0.9 μg/cm²/min	DMSO	23	0.25			ASTM F739	313
Ansell Edmont Neox; supported (lined) glove film; specified by glove film weight			good permeation rate; 6 to 50 eyedropper size drops per hour; breakthrough time: >180 minutes; permeation rate: <90 μg/cm²/min	DMSO	23				ASTM F739	313
Pioneer Industrial Products Stanzoil N-44; glove film			no permeation rate at steady state detected; breakthrough time: 0 minutes; lower detection limit: 0.004 ppm; permeation rate: 0 μg/cm²/min	DMSO					ASTM F739.85	312
Rubber, Nitrile (NBR)										
Ansell Edmont Sol-Vex 37-165; unsupported glove film; 0.54 mm thick			very good permeation rate; 1 to 5 eyedropper size drops per hour; breakthrough time: >240 minutes; permeation rate: <9 μg/cm²/min	DMSO	23				ASTM F739	313
Pioneer Industrial Products Stansolv A-14; glove film			breakthrough time: 0 minutes; lower detection limit: 0.004 ppm; permeation rate: 6 μg/cm²/min	DMSO					ASTM F739.85	312
TPE, Vinyl										
Pioneer Industrial Products Pylox V-20; glove film			no permeation rate at steady state detected; breakthrough time: 60 minutes; lower detection limit: 0.004 ppm; permeation rate: 0 μg/cm²/min	DMSO					ASTM F739.85	312

DIMETHYLACETAMIDE

Material Family / Material Note	Perm. Coefficient (cm³·mm/m²·day·atm)	VTR (g·mm/m²·day)	Non-normalized Data	Penetrant Note	Temp. (°C)	Time (days)	RH (%)	Pressure (kPa)	Test Note	Source
Rubber, Latex (NR)										
Ansell Edmont Canners and Handlers 392; unsupported glove film; 0.48 mm thick			good permeation rate; 6 to 50 eyedropper size drops per hour; breakthrough time: 15 minutes; permeation rate: <90 μg/cm²/min	DMAC	23				ASTM F739	313
Rubber, Nitrile (NBR)										
Pioneer Industrial Products Stansolv A-14; glove film			permeation rate too large to measure; breakthrough time: 28 minutes; lower detection limit: 0.001 ppm	DMAC					ASTM F739.85	312
TPE, Vinyl										
Pioneer Industrial Products Pylox V-20; glove film			permeation rate too large to measure; breakthrough time: 20 minutes; lower detection limit: 0.005 ppm	DMAC					ASTM F739.85	312

DIMETHYLFORMAMIDE

Material Family / Material Note	Perm. Coefficient (cm³·mm/m²·day·atm)	VTR (g·mm/m²·day)	Non-normalized Data	Penetrant Note	Temp. (°C)	Time (days)	RH (%)	Pressure (kPa)	Test Note	Source
Rubber, Latex (NR)										
Ansell Edmont Canners and Handlers 392; unsupported glove film; 0.48 mm thick			very good permeation rate; 1 to 5 eyedropper size drops per hour; breakthrough time: 25 minutes; permeation rate: <9 μg/cm²/min	DMF	23				ASTM F739	313
Pioneer Industrial Products L-118; glove film			breakthrough time: 67 minutes; lower detection limit: 0.1 ppm; permeation rate: 246 μg/cm²/min	DMF					ASTM F739.85	312

DIMETHYLFORMAMIDE (continued)

Material Family	Material Note	Perm. Coefficient (cm³·mm/m²·day·atm)	VTR (g·mm/m²·day)	Non-normalized Data	Penetrant Note	Temp. (°C)	Time (days)	RH (%)	Pressure (kPa)	Test Note	Source
Rubber, Neoprene (CR)	Ansell Edmont Neoprene 29-840; unsupported glove film; 0.38 mm thick			good permeation rate: 6 to 50 eyedropper size drops per hour; breakthrough time: 10 minutes; permeation rate: <90 µg/cm²/min	DMF	23				ASTM F739	313
	Ansell Edmont Neox; supported (lined) glove film; specified by glove film weight			good permeation rate: 6 to 50 eyedropper size drops per hour; breakthrough time: 60 minutes; permeation rate: <90 µg/cm²/min	DMF	23				ASTM F739	313
	Pioneer Industrial Products Stanzoil N-44; glove film			breakthrough time: 110 minutes; lower detection limit: 0.1 ppm; permeation rate: 246 µg/cm²/min	DMF					ASTM F739.85	312
Rubber, Nitrile (NBR)	Pioneer Industrial Products Stansolv A-14; glove film			breakthrough time: 35 minutes; lower detection limit: 0.2 ppm; permeation rate: 246 µg/cm²/min	DMF					ASTM F739.85	312

DIOCTYL PHTHALATE

Material Family	Material Note	Perm. Coefficient (cm³·mm/m²·day·atm)	VTR (g·mm/m²·day)	Non-normalized Data	Penetrant Note	Temp. (°C)	Time (days)	RH (%)	Pressure (kPa)	Test Note	Source
Polyvinyl Alcohol	Ansell Edmont PVA; supported (lined) glove film; specified by glove film weight			fair permeation rate: 51 to 100 eyedropper size drops per hour; PVA coating is water soluble; breakthrough time: 30 minutes; permeation rate: <900 µg/cm²/min	DOP	23				ASTM F739	313
Rubber, Neoprene (CR)	Ansell Edmont Neoprene 29-840; unsupported glove film; 0.38 mm thick			low permeation rate: 0 to 1/2 eyedropper size drops per hour; breakthrough time: >360 minutes; permeation rate: <0.9 µg/cm²/min	DOP	23				ASTM F739	313
	Ansell Edmont Neox; supported (lined) glove film; specified by glove film weight			low permeation rate: 0 to 1/2 eyedropper size drops per hour; breakthrough time: 120 minutes; permeation rate: <0.9 µg/cm²/min	DOP	23				ASTM F739	313
Rubber, Nitrile (NBR)	Ansell Edmont Sol-Vex 37-165; unsupported glove film; 0.54 mm thick			low permeation rate: 0 to 1/2 eyedropper size drops per hour; breakthrough time: >360 minutes; permeation rate: <0.9 µg/cm²/min	DOP	23				ASTM F739	313
TPE, Vinyl	Ansell Edmont Wet-Wear 550; clothing; PVC stretch outer layer bonded to a lightweight liner			breakthrough time: 11 minutes	DOP	23				ASTM F739	313
	Ansell Edmont Wet-Wear 600; clothing; nylon netting bonded between two layers of PVC			low permeation rate: 0 to 1/2 eyedropper size drops per hour; lower detection limit: 0 ppm; permeation rate: <0.9 µg/cm²/min	DOP	23	0.25			ASTM F739	313
TPE, Vinyl coated fabric	Ansell Edmont Wet-Wear 700; clothing; PVC coating with Nylon/Polyester lining			breakthrough time: 25 minutes	DOP	23				ASTM F739	313

DIOXANE

Material Family	Material Note	Perm. Coefficient (cm³·mm/m²·day·atm)	VTR (g·mm/m²·day)	Non-normalized Data	Penetrant Note	Temp. (°C)	Time (days)	RH (%)	Pressure (kPa)	Test Note	Source
Rubber, Latex (NR)	Ansell Edmont Canners and Handlers 392; unsupported glove film; 0.48 mm thick			fair permeation rate: 51 to 100 eyedropper size drops per hour; breakthrough time: 5 minutes; permeation rate: <900 µg/cm²/min		23				ASTM F739	313

DIOXANE (1,4-)

Material Family	Material Note	Perm. Coefficient (cm³·mm/m²·day·atm)	VTR (g·mm/m²·day)	Non-normalized Data	Penetrant Note	Temp. (°C)	Time (days)	RH (%)	Pressure (kPa)	Test Note	Source
Rubber, Neoprene (CR)	Pioneer Industrial Products Stanzoil N-44; glove film			breakthrough time: 28 minutes; permeation rate: 372 µg/cm²/min		23				ASTM F739.85	312
TPE, Vinyl	Pioneer Industrial Products Pylox V-20; glove film			breakthrough time: 6 minutes; permeation rate: 1500 µg/cm²/min		23				ASTM F739.85	312

ELECTROLESS COPPER

Material Family	Material Note	Perm. Coefficient (cm³·mm/m²·day·atm)	VTR (g·mm/m²·day)	Non-normalized Data	Penetrant Note	Temp. (°C)	Time (days)	RH (%)	Pressure (kPa)	Test Note	Source
Rubber, Latex (NR)	Ansell Edmont Canners and Handlers 392; unsupported glove film; 0.48 mm thick			no permeation detected during a 6 hour test; lower detection limit: 0 ppm	MacDermid 9048	23	0.25			ASTM F739	313
Rubber, Neoprene (CR)	Ansell Edmont Neoprene 29-840; unsupported glove film; 0.38 mm thick			no permeation detected during a 6 hour test; lower detection limit: 0 ppm	MacDermid 9048	23	0.25			ASTM F739	313
	Ansell Edmont Neox; supported (lined) glove film; specified by glove film weight			no permeation detected during a 6 hour test; lower detection limit: 0 ppm	MacDermid 9048	23	0.25			ASTM F739	313
Rubber, Nitrile (NBR)	Ansell Edmont Sol-Vex 37-165; unsupported glove film; 0.54 mm thick			no permeation detected during a 6 hour test; lower detection limit: 0 ppm	MacDermid 9048	23	0.25			ASTM F739	313

Material Family	Material Note	Perm. Coefficient (cm³·mm/m²·day·atm)	VTR (g·mm/m²·day)	Non-normalized Data	Penetrant Note	Temp. (°C)	Time (days)	RH (%)	Pressure (kPa)	Test Note	Source
ELECTROLESS NICKEL											
Rubber, Latex (NR)	Ansell Edmont Canners and Handlers 392; unsupported glove film; 0.48 mm thick			no permeation detected during a 6 hour test; lower detection limit: 0 ppm	MacDermid J60/61	23	0.25			ASTM F739	313
Rubber, Neoprene (CR)	Ansell Edmont Neoprene 29-840; unsupported glove film; 0.38 mm thick			no permeation detected during a 6 hour test; lower detection limit: 0 ppm	MacDermid J60/61; 90% conc.	23	0.25			ASTM F739	313
Rubber, Latex (NR)	Ansell Edmont Neox; supported (lined) glove film; specified by glove film weight			no permeation detected during a 6 hour test; lower detection limit: 0 ppm	MacDermid J60/61	23	0.25			ASTM F739	313
Rubber, Nitrile (NBR)	Ansell Edmont Sol-Vex 37-165; unsupported glove film; 0.54 mm thick			no permeation detected during a 6 hour test; lower detection limit: 0 ppm	MacDermid J60/61	23	0.25			ASTM F739	313
TPE, Vinyl	Ansell Edmont Monkey Grip; supported (lined) glove film; specified by glove film weight			no permeation detected during a 6 hour test; lower detection limit: 0 ppm	MacDermid J60/61	23	0.25			ASTM F739	313
EPICHLOROHYDRIN											
Polyvinyl Alcohol	Ansell Edmont PVA; supported (lined) glove film; specified by glove film weight			low permeation rate; 0 to 1/2 eyedropper size drops per hour; PVA coating is water soluble; breakthrough time: 300 minutes; permeation rate: <0.9 µg/cm²/min		23				ASTM F739	313
Rubber, Latex (NR)	Ansell Edmont Canners and Handlers 392; unsupported glove film; 0.48 mm thick			fair permeation rate: 51 to 100 eyedropper size drops per hour; breakthrough time: 5 minutes; permeation rate: <900 µg/cm²/min		23				ASTM F739	313
Rubber, Neoprene (CR)	Ansell Edmont Neox; supported (lined) glove film; specified by glove film weight			fair permeation rate: 51 to 100 eyedropper size drops per hour; breakthrough time: 10 minutes; permeation rate: <900 µg/cm²/min		23				ASTM F739	313
ETHANE											
Polyethylene, HDPE	Hoechst AG Hostalen	90.2				20				volume at standard temperature and pressure; useable average for all Hostalen grades	94
	Phillips Marlex; film	92.9				23				ASTM D1434	101
ETHOXYETHANOL (2-)											
Rubber, Latex (NR)	Pioneer Industrial Products L-118; glove film			breakthrough time: 10 minutes; lower detection limit: 0.03 ppm; permeation rate: 12 µg/cm²/min						ASTM F739.85	312
Rubber, Neoprene (CR)	Pioneer Industrial Products Stanzoil N-44; glove film			breakthrough time: 352 minutes; lower detection limit: 0.06 ppm; permeation rate: 18 µg/cm²/min						ASTM F739.85	312
Rubber, Nitrile (NBR)	Pioneer Industrial Products Stansolv A-14; glove film			breakthrough time: 416 minutes; lower detection limit: 0.03 ppm; permeation rate: 24 µg/cm²/min						ASTM F739.85	312
ETHOXYETHYL ACETATE (2-)											
Rubber, Latex (NR)	Pioneer Industrial Products L-118; glove film			breakthrough time: 13 minutes							312
Rubber, Neoprene (CR)	Pioneer Industrial Products Stanzoil N-44; glove film			breakthrough time: 76 minutes; lower detection limit: 0.07 ppm; permeation rate: 252 µg/cm²/min						ASTM F739.85	312
Rubber, Nitrile (NBR)	Pioneer Industrial Products Stansolv A-14; glove film			breakthrough time: 162 minutes; lower detection limit: 0.1 ppm; permeation rate: 72 µg/cm²/min						ASTM F739.85	312
ETHYL ACETATE											
Acrylonitrile Copolymer, AMA	barrier prop.: bottles			5.85% penetrant weight loss (crazing)		50	28				293
Laminar, Nylon/HDPE	DuPont Selar RB/ HDPE; solvent packaging; 18% Selar RB; barrier prop.: extrusion blow molded; one pint bottle			0.75% penetrant weight loss (11% relative improvement vs. HDPE)	oxygen containing	48.9	28				291

© *Plastics Design Library*

ETHYL ACETATE (continued)

Material Family	Material Note	Perm. Coefficient (cm³·mm/m²·day·atm)	VTR (g·mm/m²·day)	Non-normalized Data	Penetrant Note	Temp. (°C)	Time (days)	RH (%)	Pressure (kPa)	Test Note	Source
Laminar, Nylon/Olefin	DuPont Selar RB/ Polyolefin; 8% Selar RB; barrier prop.; extrusion blow molded; bottles			0.42% penetrant weight loss		50					293
Polyester, PET	DuPont Mylar; film		0.03			40	28			ASTM E96-80; modified test, permeabilities determined at the partial pressure of the vapor at the test temperature	270
Polyethylene, Fluorinated	barrier prop.; bottles			2.7% penetrant weight loss		50	28				293
Polyethylene, HDPE	bottles			4.0% penetrant weight loss		50	28				293
Polyethylene, LDPE	Dow; film		11.8 - 118			24					250
Polystyrene	Dow Styron; film			sample failed		24					250
Polyvinyl Alcohol	Ansell Edmont PVA; supported (lined) glove film; specified by glove film weight			low permeation rate; 0 to 1/2 eyedropper size drops per hour; PVA coating is water soluble; lower detection limit: 0 ppm; permeation rate: <0.9 µg/cm²/min		23	0.25			ASTM F739	313
Polyvinyl Chloride	bottles			failed		23	180				293
Rubber, Latex (NR)	Ansell Edmont Canners and Handlers 392; unsupported glove film; 0.48 mm thick			fair permeation rate; 51 to 100 eyedropper size drops per hour; breakthrough time: 5 minutes; permeation rate: <900 µg/cm²/min		23				ASTM F739	313
Rubber, Neoprene (CR)	Ansell Edmont Neoprene 29-840; unsupported glove film; 0.38 mm thick			good permeation rate; 6 to 50 eyedropper size drops per hour; breakthrough time: 15 minutes; permeation rate: <90 µg/cm²/min		23				ASTM F739	313
	Ansell Edmont Neox; supported (lined) glove film; specified by glove film weight			good permeation rate; 6 to 50 eyedropper size drops per hour; breakthrough time: 200 minutes; permeation rate: <90 µg/cm²/min		23				ASTM F739	313
	Pioneer Industrial Products Stanzoil N-44; glove film			breakthrough time: 34 minutes; lower detection limit: 0.08 ppm; permeation rate: 1068 µg/cm²/min						ASTM F739.85	312
SAN	Dow Tyril; film			sample failed		24					250

ETHYL ALCOHOL

Material Family	Material Note	Perm. Coefficient (cm³·mm/m²·day·atm)	VTR (g·mm/m²·day)	Non-normalized Data	Penetrant Note	Temp. (°C)	Time (days)	RH (%)	Pressure (kPa)	Test Note	Source
Acetal	DuPont Delrin		0.1		90%; with 10% water	23		50			201
	DuPont Delrin		0.59		70%; with 30% water	23		50			201
	DuPont Delrin		3.1		70%; with 30% water	38					201
Polyethylene, LDPE	Dow; film		0.79 - 1.6			24					250
Polystyrene	Dow Styron; film		0.39 - sample failed			24					250
Rubber, Latex (NR)	Ansell Edmont Canners and Handlers 392; unsupported glove film; 0.48 mm thick			very good permeation rate; 1 to 5 eyedropper size drops per hour; breakthrough time: 15 minutes; permeation rate: <9 µg/cm²/min		23				ASTM F739	313
	Pioneer Industrial Products L-118; glove film			no permeation rate at steady state detected; breakthrough time: >480 minutes; lower detection limit: 0.02 ppm; permeation rate: 0 µg/cm²/min	ethanol					ASTM F739.85	312
Rubber, Neoprene (CR)	Ansell Edmont Neoprene 29-840; unsupported glove film; 0.38 mm thick			very good permeation rate; 1 to 5 eyedropper size drops per hour; breakthrough time: 90 minutes; permeation rate: <9 µg/cm²/min		23				ASTM F739	313
	Ansell Edmont Neox; supported (lined) glove film; specified by glove film weight			very good permeation rate; 1 to 5 eyedropper size drops per hour; breakthrough time: 180 minutes; permeation rate: <9 µg/cm²/min		23				ASTM F739	313

ETHYL ALCOHOL

Material Family	Material Note	Perm. Coefficient (cm³·mm/m²·day·atm)	VTR (g·mm/m²·day)	Non-normalized Data	Penetrant Note	Temp. (°C)	Time (days)	RH (%)	Pressure (kPa)	Test Note	Source
Rubber, Neoprene (CR)	Pioneer Industrial Products Stanzoil N-44; glove film			no permeation rate at steady state detected; breakthrough time: >480 minutes; lower detection limit: 0.002 ppm; permeation rate: 0 µg/cm²/min	ethanol					ASTM F739.85	312
Rubber, Nitrile (NBR)	Ansell Sol-Vex 37-165; unsupported glove film; 0.54 mm thick			very good permeation rate: 1 to 5 eyedropper size drops per hour; breakthrough time: 240 minutes; permeation rate: <9 µg/cm²/min		23				ASTM F739	313
	Pioneer Industrial Products Stansolv A-14; glove film			no permeation rate at steady state detected; breakthrough time: >480 minutes; lower detection limit: 0.002 ppm; permeation rate: 0 µg/cm²/min	ethanol					ASTM F739.85	312
SAN	Dow Tyril; film			sample failed		24					250
TPE, Vinyl	Ansell Edmont Monkey Grip; supported (lined) glove film; specified by glove film weight			very good permeation rate: 1 to 5 eyedropper size drops per hour; breakthrough time: 60 minutes; permeation rate: <9 µg/cm²/min		23				ASTM F739	313
	Pioneer Industrial Products Pylox V-20; glove film			breakthrough time: 20 minutes; permeation rate: 30 µg/cm²/min	ethanol					ASTM F739.85	312

ETHYL ALCOHOL AMINE

Material Family	Material Note	Perm. Coefficient	VTR	Non-normalized Data	Penetrant Note	Temp. (°C)	Time (days)	RH (%)	Pressure (kPa)	Test Note	Source
Polyvinyl Alcohol	Ansell Edmont PVA; supported (lined) glove film; specified by glove film weight			low permeation rate; 0 to 1/2 eyedropper size drops per hour; PVA coating is water soluble; lower detection limit: 0 ppm; permeation rate: <0.9 µg/cm²/min		23	0.25			ASTM F739	313
Rubber, Latex (NR)	Ansell Edmont Canners and Handlers 392; unsupported glove film; 0.48 mm thick			low permeation rate; 0 to 1/2 eyedropper size drops per hour; breakthrough time: 50 minutes; permeation rate: <0.9 µg/cm²/min		23				ASTM F739	313
Rubber, Neoprene (CR)	Ansell Edmont Neoprene 29-840; unsupported glove film; 0.38 mm thick			low permeation rate; 0 to 1/2 eyedropper size drops per hour; lower detection limit: 0 ppm; permeation rate: <0.9 µg/cm²/min		23	0.25			ASTM F739	313
	Ansell Edmont Neox; supported (lined) glove film; specified by glove film weight			low permeation rate; 0 to 1/2 eyedropper size drops per hour; lower detection limit: 0 ppm; permeation rate: <0.9 µg/cm²/min		23	0.25			ASTM F739	313
Rubber, Nitrile (NBR)	Ansell Edmont Sol-Vex 37-165; unsupported glove film; 0.54 mm thick			low permeation rate; 0 to 1/2 eyedropper size drops per hour; lower detection limit: 0 ppm; permeation rate: <0.9 µg/cm²/min		23	0.25			ASTM F739	313
TPE, Vinyl	Ansell Edmont Monkey Grip; supported (lined) glove film; specified by glove film weight			low permeation rate; 0 to 1/2 eyedropper size drops per hour; lower detection limit: 0 ppm; permeation rate: <0.9 µg/cm²/min		23	0.25			ASTM F739	313
	Ansell Edmont Wet-Wear 550; clothing; PVC stretch outer layer bonded to a lightweight liner			no permeation detected during a 6 hour test; lower detection limit: 0 ppm		23	0.25			ASTM F739	313
	Ansell Edmont Wet-Wear 600; clothing; nylon netting bonded between two layers of PVC			no permeation detected during a 6 hour test; lower detection limit: 0 ppm		23	0.25			ASTM F739	313
TPE, Vinyl coated fabric	Ansell Edmont Wet-Wear 700; clothing; PVC coating with Nylon/Polyester lining			no permeation detected during a 6 hour test; lower detection limit: 0 ppm		23	0.25			ASTM F739	313

ETHYL ETHER

Material Family	Material Note	Perm. Coefficient	VTR	Non-normalized Data	Penetrant Note	Temp. (°C)	Time (days)	RH (%)	Pressure (kPa)	Test Note	Source
Polyvinyl Alcohol	Ansell Edmont PVA; supported (lined) glove film; specified by glove film weight			low permeation rate; 0 to 1/2 eyedropper size drops per hour; PVA coating is water soluble; lower detection limit: 0 ppm; permeation rate: <0.9 µg/cm²/min		23	0.25			ASTM F739	313
Rubber, Neoprene (CR)	Ansell Edmont Neoprene 29-840; unsupported glove film; 0.38 mm thick			good permeation rate; 6 to 50 eyedropper size drops per hour; breakthrough time: 10 minutes; permeation rate: <90 µg/cm²/min		23				ASTM F739	313
	Ansell Edmont Neox; supported (lined) glove film; specified by glove film weight			good permeation rate; 6 to 50 eyedropper size drops per hour; breakthrough time: 10 minutes; permeation rate: <90 µg/cm²/min		23				ASTM F739	313
	Pioneer Industrial Products Stanzoil N-44; glove film			breakthrough time: 18 minutes; lower detection limit: 0.1 ppm; permeation rate: 486 µg/cm²/min						ASTM F739.85	312

ETHYL ETHER (continued)

Material Family	Material Note	Perm. Coefficient (cm³·mm/m²·day·atm)	VTR (g·mm/m²·day)	Non-normalized Data	Penetrant Note	Temp. (°C)	Time (days)	RH (%)	Pressure (kPa)	Test Note	Source
Rubber, Nitrile (NBR)	Ansell Edmont Sol-Vex 37-165; unsupported glove film; 0.54 mm thick			good permeation rate; 6 to 50 eyedropper size drops per hour; breakthrough time: 120 minutes; permeation rate: <90 µg/cm²/min		23				ASTM F739	313
	Pioneer Industrial Products Stansolv A-14; glove film			breakthrough time: 64 minutes; lower detection limit: 0.1 ppm; permeation rate: 78 µg/cm²/min						ASTM F739.85	312

ETHYL GLYCOL ETHER

Material Family	Material Note	Perm. Coefficient	VTR	Non-normalized Data	Penetrant Note	Temp.	Time	RH	Pressure	Test Note	Source
Polyvinyl Alcohol	Ansell Edmont PVA; supported (lined) glove film; specified by glove film weight			good permeation rate; 6 to 50 eyedropper size drops per hour; PVA coating is water soluble; breakthrough time: 75 minutes; permeation rate: <90 µg/cm²/min		23				ASTM F739	313
Rubber, Latex (NR)	Ansell Edmont Canners and Handlers 392; unsupported glove film; 0.48 mm thick			very good permeation rate; 1 to 5 eyedropper size drops per hour; breakthrough time: 25 minutes; permeation rate: <9 µg/cm²/min		23				ASTM F739	313
Rubber, Neoprene (CR)	Ansell Edmont Neoprene 29-840; unsupported glove film; 0.38 mm thick			low permeation rate; 0 to 1/2 eyedropper size drops per hour; breakthrough time: 45 minutes; permeation rate: <0.9 µg/cm²/min		23				ASTM F739	313
	Ansell Edmont Neox; supported (lined) glove film; specified by glove film weight			low permeation rate; 0 to 1/2 eyedropper size drops per hour; breakthrough time: 240 minutes; permeation rate: <0.9 µg/cm²/min		23				ASTM F739	313
Rubber, Nitrile (NBR)	Ansell Edmont Sol-Vex 37-165; unsupported glove film; 0.54 mm thick			good permeation rate; 6 to 50 eyedropper size drops per hour; breakthrough time: 210 minutes; permeation rate: <90 µg/cm²/min		23				ASTM F739	313

ETHYLENE

Material Family	Material Note	Perm. Coefficient	VTR	Non-normalized Data	Penetrant Note	Temp.	Time	RH	Pressure	Test Note	Source
Polyethylene, HDPE	Hoechst AG Hostalen	111				20				volume at standard temperature and pressure; useable average for all Hostalen grades	94

ETHYLENE DICHLORIDE

Material Family	Material Note	Perm. Coefficient	VTR	Non-normalized Data	Penetrant Note	Temp.	Time	RH	Pressure	Test Note	Source
Polyvinyl Alcohol	Ansell Edmont PVA; supported (lined) glove film; specified by glove film weight			low permeation rate; 0 to 1/2 eyedropper size drops per hour; PVA coating is water soluble; lower detection limit: 0 ppm; permeation rate: <0.9 µg/cm²/min		23	0.25			ASTM F739	313
Rubber, Neoprene (CR)	Pioneer Industrial Products Stanzoil N-44; glove film			breakthrough time: 33 minutes; lower detection limit: 0.09 ppm; permeation rate: 1482 µg/cm²/min						ASTM F739.85	312
Rubber, Nitrile (NBR)	Pioneer Industrial Products L-118; glove film			no permeation rate at steady state detected; breakthrough time: >480 minutes; permeation rate: 0 µg/cm²/min						ASTM F739.85	312

ETHYLENE GLYCOL

Material Family	Material Note	Perm. Coefficient	VTR	Non-normalized Data	Penetrant Note	Temp.	Time	RH	Pressure	Test Note	Source
Polyvinyl Alcohol	Ansell Edmont PVA; supported (lined) glove film; specified by glove film weight			very good permeation rate; 1 to 5 eyedropper size drops per hour; PVA coating is water soluble; breakthrough time: 120 minutes; permeation rate: <9 µg/cm²/min		23				ASTM F739	313
Rubber, Latex (NR)	Ansell Edmont Canners and Handlers 392; unsupported glove film; 0.48 mm thick			low permeation rate; 0 to 1/2 eyedropper size drops per hour; lower detection limit: 0 ppm; permeation rate: <0.9 µg/cm²/min		23	0.25			ASTM F739	313
	Pioneer Industrial Products L-118; glove film			breakthrough time: 16 minutes; lower detection limit: 0.06 ppm; permeation rate: 1752 µg/cm²/min						ASTM F739.85	312
Rubber, Neoprene (CR)	Ansell Edmont Neoprene 29-840; unsupported glove film; 0.38 mm thick			low permeation rate; 0 to 1/2 eyedropper size drops per hour; permeation limit: 0 ppm; permeation rate: <0.9 µg/cm²/min		23	0.25			ASTM F739	313
	Ansell Edmont Neox; supported (lined) glove film; specified by glove film weight			low permeation rate; 0 to 1/2 eyedropper size drops per hour; permeation limit: 0 ppm; permeation rate: <0.9 µg/cm²/min		23	0.25			ASTM F739	313

ETHYLENE GLYCOL

Material Family	Material Note	Perm. Coefficient (cm³·mm/m²·day·atm)	VTR (g·mm/m²·day)	Non-normalized Data	Penetrant Note	Temp. (°C)	Time (days)	RH (%)	Pressure (kPa)	Test Note	Source
Rubber, Neoprene (CR)	Pioneer Industrial Products Stanzoil N-44; glove film			no permeation rate at steady state detected; breakthrough time: >480 minutes; permeation rate: 0 µg/cm²/min						ASTM F739.85	312
Rubber, Nitrile (NBR)	Ansell Edmont Sol-Vex 37-165; unsupported glove film; 0.54 mm thick			low permeation rate; 0 to 1/2 eyedropper size drops per hour; lower detection limit: 0 ppm; permeation rate: <0.9 µg/cm²/min		23	0.25			ASTM F739	313
	Pioneer Industrial Products Stansolv A-14; glove film			no permeation rate at steady state detected; breakthrough time: >480 minutes; permeation rate: 0 µg/cm²/min		23				ASTM F739.85	312
TPE, Vinyl	Ansell Edmont Monkey Grip; supported (lined) glove film; specified by glove film weight			low permeation rate; 0 to 1/2 eyedropper size drops per hour; lower detection limit: 0 ppm; permeation rate: <0.9 µg/cm²/min		23	0.25			ASTM F739	313
	Ansell Edmont Wet-Wear 550; clothing; PVC stretch outer layer bonded to a lightweight liner			breakthrough time: 200 minutes		23				ASTM F739	313
	Ansell Edmont Wet-Wear 600; clothing; nylon netting bonded between two layers of PVC			low permeation rate; 0 to 1/2 eyedropper size drops per hour; lower detection limit: 0 ppm; permeation rate: <0.9 µg/cm²/min		23	0.25			ASTM F739	313
	Pioneer Industrial Products Pylox V-20; glove film			no permeation rate at steady state detected; breakthrough time: >480 minutes; permeation rate: 0 µg/cm²/min		23				ASTM F739.85	312
TPE, Vinyl coated fabric	Ansell Edmont Wet-Wear 700; clothing; PVC coating with Nylon/Polyester lining			low permeation rate; 0 to 1/2 eyedropper size drops per hour; lower detection limit: 0 ppm; permeation rate: <0.9 µg/cm²/min		23	0.25			ASTM F739	313

ETHYLENE OXIDE

Material Family	Material Note	Perm. Coefficient (cm³·mm/m²·day·atm)	VTR (g·mm/m²·day)	Non-normalized Data	Penetrant Note	Temp. (°C)	Time (days)	RH (%)	Pressure (kPa)	Test Note	Source
Polyethylene, LDPE	0.920 g/cm³ density; 0.05 mm thick; 2.5 BUR; blown film; 4 g/10 min. MFI; film	2100								ASTM D1434	216
Rubber, Neoprene (CR)	Pioneer Industrial Products Stanzoil N-44; glove film			breakthrough time: 31 minutes; lower detection limit: 3 ppm; permeation rate: 60 µg/cm²/min						ASTM F739.85	312
Rubber, Nitrile (NBR)	Pioneer Industrial Products Stansolv A-14; glove film			breakthrough time: 32 minutes; lower detection limit: 0.3 ppm; permeation rate: 126 µg/cm²/min						ASTM F739.85	312
TPE, Polybutadiene	Jap. Synth. JSR RB820; 1,2-polybutadiene; 0.910 g/cm³ density; transparent; 2.5 BUR; 0.05 mm thick; blown film; 15% crystallinity; film	32,000								ASTM D1434	216
	Jap. Synth. JSR RB830; 1,2-polybutadiene; 0.910 g/cm³ density; transparent; 2.5 BUR; 0.05 mm thick; blown film; 25% crystallinity; film	25,000								ASTM D1434	216

FORMALDEHYDE

Material Family	Material Note	Perm. Coefficient (cm³·mm/m²·day·atm)	VTR (g·mm/m²·day)	Non-normalized Data	Penetrant Note	Temp. (°C)	Time (days)	RH (%)	Pressure (kPa)	Test Note	Source
Polyethylene, LDPE	Dow; film		0.79 - 2.0			24					250
Polystyrene	Dow Styron; film		1.6 - 2.0			24					250
Rubber, Latex (NR)	Ansell Edmont Canners and Handlers 392; unsupported glove film; 0.48 mm thick			good permeation rate: 6 to 50 eyedropper size drops per hour; breakthrough time: 10 minutes; permeation rate: <90 µg/cm²/min		23				ASTM F739	313
	Pioneer Industrial Products L-118; glove film			no permeation rate at steady state detected; breakthrough time: >480 minutes; lower detection limit: 0.01 ppm; permeation rate: 0 µg/cm²/min	37% conc.	23				ASTM F739.85	312
Rubber, Neoprene (CR)	Ansell Edmont Neoprene 29-840; unsupported glove film; 0.38 mm thick			low permeation rate; 0 to 1/2 eyedropper size drops per hour; breakthrough time: 120 minutes; permeation rate: <0.9 µg/cm²/min		23				ASTM F739	313
	Ansell Edmont Neox; supported (lined) glove film; specified by glove film weight			very good permeation rate; 1 to 5 eyedropper size drops per hour; breakthrough time: 120 minutes; permeation rate: <9 µg/cm²/min		23				ASTM F739	313

FORMALDEHYDE (continued)

Material Family	Material Note	Perm. Coefficient (cm³ · mm/m² · day · atm)	VTR (g · mm/m² · day)	Non-normalized Data	Penetrant Note	Temp. (°C)	Time (days)	RH (%)	Pressure (kPa)	Test Note	Source
Rubber, Neoprene (CR)	Pioneer Industrial Products Stanzoil N-44; glove film			no permeation rate at steady state detected; breakthrough time: >480 minutes; lower detection limit: 3 ppm; permeation rate: 0 µg/cm²/min	37% conc.					ASTM F739.85	312
Rubber, Nitrile (NBR)	Ansell Edmont Sol-Vex 37-165; unsupported glove film; 0.54 mm thick			low permeation rate: 0 to 1/2 eyedropper size drops per hour; lower detection limit: 0 ppm; permeation rate: <0.9 µg/cm²/min		23	0.25			ASTM F739	313
	Pioneer Industrial Products Stansolv A-14; glove film			no permeation rate at steady state detected; breakthrough time: >480 minutes; lower detection limit: 8 ppm; permeation rate: 0 µg/cm²/min	37% conc.					ASTM F739.85	312
SAN	Dow Tyril; film		2.0 - 3.9			24					250
TPE, Vinyl	Ansell Edmont Monkey Grip; supported (lined) glove film; specified by glove film weight			very good permeation rate: 1 to 5 eyedropper size drops per hour; breakthrough time: 80 minutes; permeation rate: <9 µg/cm²/min		23				ASTM F739	313
	Ansell Edmont Wet-Wear 550; clothing; PVC stretch outer layer bonded to a lightweight liner			very good permeation rate: 1 to 5 eyedropper size drops per hour; breakthrough time: 5 minutes; permeation rate: <9 µg/cm²/min		23				ASTM F739	313
	Ansell Edmont Wet-Wear 600; clothing; nylon netting bonded between two layers of PVC			good permeation rate: 6 to 50 eyedropper size drops per hour; breakthrough time: 360 minutes; permeation rate: <90 µg/cm²/min		23				ASTM F739	313
	Pioneer Industrial Products Pylox V-20; glove film			no permeation rate at steady state detected; breakthrough time: >480 minutes; lower detection limit: 0.01 ppm; permeation rate: 0 µg/cm²/min	37% conc.					ASTM F739.85	312
TPE, Vinyl coated fabric	Ansell Edmont Wet-Wear 700; clothing; PVC coating with Nylon/Polyester lining			very good permeation rate: 1 to 5 eyedropper size drops per hour; breakthrough time: 30 minutes; permeation rate: <9 µg/cm²/min		23				ASTM F739	313

FORMALIN

Material Family	Material Note	Perm. Coefficient (cm³ · mm/m² · day · atm)	VTR (g · mm/m² · day)	Non-normalized Data	Penetrant Note	Temp. (°C)	Time (days)	RH (%)	Pressure (kPa)	Test Note	Source
Rubber, Latex (NR)	Pioneer Industrial Products L-118; glove film			no permeation rate at steady state detected; breakthrough time: >480 minutes; lower detection limit: 0.01 ppm; permeation rate: 0 µg/cm²/min	solution					ASTM F739.85	312
Rubber, Neoprene (CR)	Pioneer Industrial Products Stanzoil N-44; glove film			no permeation rate at steady state detected; breakthrough time: >480 minutes; lower detection limit: 3 ppm; permeation rate: 0 µg/cm²/min	solution					ASTM F739.85	312
Rubber, Nitrile (NBR)	Pioneer Industrial Products Stansolv A-14; glove film			no permeation rate at steady state detected; breakthrough time: >480 minutes; lower detection limit: 8 ppm; permeation rate: 0 µg/cm²/min	solution					ASTM F739.85	312
TPE, Vinyl	Pioneer Industrial Products Pylox V-20; glove film			no permeation rate at steady state detected; breakthrough time: >480 minutes; lower detection limit: 0.01 ppm; permeation rate: 0 µg/cm²/min	solution					ASTM F739.85	312

FORMIC ACID

Material Family	Material Note	Perm. Coefficient (cm³ · mm/m² · day · atm)	VTR (g · mm/m² · day)	Non-normalized Data	Penetrant Note	Temp. (°C)	Time (days)	RH (%)	Pressure (kPa)	Test Note	Source
Rubber, Latex (NR)	Ansell Edmont Canners and Handlers 392; unsupported glove film; 0.48 mm thick			breakthrough time: 150 minutes	90% conc.	23				ASTM F739	313
Rubber, Neoprene (CR)	Ansell Edmont Neoprene 29-840; unsupported glove film; 0.38 mm thick			no permeation detected during a 6 hour test; lower detection limit: 0 ppm	90% conc.	23	0.25			ASTM F739	313
	Ansell Edmont Neox; supported (lined) glove film; specified by glove film weight			no permeation detected during a 6 hour test; lower detection limit: 0 ppm	90% conc.	23	0.25			ASTM F739	313
Rubber, Nitrile (NBR)	Ansell Edmont Sol-Vex 37-165; unsupported glove film; 0.54 mm thick			breakthrough time: 240 minutes	90% conc.	23				ASTM F739	313
TPE, Vinyl	Ansell Edmont Monkey Grip; supported (lined) glove film; specified by glove film weight			breakthrough time: >360 minutes	90% conc.	23				ASTM F739	313
	Ansell Edmont Wet-Wear 550; clothing; PVC stretch outer layer bonded to a lightweight liner			breakthrough time: 25 minutes	90% conc.	23				ASTM F739	313

FORMIC ACID (continued)

Material Family	Material Note	Perm. Coefficient (cm³·mm/m²·day·atm)	VTR (g·mm/m²·day)	Non-normalized Data	Penetrant Note	Temp. (°C)	Time (days)	RH (%)	Pressure (kPa)	Test Note	Source
TPE, Vinyl	Ansell Edmont Wet-Wear 600; clothing; nylon netting bonded between two layers of PVC			breakthrough time: 75 minutes	90% conc.	23				ASTM F739	313
TPE, Vinyl coated fabric	Ansell Edmont Wet-Wear 700; clothing; PVC coating with Nylon/Polyester lining			breakthrough time: 30 minutes	90% conc.	23				ASTM F739	313

FURFURAL

Material Family	Material Note	Perm. Coefficient (cm³·mm/m²·day·atm)	VTR (g·mm/m²·day)	Non-normalized Data	Penetrant Note	Temp. (°C)	Time (days)	RH (%)	Pressure (kPa)	Test Note	Source
Polyvinyl Alcohol	Ansell Edmont PVA; supported (lined) glove film; specified by glove film weight			low permeation rate; 0 to 1/2 eyedropper size drops per hour; PVA coating is water soluble; lower detection limit: 0 ppm; permeation rate: <0.9 µg/cm²/min		23	0.25			ASTM F739	313
Rubber, Latex (NR)	Ansell Edmont Canners and Handlers 392; unsupported glove film; 0.48 mm thick			very good permeation rate; 1 to 5 eyedropper size drops per hour; breakthrough time: 15 minutes; permeation rate: <9 µg/cm²/min		23				ASTM F739	313
Rubber, Neoprene (CR)	Ansell Edmont Neoprene 29-840; unsupported glove film; 0.38 mm thick			good permeation rate; 6 to 50 eyedropper size drops per hour; breakthrough time: 200 minutes; permeation rate: <90 µg/cm²/min		23				ASTM F739	313
	Ansell Edmont Neox; supported (lined) glove film; specified by glove film weight			good permeation rate; 6 to 50 eyedropper size drops per hour; breakthrough time: 120 minutes; permeation rate: <90 µg/cm²/min		23				ASTM F739	313

FREON 114

Material Family	Material Note	Perm. Coefficient (cm³·mm/m²·day·atm)	VTR (g·mm/m²·day)	Non-normalized Data	Penetrant Note	Temp. (°C)	Time (days)	RH (%)	Pressure (kPa)	Test Note	Source
Fluoropolymer, PVDF	Solvay Solef; 0.025 mm thick; cast film; film	0.25				23				ASTM D1434	125
TPE, Polyester	DuPont Hytrel 4056; 40 Shore D	3542				21.5			34.5		274
	DuPont Hytrel 5556; 55 Shore D	2419				21.5			34.5		274
	DuPont Hytrel 6346; 63 Shore D	397				21.5			34.5		274
	DuPont Hytrel 7246; 72 Shore D	233				21.5			34.5		274

FREON 115

Material Family	Material Note	Perm. Coefficient (cm³·mm/m²·day·atm)	VTR (g·mm/m²·day)	Non-normalized Data	Penetrant Note	Temp. (°C)	Time (days)	RH (%)	Pressure (kPa)	Test Note	Source
Fluoropolymer, PVDF	Solvay Solef; 0.025 mm thick; cast film; film	0.1				23				ASTM D1434	125

FREON 12

Material Family	Material Note	Perm. Coefficient (cm³·mm/m²·day·atm)	VTR (g·mm/m²·day)	Non-normalized Data	Penetrant Note	Temp. (°C)	Time (days)	RH (%)	Pressure (kPa)	Test Note	Source
Acetal	DuPont Delrin		0.08		30%; with 70% Freon 11; propellant	23		50			201
	DuPont Delrin		0.08		20%; with 80% Freon 114; propellant	23		50			201
	DuPont Delrin		0.21		30%; with 70% Freon 11; propellant	38					201
	DuPont Delrin		0.17		20%; with 80% Freon 114; propellant	38					201
Fluoropolymer, PVDF	Solvay Solef; 0.025 mm thick; cast film; film	0.16				23				ASTM D1434	125
Nylon 6/66	305 mm long, 1 mm thick, 15.9 mm ID			0.067 lb/ft · yr penetrant loss	air conditioning refrigerant	93				refrigerant at saturated vapor pressure; calculated from permeation coefficient data	275
Polyethylene, HDPE	Phillips Marlex; film	37.4				23				ASTM D1434	101

Material Family / Material Note	Perm. Coefficient (cm³·mm/m²·day·atm)	VTR (g·mm/m²·day)	Non-normalized Data	Penetrant Note	Temp. (°C)	Time (days)	RH (%)	Pressure (kPa)	Test Note	Source
FREON 12										
Rubber, Neoprene (CR)										
Pioneer Industrial Products Stansoil N-44; glove film			no permeation rate at steady state detected; breakthrough time: >480 minutes; lower detection limit: 0.03 ppm; permeation rate: 0 µg/cm²/min						ASTM F739.85	312
Rubber, Nitrile (NBR)										
Pioneer Industrial Products Stansolv A-14; glove film			no permeation rate at steady state detected; breakthrough time: >480 minutes; lower detection limit: 8 ppm; permeation rate: 0 µg/cm²/min						ASTM F739.85	312
305 long, 1 mm thick, 15.9 mm ID			0.662 lb/ft · yr penetrant loss	air conditioning refrigerant	93				refrigerant at saturated vapor pressure, calculated from permeation coefficient data	275
TPE, Polyamide										
DuPont Zytel FN 726; 15.9 mm ID, 305 mm long, 58 Shore D, 1 mm thick			0.012 lb/ft · yr penetrant loss	air conditioning refrigerant	93				refrigerant at saturated vapor pressure, calculated from permeation coefficient data	275
TPE, Polyester										
DuPont Hytrel 4056; 40 Shore D	121				21.5			34.5		274
DuPont Hytrel 5556; 55 Shore D	104				21.5			34.5		274
DuPont Hytrel 6346; 63 Shore D	104				21.5			34.5		274
DuPont Hytrel 7246; 72 Shore D	71				21.5			34.5		274
FREON 22										
TPE, Polyester										
DuPont Hytrel 4056; 40 Shore D	41				21.5			34.5		274
DuPont Hytrel 5556; 55 Shore D	51				21.5			34.5		274
DuPont Hytrel 6346; 63 Shore D	<17.3				21.5			34.5		274
FREON 318										
Fluoropolymer, PVDF										
Solvay Solef; 0.025 mm thick; cast film	0.18				23				ASTM D1434	125
FREON TF										
Polyvinyl Alcohol										
Ansell Edmont PVA; supported (lined) glove film; specified by glove film weight			low permeation rate; 0 to 1/2 eyedropper size drops per hour; PVA coating is water soluble; lower detection limit: 0 ppm; permeation rate: <0.9 µg/cm²/min		23	0.25			ASTM F739	313
Rubber, Neoprene (CR)										
Ansell Edmont Neoprene 29-840; unsupported glove film; 0.38 mm thick			low permeation rate; 0 to 1/2 eyedropper size drops per hour; breakthrough time: 240 minutes; permeation rate: <0.9 µg/cm²/min		23				ASTM F739	313
Ansell Edmont Neox; supported (lined) glove film; specified by glove film weight			very good permeation rate; 1 to 5 eyedropper size drops per hour; breakthrough time: 120 minutes; permeation rate: <9 µg/cm²/min		23				ASTM F739	313
Pioneer Industrial Products Stansoil N-44; glove film			no permeation rate at steady state detected; breakthrough time: >480 minutes; lower detection limit: 0.03 ppm; permeation rate: 0 µg/cm²/min						ASTM F739.85	312
Rubber, Nitrile (NBR)										
Ansell Edmont Sol-Vex 37-165; unsupported glove film; 0.54 mm thick			low permeation rate; 0 to 1/2 eyedropper size drops per hour; lower detection limit: 0 ppm; permeation rate: <0.9 µg/cm²/min		23	0.25			ASTM F739	313
Pioneer Industrial Products Stansolv A-14; glove film			no permeation rate at steady state detected; breakthrough time: >480 minutes; lower detection limit: 0.01 ppm; permeation rate: 0 µg/cm²/min						ASTM F739.85	312
TPE, Vinyl										
Ansell Edmont Wet-Wear 550; clothing; PVC stretch outer layer bonded to a lightweight liner			fair permeation rate; 51 to 100 eyedropper size drops per hour; breakthrough time: 6 minutes; permeation rate: <900 µg/cm²/min		23				ASTM F739	313
Ansell Edmont Wet-Wear 800; clothing; nylon netting bonded between two layers of PVC			good permeation rate; 6 to 50 eyedropper size drops per hour; breakthrough time: 60 minutes; permeation rate: <90 µg/cm²/min		23				ASTM F739	313
Pioneer Industrial Products Pylox V-20; glove film			breakthrough time: 11 minutes; permeation rate: 192 µg/cm²/min						ASTM F739.85	312

Material Family	Material Note	Perm. Coefficient (cm³·mm/m²·day·atm)	VTR (g·mm/m²·day)	Non-normalized Data	Penetrant Note	Temp (°C)	Time (days)	RH (%)	Pressure (kPa)	Test Note	Source
FREON TF (continued)											
TPE, Vinyl coated fabric	Ansell Edmont Wet-Wear 700; clothing; PVC coating with Nylon/Polyester lining			fair permeation rate; 51 to 100 eyedropper size drops per hour; breakthrough time: 15 minutes; permeation rate: <900 µg/cm²/min		23				ASTM F739	313
FREON TMC											
Polyvinyl Alcohol	Ansell Edmont PVA, supported (lined) glove film; specified by glove film weight			low permeation rate; 0 to 1/2 eyedropper size drops per hour; PVA coating is water soluble; lower detection limit: 0 ppm; permeation rate: <0.9 µg/cm²/min		23	0.25			ASTM F739	313
GAS ADDITIVE											
Laminar, Nylon/Olefin	DuPont Selar RB/ Polyolefin; 8% Selar RB; barrier prop.; extrusion blow molded; bottles			0.03% penetrant weight loss	Brakleen	50	28				293
Polyethylene, Fluorinated	barrier prop.; bottles			0.06% penetrant weight loss	Brakleen	50	28				293
Polyethylene, HDPE	bottles			10.6% penetrant weight loss	Brakleen	50	28				293
GASOLINE											
Acetal	DuPont Delrin		0.04			23		50			201
HDPE/EVAL Film	1.27/0.015 mm thick; EVAL F inside; HDPE outside		6.4								266
	1.27/0.061 mm thick; EVAL F inside; HDPE outside		1.4								266
Polyethylene, HDPE	1.27 mm thick		25.4								266
Polyvinyl Alcohol	Ansell Edmont PVA, supported (lined) glove film; specified by glove film weight			low permeation rate; 0 to 1/2 eyedropper size drops per hour; permeation rate: <0.9 µg/cm²/min	white	23	0.25			ASTM F739	313
Rubber, Neoprene (CR)	Pioneer Industrial Products Stanzoil N-44; glove film			breakthrough time: 96 minutes; lower detection limit: 0.2 ppm; permeation rate: 96 µg/cm²/min	white	50				ASTM F739.85	312
Rubber, Nitrile (NBR)	Ansell Edmont Sol-Vex 37-165; unsupported glove film; 0.54 mm thick			low permeation rate; 0 to 1/2 eyedropper size drops per hour; lower detection limit: 0 ppm; permeation rate: <0.9 µg/cm²/min	white	23	0.25			ASTM F739	313
	Pioneer Industrial Products Stansolv A-14; glove film			no permeation rate at steady state detected; breakthrough time: >480 minutes; lower detection limit: 0.1 ppm; permeation rate: 0 µg/cm²/min	white	50				ASTM F739.85	312
TPE, Vinyl	Ansell Edmont Wet-Wear 550; clothing; PVC stretch outer layer bonded to a lightweight liner			good permeation rate; 6 to 50 eyedropper size drops per hour; breakthrough time: 6 minutes; permeation rate: <90 µg/cm²/min	white	23				ASTM F739	313
	Ansell Edmont Wet-Wear 600; clothing; nylon netting bonded between two layers of PVC			fair permeation rate; 51 to 100 eyedropper size drops per hour; breakthrough time: 5 minutes; permeation rate: <900 µg/cm²/min	white	23				ASTM F739	313
TPE, Vinyl coated fabric	Ansell Edmont Wet-Wear 700; clothing; PVC coating with Nylon/Polyester lining			fair permeation rate; 51 to 100 eyedropper size drops per hour; breakthrough time: 5 minutes; permeation rate: <900 µg/cm²/min	white	23				ASTM F739	313
HAIR SPRAY											
Acetal	DuPont Delrin	0.31			various formulations	23		50			201
	DuPont Delrin	2.4			various formulations	38					201

HELIUM

Material Family	Material Note	Perm. Coefficient (cm³·mm/m²·day·atm)	VTR (g·mm/m²·day)	Non-normalized Data	Penetrant Note	Temp. (°C)	Time (days)	RH (%)	Pressure (kPa)	Test Note	Source
EVOH	Eval Co. Eval E; barrier prop.; 44% ethylene	2.6				5		0			264
	Eval Co. Eval F; barrier prop.; 32% ethylene	1.06				5		0			264
	Eval Co. Eval H; barrier prop.; 38% ethylene	1.8				5		0			264
	Eval Co. Eval E; barrier prop.; 44% ethylene	9.37				23		0			264
	Eval Co. Eval F; barrier prop.; 32% ethylene	3.7				23		0			264
	Eval Co. Eval H; barrier prop.; 38% ethylene	6.5				23		0			264
	Eval Co. Eval E; barrier prop.; 44% ethylene	14.0				35		0			264
	Eval Co. Eval F; barrier prop.; 32% ethylene	5.4				35		0			264
	Eval Co. Eval H; barrier prop.; 38% ethylene	9.4				35		0			264
Fluoroelastomer, FKM	DuPont Viton; compounded	771				24				one atmosphere @60°C; specimen size: 1 cm² x 1 cm thick	305
	DuPont Viton; compounded	15,034				121				one atmosphere @60°C; specimen size: 1 cm² x 1 cm thick	305
	DuPont Viton; compounded	57,888				204				one atmosphere @60°C; specimen size: 1 cm² x 1 cm thick	305
Fluoropolymer, CTFE	3M Kel-F 81; amorphous form; film	142				25					96
Fluoropolymer, ETFE	Ausimont Hyflon 700; high molecular weight; 4 g/10 min. MFI	591				23				ASTM D1434; activation energy = 6-8 kcal/mole	114
	Ausimont Hyflon 800; low molecular weight; 11 g/10 min. MFI	591				23				ASTM D1434; activation energy = 6-8 kcal/mole	114
	DuPont Tefzel; developmental material, 0.102 mm thick; film	354				25				ASTM D1434	205
Fluoropolymer, PVDF	Solvay Solef; 0.1 mm thick; cast film; film	86.1				23				ASTM D1434	125
Nylon 6	oriented	45.7				35		0			264
Nylon 66	DuPont Zytel 42; low flow; film	59.1				23		50			68
Polyester, PET	oriented	70.9				35		0			264
Polyethylene, HDPE	Hoechst AG Hostalen	152				20				volume at standard temperature and pressure; useable average for all Hostalen grades	94
	Phillips Marlex; film	97.2				23				ASTM D1434	101
	Hoechst AG Hostalen	213				30				volume at standard temperature and pressure; useable average for all Hostalen grades	94
	Hoechst AG Hostalen	466				50				"	94
Polyimide	Ube Upilex R; 0.025 mm thick; film	55.9				30				ASTM D1434	97
Polysulfone	Amoco Udel; transparent, amber tint	772				23		dry		ASTM D1434	15
Polyvinyl Chloride	film	639				25		0			63
Polyvinylidene Chloride	Dow Saran 5253; barrier prop.	10.8				35		0			264
Rubber, Isobutylene	gum vulcanizate			86% relative to Butyl Rubber		25					300
Rubber, Latex (NR)	gum vulcanizate			357% relative to Butyl Rubber		25					300
Rubber, Nitrile (NBR)	61% butadiene/ 39% acrylonitrile; gum vulcanizate			79% relative to Butyl Rubber		25					300

Appendix I - Permeability Sort

HELIUM (continued)

Material Family	Material Note	Perm. Coefficient (cm³·mm/m²·day·atm)	VTR (g·mm/m²·day)	Non-normalized Data	Penetrant Note	Temp. (°C)	Time (days)	RH (%)	Pressure (kPa)	Test Note	Source
Rubber, Nitrile (NBR)	68% butadiene / 32% acrylonitrile; gum vulcanizate			114% relative to Butyl Rubber		25					300
	73% butadiene / 27% acrylonitrile; gum vulcanizate			139% relative to Butyl Rubber		25					300
	80% butadiene / 20% acrylonitrile; gum vulcanizate			196% relative to Butyl Rubber		25					300
Rubber, Styrene Butadiene	hot; gum vulcanizate			264% relative to Butyl Rubber		25					300
TPE, Polyamide	Atochem Pebax 2533; 25 Shore D, 0.12 mm thick; film	1543				23		dry			287
	Atochem Pebax 3533; 35 Shore D, 0.12 mm thick; film	1142				23		dry			287
	Atochem Pebax 4033; 40 Shore D, 0.12 mm thick; film	965				23		dry			287
	Atochem Pebax 5533; 55 Shore D, 0.12 mm thick; film	460				23		dry			287
	Atochem Pebax 6333; 63 Shore D, 0.12 mm thick; film	302				23		dry			287
TPE, Polyester	DuPont Hytrel 4056; 40 Shore D	1356				21.5			34.5		274
	DuPont Hytrel 5556; 55 Shore D	855				21.5			34.5		274
	DuPont Hytrel 7246; 72 Shore D	276				21.5			34.5		274
TPE, Urethane (TPAU)	BASF Elastollan C80A; 80 Shore A	306				20					130
	BASF Elastollan C85A; 85 Shore A	263				20					130
	BASF Elastollan C90A; 90 Shore A	219				20					130
	BASF Elastollan C95A; 95 Shore A	175				20					130
TPE, Urethane (TPEU)	BASF Elastollan 1180A; 80 Shore A	438				20					130
	BASF Elastollan 1185A; 85 Shore A	350				20					130
	BASF Elastollan 1190A; 90 Shore A	175				20					130
	BASF Elastollan 1195A; 95 Shore A	263				20					130

HEPTANE

Material Family	Material Note	Perm. Coefficient (cm³·mm/m²·day·atm)	VTR (g·mm/m²·day)	Non-normalized Data	Penetrant Note	Temp. (°C)	Time (days)	RH (%)	Pressure (kPa)	Test Note	Source
Laminar, Nylon/HDPE	DuPont Selar RB/ HDPE; solvent packaging; 18% Selar RB; barrier prop.; extrusion blow molded; one pint bottle			0.25% penetrant weight loss (100% relative improvement vs. HDPE)	aliphatic	48.9	28				291
Rubber, Neoprene (CR)	Pioneer Industrial Products Stanzoil N-44; glove film			breakthrough time: 124 minutes; lower detection limit: 0.02 ppm; permeation rate: 12 µg/cm²/min						ASTM F739.85	312
Rubber, Nitrile (NBR)	Pioneer Industrial Products Stansolv A-14; glove film			breakthrough time: 2 minutes; lower detection limit: 0.01 ppm; permeation rate: 0.018 µg/cm²/min						ASTM F739.85	312

HEXAMETHYLDISILIZANE

Material Family	Material Note	Perm. Coefficient (cm³·mm/m²·day·atm)	VTR (g·mm/m²·day)	Non-normalized Data	Penetrant Note	Temp. (°C)	Time (days)	RH (%)	Pressure (kPa)	Test Note	Source
Polyvinyl Alcohol	Ansell Edmont PVA; supported (lined) glove film; specified by glove film weight			no permeation detected during a 6 hour test; PVA coating is water soluble; lower detection limit: 0 ppm		23	0.25			ASTM F739	313
Rubber, Latex (NR)	Ansell Edmont Canners and Handlers 392; unsupported glove film; 0.46 mm thick			fair permeation rate; 51 to 100 eyedropper size drops per hour; breakthrough time: 15 minutes; permeation rate: <300 µg/cm²/min		23				ASTM F739	313
Rubber, Neoprene (CR)	Ansell Edmont Neoprene 29-840; unsupported glove film; 0.38 mm thick			breakthrough time: 50 minutes		23				ASTM F739	313
	Ansell Edmont Neox; supported (lined) glove film; specified by glove film weight			breakthrough time: 60 minutes		23				ASTM F739	313
Rubber, Nitrile (NBR)	Ansell Edmont Sol-Vex 37-165; unsupported glove film; 0.54 mm thick			no permeation detected during a 6 hour test; lower detection limit: 0 ppm		23	0.25			ASTM F739	313

HEXANE

Material Family / Material Note	Perm. Coefficient (cm³·mm/m²·day·atm)	VTR (g·mm/m²·day)	Non-normalized Data	Penetrant Note	Temp. (°C)	Time (days)	RH (%)	Pressure (kPa)	Test Note	Source
Acrylonitrile Copolymer, AMA										
barrier prop.; bottles			0.93% penetrant weight loss		50	28				293
Laminar, Nylon/Olefin										
DuPont Selar RB/ Polyolefin; 8% Selar RB; barrier prop.; extrusion blow molded; bottles			0.17% penetrant weight loss		50	28				293
Polyester, PET										
DuPont Mylar film		0.05			40				ASTM E96-80; modified test; permeabilities determined at the partial pressure of the vapor at the test temperature	270
Polyethylene, Fluorinated										
barrier prop.; bottles			0.42% penetrant weight loss		50	28				293
Polyethylene, HDPE										
bottles			32.9% penetrant weight loss		50	28				293
Polyvinyl Alcohol										
Ansell Edmont PVA; supported (lined) glove film; specified by glove film weight			low permeation rate; 0 to 1/2 eyedropper size drops per hour; PVA coating is water soluble; lower detection limit: 0 ppm; permeation rate: <0.9 µg/cm²/min		23	0.25			ASTM F739	313
Polyvinyl Chloride										
bottles			6.06% penetrant weight loss		23	180				293
Rubber, Neoprene (CR)										
Ansell Edmont Neoprene 29-840; unsupported glove film; 0.38 mm thick			fair permeation rate; 51 to 100 eyedropper size drops per hour; breakthrough time: 45 minutes; permeation rate: <900 µg/cm²/min		23				ASTM F739	313
Ansell Edmont Neox; supported (lined) glove film; specified by glove film weight			good permeation rate; 6 to 50 eyedropper size drops per hour; breakthrough time: 90 minutes; permeation rate: <90 µg/cm²/min		23				ASTM F739	313
Pioneer Industrial Products Stanzoil N-44; glove film			breakthrough time: 39 minutes; lower detection limit: 0.08 ppm; permeation rate: 36 µg/cm²/min		23				ASTM F739.85	312
Rubber, Nitrile (NBR)										
Ansell Edmont Sol-Vex 37-165; unsupported glove film; 0.54 mm thick			low permeation rate; 0 to 1/2 eyedropper size drops per hour; lower detection limit: 0 ppm; permeation rate: <0.9 µg/cm²/min		23	0.25			ASTM F739	313
Pioneer Industrial Products Stansolv A-14; glove film			no permeation rate at steady state detected; breakthrough time: >480 minutes; lower detection limit: 0.08 ppm; permeation rate: 0 µg/cm²/min		23				ASTM F739.85	312

HYDRAZINE

Material Family / Material Note	Perm. Coefficient (cm³·mm/m²·day·atm)	VTR (g·mm/m²·day)	Non-normalized Data	Penetrant Note	Temp. (°C)	Time (days)	RH (%)	Pressure (kPa)	Test Note	Source
Rubber, Latex (NR)										
Ansell Edmont Canners and Handlers 392; unsupported glove film; 0.48 mm thick			very good permeation rate; 1 to 5 eyedropper size drops per hour; breakthrough time: 150 minutes; permeation rate: <9 µg/cm²/min	65% conc.	23				ASTM F739	313
Pioneer Industrial Products L-118; glove film			breakthrough time: 218 minutes; lower detection limit: 0.7 ppm; permeation rate: 12 µg/cm²/min	65% conc.	23				ASTM F739.85	312
Rubber, Neoprene (CR)										
Ansell Edmont Neoprene 29-840; unsupported glove film; 0.38 mm thick			no permeation detected during a 6 hour test; lower detection limit: 0 ppm	65% conc.	23	0.25			ASTM F739	313
Ansell Edmont Neox; supported (lined) glove film; specified by glove film weight			no permeation detected during a 6 hour test; lower detection limit: 0 ppm	65% conc.	23	0.25			ASTM F739	313
Pioneer Industrial Products Stanzoil N-44; glove film			no permeation rate at steady state detected; breakthrough time: >480 minutes; lower detection limit: 0.7 ppm; permeation rate: 0 µg/cm²/min		23				ASTM F739.85	312
Rubber, Nitrile (NBR)										
Ansell Edmont Sol-Vex 37-165; unsupported glove film; 0.54 mm thick			no permeation detected during a 6 hour test; lower detection limit: 0 ppm	65% conc.	23	0.25			ASTM F739	313
Pioneer Industrial Products Stansolv A-14; glove film			no permeation rate at steady state detected; breakthrough time: >480 minutes; lower detection limit: 0.7 ppm; permeation rate: 0 µg/cm²/min		23				ASTM F739.85	312
TPE, Vinyl										
Ansell Edmont Monkey Grip; supported (lined) glove film; specified by glove film weight			no permeation detected during a 6 hour test; lower detection limit: 0 ppm	65% conc.	23	0.25			ASTM F739	313
Ansell Edmont Wet-Wear 600; clothing; nylon netting bonded between two layers of PVC			no permeation detected during a 6 hour test; lower detection limit: 0 ppm	65% conc.	23	0.25			ASTM F739	313
Pioneer Industrial Products Pylox V-20; glove film			no permeation rate at steady state detected; breakthrough time: >480 minutes; lower detection limit: 0.7 ppm; permeation rate: 0 µg/cm²/min		23				ASTM F739.85	312
TPE, Vinyl coated fabric										
Ansell Edmont Wet-Wear 700; clothing; PVC coating with Nylon/Polyester lining			no permeation detected during a 6 hour test; lower detection limit: 0 ppm	65% conc.	23	0.25			ASTM F739	313

HYDROCHLORIC ACID

Material Family	Material Note	Perm. Coefficient (cm³·mm/m²·day·atm)	VTR (g·mm/m²·day)	Non-normalized Data	Penetrant Note	Temp. (°C)	Time (days)	RH (%)	Pressure (kPa)	Test Note	Source
Polyphenylene Sulfide	Phillips Ryton; 0.127 mm thick; baked coating; film		0.03		37	23				Die cut samples were fitted to tops of glass bottles by a rubber gasket and a lid, with a surface area of 96.8 cm². Liquids were placed in bottles, gaskets and film put in place, and the lid screwed on. Apparatus was inverted to put liquid in direct contact with film. Weight loss measurements were made at 1 week intervals throughout 4 weeks of conditioning.	102
Rubber, Latex (NR)	Ansell Edmont Canners and Handlers 392; unsupported glove film; 0.48 mm thick			breakthrough time: 290 minutes	concentrated	23				ASTM F739	313
	Ansell Edmont Canners and Handlers 392; unsupported glove film; 0.48 mm thick			no permeation detected during a 6 hour test; lower detection limit: 0 ppm	10% conc.	23	0.25			ASTM F739	313
	Pioneer Industrial Products L-118; glove film			breakthrough time: 211 minutes; lower detection limit: 4 ppm; permeation rate: 1308 µg/cm²/min	37.5% conc.					ASTM F739.85	312
Rubber, Neoprene (CR)	Ansell Edmont Neoprene 29-840; unsupported glove film; 0.38 mm thick			no permeation detected during a 6 hour test; lower detection limit: 0 ppm	concentrated	23	0.25			ASTM F739	313
	Ansell Edmont Neoprene 29-840; unsupported glove film; 0.38 mm thick			no permeation detected during a 6 hour test; lower detection limit: 0 ppm	10% conc.	23	0.25			ASTM F739	313
	Ansell Edmont Neox; supported (lined) glove film; specified by glove film weight			no permeation detected during a 6 hour test; lower detection limit: 0 ppm	concentrated	23	0.25			ASTM F739	313
	Ansell Edmont Neox; supported (lined) glove film; specified by glove film weight			no permeation detected during a 6 hour test; lower detection limit: 0 ppm	10% conc.	23	0.25			ASTM F739	313
	Pioneer Industrial Products Stanzoil N-44; glove film			no permeation rate at steady state detected; breakthrough time: >480 minutes; lower detection limit: 4 ppm; permeation rate: 0 µg/cm²/min	37.5% conc.	23				ASTM F739.85	312
Rubber, Nitrile (NBR)	Ansell Edmont Sol-Vex 37-165; unsupported glove film; 0.54 mm thick			no permeation detected during a 6 hour test; lower detection limit: 0 ppm	concentrated	23	0.25			ASTM F739	313
	Ansell Edmont Sol-Vex 37-165; unsupported glove film; 0.54 mm thick			no permeation detected during a 6 hour test; lower detection limit: 0 ppm	10% conc.	23	0.25			ASTM F739	313
	Pioneer Industrial Products Stansolv A-14; glove film			no permeation rate at steady state detected; breakthrough time: >480 minutes; lower detection limit: 0.4 ppm; permeation rate: 0 µg/cm²/min	37.5% conc.					ASTM F739.85	312
TPE, Vinyl	Ansell Edmont Monkey Grip; supported (lined) glove film; specified by glove film weight			breakthrough time: >300 minutes	concentrated	23				ASTM F739	313
	Ansell Edmont Monkey Grip; supported (lined) glove film; specified by glove film weight			no permeation detected during a 6 hour test; lower detection limit: 0 ppm	10% conc.	23	0.25			ASTM F739	313
	Ansell Edmont Wet-Wear 550; clothing; PVC stretch outer layer bonded to a lightweight liner			no permeation detected during a 6 hour test; lower detection limit: 0 ppm	10% conc.	23	0.25			ASTM F739	313
	Ansell Edmont Wet-Wear 600; clothing; nylon netting bonded between two layers of PVC			breakthrough time: 300 minutes	10% conc.	23				ASTM F739	313
	Pioneer Industrial Products Pylox V-20; glove film			no permeation rate at steady state detected; breakthrough time: >480 minutes; lower detection limit: 4 ppm; permeation rate: 0 µg/cm²/min	37.5% conc.					ASTM F739.85	312
TPE, Vinyl coated fabric	Ansell Edmont Wet-Wear 700; clothing; PVC coating with Nylon/Polyester lining			breakthrough time: 55 minutes	10% conc.	23				ASTM F739	313

HYDROFLUORIC ACID

Material Family	Material Note	Perm. Coefficient (cm³·mm/m²·day·atm)	VTR (g·mm/m²·day)	Non-normalized Data	Penetrant Note	Temp. (°C)	Time (days)	RH (%)	Pressure (kPa)	Test Note	Source
Rubber, Latex (NR)	Ansell Edmont Canners and Handlers 392; unsupported glove film; 0.48 mm thick			breakthrough time: 190 minutes	48% conc.	23				ASTM F739	313

HYDROFLUORIC ACID (continued)

Material Family	Material Note	Perm. Coefficient (cm³ · mm/m² · day · atm)	VTR (g · mm/m² · day)	Non-normalized Data	Penetrant Note	Temp. (°C)	Time (days)	RH (%)	Pressure (kPa)	Test Note	Source
Rubber, Latex (NR)	Pioneer Industrial Products L-11B; glove film			no permeation rate at steady state detected; breakthrough time: >480 minutes; lower detection limit: 1 ppm; permeation rate: 0 µg/cm²/min	48% conc.					ASTM F739.85	312
Rubber, Neoprene (CR)	Ansell Edmont Neoprene 29-840; unsupported glove film; 0.38 mm thick			breakthrough time: 60 minutes	48% conc.	23				ASTM F739	313
	Ansell Edmont Neox; supported (lined) glove film; specified by glove film weight			breakthrough time: 75 minutes	48% conc.	23				ASTM F739	313
	Pioneer Industrial Products Stanzoil N-44; glove film			no permeation rate at steady state detected; breakthrough time: >480 minutes; lower detection limit: 1 ppm; permeation rate: 0 µg/cm²/min	48% conc.					ASTM F739.85	312
Rubber, Nitrile (NBR)	Ansell Edmont Sol-Vex 37-165; unsupported glove film; 0.54 mm thick			breakthrough time: 120 minutes	48% conc.	23				ASTM F739	313
	Pioneer Industrial Products Starsolv A-14; glove film			breakthrough time: 134 minutes; lower detection limit: 0.001 ppm; permeation rate: 30 µg/cm²/min	48% conc.					ASTM F739.85	312
TPE, Vinyl	Ansell Edmont Monkey Grip; supported (lined) glove film; specified by glove film weight			breakthrough time: 40 minutes	48% conc.	23				ASTM F739	313
	Ansell Edmont Wet-Wear 550; clothing; PVC stretch outer layer bonded to a lightweight liner			breakthrough time: 75 minutes	48% conc.	23				ASTM F739	313
	Ansell Edmont Wet-Wear 600; clothing; nylon netting bonded between two layers of PVC			breakthrough time: 90 minutes	48% conc.	23				ASTM F739	313
	Pioneer Industrial Products Pylox V-20; glove film			no permeation rate at steady state detected; breakthrough time: 110 minutes; lower detection limit: 1 ppm; permeation rate: 0 µg/cm²/min	48% conc.					ASTM F739.85	312
TPE, Vinyl coated fabric	Ansell Edmont Wet-Wear 700; clothing; PVC coating with Nylon/Polyester lining			breakthrough time: 5 minutes	48% conc.	23				ASTM F739	313

HYDROGEN

Material Family	Material Note	Perm. Coefficient (cm³ · mm/m² · day · atm)	VTR (g · mm/m² · day)	Non-normalized Data	Penetrant Note	Temp. (°C)	Time (days)	RH (%)	Pressure (kPa)	Test Note	Source
ASA	BASF AG Luran S 757 R; 0.1 mm thick; blown film	507				23				DIN 53380	143
	BASF AG Luran S 776 S; 0.1 mm thick; blown film	507				23				DIN 53380	143
Epoxy		43.3				25				ASTM D1434-63T	121
Fluoropolymer, CTFE	3M Kel-F; 0.01 mm thick	5.6				-15			1724	mass spectrometry and calibrated standard gas leaks; developed by McDonnell Douglas Space Systems Caompany Chemistry Laboratory	306
	3M Kel-F; 0.01 mm thick	5.9				-12			3447	"	306
	3M Kel-F; 0.01 mm thick	5.1				-16			6895	"	306
	3M Kel-F 81; amorphous form; film	21.0				0					96
	3M Kel-F; 0.01 mm thick	35.6				25			1724	mass spectrometry and calibrated standard gas leaks; developed by McDonnell Douglas Space Systems Caompany Chemistry Laboratory	306
	3M Kel-F; 0.01 mm thick	36.2				25			3447	"	306
	3M Kel-F; 0.01 mm thick	36.2				25			6895	"	306
	3M Kel-F 81; amorphous form; film	64.4				25					96
	3M Kel-F 81; amorphous form; film	158				50					96

HYDROGEN (continued)

Material Family	Material Note	Perm. Coefficient (cm³·mm/m²·day·atm)	VTR (g·mm/m²·day)	Non-normalized Data	Penetrant Note	Temp (°C)	Time (days)	RH (%)	Pressure (kPa)	Test Note	Source
Fluoropolymer, CTFE	3M Kel-F; 0.01 mm thick	204				68			1724	mass spectrometry and calibrated standard gas leaks; developed by McDonnell Douglas Space Systems Caompany Chemistry Laboratory	306
	3M Kel-F; 0.01 mm thick	197				67			3447	"	306
	3M Kel-F; 0.01 mm thick	218				70			6895	"	306
Fluoropolymer, ECTFE	Ausimont Halar; 0.02 mm thick	10.4				-22			1724	"	306
	Ausimont Halar; 0.02 mm thick	10.3				-20			3447	"	306
	Ausimont Halar; 0.02 mm thick	10.3				-21			6895	"	306
	Ausimont Halar; 0.02 mm thick	106				25			1724	"	306
	Ausimont Halar; 0.02 mm thick	109				25			3447	"	306
	Ausimont Halar; 0.02 mm thick	108				25			6895	"	306
	Ausimont Halar; 0.02 mm thick	576				66			1724	"	306
	Ausimont Halar; 0.02 mm thick	582				67			3447	"	306
	Ausimont Halar; 0.02 mm thick	590				68			6895	"	306
Fluoropolymer, FEP	DuPont Teflon; 0.05 mm thick	79.3				-15			1724	"	306
	DuPont Teflon; 0.05 mm thick	84.4				-13			3447	"	306
	DuPont Teflon; 0.05 mm thick	76.8				-16			6895	"	306
	DuPont Teflon; 0.05 mm thick	386				25			1724	"	306
	DuPont Teflon; 0.05 mm thick	381				25			3447	"	306
	DuPont Teflon; 0.05 mm thick	385				25			6895	"	306
	DuPont Teflon; 0.05 mm thick	1637				68			1724	"	306
	DuPont Teflon; 0.05 mm thick	1550				67			3447	"	306
	DuPont Teflon; 0.05 mm thick	1576				67			6895	"	306
Fluoropolymer, PVDF	Solvay Solef; 0.1 mm thick; cast film; film	21.3				23				ASTM D1434	125
Fluoropolymer, TFE	DuPont Teflon; 0.03 mm thick	149				-16			1724	mass spectrometry and calibrated standard gas leaks; developed by McDonnell Douglas Space Systems Ceompany Chemistry Laboratory	306
	DuPont Teflon; 0.03 mm thick	143				-17			3447	"	306
	DuPont Teflon; 0.03 mm thick	139				-18			6895	"	306
	DuPont Teflon; 0.05 mm thick; carbon filled	346				-15			1724	"	306
	DuPont Teflon; 0.05 mm thick; carbon filled	395				-11			3447	"	306
	DuPont Teflon; 0.05 mm thick; carbon filled	365				-14			6895	"	306
	DuPont Teflon; 0.03 mm thick	555				25			1724	"	306
	DuPont Teflon; 0.03 mm thick	516				25			3447	"	306
	DuPont Teflon; 0.03 mm thick	520				25			6895	"	306
	DuPont Teflon; 0.05 mm thick; carbon filled	1173				25			1724	"	306
	DuPont Teflon; 0.05 mm thick; carbon filled	1112				25			3447	"	306
	DuPont Teflon; 0.05 mm thick; carbon filled	1077				25			6895	"	306

HYDROGEN (continued)

Material Family	Material Note	Perm. Coefficient (cm³·mm/m²·day·atm)	VTR (g·mm/m²·day)	Non-normalized Data	Penetrant Note	Temp. (°C)	Time (days)	RH (%)	Pressure (kPa)	Test Note	Source
Fluoropolymer, TFE	DuPont Teflon; 0.03 mm thick	1646				68			1724	mass spectrometry and calibrated standard gas leaks; developed by McDonnell Douglas Space Systems Caompany Chemistry Laboratory	306
	DuPont Teflon; 0.03 mm thick	1628				67			3447	"	306
	DuPont Teflon; 0.03 mm thick	1436				63			6895	"	306
	DuPont Teflon; 0.05 mm thick; carbon filled	3090				68			1724	"	306
	DuPont Teflon; 0.05 mm thick; carbon filled	2994				67			3447	"	306
	DuPont Teflon; 0.05 mm thick; carbon filled	2906				65			6895	"	306
Parylene	Union Carbide Parylene C; vapor phase deposition; thin film	43.3				25				ASTM D1434-63T	121
	Union Carbide Parylene D; vapor phase deposition; thin film	94.5				25				ASTM D1434-63T	121
	Union Carbide Parylene N; highly crystalline, high molecular weight, completely linear; vapor phase deposition; thin film	213				25				ASTM D1434-63T	121
Polyester, PET	DuPont Mylar; film	39.4				25				ASTM D1434-72	270
Polyethylene, HDPE	0.03 mm thick	31.9				-15			1724	mass spectrometry and calibrated standard gas leaks; developed by McDonnell Douglas Space Systems Caompany Chemistry Laboratory	306
	0.03 mm thick	30.6				-16			3447	"	306
	0.03 mm thick	27.9				-18			6895	"	306
	Hoechst AG Hostalen	223				20				volume at standard temperature and pressure; useable average for all Hostalen grades	94
	Hoechst AG Hostalen	243				25				volume at standard temperature and pressure; useable average for all Hostalen grades	94
	Phillips Marlex; film	126				23				ASTM D1434	101
	0.03 mm thick	156				25			1724	mass spectrometry and calibrated standard gas leaks; developed by McDonnell Douglas Space Systems Caompany Chemistry Laboratory	306
	0.03 mm thick	154				25			3447	"	306
	0.03 mm thick	161				25			6895	"	306
	Hoechst AG Hostalen	294				30				volume at standard temperature and pressure; useable average for all Hostalen grades	94
	Hoechst AG Hostalen	446				40				"	94
	Hoechst AG Hostalen	679				50				"	94
	0.03 mm thick	761				68			1724	mass spectrometry and calibrated standard gas leaks; developed by McDonnell Douglas Space Systems Caompany Chemistry Laboratory	306
	0.03 mm thick	748				67			3447	"	306

HYDROGEN (continued)

Material Family	Material Note	Perm. Coefficient (cm³·mm/m²·day·atm)	VTR (g·mm/m²·day)	Non-normalized Data	Penetrant Note	Temp. (°C)	Time (days)	RH (%)	Pressure (kPa)	Test Note	Source
Polyethylene, HDPE	0.03 mm thick	740				67			6895	mass spectrometry and calibrated standard gas leaks; developed by McDonnell Douglas Space Systems Caompany Chemistry Laboratory	306
Polyphenylene Sulfide	Phillips Ryton; 0.127 mm thick; baked coating; film	165								ASTM D1434, Method M	102
Polypyrrole	BASF AG Lutamer ES 9567: 0.031-0.037 mm thick, intrinsically conductive, high flexibility; w/ benzenesulfonate anions; film	69.9									182
Polysulfone	Amoco Udel; transparent, amber tint	709				23		dry		ASTM D1434	15
Rubber, Isobutylene	gum vulcanizate			87% relative to Butyl Rubber		25					300
Rubber, Latex (NR)	gum vulcanizate			667% relative to Butyl Rubber		25					300
Rubber, Neoprene (CR)	Neoprene G type; gum vulcanizate			180% relative to Butyl Rubber		25					300
Rubber, Nitrile (NBR)	61% butadiene/ 39% acrylonitrile; gum vulcanizate			100% relative to Butyl Rubber		25					300
	68% butadiene/ 32% acrylonitrile; gum vulcanizate			161% relative to Butyl Rubber		25					300
	73% butadiene/ 27% acrylonitrile; gum vulcanizate			214% relative to Butyl Rubber		25					300
	80% butadiene/ 20% acrylonitrile; gum vulcanizate			341% relative to Butyl Rubber		25					300
Rubber, Polybutadiene	hot emulsion; gum vulcanizate			575% relative to Butyl Rubber		25					300
Rubber, Styrene Butadiene	hot; gum vulcanizate			641% relative to Butyl Rubber		25					300
Silicone	gum vulcanizate	17,716		7150% relative to Butyl Rubber		25					300
TPE, Urethane (TPAU)	BASF Elastollan C80A; 80 Shore A	394				20				ASTM D1434-63T	121
	BASF Elastollan C85A; 85 Shore A	350				20					130
	BASF Elastollan C90A; 90 Shore A	263				20					130
	BASF Elastollan C95A; 95 Shore A	175				20					130
TPE, Urethane (TPEU)	BASF Elastollan 1180A; 80 Shore A	613				20					130
	BASF Elastollan 1185A; 85 Shore A	525				20					130
	BASF Elastollan 1190A; 90 Shore A	438				20					130
	BASF Elastollan 1195A; 95 Shore A	350				20					130

HYDROGEN PEROXIDE

Material Family	Material Note	Perm. Coefficient (cm³·mm/m²·day·atm)	VTR (g·mm/m²·day)	Non-normalized Data	Penetrant Note	Temp. (°C)	Time (days)	RH (%)	Pressure (kPa)	Test Note	Source
Rubber, Latex (NR)	Ansell Edmont Canners and Handlers 392; unsupported glove film; 0.48 mm thick			no permeation detected during a 6 hour test; lower detection limit: 0 ppm	30% conc.	23	0.25			ASTM F739	313
Rubber, Neoprene (CR)	Ansell Edmont Neoprene 29-840; unsupported glove film; 0.38 mm thick			breakthrough time: 5 minutes	30% conc.	23				ASTM F739	313
	Ansell Edmont Neox; supported (lined) glove film; specified by glove film weight			breakthrough time: 7 minutes	30% conc.	23				ASTM F739	313
Rubber, Nitrile (NBR)	Ansell Edmont Sol-Vex 37-165; unsupported glove film; 0.54 mm thick			no permeation detected during a 6 hour test; lower detection limit: 0 ppm	30% conc.	23	0.25			ASTM F739	313
TPE, Vinyl	Ansell Edmont Monkey Grip; supported (lined) glove film; specified by glove film weight			no permeation detected during a 6 hour test; lower detection limit: 0 ppm	30% conc.	23	0.25			ASTM F739	313

HYDROGEN SULFIDE

Material Note	Perm. Coefficient (cm³·mm/m²·day·atm)	VTR (g·mm/m²·day)	Non-normalized Data	Penetrant Note	Temp. (°C)	Time (days)	RH (%)	Pressure (kPa)	Test Note	Source
Fluoropolymer, CTFE										
3M Kel-F 81; amorphous form; film	2.3				50					96
3M Kel-F 81; amorphous form; film	13.1				75					96
Fluoropolymer, PVDF										
Solvay Solef; 0.025 mm thick; cast film; film	1.5				23				ASTM D1434	125
Polyphenylene Sulfide										
Phillips Ryton; 0.127 mm thick; baked coating; film	1.2								ASTM D1434, Method M	102

HYDROQUINONE

Material Note	Perm. Coefficient (cm³·mm/m²·day·atm)	VTR (g·mm/m²·day)	Non-normalized Data	Penetrant Note	Temp. (°C)	Time (days)	RH (%)	Pressure (kPa)	Test Note	Source
Rubber, Latex (NR)										
Ansell Edmont Canners and Handlers 392; unsupported glove film; 0.48 mm thick			low permeation rate; 0 to 1/2 eyedropper size drops per hour; lower detection limit: 0 ppm; permeation rate: <0.9 µg/cm²/min	saturated	23	0.25			ASTM F739	313
Rubber, Neoprene (CR)										
Ansell Edmont Neoprene 29-840; unsupported glove film; 0.38 mm thick			low permeation rate; 0 to 1/2 eyedropper size drops per hour; lower detection limit: 0 ppm; permeation rate: <0.9 µg/cm²/min	saturated	23	0.25			ASTM F739	313
Ansell Edmont Neox; supported (lined) glove film; specified by glove film weight			low permeation rate; 0 to 1/2 eyedropper size drops per hour; lower detection limit: 0 ppm; permeation rate: <0.9 µg/cm²/min	saturated	23	0.25			ASTM F739	313
Rubber, Nitrile (NBR)										
Ansell Edmont Sol-Vex 37-165; unsupported glove film; 0.54 mm thick			low permeation rate; 0 to 1/2 eyedropper size drops per hour; lower detection limit: 0 ppm; permeation rate: <0.9 µg/cm²/min	saturated	23	0.25			ASTM F739	313
TPE, Vinyl										
Ansell Edmont Monkey Grip; supported (lined) glove film; specified by glove film weight			low permeation rate; 0 to 1/2 eyedropper size drops per hour; lower detection limit: 0 ppm; permeation rate: <0.9 µg/cm²/min	saturated	23	0.25			ASTM F739	313

HYDROXYETHYL AMINE (BIS(2-))

Material Note	Perm. Coefficient (cm³·mm/m²·day·atm)	VTR (g·mm/m²·day)	Non-normalized Data	Penetrant Note	Temp. (°C)	Time (days)	RH (%)	Pressure (kPa)	Test Note	Source
Rubber, Latex (NR)										
Pioneer Industrial Products L-118; glove film			no permeation rate at steady state detected; breakthrough time: >480 minutes; lower detection limit: 1.1 ppm; permeation rate: 0 µg/cm²/min						ASTM F739.85	312
Rubber, Neoprene (CR)										
Pioneer Industrial Products Stanzoil N-44; glove film			no permeation rate at steady state detected; breakthrough time: >480 minutes; lower detection limit: 1.1 ppm; permeation rate: 0 µg/cm²/min						ASTM F739.85	312
Rubber, Nitrile (NBR)										
Pioneer Industrial Products Stansolv A-14; glove film			no permeation rate at steady state detected; breakthrough time: >480 minutes; lower detection limit: 1.1 ppm; permeation rate: 0 µg/cm²/min						ASTM F739.85	312
TPE, Vinyl										
Pioneer Industrial Products Pylox V-20; glove film			no permeation rate at steady state detected; breakthrough time: >480 minutes; lower detection limit: 1.1 ppm; permeation rate: 0 µg/cm²/min						ASTM F739.85	312

HCFC-22/ HCFC-124/ HFC-152A

Material Note	Perm. Coefficient (cm³·mm/m²·day·atm)	VTR (g·mm/m²·day)	Non-normalized Data	Penetrant Note	Temp. (°C)	Time (days)	RH (%)	Pressure (kPa)	Test Note	Source
Nylon 6/66										
305 mm long, 1 mm thick, 15.9 mm ID			0.178 lb/ft · yr penetrant loss	air conditioning refrigerant, ternary blend	93				refrigerant at saturated vapor pressure, calculated from permeation coefficient data	275
Rubber, Nitrile (NBR)										
305 mm long, 1 mm thick, 15.9 mm ID			0.938 lb/ft · yr penetrant loss	"	93				"	275
TPE, Polyamide										
DuPont Zytel FN 726; 15.9 mm ID, 305 mm long, 58 Shore D, 1 mm thick			0.086 lb/ft · yr penetrant loss	"	93				"	275

HCFCX-134A

Material Note	Perm. Coefficient (cm³·mm/m²·day·atm)	VTR (g·mm/m²·day)	Non-normalized Data	Penetrant Note	Temp. (°C)	Time (days)	RH (%)	Pressure (kPa)	Test Note	Source
Nylon 6/66										
305 mm long, 1 mm thick, 15.9 mm ID			0.077 lb/ft · yr penetrant loss	air conditioning refrigerant	93				refrigerant at saturated vapor pressure, calculated from permeation coefficient data	275
Rubber, Nitrile (NBR)										
305 mm long, 1 mm thick, 15.9 mm ID			0.56 lb/ft · yr penetrant loss	"	93				"	275

HCFCX-134A (continued)

Material Family	Material Note	Perm. Coefficient (cm³·mm/m²·day·atm)	VTR (g·mm/m²·day)	Non-normalized Data	Penetrant Note	Temp. (°C)	Time (days)	RH (%)	Pressure (kPa)	Test Note	Source
TPE, Polyamide	DuPont Zytel FN 726; 15.9 mm ID, 305 mm long, 58 Shore D, 1 mm thick			0.015 lb/ft · yr penetrant loss	air conditioning refrigerant	93				refrigerant at saturated vapor pressure, calculated from permeation coefficient data	275

ISOAMYL ACETATE

Material Family	Material Note	Perm. Coefficient (cm³·mm/m²·day·atm)	VTR (g·mm/m²·day)	Non-normalized Data	Penetrant Note	Temp. (°C)	Time (days)	RH (%)	Pressure (kPa)	Test Note	Source
Rubber, Neoprene (CR)	Pioneer Industrial Products Stanzoil N-44; glove film			breakthrough time: 30 minutes; permeation rate: 312 µg/cm²/min						ASTM F739.85	312
TPE, Vinyl	Pioneer Industrial Products Pylox V-20; glove film			breakthrough time: 5 minutes; permeation rate: 1602 µg/cm²/min						ASTM F739.85	312

ISOBUTYL ALCOHOL

Material Family	Material Note	Perm. Coefficient (cm³·mm/m²·day·atm)	VTR (g·mm/m²·day)	Non-normalized Data	Penetrant Note	Temp. (°C)	Time (days)	RH (%)	Pressure (kPa)	Test Note	Source
Rubber, Latex (NR)	Ansell Edmont Canners and Handlers 392; unsupported glove film; 0.48 mm thick			very good permeation rate: 1 to 5 eyedropper size drops per hour; breakthrough time: 15 minutes; permeation rate: <9 µg/cm²/min		23				ASTM F739	313
Rubber, Neoprene (CR)	Ansell Edmont Neoprene 29-840; unsupported glove film; 0.38 mm thick			low permeation rate: 0 to 1/2 eyedropper size drops per hour; breakthrough time: 10 minutes; permeation rate: <0.9 µg/cm²/min		23				ASTM F739	313
Rubber, Neoprene (CR)	Ansell Edmont Neox; supported (lined) glove film; specified by glove film weight			low permeation rate: 0 to 1/2 eyedropper size drops per hour; lower detection limit: 0 ppm; permeation rate: <0.9 µg/cm²/min		23	0.25			ASTM F739	313
Rubber, Nitrile (NBR)	Ansell Edmont Sol-Vex 37-165; unsupported glove film; 0.54 mm thick			low permeation rate: 0 to 1/2 eyedropper size drops per hour; lower detection limit: 0 ppm; permeation rate: <0.9 µg/cm²/min		23	0.25			ASTM F739	313

ISOBUTYL ALCOHOL (continued)

Material Family	Material Note	Perm. Coefficient (cm³·mm/m²·day·atm)	VTR (g·mm/m²·day)	Non-normalized Data	Penetrant Note	Temp. (°C)	Time (days)	RH (%)	Pressure (kPa)	Test Note	Source
TPE, Vinyl	Ansell Edmont Monkey Grip; supported (lined) glove film; specified by glove film weight			very good permeation rate: 1 to 5 eyedropper size drops per hour; breakthrough time: 10 minutes; permeation rate: <9 µg/cm²/min		23				ASTM F739	313
TPE, Vinyl	Ansell Edmont Wet-Wear 550; clothing; PVC stretch outer layer bonded to a lightweight liner			very good permeation rate: 1 to 5 eyedropper size drops per hour; breakthrough time: 200 minutes; permeation rate: <9 µg/cm²/min		23				ASTM F739	313
TPE, Vinyl	Ansell Edmont Wet-Wear 600; clothing; nylon netting bonded between two layers of PVC			very good permeation rate: 1 to 5 eyedropper size drops per hour; breakthrough time: 120 minutes; permeation rate: <9 µg/cm²/min		23				ASTM F739	313
TPE, Vinyl coated fabric	Ansell Edmont Wet-Wear 700; clothing; PVC coating with Nylon/Polyester lining			very good permeation rate: 1 to 5 eyedropper size drops per hour; breakthrough time: 7 minutes; permeation rate: <9 µg/cm²/min		23				ASTM F739	313

ISOOCTANE

Material Family	Material Note	Perm. Coefficient (cm³·mm/m²·day·atm)	VTR (g·mm/m²·day)	Non-normalized Data	Penetrant Note	Temp. (°C)	Time (days)	RH (%)	Pressure (kPa)	Test Note	Source
Polyvinyl Alcohol	Ansell Edmont PVA; supported (lined) glove film; specified by glove film weight			low permeation rate: 0 to 1/2 eyedropper size drops per hour; PVA coating is water soluble; lower detection limit: 0 ppm; permeation rate: <0.9 µg/cm²/min		23	0.25			ASTM F739	313
Rubber, Neoprene (CR)	Ansell Edmont Neoprene 29-840; unsupported glove film; 0.38 mm thick			good permeation rate: 6 to 50 eyedropper size drops per hour; breakthrough time: 60 minutes; permeation rate: <90 µg/cm²/min		23				ASTM F739	313
Rubber, Neoprene (CR)	Ansell Edmont Neox; supported (lined) glove film; specified by glove film weight			low permeation rate: 0 to 1/2 eyedropper size drops per hour; breakthrough time: 360 minutes; permeation rate: <0.9 µg/cm²/min		23				ASTM F739	313
Rubber, Nitrile (NBR)	Ansell Edmont Sol-Vex 37-165; unsupported glove film; 0.54 mm thick			low permeation rate: 0 to 1/2 eyedropper size drops per hour; breakthrough time: 360 minutes; permeation rate: <0.9 µg/cm²/min		23				ASTM F739	313

ISOPROPYL ACETATE

Material Family / Material Note	Perm. Coefficient (cm³·mm/m²·day·atm)	VTR (g·mm/m²·day)	Non-normalized Data	Penetrant Note	Temp. (°C)	Time (days)	RH (%)	Pressure (kPa)	Test Note	Source
Acrylonitrile Copolymer, AMA — barrier prop.; bottles			0.19% penetrant weight loss		50	28				293
Laminar, Nylon/Olefin — DuPont Selar RB/ Polyolefin; 8% Selar RB; barrier prop.; extrusion blow molded; bottles			0.03% penetrant weight loss		50	28				293
Polyethylene, Fluorinated — barrier prop.; bottles			0.62% penetrant weight loss		50	28				293
Polyethylene, HDPE — bottles			2.4% penetrant weight loss		50	28				293
Polyvinyl Chloride — bottles			failed		23	180				293

ISOPROPYL ALCOHOL

Material Family / Material Note	Perm. Coefficient (cm³·mm/m²·day·atm)	VTR (g·mm/m²·day)	Non-normalized Data	Penetrant Note	Temp. (°C)	Time (days)	RH (%)	Pressure (kPa)	Test Note	Source
Rubber, Latex (NR) — Ansell Edmont Canners and Handlers 392; unsupported glove film; 0.46 mm thick			very good permeation rate: 1 to 5 eyedropper size drops per hour; breakthrough time: 200 minutes; permeation rate: <9 µg/cm²/min		23				ASTM F739	313
Pioneer Industrial Products L-118; glove film			no permeation rate at steady state detected; breakthrough time: >60 minutes; permeation rate: 0 µg/cm²/min	isopropanol; IPA	23				ASTM F739.85	312
Rubber, Neoprene (CR) — Ansell Edmont Neoprene 29-840; unsupported glove film; 0.38 mm thick			low permeation rate; 0 to 1/2 eyedropper size drops per hour; lower detection limit: 0.9 µg/cm²/min		23	0.25			ASTM F739	313
Ansell Edmont Neox; supported (lined) glove film; specified by glove film weight			low permeation rate; 0 to 1/2 eyedropper size drops per hour; lower detection limit: <0.9 µg/cm²/min		23	0.25			ASTM F739	313
Pioneer Industrial Products Stanzoil N-44; glove film			no permeation rate at steady state detected; breakthrough time: >60 minutes; lower detection limit: 0.0006 ppm; permeation rate: 0 µg/cm²/min	isopropanol; IPA	23				ASTM F739.85	312
Rubber, Nitrile (NBR) — Ansell Edmont Sol-Vex 37-165; unsupported glove film; 0.54 mm thick			low permeation rate; 0 to 1/2 eyedropper size drops per hour; lower detection limit: 0 ppm; permeation rate: <0.9 µg/cm²/min		23	0.25			ASTM F739	313
Pioneer Industrial Products Stansolv A-14; glove film			no permeation rate at steady state detected; breakthrough time: >480 minutes; lower detection limit: 0.05 ppm; permeation rate: 0 µg/cm²/min	isopropanol; IPA	23				ASTM F739.85	312
TPE, Vinyl — Ansell Edmont Monkey Grip; supported (lined) glove film; specified by glove film weight			low permeation rate; 0 to 1/2 eyedropper size drops per hour; breakthrough time: 150 minutes; permeation rate: <0.9 µg/cm²/min		23				ASTM F739	313
Pioneer Industrial Products Pylox V-20; glove film			no permeation rate at steady state detected; breakthrough time: 208 minutes; permeation rate: 0 µg/cm²/min	isopropanol; IPA	23				ASTM F739.85	312

ISOPROPYL BENZENE

Material Family / Material Note	Perm. Coefficient (cm³·mm/m²·day·atm)	VTR (g·mm/m²·day)	Non-normalized Data	Penetrant Note	Temp. (°C)	Time (days)	RH (%)	Pressure (kPa)	Test Note	Source
Rubber, Neoprene (CR) — Pioneer Industrial Products Stanzoil N-44; glove film			breakthrough time: 41 minutes; lower detection limit: 0.4 ppm; permeation rate: 216 µg/cm²/min						ASTM F739.85	312
Rubber, Nitrile (NBR) — Pioneer Industrial Products Stansolv A-14; glove film			breakthrough time: 271 minutes; lower detection limit: 0.03 ppm; permeation rate: 48 µg/cm²/min						ASTM F739.85	312

KEROSINE

Material Family / Material Note	Perm. Coefficient (cm³·mm/m²·day·atm)	VTR (g·mm/m²·day)	Non-normalized Data	Penetrant Note	Temp. (°C)	Time (days)	RH (%)	Pressure (kPa)	Test Note	Source
EVOH — Eval Co. Eval E: 0.02 mm thick; 44% ethylene; film		<0.0009			20		65			266
Eval Co. Eval E; 0.032 mm thick; 44% ethylene; film		<0.0015			20		65			266
Eval Co. Eval EF-E; barrier prop.; 44% ethylene; film		0.00098			20					265
Eval Co. Eval EF-F; barrier prop.; 32% ethylene; film		>0.0004			20					265
Eval Co. Eval EF-XL; 0.015 mm thick; biaxially oriented; film		<0.0007			20		65			266

KEROSINE (continued)

Material Family	Material Note	Perm. Coefficient (cm³·mm/m²·day·atm)	VTR (g·mm/m²·day)	Non-normalized Data	Penetrant Note	Temp. (°C)	Time (days)	RH (%)	Pressure (kPa)	Test Note	Source
EVOH	Eval Co. Eval EF-XL; barrier prop.; biaxially oriented; film		0.0004			20					265
	Eval Co. Eval F: 0.02 mm thick; 32% ethylene; film		<0.0009			20		65			266
	Eval Co. Eval F: 0.032 mm thick; 32% ethylene; film		<0.0015			20		65			266
Laminar, Nylon/Olefin	DuPont Selar RB/ Polyolefin; 8% Selar RB; barrier prop.; extrusion blow molded; bottles			0.01% penetrant weight loss		50	28				293
LDPE/ EVAL Film	0.06/0.015 mm thick; EVAL EF-F inside; LDPE outside; film		<0.0035			20		65			266
	0.06/0.015 mm thick; EVAL EF-XL inside; LDPE outside; film		<0.0035			20		65			266
	0.06/0.025 mm thick; EVAL EF-E inside; LDPE outside; film		<0.004			20		65			266
Nylon	0.0254 mm thick, oriented; film		0.01			20		65			266
	oriented, 0.015 mm thick; PVDC coated; film		<0.0007			20		65			266
Nylon 66	DuPont Zytel 42; low flow; 2.54 mm thick; bottles		0.08								68
Polyester, PET	0.0254 mm thick; film		0.01			20		65			266
Polyethylene, Fluorinated	barrier prop.; bottles			0.04% penetrant weight loss		50	28				293
Polyethylene, HDPE	bottles			2.3% penetrant weight loss		50	28				293
Polyethylene, LDPE	0.051 mm thick; film		3.9			20		65			266
	film		3.8			20					265
Polypropylene	0.02 mm thick, oriented; film		1.1			20		65			266
	biaxially oriented; film		1.1			20					265
Polyvinyl Alcohol	Ansell Edmont PVA; supported (lined) glove film; specified by glove film weight			low permeation rate; 0 to 1/2 eyedropper size drops per hour; PVA coating is water soluble; lower detection limit: 0 ppm; permeation rate: <0.9 µg/cm²/min		23	0.25			ASTM F739	313
Rubber, Neoprene (CR)	Ansell Edmont Neoprene 29-840; unsupported glove film; 0.38 mm thick			low permeation rate; 0 to 1/2 eyedropper size drops per hour; breakthrough time: >360 minutes; permeation rate: <0.9 µg/cm²/min		23				ASTM F739	313
	Ansell Edmont Neox; supported (lined) glove film; specified by glove film weight			low permeation rate; 0 to 1/2 eyedropper size drops per hour; lower detection limit: 0 ppm; permeation rate: <0.9 µg/cm²/min		23	0.25			ASTM F739	313
	Pioneer Industrial Products Stanzoil N-44; glove film			no permeation rate at steady state detected; breakthrough time: >480 minutes; lower detection limit: 5.0e-05 ppm; permeation rate: 0 µg/cm²/min		23				ASTM F739.85	312
Rubber, Nitrile (NBR)	Ansell Edmont Sol-Vex 37-165; unsupported glove film; 0.54 mm thick			low permeation rate; 0 to 1/2 eyedropper size drops per hour; lower detection limit: 0 ppm; permeation rate: <0.9 µg/cm²/min		23	0.25			ASTM F739	313
	Pioneer Industrial Products Stansolv A-14; glove film			no permeation rate at steady state detected; breakthrough time: >480 minutes; lower detection limit: 0.007 ppm; permeation rate: 0 µg/cm²/min		23				ASTM F739.85	312
TPE, Vinyl	Ansell Edmont Monkey Grip; supported (lined) glove film; specified by glove film weight			low permeation rate; 0 to 1/2 eyedropper size drops per hour; breakthrough time: >360 minutes; permeation rate: <0.9 µg/cm²/min		23				ASTM F739	313
	Ansell Edmont Wet-Wear 550; clothing; PVC stretch outer layer bonded to a lightweight liner			very good permeation rate; 1 to 5 eyedropper size drops per hour; breakthrough time: 10 minutes; permeation rate: <9 µg/cm²/min		23				ASTM F739	313
	Ansell Edmont Wet-Wear 600; clothing; nylon netting bonded between two layers of PVC			good permeation rate; 6 to 50 eyedropper size drops per hour; breakthrough time: 180 minutes; permeation rate: <90 µg/cm²/min		23				ASTM F739	313
TPE, Vinyl coated fabric	Ansell Edmont Wet-Wear 700; clothing; PVC coating with Nylon/Polyester lining			good permeation rate; 6 to 50 eyedropper size drops per hour; breakthrough time: 75 minutes; permeation rate: <90 µg/cm²/min		23				ASTM F739	313

Material Family	Material Note	Permeability Data — Perm. Coefficient (cm³·mm/m²·day·atm)	VTR (g·mm/m²·day)	Non-normalized Data	Penetrant Note	Temp. (°C)	Time (days)	RH (%)	Pressure (kPa)	Test Note	Source

LACTIC ACID

Material Family	Material Note			Non-normalized Data	Penetrant Note	Temp.	Time			Test Note	Source
Polyvinyl Alcohol	Ansell Edmont PVA; supported (lined) glove film; specified by glove film weight			low permeation rate; 0 to 1/2 eyedropper size drops per hour; PVA coating is water soluble; lower detection limit: 0 ppm; permeation rate: <0.9 µg/cm²/min	85% conc.	23	0.25			ASTM F739	313
Rubber, Latex (NR)	Ansell Edmont Canners and Handlers 392; unsupported glove film; 0.48 mm thick			no permeation detected during a 6 hour test; lower detection limit: 0 ppm	85% conc.	23	0.25			ASTM F739	313
Rubber, Neoprene (CR)	Ansell Edmont Neoprene 29-840; unsupported glove film; 0.38 mm thick			low permeation rate; 0 to 1/2 eyedropper size drops per hour; lower detection limit: 0 ppm; permeation rate: <0.9 µg/cm²/min	85% conc.	23	0.25			ASTM F739	313
	Ansell Edmont Neox; supported (lined) glove film; specified by glove film weight			low permeation rate; 0 to 1/2 eyedropper size drops per hour; lower detection limit: 0 ppm; permeation rate: <0.9 µg/cm²/min	85% conc.	23	0.25			ASTM F739	313
Rubber, Nitrile (NBR)	Ansell Edmont Sol-Vex 37-165; unsupported glove film; 0.54 mm thick			low permeation rate; 0 to 1/2 eyedropper size drops per hour; lower detection limit: 0 ppm; permeation rate: <0.9 µg/cm²/min	85% conc.	23	0.25			ASTM F739	313
TPE, Vinyl	Ansell Edmont Monkey Grip; supported (lined) glove film; specified by glove film weight			low permeation rate; 0 to 1/2 eyedropper size drops per hour; lower detection limit: 0 ppm; permeation rate: <0.9 µg/cm²/min	85% conc.	23	0.25			ASTM F739	313

LAURIC ACID

Material Family	Material Note			Non-normalized Data	Penetrant Note	Temp.	Time			Test Note	Source
Rubber, Latex (NR)	Ansell Edmont Canners and Handlers 392; unsupported glove film; 0.48 mm thick			no permeation detected during a 6 hour test; lower detection limit: 0 ppm	with ethylene oxide; 36% conc.	23	0.25			ASTM F739	313
Rubber, Neoprene (CR)	Ansell Edmont Neoprene 29-840; unsupported glove film; 0.38 mm thick			no permeation detected during a 6 hour test; lower detection limit: 0 ppm	with ethylene oxide; 36% conc.	23	0.25			ASTM F739	313
	Ansell Edmont Neox; supported (lined) glove film; specified by glove film weight			no permeation detected during a 6 hour test; lower detection limit: 0 ppm	with ethylene oxide; 36% conc.	23	0.25			ASTM F739	313
Rubber, Nitrile (NBR)	Ansell Edmont Sol-Vex 37-165; unsupported glove film; 0.54 mm thick			no permeation detected during a 6 hour test; lower detection limit: 0 ppm	with ethylene oxide; 36% conc.	23	0.25			ASTM F739	313
TPE, Vinyl	Ansell Edmont Monkey Grip; supported (lined) glove film; specified by glove film weight			breakthrough time: 15 minutes	with ethylene oxide; 36% conc.	23					313

MALEIC ACID

Material Family	Material Note			Non-normalized Data	Penetrant Note	Temp.	Time			Test Note	Source
Rubber, Latex (NR)	Ansell Edmont Canners and Handlers 392; unsupported glove film; 0.48 mm thick			no permeation detected during a 6 hour test; lower detection limit: 0 ppm	saturated	23	0.25			ASTM F739	313
Rubber, Neoprene (CR)	Ansell Edmont Neoprene 29-840; unsupported glove film; 0.38 mm thick			no permeation detected during a 6 hour test; lower detection limit: 0 ppm	saturated	23	0.25			ASTM F739	313
	Ansell Edmont Neox; supported (lined) glove film; specified by glove film weight			no permeation detected during a 6 hour test; lower detection limit: 0 ppm	saturated	23	0.25			ASTM F739	313
Rubber, Nitrile (NBR)	Ansell Edmont Sol-Vex 37-165; unsupported glove film; 0.54 mm thick			no permeation detected during a 6 hour test; lower detection limit: 0 ppm	saturated	23	0.25			ASTM F739	313
TPE, Vinyl	Ansell Edmont Monkey Grip; supported (lined) glove film; specified by glove film weight			no permeation detected during a 6 hour test; lower detection limit: 0 ppm	saturated	23	0.25			ASTM F739	313
	Ansell Edmont Wet-Wear 550; clothing; PVC stretch outer layer bonded to a lightweight liner			no permeation detected during a 6 hour test; lower detection limit: 0 ppm	saturated	23	0.25			ASTM F739	313
	Ansell Edmont Wet-Wear 600; clothing; nylon netting bonded between two layers of PVC			no permeation detected during a 6 hour test; lower detection limit: 0 ppm	saturated	23	0.25			ASTM F739	313
TPE, Vinyl coated fabric	Ansell Edmont Wet-Wear 700; clothing; PVC coating with Nylon/Polyester lining			no permeation detected during a 6 hour test; lower detection limit: 0 ppm	saturated	23	0.25			ASTM F739	313

METHANE

Material Family	Material Note	Perm. Coefficient (cm³·mm/m²·day·atm)	VTR (g·mm/m²·day)	Non-normalized Data	Penetrant Note	Temp. (°C)	Time (days)	RH (%)	Pressure (kPa)	Test Note	Source
ASA	BASF AG Luran S 757 R; 0.1 mm thick; blown film	10.1				23				DIN 53380	143
	BASF AG Luran S 776 S; 0.1 mm thick; blown film	11.1				23				DIN 53380	143
Fluoropolymer, ETFE	Ausimont Hylon 700; high molecular weight; 4 g/10 min. MFI	7.9				23				ASTM D1434; activation energy = 6-8 kcal/mole	114
	Ausimont Hylon 800; low molecular weight; 11 g/10 min. MFI	7.9				23				ASTM D1434; activation energy = 6-8 kcal/mole	114
Polyethylene, HDPE	Hoechst AG Hostalen	56.7				20				volume at standard temperature and pressure; useable average for all Hostalen grades	94
Polypyrrole	BASF AG Lutamer ES 9567; 0.031-0.037 mm thick, intrinsically conductive, high flexibility; w/ benzenesulfonate anions; film	0.41									182
Polysulfone	Amoco Udel; transparent, amber tint	14.8				23		dry		ASTM D1434	15
TPE, Urethane (TPAU)	BASF Elastollan C90A; 80 Shore A	96.3				20					130
	BASF Elastollan C85A; 85 Shore A	52.5				20					130
	BASF Elastollan C90A; 90 Shore A	35.0				20					130
	BASF Elastollan C95A; 95 Shore A	17.5				20					130
TPE, Urethane (TPEU)	BASF Elastollan 1180A; 80 Shore A	158				20					130
	BASF Elastollan 1185A; 85 Shore A	123				20					130
	BASF Elastollan 1190A; 90 Shore A	78.8				20					130
	BASF Elastollan 1195A; 95 Shore A	43.8				20					130

METHANE DICHLORIDE

Material Family	Material Note	Perm. Coefficient	VTR	Non-normalized Data	Penetrant Note	Temp. (°C)	Time (days)	RH (%)	Pressure (kPa)	Test Note	Source
Rubber, Neoprene (CR)	Pioneer Industrial Products Stanzoil N-44; glove film			breakthrough time: 6 minutes; permeation rate: 1434 µg/cm²/min						ASTM F739.85	312

METHYL ALCOHOL

Material Family	Material Note	Perm. Coefficient	VTR	Non-normalized Data	Penetrant Note	Temp. (°C)	Time (days)	RH (%)	Pressure (kPa)	Test Note	Source
Laminar, EVOH/Olefin	DuPont Selar RB/Polyolefin; 15% Selar RB; barrier prop., laminar technology; extrusion blow molded; 1 L bottle			0.74% penetrant weight loss	with 25% xylene	23	180				293
	DuPont Selar RB/Polyolefin; 15% Selar RB; barrier prop., laminar technology; extrusion blow molded; 1 L bottle			0.3% penetrant weight loss		23	180				293
Laminar, Nylon/HDPE	DuPont Selar RB/HDPE; solvent packaging; 18% Selar RB; barrier prop.; extrusion blow molded; one pint bottle			0.75% penetrant weight loss (1% relative improvement vs. HDPE)	oxygen containing	48.9	28				291
Laminar, Nylon/Olefin	DuPont Selar RB/Polyolefin; 8% Selar RB; barrier prop., laminar technology; extrusion blow molded; 1 L bottle			3.56% penetrant weight loss	with 25% xylene	23	180				293
	DuPont Selar RB/Polyolefin; 8% Selar RB; barrier prop., laminar technology; extrusion blow molded; 1 L bottle			0.28% penetrant weight loss		23	180				293
Polyethylene, HDPE	DuPont; 1 L bottle			4.90% penetrant weight loss	with 25% xylene	23	180				293
	DuPont; 1 L bottle			0.29% penetrant weight loss		23	180				293
Polyethylene, LDPE	Dow; film		2.4 - 3.2			24					250

METHYL ALCOHOL (continued)

Material Family	Material Note	Perm. Coefficient (cm³·mm/m²·day·atm)	VTR (g·mm/m²·day)	Non-normalized Data	Per- metrant Note	Temp. (°C)	Time (days)	RH (%)	Pressure (kPa)	Test Note	Source
Polyphenylene Sulfide	Phillips Ryton; 0.127 mm thick; baked coating; film		0.12			23				Die cut samples were fitted to tops of glass bottles by a rubber gasket and a lid, with a surface area of 96.8 cm²; Liquids were placed in bottles, gaskets and film put in place, and the lid screwed on. Apparatus was inverted to put liquid in direct contact with film. Weight loss measurements were made at 1 week intervals throughout 4 weeks of conditioning.	102
Polystyrene	Dow Styron; film		0.39 - 2.4			24					250
Rubber, Latex (NR)	Ansell Canners and Handlers 392; unsupported glove film; 0.48 mm thick			very good permeation rate; 1 to 5 eyedropper size drops per hour; breakthrough time: 200 minutes; permeation rate: <9 µg/cm²/min		23				ASTM F739	313
	Pioneer Industrial Products L-118; glove film			no permeation rate at steady state detected; breakthrough time: >60 minutes; permeation rate: 0 µg/cm²/min	methanol					ASTM F739.85	312
Rubber, Neoprene (CR)	Ansell Edmont Neoprene 29-840; unsupported glove film; 0.38 mm thick			low permeation rate; 0 to 1/2 eyedropper size drops per hour; breakthrough time: 60 minutes; permeation rate: <0.9 µg/cm²/min		23				ASTM F739	313
	Ansell Edmont Neox; supported (lined) glove film; specified by glove film weight			low permeation rate; 0 to 1/2 eyedropper size drops per hour; breakthrough time: 15 minutes; permeation rate: <0.9 µg/cm²/min		23				ASTM F739	313
	Pioneer Industrial Products Stanzoil N-44; glove film			no permeation rate at steady state detected; breakthrough time: >60 minutes; permeation rate: 0 µg/cm²/min	methanol					ASTM F739.85	312
Rubber, Nitrile (NBR)	Ansell Edmont Sol-Vex 37-165; unsupported glove film; 0.54 mm thick			fair permeation rate; 51 to 100 eyedropper size drops per hour; breakthrough time: 11 minutes; permeation rate: <900 µg/cm²/min		23				ASTM F739	313
	Pioneer Industrial Products Starsolv A-14; glove film			breakthrough time: 118 minutes; lower detection limit: 0.08 ppm; permeation rate: 18 µg/cm²/min	methanol					ASTM F739.85	312
SAN	Dow Tyril; film			sample failed		24					250
TPE, Vinyl	Ansell Edmont Monkey Grip; supported (lined) glove film; specified by glove film weight			good permeation rate; 6 to 50 eyedropper size drops per hour; breakthrough time: 45 minutes; permeation rate: <90 µg/cm²/min		23				ASTM F739	313
	Ansell Edmont Wet-Wear 550; clothing; PVC stretch outer layer bonded to a lightweight liner			good permeation rate; 6 to 50 eyedropper size drops per hour; breakthrough time: 5 minutes; permeation rate: <90 µg/cm²/min		23				ASTM F739	313
	Ansell Edmont Wet-Wear 600; clothing; nylon netting bonded between two layers of PVC			good permeation rate; 6 to 50 eyedropper size drops per hour; breakthrough time: 30 minutes; permeation rate: <90 µg/cm²/min		23				ASTM F739	313
	Pioneer Industrial Products Pylox V-20; glove film			breakthrough time: 3 minutes; permeation rate: 18 µg/cm²/min	methanol					ASTM F739.85	312
TPE, Vinyl coated fabric	Ansell Edmont Wet-Wear 700; clothing; PVC coating with Nylon/Polyester lining			very good permeation rate; 1 to 5 eyedropper size drops per hour; breakthrough time: 200 minutes; permeation rate: <9 µg/cm²/min		23				ASTM F739	313

METHYL CELLOSOLVE

Material Family	Material Note	Perm. Coefficient (cm³·mm/m²·day·atm)	VTR (g·mm/m²·day)	Non-normalized Data	Per- metrant Note	Temp. (°C)	Time (days)	RH (%)	Pressure (kPa)	Test Note	Source
Polyvinyl Alcohol	Ansell Edmont PVA; supported (lined) glove film; specified by glove film weight			good permeation rate; 6 to 50 eyedropper size drops per hour; PVA coating is water soluble; breakthrough time: 30 minutes; permeation rate: <90 µg/cm²/min		23				ASTM F739	313
Rubber, Latex (NR)	Ansell Edmont Canners and Handlers 392; unsupported glove film; 0.48 mm thick			very good permeation rate; 1 to 5 eyedropper size drops per hour; breakthrough time: 200 minutes; permeation rate: <9 µg/cm²/min		23				ASTM F739	313

METHYL CELLOSOLVE (continued)

Material Family / Material Note	Perm. Coefficient (cm³·mm/m²·day·atm)	VTR (g·mm/m²·day)	Non-normalized Data	Penetrant Note	Temp. (°C)	Time (days)	RH (%)	Pressure (kPa)	Test Note	Source
Rubber, Neoprene (CR)										
Ansell Edmont Neoprene 29-840; unsupported glove film; 0.38 mm thick			good permeation rate: 6 to 50 eyedropper size drops per hour; breakthrough time: 25 minutes; permeation rate: <90 µg/cm²/min		23				ASTM F739	313
Ansell Edmont Neox; supported (lined) glove film; specified by glove film weight			very good permeation rate; 1 to 5 eyedropper size drops per hour; breakthrough time: 70 minutes; permeation rate: <9 µg/cm²/min		23				ASTM F739	313
Rubber, Nitrile (NBR)										
Ansell Edmont Sol-Vex 37-165; unsupported glove film; 0.54 mm thick			good permeation rate: 6 to 50 eyedropper size drops per hour; breakthrough time: 11 minutes; permeation rate: <90 µg/cm²/min		23				ASTM F739	313

METHYL CHLOROFORM

Material Family / Material Note	Perm. Coefficient (cm³·mm/m²·day·atm)	VTR (g·mm/m²·day)	Non-normalized Data	Penetrant Note	Temp. (°C)	Time (days)	RH (%)	Pressure (kPa)	Test Note	Source
Rubber, Neoprene (CR)										
Pioneer Industrial Products Stanzoil N-44; glove film			breakthrough time: 27 minutes; permeation rate: 1182 µg/cm²/min						ASTM F739.65	312
Rubber, Nitrile (NBR)										
Pioneer Industrial Products Stansolv A-14; glove film			breakthrough time: 131 minutes; lower detection limit: 0.05 ppm; permeation rate: 264 µg/cm²/min						ASTM F739.65	312

METHYL CYANIDE

Material Family / Material Note	Perm. Coefficient (cm³·mm/m²·day·atm)	VTR (g·mm/m²·day)	Non-normalized Data	Penetrant Note	Temp. (°C)	Time (days)	RH (%)	Pressure (kPa)	Test Note	Source
Rubber, Neoprene (CR)										
Pioneer Industrial Products Stanzoil N-44; glove film			breakthrough time: 40 minutes; permeation rate: 42 µg/cm²/min						ASTM F739.85	312

METHYL ETHYL KETONE

Material Family / Material Note	Perm. Coefficient (cm³·mm/m²·day·atm)	VTR (g·mm/m²·day)	Non-normalized Data	Penetrant Note	Temp. (°C)	Time (days)	RH (%)	Pressure (kPa)	Test Note	Source
Acrylonitrile Copolymer, AMA										
barrier prop.; bottles			failed		50	28				293
EVOH										
Eval Co. Eval E; 0.02 mm thick; 44% ethylene; film		0.1			20		65			266
Eval Co. Eval E; 0.032 mm thick; 44% ethylene; film		0.01			20		65			266
Eval Co. Eval EF-XL; 0.015 mm thick, biaxially oriented; film		0.0047			20		65			266
Eval Co. Eval F; 0.02 mm thick; 32% ethylene; film		0.08			20		65			266
Eval Co. Eval F; 0.032 mm thick; 32% ethylene; film		0.01			20		65			266
Laminar, Nylon/HDPE										
DuPont Selar RB/ HDPE; solvent packaging; 18% Selar RB; barrier prop.; extrusion blow molded; one pint bottle			0.65% penetrant weight loss (7% relative improvement vs. HDPE)	oxygen containing	48.9	28				291
Laminar, Nylon/Olefin										
DuPont Selar RB/ Polyolefin; 8% Selar RB; barrier prop.; extrusion blow molded; bottles			0.97% penetrant weight loss		50	28				293
LDPE/ EVAL Film										
0.06/ 0.015 mm thick; EVAL EF-F inside; LDPE outside; film		0.01			20		65			266
0.06/ 0.015 mm thick; EVAL EF-XL inside; LDPE outside; film		0.0035			20		65			266
0.06/ 0.025 mm thick; EVAL EF-E inside; LDPE outside; film		0.01			20		65			266
Nylon										
0.0254 mm thick, oriented; film		0.07			20		65			266
Polyester, PET										
oriented, 0.015 mm thick; PVDC coated; film		0.02			20		65			266
0.0254 mm thick; film		0.04			20		65			266
Polyethylene, Fluorinated										
barrier prop.; bottles			2.7% penetrant weight loss		50	28				293
Polyethylene, HDPE										
bottles			2.8% penetrant weight loss		50	28				293

METHYL ETHYL KETONE (continued)

Material Family	Material Note	Perm. Coefficient (cm³·mm/m²·day·atm)	VTR (g·mm/m²·day)	Non-normalized Data	Penetrant Note	Temp. (°C)	Time (days)	RH (%)	Pressure (kPa)	Test Note	Source
Polyethylene, LDPE	0.051 mm thick; film		3.8			20		65			266
Polypropylene	0.02 mm thick; oriented; film		0.24			20		65			266
Polyvinyl Alcohol	Ansell Edmont PVA; supported (lined) glove film; specified by glove film weight			very good permeation rate; 1 to 5 eyedropper size drops per hour; PVA coating is water soluble; breakthrough time: 90 minutes; permeation rate: <9 µg/cm²/min	MEK	23				ASTM F739	313
Polyvinyl Chloride	bottles			failed		23	180				293
Rubber, Latex (NR)	Ansell Edmont Canners and Handlers 392; unsupported glove film; 0.48 mm thick			fair permeation rate; 51 to 100 eyedropper size drops per hour; breakthrough time: 5 minutes; permeation rate: <300 µg/cm²/min	MEK	23				ASTM F739	313
Rubber, Neoprene (CR)	Pioneer Industrial Products L-118; glove film			breakthrough time: 6 minutes; permeation rate: 522 µg/cm²/min	MEK					ASTM F739.85	312
	Pioneer Industrial Products Stanzoil N-44; glove film			breakthrough time: 22 minutes; permeation rate: 930 µg/cm²/min	MEK					ASTM F739.85	312
Rubber, Nitrile (NBR)	Pioneer Industrial Products Stansolv A-14; glove film			breakthrough time: 6 minutes; permeation rate: 522 µg/cm²/min	MEK					ASTM F739.85	312
TPE, Vinyl	Pioneer Industrial Products Pylox V-20; glove film			permeation rate too large to measure; breakthrough time: 1 minutes	MEK					ASTM F739.85	312

METHYL GLYCOL ETHER

Material Family	Material Note	Perm. Coefficient (cm³·mm/m²·day·atm)	VTR (g·mm/m²·day)	Non-normalized Data	Penetrant Note	Temp. (°C)	Time (days)	RH (%)	Pressure (kPa)	Test Note	Source
Polyvinyl Alcohol	Ansell Edmont PVA; supported (lined) glove film; specified by glove film weight			good permeation rate; 6 to 50 eyedropper size drops per hour; PVA coating is water soluble; breakthrough time: 30 minutes; permeation rate: <90 µg/cm²/min		23				ASTM F739	313
Rubber, Latex (NR)	Ansell Edmont Canners and Handlers 392; unsupported glove film; 0.48 mm thick			very good permeation rate; 1 to 5 eyedropper size drops per hour; breakthrough time: 200 minutes; permeation rate: <9 µg/cm²/min		23				ASTM F739	313
Rubber, Neoprene (CR)	Ansell Edmont Neoprene 29-840; unsupported glove film; 0.38 mm thick			good permeation rate; 6 to 50 eyedropper size drops per hour; breakthrough time: 25 minutes; permeation rate: <90 µg/cm²/min		23				ASTM F739	313
	Ansell Edmont Neox; supported (lined) glove film; specified by glove film weight			very good permeation rate; 1 to 5 eyedropper size drops per hour; breakthrough time: 70 minutes; permeation rate: <9 µg/cm²/min		23				ASTM F739	313
Rubber, Nitrile (NBR)	Ansell Edmont Sol-Vex 37-165; unsupported glove film; 0.54 mm thick			good permeation rate; 6 to 50 eyedropper size drops per hour; breakthrough time: 11 minutes; permeation rate: <90 µg/cm²/min		23				ASTM F739	313

METHYL IODIDE

Material Family	Material Note	Perm. Coefficient (cm³·mm/m²·day·atm)	VTR (g·mm/m²·day)	Non-normalized Data	Penetrant Note	Temp. (°C)	Time (days)	RH (%)	Pressure (kPa)	Test Note	Source
Polyvinyl Alcohol	Ansell Edmont PVA; supported (lined) glove film; specified by glove film weight			low permeation rate; 0 to 1/2 eyedropper size drops per hour; PVA coating is water soluble; lower detection limit: 0 ppm; permeation rate: <0.9 µg/cm²/min		23	0.25			ASTM F739	313
Rubber, Neoprene (CR)	Pioneer Industrial Products Stanzoil N-44; glove film			breakthrough time: 12 minutes; permeation rate: 3702 µg/cm²/min						ASTM F739.85	312
TPE, Vinyl	Pioneer Industrial Products Pylox V-20; glove film			permeation rate too large to measure; breakthrough time: 1 minutes						ASTM F739.85	312

METHYL ISOBUTYL KETONE

Material Family	Material Note	Perm. Coefficient (cm³·mm/m²·day·atm)	VTR (g·mm/m²·day)	Non-normalized Data	Penetrant Note	Temp. (°C)	Time (days)	RH (%)	Pressure (kPa)	Test Note	Source
Acrylonitrile Copolymer, AMA	barrier prop.; bottles			0.09% penetrant weight loss		50	28				293
Laminar, Nylon/Olefin	DuPont Selar RB/ Polyolefin; 8% Selar RB; barrier prop.; extrusion blow molded; bottles			0.04% penetrant weight loss		50	28				293
Polyethylene, Fluorinated	barrier prop.; bottles			0.56% penetrant weight loss		50	28				293
Polyethylene, HDPE	bottles			1.8% penetrant weight loss		50	28				293

Material Family	Material Note	Permeability Data			Test Conditions						Source
		Perm. Coefficient (cm³·mm/m²·day·atm)	VTR (g·mm/m²·day)	Non-normalized Data	Penetrant Note	Temp. (°C)	Time (days)	RH (%)	Pressure (kPa)	Test Note	
METHYL ISOBUTYL KETONE (continued)											
Polyvinyl Alcohol	Ansell Edmont PVA; supported (lined) glove film; specified by glove film weight			low permeation rate; 0 to 1/2 eyedropper size drops per hour; PVA coating is water soluble; lower detection limit: 0 ppm; permeation rate: <0.9 µg/cm²/min	MIBK	23	0.25			ASTM F739	313
Polyvinyl Chloride	bottles			failed		23	180				293
METHYL METHACRYLATE											
Polyvinyl Alcohol	Ansell Edmont PVA; supported (lined) glove film; specified by glove film weight			low permeation rate; 0 to 1/2 eyedropper size drops per hour; PVA coating is water soluble; lower detection limit: 0 ppm; permeation rate: <0.9 µg/cm²/min		23	0.25			ASTM F739	313
METHYL SALICYLATE											
Acetal	DuPont Delrin		0.12			23		50			201
Acrylonitrile Copolymer, AMA	barrier prop.; bottles			0.11% penetrant weight loss		50	28				293
Laminar, Nylon/Olefin	DuPont Selar RB/ Polyolefin: 8% Selar RB; barrier prop.; extrusion blow molded; bottles			0.01% penetrant weight loss		50	28				293
Nylon 66	DuPont Zytel 42; low flow, 2.54 mm thick; bottles		0.08								68
Polyethylene, Fluorinated	barrier prop.; bottles			0.03% penetrant weight loss		50	28				293
Polyethylene, HDPE	bottles			1.02% penetrant weight loss		50	28				293
Polyvinyl Chloride	bottles			failed		23	180				293
METHYL TERTIARY BUTYL ETHER											
Polyvinyl Alcohol	Ansell Edmont PVA; supported (lined) glove film; specified by glove film weight			low permeation rate; 0 to 1/2 eyedropper size drops per hour; PVA coating is water soluble; lower detection limit: 0 ppm; permeation rate: <0.9 µg/cm²/min	MTBE	23	0.25			ASTM F739	313
Rubber, Nitrile (NBR)	Ansell Edmont Sol-Vex 37-165; unsupported glove film; 0.54 mm thick			low permeation rate; 0 to 1/2 eyedropper size drops per hour; lower detection limit: 0 ppm; permeation rate: <0.9 µg/cm²/min	MTBE	23	0.25			ASTM F739	313
METHYL-2-PYRROLIDONE (N-)											
Rubber, Latex (NR)	Ansell Edmont Canners and Handlers 392; unsupported glove film; 0.48 mm thick			very good permeation rate; 1 to 5 eyedropper size drops per hour; breakthrough time: 75 minutes; permeation rate: <9 µg/cm²/min	NMP	23				ASTM F739	313
METHYLAMINE											
Rubber, Latex (NR)	Ansell Edmont Canners and Handlers 392; unsupported glove film; 0.48 mm thick			very good permeation rate; 1 to 5 eyedropper size drops per hour; breakthrough time: 55 minutes; permeation rate: <9 µg/cm²/min		23				ASTM F739	313
Rubber, Neoprene (CR)	Ansell Edmont Neoprene 29-840; unsupported glove film; 0.38 mm thick			good permeation rate; 6 to 50 eyedropper size drops per hour; breakthrough time: 270 minutes; permeation rate: <90 µg/cm²/min		23				ASTM F739	313
	Ansell Edmont Neox; supported (lined) glove film; specified by glove film weight			low permeation rate; 0 to 1/2 eyedropper size drops per hour; breakthrough time: 360 minutes; permeation rate: <0.9 µg/cm²/min		23				ASTM F739	313

Material Family	Material Note	Perm. Coefficient (cm³·mm/m²·day·atm)	VTR (g·mm/m²·day)	Non-normalized Data	Penetrant Note	Temp. (°C)	Time (days)	RH (%)	Pressure (kPa)	Test Note	Source

METHYLAMINE (continued)

Material Family	Material Note	Non-normalized Data	Penetrant Note	Temp. (°C)	Time (days)	Test Note	Source
Rubber, Nitrile (NBR)	Ansell Edmont Sol-Vex 37-165; unsupported glove film; 0.54 mm thick	low permeation rate; 0 to 1/2 eyedropper size drops per hour; lower detection limit: 0 ppm; permeation rate: <0.9 µg/cm²/min		23	0.25	ASTM F739	313
TPE, Vinyl	Ansell Edmont Monkey Grip; supported (lined) glove film; specified by glove film weight	very good permeation rate; 1 to 5 eyedropper size drops per hour; breakthrough time: 135 minutes; permeation rate: <9 µg/cm²/min		23		ASTM F739	313
	Ansell Edmont Wet-Wear 550; clothing; PVC stretch outer layer bonded to a lightweight liner	low permeation rate; 0 to 1/2 eyedropper size drops per hour; breakthrough time: 5 minutes; permeation rate: <0.9 µg/cm²/min		23		ASTM F739	313
	Ansell Edmont Wet-Wear 600; clothing; nylon netting bonded between two layers of PVC	very good permeation rate; 1 to 5 eyedropper size drops per hour; breakthrough time: 60 minutes; permeation rate: <9 µg/cm²/min		23		ASTM F739	313
TPE, Vinyl coated fabric	Ansell Edmont Wet-Wear 700; clothing; PVC coating with Nylon/Polyester lining	good permeation rate; 6 to 50 eyedropper size drops per hour; breakthrough time: 30 minutes; permeation rate: <90 µg/cm²/min		23		ASTM F739	313

METHYLENE BROMIDE

Material Family	Material Note	Non-normalized Data	Penetrant Note	Temp. (°C)	Time (days)	Test Note	Source
Polyvinyl Alcohol	Ansell Edmont PVA; supported (lined) glove film; specified by glove film weight	low permeation rate; 0 to 1/2 eyedropper size drops per hour; PVA coating is water soluble; lower detection limit: 0 ppm; permeation rate: <0.9 µg/cm²/min		23	0.25	ASTM F739	313

METHYLENE CHLORIDE

Material Family	Material Note	Non-normalized Data	Penetrant Note	Temp. (°C)	Time (days)	Test Note	Source
	Ansell Edmont PVA; supported (lined) glove film; specified by glove film weight	low permeation rate; 0 to 1/2 eyedropper size drops per hour; PVA coating is water soluble; lower detection limit: 0 ppm; permeation rate: <0.9 µg/cm²/min		23	0.25	ASTM F739	313
Rubber, Neoprene (CR)	Pioneer Industrial Products Stanzoil N-44; glove film	breakthrough time: 6 minutes; permeation rate: 1434 µg/cm²/min				ASTM F739.85	312

METHYLPHENOL (3-)

Material Family	Material Note	Non-normalized Data	Penetrant Note	Test Note	Source
Rubber, Latex (NR)	Pioneer Industrial Products L-118; glove film	breakthrough time: 150 minutes; lower detection limit: 5 ppm; permeation rate: 12 µg/cm²/min	m-methylphenol	ASTM F739.85	312
Rubber, Neoprene (CR)	Pioneer Industrial Products Stanzoil N-44; glove film	no permeation rate at steady state detected; breakthrough time: >480 minutes; lower detection limit: 5 ppm; permeation rate: 0 µg/cm²/min	m-methylphenol	ASTM F739.85	312
Rubber, Nitrile (NBR)	Pioneer Industrial Products Stansolv A-14; glove film	breakthrough time: 210 minutes; lower detection limit: 5 ppm; permeation rate: 126 µg/cm²/min	m-methylphenol	ASTM F739.85	312
TPE, Vinyl	Pioneer Industrial Products Pylox V-20; glove film	breakthrough time: 150 minutes; lower detection limit: 5 ppm; permeation rate: 36 µg/cm²/min	m-methylphenol	ASTM F739.85	312

METHYLPHENOL (m-)

Material Family	Material Note	Non-normalized Data	Penetrant Note	Test Note	Source
Rubber, Latex (NR)	Pioneer Industrial Products L-118; glove film	breakthrough time: 150 minutes; lower detection limit: 5 ppm; permeation rate: 12 µg/cm²/min	3-methylphenol	ASTM F739.85	312
Rubber, Neoprene (CR)	Pioneer Industrial Products Stanzoil N-44; glove film	no permeation rate at steady state detected; breakthrough time: >480 minutes; lower detection limit: 5 ppm; permeation rate: 0 µg/cm²/min	3-methylphenol	ASTM F739.85	312
Rubber, Nitrile (NBR)	Pioneer Industrial Products Stansolv A-14; glove film	breakthrough time: 210 minutes; lower detection limit: 5 ppm; permeation rate: 126 µg/cm²/min	3-methylphenol	ASTM F739.85	312
TPE, Vinyl	Pioneer Industrial Products Pylox V-20; glove film	breakthrough time: 150 minutes; lower detection limit: 5 ppm; permeation rate: 36 µg/cm²/min	3-methylphenol	ASTM F739.85	312

MINERAL OILS

Material Family	Material Note	Perm. Coefficient (cm³·mm/m²·day·atm)	VTR (g·mm/m²·day)	Non-normalized Data	Penetrant Note	Temp. (°C)	Time (days)	RH (%)	Pressure (kPa)	Test Note	Source
Acetal	DuPont Delrin		0								201
	DuPont Delrin		0			38		50			201

MINERAL SPIRITS

Material Family	Material Note	Perm. Coefficient (cm³·mm/m²·day·atm)	VTR (g·mm/m²·day)	Non-normalized Data	Penetrant Note	Temp. (°C)	Time (days)	RH (%)	Pressure (kPa)	Test Note	Source
Laminar, Nylon/Olefin	DuPont Selar RB/ Polyolefin; 8% Selar RB; barrier prop.; extrusion blow molded; bottles			0.01% penetrant weight loss		50	28				293
Polyethylene, Fluorinated	barrier prop.; bottles			0.02% penetrant weight loss		50	28				293
Polyethylene, HDPE	bottles			0.8% penetrant weight loss		50	28				293
Polyvinyl Alcohol	Ansell Edmont PVA; supported (lined) glove film; specified by glove film weight			low permeation rate: 0 to 1/2 eyedropper size drops per hour; PVA coating is water soluble; breakthrough time: 90 minutes; permeation rate: <0.9 µg/cm²/min	Rule 66	23	0.25			ASTM F739	313
Rubber, Neoprene (CR)	Ansell Edmont Neoprene 29-840; unsupported glove film; 0.38 mm thick			very good permeation rate: 1 to 5 eyedropper size drops per hour; breakthrough time: 90 minutes; permeation rate: <9 µg/cm²/min	Rule 66	23				ASTM F739	313
	Ansell Edmont Neox; supported (lined) glove film; specified by glove film weight			low permeation rate: 0 to 1/2 eyedropper size drops per hour; lower detection limit: 0 ppm; permeation rate: <0.9 µg/cm²/min	Rule 66	23	0.25			ASTM F739	313
	Pioneer Industrial Products Stanzoil N-44; glove film			breakthrough time: 126 minutes; lower detection limit: 0.002 ppm; permeation rate: 12 µg/cm²/min						ASTM F739.85	312
Rubber, Nitrile (NBR)	Ansell Sol-Vex 37-165; unsupported glove film; 0.54 mm thick			low permeation rate: 0 to 1/2 eyedropper size drops per hour; lower detection limit: 0 ppm; permeation rate: <0.9 µg/cm²/min	Rule 66	23	0.25			ASTM F739	313
	Pioneer Industrial Products Stansolv A-14; glove film			no permeation rate at steady state detected; breakthrough time: >480 minutes; lower detection limit: 0.02 ppm; permeation rate: 0 µg/cm²/min						ASTM F739.85	312
TPE, Vinyl	Ansell Edmont Monkey Grip; supported (lined) glove film; specified by glove film weight			very good permeation rate: 1 to 5 eyedropper size drops per hour; breakthrough time: 150 minutes; permeation rate: <9 µg/cm²/min	Rule 66	23				ASTM F739	313

MORPHOLINE

Material Family	Material Note	Perm. Coefficient (cm³·mm/m²·day·atm)	VTR (g·mm/m²·day)	Non-normalized Data	Penetrant Note	Temp. (°C)	Time (days)	RH (%)	Pressure (kPa)	Test Note	Source
Polyvinyl Alcohol	Ansell Edmont PVA; supported (lined) glove film; specified by glove film weight			good permeation rate: 6 to 50 eyedropper size drops per hour; PVA coating is water soluble; breakthrough time: 90 minutes; permeation rate: <90 µg/cm²/min		23				ASTM F739	313
Rubber, Latex (NR)	Ansell Edmont Canners and Handlers 392; unsupported glove film; 0.48 mm thick			good permeation rate: 6 to 50 eyedropper size drops per hour; breakthrough time: 200 minutes; permeation rate: <90 µg/cm²/min		23				ASTM F739	313

MOTOR OILS

Material Family	Material Note	Perm. Coefficient (cm³·mm/m²·day·atm)	VTR (g·mm/m²·day)	Non-normalized Data	Penetrant Note	Temp. (°C)	Time (days)	RH (%)	Pressure (kPa)	Test Note	Source
Acetal	DuPont Delrin		0			23		50			201
	DuPont Delrin		0			38		50			201
Laminar, Nylon/Olefin	DuPont Selar RB/ Polyolefin; 8% Selar RB; barrier prop.; extrusion blow molded; bottles			0.02% penetrant weight loss	2 cycle	50	28				293
Nylon 66	DuPont Zytel 42; low flow; 2.54 mm thick; bottles		0.08		SAE 10						68
Polyethylene, Fluorinated	barrier prop.; bottles			0.07% penetrant weight loss	2 cycle	50	28				293
Polyethylene, HDPE	bottles			0.4% penetrant weight loss	2 cycle	50	28				293

MURIATIC ACID

Material Family	Material Note	Perm. Coefficient (cm³·mm/m²·day·atm)	VTR (g·mm/m²·day)	Non-normalized Data	Penetrant Note	Temp (°C)	Time (days)	RH (%)	Pressure (kPa)	Test Note	Source
Rubber, Latex (NR)	Ansell Edmont Canners and Handlers 392; unsupported glove film; 0.48 mm thick			breakthrough time: 290 minutes		23				ASTM F739	313
	Pioneer Industrial Products L-118; glove film			breakthrough time: 211 minutes; lower detection limit: 4 ppm; permeation rate: 1308 μg/cm²/min						ASTM F739.85	312
Rubber, Neoprene (CR)	Ansell Edmont Neoprene 29-840; unsupported glove film; 0.38 mm thick			no permeation detected during a 6 hour test; lower detection limit: 0 ppm	10% conc.	23	0.25			ASTM F739	313
	Ansell Edmont Neox; supported (lined) glove film; specified by glove film weight			no permeation detected during a 6 hour test; lower detection limit: 0 ppm		23	0.25			ASTM F739	313
	Pioneer Industrial Products Stanzoil N-44; glove film			no permeation rate at steady state detected; breakthrough time: >480 minutes; lower detection limit: 4 ppm; permeation rate: 0 μg/cm²/min						ASTM F739.85	312
Rubber, Nitrile (NBR)	Ansell Edmont Sol-Vex 37-165; unsupported glove film; 0.54 mm thick			no permeation detected during a 6 hour test; lower detection limit: 0 ppm	10% conc.	23	0.25			ASTM F739	313
	Pioneer Industrial Products Starsolv A-14; glove film			no permeation rate at steady state detected; breakthrough time: >480 minutes; lower detection limit: 0.4 ppm; permeation rate: 0 μg/cm²/min						ASTM F739.85	312
TPE, Vinyl	Ansell Edmont Monkey Grip; supported (lined) glove film; specified by glove film weight			breakthrough time: >300 minutes		23				ASTM F739	313
	Pioneer Industrial Products Pylox V-20; glove film			no permeation rate at steady state detected; breakthrough time: >480 minutes; lower detection limit: 4 ppm; permeation rate: 0 μg/cm²/min						ASTM F739.85	312

N-HEPTANE

Material Family	Material Note	Perm. Coefficient (cm³·mm/m²·day·atm)	VTR (g·mm/m²·day)	Non-normalized Data	Penetrant Note	Temp (°C)	Time (days)	RH (%)	Pressure (kPa)	Test Note	Source
Polyethylene, LDPE	Dow, film		118 - 197			24					250
Polystyrene	Dow Styron; film			sample failed		24					250
SAN	Dow Tyril; film		0.79 - 7.9			24					250

NAPHTHA

Material Family	Material Note	Perm. Coefficient (cm³·mm/m²·day·atm)	VTR (g·mm/m²·day)	Non-normalized Data	Penetrant Note	Temp (°C)	Time (days)	RH (%)	Pressure (kPa)	Test Note	Source
Laminar, Nylon/HDPE	DuPont Selar RB/ HDPE; solvent packaging; 18% Selar RB; barrier prop.; extrusion blow molded; one pint bottle			0.15% penetrant weight loss (140% relative improvement vs. HDPE)	aromatic	48.9	28				291
Laminar, Nylon/Olefin	DuPont Selar RB/ Polyolefin; 8% Selar RB; barrier prop.; extrusion blow molded; bottles			0.03% penetrant weight loss		50	28				293
Nylon 66	DuPont Zytel 42; low flow, 2.54 mm thick; bottles		2.4		VMP naphtha						68
Polyethylene, Fluorinated	barrier prop.; bottles			0.06% penetrant weight loss		50	28				293
Polyethylene, HDPE	bottles			8.8% penetrant weight loss		50	28				293
Polyvinyl Alcohol	Ansell Edmont PVA; supported (lined) glove film; specified by glove film weight			low permeation rate; 0 to 1/2 eyedropper size drops per hour; PVA coating is water soluble; breakthrough time: >420 minutes; permeation rate: <0.9 μg/cm²/min	VM&P	23				ASTM F739	313
Rubber, Neoprene (CR)	Ansell Edmont Neoprene 29-840; unsupported glove film; 0.38 mm thick			fair permeation rate; 51 to 100 eyedropper size drops per hour; breakthrough time: 15 minutes; permeation rate: <900 μg/cm²/min	VM&P	23				ASTM F739	313
	Ansell Edmont Neox; supported (lined) glove film; specified by glove film weight			low permeation rate; 0 to 1/2 eyedropper size drops per hour; lower detection limit: 0 ppm; permeation rate: <0.9 μg/cm²/min	VM&P	23	0.25			ASTM F739	313
	Pioneer Industrial Products Stanzoil N-44; glove film			breakthrough time: 126 minutes; lower detection limit: 0.002 ppm; permeation rate: 12 μg/cm²/min						ASTM F739.85	312
Rubber, Nitrile (NBR)	Ansell Edmont Sol-Vex 37-165; unsupported glove film; 0.54 mm thick			low permeation rate; 0 to 1/2 eyedropper size drops per hour; lower detection limit: 0 ppm; permeation rate: <0.9 μg/cm²/min	VM&P	23	0.25			ASTM F739	313

NAPHTHA (continued)

Material Family	Material Note	Perm. Coefficient (cm³·mm/m²·day·atm)	VTR (g·mm/m²·day)	Non-normalized Data	Penetrant Note	Temp. (°C)	Time (days)	RH (%)	Pressure (kPa)	Test Note	Source
Rubber, Nitrile (NBR)	Pioneer Industrial Products Stansolv A-14; glove film			no permeation rate at steady state detected; breakthrough time: >480 minutes; lower detection limit: 0.02 ppm; permeation rate: 0 µg/cm²/min						ASTM F739.85	312
TPE, Vinyl	Ansell Edmont Monkey Grip; supported (lined) glove film; specified by glove film weight			very good permeation rate: 1 to 5 eyedropper size drops per hour; breakthrough time: 120 minutes; permeation rate: <9 µg/cm²/min	VM&P	23				ASTM F739	313
	Ansell Edmont Wet-Wear 550; clothing; PVC stretch outer layer bonded to a lightweight liner			very good permeation rate: 1 to 5 eyedropper size drops per hour; breakthrough time: 6 minutes; permeation rate: <9 µg/cm²/min	VM&P	23				ASTM F739	313
	Ansell Edmont Wet-Wear 600; clothing; nylon netting bonded between two layers of PVC			good permeation rate: 6 to 50 eyedropper size drops per hour; breakthrough time: 200 minutes; permeation rate: <90 µg/cm²/min	VM&P	23				ASTM F739	313
TPE, Vinyl coated fabric	Ansell Edmont Wet-Wear 700; clothing; PVC coating with Nylon/Polyester lining			good permeation rate: 6 to 50 eyedropper size drops per hour; breakthrough time: 9 minutes; permeation rate: <90 µg/cm²/min	VM&P	23				ASTM F739	313

NATURAL GAS

Material Family	Material Note	Perm. Coefficient (cm³·mm/m²·day·atm)	VTR (g·mm/m²·day)	Non-normalized Data	Penetrant Note	Temp. (°C)	Time (days)	RH (%)	Pressure (kPa)	Test Note	Source
Polyethylene, HDPE	Phillips Marlex; film	44.5				23				ASTM D1434	101

NITRIC ACID

Material Family	Material Note	Perm. Coefficient (cm³·mm/m²·day·atm)	VTR (g·mm/m²·day)	Non-normalized Data	Penetrant Note	Temp. (°C)	Time (days)	RH (%)	Pressure (kPa)	Test Note	Source
Rubber, Latex (NR)	Ansell Edmont Canners and Handlers 392; unsupported glove film; 0.48 mm thick			no permeation detected during a 6 hour test; lower detection limit: 0 ppm	10% conc.	23	0.25			ASTM F739	313
	Pioneer Industrial Products L-118; glove film			permeation rate too large to measure; breakthrough time: 233 minutes; lower detection limit: 0.1 ppm	50% conc.	23				ASTM F739.85	312
Rubber, Neoprene (CR)	Ansell Edmont Neoprene 29-840; unsupported glove film; 0.38 mm thick			no permeation detected during a 6 hour test; lower detection limit: 0 ppm	10% conc.	23	0.25			ASTM F739	313
	Ansell Edmont Neoprene 29-840; unsupported glove film; 0.38 mm thick			breakthrough time: 140 minutes	70% conc.	23				ASTM F739	313
	Ansell Edmont Neox; supported (lined) glove film; specified by glove film weight			no permeation detected during a 6 hour test; lower detection limit: 0 ppm	10% conc.	23	0.25			ASTM F739	313
	Ansell Edmont Neox; supported (lined) glove film; specified by glove film weight			no permeation detected during a 6 hour test; lower detection limit: 0 ppm	70% conc.	23	0.25			ASTM F739	313
	Pioneer Industrial Products Stanzoil N-44; glove film			no permeation rate at steady state detected; breakthrough time: >480 minutes; lower detection limit: 0.0003 ppm; permeation rate: 0 µg/cm²/min	50% conc.	23				ASTM F739.85	312
Rubber, Nitrile (NBR)	Ansell Sol-Vex 37-165; unsupported glove film; 0.54 mm thick			no permeation detected during a 6 hour test; lower detection limit: 0 ppm	10% conc.	23	0.25			ASTM F739	313
	Pioneer Industrial Products Stansolv A-14; glove film			breakthrough time: 72 minutes; lower detection limit: 0.07 ppm; permeation rate: 1206 µg/cm²/min	50% conc.	23				ASTM F739.85	312
TPE, Vinyl	Ansell Edmont Monkey Grip; supported (lined) glove film; specified by glove film weight			no permeation detected during a 6 hour test; lower detection limit: 0 ppm	10% conc.	23	0.25			ASTM F739	313
	Ansell Edmont Monkey Grip; supported (lined) glove film; specified by glove film weight			breakthrough time: 345 minutes	70% conc.	23				ASTM F739	313
	Ansell Edmont Wet-Wear 550; clothing; PVC stretch outer layer bonded to a lightweight liner			no permeation detected during a 6 hour test; lower detection limit: 0 ppm	10% conc.	23	0.25			ASTM F739	313
	Ansell Edmont Wet-Wear 550; clothing; PVC stretch outer layer bonded to a lightweight liner			breakthrough time: 45 minutes	70% conc.	23				ASTM F739	313
TPE, Vinyl	Ansell Edmont Wet-Wear 600; clothing; nylon netting bonded between two layers of PVC			breakthrough time: 285 minutes	10% conc.	23				ASTM F739	313
	Ansell Edmont Wet-Wear 600; clothing; nylon netting bonded between two layers of PVC			breakthrough time: 160 minutes	70% conc.	23				ASTM F739	313

Material Family	Material Note	Permeability Data			Test Conditions						Source
		Perm. Coefficient (cm³·mm/m²·day·atm)	VTR (g·mm/m²·day)	Non-normalized Data	Penetrant Note	Temp. (°C)	Time (days)	RH (%)	Pressure (kPa)	Test Note	

NITRIC ACID (continued)

Material Family	Material Note	Perm. Coeff.	VTR	Non-normalized Data	Penetrant Note	Temp.	Time	RH	Press.	Test Note	Source
TPE, Vinyl	Pioneer Industrial Products Pylox V-20; glove film			no permeation rate at steady state detected; breakthrough time: 114 minutes; lower detection limit: 0.08 ppm; permeation rate: 0 μg/cm²/min	50% conc.					ASTM F739.85	312
TPE, Vinyl coated fabric	Ansell Edmont Wet-Wear 700; clothing; PVC coating with Nylon/Polyester lining			no permeation detected during a 6 hour test; lower detection limit: 0 ppm	10% conc.	23	0.25			ASTM F739	313
	Ansell Edmont Wet-Wear 700; clothing; PVC coating with Nylon/Polyester lining			breakthrough time: 6 minutes	70% conc.	23				ASTM F739	313

NITROBENZENE

Material Family	Material Note	Perm. Coeff.	VTR	Non-normalized Data	Penetrant Note	Temp.	Time	RH	Press.	Test Note	Source
Polyvinyl Alcohol	Ansell Edmont PVA; supported (lined) glove film; specified by glove film weight			low permeation rate; 0 to 1/2 eyedropper size drops per hour; PVA coating is water soluble; lower detection limit: 0 ppm; permeation rate: <0.9 μg/cm²/min		23	0.25			ASTM F739	313
Rubber, Latex (NR)	Ansell Edmont Canners and Handlers 392; unsupported glove film; 0.46 mm thick			good permeation rate; 6 to 50 eyedropper size drops per hour; breakthrough time: 15 minutes; permeation rate: <90 μg/cm²/min		23				ASTM F739	313
Rubber, Neoprene (CR)	Pioneer Industrial Products Stanzoil N-44; glove film			breakthrough time: 60 minutes; lower detection limit: 1 ppm; permeation rate: 120 μg/cm²/min						ASTM F739.85	312
Rubber, Nitrile (NBR)	Pioneer Industrial Products Stansolv A-14; glove film			breakthrough time: 60 minutes; lower detection limit: 1 ppm; permeation rate: 90 μg/cm²/min						ASTM F739.85	312

NITROGEN

Material Family	Material Note	Perm. Coeff.	VTR	Non-normalized Data	Penetrant Note	Temp.	Time	RH	Press.	Test Note	Source
ABS	BASF AG Terluran 877 M; moderate flow; 0.1 mm thick; film	10.1				23				DIN 53380; Values for permeability depend on the conditions under which the film was produced. Figures determined may differ by as much as 50% from those given.	137
	BASF AG Terluran 967 K; moderate flow, 0.1 mm thick; w/ butadiene acrylic rubber; film	10.1				23				DIN 53380; Values for permeability depend on the conditions under which the film was produced. Figures determined may differ by as much as 50% from those given.	137
	BASF AG Terluran 997 VE; 0.1 mm thick, low flow; w/ butadiene acrylic rubber; film	20.3				23				DIN 53380; Values for permeability depend on the conditions under which the film was produced. Figures determined may differ by as much as 50% from those given.	137
	Dow; low nitrile content; film	9.8 - 13.8				24					250
	Dow; medium nitrile content; film	3.9 - 5.9				24					250
Acetal	DuPont Delrin		0.02		at 620 kPa (90 psi)	23		50			201
Acetal Copolymer	Hoechst Cel. Celcon M25; high molecular weight, 0.15 mm thick; 2.5 g/10 min. MFI; film	0.87 - 1.3									210
	Hoechst Cel. Celcon M270; low molecular weight, high flow, 0.15 mm thick; 27.0 g/10 min. MFI; film	0.87 - 1.3									210
	Hoechst Cel. Celcon M90; gen. purp. grade, 0.15 mm thick; 9.0 g/10 min. MFI; film	0.87 - 1.3									210
Acrylonitrile Copolymer, AMA	BP Chem. Barex 210; packaging; impact modified; barrier prop.	0.08				22.8		100			296
	BP Chem. Barex 218; packaging; impact modified, high impact, barrier prop.	0.16				22.8		100			296

NITROGEN (continued)

Material Family	Material Note	Perm. Coefficient (cm³·mm/m²·day·atm)	VTR (g·mm/m²·day)	Non-normalized Data	Penetrant Note	Temp. (°C)	Time (days)	RH (%)	Pressure (kPa)	Test Note	Source
ASA	BASF AG Luran S 757 R; 0.1 mm thick; blown film	6.1				23				DIN 53380	143
	BASF AG Luran S 776 S; 0.1 mm thick; blown film	7.1				23				DIN 53380	143
	BASF AG Luran S 776 S; 0.1 mm thick; film	10.1				23				DIN 53380 part 2 method M; Values for permeability depend on the conditions under which the film was produced. Figures determined may differ by as much as 50% from those given.	142
	BASF AG Luran S 797 S; 0.1 mm thick; film	7.6				23					142
AU (Polyester Urethane)		1382				21.1				applies to general class of base polymer	309
Epoxy		1.6				25				ASTM D1434-63T	121
EU (Polyether Urethane)		82.1				21.1				applies to general class of base polymer	309
EVOH	Eval Co. Eval E; barrier prop.; 44% ethylene	0.0031				23		0			264
	Eval Co. Eval F; barrier prop.; 32% ethylene	0.00039				23		0			264
	Eval Co. Eval H; barrier prop.; 38% ethylene	0.0016				23		0			264
	Eval Co. Eval E; barrier prop.; 44% ethylene	0.01				35		0			264
	Eval Co. Eval F; barrier prop.; 32% ethylene	0.00079				35		0			264
	Eval Co. Eval H; barrier prop.; 38% ethylene	0.0031				35		0			264
Fluoroelastomer, FKM	DuPont Viton; compounded	4.7				24				one atmosphere @80°C; specimen size: 1 cm² x 1 cm thick	305
		17.3				21.1				applies to general class of base polymer	309
Fluoropolymer, CTFE	3M Kel-F; 0.01 mm thick	0.02				25			1724	mass spectrometry and calibrated standard gas leaks; developed by McDonnell Douglas Space Systems Caompany Chemistry Laboratory	306
	3M Kel-F; 0.01 mm thick	0.02				25			3447	"	306
	3M Kel-F; 0.01 mm thick	0.02				25			6895	"	306
	3M Kel-F 81; amorphous form; film	0.33				25					96
	Allied Sig. Aclar 22A; transparent; film	0.98				25				STP conditions	138
	Allied Sig. Aclar 22C; transparent; film	0.98				25				STP conditions	138
	3M Kel-F 81; amorphous form; film	2.0				50					96
	3M Kel-F; 0.01 mm thick	3.6				68			1724	mass spectrometry and calibrated standard gas leaks; developed by McDonnell Douglas Space Systems Caompany Chemistry Laboratory	306
	3M Kel-F; 0.01 mm thick	3.8				69			3447	"	306
	3M Kel-F; 0.01 mm thick	3.9				70			6895	"	306
	3M Kel-F 81; amorphous form; film	6.0				75					96
Fluoropolymer, ECTFE	Ausimont Halar; 0.02 mm thick	0.48				11			1724	mass spectrometry and calibrated standard gas leaks; developed by McDonnell Douglas Space Systems Caompany Chemistry Laboratory	306

NITROGEN (continued)

Material Family	Material Note	Perm. Coefficient (cm³·mm/m²·day·atm)	VTR (g·mm/m²·day)	Non-normalized Data	Penetrant Note	Temp. (°C)	Time (days)	RH (%)	Pressure (kPa)	Test Note	Source
Fluoropolymer, ECTFE	Ausimont Halar; 0.02 mm thick	0.48				10			3447	mass spectrometry and calibrated standard gas leaks; developed by McDonnell Douglas Space Systems Caompany Chemistry Laboratory	306
	Ausimont Halar; 0.02 mm thick	0.53				10			6895	"	306
	Ausimont Halar; 0.02 mm thick	1.1				25			1724	"	306
	Ausimont Halar; 0.02 mm thick	1.3				25			3447	"	306
	Ausimont Halar; 0.02 mm thick	1.2				25			6895	"	306
	Ausimont Halar; 0.02 mm thick	21.3				71			1724	"	306
	Ausimont Halar; 0.02 mm thick	37.4				72			3447	"	306
	Ausimont Halar; 0.02 mm thick	21.7				68			6895	"	306
Fluoropolymer, ETFE	Ausimont Hylfon 700; high molecular weight; 4 g/10 min. MFI	21.7				23				ASTM D1434; activation energy = 6-8 kcal/mole	114
	Ausimont Hyflon 800; low molecular weight; 11 g/10 min. MFI	21.7				23				ASTM D1434; activation energy = 6-8 kcal/mole	114
	DuPont Tefzel; developmental material, 0.102 mm thick; film	11.8				25				ASTM D1434	205
Fluoropolymer, FEP	DuPont Teflon; 0.05 mm thick	4.4				-9			1724	mass spectrometry and calibrated standard gas leaks; developed by McDonnell Douglas Space Systems Caompany Chemistry Laboratory	306
	DuPont Teflon; 0.05 mm thick	4.9				-7			3447	"	306
	DuPont Teflon; 0.05 mm thick	5.6				-5			6895	"	306
	DuPont Teflon; 0.05 mm thick	33.8				25			3447	"	306
	DuPont Teflon; 0.05 mm thick	33.7				25			6895	"	306
	DuPont Teflon; 0.05 mm thick	33.3				25			1724	"	306
	DuPont Teflon; 0.05 mm thick	126				25				STP conditions	138
	DuPont Teflon; 0.05 mm thick	332				71			1724	mass spectrometry and calibrated standard gas leaks; developed by McDonnell Douglas Space Systems Caompany Chemistry Laboratory	306
	DuPont Teflon; 0.05 mm thick	337				66			3447	mass spectrometry and calibrated standard gas leaks; developed by McDonnell Douglas Space Systems Caompany Chemistry Laboratory	306
	DuPont Teflon; 0.05 mm thick	333				68			6895		306
Fluoropolymer, PFA	DuPont Teflon PFA; film	115				25				ASTM D1434	39
Fluoropolymer, PVDF	Solvay Solef 1008; 0.1 mm thick, translucent; film	3.0				23				ASTM D1434	125
		3.5				25				STP conditions	138
Fluoropolymer, PVF		0.1				25				STP conditions	138

NITROGEN (continued)

NITROGEN (continued)

Material Family	Material Note	Perm. Coefficient (cm³·mm/m²·day·atm)	VTR (g·mm/m²·day)	Non-normalized Data	Penetrant Note	Temp (°C)	Time (days)	RH (%)	Pressure (kPa)	Test Note	Source
Fluoropolymer, TFE	DuPont Teflon; 0.03 mm thick	8.3				-23			1724	mass spectrometry and calibrated standard gas leaks; developed by McDonnell Douglas Space Systems Company Chemistry Laboratory	306
	DuPont Teflon; 0.03 mm thick	7.8				-25			3447	"	306
	DuPont Teflon; 0.03 mm thick	8.3				-23			6895	"	306
	DuPont Teflon; 0.05 mm thick; carbon filled	21.9				-14			1724	"	306
	DuPont Teflon; 0.05 mm thick; carbon filled	20.5				-17			3447	"	306
	DuPont Teflon; 0.05 mm thick; carbon filled	20.5				-17			6895	"	306
	DuPont Teflon; 0.03 mm thick	68.9				25			1724	"	306
	DuPont Teflon; 0.03 mm thick	69.0				25			3447	"	306
	DuPont Teflon; 0.03 mm thick	68.6				25			6895	"	306
	DuPont Teflon; 0.05 mm thick; carbon filled	128				25			1724	"	306
	DuPont Teflon; 0.05 mm thick; carbon filled	133				25			3447	"	306
	DuPont Teflon; 0.05 mm thick; carbon filled	124				25			6895	"	306
	DuPont Teflon; 0.03 mm thick	254				71			1724	"	306
	DuPont Teflon; 0.03 mm thick	253				71			3447	"	306
	DuPont Teflon; 0.03 mm thick	251				70			6895	"	306
	DuPont Teflon; 0.05 mm thick; carbon filled	462				68			1724	"	306
	DuPont Teflon; 0.05 mm thick; carbon filled	466				68			3447	"	306
	DuPont Teflon; 0.05 mm thick; carbon filled	418				67			6895	"	306
HDPE/ EVA/ PVDC/ EVA Film	Dow Saranex 25; form-fill-seal pouch, bag-in-box; neutral, barrier prop., 0.051 mm thick; high slip; coextruded film	0.06				23		10	6895	ASTM D1434	257
LDPE/ EVA/ PVDC/ EVA Film	Dow Saranex 23; laminations; neutral, barrier prop., 0.051 mm thick; medium slip; coextruded film	0.12				23		10	6895	ASTM D1434	257
LDPE/ EVA/ PVDC/ EVA/ LDPE Film	Dow Saranex 11; laminations; neutral, barrier prop., 0.038 mm thick; medium slip; coextruded film	0.09				23		10	6895	ASTM D1434	257
	Dow Saranex 14; form-fill-seal pouch; neutral, barrier prop., 0.051 mm thick; medium slip; coextruded film	0.06				23		10	6895	ASTM D1434	257
	Dow Saranex 15; form-fill-seal pouch; neutral, barrier prop., 0.076 mm thick; medium slip; coextruded film	0.08				23		10	6895	ASTM D1434	257
	Dow Saranex 15; form-fill-seal pouch; neutral, barrier prop., 0.102 mm thick; medium slip; coextruded film	0.11				23		10	6895	ASTM D1434	257
Nylon 6	Allied Sig. Capran 77C; 0.0254 mm thick; film	0.08				0		0		STP conditions	285
	Allied Sig. Capran 77C; 0.0254 mm thick; film	0.35				23		0		STP conditions	285
	Allied Sig. Capran 77K; 0.0254 mm thick; PVDC coated; film; oriented	0.04				23		0		STP conditions	285
	Allied Sig. Capran 77C; 0.0254 mm thick; film	0.28				23		0			264
	Allied Sig. Capran 77C; 0.0254 mm thick; film	4.7				50		0		STP conditions	285
Nylon 66	DuPont Can. Dartek; transparent, 0.0254 mm thick; film	0.28				23		0			276
	DuPont Zytel 42; low flow; film	0.28				23		50		isotactic gas permeability cell	68

NITROGEN (continued)

Material Family	Material Note	Perm. Coefficient (cm³·mm/m²·day·atm)	VTR (g·mm/m²·day)	Non-normalized Data	Penetrant Note	Temp. (°C)	Time (days)	RH (%)	Pressure (kPa)	Test Note	Source
Nylon 66/610	Emser Grilon XE3303; transparent, barrier prop., 0.05 mm thick; film	0.61				23		50			307
Nylon 6/66	Allied Sig. Capran; 0.0254 mm thick; film	0.2				23				ASTM D1435; Dow Cell	284
Nylon, Amorphous	Emser Grivory G21; transparent, barrier prop., 0.05 mm thick; film	0.51				23		50		DIN 53380	307
Parylene	Union Carbide Parylene C; vapor phase deposition; thin film	0.39				25				ASTM D1434-63T	121
	Union Carbide Parylene D; vapor phase deposition; thin film	1.8				25				ASTM D1434-63T	121
	Union Carbide Parylene N; highly crystalline, high molecular weight, completely linear; vapor phase deposition; thin film	3.03				25				ASTM D1434-63T	121
PE/ PVC-PVDC Copolymer Multilayer Film	Dow Saranex; multilayer film	0.04 - 0.2				24					250
Polybutylene	Shell Duraflex 1600; peelable seals; FDA grade, 0.051 mm thick, heat sealable; blown film; 5 phr zinc oxide, slip and antiblock formulations; film	43.3				22.8		50		ASTM D1434, method M	304
	Shell Duraflex 1710; peelable seals; 0.910 g/cm³ density, 0.909 g/cm³ density; FDA grade, heat sealable, 0.051 mm thick; blown film; slip and antiblock formulations, antiblock formulations; film	43.3				22.8		50		ASTM D1434, method M	304
Polycarbonate	Bayer Makrolon; 0.1 mm thick; film	11.2								DIN 53380, pt. 3	289
	Dow Calibre 300-15; transparent, gen. purp. grade; 15 g/10 min. MFI	10.6								ASTM D2752	78
	Dow Calibre 300-4; transparent, gen. purp. grade; 4 g/10 min. MFI	12.2								ASTM D2752	78
	Dow Calibre 800-6; transparent, flame retardant; 6 g/10 min. MFI	22.4								ASTM D2752	78
Polyester, PBT	BASF AG Ultradur B4550; 0.25 mm thick	3.0				23		50		DIN 53380; measured in standard laboratory atmosphere	180
Polyester, PCTG	Eastman Kodar PCTG 5445; transparent, 0.25 mm thick; film	3				23				ASTM D1434	166
Polyester, PET	DuPont Mylar; film	0.39				25				ASTM D1434-72	270
	oriented	0.18				23			0		264
	oriented	0.28 - 0.39				25				STP conditions	138
Polyester, PETG	Eastman Kodar PETG 6763; transparent, amorphous, 0.25 mm thick; film	3.9				23				ASTM D1434	165
Polyethylene, HDPE	0.03 mm thick	1.6				-10			1724	mass spectrometry and calibrated standard gas leaks; developed by McDonnell Douglas Space Systems Caompany Chemistry Laboratory	306
	0.03 mm thick	0.95				-19			3447	"	306
	0.03 mm thick	0.99				-17			6895	"	306
	Dow; film	15.8 - 23.6				24				volume at standard temperature and pressure; useable average for all Hostalen grades	250
	Hoechst AG Hostalen	18.2				20					94

NITROGEN (continued)

Material Family	Material Note	Perm. Coefficient (cm³·mm/m²·day·atm)	VTR (g·mm/m²·day)	Non-normalized Data	Penetrant Note	Temp. (°C)	Time (days)	RH (%)	Pressure (kPa)	Test Note	Source
Polyethylene, HDPE	Hoechst AG Hostalen	21.3				25				volume at standard temperature and pressure; useable average for all Hostalen grades	94
	Phillips Marlex; film	20.9				23				ASTM D1434	101
	0.03 mm thick	15.5				25			1724	mass spectrometry and calibrated standard gas leaks; developed by McDonnell Douglas Space Systems Caompany Chemistry Laboratory	306
	0.03 mm thick	14.0				25			3447	"	306
	0.03 mm thick	14.7				25			6895	"	306
		16.5				25				STP conditions	138
	Hoechst AG Hostalen	29.4				30				volume at standard temperature and pressure; useable average for all Hostalen grades	94
	Hoechst AG Hostalen	48.6				40				"	94
	Hoechst AG Hostalen	85.1				50				"	94
	0.03 mm thick	173.3				72			1724	mass spectrometry and calibrated standard gas leaks; developed by McDonnell Douglas Space Systems Caompany Chemistry Laboratory	306
	0.03 mm thick	128				69			3447	"	306
	0.03 mm thick	150				68			6895	"	306
Polyethylene, LDPE	Dow; film	39.4 - 78.7				24					250
	film	70.9				25				STP conditions	138
Polyethylene, LLDPE	DuPont Can, Sclairfilm SL1; laminations; 0.918 g/cm³ density; 0.0254 mm thick; film	3.8								ASTM D3985; approximate values	278
	DuPont Can, Sclairfilm SL3; laminations; 0.918 g/cm³ density; 0.0254 mm thick; film	3.8								ASTM D3985; approximate values	278
Polyethylene, MDPE		33.5 - 124				25				STP conditions	138
Polyimide	Ube Upilex R; 0.025 mm thick; film	0.76				30				ASTM D1434	97
Polyisoprene Rubber		529				21.1				applies to general class of base polymer	309
Polymethylpentene	Mitsui TPX X-22; 0.835 g/cm³ density; transparent, 0.05 mm thick; film	475				23					302
	Mitsui TPX X-44; 0.834 g/cm³ density; transparent, 0.05 mm thick; film	470				23					302
	Mitsui TPX X-88; 0.835 g/cm³ density; transparent, food grade, 0.05 mm thick; film	400				23					302
Polypropylene	oriented; film	16.5				25		0			63
Polypyrrole	BASF AG Lutamer ES 9567; 0.031-0.037 mm thick, intrinsically conductive, high flexibility; w/ benzensulfonate anions; film	0.41									182
Polystyrene	Dow Styron; film	15.8 - 19.7				24					250
	Dow Trycite; film	19.7 - 23.6				24					250
Polystyrene, GP	BASF AG Polystyrol 168 N; transparent, 0.1 mm thick; film	25.3				23				DIN 53380	26
	Dow Styron; injection molding	15.8 - 19.7				23				ASTM D1434	263

NITROGEN (continued)

Material Family	Material Note	Permeability Data			Test Conditions						Source
		Perm. Coefficient (cm³·mm/m²·day·atm)	VTR (g·mm/m²·day)	Non-normalized Data	Penetrant Note	Temp. (°C)	Time (days)	RH (%)	Pressure (kPa)	Test Note	
Polystyrene, GP	Dow Styron; oriented; sheet	19.7 - 23.6				23				ASTM D1434	263
Polystyrene, IPS	BASF AG Polystyrol 476 L; 0.1 mm thick; film	40.5				23				DIN 53380	26
	Dow Styron; injection molding	15.7 - 19.7				23				ASTM D1434	262
Polysulfone	Amoco Udel; transparent; amber tint	15.8				23		dry		ASTM D1434	15
Polyurethane	film	31.5				25				ASTM D1434-63T	121
Polyvinyl Chloride	film	70.9				25		0			63
Polyvinylidene Chloride	Dow Saran; film	0.04 - 0.08				24					250
	Dow Saran 5253; barrier prop.	0.0047				23		0			264
	Dow Saran Wrap 16; chub packaging; laminations; transparent; barrier prop., 0.019 mm thick, biaxially oriented; monolayer film	0.07				23		10		ASTM D1434	256
	Dow Saran Wrap 18L; laminations; transparent; barrier prop., 0.019 mm thick, preshrunk; biaxially oriented; monolayer film	0.09				23		10		ASTM D1434	256
	Dow Saran Wrap 19; chub packaging; transparent; barrier prop., 0.0254 mm thick, biaxially oriented; monolayer film	0.07				23		10		ASTM D1434	256
	Dow Saran Wrap 28; chub packaging; transparent; barrier prop., 0.0254 mm thick, biaxially oriented; monolayer film	0.12				23		10		ASTM D1434	256
	Dow Saran Wrap 560; unit packaging; transparent; barrier prop., 0.152 mm thick, biaxially oriented; coextruded film	0.02				23		10		ASTM D1434	256
PVC-PVDC Copolymer		0.05 - 0.59				25				STP conditions	138
Rubber, Butyl (IIR)		21.6				21.1				applies to general class of base polymer	309
Rubber, Chlorohydrin (CO)		14.7				21.1				applies to general class of base polymer	309
Rubber, Chlorohydrin (ECO)		57				21.1				applies to general class of base polymer	309
Rubber, Chlorosulf. PE (CSM)	Hypalon	60.5 - 77.8				21.1				applies to general class of base polymer	309
Rubber, EACM		76				21.1				applies to general class of base polymer	309
Rubber, EPDM	DuPont Nordel 1040	533 - 648				23				STP conditions	311
	gum vulcanizate	553		1600% relative to Butyl Rubber		25					300
Rubber, EPM		553				21.1				applies to general class of base polymer	309
Rubber, Isobutylene	gum vulcanizate			72% relative to Butyl Rubber		25					300
Rubber, Latex (NR)	gum vulcanizate	529		2000% relative to Butyl Rubber		25					300
Rubber, Neoprene (CR)	Neoprene G type; gum vulcanizate	77		280% relative to Butyl Rubber		25					300
Rubber, Nitrile (NBR)	34% acrylonitrile	39.7				21.1				applies to general class of base polymer	309
	38% acrylonitrile	15.6				21.1				applies to general class of base polymer	309

NITROGEN (continued)

Material Family	Material Note	Perm. Coefficient (cm³·mm/m²·day·atm)	VTR (g·mm/m²·day)	Non-normalized Data	Penetrant Note	Temp. (°C)	Time (days)	RH (%)	Pressure (kPa)	Test Note	Source
Rubber, Nitrile (NBR)	61% butadiene/ 39% acrylonitrile; gum vulcanizate			58% relative to Butyl Rubber		25					300
	68% butadiene/ 32% acrylonitrile; gum vulcanizate			150% relative to Butyl Rubber		25					300
	73% butadiene/ 27% acrylonitrile; gum vulcanizate			260% relative to Butyl Rubber		25					300
	80% butadiene/ 20% acrylonitrile; gum vulcanizate			620% relative to Butyl Rubber		25					300
Rubber, Polybutadiene	film	787				24					250
	hot emulsion; gum vulcanizate	1728		1600% relative to Butyl Rubber		25					300
						21.1				applies to general class of base polymer	309
Rubber, Polysulfide (T)		57				21.1				applies to general class of base polymer	309
Rubber, Propylene Oxide (PO)		77.8				21.1				applies to general class of base polymer	309
Rubber, Styrene Butadiene	hot; gum vulcanizate	415		1560% relative to Butyl Rubber		25					300
						21.1				applies to general class of base polymer	309
SAN	Dow Tyril; low nitrile content; film	3.9				24					250
Silicone	gum vulcanizate	17,280		66,000% relative to Butyl Rubber		25					300
						21.1				applies to general class of base polymer	309
Silicone, FVMQ		14,256				21.1				applies to general class of base polymer	309
Styrene-Butad Block Cop	BASF AG Styrolux 637 D; 0.1 mm thick; film	45.6				23					29
	BASF AG Styrolux 656 C; high flow, 0.1 mm thick; film	35				23					29
	BASF AG Styrolux 684 D; 0.1 mm thick; film	70.9				23					29
TPE, Olefinic	Adv. Elast. Santoprene 201-73; 73 Shore A, 0.5 mm thick	194			gas at 0°C and 0.11 Mpa	23			110	ASTM D1434	282
	Adv. Elast. Santoprene 201-87; 87 Shore A, 0.5 mm thick	264			gas at 0°C and 0.11 Mpa	23			110	ASTM D1434	282
	Adv. Elast. Santoprene 203-50; 50 Shore D, 0.5 mm thick	93			gas at 0°C and 0.11 Mpa	23			110	ASTM D1434	282
TPE, Polyamide	Atochem Pebax 2533; 25 Shore D, 0.12 mm thick; film	1116				23		dry			287
	Atochem Pebax 3533; 35 Shore D, 0.12 mm thick; film	657				23		dry			287
	Atochem Pebax 4033; 40 Shore D, 0.12 mm thick; film	256				23		dry			287
	Atochem Pebax 5533; 55 Shore D, 0.12 mm thick; film	72				23		dry			287
	Atochem Pebax 6333; 63 Shore D, 0.12 mm thick; film	33				23		dry			287
TPE, Polyester	DuPont Hytrel 4056; 40 Shore D	147				21.5			34.5		274
	DuPont Hytrel 5556; 55 Shore D	121				21.5			34.5		274
	Eastman Ecdel; copolyester ether; transparent, crystalline, 0.11 - 0.14 mm thick; film	25.6				23				ASTM D1434	60
TPE, Urethane (TPAU)	BASF Elastollan C80A; 80 Shore A	35.0				20					130
	BASF Elastollan C85A; 85 Shore A	26.3				20					130

NITROGEN (continued)

Material Family	Material Note	Perm. Coefficient (cm³ · mm/m² · day · atm)	VTR (g · mm/m² · day)	Non-normalized Data	Penetrant Note	Temp. (°C)	Time (days)	RH (%)	Pressure (kPa)	Test Note	Source
TPE, Urethane (TPAU)	BASF Elastollan C90A; 90 Shore A	17.5				20					130
	BASF Elastollan C95A; 95 Shore A	8.8				20					130
TPE, Urethane (TPEU)	BASF Elastollan 1180A; 80 Shore A	52.5				20					130
	BASF Elastollan 1185A; 85 Shore A	43.8				20					130
	BASF Elastollan 1190A; 90 Shore A	35.0				20					130
	BASF Elastollan 1195A; 95 Shore A	26.3				20					130

NITROMETHANE

Material Family	Material Note	Perm. Coefficient (cm³ · mm/m² · day · atm)	VTR (g · mm/m² · day)	Non-normalized Data	Penetrant Note	Temp. (°C)	Time (days)	RH (%)	Pressure (kPa)	Test Note	Source
Polyvinyl Alcohol	Ansell Edmont PVA; supported (lined) glove film; specified by glove film weight			low permeation rate; 0 to 1/2 eyedropper size drops per hour; PVA coating is water soluble; lower detection limit: 0 ppm; permeation rate: <0.9 µg/cm²/min	95.5% conc.	23	0.25			ASTM F739	313
Rubber, Latex (NR)	Ansell Edmont Canners and Handlers 392; unsupported glove film; 0.48 mm thick			good permeation rate; 6 to 50 eyedropper size drops per hour; breakthrough time: 10 minutes; permeation rate: <90 µg/cm²/min	95.5% conc.	23				ASTM F739	313
Rubber, Neoprene (CR)	Ansell Edmont Neoprene 29-840; unsupported glove film; 0.38 mm thick			very good permeation rate; 1 to 5 eyedropper size drops per hour; breakthrough time: 60 minutes; permeation rate: <9 µg/cm²/min	95.5% conc.	23				ASTM F739	313
	Ansell Edmont Neoc; supported (lined) glove film; specified by glove film weight			low permeation rate; 0 to 1/2 eyedropper size drops per hour; breakthrough time: 90 minutes; permeation rate: <0.9 µg/cm²/min	95.5% conc.	23				ASTM F739	313
Rubber, Nitrile (NBR)	Ansell Edmont Sol-Vex 37-165; unsupported glove film; 0.54 mm thick			fair permeation rate; 51 to 100 eyedropper size drops per hour; breakthrough time: 30 minutes; permeation rate: <900 µg/cm²/min	95.5% conc.	23				ASTM F739	313

NITROPROPANE

Material Family	Material Note	Perm. Coefficient (cm³ · mm/m² · day · atm)	VTR (g · mm/m² · day)	Non-normalized Data	Penetrant Note	Temp. (°C)	Time (days)	RH (%)	Pressure (kPa)	Test Note	Source
Polyvinyl Alcohol	Ansell Edmont PVA; supported (lined) glove film; specified by glove film weight			low permeation rate; 0 to 1/2 eyedropper size drops per hour; PVA coating is water soluble; breakthrough time: >360 minutes; permeation rate: <0.9 µg/cm²/min	95.5% conc.	23				ASTM F739	313
Rubber, Latex (NR)	Ansell Edmont Canners and Handlers 392; unsupported glove film; 0.48 mm thick			good permeation rate; 6 to 50 eyedropper size drops per hour; breakthrough time: 5 minutes; permeation rate: <90 µg/cm²/min	95.5% conc.	23				ASTM F739	313
Rubber, Neoprene (CR)	Ansell Edmont Neoprene 29-840; unsupported glove film; 0.38 mm thick			fair permeation rate; 51 to 100 eyedropper size drops per hour; breakthrough time: 5 minutes; permeation rate: <300 µg/cm²/min	95.5% conc.	23				ASTM F739	313
	Ansell Edmont Neoc; supported (lined) glove film; specified by glove film weight			good permeation rate; 6 to 50 eyedropper size drops per hour; breakthrough time: 60 minutes; permeation rate: <90 µg/cm²/min	95.5% conc.	23				ASTM F739	313
TPE, Vinyl coated fabric	Ansell Edmont Wet-Wear 700; clothing; PVC coating with Nylon/Polyester lining			good permeation rate; 6 to 50 eyedropper size drops per hour; breakthrough time: <5 minutes; permeation rate: <90 µg/cm²/min	95.5% conc.	23				ASTM F739	313

NITROUS OXIDE

Material Family	Material Note	Perm. Coefficient (cm³ · mm/m² · day · atm)	VTR (g · mm/m² · day)	Non-normalized Data	Penetrant Note	Temp. (°C)	Time (days)	RH (%)	Pressure (kPa)	Test Note	Source
Fluoropolymer, PVDF	Solvay Solef; 0.025 mm thick; cast film	22.8				23				ASTM D1434	125

OCTYL ALCOHOL

Material Family	Material Note	Perm. Coefficient (cm³ · mm/m² · day · atm)	VTR (g · mm/m² · day)	Non-normalized Data	Penetrant Note	Temp. (°C)	Time (days)	RH (%)	Pressure (kPa)	Test Note	Source
Polyvinyl Alcohol	Ansell Edmont PVA; supported (lined) glove film; specified by glove film weight			low permeation rate; 0 to 1/2 eyedropper size drops per hour; PVA coating is water soluble; lower detection limit: 0 ppm; permeation rate: <0.9 µg/cm²/min	95.5% conc.	23	0.25			ASTM F739	313
Rubber, Latex (NR)	Ansell Edmont Canners and Handlers 392; unsupported glove film; 0.48 mm thick			very good permeation rate; 1 to 5 eyedropper size drops per hour; breakthrough time: 30 minutes; permeation rate: <9 µg/cm²/min	95.5% conc.	23				ASTM F739	313

OCTYL ALCOHOL (continued)

Material Family	Material Note	Perm. Coefficient (cm³·mm/m²·day·atm)	VTR (g·mm/m²·day)	Non-normalized Data	Penetrant Note	Temp. (°C)	Time (days)	RH (%)	Pressure (kPa)	Test Note	Source
Rubber, Neoprene (CR)	Ansell Edmont Neoprene 29-840; unsupported glove film; 0.38 mm thick			low permeation rate; 0 to 1/2 eyedropper size drops per hour; breakthrough time: 420 minutes; permeation rate: <0.9 µg/cm²/min		23				ASTM F739	313
	Ansell Edmont Neox; supported (lined) glove film; specified by glove film weight			low permeation rate; 0 to 1/2 eyedropper size drops per hour; breakthrough time: >420 minutes; permeation rate: <0.9 µg/cm²/min		23					313
Rubber, Nitrile (NBR)	Ansell Edmont Sol-Vex 37-165; unsupported glove film; 0.54 mm thick			low permeation rate; 0 to 1/2 eyedropper size drops per hour; lower detection limit: 0 ppm; permeation rate: <0.9 µg/cm²/min		23	0.25			ASTM F739	313
TPE, Vinyl	Ansell Edmont Monkey Grip; supported (lined) glove film; specified by glove film weight			low permeation rate; 0 to 1/2 eyedropper size drops per hour; breakthrough time: >300 minutes; permeation rate: <0.9 µg/cm²/min		23				ASTM F739	313
	Ansell Edmont Wet-Wear 550; clothing; PVC stretch outer layer bonded to a lightweight liner			low permeation rate; 0 to 1/2 eyedropper size drops per hour; breakthrough time: 45 minutes; permeation rate: <0.9 µg/cm²/min		23				ASTM F739	313
	Ansell Edmont Wet-Wear 600; clothing; nylon netting bonded between two layers of PVC			low permeation rate; 0 to 1/2 eyedropper size drops per hour; breakthrough time: 9 minutes; permeation rate: <0.9 µg/cm²/min		23				ASTM F739	313
TPE, Vinyl coated fabric	Ansell Edmont Wet-Wear 700; clothing; PVC coating with Nylon/Polyester lining			low permeation rate; 0 to 1/2 eyedropper size drops per hour; lower detection limit: 0 ppm; permeation rate: <0.9 µg/cm²/min		23	0.25			ASTM F739	313

OLEIC ACID

Material Family	Material Note	Perm. Coefficient (cm³·mm/m²·day·atm)	VTR (g·mm/m²·day)	Non-normalized Data	Penetrant Note	Temp. (°C)	Time (days)	RH (%)	Pressure (kPa)	Test Note	Source
Polyvinyl Alcohol	Ansell Edmont PVA; supported (lined) glove film; specified by glove film weight			low permeation rate; 0 to 1/2 eyedropper size drops per hour; PVA coating is water soluble; breakthrough time: 60 minutes; permeation rate: <0.9 µg/cm²/min		23				ASTM F739	313
Rubber, Latex (NR)	Ansell Edmont Canners and Handlers 392; unsupported glove film; 0.48 mm thick			no permeation detected during a 6 hour test; lower detection limit: 0 ppm		23	0.25			ASTM F739	313
Rubber, Neoprene (CR)	Ansell Edmont Neoprene 29-840; unsupported glove film; 0.38 mm thick			very good permeation rate; 1 to 5 eyedropper size drops per hour; breakthrough time: 60 minutes; permeation rate: <9 µg/cm²/min		23				ASTM F739	313
	Ansell Edmont Neox; supported (lined) glove film; specified by glove film weight			low permeation rate; 0 to 1/2 eyedropper size drops per hour; breakthrough time: 150 minutes; permeation rate: <0.9 µg/cm²/min		23				ASTM F739	313
Rubber, Nitrile (NBR)	Ansell Edmont Sol-Vex 37-165; unsupported glove film; 0.54 mm thick			low permeation rate; 0 to 1/2 eyedropper size drops per hour; lower detection limit: 0 ppm; permeation rate: <0.9 µg/cm²/min		23	0.25			ASTM F739	313
TPE, Vinyl	Ansell Edmont Monkey Grip; supported (lined) glove film; specified by glove film weight			very good permeation rate; 1 to 5 eyedropper size drops per hour; breakthrough time: 90 minutes; permeation rate: <9 µg/cm²/min		23				ASTM F739	313

OXALIC ACID

Material Family	Material Note	Perm. Coefficient (cm³·mm/m²·day·atm)	VTR (g·mm/m²·day)	Non-normalized Data	Penetrant Note	Temp. (°C)	Time (days)	RH (%)	Pressure (kPa)	Test Note	Source
Rubber, Latex (NR)	Ansell Edmont Canners and Handlers 392; unsupported glove film; 0.48 mm thick			no permeation detected during a 6 hour test; lower detection limit: 0 ppm	saturated	23	0.25			ASTM F739	313
Rubber, Neoprene (CR)	Ansell Edmont Neoprene 29-840; unsupported glove film; 0.38 mm thick			no permeation detected during a 6 hour test; lower detection limit: 0 ppm	saturated	23	0.25			ASTM F739	313
	Ansell Edmont Neox; supported (lined) glove film; specified by glove film weight			no permeation detected during a 6 hour test; lower detection limit: 0 ppm	saturated	23	0.25			ASTM F739	313
Rubber, Nitrile (NBR)	Ansell Edmont Sol-Vex 37-165; unsupported glove film; 0.54 mm thick			no permeation detected during a 6 hour test; lower detection limit: 0 ppm	saturated	23	0.25			ASTM F739	313
TPE, Vinyl	Ansell Edmont Monkey Grip; supported (lined) glove film; specified by glove film weight			no permeation detected during a 6 hour test; lower detection limit: 0 ppm	saturated	23	0.25			ASTM F739	313

OXYGEN

Material Family	Material Note	Perm. Coefficient (cm³·mm/m²·day·atm)	VTR (g·mm/m²·day)	Non-normalized Data	Penetrant Note	Temp. (°C)	Time (days)	RH (%)	Pressure (kPa)	Test Note	Source
ABS	BASF AG Terluran 877 M; moderate flow, 0.1 mm thick; film	45.6				23				DIN 53380; Values for permeability depend on the conditions under which the film was produced. Figures determined may differ by as much as 50% from those given.	137
	BASF AG Terluran 967 K; moderate flow, 0.1 mm thick; w/ butadiene acrylic rubber; film	50.7				23				"	137
	BASF AG Terluran 997 VE; 0.1 mm thick, low flow, w/ butadiene acrylic rubber; film	81.1				23				"	137
	Dow; low nitrile content; film	78.7 - 102				24					250
	Dow; medium nitrile content; film	47.2 - 55.1				24					250
Acetal	DuPont Delrin	4.7 - 6.7				23		50			201
Acetal Copolymer	Hoechst Cel. Celcon M25; high molecular weight, 0.15 mm thick; 2.5 g/10 min. MFI; film	2.0 - 2.9									210
	Hoechst Cel. Celcon M270; low molecular weight, high flow, 0.15 mm thick; 27.0 g/10 min. MFI; film	2.0 - 2.9									210
	Hoechst Cel. Celcon M90; gen. purp. grade, 0.15 mm thick; 9.0 g/10 min. MFI; film	2.0 - 2.9									210
Acrylonitrile	film	0.02				24					250
Acrylonitrile Copolymer, AMA	film	0.08				24					250
	BP Chem. Barex 210; barrier prop.	0.06				5		0			264
	BP Chem. Barex 210; barrier prop.	0.31				23		0			264
	BP Chem. Barex 210; packaging; impact modified, barrier prop.	0.32				22.8		100		ASTM D3985	296
	BP Chem. Barex 210; packaging; impact modified, barrier prop.	0.32				22.8		0		ASTM D1434	296
	BP Chem. Barex 210; packaging; impact modified, barrier prop.	0.32				22.8		0		ASTM D3885	296
	BP Chem. Barex 210; packaging; impact modified, barrier prop.	0.32				22.8		90		ASTM D3885	296
	BP Chem. Barex 218; packaging; impact modified, high impact, barrier prop.	0.64				22.8		100		ASTM D3985	296
	BP Chem. Barex 218; packaging; impact modified, high impact, barrier prop.	0.64				22.8		0		ASTM D1434	296
	BP Chem. Barex 210; barrier prop.	0.79				35		0			264
	BP Chem. Barex 210; barrier prop.	2.41				50		0			264
ASA	BASF AG Luran S 757 R; 0.1 mm thick; blown film	15.2				23				DIN 53380	143
	BASF AG Luran S 776 S; 0.1 mm thick; blown film	18.2				23				DIN 53380	143
	BASF AG Luran S 776 S; 0.1 mm thick; film	55.7				23				DIN 53380 part 2 method M; Values for permeability depend on the conditions under which the film was produced. Figures determined may differ by as much as 50% from those given.	142
	BASF AG Luran S 797 S; 0.1 mm thick; film	50.7				23					142
Cellulosic Plastic	Cellophane; 0.023 mm thick; PVDC coated; film	0.1				20		65		ASTM D3985	268
	Cellophane; 0.023 mm thick; PVDC coated; film	0.28				20		85		ASTM D3985	268

OXYGEN (continued)

Material Family	Material Note	Perm. Coefficient (cm³·mm/m²·day·atm)	VTR (g·mm/m²·day)	Non-normalized Data	Penetrant Note	Temp (°C)	Time (days)	RH (%)	Pressure (kPa)	Test Note	Source
Cellulosic Plastic	Cellophane; 0.023 mm thick; PVDC coated; film	0.81				20		100		ASTM D3985	268
	Cellophane; 0.023 mm thick; PVDC coated; film	0.03				35		0		JIS Z1707	268
Epoxy		2.0 - 3.9				25				ASTM D1434-63T	121
EVAL/ LDPE Film	EVAL EF-XL barrier; 0.015/ 0.051 mm thick; LDPE inside	0.02				20				RH: 65% outside; 100% inside	265
	EVAL EF-XL barrier; 0.015/ 0.051 mm thick; LDPE inside	0.04				20				RH: 80% outside; 100% inside	265
	EVAL EF-XL barrier; 0.015/ 0.051 mm thick; LDPE inside	0.02				20				RH: 65% outside; 10% inside	265
	EVAL EF-XL barrier; 0.015/ 0.051 mm thick; LDPE inside	0.04				20				RH: 80% outside; 10% inside	265
EVOH	Eval Co. Eval E; barrier prop.; 44% ethylene	0.01				5		0			264
	Eval Co. Eval F; barrier prop.; 32% ethylene	0.0012				5		0			264
	Eval Co. Eval G; barrier prop.; 46% ethylene	0.03				5		0			264
	Eval Co. Eval H; barrier prop.; 38% ethylene	0.0024				5		0			264
	Eval Co. Eval K; barrier prop.; 38% ethylene	0.0024				5		0			264
	Eval Co. Eval L; barrier prop.; 27 % ethylene	0.00039				5		0			264
	Eval Co. Eval E; 110°C chill roll temp.; no heat treatment; 44% ethylene; not oriented	0.026				20		0			264
	Eval Co. Eval E; 110°C chill roll temp.; no heat treatment; 44% ethylene; not oriented	0.24				20		100			264
	Eval Co. Eval E; 50°C chill roll temp.; 140°C heat treatment; 44% ethylene; biaxially oriented	0.024				20		0			264
	Eval Co. Eval E; 50°C chill roll temp.; 140°C heat treatment; 44% ethylene; biaxially oriented	0.06				20		100			264
	Eval Co. Eval E; 50°C chill roll temp.; 140°C heat treatment; 44% ethylene; not oriented	0.024				20		0			264
	Eval Co. Eval E; 50°C chill roll temp.; 140°C heat treatment; 44% ethylene; not oriented	0.16				20		100			264
	Eval Co. Eval E; 50°C chill roll temp.; 140°C heat treatment; 44% ethylene; uniaxially oriented	0.024				20		0			264
	Eval Co. Eval E; 50°C chill roll temp.; 140°C heat treatment; 44% ethylene; uniaxially oriented	0.079				20		100			264
	Eval Co. Eval E; 50°C chill roll temp.; no heat treatment; 44% ethylene; biaxially oriented	0.028				20		0			264
	Eval Co. Eval E; 50°C chill roll temp.; no heat treatment; 44% ethylene; biaxially oriented	0.28				20		100			264
	Eval Co. Eval E; 50°C chill roll temp.; no heat treatment; 44% ethylene; not oriented	0.03				20		0			264
	Eval Co. Eval E; 50°C chill roll temp.; no heat treatment; 44% ethylene; not oriented	0.30				20		100			264
	Eval Co. Eval E; 50°C chill roll temp.; no heat treatment; 44% ethylene; uniaxially oriented	0.028				20		0			264
	Eval Co. Eval E; 50°C chill roll temp.; no heat treatment; 44% ethylene; uniaxially oriented	0.28				20		100			264
	Eval Co. Eval E; barrier EF-E; 0.02 mm thick; 44% ethylene	0.02				23		0			264
	Eval Co. Eval EF-E; 0.02 mm thick; 44% ethylene; film	0.04				20		65		ASTM D3985	268
	Eval Co. Eval EF-E; 0.02 mm thick; 44% ethylene; film	0.08				20		85		ASTM D3985	268

OXYGEN (continued)

Material Family	Material Note	Perm. Coefficient (cm³·mm/m²·day·atm)	VTR (g·mm/m²·day)	Non-normalized Data	Penetrant Note	Temp. (°C)	Time (days)	RH (%)	Pressure (kPa)	Test Note	Source
EVOH	Eval Co. Eval EF-E; 0.02 mm thick; 44% ethylene; film	0.26				20		100		ASTM D3985	268
	Eval Co. Eval EF-E; barrier prop.; 44% ethylene; film	0.03				20		65			265
	Eval Co. Eval EF-E; barrier prop.; 44% ethylene; film	0.07				20		85			265
	Eval Co. Eval EF-E; barrier prop.; 44% ethylene; film	0.2				20		100			265
	Eval Co. Eval EF-E; barrier prop.; 44% ethylene; film	0.06				20		0			265
	Eval Co. Eval EF-F; 0.015 mm thick; 32% ethylene; film	0.01				20		65		ASTM D3985	268
	Eval Co. Eval EF-F; 0.015 mm thick; 32% ethylene; film	0.05				20		85		ASTM D3985	268
	Eval Co. Eval EF-F; 0.015 mm thick; 32% ethylene; film	0.63				20		100		ASTM D3985	268
	Eval Co. Eval EF-F; barrier prop.; 32% ethylene; film	0.01				20		65			265
	Eval Co. Eval EF-F; barrier prop.; 32% ethylene; film	0.03				20		85			265
	Eval Co. Eval EF-F; barrier prop.; 32% ethylene; film	0.39				20		100			265
	Eval Co. Eval EF-F; barrier prop.; 32% ethylene; film	0.01				20		0			265
	Eval Co. Eval EF-XL; barrier prop., biaxially oriented; film	0.004				20		65			265
	Eval Co. Eval EF-XL; barrier prop., biaxially oriented; film	0.02				20		85			265
	Eval Co. Eval EF-XL; barrier prop., biaxially oriented; film	0.09				20		100			265
	Eval Co. Eval EF-XL; barrier prop., biaxially oriented; film	0.01				20		0			265
	Eval Co. Eval EF-XL; biaxially oriented, 0.015 mm thick; film	0.01				20		65		ASTM D3985	268
	Eval Co. Eval EF-XL; biaxially oriented, 0.015 mm thick; film	0.03				20		85		ASTM D3985	268
	Eval Co. Eval EF-XL; biaxially oriented, 0.015 mm thick; film	0.15				20		100		ASTM D3985	268
	Eval Co. Eval F; 110°C chill roll temp.; no heat treatment; 32% ethylene; not oriented	0.003				20		0			264
	Eval Co. Eval F; 110°C chill roll temp.; no heat treatment; 32% ethylene; not oriented	0.87				20		100			264
	Eval Co. Eval F; 50°C chill roll temp.; 140°C heat treatment; 32% ethylene; biaxially oriented	0.0024				20		0			264
	Eval Co. Eval F; 50°C chill roll temp.; 140°C heat treatment; 32% ethylene; biaxially oriented	0.06				20		100			264
	Eval Co. Eval F; 50°C chill roll temp.; 140°C heat treatment; 32% ethylene; not oriented	0.0026				20		0			264
	Eval Co. Eval F; 50°C chill roll temp.; 140°C heat treatment; 32% ethylene; not oriented	0.28				20		100			264
	Eval Co. Eval F; 50°C chill roll temp.; 140°C heat treatment; 32% ethylene; uniaxially oriented	0.0002				20		0			264
	Eval Co. Eval F; 50°C chill roll temp.; 140°C heat treatment; 32% ethylene; uniaxially oriented	0.1				20		100			264

OXYGEN (continued)

Material Family	Material Note	Permeability Data			Test Conditions					Test Note	Source
		Perm. Coefficient (cm³·mm/m²·day·atm)	VTR (g·mm/m²·day)	Non-normalized Data	Penetrant Note	Temp. (°C)	Time (days)	RH (%)	Pressure (kPa)		
EVOH	Eval Co. Eval F; 50°C chill roll temp.; no heat treatment; 32% ethylene; biaxially oriented	0.003				20		0			264
	Eval Co. Eval F; 50°C chill roll temp.; no heat treatment; 32% ethylene; biaxially oriented	0.79				20		100			264
	Eval Co. Eval F; 50°C chill roll temp.; no heat treatment; 32% ethylene; not oriented	0.0031				20		0			264
	Eval Co. Eval F; 50°C chill roll temp.; no heat treatment; 32% ethylene; not oriented	1.02				20		100			264
	Eval Co. Eval F; 50°C chill roll temp.; no heat treatment; 32% ethylene; uniaxially oriented	0.003				20		0			264
	Eval Co. Eval F; 50°C chill roll temp.; no heat treatment; 32% ethylene; uniaxially oriented	0.83				20		100			264
	Eval Co. Eval F; barrier prop.; 32% ethylene	0.01				23		0			264
	Eval Co. Eval G; barrier prop.; 48% ethylene	0.05				23		0			264
	Eval Co. Eval H; barrier prop.; 38% ethylene	0.01				23		0			264
	Eval Co. Eval K; barrier prop.; 38% ethylene	0.01				23		0			264
	Eval Co. Eval L; barrier prop.; 27 % ethylene	0.0024				23		0			264
		0.01				22.8		90		ASTM D3895	296
		0.39				22.8		90		ASTM D3895	296
	DuPont Selar OH 3003; packaging; barrier prop.; 32% ethylene; 3 g/10 min. MFI	0.01				30		dry			295
	DuPont Selar OH 3003; packaging; barrier prop.; 32% ethylene; 3 g/10 min. MFI	0.05				30		80			295
	DuPont Selar OH 4416; packaging; barrier prop.; 44% ethylene; 12 g/10 min. MFI	0.06				30		dry			295
	DuPont Selar OH 4416; packaging; barrier prop.; 44% ethylene; 12 g/10 min. MFI	0.13				30		80			295
	DuPont Selar OH BX220; packaging; barrier prop., heat stabilized; 32% ethylene; 3 g/10 min. MFI	0.01				30		dry			295
	DuPont Selar OH BX220; packaging; barrier prop., heat stabilized; 32% ethylene; 3 g/10 min. MFI	0.05				30		80			295
	DuPont Selar OH BX228; packaging; barrier prop.; 44% ethylene; 11 g/10 min. MFI	0.06				30		dry			295
	DuPont Selar OH BX228; packaging; barrier prop.; 44% ethylene; 11 g/10 min. MFI	0.13				30		80			295
	DuPont Selar OH BX250; packaging; barrier prop., heat stabilized, formability; 32% ethylene; 1.5 g/10 min. MFI	0.01				30		dry			295
	DuPont Selar OH BX250; packaging; barrier prop., heat stabilized, formability; 32% ethylene; 1.5 g/10 min. MFI	0.05				30		80			295
	Eval Co. Eval E; barrier prop.; 44% ethylene	0.05				35		0			264
	Eval Co. Eval E; barrier prop.; 44% ethylene	0.14				50		0			264
	Eval Co. Eval EF-E; 0.02 mm thick; 44% ethylene; film	0.08				35		0		JIS Z1707	268
	Eval Co. Eval EF-F; 0.015 mm thick; 32% ethylene; film	0.01				35		0		JIS Z1707	268
	Eval Co. Eval EF-XL; biaxially oriented, 0.015 mm thick; film	0.01				35		0		JIS Z1707	268
	Eval Co. Eval F; barrier prop.; 32% ethylene	0.01				35		0			264
	Eval Co. Eval F; barrier prop.; 32% ethylene	0.03				50		0			264
	Eval Co. Eval G; barrier prop.; 48% ethylene	0.07				35		0			264

OXYGEN (continued)

Material Family	Material Note	Perm. Coefficient (cm³·mm/m²·day·atm)	VTR (g·mm/m²·day)	Non-normalized Data	Penetrant Note	Temp. (°C)	Time (days)	RH (%)	Pressure (kPa)	Test Note	Source
EVOH	Eval Co. Eval G; barrier prop.: 48% ethylene	0.16				50		0			264
	Eval Co. Eval H; barrier prop.; 38% ethylene	0.02				35		0			264
	Eval Co. Eval H; barrier prop.; 38% ethylene	0.07				50		0			264
	Eval Co. Eval K; barrier prop.; 38% ethylene	0.02				35		0			264
	Eval Co. Eval K; barrier prop.: 38% ethylene	0.07				50		0			264
	Eval Co. Eval L; barrier prop.; 27 % ethylene	0.01				35		0			264
	Eval Co. Eval L; barrier prop.; 27 % ethylene	0.02				50		0			264
Fluoroelastomer, FKM	DuPont Viton; compounded	95.0				30				one atmosphere @80°C; specimen size: 1 cm² x 1 cm thick	305
Fluoropolymer, CTFE	3M Kel-F 81; amorphous form; film	0.46				0					96
	3M Kel-F; 0.01 mm thick	0.26				25			1724	mass spectrometry and calibrated standard gas leaks; developed by McDonnell Douglas Space Systems Caompany Chemistry Laboratory	306
	3M Kel-F; 0.01 mm thick	0.25				25			3447	"	306
	3M Kel-F 81; amorphous form; film	2.6				25				STP conditions	96
	Allied Sig. Aclar 22A; transparent; film	4.7				25				STP conditions	138
	Allied Sig. Aclar 22C; transparent; film	5.9				25				STP conditions	138
Fluoropolymer, CTFE	Allied Sig. Aclar 33C; transparent; film	2.8				25					138
	3M Kel-F 81; amorphous form; film	9.2				50				STP conditions	96
	3M Kel-F; 0.01 mm thick	8.2				52			1724	mass spectrometry and calibrated standard gas leaks; developed by McDonnell Douglas Space Systems Caompany Chemistry Laboratory	306
	3M Kel-F; 0.01 mm thick	8.2				52			3447	mass spectrometry and calibrated standard gas leaks; developed by McDonnell Douglas Space Systems Caompany Chemistry Laboratory	306
	3M Kel-F 81; amorphous form; film	37.4				75					96
Fluoropolymer, ECTFE	Ausimont Halar; 0.02 mm thick	0.48				-18			1724	mass spectrometry and calibrated standard gas leaks; developed by McDonnell Douglas Space Systems Caompany Chemistry Laboratory	306
	Ausimont Halar; 0.02 mm thick	0.5				-15			3447	"	306
	Ausimont Halar; 0.02 mm thick	10.2				25			1724	"	306
	Ausimont Halar; 0.02 mm thick	9.6				25			3447	"	306
	Ausimont Halar; 0.02 mm thick	45.2				55			1724	"	306
	Ausimont Halar; 0.02 mm thick	46.1				56			3447	"	306
Fluoropolymer, ETFE	Ausimont Hyflon 700; high molecular weight; 4 g/10 min. MFI	62.6				23				ASTM D1434; activation energy = 6-8 kcal/mole	114
	Ausimont Hyflon 800; low molecular weight; 11 g/10 min. MFI	62.6				23				ASTM D1434; activation energy = 6-8 kcal/mole	114
	DuPont Tefzel; developmental material. 0.102 mm thick; film	39.4				25				ASTM D1434	205

OXYGEN (continued)

Material Family	Material Note	Permeability Data					Test Conditions						Source
		Perm. Coefficient (cm³ · mm/m² · day · atm)	VTR (g · mm/m² · day)	Non-normalized Data			Penetrant Note	Temp. (°C)	Time (days)	RH (%)	Pressure (kPa)	Test Note	
Fluoropolymer, FEP	DuPont Teflon; 0.05 mm thick	9.1						-16			1724	mass spectrometry and calibrated standard gas leaks; developed by McDonnell Douglas Space Systems Caompany Chemistry Laboratory	306
	DuPont Teflon; 0.05 mm thick	9.0						-16			3447	"	306
	DuPont Teflon; 0.05 mm thick	116						25			1724	"	306
	DuPont Teflon; 0.05 mm thick	101						25			3447	"	306
	DuPont Teflon; 0.05 mm thick	295						25				STP conditions	138
	DuPont Teflon; 0.05 mm thick	452						52			1724	mass spectrometry and calibrated standard gas leaks; developed by McDonnell Douglas Space Systems Caompany Chemistry Laboratory	306
Fluoropolymer, PFA	DuPont Teflon PFA; film	465						53			3447		306
Fluoropolymer, PVDF	Solvay Solef 1008; 0.1 mm thick, transluscent; film	347						25				ASTM D1434	39
		2.1						23				ASTM D1434	125
	Atochem Foraflon; 0.037 mm thick; extruded film	0.55						25				STP conditions	138
		5.2						30				ISO 2556	89
Fluoropolymer, PVF	DuPont Teflon; 0.03 mm thick	1.2						25				STP conditions	138
Fluoropolymer, TFE	DuPont Teflon; 0.03 mm thick	46.1						-17			1724	mass spectrometry and calibrated standard gas leaks; developed by McDonnell Douglas Space Systems Caompany Chemistry Laboratory	306
	DuPont Teflon; 0.03 mm thick	39.8						-17			3447	"	306
	DuPont Teflon; 0.05 mm thick; carbon filled	81.2						-16			1724	"	306
	DuPont Teflon; 0.05 mm thick; carbon filled	83.7						-15			3447	"	306
	DuPont Teflon; 0.03 mm thick	223						25			1724	"	306
	DuPont Teflon; 0.03 mm thick	222						25			3447	"	306
	DuPont Teflon; 0.05 mm thick; carbon filled	442						25			1724	"	306
	DuPont Teflon; 0.05 mm thick; carbon filled	451						25			3447	"	306
	DuPont Teflon; 0.03 mm thick	471						51			1724	"	306
	DuPont Teflon; 0.03 mm thick	478						51			3447	"	306
	DuPont Teflon; 0.05 mm thick; carbon filled	1016						55			1724	"	306
	DuPont Teflon; 0.05 mm thick; carbon filled	884						53			3447	"	306
HDPE/ EAA/ Nylon/ EAA Film	Dow Nylopak 570; form-fill-seal pouch; neutral, barrier prop., 0.051 mm thick; coextruded film	9.5						23		10		ASTM D3985	258
HDPE/ EVA/ PVDC/ EVA Film	Dow Saranex 25; form-fill-seal pouch, bag-in-box; neutral, barrier prop., 0.051 mm thick; high slip; coextruded film	0.4						23		10		ASTM D3985	257

OXYGEN (continued)

Material Family	Material Note	Permeability Data			Test Conditions						Source
		Perm. Coefficient (cm³·mm/m²·day·atm)	VTR (g·mm/m²·day)	Non-normalized Data	Penetrant Note	Temp. (°C)	Time (days)	RH (%)	Pressure (kPa)	Test Note	
HDPE/ EVAL/ LDPE Film	EVAL EP-E barrier; 0.6/ 0.025/ 0.15 mm thick; LDPE inside; HDPE outside	0.6				20				RH: 65% outside; 10 (wet)% inside; 14.5% barrier	264
	EVAL EP-E barrier; 0.6/ 0.025/ 0.15 mm thick; LDPE inside; HDPE outside	0.6				20				RH: 75% outside; 10 (wet)% inside; 15.3% barrier	264
HDPE/ EVAL/ PP Film	EVAL EP-F barrier; 0.15/ 0.025/ 0.6 mm thick; PP inside; HDPE outside	0.42				20				RH: 65% outside; 100 (wet)% inside; 75.9% barrier	264
	EVAL EP-F barrier; 0.15/ 0.025/ 0.6 mm thick; PP inside; HDPE outside	0.71				20				RH: 75% outside; 100 (wet)% inside; 82.8% barrier	264
HDPE/ PVDC/ LDPE Film	Saran VC barrier; 0.6/ 0.025/ 0.15 mm thick; LDPE inside; HDPE outside	1.8				20				RH: 65% outside; 10 (wet)% inside; 14.5% barrier	264
	Saran VC barrier; 0.6/ 0.025/ 0.15 mm thick; LDPE inside; HDPE outside	1.8				20				RH: 75% outside; 10 (wet)% inside; 15.3% barrier	264
HDPE/ PVDC/ PP Film	Saran VC barrier; 0.15/ 0.025/ 0.6 mm thick; PP inside; HDPE outside	1.4				20				RH: 75% outside; 100 (wet)% inside; 82.8% barrier	264
	Saran VC barrier; 0.15/ 0.025/ 0.6 mm thick; PP inside; HDPE outside	1.4				20				RH: 65% outside; 100 (wet)% inside; 75.9% barrier	264
Ionomer	DuPont Surlyn 1601; 0.94 g/cm³ density; sodium ion, 0.051 mm thick; blown film; 1.3 g/10 min. MFI	209									280
	DuPont Surlyn 1603; 0.94 g/cm³ density; sodium ion, 0.051 mm thick; blown film; 1.7 g/10 min. MFI	150									280
	DuPont Surlyn 1650; 0.95 g/cm³ density; sodium ion, 0.051 mm thick; blown film; 1.6 g/10 min. MFI	174									280
	DuPont Surlyn 1652; 0.94 g/cm³ density; zinc ion, 0.051 mm thick; blown film; 5.0 g/10 min. MFI	142									280
	DuPont Surlyn 1702; 0.94 g/cm³ density; zinc ion, 0.051 mm thick; blown film; 14.0 g/10 min. MFI	138									280
	DuPont Surlyn 1705; 0.95 g/cm³ density; zinc ion, 0.051 mm thick; blown film; 5.5 g/10 min. MFI	134									280
	DuPont Surlyn 1707; 0.95 g/cm³ density; sodium ion, 0.051 mm thick; blown film; 0.9 g/10 min. MFI	130									280
	DuPont Surlyn F1605; 0.95 g/cm³ density; sodium ion, 0.051 mm thick; blown film; 2.8 g/10 min. MFI	158									280
	DuPont Surlyn F1706; 0.96 g/cm³ density; zinc ion, 0.051 mm thick; blown film; 0.7 g/10 min. MFI	146									280
	DuPont Surlyn F1801; 0.96 g/cm³ density; zinc ion, 0.051 mm thick; blown film; 1.0 g/10 min. MFI	170									280
	DuPont Surlyn F1855; 0.96 g/cm³ density; zinc ion, 0.051 mm thick; blown film; 1.0 g/10 min. MFI	233.2									280
	DuPont Surlyn F1856; 0.95 g/cm³ density; sodium ion, 0.051 mm thick; blown film; 1.0 g/10 min. MFI	229									280
Laminar, EVOH/HDPE	DuPont Selar RB 421/ HDPE; 15% Selar RB 421; barrier prop.; extrusion blow molded	0.42				23		75			293

Appendix I - Permeability Sort

OXYGEN (continued)

Material Family	Material Note	Perm. Coefficient (cm³·mm/m²·day·atm)	VTR (g·mm/m²·day)	Non-normalized Data	Penetrant Note	Temp. (°C)	Time (days)	RH (%)	Pressure (kPa)	Test Note	Source
Laminar, Nylon/HDPE	DuPont Selar RB 215/ HDPE; 10% Selar RB 215; barrier prop.; extrusion blow molded	15.0				23		75			293
	DuPont Selar RB 300/ HDPE; 10% Selar RB 300; barrier prop.; extrusion blow molded	13.4				23		75			293
Laminar, Nylon/LDPE	DuPont Selar RB 215/ LDPE; 15% Selar RB 215; barrier prop.; extrusion blow molded	35.4				23		75			293
Laminar, Nylon/PP	DuPont Selar RB 421/ PP; 10% Selar RB 240; barrier prop.; extrusion blow molded	14.2				23		75			293
LDPE/ EVA/ PVDC/ EVA Film	Dow Saranex 23; laminations; neutral, barrier prop., 0.051 mm thick; medium slip; coextruded film	0.4				23		10		ASTM D3985	257
LDPE/ EVA/ PVDC/ EVA/ LDPE Film	Dow Saranex 11; laminations; neutral, barrier prop., 0.038 mm thick; medium slip; coextruded film	0.88				23		10		ASTM D3985	257
	Dow Saranex 12; barrier prop., 0.051 mm thick; film	0.39				20		65			265
	Dow Saranex 12; barrier prop., 0.051 mm thick; film	0.39				20		85			265
	Dow Saranex 12; barrier prop., 0.051 mm thick; film	0.39				20		100			265
	Dow Saranex 14; barrier prop., 0.051 mm thick; film	0.18				20		65			265
	Dow Saranex 14; barrier prop., 0.051 mm thick; film	0.18				20		85			265
	Dow Saranex 14; barrier prop., 0.051 mm thick; film	0.18				20		100			265
	Dow Saranex 14; form-fill-seal pouch; neutral, barrier prop., 0.051 mm thick; medium slip; coextruded film	0.4				23		10		ASTM D3985	257
	Dow Saranex 15; form-fill-seal pouch; neutral, barrier prop., 0.076 mm thick; medium slip; coextruded film	0.59				23		10		ASTM D3985	257
	Dow Saranex 15; form-fill-seal pouch; neutral, barrier prop., 0.102 mm thick; medium slip; coextruded film	0.79				23		10		ASTM D3985	257
LDPE/ EVAL/ PP Film	EVAL EP-E barrier; 0.15/ 0.025/ 0.6 mm thick; PP inside; LDPE outside	1.1				20				RH: 65% outside; 100 (wet)% inside; 69.8% barrier	264
	EVAL EP-E barrier; 0.15/ 0.025/ 0.6 mm thick; PP inside; LDPE outside	1.4				20				RH: 75% outside; 100 (wet)% inside; 78.4% barrier	264
LDPE/ PVDC/ PP Film	Saran VC barrier; 0.15/ 0.025/ 0.6 mm thick; PP inside; LDPE outside	1.4				20				RH: 65% outside; 100 (wet)% inside; 69.8% barrier	264
	Saran VC barrier; 0.15/ 0.025/ 0.6 mm thick; PP inside; LDPE outside	1.4				20				RH: 75% outside; 100 (wet)% inside; 78.4% barrier	264
Liquid Crystal Polymer	Hoechst AG Vectra A950; 0.019 mm thick; film	0.03				23		0		test area: 5 cm²	70
	"	0.02				23		100		test area: 5 cm²	70
	"	0.14				38		0		test area: 5 cm²	70
	"	0.06				38		100		test area: 5 cm²	70
Multilayer, LIM Structure	10% nylon MXD6; SMA; barrier prop., clarity, hot fill; SAN; lamellar IM	0.39									253
	2.5% nylon MXD6; SMA; barrier prop., clarity, hot fill; SAN; lamellar IM	1.8									253

OXYGEN (continued)

Material Family	Material Note	Permeability Data: Perm. Coefficient (cm³·mm/m²·day·atm)	VTR (g·mm/m²·day)	Non-normalized Data	Penetrant Note	Temp. (°C)	Time (days)	RH (%)	Pressure (kPa)	Test Note	Source
Multilayer, LIM Structure	20% nylon MXD6; SMA; barrier prop., hot fill, clarity; SAN; lamellar IM	non-detectable									253
	5% nylon MXD6; SMA; barrier prop., hot fill, clarity; SAN; lamellar IM	0.91									253
	EVOH; EVA; barrier prop.; PET; lamellar IM	0.24									253
	EVOH; PE-g-maleic anhydride; high impact, barrier prop., hot fill; HDPE; lamellar IM	0.24									253
	EVOH; PP-g-maleic anhydride; high impact, barrier prop., retort; PP; lamellar IM	0.24									253
	nylon MXD6; high impact, barrier prop., compatibilizer not required, clarity, TPU; lamellar IM	0.16									253
	nylon MXD6; high impact, transparent, barrier prop., clarity, compatibilizer not required, hot fill; IPS; lamellar IM	0.39									253
	nylon MXD6; SMA; barrier prop., clarity, hot fill; PS; lamellar IM	0.39									253
Nylon	oriented, 0.015 mm thick; film	1.3				20		65		ASTM D3985	268
	oriented, 0.015 mm thick; film	3.6				20		85		ASTM D3985	268
	oriented, 0.015 mm thick; film	12.4				20		100		ASTM D3985	268
	oriented, 0.017 mm thick; PVDC coated; film	0.2				20		65		ASTM D3985	268
	oriented, 0.017 mm thick; PVDC coated; film	0.2				20		85		ASTM D3985	268
	oriented, 0.017 mm thick; PVDC coated; film	0.2				20		100		ASTM D3985	268
	oriented; film	0.76				20		65			265
	oriented; film	2.1				20		85			265
	oriented; film	7.5				20		100			265
	oriented; film	0.98				20		0			265
	oriented; PVDC coated; film	0.14				20		65			265
	oriented; PVDC coated; film	0.14				20		85			265
	oriented; PVDC coated; film	0.14				20		100			265
	oriented; PVDC coated; film	0.28				20		0			265
	oriented, 0.015 mm thick; film	1.6				35		0		JIS Z1707	268
	oriented, 0.017 mm thick; PVDC coated; film	0.41				35		0		JIS Z1707	268
Nylon 6	Allied Sig. Capran 77C; 0.0254 mm thick; film	0.2				0		0		STP conditions	285
	oriented	0.19				5		0			264
	oriented	0.57				5		0			264
	Allied Sig. Capran; 0.0254 mm thick; film	1.02				23		0		permeability cell; STP conditions	284
	Allied Sig. Capran 77C; 0.019 mm thick; film	0.94				23		0		STP conditions	285
	Allied Sig. Capran 77C; 0.0254 mm thick; film	1.02				23		0		STP conditions	285
	Allied Sig. Capran 77K; 0.0254 mm thick; PVDC coated; film	0.2				23		0		STP conditions	285
	BASF Ultramid B36F; moderate flow, clarity; flat film, tubular film	0.61 - 0.71				23		40		DIN 53380	93
	BASF Ultramid B4; 0.02-0.05 mm thick, unstretched	0.57 - 0.8				20		40		DIN 53380	252
	BASF Ultramid B4; biaxially stretched, 0.02 mm thick	0.24 - 0.3				20		40		DIN 53380	252
	BASF Ultramid B4; moderate flow, 0.02-0.1 mm thick; flat film, tubular film	0.61 - 0.71				23		40		DIN 53380	93
	barrier prop.; film	1.4				22.8		0			294
	barrier prop.; film	2.8				22.8		80			294

OXYGEN (continued)

Material Family	Material Note	Perm. Coefficient (cm³·mm/m²·day·atm)	VTR (g·mm/m²·day)	Non-normalized Data	Penetrant Note	Temp. (°C)	Time (days)	RH (%)	Pressure (kPa)	Test Note	Source
Nylon 6	oriented	0.7				23		0			264
		2				23		0			264
		1.2				22.8		0		ASTM D1434	296
	Allied Sig. Capran 77C; 0.0254 mm thick; film	5.5				50		0		STP conditions	285
	oriented	1.3				35		0			264
		3.94				35		0			264
Nylon 6 / LDPE Film	BASF Ultramid B4/ Lupolen 3020 D; 0.03 mm/ 0.05 mm thick	1.6 - 2.0				20		40		DIN 53380	252
Nylon 66	BASF Ultramid A5; 0.02-0.1 mm thick, low flow; tubular film	0.3 - 0.41				23		40		DIN 53380	93
	BASF Ultramid A5; 0.05 mm thick; blown film	0.76				20		40		DIN 53380	252
	BASF Ultramid A5; low flow, 0.02-0.1 mm thick; flat film	0.61 - 0.71				23		40		DIN 53380	93
	DuPont Can. Dartek; transparent, 0.0254 mm thick; film	1.4				23		0		ASTM D1434-66, method V	276
	DuPont Can. Dartek; transparent, 0.0254 mm thick; film	6.3				23		100		ASTM D1434-66, method V	276
	DuPont Can. Dartek B-601; barrier prop., 0.0254 mm thick; PVDC coated; film	0.2				23		0		ASTM D1434-66	276
	DuPont Can. Dartek B-602; barrier prop., 0.038 mm thick PVDC coated; film	0.29				23		0		ASTM D1434-66	276
	DuPont Zytel 42; low flow; film	0.79				23		50			68
Nylon 66/610	Emser Grilon XE3303; transparent, barrier prop., 0.05 mm thick; film	2.8				23		50			307
	Emser Grilon XE3303; transparent, barrier prop., 0.05 mm thick; film	3.8				23		85			307
	Emser Grilon XE3303; transparent, barrier prop., 0.05 mm thick; film	3.8						85		pre-test exposure: pasteurization	307
	Emser Grilon XE3303; transparent, barrier prop., 0.05 mm thick; film	3.8						85		pre-test exposure: steam sterilization	307
	Emser Grilon XE3303; transparent, barrier prop., 0.05 mm thick; film	3.8				23		100			307
Nylon 6/66	Allied Sig. Capran; 0.0254 mm thick; film	0.94				23		0 (dry)		ASTM D3985	284
	Allied Sig. Capran; 0.0254 mm thick; film	5.9				23		90 (wet)		permeability cell	284
	BASF Ultramid C35; mod-high flow, 0.02-0.1 mm thick; flat film, tubular film	0.81 - 0.91				23		40		DIN 53380	93
Nylon MXD6	barrier prop.	0.02				5		0			264
	barrier prop.	0.06				23		0			264
		0.07				22.8		0		ASTM D3985	296
		0.32				22.8		90		ASTM D3985	296
	barrier prop.	0.11				35		0			264
	barrier prop.	0.36				50		0			264
Nylon, Amorphous	DuPont Selar PA; barrier prop.; film	0.98				22.8		0			294
	DuPont Selar PA; barrier prop.; film	0.47				22.8		80			294
	DuPont Selar PT; food packaging; 0.23 mm thick, 80 mm height, 60 mm diameter, retort, hot fill, thermoformed; cup			0.204 cm³/pkg/day · atm						container before retort	290
	DuPont Selar PT; food packaging; 0.23 mm thick, 80 mm height, 60 mm diameter, retort, hot fill, thermoformed; cup			0.228 cm³/pkg/day · atm						container after retort; retort temperature: 121°C; retort time: 40 minutes	290

OXYGEN (continued)

Material Family	Material Note	Permeability Data — Perm. Coefficient (cm³·mm/m²·day·atm)	VTR (g·mm/m²·day)	Non-normalized Data	Test Conditions — Penetrant Note	Temp (°C)	Time (days)	RH (%)	Pressure (kPa)	Test Note	Source
Nylon, Amorphous	DuPont Selar PT; food packaging; transparent, heat stabilized, hot fill, retort; thermoformed; container	1.97				25					290
	Emser Grivory G21; transparent, barrier prop., 0.05 mm thick; film	1.5				23		50		DIN 53380	307
	Emser Grivory G21; transparent, barrier prop., 0.05 mm thick; film	0.41				23		100		DIN 53380	307
Nylon/ EVAL/ LDPE Film	EVAL EF-E barrier; 0.02/0.02/0.051 mm thick; LDPE inside; coated nylon outside	0.14				20				RH: 65% outside; 100% inside; 65% barrier	265
	EVAL EF-E barrier; 0.02/0.02/0.051 mm thick; LDPE inside; coated nylon outside	0.2				20				RH: 80% outside; 100% inside; 64% barrier	265
	EVAL EF-E barrier; 0.02/0.02/0.051 mm thick; LDPE inside; coated nylon outside	0.14				20				RH: 65% outside; 10% inside; 80% barrier	265
	EVAL EF-E barrier; 0.02/0.02/0.051 mm thick; LDPE inside; coated nylon outside	0.21				20				RH: 80% outside; 10% inside; 78% barrier	265
	EVAL EF-F barrier; 0.015/0.015/0.051 mm thick; LDPE inside; oriented nylon outside	0.05				20				RH - 65% outside; 100% inside; 66% barrier	265
Nylon/ EVAL/ LDPE Film	EVAL EF-F barrier; 0.015/0.015/0.051 mm thick; LDPE inside; oriented nylon outside	0.1				20				RH: 80% outside; 100% inside; 63% barrier	265
	EVAL EF-F barrier; 0.015/0.015/0.051 mm thick; LDPE inside; oriented nylon outside	0.04				20				RH: 65% outside; 10% inside; 81% barrier	265
	EVAL EF-F barrier; 0.015/0.015/0.051 mm thick; LDPE inside; oriented nylon outside	0.08				20				RH: 80% outside; 10% inside; 77% barrier	265
Nylon/ EVAL/ PP Film	EVAL EP-E barrier; 0.15/0.025/0.6 mm thick; PP inside; nylon outside	1				20				RH: 65% outside; 100 (wet)% inside; 66.4% barrier	264
	EVAL EP-E barrier; 0.15/0.025/0.6 mm thick; PP inside; nylon outside	1.3				20				RH: 75% outside; 100 (wet)% inside; 76% barrier	264
Nylon/ PVDC/ PP Film	Saran VC barrier; 0.15/0.025/0.6 mm thick; PP inside; nylon outside	1.4				20				RH: 65% outside; 100 (wet)% inside; 66.4% barrier	264
	Saran VC barrier; 0.15/0.025/0.6 mm thick; PP inside; nylon outside	1.4				20				RH: 75% outside; 100 (wet)% inside; 76% barrier	264
Parylene	Union Carbide Parylene C; vapor phase deposition; thin film	2.8				25				ASTM D1434-63T	121
	Union Carbide Parylene D; vapor phase deposition; thin film	12.6				25				ASTM D1434-63T	121
	Union Carbide Parylene N; highly crystalline, high molecular weight, completely linear; vapor phase deposition; thin film	15.4				25				ASTM D1434-63T	121
PC/ EVAL/ PP Film	EVAL EP-E barrier; 0.15/0.025/0.6 mm thick; PP inside; PC outside	0.97				20				RH: 65% outside; 100 (wet)% inside; 65.9% barrier	264
	EVAL EP-E barrier; 0.15/0.025/0.6 mm thick; PP inside; PC outside	1.3				20				RH: 75% outside; 100 (wet)% inside; 75.6% barrier	264
	EVAL EP-F barrier; 0.15/0.025/0.6 mm thick; PP inside; PC outside	0.26				20				RH: 65% outside; 100 (wet)% inside; 65.7% barrier	264
	EVAL EP-F barrier; 0.15/0.025/0.6 mm thick; PP inside; PC outside	0.42				20				RH: 75% outside; 100 (wet)% inside; 75.5% barrier	264
PC/ PVDC/ PP Film	Saran VC barrier; 0.15/0.025/0.6 mm thick; PP inside; PC outside	1.4				20				RH: 65% outside; 100 (wet)% inside; 65.7% barrier	264

OXYGEN (continued)

Material Family	Material Note	Perm. Coefficient (cm³·mm/m²·day·atm)	VTR (g·mm/m²·day)	Non-normalized Data	Penetrant Note	Temp. (°C)	Time (days)	RH (%)	Pressure (kPa)	Test Note	Source
PC/PVDC/PP Film	Saran VC barrier: 0.15/0.025/0.6 mm thick; PP inside; PC outside	1.4				20				RH: 65% outside; 100 (wet)% inside; 65.9% barrier	264
	Saran VC barrier: 0.15/0.025/0.6 mm thick; PP inside; PC outside	1.4				20				RH: 75% outside; 100 (wet)% inside; 75.5% barrier	264
	Saran VC barrier: 0.15/0.025/0.6 mm thick; PP inside; PC outside	1.4				20				RH: 75% outside; 100 (wet)% inside; 75.6% barrier	264
PE Ionomer Copolymer	BASF AG Lucalen H300MX; 0.1 mm thick; 8 wt.% acrylic acid; 7 g/10 min. MFI	159				23				DIN 53380	25
PE-Acrylic Acid Copolymer	BASF AG Lucalen A2710H; 0.1 mm thick; 17 wt.% acrylic acid; 1.7 g/10 min. MFI	550				23				DIN 53380	25
	BASF AG Lucalen A2910M; 0.1 mm thick; 11 wt.% acrylic acid; 7 g/10 min. MFI	243				23				DIN 53380	25
	BASF AG Lucalen A3710MX; 0.1 mm thick; 8 wt.% acrylic acid; 7 g/10 min. MFI	178				23				DIN 53380	25
PE/PVC-PVDC Copolymer Multilayer Film	Dow Saranex; multilayer film	0.2 - 0.79				24					250
PET/EVAL/LDPE Film	EVAL EF-F barrier: 0.013/0.015/0.051 mm thick; LDPE inside; PET outside	0.05				20				RH: 65% outside; 100% inside; 70% barrier	265
	EVAL EF-F barrier: 0.013/0.015/0.051 mm thick; LDPE inside; PET outside	0.12				20				RH: 80% outside; 100% inside; 57% barrier	265
PET/EVAL/LDPE Film	EVAL EF-F barrier: 0.013/0.015/0.051 mm thick; LDPE inside; PET outside	0.02				20				RH: 65% outside; 10% inside; 83% barrier	265
	EVAL EF-F barrier: 0.013/0.015/0.051 mm thick; LDPE inside; PET outside	0.05				20				RH: 80% outside; 10% inside; 69% barrier	265
PET/EVAL/PP Film	EVAL EP-F barrier: 0.15/0.025/0.6 mm thick; PP inside; PET outside	0.18				20				RH: 65% outside; 100 (wet)% inside; 69.2% barrier	264
	EVAL EP-F barrier: 0.15/0.025/0.6 mm thick; PP inside; PET outside	0.52				20				RH: 75% outside; 100 (wet)% inside; 78% barrier	264
PET/PVDC/PP Film	Saran VC barrier: 0.15/0.025/0.6 mm thick; PP inside; PET outside	1.4				20				RH: 65% outside; 100 (wet)% inside; 69.2% barrier	264
	Saran VC barrier: 0.15/0.025/0.6 mm thick; PP inside; PET outside	1.4				20				RH: 75% outside; 100 (wet)% inside; 78% barrier	264
Polybutylene	Shell Duraflex 1600; peelable seals; FDA grade, 0.051 mm thick, heat sealable; blown film; 5 phr zinc oxide, slip and antiblock formulations; film	152				22.8		50		ASTM D1434, method M	304
	Shell Duraflex 1710; peelable seals; 0.910 g/cm³ density; FDA grade, heat sealable, 0.051 mm thick; blown film; slip and antiblock formulations, antiblock formulations; film	157				22.8		50		ASTM D1434, method M	304
Polycarbonate	Bayer Makrolon; 0.1 mm thick; film	67.9								DIN 53380, pt. 3	289
	Dow Calibre 300-15; transparent, gen. purp. grade; 15 g/10 min. MFI	90.6								ASTM D2752	78
	Dow Calibre 300-4; transparent, gen. purp. grade; 4 g/10 min. MFI	102								ASTM D2752	78
	Dow Calibre 600-6; transparent, flame retardant; 6 g/10 min. MFI	124								ASTM D2752	78
	film	102				22.8		0			294

OXYGEN (continued)

Material Family	Material Note	Perm. Coefficient (cm³·mm/m²·day·atm)	VTR (g·mm/m²·day)	Non-normalized Data	Penetrant Note	Temp. (°C)	Time (days)	RH (%)	Pressure (kPa)	Test Note	Source
Polyester, PBT	BASF AG Ultradur B4550; 0.25 mm thick	15.2				23		50		DIN 53380; measured in standard laboratory atmosphere	180
Polyester, PCTG	Eastman Kodar PCTG 5445; transparent, 0.25 mm thick; film	10				23		50		ASTM D1434	166
Polyester, PET	oriented	0.26				5		0			264
	DuPont Mylar; film	2.4				25		0		ASTM D1434-72	270
	Shell Cleartuf; packaging; transparent, barrier prop., oriented	2.0				25		0		ASTM D1434	297
	Shell Cleartuf; packaging; transparent, barrier prop., unoriented	3.9				25		0		ASTM D1434	297
	0.012 mm thick; film	2.3				20		65		ASTM D3985	268
	0.012 mm thick; film	2.3				20		85		ASTM D3985	268
	0.012 mm thick; film	2.3				20		100		ASTM D3985	268
	film	1.1				20		65			265
	film	1.1				20		85			265
	film	1.1				20		100			265
	film	2.5				20		0			265
	oriented	0.91				23		0			264
	oriented	1.2 - 2.4				25				STP conditions	138
	oriented, 0.014 mm thick; PVDC coated; film	0.2				20		65		ASTM D3985	268
	oriented, 0.014 mm thick; PVDC coated; film	0.2				20		85		ASTM D3985	268
	oriented, 0.014 mm thick; PVDC coated; film	0.2				20		100		ASTM D3985	268
	0.012 mm thick; film	2.8				22.8		0		ASTM D1434	296
	oriented	5.1				35		0		JIS Z1707	268
	oriented	2.01				35		0			264
	oriented	6.61				50		0			264
	oriented, 0.014 mm thick; PVDC coated; film	0.43				35		0		JIS Z1707	268
Polyester, PETG	Eastman Kodar PETG 6763; transparent, amorphous, 0.25 mm thick; film	9.8				23		0		ASTM D1434	165
		10.0				22.8		0		ASTM D1434	296
Polyethylene, HDPE	0.03 mm thick	5.0				-16			1724	mass spectrometry and calibrated standard gas leaks; developed by McDonnell Douglas Space Systems Caompany Chemistry Laboratory	306
	0.03 mm thick	5.2				-15				"	306
	Dow; film	39.4 - 78.7				24			3447		250
	DuPont Can. Sclair 15A; merchandising bags; 0.941 g/cm³ density; 59 Shore D, 0.0254 mm thick; 0.35 g/10 min. MFI; blown film	66.0				23				ASTM D1434	277
	DuPont Can. Sclair 16A; merchandising bags; 0.945 g/cm³ density; 60 Shore D, 0.0254 mm thick; 0.28 g/10 min. MFI; blown film	55.9				23				ASTM D1434	277
	DuPont Can. Sclair 19A; laminations, coextrusion; 0.96 g/cm³ density; 65 Shore D, 0.0254 mm thick; 0.75 g/10 min. MFI; blown film	40.6				23				ASTM D1434	277
	Hoechst AG Hostalen	73.0				20				volume at standard temperature and pressure; useable average for all Hostalen grades	94

OXYGEN (continued)

Material Family	Material Note	Permeability Data			Test Conditions						Source
		Perm. Coefficient (cm³·mm/m²·day·atm)	VTR (g·mm/m²·day)	Non-normalized Data	Penetrant Note	Temp. (°C)	Time (days)	RH (%)	Pressure (kPa)	Test Note	
Polyethylene, HDPE	Hoechst AG Hostalen	77.0				25					94
	Phillips Marlex; film	43.7				23					ASTM D1434	101
	0.03 mm thick	50.3				25			1724	mass spectrometry and calibrated standard gas leaks; developed by McDonnell Douglas Space Systems Caompany Chemistry Laboratory	306
	0.03 mm thick	49.4				25			3447		306
		>75.8				22.8		0		ASTM D1434	296
		72.8				25				STP conditions	138
		49.6				23		75			293
		59.1				23		0			264
	Hoechst AG Hostalen	93.2				30				volume at standard temperature and pressure; useable average for all Hostalen grades	94
	Hoechst AG Hostalen	142				40				"	94
	Hoechst AG Hostalen	233				50				"	94
	0.03 mm thick	113				35		0			264
	0.03 mm thick	218				51			1724	mass spectrometry and calibrated standard gas leaks; developed by McDonnell Douglas Space Systems Caompany Chemistry Laboratory	306
	0.03 mm thick	178				52			3447		306
Polyethylene, LDPE	Dow, film	98.4 - 138				24					250
	0.05 mm thick; film	68.5				20		65		ASTM D3985	268
	0.05 mm thick; film	68.5				20		85		ASTM D3985	268
	0.05 mm thick; film	68.5				20		100		ASTM D3985	268
	0.920 g/cm³ density; 0.05 mm thick, 2.5 BUR; blown film; 4 g/10 min. MFI; film	150								ASTM D1434	216
		197				25				STP conditions	138
		177				23		75			293
		218				23		0			264
	0.05 mm thick; film	152				35		0		JIS Z1707	268
		293				35		0			264
Polyethylene, LLDPE	Dow Dowlex 2045; 0.920 g/cm³ density; 0.0254 mm thick; blown film; 1.0 g/10 min. MFI	207				23				RH: <1% (dry test); ASTM D3985-81	11
	DuPont Can. Sclair 11F9; blending resin, multi-purpose bags; 0.921 g/cm³ density; 0.0254 mm thick; 0.75 g/10 min. MFI; blown film	132				23				ASTM D1434	277
	DuPont Can. Sclairfilm S1.1; laminations; 0.918 g/cm³ density; 0.038 mm thick; film	236								ASTM D3985	278
	DuPont Can. Sclairfilm S1.1; laminations; 0.918 g/cm³ density; 0.051 mm thick; film	199								ASTM D3985	278
	DuPont Can. Sclairfilm S1.1; laminations; 0.918 g/cm³ density; 0.076 mm thick; film	236								ASTM D3985	278
	DuPont Can. Sclairfilm S1.3; laminations; 0.918 g/cm³ density; 0.038 mm thick; film	236								ASTM D3985	278

OXYGEN (continued)

Material Family	Material Note	Perm. Coefficient (cm³·mm/m²·day·atm)	VTR (g·mm/m²·day)	Non-normalized Data	Penetrant Note	Temp. (°C)	Time (days)	RH (%)	Pressure (kPa)	Test Note	Source
Polyethylene, LLDPE	DuPont Can. Sclairfilm SL3; laminations; 0.918 g/cm³ density; 0.051 mm thick; film	199								ASTM D3985	278
	DuPont Can. Sclairfilm SL3; laminations; 0.918 g/cm³ density; 0.076 mm thick; film	236								ASTM D3985	278
Polyethylene, MDPE	DuPont Can. Sclair 14D; merchandising bags; 0.935 g/cm³ density; 59 Shore D, 0.0254 mm thick; 0.28 g/10 min. MFI; blown film	78.7				23				ASTM D1434	277
		98.4 - 211				25				STP conditions	138
Polyethylene, PE/EVA Cop	DuPont Elvax 3120; 0.930 g/cm³ density; 0.051 mm thick; 7.5% VA; 1.2 g/10 min. MFI; antiblock additive, slip additive; blown film	237								ASTM D3985	281
	DuPont Elvax 3121A; 0.930 g/cm³ density; 0.051 mm thick; 7.5% VA; 0.35 g/10 min. MFI; antiblock additive, slip additive; blown film	257								ASTM D3985	281
	DuPont Elvax 3128; 0.930 g/cm³ density; 0.051 mm thick; 8.9% VA; 2.0 g/10 min. MFI; blown film	300								ASTM D3985	281
	DuPont Elvax 3128SB; 0.930 g/cm³ density; 0.051 mm thick; 8.9% VA; 2.0 g/10 min. MFI; antiblock additive, slip additive; blown film	300								ASTM D3985	281
	DuPont Elvax 3130; 0.94 g/cm³ density; 0.051 mm thick; 12.0% VA; 2.5 g/10 min. MFI; blown film	356								ASTM D3985	281
	DuPont Elvax 3130SB; 0.94 g/cm³ density; 0.051 mm thick; 12.0% VA; 2.5 g/10 min. MFI; antiblock additive, slip additive; blown film	356								ASTM D3985	281
	DuPont Elvax 3135; 0.94 g/cm³ density; 0.051 mm thick; 12.0% VA; 0.25-0.35 g/10 min. MFI; blown film	356								ASTM D3985	281
	DuPont Elvax 3135SB; 0.94 g/cm³ density; 0.051 mm thick; 12.0% VA; 0.25-0.35 g/10 min. MFI; antiblock additive, slip additive; blown film	356								ASTM D3985	281
	DuPont Elvax 3135X; 0.94 g/cm³ density; 0.051 mm thick; 12.0% VA; 0.25-0.35 g/10 min. MFI; blown film	356								ASTM D3985	281
	DuPont Elvax 3137; 0.94 g/cm³ density; 0.051 mm thick; 12.0% VA; 0.3 g/10 min. MFI; blown film	435								ASTM D3985	281
	DuPont Elvax 3159; 0.94 g/cm³ density; 0.051 mm thick; 15.0% VA; 0.5 g/10 min. MFI; blown film	419								ASTM D3985	281
	DuPont Elvax 3165; 0.94 g/cm³ density; 0.051 mm thick; 18.0% VA; 0.7 g/10 min. MFI; blown film	470								ASTM D3985	281
	DuPont Elvax 3170; 0.94 g/cm³ density; 0.051 mm thick; 18.0% VA; 2.5 g/10 min. MFI; blown film	435								ASTM D3985	281
	DuPont Elvax 3170SB; 0.94 g/cm³ density; 0.051 mm thick; 18.0% VA; 2.5 g/10 min. MFI; antiblock additive, slip additive; blown film	435								ASTM D3985	281
	DuPont Elvax 3170SHB; 0.94 g/cm³ density; 0.051 mm thick; 18.0% VA; 2.5 g/10 min. MFI; high antiblock, slip additive; blown film	435								ASTM D3985	281
	0.930 g/cm³ density; 0.05 mm thick, 2.5 BUR; blown film; 12.0% VA; film	180								ASTM D1434	216

OXYGEN (continued)

Material Family	Material Note	Perm. Coefficient (cm³·mm/m²·day·atm)	VTR (g·mm/m²·day)	Non-normalized Data	Penetrant Note	Temp. (°C)	Time (days)	RH (%)	Pressure (kPa)	Test Note	Source
Polyethylene, ULDPE	Dow Attane 4001; 0.905 g/cm³ density; 0.0254 mm thick; blown film; 1.0 g/10 min. MFI	256								RH: <1% (dry test); ASTM D3985-81	11
	Dow Attane 4003; 0.912 g/cm³ density; 0.0254 mm thick; blown film; 0.8 g/10 min. MFI	378								RH: <1% (dry test); ASTM D3985-81	11
Polyimide	DuPont Kapton Type-E; 0.076 mm thick; film	0.3									273
	DuPont Kapton Type-K; 0.076 mm thick; film	8.0									273
	DuPont Kapton Type-V; 0.076 mm thick; film	8.7									273
	Ube Upilex R; 0.025 mm thick; film	2.5				30				ASTM D1434	97
	Ube Upilex S; 0.025 mm thick; film	0.02				30				ASTM D1434	97
Polymethylpentene	Mitsui TPX X-22; 0.835 g/cm³ density; transparent, 0.05 mm thick; film	2000				23					302
	Mitsui TPX X-44; 0.834 g/cm³ density; transparent, 0.05 mm thick; film	1900				23					302
	Mitsui TPX X-88; 0.835 g/cm³ density; transparent, food grade, 0.05 mm thick; film	1600				23					302
Polyphenylene Sulfide	Phillips Ryton; 0.127 mm thick; baked coating; film	11.8								ASTM D1434, Method M	102
	Phillips Ryton; 0.875 mm thick; baked coating; film	5.9 - 7.9								ASTM D1434, Method M	102
Polypropylene	biaxially oriented; PVDC coated; film	0.22				20		65			265
	biaxially oriented; PVDC coated; film	0.22				20		85			265
	biaxially oriented; PVDC coated; film	0.22				20		100			265
	biaxially oriented; PVDC coated; film	0.43				20		0			265
	film	107.1				23				ASTM D1434	101
	oriented	64.2				23		0			264
	oriented, 0.02 mm thick; film	53.2				20		65		ASTM D3985	268
	oriented, 0.02 mm thick; film	53.2				20		85		ASTM D3985	268
	oriented, 0.02 mm thick; film	53.2				20		100		ASTM D3985	268
	oriented, 0.022 mm thick; PVDC coated; film	0.26				20		65		ASTM D3985	268
	oriented, 0.022 mm thick; PVDC coated; film	0.26				20		85		ASTM D3985	268
	oriented, 0.022 mm thick; PVDC coated; film	0.26				20		100		ASTM D3985	268
	oriented; film	68.5				25		0			63
	oriented; film	78.7				23		75			293
	oriented	>99.7				22.8		0		ASTM D1434	296
	oriented, 0.02 mm thick; film	79.9				35		0			264
	oriented, 0.02 mm thick; film	89.0				35		0		JIS Z1707	268
	oriented, 0.022 mm thick; PVDC coated; film	0.48				35		0		JIS Z1707	268
Polypyrrole	BASF AG Lutamer ES 8567; 0.031-0.037 mm thick, intrinsically conductive, high flexibility; w/ benzenesulfonate anions; film	2.3									182
Polystyrene	Dow Styron; film	118 - 157				24					250
	Dow Trycite; film	98.4 - 138				24					250
		102				23		0			264
		>140				22.8		0		ASTM D1434	296
Polystyrene, GP	BASF AG Polystyrol 168 N; transparent, 0.1 mm thick; film	101				23				DIN 53380	26
	Dow Styron; injection molding	118 - 157				23				ASTM D1434	263
	Dow Styron; oriented; sheet	98.4 - 138				23				ASTM D1434	263

OXYGEN (continued)

Material Family	Material Note	Perm. Coefficient (cm³·mm/m²·day·atm)	VTR (g·mm/m²·day)	Non-normalized Data	Penetrant Note	Temp. (°C)	Time (days)	RH (%)	Pressure (kPa)	Test Note	Source
Polystyrene, IPS	BASF AG Polystyrol 476 L; 0.1 mm thick; film	162				23				DIN 53390	26
	Dow Styron; injection molding	118 - 157				23				ASTM D1434	262
Polysulfone	Amoco Udel; transparent, amber tint	90.6				23		dry		ASTM D1434	15
Polyurethane		78.7				25				ASTM D1434-63T	121
Polyvinyl Alcohol	Air Products Vinex 1003; 0.03 mm thick; 5-7 g/10 min. MFI; 1.25 density	0.02						0		Pressure: 760 mm Hg	283
	Air Products Vinex 1003; 0.03 mm thick; 5-7 g/10 min. MFI; 1.25 density	0.56						50		Pressure: 760 mm Hg	283
	Air Products Vinex 2144; 0.033 mm thick; 6-8 g/10 min. MFI; 1.25 density	0.02						0		Pressure: 760 mm Hg	283
	Air Products Vinex 2144; 0.033 mm thick; 6-8 g/10 min. MFI; 1.25 density	0.74						50		Pressure: 760 mm Hg	283
	Air Products Vinex 5030; 0.038 mm thick; 9-11 g/10 min. MFI; 1.25 density	0.02						0		Pressure: 760 mm Hg	283
	Air Products Vinex 5030; 0.038 mm thick; 9-11 g/10 min. MFI; 1.25 density	0.88						50		Pressure: 760 mm Hg	283
	0.014 mm thick; film	0.06				20		65		ASTM D3985	268
	0.014 mm thick; film	0.35				20		85		ASTM D3985	268
	0.014 mm thick; film	15.2				20		100		ASTM D3985	268
	film	0.02				24		0			250
	film	0.09				24		75			250
	0.014 mm thick; film	0.01				35		0		JIS Z1707	268
Polyvinyl Chloride	film	4.8				20		dry			63
	film	5.2				23		dry			63
	film	274				25		0			63
	unplasticized; film	2.0 - 7.9				24		0			250
		3.2				22.8		0		ASTM D1434	296
	film	7.4				35		dry			63
Polyvinylidene Chloride	Dow Saran MA; VDC methyl acrylate; barrier prop.	0.0024				5		0			264
	Dow Saran VC; VDC vinyl chloride; barrier prop.	0.0047				5		0			264
	Dow Saran; film	0.31 - 0.43				24					250
	Dow Saran; VDC vinyl chloride; barrier prop.; film	0.06				23		75			255
	Dow Saran; VDC vinyl chloride; barrier prop.; film	0.06				23		100			255
	Dow Saran 313; VDC vinyl chloride; barrier prop.; extrusion; film; 1.69 spec. grav.	0.47				23		75		ASTM D3985	254
	Dow Saran 416; VDC vinyl chloride; barrier prop.; extrusion; film; 1.73 spec. grav.	0.03				23		75		ASTM D3985	254
	Dow Saran 469; VDC vinyl chloride; barrier prop.; extrusion; 43,000 mol. wgt.; film; 1.76 spec. grav.	0.04				23		75		ASTM D3985	254
	Dow Saran 516; VDC vinyl chloride; barrier prop.; extrusion; film; 1.73 spec. grav.	0.04				23		75		ASTM D3985	254
	Dow Saran 525; VDC vinyl chloride; barrier prop.; extrusion; film; 1.73 spec. grav.	0.04				23		75		ASTM D3985	254
	Dow Saran 866; VDC vinyl chloride; barrier prop.; extrusion; film; 1.70 spec. grav.	0.43				23		75		ASTM D3985	254
	Dow Saran F-239; VDC vinyl chloride; barrier prop.; 0.003 mm thick; coating; film	0.01				23		75		ASTM D3985	254

OXYGEN (continued)

Material Family	Material Note	Perm. Coefficient (cm³·mm/m²·day·atm)	VTR (g·mm/m²·day)	Non-normalized Data	Penetrant Note	Temp. (°C)	Time (days)	RH (%)	Pressure (kPa)	Test Note	Source
Polyvinylidene Chloride	Dow Saran F-278; VDC vinyl chloride; barrier prop., 0.003 mm thick; coating; film	0.01				23		75		ASTM D3985	254
	Dow Saran F-310; VDC vinyl chloride; barrier prop., 0.003 mm thick; coating; film	0.03				23		75		ASTM D3985	254
	Dow Saran HB; barrier prop.; film	0.03				20		65			265
	Dow Saran HB; barrier prop.; film	0.03				20		85			265
	Dow Saran HB; barrier prop.; film	0.03				20		100			265
	Dow Saran MA; VDC methyl acrylate; barrier prop.	0.03				23		0			264
	Dow Saran MA; VDC methyl acrylate; barrier prop.; film	0.03				23		75			255
	Dow Saran MA; VDC methyl acrylate; barrier prop.; film	0.03				23		100			255
	Dow Saran MA 119; VDC methyl acrylate; barrier prop.; extrusion; 90,000 mol. wgt.; film; 1.78 spec. grav.	0.03				23		75		ASTM D3985	254
	Dow Saran VC; VDC vinyl chloride; barrier prop.	0.06				23		0			264
	Dow Saran Wrap 18; chub packaging, laminations; transparent, barrier prop., 0.019 mm thick, biaxially oriented; monolayer film	0.47				23		10		ASTM D3985	256
	Dow Saran Wrap 18L, laminations; transparent, barrier prop., 0.019 mm thick, preshrunk, biaxially oriented; monolayer film	0.63				23		10		ASTM D3985	256
	Dow Saran Wrap 19; chub packaging, transparent, barrier prop., 0.0254 mm thick, biaxially oriented; monolayer film	0.47				23		10		ASTM D3985	256
	Dow Saran Wrap 28; chub packaging, transparent, barrier prop., 0.0254 mm thick, biaxially oriented; monolayer film	0.71				23		10		ASTM D3985	256
	Dow Saran Wrap 560; unit packaging, transparent, barrier prop., 0.152 mm thick, biaxially oriented; coextruded film	0.1				23		10		ASTM D3985	256
		0.03				22.8		0		ASTM D3985	296
		0.03				22.8		90		ASTM D3985	296
	Dow Saran MA; VDC methyl acrylate; barrier prop.	0.11				35		0			264
	Dow Saran MA; VDC methyl acrylate; barrier prop.	0.5				50		0			264
	Dow Saran VC; VDC vinyl chloride; barrier prop.	0.17				35		0			264
	Dow Saran VC; VDC vinyl chloride; barrier prop.	0.75				50		0			264
PP/ EVAL/ HDPE Film	EVAL EP-F barrier 0.6/ 0.025/ 0.15 mm thick; HDPE inside; PP outside	0.16				20				RH: 65% outside; 10 (wet)% inside; 29.1% barrier	264
	EVAL EP-F barrier 0.6/ 0.025/ 0.15 mm thick; HDPE inside; PP outside	0.16				20				RH: 75% outside; 10 (wet)% inside; 31.7% barrier	264
PP/ EVAL/ LDPE Film	EVAL EF-F barrier: 0.02/ 0.015/ 0.051 mm thick; LDPE inside; OPP outside	0.15				20				RH: 65% outside; 100% inside; 84% barrier	265
	EVAL EF-F barrier: 0.02/ 0.015/ 0.051 mm thick; LDPE inside; OPP outside	0.32				20				RH: 80% outside; 100% inside; 34% barrier	265
	EVAL EF-F barrier: 0.02/ 0.015/ 0.051 mm thick; LDPE inside; OPP outside	0.01				20				RH: 65% outside; 10% inside; 91% barrier	265
	EVAL EF-F barrier: 0.02/ 0.015/ 0.051 mm thick; LDPE inside; OPP outside	0.03				20				RH: 80% outside; 10% inside; 40% barrier	265

OXYGEN (continued)

Material Family	Material Note	Permeability Data			Penetrant Note	Test Conditions					Source
		Perm. Coefficient (cm³·mm/m²·day·atm)	VTR (g·mm/m²·day)	Non-normalized Data		Temp (°C)	Time (days)	RH (%)	Pressure (kPa)	Test Note	
PP/EVAL/LDPE Film	EVAL EP-F barrier; 0.6/ 0.025/ 0.15 mm thick; LDPE inside; PP outside	0.15				20				RH: 75% outside; 10 (wet)% inside; 22.5% barrier	264
	EVAL EP-F barrier; 0.6/ 0.025/ 0.15 mm thick; LDPE inside; PP outside	0.16				20				RH: 65% outside; 10 (wet)% inside; 21.4% barrier	264
PP/EVAL/PC Film	EVAL EP-E barrier; 0.6/ 0.025/ 0.15 mm thick; PC inside; PP outside	0.6				20				RH: 65% outside; 10 (wet)% inside; 21% barrier	264
	EVAL EP-F barrier; 0.6/ 0.025/ 0.15 mm thick; PC inside; PP outside	0.62				20				RH: 75% outside; 10 (wet)% inside; 21.3% barrier	264
PP/EVAL/PET Film	EVAL EP-F barrier; 0.6/ 0.025/ 0.15 mm thick; PET inside; PP outside	0.15				20				RH: 75% outside; 10 (wet)% inside; 17.8% barrier	264
	EVAL EP-F barrier; 0.6/ 0.025/ 0.15 mm thick; PET inside; PP outside	0.15				20				RH: 65% outside; 10 (wet)% inside; 16.6% barrier	264
PP/EVAL/PP Film	EVAL EF-F barrier; 0.02/ 0.015/ 0.051 mm thick; OPP outside, coated PP outside	0.11				20				RH: 65% outside; 100% inside; 80% barrier	265
	EVAL EF-F barrier; 0.02/ 0.015/ 0.051 mm thick; OPP outside, coated PP outside	0.25				20				RH: 80% outside; 100% inside; 41% barrier	265
	EVAL EF-F barrier; 0.02/ 0.015/ 0.051 mm thick; OPP outside, coated PP outside	0.03				20				RH: 65% outside; 10% inside; 89% barrier	265
	EVAL EF-F barrier; 0.02/ 0.015/ 0.051 mm thick; OPP outside, coated PP outside	0.03				20				RH: 80% outside; 10% inside; 49% barrier	265
	EVAL EP-E barrier; 0.15/ 0.025/ 0.6 mm thick; PP inside; PP outside	1.2				20				RH: 65% outside; 100 (wet)% inside; 72.2% barrier	264
	EVAL EP-E barrier; 0.15/ 0.025/ 0.6 mm thick; PP inside; PP outside	1.6				20				RH: 75% outside; 100 (wet)% inside; 80.1% barrier	264
	EVAL EP-F barrier; 0.15/ 0.025/ 0.6 mm thick; PP inside; PP outside	0.34				20				RH: 65% outside; 100 (wet)% inside; 72% barrier	264
	EVAL EP-F barrier; 0.15/ 0.025/ 0.6 mm thick; PP inside; PP outside	0.58				20				RH: 75% outside; 100 (wet)% inside; 80% barrier	264
	EVAL EP-F barrier; 0.6/ 0.025/ 0.15 mm thick; PP inside; PP outside	0.16				20				RH: 65% outside; 10 (wet)% inside; 24.2% barrier	264
	EVAL EP-F barrier; 0.6/ 0.025/ 0.15 mm thick; PP inside; PP outside	0.16				20				RH: 75% outside; 10 (wet)% inside; 25.9% barrier	264
PP/EVAL/PS Film	EVAL EP-F barrier; 0.6/ 0.025/ 0.15 mm thick; PS inside; PP outside	0.15				20				RH: 75% outside; 10 (wet)% inside; 17.1% barrier	264
	EVAL EP-F barrier; 0.6/ 0.025/ 0.15 mm thick; PS inside; PP outside	0.15				20				RH: 65% outside; 10 (wet)% inside; 16.9% barrier	264
PP/PVDC/HDPE Film	Saran VC barrier; 0.6/ 0.025/ 0.15 mm thick; HDPE inside; PP outside	1.8				20				RH: 65% outside; 10 (wet)% inside; 29.1% barrier	264
	Saran VC barrier; 0.6/ 0.025/ 0.15 mm thick; HDPE inside; PP outside	1.8				20				RH: 75% outside; 10 (wet)% inside; 31.7% barrier	264
PP/PVDC/LDPE Film	Saran VC barrier; 0.6/ 0.025/ 0.15 mm thick; LDPE inside; PP outside	1.8				20				RH: 75% outside; 10 (wet)% inside; 22.5% barrier	264

OXYGEN (continued)

Material Family	Material Note	Permeability Data			Test Conditions						Source
		Perm. Coefficient (cm³·mm/m²·day·atm)	VTR (g·mm/m²·day)	Non-normalized Data	Penetrant Note	Temp. (°C)	Time (days)	RH (%)	Pressure (kPa)	Test Note	
PP/PVDC/LDPE Film	Saran VC barrier; 0.6/0.025/0.15 mm thick; LDPE inside; PP outside	1.8				20				RH: 65% outside; 10 (wet)% inside; 21.4% barrier	264
PP/PVDC/PC Film	Saran VC barrier; 0.6/0.025/0.15 mm thick; PC inside; PP outside	1.8				20				RH: 65% outside; 10 (wet)% inside; 21% barrier	264
	Saran VC barrier; 0.6/0.025/0.15 mm thick; PC inside; PP outside	1.8				20				RH: 75% outside; 10 (wet)% inside; 21.3% barrier	264
PP/PVDC/PET Film	Saran VC barrier; 0.6/0.025/0.15 mm thick; PET inside; PP outside	1.8				20				RH: 75% outside; 10 (wet)% inside; 17.8% barrier	264
	Saran VC barrier; 0.6/0.025/0.15 mm thick; PET inside; PP outside	1.8				20				RH: 65% outside; 10 (wet)% inside; 16.6% barrier	264
PP/PVDC/PP Film	Saran VC barrier; 0.15/0.025/0.6 mm thick; PP inside; PP outside	1.4				20				RH: 65% outside; 100 (wet)% inside; 72% barrier	264
	Saran VC barrier; 0.15/0.025/0.6 mm thick; PP inside; PP outside	1.4				20				RH: 65% outside; 100 (wet)% inside; 72.2% barrier	264
	Saran VC barrier; 0.15/0.025/0.6 mm thick; PP inside; PP outside	1.4				20				RH: 75% outside; 100 (wet)% inside; 80% barrier	264
	Saran VC barrier; 0.15/0.025/0.6 mm thick; PP inside; PP outside	1.4				20				RH: 75% outside; 100 (wet)% inside; 80.1% barrier	264
	Saran VC barrier; 0.6/0.025/0.15 mm thick; PP inside; PP outside	1.8				20				RH: 75% outside; 10 (wet)% inside; 25.9% barrier	264
	Saran VC barrier; 0.6/0.025/0.15 mm thick; PP inside; PP outside	1.8				20				RH: 65% outside; 10 (wet)% inside; 24.2% barrier	264
PP/PVDC/PS Film	Saran VC barrier; 0.6/0.025/0.15 mm thick; PS inside; PP outside	1.8				20				RH: 65% outside; 10 (wet)% inside; 16.9% barrier	264
	Saran VC barrier; 0.6/0.025/0.15 mm thick; PS inside; PP outside	1.8				20				RH: 75% outside; 10 (wet)% inside; 17.1% barrier	264
PS/EVAL/PP Film	EVAL EP-F barrier; 0.15/0.025/0.6 mm thick; PP inside; PS outside	0.26				20				RH: 65% outside; 100 (wet)% inside; 65.8% barrier	264
	EVAL EP-F barrier; 0.15/0.025/0.6 mm thick; PP inside; PS outside	0.42				20				RH: 75% outside; 100 (wet)% inside; 75.6% barrier	264
PS/PVDC/PP Film	Saran VC barrier; 0.15/0.025/0.6 mm thick; PP inside; PS outside	1.4				20				RH: 65% outside; 100 (wet)% inside; 65.8% barrier	264
	Saran VC barrier; 0.15/0.025/0.6 mm thick; PP inside; PS outside	1.4				20				RH: 75% outside; 100 (wet)% inside; 75.6% barrier	264
PVC-PVDC Copolymer		0.31 - 2.7				25				STP conditions	138
Rubber, EPDM	DuPont Nordel 1040	1641 - 1901				23				STP conditions	311
Rubber, Isobutylene	gum vulcanizate			1570% relative to Butyl Rubber		25					300
Rubber, Latex (NR)	gum vulcanizate			91% relative to Butyl Rubber		25					300
	gum vulcanizate			1780% relative to Butyl Rubber		25					300
Rubber, Neoprene (CR)	Neoprene G type; gum vulcanizate			302% relative to Butyl Rubber		25					300

OXYGEN (continued)

Material Family	Material Note	Perm. Coefficient (cm³·mm/m²·day·atm)	VTR (g·mm/m²·day)	Non-normalized Data	Penetrant Note	Temp (°C)	Time (days)	RH (%)	Pressure (kPa)	Test Note	Source
Rubber, Nitrile (NBR)	61% butadiene/39% acrylonitrile; gum vulcanizate			73% relative to Butyl Rubber		25					300
	68% butadiene/32% acrylonitrile; gum vulcanizate			178% relative to Butyl Rubber		25					300
	73% butadiene/27% acrylonitrile; gum vulcanizate			302% relative to Butyl Rubber		25					300
	80% butadiene/20% acrylonitrile; gum vulcanizate			622% relative to Butyl Rubber		25					300
Rubber, Polybutadiene	hot emulsion; gum vulcanizate			1430% relative to Butyl Rubber		25					300
	film	1181									250
Rubber, Styrene Butadiene	hot, gum vulcanizate			1290% relative to Butyl Rubber		25					300
SAN	BASF AG Luran 358 N; transparent, high flow, 0.1 mm thick; film	20.3 - 50.7				23				DIN 53380	30
	BASF AG Luran 368 R; transparent, 0.1 mm thick, moderate to low flow; film	20.3 - 50.7				23				DIN 53380	30
	BASF AG Luran 378 P; transparent, mod-high flow, 0.1 mm thick; film	20.3 - 30.4				23				DIN 53380	30
	BASF AG Luran 388 S; transparent, 0.1 mm thick, low flow; film	20.3 - 30.4				23				DIN 53380	30
	Dow Tyril; low nitrile content; film	31.5 - 39.4				24					250
	Dow Tyril; medium nitrile content; film	15.8 - 27.6				24					250
Silicone	gum vulcanizate			3920% relative to Butyl Rubber		25					300
		19,685				25				ASTM D1434-63T	121
Styrene-Butadiene Block Copol.	BASF AG Styrolux 637 D; 0.1 mm thick; film	192				23					29
	BASF AG Styrolux 656 C; high flow, 0.1 mm thick; film	162				23					29
	BASF AG Styrolux 684 D; 0.1 mm thick; film	263				23					29
TPE, Olefinic	Adv. Elast. Santoprene 201-73; 73 Shore A, 0.5 mm thick	504			gas at 0°C and 0.11 Mpa	23			110	ASTM D1434	282
	Adv. Elast. Santoprene 201-87; 87 Shore A, 0.5 mm thick	589			gas at 0°C and 0.11 Mpa	23			110	ASTM D1434	282
	Adv. Elast. Santoprene 203-50; 50 Shore D, 0.5 mm thick	279			gas at 0°C and 0.11 Mpa	23			110	ASTM D1434	282
TPE, Polyamide	Atochem Pebax 2533; 25 Shore D, 0.12 mm thick; film	985				23		dry			287
	Atochem Pebax 3533; 35 Shore D, 0.12 mm thick; film	860				23		dry			287
	Atochem Pebax 4033; 40 Shore D, 0.12 mm thick; film	387				23		dry			287
	Atochem Pebax 5533; 55 Shore D, 0.12 mm thick; film	230				23		dry			287
	Atochem Pebax 6533; 63 Shore D, 0.12 mm thick; film	204				23		dry			287
TPE, Polybutadiene	Jap. Synth. JSR RB820; 1,2-polybutadiene; 0.910 g/cm³ density; transparent, 2.5 BUR, 0.05 mm thick; blown film; 15% crystallinity; film	550								ASTM D1434	216
	Jap. Synth. JSR RB830; 0.910 g/cm³ density; stretch film, 0.018 mm thick, 4.8 BUR; blown film; 29% crystallinity; film	550								ASTM D1434	216
	Jap. Synth. JSR RB830; 1,2-polybutadiene; 0.910 g/cm³ density; transparent, 2.5 BUR, 0.05 mm thick; blown film; 25% crystallinity; film	550								ASTM D1434	216

OXYGEN (continued)

Material Family	Material Note	Permeability Data Perm. Coefficient (cm³·mm/m²·day·atm)	VTR (g·mm/m²·day)	Non-normalized Data	Test Conditions Penetrant Note	Temp. (°C)	Time (days)	RH (%)	Pressure (kPa)	Test Note	Source
TPE, Polyester	Eastman Ecdel; copolyester ether; transparent, crystalline, 0.11 - 0.14 mm thick; film	127				23				ASTM D1434	60
TPE, Styrenic	Shell Kraton D 1101; SBS; FDA grade, 71 Shore A; unsaturated; 31% styrene/ 69% rubber, neat rubber	1717				23				ASTM D1434-82; area: 50 cm²; gradient: 100% gas at 740 mm Hg	303
	Shell Kraton D 1107; SIS; FDA grade, 37 Shore A; unsaturated; 14% styrene/ 86% rubber, neat rubber	1248				23				"	303
	Shell Kraton D 2103; SBS; FDA grade, 70 Shore A; unsaturated; ready to use compound	1902				23				"	303
	Shell Kraton D 2104; SBS; FDA grade, 27 Shore A; unsaturated; ready to use compound	6831				23				"	303
	Shell Kraton D 2109; SBS; FDA grade, 44 Shore A; unsaturated; ready to use compound	3543				23				"	303
	Shell Kraton G 1650; SEBS; FDA grade, 75 Shore A; saturated; 29% styrene/ 71% rubber, neat rubber	909				23				"	303
	Shell Kraton G 1651; SEBS; FDA grade, 75 Shore A; saturated; 29% styrene/ 71% rubber, neat rubber	850				23				"	303
	Shell Kraton G 1652; SEBS; FDA grade, 75 Shore A; saturated; 29% styrene/ 71% rubber, neat rubber	1059				23				"	303
	Shell Kraton G 2701; SEBS; FDA grade, 67 Shore A; saturated; ready to use compound	1102				23				"	303
	Shell Kraton G 2705; SEBS; FDA grade, 55 Shore A; saturated; ready to use compound	1646				23				"	303
	Shell Kraton G 2706; SEBS; FDA grade, 28 Shore A; saturated; ready to use compound	2177				23				"	303
TPE, Urethane (TPAU)	BASF Elastollan C90A; 90 Shore A	123				20					130
	BASF Elastollan C85A; 85 Shore A	87.5				20					130
	BASF Elastollan C90A; 90 Shore A	61.3				20					130
	BASF Elastollan C95A; 95 Shore A	35.0				20					130
TPE, Urethane (TPEU)	BASF Elastollan 1180A; 80 Shore A	184				20					130
	BASF Elastollan 1185A; 85 Shore A	140				20					130
	BASF Elastollan 1190A; 90 Shore A	105				20					130
	BASF Elastollan 1195A; 95 Shore A	70.0				20					130
TPE, Vinyl	1.23 - 1.31 g/cm³ density; stretch film, 0.018 mm thick, 4.8 BUR; blown film; film	190 - 360									ASTM D1434	216
	1.26 g/cm³ density; clear, 2.5 BUR, 0.05 mm thick; blown film; 50 phr plasticizer; film	93									ASTM D1434	216
	plasticized; film	11.8 - 787				24						250

PAINT THINNER

Material Family	Material Note	Perm. Coefficient	VTR	Non-normalized Data	Penetrant Note	Temp. (°C)	Time (days)	RH (%)	Pressure (kPa)	Test Note	Source
Laminar, Nylon/Olefin	DuPont Selar RB/ Polyolefin; 8% Selar RB; barrier prop.; extrusion blow molded; bottles			0.06% penetrant weight loss		50	28				293
Polyethylene, Fluorinated	barrier prop.; bottles			0.08% penetrant weight loss		50	28				293
Polyethylene, HDPE	bottles			10.3% penetrant weight loss		50	28				293

Material Family	Material Note	Permeability Data — Perm. Coefficient (cm³·mm/m²·day·atm)	VTR (g·mm/m²·day)	Non-normalized Data	Penetrant Note	Temp. (°C)	Time (days)	RH (%)	Pressure (kPa)	Test Note	Source

PALMITIC ACID

Material Family	Material Note	Perm. Coefficient	VTR	Non-normalized Data	Penetrant Note	Temp. (°C)	Time (days)	RH (%)	Pressure (kPa)	Test Note	Source
Rubber, Latex (NR)	Ansell Edmont Canners and Handlers 392; unsupported glove film; 0.48 mm thick			breakthrough time: 5 minutes	saturated	23				ASTM F739	313
Rubber, Neoprene (CR)	Ansell Edmont Neoprene 29-840; unsupported glove film; 0.38 mm thick			no permeation detected during a 6 hour test; lower detection limit: 0 ppm	saturated	23	0.25			ASTM F739	313
	Ansell Edmont Neox; supported (lined) glove film; specified by glove film weight			no permeation detected during a 6 hour test; lower detection limit: 0 ppm	saturated	23	0.25			ASTM F739	313
Rubber, Nitrile (NBR)	Ansell Edmont Sol-Vex 37-165; unsupported glove film; 0.54 mm thick			breakthrough time: 30 minutes	saturated	23				ASTM F739	313
TPE, Vinyl	Ansell Edmont Monkey Grip; supported (lined) glove film; specified by glove film weight			breakthrough time: 75 minutes	saturated	23				ASTM F739	313

PENTACHLOROPHENOL

Material Family	Material Note	Perm. Coefficient	VTR	Non-normalized Data	Penetrant Note	Temp. (°C)	Time (days)	RH (%)	Pressure (kPa)	Test Note	Source
Polyvinyl Alcohol	Ansell Edmont PVA; supported (lined) glove film; specified by glove film weight			fair permeation rate; 51 to 100 eyedropper size drops per hour; PVA coating is water soluble; breakthrough time: 5 minutes; permeation rate: <900 µg/cm²/min		23				ASTM F739	313
Rubber, Neoprene (CR)	Ansell Edmont Neoprene 29-840; unsupported glove film; 0.38 mm thick			low permeation rate; 0 to 1/2 eyedropper size drops per hour; breakthrough time: 6 minutes; permeation rate: <0.9 µg/cm²/min		23				ASTM F739	313
	Ansell Edmont Neox; supported (lined) glove film; specified by glove film weight			low permeation rate; 0 to 1/2 eyedropper size drops per hour; breakthrough time: 6 minutes; permeation rate: <0.9 µg/cm²/min		23				ASTM F739	313
Rubber, Nitrile (NBR)	Ansell Edmont Sol-Vex 37-165; unsupported glove film; 0.54 mm thick			low permeation rate; 0 to 1/2 eyedropper size drops per hour; lower detection limit: 0 ppm; permeation rate: <0.9 µg/cm²/min		23	0.25			ASTM F739	313
TPE, Vinyl	Ansell Edmont Monkey Grip; supported (lined) glove film; specified by glove film weight			low permeation rate; 0 to 1/2 eyedropper size drops per hour; breakthrough time: 180 minutes; permeation rate: <0.9 µg/cm²/min		23				ASTM F739	313

PENTANE

Material Family	Material Note	Perm. Coefficient	VTR	Non-normalized Data	Penetrant Note	Temp. (°C)	Time (days)	RH (%)	Pressure (kPa)	Test Note	Source
Polyvinyl Alcohol	Ansell Edmont PVA; supported (lined) glove film; specified by glove film weight			low permeation rate; 0 to 1/2 eyedropper size drops per hour; PVA coating is water soluble; lower detection limit: 0 ppm; permeation rate: <0.9 µg/cm²/min		23	0.25			ASTM F739	313
Rubber, Neoprene (CR)	Ansell Edmont Neoprene 29-840; unsupported glove film; 0.38 mm thick			fair permeation rate; 51 to 100 eyedropper size drops per hour; breakthrough time: 30 minutes; permeation rate: <900 µg/cm²/min		23				ASTM F739	313
	Ansell Edmont Neox; supported (lined) glove film; specified by glove film weight			very good permeation rate; 1 to 5 eyedropper size drops per hour; breakthrough time: 45 minutes; permeation rate: <9 µg/cm²/min		23				ASTM F739	313
	Pioneer Industrial Products Stanzoil N-44; glove film			breakthrough time: 38 minutes; permeation rate: 18 µg/cm²/min						ASTM F739.85	312
Rubber, Nitrile (NBR)	Ansell Edmont Sol-Vex 37-165; unsupported glove film; 0.54 mm thick			low permeation rate; 0 to 1/2 eyedropper size drops per hour; lower detection limit: 0 ppm; permeation rate: <0.9 µg/cm²/min		23	0.25			ASTM F739	313
TPE, Vinyl	Ansell Edmont Wet-Wear 550; clothing; PVC stretch outer layer bonded to a lightweight liner			good permeation rate; 6 to 50 eyedropper size drops per hour; breakthrough time: 4 minutes; permeation rate: <90 µg/cm²/min		23				ASTM F739	313
	Pioneer Industrial Products Pylox V-20; glove film			breakthrough time: 9 minutes; permeation rate: 102 µg/cm²/min						ASTM F739.85	312
TPE, Vinyl coated fabric	Ansell Edmont Wet-Wear 700; clothing; PVC coating with Nylon/Polyester lining			good permeation rate; 6 to 50 eyedropper size drops per hour; breakthrough time: 11 minutes; permeation rate: <90 µg/cm²/min		23				ASTM F739	313

Material Family	Material Note	Perm. Coefficient (cm³·mm/m²·day·atm)	VTR (g·mm/m²·day)	Non-normalized Data	Penetrant Note	Temp. (°C)	Time (days)	RH (%)	Pressure (kPa)	Test Note	Source

PEPPERMINT (MENTHOL)

Material Family	Material Note	Perm. Coefficient	VTR	Non-normalized Data	Penetrant Note	Temp. (°C)	Time (days)	RH (%)	Pressure (kPa)	Test Note	Source
EVAL/ PE Film	0.013/ 0.051 mm thick; PE inside; EVAL EF-XL outside; film			>30 days to leakage							265
Nylon/ EVAL Film	0.015/ 0.015 mm thick; EVAL EF-F inside; oriented nylon outside; film			>30 days to leakage							265
Nylon/ PE Film	0.015/ 0.051 mm thick; PE inside; oriented nylon outside; film			20 days to leakage							265
PET/ EVAL Film	0.013/ 0.015 mm thick; EVAL EF-F inside; PET outside; film			>30 days to leakage							265
PET/ EVAL/ LDPE Film	EVAL EF-F barrier; 0.013/ 0.015/ 0.051 mm thick; LDPE inside; PET outside			25 days to leakage							265
PET/ PE Film	0.013/ 0.051 mm thick; PE inside; PET outside; film			16 days to leakage							265
PP/ PE Film	0.018/ 0.051 mm thick; PE inside; PVDC coated BOPP outside; film			2 days to leakage							265

PERCHLORIC ACID

Material Family	Material Note	Perm. Coefficient	VTR	Non-normalized Data	Penetrant Note	Temp. (°C)	Time (days)	RH (%)	Pressure (kPa)	Test Note	Source
Rubber, Latex (NR)	Ansell Edmont Canners and Handlers 392; unsupported glove film; 0.48 mm thick			no permeation detected during a 6 hour test; lower detection limit: 0 ppm	60% conc.	23	0.25			ASTM F739	313
Rubber, Neoprene (CR)	Ansell Edmont Neoprene 29-840; unsupported glove film; 0.38 mm thick			no permeation detected during a 6 hour test; lower detection limit: 0 ppm	60% conc.	23	0.25			ASTM F739	313
	Ansell Edmont Neor; supported (lined) glove film; specified by glove film weight			no permeation detected during a 6 hour test; lower detection limit: 0 ppm	60% conc.	23	0.25			ASTM F739	313
Rubber, Nitrile (NBR)	Ansell Edmont Sol-Vex 37-165; unsupported glove film; 0.54 mm thick			no permeation detected during a 6 hour test; lower detection limit: 0 ppm	60% conc.	23	0.25			ASTM F739	313
TPE, Vinyl	Ansell Edmont Monkey Grip; supported (lined) glove film; specified by glove film weight			no permeation detected during a 6 hour test; lower detection limit: 0 ppm	60% conc.	23	0.25			ASTM F739	313

PERCHLOROETHYLENE

Material Family	Material Note	Perm. Coefficient	VTR	Non-normalized Data	Penetrant Note	Temp. (°C)	Time (days)	RH (%)	Pressure (kPa)	Test Note	Source
Acetal	DuPont Delrin		0.08			23		50			201
Polyvinyl Alcohol	Ansell Edmont PVA; supported (lined) glove film; specified by glove film weight			low permeation rate; 0 to 1/2 eyedropper size drops per hour; PVA coating is water soluble; lower detection limit: 0 ppm; permeation rate: <0.9 µg/cm²/min		23	0.25			ASTM F739	313
Rubber, Neoprene (CR)	Pioneer Industrial Products Stancoil N-44; glove film			breakthrough time: 28 minutes; lower detection limit: 0.0002 ppm; permeation rate: 453 µg/cm²/min						ASTM F739.85	312
Rubber, Nitrile (NBR)	Ansell Edmont Sol-Vex 37-165; unsupported glove film; 0.54 mm thick			very good permeation rate; 1 to 5 eyedropper size drops per hour; breakthrough time: 300 minutes; permeation rate: <9 µg/cm²/min		23				ASTM F739	313
	Pioneer Industrial Products Stansolv A-14; glove film			breakthrough time: 373 minutes; lower detection limit: 0.0002 ppm; permeation rate: 27 µg/cm²/min						ASTM F739.85	312
TPE, Vinyl	Ansell Edmont Wet-Wear 550; clothing; PVC stretch outer layer bonded to a lightweight liner			good permeation rate; 6 to 50 eyedropper size drops per hour; breakthrough time: 4 minutes; permeation rate: <90 µg/cm²/min		23				ASTM F739	313
TPE, Vinyl coated fabric	Ansell Edmont Wet-Wear 700; clothing; PVC coating with Nylon/Polyester lining			fair permeation rate; 51 to 100 eyedropper size drops per hour; breakthrough time: 5 minutes; permeation rate: <900 µg/cm²/min		23				ASTM F739	313

PETROLEUM ETHER

Material Family	Material Note	Perm. Coefficient	VTR	Non-normalized Data	Penetrant Note	Temp. (°C)	Time (days)	RH (%)	Pressure (kPa)	Test Note	Source
Rubber, Neoprene (CR)	Pioneer Industrial Products Stancoil N-44; glove film			breakthrough time: 126 minutes; lower detection limit: 0.002 ppm; permeation rate: 12 µg/cm²/min						ASTM F739.85	312
Rubber, Nitrile (NBR)	Pioneer Industrial Products Stansolv A-14; glove film			no permeation rate at steady state detected; breakthrough time: >480 minutes; lower detection limit: 0.02 ppm; permeation rate: 0 µg/cm²/min						ASTM F739.85	312

PHENOL

Material Family	Material Note	Perm. Coefficient (cm³·mm/m²·day·atm)	VTR (g·mm/m²·day)	Non-normalized Data	Penetrant Note	Temp. (°C)	Time (days)	RH (%)	Pressure (kPa)	Test Note	Source
Polyvinyl Alcohol	Ansell Edmont PVA; supported (lined) glove film; specified by glove film weight			low permeation rate: 0 to 1/2 eyedropper size drops per hour; PVA coating is water soluble; lower detection limit: 0 ppm; permeation rate: <0.9 µg/cm²/min		23	0.25			ASTM F739	313
Rubber, Latex (NR)	Ansell Edmont Canners and Handlers 392; unsupported glove film; 0.48 mm thick			breakthrough time: 90 minutes		23				ASTM F739	313
	Pioneer Industrial Products L-118; glove film			no permeation rate at steady state detected; breakthrough time: >480 minutes; lower detection limit: 0.06 ppm; permeation rate: 0 µg/cm²/min	saturated					ASTM F739.85	312
Rubber, Neoprene (CR)	Ansell Edmont Neoprene 29-840; unsupported glove film; 0.38 mm thick			good permeation rate; 6 to 50 eyedropper size drops per hour; breakthrough time: 180 minutes; permeation rate: <90 µg/cm²/min		23				ASTM F739	313
	Ansell Edmont Neox; supported (lined) glove film; specified by glove film weight			low permeation rate; 0 to 1/2 eyedropper size drops per hour; breakthrough time: 390 minutes; permeation rate: <0.9 µg/cm²/min		23				ASTM F739	313
	Pioneer Industrial Products Stanzoil N-44; glove film			no permeation rate at steady state detected; breakthrough time: >480 minutes; lower detection limit: 0.2 ppm; permeation rate: 0 µg/cm²/min	saturated					ASTM F739.85	312
Rubber, Nitrile (NBR)	Pioneer Industrial Products Stansolv A-14; glove film			no permeation rate at steady state detected; breakthrough time: >480 minutes; permeation rate: 0 µg/cm²/min	saturated					ASTM F739.85	312
TPE, Vinyl	Ansell Edmont Monkey Grip (lined) glove film; specified by glove film weight			very good permeation rate; 1 to 5 eyedropper size drops per hour; breakthrough time: 75 minutes; permeation rate: <9 µg/cm²/min		23				ASTM F739	313
	Ansell Edmont Wet-Wear 550; clothing; PVC stretch outer layer bonded to a lightweight liner			very good permeation rate; 1 to 5 eyedropper size drops per hour; breakthrough time: 35 minutes; permeation rate: <9 µg/cm²/min		23				ASTM F739	313
	Ansell Edmont Wet-Wear 600; clothing; nylon netting bonded between two layers of PVC			very good permeation rate; 1 to 5 eyedropper size drops per hour; breakthrough time: 90 minutes; permeation rate: <9 µg/cm²/min		23				ASTM F739	313
	Pioneer Industrial Products Pylox V-20; glove film			breakthrough time: 32 minutes; permeation rate: 78 µg/cm²/min	saturated					ASTM F739.85	312
TPE, Vinyl coated fabric	Ansell Edmont Wet-Wear 700; clothing; PVC coating with Nylon/Polyester lining			very good permeation rate; 1 to 5 eyedropper size drops per hour; breakthrough time: 50 minutes; permeation rate: <9 µg/cm²/min		23				ASTM F739	313

PHOSPHORIC ACID

Material Family	Material Note	Perm. Coefficient (cm³·mm/m²·day·atm)	VTR (g·mm/m²·day)	Non-normalized Data	Penetrant Note	Temp. (°C)	Time (days)	RH (%)	Pressure (kPa)	Test Note	Source
Rubber, Latex (NR)	Ansell Edmont Canners and Handlers 392; unsupported glove film; 0.48 mm thick			no permeation detected during a 6 hour test; lower detection limit: 0 ppm	concentrated	23	0.25			ASTM F739	313
	Pioneer Industrial Products L-118; glove film			no permeation rate at steady state detected; breakthrough time: >480 minutes; lower detection limit: 0.04 ppm; permeation rate: 0 µg/cm²/min	85% conc.	23				ASTM F739.85	312
Rubber, Neoprene (CR)	Ansell Edmont Neoprene 29-840; unsupported glove film; 0.38 mm thick			no permeation detected during a 6 hour test; lower detection limit: 0 ppm	concentrated	23	0.25			ASTM F739	313
	Ansell Edmont Neox; supported (lined) glove film; specified by glove film weight			no permeation detected during a 6 hour test; lower detection limit: 0 ppm	concentrated	23	0.25			ASTM F739	313
	Pioneer Industrial Products Stanzoil N-44; glove film			no permeation rate at steady state detected; breakthrough time: >480 minutes; lower detection limit: 0.04 ppm; permeation rate: 0 µg/cm²/min	85% conc.	23				ASTM F739.85	312
Rubber, Nitrile (NBR)	Ansell Edmont Sol-Vex 37-165; unsupported glove film; 0.54 mm thick			no permeation detected during a 6 hour test; lower detection limit: 0 ppm	concentrated	23	0.25			ASTM F739	313
	Pioneer Industrial Products Stansolv A-14; glove film			no permeation rate at steady state detected; breakthrough time: >480 minutes; lower detection limit: 0.04 ppm; permeation rate: 0 µg/cm²/min	85% conc.	23				ASTM F739.85	312
TPE, Vinyl	Ansell Edmont Monkey Grip; supported (lined) glove film; specified by glove film weight			no permeation detected during a 6 hour test; lower detection limit: 0 ppm	concentrated	23	0.25			ASTM F739	313
	Ansell Edmont Wet-Wear 550; clothing; PVC stretch outer layer bonded to a lightweight liner			no permeation detected during a 6 hour test; lower detection limit: 0 ppm	concentrated	23	0.25			ASTM F739	313

Material Family	Material Note	Permeability Data			Test Conditions						Source
		Perm. Coefficient (cm³·mm/m²·day·atm)	VTR (g·mm/m²·day)	Non-normalized Data	Penetrant Note	Temp. (°C)	Time (days)	RH (%)	Pressure (kPa)	Test Note	

PHOSPHORIC ACID (continued)

Material Family	Material Note			Non-normalized Data	Penetrant Note	Temp.	Time			Test Note	Source
TPE, Vinyl	Ansell Edmont Wet-Wear 600; clothing; nylon netting bonded between two layers of PVC			no permeation detected during a 6 hour test; lower detection limit: 0 ppm	concentrated	23	0.25			ASTM F739	313
	Pioneer Industrial Products Pylox V-20; glove film			no permeation rate at steady state detected; breakthrough time: >480 minutes; lower detection limit: 0.04 ppm; permeation rate: 0 μg/cm²/min	85% conc.					ASTM F739.85	312
TPE, Vinyl coated fabric	Ansell Edmont Wet-Wear 700; clothing; PVC coating with Nylon/Polyester lining			no permeation detected during a 6 hour test; lower detection limit: 0 ppm	concentrated	23	0.25			ASTM F739	313

PICRIC ACID

Material Family	Material Note			Non-normalized Data	Penetrant Note	Temp.				Test Note	Source
Rubber, Neoprene (CR)	Ansell Edmont Neoprene 29-840; unsupported glove film; 0.38 mm thick			very good permeation rate; 1 to 5 eyedropper size drops per hour; breakthrough time: 150 minutes; permeation rate: <9 μg/cm²/min	saturated; with ethylene oxide	23				ASTM F739	313
	Ansell Edmont Neox; supported (lined) glove film; specified by glove film weight			very good permeation rate; 1 to 5 eyedropper size drops per hour; breakthrough time: 180 minutes; permeation rate: <9 μg/cm²/min	saturated; with ethylene oxide	23				ASTM F739	313
Rubber, Nitrile (NBR)	Ansell Edmont Sol-Vex 37-165; unsupported glove film; 0.54 mm thick			very good permeation rate; 1 to 5 eyedropper size drops per hour; breakthrough time: 160 minutes; permeation rate: <9 μg/cm²/min	saturated; with ethylene oxide	23				ASTM F739	313
TPE, Vinyl	Ansell Edmont Monkey Grip; supported (lined) glove film; specified by glove film weight			very good permeation rate; 1 to 5 eyedropper size drops per hour; breakthrough time: 40 minutes; permeation rate: <9 μg/cm²/min	saturated; with ethylene oxide	23				ASTM F739	313

PINE OIL

Material Family	Material Note			Non-normalized Data	Penetrant Note	Temp.	Time			Test Note	Source
Laminar, Nylon/Olefin	DuPont Selar RB/ Polyolefin; 8% Selar RB; barrier prop.; extrusion blow molded; bottles			0.27% penetrant weight loss	cleaner	50	28				293
Polyethylene, Fluorinated	barrier prop.; bottles			0.10% penetrant weight loss	cleaner	50	28				293
Polyethylene, HDPE	bottles			1.7% penetrant weight loss (oily surface)	cleaner	50	28				293

PIPERONOL (HELIOTROPIN)

Material Family	Material Note			Non-normalized Data							Source
EVAL/ PE Film	0.013/ 0.051 mm thick; PE inside; EVAL EF-XL outside; film			>30 days to leakage							265
Nylon/ EVAL Film	0.015/ 0.015 mm thick; EVAL EF-F inside; oriented nylon outside; film			27 days to leakage							265
Nylon/ PE Film	0.015/ 0.051 mm thick; PE inside; oriented nylon outside; film			5 days to leakage							265
PET/ EVAL Film	0.013/ 0.015 mm thick; EVAL EF-F inside; PET outside; film			30 days to leakage							265
PET/ EVAL/ LDPE Film	EVAL EF-F barrier; 0.013/ 0.015/ 0.051 mm thick; LDPE inside; PET outside			27 days to leakage							265
PET/ PE Film	0.013/ 0.051 mm thick; PE inside; PET outside; film			5 days to leakage							265
PP/ PE Film	0.018/ 0.051 mm thick; PE inside; PVDC coated BOPP outside; film			1 days to leakage							265

POLYCHLORINATED BIPHENYLS

Material Family	Material Note			Non-normalized Data	Penetrant Note					Test Note	Source
Rubber, Neoprene (CR)	Pioneer Industrial Products Stanzoil N-44; glove film			no permeation rate at steady state detected; breakthrough time: >480 minutes; lower detection limit: 1 ppm; permeation rate: 0 μg/cm²/min	PCBs; 50% TCB					ASTM F739.85	312
Rubber, Nitrile (NBR)	Pioneer Industrial Products Stansolv A-14; glove film			breakthrough time: 343 minutes; lower detection limit: 1 ppm; permeation rate: 216 μg/cm²/min	PCBs; 50% TCB					ASTM F739.85	312

POTASSIUM HYDROXIDE

Material Family	Material Note	Perm. Coefficient (cm³ · mm/m² · day · atm)	VTR (g · mm/m² · day)	Non-normalized Data	Penetrant Note	Temp. (°C)	Time (days)	RH (%)	Pressure (kPa)	Test Note	Source
Rubber, Latex (NR)	Ansell Edmont Canners and Handlers 392; unsupported glove film; 0.48 mm thick			no permeation detected during a 6 hour test; lower detection limit: 0 ppm	KOH; 50% conc.	23	0.25			ASTM F739	313
	Pioneer Industrial Products L-118; glove film			no permeation rate at steady state detected; breakthrough time: >480 minutes; lower detection limit: 0.3 ppm; permeation rate: 0 µg/cm²/min	KOH; 50% conc.					ASTM F739.85	312
Rubber, Neoprene (CR)	Ansell Edmont Neoprene 29-840; unsupported glove film; 0.38 mm thick			no permeation detected during a 6 hour test; lower detection limit: 0 ppm	KOH; 50% conc.	23	0.25			ASTM F739	313
	Ansell Edmont Neox; supported (lined) glove film; specified by glove film weight			no permeation detected during a 6 hour test; lower detection limit: 0 ppm	KOH; 50% conc.	23	0.25			ASTM F739	313
	Pioneer Industrial Products Stanzoil N-44; glove film			no permeation rate at steady state detected; breakthrough time: >480 minutes; lower detection limit: 0.3 ppm; permeation rate: 0 µg/cm²/min	KOH; 50% conc.					ASTM F739.85	312
Rubber, Nitrile (NBR)	Ansell Edmont Sol-Vex 37-165; unsupported glove film; 0.54 mm thick			no permeation detected during a 6 hour test; lower detection limit: 0 ppm	KOH; 50% conc.	23	0.25			ASTM F739	313
	Pioneer Industrial Products Stansolv A-14; glove film			no permeation rate at steady state detected; breakthrough time: >480 minutes; lower detection limit: 0.4 ppm; permeation rate: 0 µg/cm²/min	KOH; 50% conc.					ASTM F739.85	312
TPE, Vinyl	Ansell Edmont Monkey Grip; supported (lined) glove film; specified by glove film weight			no permeation detected during a 6 hour test; lower detection limit: 0 ppm	KOH; 50% conc.	23	0.25			ASTM F739	313
	Ansell Edmont Wet-Wear 550; clothing; PVC stretch outer layer bonded to a lightweight liner			no permeation detected during a 6 hour test; lower detection limit: 0 ppm	KOH; 50% conc.	23	0.25			ASTM F739	313
	Ansell Edmont Wet-Wear 600; clothing; nylon netting bonded between two layers of PVC			no permeation detected during a 6 hour test; lower detection limit: 0 ppm	KOH; 50% conc.	23	0.25			ASTM F739	313
	Pioneer Industrial Products Pylox V-20; glove film			no permeation rate at steady state detected; breakthrough time: >480 minutes; lower detection limit: 0.4 ppm; permeation rate: 0 µg/cm²/min	KOH; 50% conc.					ASTM F739.85	312
TPE, Vinyl coated fabric	Ansell Edmont Wet-Wear 700; clothing; PVC coating with Nylon/Polyester lining			no permeation detected during a 6 hour test; lower detection limit: 0 ppm	KOH; 50% conc.	23	0.25			ASTM F739	313

PROPANE

Material Family	Material Note	Perm. Coefficient (cm³ · mm/m² · day · atm)	VTR (g · mm/m² · day)	Non-normalized Data	Penetrant Note	Temp. (°C)	Time (days)	RH (%)	Pressure (kPa)	Test Note	Source
Polyethylene, HDPE	Hoechst AG Hostalen	35.5			gas at 0°C and 0.11 Mpa	20				volume at standard temperature and pressure; useable average for all Hostalen grades	94
TPE, Olefinic	Adv. Elast. Santoprene 201-73; 73 Shore A, 0.5 mm thick	1162			gas at 0°C and 0.11 Mpa	23			110	ASTM D1434	282
	Adv. Elast. Santoprene 201-87; 87 Shore A, 0.5 mm thick	3332			gas at 0°C and 0.11 Mpa	23			110	ASTM D1434	282
	Adv. Elast. Santoprene 203-50; 50 Shore D, 0.5 mm thick	1938			gas at 0°C and 0.11 Mpa	23			110	ASTM D1434	282
TPE, Polyamide	Atochem Pebax 5533; 55 Shore D, 0.12 mm thick; film	789				23		dry			287
	Atochem Pebax 6333; 63 Shore D, 0.12 mm thick; film	236				23		dry			287
TPE, Polyester	DuPont Hytrel 4056; 40 Shore D	<17.3				21.5			34.5		274
	DuPont Hytrel 5556; 55 Shore D	<17.3				21.5			34.5		274
	DuPont Hytrel 6346; 63 Shore D	<17.3				21.5			34.5		274

PROPYL ACETATE

Material Family	Material Note	Perm. Coefficient (cm³ · mm/m² · day · atm)	VTR (g · mm/m² · day)	Non-normalized Data	Penetrant Note	Temp. (°C)	Time (days)	RH (%)	Pressure (kPa)	Test Note	Source
Polyvinyl Alcohol	Ansell Edmont PVA; supported (lined) glove film; specified by glove film weight			very good permeation rate; 1 to 5 eyedropper size drops per hour; PVA coating is water soluble; breakthrough time: 120 minutes; permeation rate: <9 µg/cm²/min		23				ASTM F739	313

PROPYL ACETATE (continued)

Material Family	Material Note	Perm. Coefficient (cm³·mm/m²·day·atm)	VTR (g·mm/m²·day)	Non-normalized Data	Penetrant Note	Temp. (°C)	Time (days)	RH (%)	Pressure (kPa)	Test Note	Source
Rubber, Nitrile (NBR)	Ansell Edmont Sol-Vex 37-165; unsupported glove film; 0.54 mm thick			good permeation rate; 6 to 50 eyedropper size drops per hour; breakthrough time: 200 minutes; permeation rate: <90 µg/cm²/min		23				ASTM F739	313

PROPYL ALCHOHOL

Material Family	Material Note	Perm. Coefficient	VTR	Non-normalized Data	Penetrant Note	Temp. (°C)	Time (days)	RH (%)	Pressure (kPa)	Test Note	Source
Laminar, EVOH/Olefin	DuPont Selar RB/ Polyolefin, 15% Selar RB; barrier prop., laminar technology; extrusion blow molded; 1 L bottle			0.02% penetrant weight loss	with 25% xylene	50	28				293
	DuPont Selar RB/ Polyolefin, 15% Selar RB; barrier prop., laminar technology; extrusion blow molded; 1 L bottle			0.01% penetrant weight loss		50	28				293
Laminar, Nylon/Olefin	DuPont Selar RB/ Polyolefin, 8% Selar RB; barrier prop., laminar technology; extrusion blow molded; 1 L bottle			1.44% penetrant weight loss	with 25% xylene	50	28				293
	DuPont Selar RB/ Polyolefin, 8% Selar RB; barrier prop., laminar technology; extrusion blow molded; 1 L bottle			0.14% penetrant weight loss		50	28				293
Polyethylene, HDPE	DuPont; 1 L bottle			4.71% penetrant weight loss	with 25% xylene	50	28				293
	DuPont; 1 L bottle			0.15% penetrant weight loss		50	28				293
Rubber, Latex (NR)	Ansell Edmont Canners and Handlers 392; unsupported glove film; 0.48 mm thick			very good permeation rate: 1 to 5 eyedropper size drops per hour; breakthrough time: 200 minutes; permeation rate: <9 µg/cm²/min		23				ASTM F739	313
Rubber, Neoprene (CR)	Ansell Edmont Neoprene 29-840; unsupported glove film; 0.38 mm thick			low permeation rate: 0 to 1/2 eyedropper size drops per hour; breakthrough time: 150 minutes; permeation rate: <0.9 µg/cm²/min		23				ASTM F739	313
	Ansell Edmont Neox; supported (lined) glove film; specified by glove film weight			low permeation rate: 0 to 1/2 eyedropper size drops per hour, lower detection limit: 0 ppm; permeation rate: <0.9 µg/cm²/min		23	0.25			ASTM F739	313
Rubber, Nitrile (NBR)	Ansell Edmont Sol-Vex 37-165; unsupported glove film; 0.54 mm thick			low permeation rate: 0 to 1/2 eyedropper size drops per hour, lower detection limit: 0 ppm; permeation rate: <0.9 µg/cm²/min		23	0.25			ASTM F739	313
TPE, Vinyl	Ansell Edmont Monkey Grip; supported (lined) glove film; specified by glove film weight			very good permeation rate: 1 to 5 eyedropper size drops per hour; breakthrough time: 90 minutes; permeation rate: <9 µg/cm²/min		23				ASTM F739	313

PROPYLENE

Material Family	Material Note	Perm. Coefficient	VTR	Non-normalized Data	Penetrant Note	Temp. (°C)	Time (days)	RH (%)	Pressure (kPa)	Test Note	Source
Polyethylene, HDPE	Hoechst AG Hostalen	77.0				20				volume at standard temperature and pressure; useable average for all Hostalen grades	94

PROPYLENE GLYCOL

Material Family	Material Note	Perm. Coefficient	VTR	Non-normalized Data	Penetrant Note	Temp. (°C)	Time (days)	RH (%)	Pressure (kPa)	Test Note	Source
TPE, Vinyl	Ansell Edmont Wet-Wear 550; clothing; PVC stretch outer layer bonded to a lightweight liner			low permeation rate: 0 to 1/2 eyedropper size drops per hour; lower detection limit: 0 ppm; permeation rate: <0.9 µg/cm²/min		23	0.25			ASTM F739	313
	Ansell Edmont Wet-Wear 600; clothing; nylon netting bonded between two layers of PVC			no permeation detected during a 6 hour test; lower detection limit: 0 ppm		23	0.25			ASTM F739	313
TPE, Vinyl coated fabric	Ansell Edmont Wet-Wear 700; clothing; PVC coating with Nylon/Polyester lining			no permeation detected during a 6 hour test; lower detection limit: 0 ppm		23	0.25			ASTM F739	313

PROPYLENE OXIDE

Material Family	Material Note	Perm. Coefficient	VTR	Non-normalized Data	Penetrant Note	Temp. (°C)	Time (days)	RH (%)	Pressure (kPa)	Test Note	Source
Polyvinyl Alcohol	Ansell Edmont PVA; supported (lined) glove film; specified by glove film weight			good permeation rate; 6 to 50 eyedropper size drops per hour; PVA coating is water soluble; breakthrough time: 35 minutes; permeation rate: <90 µg/cm²/min		23				ASTM F739	313

Material Family	Material Note	Perm. Coefficient (cm³·mm/m²·day·atm)	VTR (g·mm/m²·day)	Non-normalized Data	Penetrant Note	Temp. (°C)	Time (days)	RH (%)	Pressure (kPa)	Test Note	Source
PYRIDINE											
Polyvinyl Alcohol	Ansell Edmont PVA; supported (lined) glove film; specified by glove film weight			fair permeation rate; 51 to 100 eyedropper size drops per hour; PVA coating is water soluble; breakthrough time: 10 minutes; permeation rate: <900 µg/cm²/min		23				ASTM F739	313
Rubber, Latex (NR)	Ansell Edmont Canners and Handlers 392; unsupported glove film; 0.48 mm thick			fair permeation rate; 51 to 100 eyedropper size drops per hour; breakthrough time: 10 minutes; permeation rate: <900 µg/cm²/min		23				ASTM F739	313
Rubber, Neoprene (CR)	Pioneer Industrial Products Stanzoil N-44; glove film			breakthrough time: 26 minutes; permeation rate: 702 µg/cm²/min						ASTM F739.85	312
TPE, Vinyl	Pioneer Industrial Products Pylox V-20; glove film			permeation rate too large to measure; breakthrough time: 1 minutes						ASTM F739.85	312
PCBS											
Rubber, Neoprene (CR)	Pioneer Industrial Products Stanzoil N-44; glove film			no permeation rate at steady state detected; breakthrough time: >480 minutes; lower detection limit: 1 ppm; permeation rate: 0 µg/cm²/min	polychlorinated biphenyls; 50% TCB					ASTM F739.85	312
Rubber, Nitrile (NBR)	Pioneer Industrial Products Stansolv A-14; glove film			breakthrough time: 343 minutes; lower detection limit: 1 ppm; permeation rate: 216 µg/cm²/min	.					ASTM F739.85	312
ROAD OIL REMOVER											
Acetal	DuPont Delrin		0.01			23		50			201
Acetal	DuPont Delrin		0.07			38					201
RUBBER SOLVENT											
Polyvinyl Alcohol	Ansell Edmont PVA; supported (lined) glove film; specified by glove film weight			low permeation rate: 0 to 1/2 eyedropper size drops per hour; PVA coating is water soluble; lower detection limit: 0 ppm; permeation rate: <0.9 µg/cm²/min		23	0.25			ASTM F739	313
Rubber, Neoprene (CR)	Ansell Edmont Neoprene 29-840; unsupported glove film; 0.38 mm thick			good permeation rate: 6 to 50 eyedropper size drops per hour; breakthrough time: 30 minutes; permeation rate: <90 µg/cm²/min		23				ASTM F739	313
Rubber, Nitrile (NBR)	Ansell Edmont Neox; supported (lined) glove film; specified by glove film weight			good permeation rate: 6 to 50 eyedropper size drops per hour; breakthrough time: 60 minutes; permeation rate: <90 µg/cm²/min		23				ASTM F739	313
Rubber, Nitrile (NBR)	Ansell Edmont Sol-Vex 37-165; unsupported glove film; 0.54 mm thick			low permeation rate: 0 - 1/2 eyedropper drops per hr.; lower detection limit: 0 ppm; permeation rate: <0.9 µg/cm²/min		23	0.25			ASTM F739	313
RULE 66											
Rubber, Latex (NR)	Pioneer Industrial Products L-118; glove film			breakthrough time: 9 minutes; lower detection limit: 0.002 ppm; permeation rate: 126 µg/cm²/min	solvent					ASTM F739.85	312
Rubber, Neoprene (CR)	Pioneer Industrial Products Stanzoil N-44; glove film			no permeation rate at steady state detected; breakthrough time: 332 minutes; lower detection limit: 0.03 ppm; permeation rate: 0 µg/cm²/min	solvent					ASTM F739.85	312
Rubber, Nitrile (NBR)	Pioneer Industrial Products Stansolv A-14; glove film			no permeation rate at steady state detected; breakthrough time: >480 minutes; lower detection limit: 0.001 ppm; permeation rate: 0 µg/cm²/min	solvent					ASTM F739.85	312
TPE, Vinyl	Pioneer Industrial Products Pylox V-20; glove film			breakthrough time: 42 minutes; lower detection limit: 0.02 ppm; permeation rate: 60 µg/cm²/min	solvent					ASTM F739.85	312
SHAMPOO											
Acetal	DuPont Delrin		0.94		various formulations	23		50			201

SHAMPOO (continued)

Material Family	Material Note	Permeability Data			Penetrant Note	Test Conditions					Source
		Perm. Coefficient (cm³·mm/m²·day·atm)	VTR (g·mm/m²·day)	Non-normalized Data		Temp. (°C)	Time (days)	RH (%)	Pressure (kPa)	Test Note	
Acetal	DuPont Delrin		3.4		various formulations	38					201

SILICON ETCH

Material Family	Material Note	Perm. Coefficient	VTR	Non-normalized Data	Penetrant Note	Temp. (°C)	Time (days)	RH (%)	Pressure (kPa)	Test Note	Source
Rubber, Neoprene (CR)	Ansell Edmont Neoprene 29-840; unsupported glove film; 0.38 mm thick			no permeation detected during a 6 hour test; lower detection limit: 0 ppm		23	0.25			ASTM F739	313
	Ansell Edmont Neox; supported (lined) glove film; specified by glove film weight			no permeation detected during a 6 hour test; lower detection limit: 0 ppm		23	0.25			ASTM F739	313
TPE, Vinyl	Ansell Edmont Monkey Grip; supported (lined) glove film; specified by glove film weight			breakthrough time: 150 minutes		23				ASTM F739	313

SODIUM HYDROXIDE

Material Family	Material Note	Perm. Coefficient	VTR	Non-normalized Data	Penetrant Note	Temp. (°C)	Time (days)	RH (%)	Pressure (kPa)	Test Note	Source
Rubber, Latex (NR)	Ansell Edmont Canners and Handlers 392; unsupported glove film; 0.48 mm thick			no permeation detected during a 6 hour test; lower detection limit: 0 ppm	NaOH; 50% conc.	23	0.25			ASTM F739	313
	Pioneer Industrial Products L-118; glove film			no permeation rate at steady state detected; breakthrough time: >480 minutes; lower detection limit: 0.1 ppm; permeation rate: 0 µg/cm²/min	NAOH; 50% conc.	23				ASTM F739.85	312
Rubber, Neoprene (CR)	Ansell Edmont Neoprene 29-840; unsupported glove film; 0.38 mm thick			no permeation detected during a 6 hour test; lower detection limit: 0 ppm	NaOH; 50% conc.	23	0.25			ASTM F739	313
	Ansell Edmont Neox; supported (lined) glove film; specified by glove film weight			no permeation detected during a 6 hour test; lower detection limit: 0 ppm	NaOH; 50% conc.	23	0.25			ASTM F739	313
	Pioneer Industrial Products Stanzoil N-44; glove film			no permeation rate at steady state detected; breakthrough time: >480 minutes; lower detection limit: 0.1 ppm; permeation rate: 0 µg/cm²/min	NAOH; 50% conc.	23				ASTM F739.85	312
Rubber, Nitrile (NBR)	Ansell Edmont Sol-Vex 37-165; unsupported glove film; 0.54 mm thick			no permeation detected during a 6 hour test; lower detection limit: 0 ppm	NaOH; 50% conc.	23	0.25			ASTM F739	313
	Pioneer Industrial Products Stanoly A-14; glove film			no permeation rate at steady state detected; breakthrough time: >480 minutes; lower detection limit: 0.1 ppm; permeation rate: 0 µg/cm²/min	NAOH; 50% conc.	23				ASTM F739.85	312
TPE, Vinyl	Ansell Edmont Monkey Grip; supported (lined) glove film; specified by glove film weight			no permeation detected during a 6 hour test; lower detection limit: 0 ppm	NaOH; 50% conc.	23	0.25			ASTM F739	313
	Ansell Edmont Wet-Wear 550; clothing; PVC stretch outer layer bonded to a lightweight liner			no permeation detected during a 6 hour test; lower detection limit: 0 ppm	NaOH; 50% conc.	23	0.25			ASTM F739	313
	Ansell Edmont Wet-Wear 600; clothing; nylon netting bonded between two layers of PVC			no permeation detected during a 6 hour test; lower detection limit: 0 ppm	NaOH; 50% conc.	23	0.25			ASTM F739	313
	Pioneer Industrial Products Pylox V-20; glove film			no permeation rate at steady state detected; breakthrough time: >480 minutes; lower detection limit: 0.04 ppm; permeation rate: 0 µg/cm²/min	NAOH; 50% conc.	23				ASTM F739.85	312
TPE, Vinyl coated fabric	Ansell Edmont Wet-Wear 700; clothing; PVC coating with Nylon/Polyester lining			no permeation detected during a 6 hour test; lower detection limit: 0 ppm	NaOH; 50% conc.	23	0.25			ASTM F739	313

STODDARD SOLVENTS

Material Family	Material Note	Perm. Coefficient	VTR	Non-normalized Data	Penetrant Note	Temp. (°C)	Time (days)	RH (%)	Pressure (kPa)	Test Note	Source
Polyvinyl Alcohol	Ansell Edmont PVA; supported (lined) glove film; specified by glove film weight			low permeation rate; 0 to 1/2 eyedropper size drops per hour; PVA coating is water soluble; lower detection limit: 0 ppm; permeation rate: <0.9 µg/cm²/min		23	0.25			ASTM F739	313
Rubber, Neoprene (CR)	Ansell Edmont Neoprene 29-840; unsupported glove film; 0.38 mm thick			very good permeation rate; 1 to 5 eyedropper size drops per hour; breakthrough time: 180 minutes; permeation rate: <9 µg/cm²/min		23				ASTM F739	313

STODDARD SOLVENTS (continued)

Material Family / Material Note	Perm. Coefficient (cm³·mm/m²·day·atm)	VTR (g·mm/m²·day)	Non-normalized Data	Penetrant Note	Temp. (°C)	Time (days)	RH (%)	Pressure (kPa)	Test Note	Source
Rubber, Neoprene (CR) — Ansell Edmont Neox; supported (lined) glove film; specified by glove film weight			low permeation rate; 0 to 1/2 eyedropper size drops per hour; lower detection limit: 0 ppm; permeation rate: <0.9 µg/cm²/min		23	0.25			ASTM F739	313
Rubber, Nitrile (NBR) — Ansell Edmont Sol-Vex 37-165; unsupported glove film; 0.54 mm thick			low permeation rate; 0 to 1/2 eyedropper size drops per hour; lower detection limit: 0 ppm; permeation rate: <0.9 µg/cm²/min		23	0.25			ASTM F739	313
TPE, Vinyl — Ansell Edmont Monkey Grip; supported (lined) glove film; specified by glove film weight			low permeation rate; 0 to 1/2 eyedropper size drops per hour; breakthrough time: 360 minutes; permeation rate: <0.9 µg/cm²/min		23				ASTM F739	313

STYRENE

Material Family / Material Note	Perm. Coefficient	VTR	Non-normalized Data	Penetrant Note	Temp. (°C)	Time (days)	RH (%)	Pressure (kPa)	Test Note	Source
Polyvinyl Alcohol — Ansell Edmont PVA; supported (lined) glove film; specified by glove film weight			low permeation rate; 0 to 1/2 eyedropper size drops per hour; PVA coating is water soluble; lower detection limit: 0 ppm; permeation rate: <0.9 µg/cm²/min		23	0.25			ASTM F739	313

SULFUR DIOXIDE

Material Family / Material Note	Perm. Coefficient	VTR	Non-normalized Data	Penetrant Note	Temp. (°C)	Time (days)	RH (%)	Pressure (kPa)	Test Note	Source
Fluoropolymer, PVDF — Solvay Solef; 0.025 mm thick; cast film; film	1.5				23				ASTM D1434	125
Polyethylene, HDPE — Hoechst AG Hostalen	436				20				volume at standard temperature and pressure; useable average for all Hostalen grades	94

SULFUR HEXAFLUORIDE

Material Family / Material Note	Perm. Coefficient	VTR	Non-normalized Data	Penetrant Note	Temp. (°C)	Time (days)	RH (%)	Pressure (kPa)	Test Note	Source
Polysulfone — Amoco Udel; transparent, amber tint	0.71				23		dry		ASTM D1434	15

SULFURIC ACID

Material Family / Material Note	Perm. Coefficient	VTR	Non-normalized Data	Penetrant Note	Temp. (°C)	Time (days)	RH (%)	Pressure (kPa)	Test Note	Source
Rubber, Latex (NR) — Ansell Edmont Canners and Handlers 392; unsupported glove film; 0.46 mm thick			no permeation detected during a 6 hour test; lower detection limit: 0 ppm	battery acid: 47% conc.	23	0.25			ASTM F739	313
Pioneer Industrial Products L-118; glove film			no permeation rate at steady state detected; breakthrough time: >480 minutes; lower detection limit: 0.1 ppm; permeation rate: 0 µg/cm²/min	50% conc.					ASTM F739.85	312
Rubber, Neoprene (CR) — Ansell Edmont Neoprene 29-840; unsupported glove film; 0.38 mm thick			no permeation detected during a 6 hour test; lower detection limit: 0 ppm	battery acid: 47% conc.	23	0.25			ASTM F739	313
Ansell Edmont Neoprene 29-840; unsupported glove film; 0.38 mm thick			breakthrough time: 180 minutes	95% conc.	23				ASTM F739	313
Ansell Edmont Neox; supported (lined) glove film; specified by glove film weight			no permeation detected during a 6 hour test; lower detection limit: 0 ppm	battery acid: 47% conc.	23	0.25			ASTM F739	313
Ansell Edmont Neox; supported (lined) glove film; specified by glove film weight			breakthrough time: >360 minutes	95% conc.	23				ASTM F739	313
Pioneer Industrial Products Starzoil N-44; glove film			no permeation rate at steady state detected; breakthrough time: >480 minutes; lower detection limit: 0.04 ppm; permeation rate: 0 µg/cm²/min	50% conc.					ASTM F739.85	312
Rubber, Nitrile (NBR) — Ansell Edmont Sol-Vex 37-165; unsupported glove film; 0.54 mm thick			no permeation detected during a 6 hour test; lower detection limit: 0 ppm	battery acid: 47% conc.	23	0.25			ASTM F739	313
Pioneer Industrial Products Stansolv A-14; glove film			no permeation rate at steady state detected; breakthrough time: >480 minutes; lower detection limit: 0.04 ppm; permeation rate: 0 µg/cm²/min	50% conc.					ASTM F739.85	312
TPE, Vinyl — Ansell Edmont Monkey Grip; supported (lined) glove film; specified by glove film weight			no permeation detected during a 6 hour test; lower detection limit: 0 ppm	battery acid: 47% conc.	23	0.25			ASTM F739	313
Ansell Edmont Monkey Grip; supported (lined) glove film; specified by glove film weight			PVC is not recommended for hot sulfuric acid or fuming oleum; breakthrough time: 220 minutes	95% conc.	23				ASTM F739	313

SULFURIC ACID (continued)

Material Family	Material Note	Perm. Coefficient (cm³·mm/m²·day·atm)	VTR (g·mm/m²·day)	Non-normalized Data	Penetrant Note	Temp. (°C)	Time (days)	RH (%)	Pressure (kPa)	Test Note	Source
TPE, Vinyl	Ansell Edmont Wet-Wear 550; clothing; PVC stretch outer layer bonded to a lightweight liner			breakthrough time: 5 minutes	95% conc.	23				ASTM F739	313
	Ansell Edmont Wet-Wear 550; clothing; PVC stretch outer layer bonded to a lightweight liner			breakthrough time: 60 minutes	10% conc.	23				ASTM F739	313
	Ansell Edmont Wet-Wear 600; clothing; nylon netting bonded between two layers of PVC			breakthrough time: 60 minutes	95% conc.	23				ASTM F739	313
	Ansell Edmont Wet-Wear 600; clothing; nylon netting bonded between two layers of PVC			breakthrough time: 140 minutes	10% conc.	23				ASTM F739	313
	Pioneer Industrial Products Pylox V-20; glove film			no permeation rate at steady state detected; breakthrough time: >480 minutes; lower detection limit: 0.04 ppm; permeation rate: 0 µg/cm²/min	50% conc.					ASTM F739.85	312
TPE, Vinyl coated fabric	Ansell Edmont Wet-Wear 700; clothing; PVC coating with Nylon/Polyester lining			breakthrough time: 37 minutes	95% conc.	23				ASTM F739	313
	Ansell Edmont Wet-Wear 700; clothing; PVC coating with Nylon/Polyester lining			breakthrough time: 120 minutes	10% conc.	23				ASTM F739	313

STP GAS TREATMENT

Material Family	Material Note	Perm. Coefficient (cm³·mm/m²·day·atm)	VTR (g·mm/m²·day)	Non-normalized Data	Penetrant Note	Temp. (°C)	Time (days)	RH (%)	Pressure (kPa)	Test Note	Source
Laminar, Nylon/Olefin	DuPont Selar RB/ Polyolefin; 8% Selar RB; barrier prop.; extrusion blow molded; bottles			0.07% penetrant weight loss		50	28				293
Polyethylene, Fluorinated	barrier prop.; bottles			0.12% penetrant weight loss		50	28				293
Polyethylene, HDPE	bottles			16.4% penetrant weight loss		50	28				293

TANNIC ACID

Material Family	Material Note	Perm. Coefficient (cm³·mm/m²·day·atm)	VTR (g·mm/m²·day)	Non-normalized Data	Penetrant Note	Temp. (°C)	Time (days)	RH (%)	Pressure (kPa)	Test Note	Source
Rubber, Latex (NR)	Ansell Edmont Canners and Handlers 392; unsupported glove film; 0.48 mm thick			no permeation detected during a 6 hour test; lower detection limit: 0 ppm	65% conc.	23	0.25			ASTM F739	313
Rubber, Neoprene (CR)	Ansell Edmont Neoprene 29-840; unsupported glove film; 0.38 mm thick			low permeation rate; 0 to 1/2 eyedropper size drops per hour; lower detection limit: 0 ppm; permeation rate: <0.9 µg/cm²/min	65% conc.	23	0.25			ASTM F739	313
	Ansell Edmont Neox; supported (lined) glove film; specified by glove film weight			low permeation rate; 0 to 1/2 eyedropper size drops per hour; lower detection limit: 0 ppm; permeation rate: <0.9 µg/cm²/min	65% conc.	23	0.25			ASTM F739	313
Rubber, Nitrile (NBR)	Ansell Edmont Sol-Vex 37-165; unsupported glove film; 0.54 mm thick			low permeation rate; 0 to 1/2 eyedropper size drops per hour; lower detection limit: 0 ppm; permeation rate: <0.9 µg/cm²/min	65% conc.	23	0.25			ASTM F739	313
TPE, Vinyl	Ansell Edmont Monkey Grip; supported (lined) glove film; specified by glove film weight			low permeation rate; 0 to 1/2 eyedropper size drops per hour; lower detection limit: 0 ppm; permeation rate: <0.9 µg/cm²/min	65% conc.	23	0.25			ASTM F739	313

TAR REMOVER

Material Family	Material Note	Perm. Coefficient (cm³·mm/m²·day·atm)	VTR (g·mm/m²·day)	Non-normalized Data	Penetrant Note	Temp. (°C)	Time (days)	RH (%)	Pressure (kPa)	Test Note	Source
Acetal	DuPont Delrin		0.01			23		50			201
	DuPont Delrin		0.07			38					201

TETRACHLOROETHANE (1,1,2,2-)

Material Family	Material Note	Perm. Coefficient (cm³·mm/m²·day·atm)	VTR (g·mm/m²·day)	Non-normalized Data	Penetrant Note	Temp. (°C)	Time (days)	RH (%)	Pressure (kPa)	Test Note	Source
Rubber, Neoprene (CR)	Pioneer Industrial Products Stanzoil N-44; glove film			breakthrough time: 18 minutes; permeation rate: 1398 µg/cm²/min						ASTM F739.85	312
TPE, Vinyl	Pioneer Industrial Products Pylox V-20; glove film			breakthrough time: 6 minutes; permeation rate: 2502 µg/cm²/min						ASTM F739.85	312

© Plastics Design Library

Material Family / Material Note	Perm. Coefficient (cm³·mm/m²·day·atm)	VTR (g·mm/m²·day)	Non-normalized Data	Penetrant Note	Temp. (°C)	Time (days)	RH (%)	Pressure (kPa)	Test Note	Source
TETRACHLOROETHENE										
Polyvinyl Alcohol — Ansell Edmont PVA; supported (lined) glove film; specified by glove film weight			low permeation rate; 0 to 1/2 eyedropper size drops per hour; PVA coating is water soluble; lower detection limit: 0 ppm; permeation rate: <0.9 µg/cm²/min		23	0.25			ASTM F739	313
Rubber, Nitrile (NBR) — Ansell Edmont Sol-Vex 37-165; unsupported glove film; 0.54 mm thick			very good permeation rate; 1 to 5 eyedropper size drops per hour; breakthrough time: 300 minutes; permeation rate: <9 µg/cm²/min		23				ASTM F739	313
TETRACHLOROETHYLENE										
Polyethylene, LDPE — Dow; film		197 - 295			24					250
Polystyrene — Dow Styron; film			sample failed		24					250
SAN — Dow Tyril; film			sample failed		24					250
TETRACHLOROETHYLENE (1,1,2,2-)										
Rubber, Neoprene (CR) — Pioneer Industrial Products Stanzoil N-44; glove film			breakthrough time: 28 minutes; lower detection limit: 0.0002 ppm; permeation rate: 453 µg/cm²/min						ASTM F739.85	312
Rubber, Nitrile (NBR) — Pioneer Industrial Products Stansolv A-14; glove film			breakthrough time: 373 minutes; lower detection limit: 0.0002 ppm; permeation rate: 27 µg/cm²/min						ASTM F739.85	312
TETRAHYDROFURAN										
Acrylonitrile Copolymer, AMA — barrier prop.; bottles			0.01% penetrant weight loss (crazing)		23	180				293
Laminar, Nylon/Olefin — DuPont Selar RB/ Polyolefin; 8% Selar RB; barrier prop.; extrusion blow molded; bottles			0.44% penetrant weight loss		23	180				293
Polyethylene, Fluorinated — barrier prop.; bottles			8.89% penetrant weight loss		23	180				293
Polyethylene, HDPE — bottles			29.19% penetrant weight loss		23	180				293
Polyvinyl Alcohol — Ansell Edmont PVA; supported (lined) glove film; specified by glove film weight			good permeation rate; 6 to 50 eyedropper size drops per hour; PVA coating is water soluble; breakthrough time: 90 minutes; permeation rate: <90 µg/cm²/min	THF	23				ASTM F739	313
Polyvinyl Chloride — bottles			failed		23	180				293
Rubber, Neoprene (CR) — Pioneer Industrial Products Stanzoil N-44; glove film			breakthrough time: 11 minutes; lower detection limit: 0.08 ppm; permeation rate: 4026 µg/cm²/min	THF					ASTM F739.85	312
Rubber, Nitrile (NBR) — Pioneer Industrial Products Stansolv A-14; glove film			breakthrough time: 17 minutes; lower detection limit: 0.08 ppm; permeation rate: 4026 µg/cm²/min	THF					ASTM F739.85	312
TPE, Vinyl — Pioneer Industrial Products Pylox V-20; glove film			permeation rate too large to measure; breakthrough time: 1 minutes	THF					ASTM F739.85	312
TOLUENE										
Acetal — DuPont Delrin	0.24				23		50			201
Acrylonitrile Copolymer, AMA — barrier prop.; bottles			0.07% penetrant weight loss		50	28				293
Laminar, Nylon/HDPE — DuPont Selar RB/ HDPE; solvent packaging; 18% Selar RB; barrier prop.; extrusion blow molded; one pint bottle			0.43% penetrant weight loss (60% relative improvement vs. HDPE)	aromatic	48.9	28				291
Laminar, Nylon/Olefin — DuPont Selar RB/ Polyolefin; 8% Selar RB; barrier prop.; extrusion blow molded; bottles			0.3% penetrant weight loss		50	28				293
Nylon 66 — DuPont Zytel 42; low flow; 2.54 mm thick; bottles	0.08									68
Polyethylene, Fluorinated — barrier prop.; bottles			0.6% penetrant weight loss		50	28				293

TOLUENE (continued)

Material Family	Material Note	Perm. Coefficient (cm³·mm/m²·day·atm)	VTR (g·mm/m²·day)	Non-normalized Data	Penetrant Note	Temp. (°C)	Time (days)	RH (%)	Pressure (kPa)	Test Note	Source
Polyethylene, HDPE	bottles			45.1% penetrant weight loss		50	28				293
Polyvinyl Alcohol	Ansell Edmont PVA; supported (lined) glove film; specified by glove film weight			low permeation rate; 0 to 1/2 eyedropper size drops per hour; PVA coating is water soluble; lower detection limit: 0 ppm; permeation rate: <0.9 µg/cm²/min	toluol	23	0.25			ASTM F739	313
Polyvinyl Chloride	bottles			failed		23	180				293
Rubber, Neoprene (CR)	Pioneer Industrial Products Stanzoil N-44; glove film			breakthrough time: 14 minutes; permeation rate: 2256 µg/cm²/min	toluol	23				ASTM F739.85	312
Rubber, Nitrile (NBR)	Ansell Edmont Sol-Vex 37-165; unsupported glove film; 0.54 mm thick			fair permeation rate; 51 to 100 eyedropper size drops per hour; breakthrough time: 10 minutes; permeation rate: <900 µg/cm²/min	toluol	23				ASTM F739	313
	Pioneer Industrial Products Stansolv A-14; glove film			breakthrough time: 28 minutes; lower detection limit: 0.002 ppm; permeation rate: 150 µg/cm²/min	toluol					ASTM F739.85	312
TPE, Vinyl	Pioneer Industrial Products Pylox V-20; glove film			breakthrough time: 3 minutes; permeation rate: 2100 µg/cm²/min	toluol					ASTM F739.85	312

TOLUENE DIISOCYANATE

Material Family	Material Note	Perm. Coefficient (cm³·mm/m²·day·atm)	VTR (g·mm/m²·day)	Non-normalized Data	Penetrant Note	Temp. (°C)	Time (days)	RH (%)	Pressure (kPa)	Test Note	Source
Polyvinyl Alcohol	Ansell Edmont PVA; supported (lined) glove film; specified by glove film weight			low permeation rate; 0 to 1/2 eyedropper size drops per hour; PVA coating is water soluble; lower detection limit: 0 ppm; permeation rate: <0.9 µg/cm²/min	TDI	23	0.25			ASTM F739	313
Rubber, Latex (NR)	Ansell Edmont Canners and Handlers 392; unsupported glove film; 0.48 mm thick			good permeation rate; 6 to 50 eyedropper size drops per hour; breakthrough time: 7 minutes; permeation rate: <90 µg/cm²/min	TDI	23				ASTM F739	313
	Pioneer Industrial Products L-118; glove film			no permeation rate at steady state detected; breakthrough time: >480 minutes; lower detection limit: 0.1 ppm; permeation rate: 0 µg/cm²/min	TDI					ASTM F739.85	312
Rubber, Nitrile (NBR)	Pioneer Industrial Products Stansolv A-14; glove film			no permeation rate at steady state detected; breakthrough time: >480 minutes; lower detection limit: 0.3 ppm; permeation rate: 0 µg/cm²/min	TDI					ASTM F739.85	312
TPE, Vinyl	Pioneer Industrial Products Pylox V-20; glove film			no permeation rate at steady state detected; breakthrough time: >480 minutes; lower detection limit: 0.06 ppm; permeation rate: 0 µg/cm²/min	TDI					ASTM F739.85	312

TRICHLOROETHANE

Material Family	Material Note	Perm. Coefficient (cm³·mm/m²·day·atm)	VTR (g·mm/m²·day)	Non-normalized Data	Penetrant Note	Temp. (°C)	Time (days)	RH (%)	Pressure (kPa)	Test Note	Source
Laminar, Nylon/HDPE	DuPont Selar RB/ HDPE; solvent packaging; 18% Selar RB; barrier prop.; extrusion blow molded; one pint bottle			0.18% penetrant weight loss (130% relative improvement vs. HDPE)	halogenated	48.9	28				291
Polyvinyl Alcohol	Ansell Edmont PVA; supported (lined) glove film; specified by glove film weight			low permeation rate; 0 to 1/2 eyedropper size drops per hour; PVA coating is water soluble; lower detection limit: 0 ppm; permeation rate: <0.9 µg/cm²/min	1,1,1-trichloroethane	23	0.25			ASTM F739	313
Rubber, Neoprene (CR)	Pioneer Industrial Products Stanzoil N-44; glove film			breakthrough time: 27 minutes; permeation rate: 1182 µg/cm²/min	1,1,1-trichloroethane					ASTM F739.85	312
Rubber, Nitrile (NBR)	Ansell Edmont Sol-Vex 37-165; unsupported glove film; 0.54 mm thick			poor permeation rate; 501 to 5000 eyedropper size drops per hour; breakthrough time: 90 minutes; permeation rate: <9000 µg/cm²/min	1,1,1-trichloroethane	23				ASTM F739	313
	Pioneer Industrial Products Stansolv A-14; glove film			breakthrough time: 131 minutes; lower detection limit: 0.05 ppm; permeation rate: 264 µg/cm²/min	1,1,1-trichloroethane					ASTM F739.85	312
TPE, Vinyl	Ansell Edmont Wet-Wear 550; clothing; PVC stretch outer layer bonded to a lightweight liner			good permeation rate; 6 to 50 eyedropper size drops per hour; breakthrough time: 4 minutes; permeation rate: <90 µg/cm²/min	1,1,1-trichloroethane	23				ASTM F739	313
	Ansell Edmont Wet-Wear 600; clothing; nylon netting bonded between two layers of PVC			fair permeation rate; 51 to 100 eyedropper size drops per hour; breakthrough time: 2 minutes; permeation rate: <300 µg/cm²/min	1,1,1-trichloroethane	23				ASTM F739	313
TPE, Vinyl coated fabric	Ansell Edmont Wet-Wear 700; clothing; PVC coating with nylon/Polyester lining			poor permeation rate; 501 to 5000 eyedropper size drops per hour; breakthrough time: 6 minutes; permeation rate: <9000 µg/cm²/min	1,1,1-trichloroethane	23				ASTM F739	313

TRICHLOROETHENE

Material Family	Material Note	Perm. Coefficient (cm³·mm/m²·day·atm)	VTR (g·mm/m²·day)	Non-normalized Data	Penetrant Note	Temp. (°C)	Time (days)	RH (%)	Pressure (kPa)	Test Note	Source
Acrylonitrile Copolymer, AMA	barrier prop.; bottles			1.75% penetrant weight loss		50	28				293
Laminar, Nylon/Olefin	DuPont Selar RB/ Polyolefin; 8% Selar RB; barrier prop.; extrusion blow molded; bottles			0.36% penetrant weight loss		50	28				293
Polyethylene, Fluorinated	barrier prop.; bottles			0.58% penetrant weight loss		50	28				293
Polyethylene, HDPE	bottles			15.0% penetrant weight loss		50	28				293
Polyvinyl Chloride	bottles			failed		23	180				293

TRICHLOROETHYLENE

Material Family	Material Note	Perm. Coefficient (cm³·mm/m²·day·atm)	VTR (g·mm/m²·day)	Non-normalized Data	Penetrant Note	Temp. (°C)	Time (days)	RH (%)	Pressure (kPa)	Test Note	Source
Acetal	DuPont Delrin		9.8			23		50			201
	DuPont Delrin		22.0			38					201
Polyvinyl Alcohol	Ansell Edmont PVA; supported (lined) glove film; specified by glove film weight			low permeation rate; 0 to 1/2 eyedropper size drops per hour; PVA coating is water soluble; lower detection limit: 0 ppm; permeation rate: <0.9 µg/cm²/min	TCE	23	0.25			ASTM F739	313
Rubber, Neoprene (CR)	Pioneer Industrial Products Stanzoil N-44; glove film			breakthrough time: 11 minutes; permeation rate: 3966 µg/cm²/min	TCE					ASTM F739.85	312
Rubber, Nitrile (NBR)	Pioneer Industrial Products Stansolv A-14; glove film			breakthrough time: 9 minutes; lower detection limit: 0.002 ppm; permeation rate: 372 µg/cm²/min	TCE					ASTM F739.85	312

TRICHLOROTRIFLUOROETHANE

Material Family	Material Note	Perm. Coefficient (cm³·mm/m²·day·atm)	VTR (g·mm/m²·day)	Non-normalized Data	Penetrant Note	Temp. (°C)	Time (days)	RH (%)	Pressure (kPa)	Test Note	Source
Rubber, Neoprene (CR)	Pioneer Industrial Products Stanzoil N-44; glove film			no permeation rate at steady state detected; breakthrough time: >480 minutes; lower detection limit: 0.2 ppm; permeation rate: 0 µg/cm²/min						ASTM F739.85	312
Rubber, Nitrile (NBR)	Pioneer Industrial Products Stansolv A-14; glove film			no permeation rate at steady state detected; breakthrough time: >480 minutes; lower detection limit: 0.01 ppm; permeation rate: 0 µg/cm²/min						ASTM F739.85	312
TPE, Vinyl	Pioneer Industrial Products Pylox V-20; glove film			breakthrough time: 11 minutes; permeation rate: 192 µg/cm²/min						ASTM F739.85	312

TRICRESYL PHOSPHATE

Material Family	Material Note	Perm. Coefficient (cm³·mm/m²·day·atm)	VTR (g·mm/m²·day)	Non-normalized Data	Penetrant Note	Temp. (°C)	Time (days)	RH (%)	Pressure (kPa)	Test Note	Source
Polyvinyl Alcohol	Ansell Edmont PVA; supported (lined) glove film; specified by glove film weight			low permeation rate; 0 to 1/2 eyedropper size drops per hour; PVA coating is water soluble; lower detection limit: 0 ppm; permeation rate: <0.9 µg/cm²/min	TCP	23	0.25			ASTM F739	313
Rubber, Latex (NR)	Ansell Edmont Canners and Handlers 392; unsupported glove film; 0.46 mm thick			low permeation rate; 0 to 1/2 eyedropper size drops per hour; breakthrough time: 45 minutes; permeation rate: <0.9 µg/cm²/min	TCP	23				ASTM F739	313
Rubber, Neoprene (CR)	Ansell Edmont Neoprene 29-840; unsupported glove film; 0.38 mm thick			low permeation rate; 0 to 1/2 eyedropper size drops per hour; lower detection limit: 0 ppm; permeation rate: <0.9 µg/cm²/min	TCP	23	0.25			ASTM F739	313
Rubber, Neoprene (CR)	Ansell Edmont Neox; supported (lined) glove film; specified by glove film weight			low permeation rate; 0 to 1/2 eyedropper size drops per hour; lower detection limit: 0 ppm; permeation rate: <0.9 µg/cm²/min	TCP	23	0.25			ASTM F739	313
Rubber, Nitrile (NBR)	Ansell Edmont Sol-Vex 37-165; unsupported glove film; 0.54 mm thick			low permeation rate; 0 to 1/2 eyedropper size drops per hour; lower detection limit: 0 ppm; permeation rate: <0.9 µg/cm²/min	TCP	23	0.25			ASTM F739	313
TPE, Vinyl	Ansell Edmont Monkey Grip; supported (lined) glove film; specified by glove film weight			low permeation rate; 0 to 1/2 eyedropper size drops per hour; lower detection limit: 0 ppm; permeation rate: <0.9 µg/cm²/min	TCP	23	0.25			ASTM F739	313
	Ansell Edmont Wet-Wear 550; clothing; PVC stretch outer layer bonded to a lightweight liner			breakthrough time: 200 minutes	TCP	23				ASTM F739	313

TRICRESYL PHOSPHATE

Material Family	Material Note	Perm. Coefficient (cm³·mm/m²·day·atm)	VTR (g·mm/m²·day)	Non-normalized Data	Penetrant Note	Temp. (°C)	Time (days)	RH (%)	Pressure (kPa)	Test Note	Source
TPE, Vinyl	Ansell Edmont Wet-Wear 600; clothing; nylon netting bonded between two layers of PVC			no permeation detected during a 6 hour test; lower detection limit: 0 ppm	TCP	23	0.25			ASTM F739	313
TPE, Vinyl coated fabric	Ansell Edmont Wet-Wear 700; clothing; PVC coating with Nylon/Polyester lining			no permeation detected during a 6 hour test; lower detection limit: 0 ppm	TCP	23	0.25			ASTM F739	313

TRIETHANOLAMINE

Material Family	Material Note	Perm. Coefficient (cm³·mm/m²·day·atm)	VTR (g·mm/m²·day)	Non-normalized Data	Penetrant Note	Temp. (°C)	Time (days)	RH (%)	Pressure (kPa)	Test Note	Source
Polyvinyl Alcohol	Ansell Edmont PVA; supported (lined) glove film; specified by glove film weight			low permeation rate; 0 to 1/2 eyedropper size drops per hour; PVA coating is water soluble; lower detection limit: 0 ppm; permeation rate: <0.9 µg/cm²/min	TEA; 85% conc.	23	0.25			ASTM F739	313
Rubber, Latex (NR)	Ansell Edmont Canners and Handlers 392; unsupported glove film; 0.48 mm thick			low permeation rate; 0 to 1/2 eyedropper size drops per hour; lower detection limit: 0 ppm; permeation rate: <0.9 µg/cm²/min	TEA; 85% conc.	23	0.25			ASTM F739	313
	Pioneer Industrial Products L-118; glove film			breakthrough time: >480 minutes; lower detection limit: 5 ppm; permeation rate: 0 µg/cm²/min	TEA					ASTM F739.85	312
Rubber, Neoprene (CR)	Ansell Edmont Neoprene 29-840; unsupported glove film; 0.38 mm thick			low permeation rate; 0 to 1/2 eyedropper size drops per hour; lower detection limit: 0 ppm; permeation rate: <0.9 µg/cm²/min	TEA; 85% conc.	23	0.25			ASTM F739	313
	Ansell Edmont Neox; supported (lined) glove film; specified by glove film weight			low permeation rate; 0 to 1/2 eyedropper size drops per hour; lower detection limit: 0 ppm; permeation rate: <0.9 µg/cm²/min	TEA; 85% conc.	23	0.25			ASTM F739	313
	Pioneer Industrial Products Stanzoil N-44; glove film			breakthrough time: >480 minutes; lower detection limit: 5 ppm; permeation rate: 0 µg/cm²/min	TEA	23				ASTM F739.85	312
Rubber, Nitrile (NBR)	Ansell Edmont Sol-Vex 37-165; unsupported glove film; 0.54 mm thick			low permeation rate; 0 to 1/2 eyedropper size drops per hour; lower detection limit: 0 ppm; permeation rate: <0.9 µg/cm²/min	TEA; 85% conc.	23	0.25			ASTM F739	313
	Pioneer Industrial Products Stansolv A-14; glove film			breakthrough time: >480 minutes; lower detection limit: 5 ppm; permeation rate: 0 µg/cm²/min	TEA					ASTM F739.85	312
TPE, Vinyl	Ansell Edmont Monkey Grip; supported (lined) glove film; specified by glove film weight			low permeation rate; 0 to 1/2 eyedropper size drops per hour; lower detection limit: 0 ppm; permeation rate: <0.9 µg/cm²/min	TEA; 85% conc.	23	0.25			ASTM F739	313
	Pioneer Industrial Products Pylox V-20; glove film			breakthrough time: >480 minutes; lower detection limit: 5 ppm; permeation rate: 0 µg/cm²/min	TEA					ASTM F739.85	312

TRIFLUOROETHANOL

Material Family	Material Note	Perm. Coefficient (cm³·mm/m²·day·atm)	VTR (g·mm/m²·day)	Non-normalized Data	Penetrant Note	Temp. (°C)	Time (days)	RH (%)	Pressure (kPa)	Test Note	Source
Rubber, Neoprene (CR)	Pioneer Industrial Products Stanzoil N-44; glove film			no permeation rate at steady state detected; breakthrough time: >60 minutes; permeation rate: 0 µg/cm²/min						ASTM F739.85	312
TPE, Vinyl	Pioneer Industrial Products Pylox V-20; glove film			breakthrough time: 15 minutes; permeation rate: 1302 µg/cm²/min						ASTM F739.85	312

TRIHYDROXYTRIETHYLAMINE

Material Family	Material Note	Perm. Coefficient (cm³·mm/m²·day·atm)	VTR (g·mm/m²·day)	Non-normalized Data	Penetrant Note	Temp. (°C)	Time (days)	RH (%)	Pressure (kPa)	Test Note	Source
Rubber, Latex (NR)	Pioneer Industrial Products L-118; glove film			no permeation rate at steady state detected; breakthrough time: >480 minutes; lower detection limit: 5 ppm; permeation rate: 0 µg/cm²/min						ASTM F739.85	312
Rubber, Neoprene (CR)	Pioneer Industrial Products Stanzoil N-44; glove film			no permeation rate at steady state detected; breakthrough time: >480 minutes; lower detection limit: 5 ppm; permeation rate: 0 µg/cm²/min						ASTM F739.85	312
Rubber, Nitrile (NBR)	Pioneer Industrial Products Stansolv A-14; glove film			no permeation rate at steady state detected; breakthrough time: >480 minutes; lower detection limit: 5 ppm; permeation rate: 0 µg/cm²/min						ASTM F739.85	312

TRIHYDROXYTRIETHYLAMINE

Material Family	Material Note	Perm. Coefficient (cm³·mm/m²·day·atm)	VTR (g·mm/m²·day)	Non-normalized Data	Penetrant Note	Temp. (°C)	Time (days)	RH (%)	Pressure (kPa)	Test Note	Source
TPE, Vinyl	Pioneer Industrial Products Pylox V-20; glove film			no permeation rate at steady state detected; breakthrough time: >480 minutes; lower detection limit: 6 ppm; permeation rate: 0 µg/cm²/min						ASTM F739.85	312

TURPENTINE

Material Family	Material Note	Perm. Coefficient (cm³·mm/m²·day·atm)	VTR (g·mm/m²·day)	Non-normalized Data	Penetrant Note	Temp. (°C)	Time (days)	RH (%)	Pressure (kPa)	Test Note	Source
Laminar, Nylon/Olefin	DuPont Selar RB/ Polyolefin; 8% Selar RB; barrier prop.; extrusion blow molded; bottles			0.03% penetrant weight loss		50	28				293
Polyethylene, Fluorinated	barrier prop.; bottles			0.06% penetrant weight loss		50	28				293
Polyethylene, HDPE	bottles			2.4% penetrant weight loss		50	28				293
Polyvinyl Alcohol	Ansell Edmont PVA; supported (lined) glove film; specified by glove film weight			low permeation rate; 0 to 1/2 eyedropper size drops per hour; PVA coating is water soluble; lower detection limit: 0 ppm; permeation rate: <0.9 µg/cm²/min		23	0.25			ASTM F739	313
Rubber, Neoprene (CR)	Pioneer Industrial Products Stanzoil N-44; glove film			no permeation rate at steady state detected; breakthrough time: >480 minutes; lower detection limit: 0.02 ppm; permeation rate: 0 µg/cm²/min						ASTM F739.85	312
Rubber, Nitrile (NBR)	Ansell Edmont Sol-Vex 37-165; unsupported glove film; 0.54 mm thick			low permeation rate; 0 to 1/2 eyedropper size drops per hour; breakthrough time: 30 minutes; permeation rate: <0.9 µg/cm²/min		23				ASTM F739	313
	Pioneer Industrial Products StansolN A-14; glove film			no permeation rate at steady state detected; breakthrough time: >480 minutes; lower detection limit: 0.0009 ppm; permeation rate: 0 µg/cm²/min						ASTM F739.85	312
TPE, Vinyl	Ansell Edmont Wet-Wear 550; clothing; PVC stretch outer layer bonded to a lightweight liner			very good permeation rate; 1 to 5 eyedropper size drops per hour; breakthrough time: 10 minutes; permeation rate: <9 µg/cm²/min		23				ASTM F739	313
	Ansell Edmont Wet-Wear 600; clothing; nylon netting bonded between two layers of PVC			good permeation rate; 6 to 50 eyedropper size drops per hour; breakthrough time: 6 minutes; permeation rate: <90 µg/cm²/min		23				ASTM F739	313
TPE, Vinyl coated fabric	Ansell Edmont Wet-Wear 700; clothing; PVC coating with Nylon/Polyester lining			good permeation rate; 6 to 50 eyedropper size drops per hour; breakthrough time: 16 minutes; permeation rate: <90 µg/cm²/min		23				ASTM F739	313

VANILLA (VANILLIN)

Material Family	Material Note	Perm. Coefficient (cm³·mm/m²·day·atm)	VTR (g·mm/m²·day)	Non-normalized Data	Penetrant Note	Temp. (°C)	Time (days)	RH (%)	Pressure (kPa)	Test Note	Source
EVAL/ PE Film	0.013/ 0.051 mm thick; PE inside; EVAL EF-XL outside; film			>30 days to leakage							265
Nylon/ EVAL Film	0.015/ 0.015 mm thick; EVAL EF-F inside; oriented nylon outside; film			2 days to leakage							265
Nylon/ PE Film	0.015/ 0.051 mm thick; PE inside; oriented nylon outside; film			2 days to leakage							265
PET/ EVAL Film	0.013/ 0.015 mm thick; EVAL EF-F inside; PET outside; film			>30 days to leakage							265
PET/ EVAL/ LDPE Film	EVAL EF-F barrier; 0.013/ 0.015/ 0.051 mm thick; LDPE inside; PET outside			15 days to leakage							265
PET/ PE Film	0.013/ 0.051 mm thick; PE inside; PET outside; film			2 days to leakage							265
PP/ PE Film	0.018/ 0.051 mm thick; PE inside; PVDC coated BOPP outside; film			6 days to leakage							265

VEGETABLE OILS

Material Family	Material Note	Perm. Coefficient (cm³·mm/m²·day·atm)	VTR (g·mm/m²·day)	Non-normalized Data	Penetrant Note	Temp. (°C)	Time (days)	RH (%)	Pressure (kPa)	Test Note	Source
Acetal	DuPont Delrin	0				23		50			201
	DuPont Delrin	0				38					201

Appendix I - Permeability Sort

Appendix I - Permeability Sort

Material Family	Material Note	Perm. Coefficient (cm³·mm/m²·day·atm)	VTR (g·mm/m²·day)	Non-normalized Data	Penetrant Note	Temp. (°C)	Time (days)	RH (%)	Pressure (kPa)	Test Note	Source
VINYL ACETATE											
Rubber, Neoprene (CR)	Pioneer Industrial Products Stanzoil N-44; glove film			breakthrough time: 30 minutes; lower detection limit: 0.03 ppm; permeation rate: 198 µg/cm²/min						ASTM F739.85	312
Rubber, Nitrile (NBR)	Pioneer Industrial Products Starsotv A-14; glove film			breakthrough time: 30 minutes; lower detection limit: 0.08 ppm; permeation rate: 402 µg/cm²/min						ASTM F739.85	312
WATER											
Laminar, Nylon/HDPE	DuPont Selar RB/ HDPE; solvent packaging; 18% Selar RB; barrier prop.; extrusion blow molded; one pint bottle			0.08% penetrant weight loss (1% relative improvement vs. HDPE)	oxygen containing	48.9	28				291
Nylon 66	DuPont Zytel 42; low flow, 2.54 mm thick; bottles		1.2 - 2.4								68
Polyphenylene Sulfide	Phillips Ryton; 0.127 mm thick; baked coating; film		0.32			23				Die cut samples were fitted to tops of glass bottles by a rubber gasket and a lid, with a surface area of 96.8 cm²; Liquids were placed in bottles, gaskets and film put in place, and the lid screwed on. Apparatus was inverted to put liquid in direct contact with film. Weight loss measurements were made at 1 week intervals throughout 4 weeks of conditioning.	102
TPE, Polyester	DuPont Hytrel 4056; 40 Shore D	267,840				25		90	34.5	assuming that permeability laws hold for water	274
	DuPont Hytrel 5556; 55 Shore D	207,360				25		90	34.5	assuming that permeability laws hold for water	274
WATER VAPOR											
ABS	BASF AG Terluran 877 M; moderate flow, 0.1 mm thick; film		3.1			23				RH: 85%-0% gradient; DIN 53122; Values for permeability depend on the conditions under which the film was produced. Figures determined may differ by as much as 50% from those given.	137
	BASF AG Terluran 967 K; moderate flow, 0.1 mm thick; w/ butadiene acrylic rubber; film		2.7			23				"	137
	BASF AG Terluran 997 VE; 0.1 mm thick, low flow; w/ butadiene acrylic rubber; film		2.7			23				"	137
ABS	Dow; low nitrile content; film		2.0 - 6.3								250
Acrylonitrile	film		0.24			24					250
Acrylonitrile Co. (AMA)	film		0.35			24					250
Acrylonitrile Copolymer, AMA	BP Chem. Barex 210; packaging; impact modified; barrier prop.		2.0			22.8		100		ASTM F1249	296
	BP Chem. Barex 218; packaging; impact modified, high impact; barrier prop.		3.0			22.8		100		ASTM F1249	296
	BP Chem. Barex 210; barrier prop.		2.4			40		90			264
	BP Chem. Barex 210; packaging; impact modified, barrier prop.		2.2			37.8		90		ASTM F1249	296
	BP Chem. Barex 218; packaging; impact modified, high impact; barrier prop.		3.0			37.8		90		ASTM F1249	296

WATER VAPOR (continued)

Material Family	Material Note	Perm. Coefficient (cm³·mm/m²·day·atm)	VTR (g·mm/m²·day)	Non-normalized Data	Penetrant Note	Temp. (°C)	Time (days)	RH (%)	Pressure (kPa)	Test Note	Source
ASA	BASF AG Luran S 757 R; 0.1 mm thick; blown film		3			23			19.86 mbar	RH: 85%-0% gradient; DIN 53122	143
	BASF AG Luran S 776 S; 0.1 mm thick; film		3.5			23			23.87 mbar	RH: 85%-0% gradient; DIN 53122; Values for permeability depend on the conditions under which the film was produced. Figures determined may differ by as much as 50% from those given.	142
	BASF AG Luran S 797 S; 0.1 mm thick; film		3			23			23.87 mbar		142
Cellulosic Plastic	Cellophane; 0.023 mm thick; PVDC coated; film		0.39			40		90		JIS Z0208	268
CTFE/ PE/ PVC Film	Allied Sig. Aclar 22C/ PE/ PVC; 0.038/ 0.051/ 0.19 mm thick; lamination		0.09			37.8		90		ASTM F372-78; Mocon Permatran	138
	Allied Sig. Aclar 88A/ PE/ PVC; 0.019/ 0.051/ 0.19 mm thick; lamination		0.12			37.8		90		ASTM F372-78; Mocon Permatran	138
CTFE/ PVC Film	Allied Sig. Aclar 33C/ PVC; 0.019/ 0.254 mm thick; lamination		0.08		.	37.8		90		ASTM F372-78; Mocon Permatran	138
Epoxy			0.7 - 0.94			37		90		ASTM E96-63T	121
EVOH	DuPont Selar OH 3003; packaging; barrier prop.; 32% ethylene; 3 g/10 min. MFI		1.2			40		80			295
	DuPont Selar OH 4416; packaging; barrier prop.. 44% ethylene; 12 g/10 min. MFI		0.55			40		80			295
	DuPont Selar OH BX220; packaging; barrier prop.; heat stabilized; 32% ethylene; 3 g/10 min. MFI		1.2			40		80			295
	DuPont Selar OH BX228; packaging; barrier prop.. heat stabilized; 44% ethylene; 11 g/10 min. MFI		0.55			40		80			295
	DuPont Selar OH BX230; packaging; barrier prop.. heat stabilized, formability; 32% ethylene; 1.5 g/10 min. MFI		0.55			40		80			295
	Eval Co. Eval E; barrier prop.; 44% ethylene		0.55			40		90			264
	Eval Co. Eval EF-E; 0.02 mm thick; 44% ethylene; film		0.79			40		90		JIS Z0208	268
	Eval Co. Eval EF-F; 0.015 mm thick; 32% ethylene; film		2.4			40		90		JIS Z0208	268
	Eval Co. Eval EF-XL; biaxially oriented, 0.015 mm thick; film		1.2			40		90		JIS Z0208	268
	Eval Co. Eval F; barrier prop.; 32% ethylene		1.5			40		90			264
	Eval Co. Eval G; barrier prop.; 48% ethylene		0.55			40		90			264
	Eval Co. Eval H; barrier prop.; 38% ethylene		0.83			40		90			264
	Eval Co. Eval K; barrier prop.; 38% ethylene		0.83			40		90			264
	Eval Co. Eval L; barrier prop.; 27 % ethylene		3.15			40		90			264
Fluoropolymer, CTFE	3M Kel-F 81; amorphous form; film		0.005			25					96
	3M Kel-F 81; amorphous form; film		0.043			50					96
	Allied Sig. Aclar 22A; transparent, 0.038 mm thick; film	0.0122 - 0.0236				37.8		90		ASTM E96, method E; measured on sealed pouches	138
	Allied Sig. Aclar 22C; transparent, 0.0254 mm thick; film	0.0119 - 0.0236				37.8		90		ASTM E96, method E; measured on sealed pouches	138
	Allied Sig. Aclar 22C; transparent, 0.051 mm thick; film	0.0122 - 0.0316				37.8		90		ASTM E96, method E; measured on sealed pouches	138
	Allied Sig. Aclar 22C; transparent, 0.19 mm thick; film	0.0171 - 0.0247				37.8		90		ASTM E96, method E; measured on sealed pouches	138

WATER VAPOR (continued)

Material Family	Material Note	Perm. Coefficient (cm³·mm/m²·day·atm)	VTR (g·mm/m²·day)	Non-normalized Data	Penetrant Note	Temp. (°C)	Time (days)	RH (%)	Pressure (kPa)	Test Note	Source
Fluoropolymer, CTFE	Allied Sig. Aclar 33C; transparent, 0.019 mm thick; film		0.0082 - 0.0112			37.8		90		ASTM E96, method E; measured on sealed pouches	138
	Allied Sig. Aclar 33C; transparent, 0.051 mm thick; film		0.0077 - 0.0158			37.8		90		ASTM E96, method E; measured on sealed pouches	138
	Allied Sig. Aclar 88A; transparent, 0.019 mm thick; film		0.0133 - 0.0163			37.8		90		ASTM E96, method E; measured on sealed pouches	138
	3M Kel-F 81; amorphous form; film		0.12			75					96
	3M Kel-F 81; amorphous form; film		0.39			100					96
Fluoropolymer, ETFE	DuPont Tefzel; developmental material, 0.102 mm thick; film		0.65			25				ASTM E96	205
Fluoropolymer, FEP			0.16			37.8		90			138
Fluoropolymer, PVDF			1.02			23		90			138
	Atochem Foraflon; 0.02 mm thick; extruded film		0.68			38				NFH 00044	89
	Atochem Foraflon; 0.028 mm thick; extruded film		0.62			38				NFH 00044	89
	Atochem Foraflon; 0.04 mm thick; extruded film		0.64			38				NFH 00044	89
	Solvay Solef 1006; 0.1 mm thick, transluscent; film		0.75			38				ASTM E96, proc. E	125
Fluoropolymer, PVF			1.3			37.8		90			138
HDPE/ EAA/ Nylon/ EAA Film	Dow Nylopak 570; form-fill-seal pouch; neutral, barrier prop., 0.051 mm thick; coextruded film		0.24			38		90		Permatran W	258
HDPE/ EVA/ PVDC/ EVA Film	Dow Saranex 25; form-fill-seal pouch, bag-in-box; neutral, barrier prop., 0.051 mm thick; high slip; coextruded film		0.06			38		90		Permatran W	257
Ionomer	DuPont Surlyn 1601; 0.94 g/cm³ density; 0.051 mm thick; blown film; 1.3 g/10 min MFI		0.63								280
	DuPont Surlyn 1603; 0.94 g/cm³ density; 0.051 mm thick; blown film; 1.7 g/10 min. MFI		0.51								280
	DuPont Surlyn 1650; 0.95 g/cm³ density; 0.051 mm thick; blown film; 1.6 g/10 min. MFI		0.59								280
	DuPont Surlyn 1652; 0.94 g/cm³ density; zinc ion, 0.051 mm thick; blown film; 5.0 g/10 min. MFI		0.47								280
	DuPont Surlyn 1702; 0.94 g/cm³ density; zinc ion, 0.051 mm thick; blown film; 14.0 g/10 min. MFI		0.55								280
	DuPont Surlyn 1705; 0.95 g/cm³ density; zinc ion, 0.051 mm thick; blown film; 5.5 g/10 min. MFI		0.55								280
	DuPont Surlyn 1707; 0.95 g/cm³ density; sodium ion, 0.051 mm thick; blown film; 0.9 g/10 min. MFI		0.63								280
	DuPont Surlyn F1605; 0.95 g/cm³ density; sodium ion, 0.051 mm thick; blown film; 2.8 g/10 min. MFI		0.63								280
	DuPont Surlyn F1706; 0.96 g/cm³ density; zinc ion, 0.051 mm thick; blown film; 0.7 g/10 min. MFI		0.55								280
	DuPont Surlyn F1801; 0.96 g/cm³ density; zinc ion, 0.051 mm thick; blown film; 1.0 g/10 min. MFI		0.55								280

WATER VAPOR (continued)

Material Family	Material Note	Perm. Coefficient (cm³·mm/m²·day·atm)	VTR (g·mm/m²·day)	Non-normalized Data	Penetrant Note	Temp (°C)	Time (days)	RH (%)	Pressure (kPa)	Test Note	Source
Ionomer	DuPont Surlyn F1855; 0.96 g/cm³ density; zinc ion, 0.051 mm thick; blown film; 1.0 g/10 min. MFI		0.79								280
	DuPont Surlyn F1856; 0.95 g/cm³ density; sodium ion, 0.051 mm thick; blown film; 1.0 g/10 min. MFI		0.95								280
LDPE/ EVA/ PVDC/ EVA Film	Dow Saranex 23; laminations; neutral, barrier prop., 0.051 mm thick; coextruded film		0.1			38		90		Permatran W	257
LDPE/ EVA/ PVDC/ EVA/ LDPE Film	Dow Saranex 11; laminations; neutral, barrier prop., 0.038 mm thick; medium slip; coextruded film		0.16			38		90		Permatran W	257
	Dow Saranex 14; form-fill-seal pouch; neutral, barrier prop., 0.051 mm thick; medium slip; coextruded film		0.08			38		90		Permatran W	257
	Dow Saranex 15; form-fill-seal pouch; neutral, barrier prop., 0.076 mm thick; medium slip; coextruded film		0.06			38		90		Permatran W	257
	Dow Saranex 15; form-fill-seal pouch; neutral, barrier prop., 0.102 mm thick; medium slip; coextruded film		0.04			38		90		Permatran W	257
Liquid Crystal Polymer	Hoechst AG Vectra A950; 0.019 mm thick; film		0.03			23		100		test area: 5 cm²	70
	Hoechst AG Vectra A950; 0.019 mm thick; film		0.05			38		100		test area: 5 cm²	70
Nylon	oriented, 0.015 mm thick; film		6.7			40		90		JIS Z0208	268
	oriented, 0.017 mm thick; PVDC coated; film		0.39			40		90		JIS Z0208	268
Nylon 6	Allied Sig. Capran 77C; 0.019 mm thick; film		0.24			23		50		pouch method	285
	Allied Sig. Capran 77C; 0.0254 mm thick; film		0.24			23		50		pouch method	285
	BASF Ultramid B36F; moderate flow, clarity; flat film, tubular film		1.5 - 1.6			23				RH: 85%–0% gradient; DIN 53122	93
	BASF Ultramid B4; 0.02-0.05 mm thick, unstretched		1.1 - 1.8			20				RH: 85%–0% gradient; DIN 53122	252
	BASF Ultramid B4; biaxially stretched, 0.02 mm thick		0.8 - 1.2			20				RH: 85%–0% gradient; DIN 53122	252
	BASF Ultramid B4; moderate flow, 0.02-0.1 mm thick; flat film, tubular film		1.5 - 1.6			23				RH: 85%–0% gradient; DIN 53122	93
	Allied Sig. Capran; 0.0254 mm thick; film		7.5 - 7.9			37.8		90		pouch method	284
	Allied Sig. Capran 77C; 0.019 mm thick; film		7.1 - 7.7			37.8		90		pouch method	285
	Allied Sig. Capran 77C; 0.019 mm thick; film		5.6 - 5.9			37.8		90		pouch method	285
	Allied Sig. Capran 77K; 0.0254 mm thick; PVDC coated; film		0.08			37.8		90		pouch method	285
	barrier prop.; film		9.8			37.8		90			294
	biaxially oriented		4.02			40		90			264
			9.2			37.8		90		ASTM F1249	296
Nylon 6/ LDPE Film	BASF Ultramid B4/ Lupolen 3020 D; 0.03 mm/ 0.05 mm thick		0.21			20				RH: 85%–0% gradient; DIN 53122	252
Nylon 66	BASF Ultramid A5; 0.02-0.1 mm thick, low flow; tubular film		0.8			23				RH: 85%–0% gradient; DIN 53122	93
	BASF Ultramid A5; 0.05 mm thick; blown film		1.5			20				RH: 85%–0% gradient; DIN 53122	252
	BASF Ultramid A5; low flow, 0.02-0.1 mm thick; flat film		1.1 - 1.2			23				RH: 85%–0% gradient; DIN 53122	93

WATER VAPOR (continued)

Material Family	Material Note	Perm. Coefficient (cm³·mm/m²·day·atm)	VTR (g·mm/m²·day)	Non-normalized Data	Penetrant Note	Temp. (°C)	Time (days)	RH (%)	Pressure (kPa)	Test Note	Source
Nylon 66	DuPont Can. Dartek; transparent, 0.0254 mm thick; film		7.5			23		100		ASTM E398-70; Honeywell MVTR tester	276
	DuPont Zytel 42; low flow; film		0.39			23		50			68
	DuPont Zytel 42; low flow; film		7.9			23		100			68
	DuPont Can. Dartek B-601; barrier prop., 0.0254 mm thick; PVDC coated; film		0.23			38		90		ASTM F372	276
	DuPont Can. Dartek B-602; barrier prop., 0.038 mm thick; PVDC coated; film		0.34			38		90		ASTM F372	276
Nylon 66/610	Emser Grilon XE3303; transparent, barrier prop., 0.05 mm thick; film		0.7			23					307
Nylon 6/66	BASF Ultramid C35; mod-high flow, 0.02-0.1 mm thick; flat film; tubular film		1.5 - 1.8			23				RH: 85%-0% gradient; DIN 53122	93
	Allied Sig. Capran; 0.0254 mm thick; film		8.7			37.8		90		cup method	284
Nylon MXD6	barrier prop.		1.3			40		90			264
Nylon, Amorphous	DuPont Selar PT; food packaging; transparent, heat stabilized, hot fill, retort; thermoformed; container		0.51			25		50			290
	Emser Grivory G21; transparent, barrier prop., 0.05 mm thick; film		0.35							DIN 53122	307
	DuPont Selar PA; barrier prop.		0.55			40		90			264
	DuPont Selar PA; barrier prop.; film		0.47			37.8		90			294
Parylene	Union Carbide Parylene C; vapor phase deposition; thin film		0.08			37		90		ASTM E96-63T	121
	Union Carbide Parylene D; vapor phase deposition; thin film		0.1			37		90		ASTM E96-63T	121
	Union Carbide Parylene N; highly crystalline, high molecular weight, completely linear; vapor phase deposition; thin film		0.59			37		90		ASTM E96-63T	121
PE Ionomer Copolymer	BASF AG Lucalen I4300MX; 0.1 mm thick; 8 wt.% acrylic acid; 7 g/10 min. MFI		0.08			23				RH: 85%-0% gradient; DIN 53122	25
PE-Acrylic Acid Copolymer	BASF AG Lucalen A2710H; 0.1 mm thick; 17 wt.% acrylic acid; 1.7 g/10 min. MFI		0.68			23				RH: 85%-0% gradient; DIN 53122	25
	BASF AG Lucalen A2910M; 0.1 mm thick; 11 wt.% acrylic acid; 7 g/10 min. MFI		0.23			23				RH: 85%-0% gradient; DIN 53122	25
	BASF AG Lucalen A3710MX; 0.1 mm thick; 8 wt.% acrylic acid; 7 g/10 min. MFI		0.08			23				RH: 85%-0% gradient; DIN 53122	25
PE/ PVC-PVDC Copolymer Multilayer Film	Dow Saranex; multilayer film		0.06 - 0.16			24					250
PE/PS Alloy	BASF AG Styroblend WS KR 2773			3 g/m² · day; conversion not possible without thickness							182
	BASF AG Styroblend WS KR 2774			3 g/m² · day; conversion not possible without thickness							182
	BASF AG Styroblend WS KR 2775			3 g/m² · day; conversion not possible without thickness							182
	BASF AG Styroblend WS KR 2776			3 g/m² · day; conversion not possible without thickness							182
	BASF AG Styroblend WS KR 2777			5 g/m² · day; conversion not possible without thickness							182
Polybutylene	Shell Duraflex 1600; peelable seals; FDA grade, 0.051 mm thick, heat sealable; blown film; 5 phr zinc oxide, slip and antiblock formulations; film		0.47			37.8		90		ASTM D96, method E	304
	Shell Duraflex 1710; peelable seals; 0.910 g/cm³ density, 0.909 g/cm³ density; FDA grade, heat sealable, 0.051 mm thick; blown film; slip and antiblock formulations; antiblock formulations; film		0.74			37.8		90		ASTM D96, method E	304
Polycarbonate	Bayer Makrolon; 0.1 mm thick; film		1.5			23		85		DIN 53122	289

WATER VAPOR (continued)

Material Family	Material Note	Perm. Coefficient (cm³·mm/m²·day·atm)	VTR (g·mm/m²·day)	Non-normalized Data	Penetrant Note	Temp. (°C)	Time (days)	RH (%)	Pressure (kPa)	Test Note	Source
Polycarbonate	film		3.8			37.8		90			294
			4.33			40		90			264
Polyester, PBT	BASF AG Ultradur B4550; 0.25 mm thick		2.5			23				RH: 85%-0% gradient; DIN 53122; measured in standard laboratory atmosphere	180
Polyester, PCTG	Eastman Kodar PCTG 5445; transparent, 0.25 mm thick; film		1.8							ASTM E96E	166
Polyester, PET	DuPont Mylar; film		0.71			37.8		90		ASTM E96-80; modified test, permeabilities determined at the partial pressure of the vapor at the test temperature	270
	Shell Cleartuf; packaging; transparent; barrier prop., oriented		0.8			37.8		100		ASTM E96	297
	Shell Cleartuf; packaging; transparent; barrier prop., unoriented		1.6			37.8		100		ASTM E96	297
	0.012 mm thick; film		1.2			40		90		JIS Z0208	268
	biaxially oriented		0.47			40		90			264
	oriented		0.39 - 0.51			37.8		90			138
	oriented, 0.014 mm thick; PVDC coated; film		0.39			40		90		JIS Z0208	268
			0.51			40		90			264
			1.7			37.8		90		ASTM F1249	296
Polyester, PETG	Eastman Kodar PETG 6763; transparent, amorphous, 0.25 mm thick; film		1.5							ASTM E96E	165
			1.6			37.8		90		ASTM F1249	296
Polyethylene, HDPE	Dow; film		0.16			24					250
	Hoechst AG Hostalen		0.03			20				useable average for all Hostalen grades	94
	Hoechst AG Hostalen		0.04			25				useable average for all Hostalen grades	94
	DuPont Can. Solair 15A; merchandising bags; 0.941 g/cm³ density; 59 Shore D, 0.0254 mm thick; 0.35 g/10 min. MFI; blown film		0.19			38		90		values are cooling rate dependent; Honeywell model 825 apparatus	277
	DuPont Can. Solair 16A; merchandising bags; 0.945 g/cm³ density; 60 Shore D, 0.0254 mm thick; 0.28 g/10 min. MFI; blown film		0.17			38		90		values are cooling rate dependent; Honeywell model 825 apparatus	277
	DuPont Can. Solair 19A; laminations, coextrusion; 0.96 g/cm³ density; 65 Shore D, 0.0254 mm thick; 0.75 g/10 min. MFI; blown film		0.13			38		90		values are cooling rate dependent; Honeywell model 825 apparatus	277
	Hoechst AG Hostalen		0.07			30				useable average for all Hostalen grades	94
	Hoechst AG Hostalen		0.14			40				useable average for all Hostalen grades	94
	Hoechst AG Hostalen		0.32			50				useable average for all Hostalen grades	94
	Phillips Marlex; film		0.12			37.8		90		ASTM D96	101
			0.1			37.8		90		ASTM F1249	296
			0.12			37.8		90			138
			0.15			40		90			264
Polyethylene, LDPE	Dow; film		0.39 - 0.59			24					250
	Dow LDPE 4005; 0.916 g/cm³ density; 0.015 mm thick; extrusion coating; 5.5 g/10 min. MFI		0.46							ASTM F1249	254

WATER VAPOR (continued)

Material Family	Material Note	Perm. Coefficient (cm³·mm/m²·day·atm)	VTR (g·mm/m²·day)	Non-normalized Data	Penetrant Note	Temp. (°C)	Time (days)	RH (%)	Pressure (kPa)	Test Note	Source
Polyethylene, LDPE	Dow LDPE 4012; 0.916 g/cm³ density; 0.01 mm min. thick; extrusion coating; 12 g/10 min. MFI		0.31							ASTM F1249	254
	Dow LDPE 5004I; 0.923 g/cm³ density; 0.01 mm min. thick; extrusion coating; 4 g/10 min. MFI		0.23							ASTM F1249	254
	Dow LDPE 722; 0.916 g/cm² density; 0.01 mm min. thick; extrusion coating; 8 g/10 min. MFI		0.26							ASTM F1249	254
	0.920 g/cm³ density; 0.05 mm thick, 2.5 BUR; blown film; 4 g/10 min. MFI; film		2.5							JIS Z0208	216
	0.05 mm thick; film		0.39			40		90		JIS Z0208	268
			0.45			40		90			264
			0.39 - 0.59			37.8		90			138
Polyethylene, LLDPE	Dow Dowlex 2045; 0.920 g/cm³ density; 0.0254 mm thick; blown film; 1.0 g/10 min. MFI		0.28					100		Mocon Test Method; Mocon Permatron W-1	11
	DuPont Can. Sclairfilm SL1; laminations; 0.918 g/cm³ density; 0.038 mm thick; film		0.47							ASTM F372	278
	DuPont Can. Sclairfilm SL1; laminations; 0.918 g/cm³ density; 0.051 mm thick; film		0.47							ASTM F372	278
	DuPont Can. Sclairfilm SL1; laminations; 0.918 g/cm³ density; 0.076 mm thick; film		0.36							ASTM F372	278
	DuPont Can. Sclairfilm SL3; laminations; 0.918 g/cm³ density; 0.038 mm thick; film		0.47							ASTM F372	278
	DuPont Can. Sclairfilm SL3; laminations; 0.918 g/cm³ density; 0.051 mm thick; film		0.47							ASTM F372	278
	DuPont Can. Sclairfilm SL3; laminations; 0.918 g/cm³ density; 0.076 mm thick; film		0.36							ASTM F372	278
	DuPont Can. Sclair 11F9; blending resin, multi-purpose bags; 0.921 g/cm³ density; 0.0254 mm thick; 0.75 g/10 min. MFI; blown film		0.38			38		90		values are cooling rate dependent; Honeywell model 825 apparatus	277
	DuPont Can. Sclair 11H4; blending resin; 0.921 g/cm³ density; 0.0254 mm thick; 1.2 g/10 min. MFI; blown film		0.46			38		90		"	277
	DuPont Can. Sclair 11R4; blending resin, multi-purpose bags; 0.921 g/cm³ density; 0.0254 mm thick; 1.6 g/10 min. MFI; blown film		0.51			38		90		values are cooling rate dependent; Honeywell model 825 apparatus	277
Polyethylene, MDPE	DuPont Can. Sclair 14D; merchandising bags; 0.935 g/cm³ density; 59 Shore D, 0.0254 mm thick; 0.28 g/10 min. MFI; blown film		0.22			38		90		"	277
			0.28			37.8		90		"	138
Polyethylene, PE-EVA Copol.	DuPont Elvax 3120; 0.930 g/cm³ density; 0.051 mm thick; 7.5% VA; 1.2 g/10 min. MFI; antiblock additive, slip additive; blown film		1.4							ASTM E96-E	281
	DuPont Elvax 3121A; 0.930 g/cm³ density; 0.051 mm thick; 7.5% VA; 0.35 g/10 min. MFI; antiblock additive, slip additive; blown film		1.4							ASTM E96-E	281
	DuPont Elvax 3128; 0.930 g/cm³ density; 0.051 mm thick; 8.9% VA; 2.0 g/10 min. MFI; blown film		1.4							ASTM E96-E	281
	DuPont Elvax 3128SB; 0.930 g/cm³ density; 0.051 mm thick; 8.9% VA; 2.0 g/10 min. MFI; antiblock additive, slip additive; blown film		1.4							ASTM E96-E	281
	DuPont Elvax 3130; 0.94 g/cm³ density; 0.051 mm thick; 12.0% VA; 2.5 g/10 min. MFI; blown film		1.6							ASTM E96-E	281

WATER VAPOR (continued)

Material Family	Material Note	Permeability Data			Test Conditions						Source
		Perm. Coefficient (cm³·mm/m²·day·atm)	VTR (g·mm/m²·day)	Non-normalized Data	Penetrant Note	Temp. (°C)	Time (days)	RH (%)	Pressure (kPa)	Test Note	
Polyethylene, PE-EVA Copol.	DuPont Elvax 3130SB; 0.94 g/cm³ density; 0.051 mm thick; 12.0% VA; 2.5 g/10 min. MFI; antiblock additive, slip additive; blown film		1.6							ASTM E96-E	281
	DuPont Elvax 3135; 0.94 g/cm³ density; 0.051 mm thick; 12.0% VA; 0.25-0.35 g/10 min. MFI; blown film		1.7							ASTM E96-E	281
	DuPont Elvax 3135SB; 0.94 g/cm³ density; 0.051 mm thick; 12.0% VA; 0.25-0.35 g/10 min. MFI; antiblock additive, slip additive; blown film		1.7							ASTM E96-E	281
	DuPont Elvax 3135X; 0.94 g/cm³ density; 0.051 mm thick; 12.0% VA; 0.25-0.35 g/10 min. MFI; blown film		1.7							ASTM E96-E	281
	DuPont Elvax 3137; 0.94 g/cm³ density; 0.051 mm thick; 12.0% VA; 0.3 g/10 min. MFI; blown film		1.6							ASTM E96-E	281
	DuPont Elvax 3159; 0.94 g/cm³ density; 0.051 mm thick; 15.0% VA; 0.5 g/10 min. MFI; blown film		3.8							ASTM E96-E	281
	DuPont Elvax 3165; 0.94 g/cm³ density; 0.051 mm thick; 18.0% VA; 0.7 g/10 min. MFI; blown film		4.7							ASTM E96-E	281
	DuPont Elvax 3170; 0.94 g/cm³ density; 0.051 mm thick; 18.0% VA; 2.5 g/10 min. MFI; blown film		4.7							ASTM E96-E	281
	DuPont Elvax 3170SB; 0.94 g/cm³ density; 0.051 mm thick; 18.0% VA; 2.5 g/10 min. MFI; antiblock additive, slip additive; blown film		4.7							ASTM E96-E	281
	DuPont Elvax 3170SHB; 0.94 g/cm³ density; 0.051 mm thick; 18.0% VA; 2.5 g/10 min. MFI; high antiblock, slip additive; blown film		4.7							ASTM E96-E	281
	0.930 g/cm³ density; 0.05 mm thick, 2.5 BUR; blown film; 12.0% VA; film		4.5							JIS Z0208	216
Polyethylene, ULDPE	Dow Attane 4001; 0.905 g/cm³ density; 0.0254 mm thick; blown film; 1.0 g/10 min. MFI		0.31					100		Mocon Test Method; Mocon Permatron W-1	11
	Dow Attane 4003; 0.912 g/cm³ density; 0.0254 mm thick; blown film; 0.8 g/10 min. MFI		0.47					100		Mocon Test Method; Mocon Permatron W-1	11
Polyimide	DuPont Kapton Type-E; 0.076 mm thick; film		0.3								273
	DuPont Kapton Type-K; 0.076 mm thick; film		2.13								273
	DuPont Kapton Type-V; 0.076 mm thick; film		1.67								273
	Ube Upilex R; 0.025 mm thick; film		0.56			38		90		ASTM E96	97
	Ube Upilex S; 0.025 mm thick; film		0.04			38		90		ASTM E96	97
Polymethylpentene	Mitsui TPX X-22; 0.835 g/cm³ density; transparent, 0.05 mm thick; film		3.25			23					302
	Mitsui TPX X-44; 0.834 g/cm³ density; transparent, 0.05 mm thick; film		3			23					302
	Mitsui TPX X-88; 0.835 g/cm³ density; transparent, food grade, 0.05 mm thick; film		2.5			23					302
Polyphenylene Sulfide	Phillips Ryton; 0.127 mm thick; baked coating; film		0.65							ASTM E96, condition E	102
Polypropylene	biaxially oriented		0.15			40		90			264
	film		0.59			37.8		90		ASTM D96	101
	oriented, 0.02 mm thick; film		<0.39			40		90		JIS Z0208	268
	oriented, 0.022 mm thick; PVDC coated; film		<0.39			40		90		JIS Z0208	268

WATER VAPOR (continued)

Material Family	Material Note	Permeability Data: Perm. Coefficient (cm³·mm/m²·day·atm)	VTR (g·mm/m²·day)	Non-normalized Data	Penetrant Note	Temp. (°C)	Time (days)	RH (%)	Pressure (kPa)	Test Note	Source
Polypropylene			0.27								264
Polypyrrole			0.4			40		90			296
	BASF AG Lutaner ES 9567; 0.031-0.037 mm thick, intrinsically conductive, high flexibility, w/ benzenesulfonate anions; film		5.4			37.8		90		ASTM F1249	182
Polystyrene	Dow Styron; film		0.79 - 3.9								250
	Dow Trycite; film		3.5			24					250
			3.35			40		90			264
			4.0			37.8		90		ASTM F1249	296
Polystyrene, GP	BASF AG Polystyrol 168 N; transparent, 0.1 mm thick; film		1.2			23				RH: 85%-0% gradient; DIN 53122	26
	Dow Styron; injection molding		0.79 - 3.9							ASTM E96	263
	Dow Styron; oriented; sheet		3.5							ASTM E96	263
Polystyrene, IPS	BASF AG Polystyrol 476 L; 0.1 mm thick; film		1.3			23				RH: 85%-0% gradient; DIN 53122	26
	Dow Styron; injection molding		0.79 - 3.9							ASTM E96	262
Polysulfone	Amoco Udel; transparent; slot cast thin film		7.1			38		90		ASTM E96	15
	Amoco Udel; transparent; slot cast thin film		27.2			71		100		ASTM E96	15
Polyurethane	0.014 mm thick; film		0.94 - 3.4			37		90		ASTM E96-63T	121
Polyvinyl Alcohol			27.9			40		90		JIS Z0208	268
Polyvinyl Chloride	unplasticized; film		1.18			38		90			250
			1.18			40		90			264
			1.7			37.8		90		ASTM F1249	296
Polyvinylidene Chloride	Dow Saran; film		0.1 - 0.12			24					250
	Dow Saran 313; VDC vinyl chloride; barrier prop.; extrusion; film; 1.69 spec. grav.			4.19 g/m² · day; conversion not possible without thickness		30				ASTM E96	254
	Dow Saran 416; VDC vinyl chloride; barrier prop.; extrusion; film; 1.73 spec. grav.			1.55 g/m² · day; conversion not possible without thickness		30				ASTM E96	254
	Dow Saran 469; VDC vinyl chloride; barrier prop.; extrusion; 43,000 mol. wgt.; film; 1.76 spec. grav.			2.02 g/m² · day; conversion not possible without thickness		30				ASTM E96	254
	Dow Saran 516; VDC vinyl chloride; barrier prop.; extrusion; film; 1.73 spec. grav.			2.02 g/m² · day; conversion not possible without thickness		30				ASTM E96	254
	Dow Saran 525; VDC vinyl chloride; barrier prop.; extrusion; film; 1.73 spec. grav.			2.02 g/m² · day; conversion not possible without thickness		30				ASTM E96	254
	Dow Saran 5253; barrier prop.		0.09			40		90			264
	Dow Saran 866; VDC vinyl chloride; barrier prop.; extrusion; film; 1.70 spec. grav.			3.1 g/m² · day; conversion not possible without thickness		30				ASTM E96	254
	Dow Saran F-239; VDC vinyl chloride; barrier prop.; 0.003 mm thick; coating; film		0.01			30				ASTM E96	254
	Dow Saran F-279; VDC vinyl chloride; barrier prop.; 0.003 mm thick; coating; film		0.01			30				ASTM E96	254
	Dow Saran F-310; VDC vinyl chloride; barrier prop.; 0.003 mm thick; coating; film		0.04			30				ASTM E96	254
	Dow Saran MA 119; VDC methyl acrylate; barrier prop.; extrusion; 90,000 mol. wgt.; film; 1.78 spec. grav.			0.78 g/m² · day; conversion not possible without thickness		30				ASTM E96	254
	Dow Saran Wrap 18; cling packaging, laminations; transparent; barrier prop.; 0.019 mm thick; biaxially oriented; monolayer film		0.11			38		90		Permatran W	256

WATER VAPOR (continued)

Material Family	Material Note	Perm. Coefficient (cm³·mm/m²·day·atm)	VTR (g·mm/m²·day)	Non-normalized Data	Penetrant Note	Temp. (°C)	Time (days)	RH (%)	Pressure (kPa)	Test Note	Source
Polyvinylidene Chloride	Dow Saran Wrap 18L; laminations; transparent, barrier prop., 0.019 mm thick, preshrunk, biaxially oriented; monolayer film		0.12			38		90		Permatran W	256
	Dow Saran Wrap 19; chub packaging; transparent, barrier prop., 0.0254 mm thick, biaxially oriented; monolayer film		0.1			38		90		Permatran W	256
	Dow Saran Wrap 28; chub packaging; transparent, barrier prop., 0.0254 mm thick, biaxially oriented; monolayer film		0.16			38		90		Permatran W	256
	Dow Saran Wrap 560; unit packaging; transparent, barrier prop., 0.152 mm thick, biaxially oriented; coextruded film		0.02			38		90		Permatran W	256
	Dow Saran XU-32004		0.02			40		90			264
PVC-PVDC Copolymer			0.08 - 0.24			37.8		90			138
Rubber, EPDM	DuPont Nordel 1040		0.06			23				STP conditions	311
Rubber, Latex (NR)	0.51 mm thick; 50 phr SRF black, 5 phr paraffinic oil			2425% relative to Butyl Rubber						Tappi Standard T464 M-45	300
Rubber, Polybutadiene	0.51 mm thick; 50 phr SRF black, 5 phr paraffinic oil film			5125% relative to Butyl Rubber						Tappi Standard T464 M-45	300
	film		17.7			39					250
Rubber, Styrene Butadiene	Exxon SBR 1500; cis 1,4-polybutadiene; 0.51 mm thick; 50 phr SRF black, 5 phr paraffinic oil			1875% relative to Butyl Rubber						Tappi Standard T464 M-45	300
SAN	BASF AG Luran 358 N; transparent, high flow, 0.1 mm thick; film		2 - 2.5			23				RH: 85%-0% gradient; DIN 53122	30
	BASF AG Luran 368 R; transparent, 0.1 mm thick, moderate to low flow; film		2 - 2.5			23				RH: 85%-0% gradient; DIN 53122	30
	BASF AG Luran 378 P; transparent, mod-high flow, 0.1 mm thick; film		2 - 2.5			23				RH: 85%-0% gradient; DIN 53122	30
	BASF AG Luran 388 S; transparent, 0.1 mm thick, low flow; film		2 - 2.5			23				RH: 85%-0% gradient; DIN 53122	30
	Dow Tyril; low nitrile content; film		1.97 - 5.51								250
Silicone			1.7 - 3.1			37		90		ASTM E96-63T	121
Styrene-Butadiene Block Copol.	BASF AG Styrolux 637 D; 0.1 mm thick; film		1.3			23					29
	BASF AG Styrolux 656 C; high flow, 0.1 mm thick; film		1.1			23					29
	BASF AG Styrolux 684 D; 0.1 mm thick; film		1.4			23					29
TPE, Olefinic	Adv. Elast. Santoprene 201-73; 73 Shore A, 0.5 mm thick		0.49			25				ASTM E96, procedure A; saturated vapor over liquid water contacts the rubber, with 25% RH on the opposite side	282
	Adv. Elast. Santoprene 201-73; 73 Shore A, 0.5 mm thick		0.23			25				ASTM E96, procedure BW; liquid water contacts the rubber, with 75% RH on the opposite side	282
	Adv. Elast. Santoprene 201-87; 87 Shore A, 0.5 mm thick		0.16			25				ASTM E96, procedure A; saturated vapor over liquid water contacts the rubber, with 25% RH on the opposite side	282
	Adv. Elast. Santoprene 201-87; 87 Shore A, 0.5 mm thick		0.23			25				ASTM E96, procedure BW; liquid water contacts the rubber, with 75% RH on the opposite side	282

WATER VAPOR (continued)

Material Family	Material Note	Permeability Data			Test Conditions						Source
		Perm. Coefficient (cm³·mm/m²·day·atm)	VTR (g·mm/m²·day)	Non-normalized Data	Penetrant Note	Temp. (°C)	Time (days)	RH (%)	Pressure (kPa)	Test Note	
TPE, Olefinic											
	Adv. Elast. Santoprene 203-50; 50 Shore D, 0.5 mm thick		0.23			25				ASTM E96, procedure A; saturated vapor over liquid water contacts the rubber, with 25% RH on the opposite side	282
	Adv. Elast. Santoprene 203-50; 50 Shore D, 0.5 mm thick		0.81			25				ASTM E96; procedure BW; liquid water contacts the rubber, with 75% RH on the opposite side	282
TPE, Polyamide											
	Atochem Pébax 2533; 25 Shore D, 0.12 mm thick; film		89			38		100			287
	Atochem Pébax 3533; 35 Shore D, 0.12 mm thick; film		67			38		100			287
	Atochem Pébax 4033; 40 Shore D, 0.12 mm thick; film		38			38		100			287
	Atochem Pébax 5533; 55 Shore D, 0.12 mm thick; film		34			38		100			287
	Atochem Pébax 6333; 63 Shore D, 0.12 mm thick; film		31			38		100			287
TPE, Polybutadiene											
	Jap. Synth. JSR RB820; 1,2-polybutadiene; 0.910 g/cm³ density; transparent, 2.5 BUR, 0.05 mm thick; blown film; 15% crystallinity; film		9.8							JIS Z0208	216
	Jap. Synth. JSR RB830; 0.910 g/cm³ density; stretch film, 0.018 mm thick, 4.8 BUR; blown film; 29% crystallinity; film		7							JIS Z0208	216
	Jap. Synth. JSR RB830; 1,2-polybutadiene; 0.910 g/cm³ density; transparent, 2.5 BUR, 0.05 mm thick; blown film; 25% crystallinity; film		7							JIS Z0208	216
TPE, Polyester											
	Eastman Ecdel; copolyester ether; transparent, crystalline, 0.11 - 0.14 mm thick; film		19.4			38		90		ASTM F372; Mocon value, confirmed by ASTM E96E	60
TPE, Styrenic											
	Shell Kraton D 1101; SBS; FDA grade, 71 Shore A; unsaturated; 31% styrene/ 69% rubber, neat rubber		257			23				RH: 90% gradient; ASTM E96-80, procedure E; area: 50 cm²	303
	Shell Kraton D 1107; SIS; FDA grade, 37 Shore A; unsaturated; 14% styrene/ 86% rubber, neat rubber		209			23				"	303
	Shell Kraton D 2103; SBS; FDA grade, 70 Shore A; unsaturated; ready to use compound		279			23				"	303
	Shell Kraton D 2104; SBS; FDA grade, 27 Shore A; unsaturated; ready to use compound		579			23				"	303
	Shell Kraton D 2109; SBS; FDA grade, 44 Shore A; unsaturated; ready to use compound		148			23				"	303
	Shell Kraton G 1650; SEBS; FDA grade, 75 Shore A; saturated; 29% styrene/71% rubber, neat rubber		54.8			23				"	303
	Shell Kraton G 1651; SEBS; FDA grade, 75 Shore A; saturated; 29% styrene/71% rubber, neat rubber		62.4			23				"	303
	Shell Kraton G 1652; SEBS; FDA grade, 75 Shore A; saturated; 29% styrene/71% rubber, neat rubber		83.2			23				"	303
	Shell Kraton G 2701; SEBS; FDA grade, 67 Shore A; saturated; ready to use compound		47			23				"	303
	Shell Kraton G 2705; SEBS; FDA grade, 55 Shore A; saturated; ready to use compound		66.1			23				ASTM E96-80, procedure E; area 50 cm²	303

WATER VAPOR (continued)

Material Family	Material Note	Perm. Coefficient (cm³·mm/m²·day·atm)	VTR (g·mm/m²·day)	Non-normalized Data	Penetrant Note	Temp. (°C)	Time (days)	RH (%)	Pressure (kPa)	Test Note	Source
TPE, Styrenic	Shell Kraton G 2706; SEBS; FDA grade, 28 Shore A; saturated; ready to use compound		86.0			23				--	303
TPE, Urethane (TPAU)	BASF Elastollan C64D; 64 Shore D			2 g/m²·day; conversion not possible without thickness		23				RH: 93% differential	130
	BASF Elastollan C80A; 80 Shore A			14 g/m²·day; conversion not possible without thickness		23				RH: 93% differential	130
	BASF Elastollan C95A; 95 Shore A			6 g/m²·day; conversion not possible without thickness		23				RH: 93% differential	130
TPE, Urethane (TPEU)	BASF Elastollan 1164D; 64 Shore D			3 g/m²·day; conversion not possible without thickness		23				RH: 93% differential	130
	BASF Elastollan 1180A; 80 Shore A			18 g/m²·day; conversion not possible without thickness		23				RH: 93% differential	130
	BASF Elastollan 1195A; 95 Shore A			9 g/m²·day; conversion not possible without thickness		23				RH: 93% differential	130
TPE, Vinyl	1.23 - 1.31 g/cm³ density; stretch film, 0.018 mm thick, 4.8 BUR; blown film; film		7.9 - 12.9							JIS Z0208	216
	1.26 g/cm³ density; transparent, 2.5 BUR, 0.05 mm thick; blown film; 50 phr plasticizer; film		10							JIS Z0208	216

XYLENE

Material Family	Material Note	Perm. Coefficient (cm³·mm/m²·day·atm)	VTR (g·mm/m²·day)	Non-normalized Data	Penetrant Note	Temp. (°C)	Time (days)	RH (%)	Pressure (kPa)	Test Note	Source
Acrylonitrile Copolymer, AMA	barrier prop.; bottles			0.06% penetrant weight loss		50	28				293
EVOH	Eval Co. Eval E; 0.02 mm thick; 44% ethylene; film	0.028				20	28	65			266
	Eval Co. Eval E; 0.032 mm thick; 44% ethylene; film	0.02				20		65			266
	Eval Co. Eval EF-E; barrier prop.; 44% ethylene; film	0.03				20					265
	Eval Co. Eval EF-F; barrier prop.; 32% ethylene; film	0.02				20					265
	Eval Co. Eval EF-XL; 0.015 mm thick, biaxially oriented; film	0.007				20		65			266
	Eval Co. Eval EF-XL; barrier prop., biaxially oriented; film	0.01				20					265
	Eval Co. Eval F; 0.02 mm thick; 32% ethylene; film	0.022				20		65			266
	Eval Co. Eval F; 0.032 mm thick; 32% ethylene; film	<0.0015				20		65			266
Laminar, EVOH/HDPE	DuPont Selar RB 421/ HDPE; 15% Selar RB 421; barrier prop.; extrusion blow molded	0.79				60	14				293
Laminar, EVOH/Olefin	DuPont Selar RB/ Polyolefin; 15% Selar RB; barrier prop., laminar technology; extrusion blow molded; 1 L bottle			2.27% penetrant weight loss	with 25% methyl alcohol	23	180				293
	DuPont Selar RB/ Polyolefin; 15% Selar RB; barrier prop., laminar technology; extrusion blow molded; 1 L bottle			1.51% penetrant weight loss	with 50% methyl alcohol	23	180				293
	DuPont Selar RB/ Polyolefin; 15% Selar RB; barrier prop., laminar technology; extrusion blow molded; 1 L bottle			0.11% penetrant weight loss		50	28				293
	DuPont Selar RB/ Polyolefin; 15% Selar RB; barrier prop., laminar technology; extrusion blow molded; 1 L bottle			0.16% penetrant weight loss	with 25% propyl alcohol	50	28				293
	DuPont Selar RB/ Polyolefin; 15% Selar RB; barrier prop., laminar technology; extrusion blow molded; 1 L bottle			0.11% penetrant weight loss	with 50% propyl alcohol	50	28				293
Laminar, Nylon/HDPE	DuPont Selar RB/ HDPE; solvent packaging; 18% Selar RB; barrier prop.; extrusion blow molded; one pint bottle			0.3% penetrant weight loss (80% relative improvement vs. HDPE)	aromatic	48.9	28				291

XYLENE (continued)

Material Family	Material Note	Perm. Coefficient (cm³·mm/m²·day·atm)	VTR (g·mm/m²·day)	Non-normalized Data	Penetrant Note	Temp. (°C)	Time (days)	RH (%)	Pressure (kPa)	Test Note	Source
Laminar, Nylon/HDPE	DuPont Selar RB 215/ HDPE; 10% Selar RB 215; barrier prop.; extrusion blow molded		0.39			60	14				293
Laminar, Nylon/LDPE	DuPont Selar RB 300/ HDPE; 10% Selar RB 300; barrier prop.; extrusion blow molded		3.5			60	14				293
	DuPont Selar RB 215/ LDPE; 15% Selar RB 215; barrier prop.; extrusion blow molded		4.7			60	14				293
Laminar, Nylon/Olefin	DuPont Selar RB/ Polyolefin; 8% Selar RB; barrier prop., laminar technology; extrusion blow molded; 1 L bottle			14.1% penetrant weight loss	with 25% methyl alcohol	23	180				293
	DuPont Selar RB/ Polyolefin; 8% Selar RB; barrier prop., laminar technology; extrusion blow molded; 1 L bottle			10.60% penetrant weight loss	with 50% methyl alcohol	23	180				293
	DuPont Selar RB/ Polyolefin; 8% Selar RB; barrier prop., laminar technology; extrusion blow molded; 1 L bottle			0.12% penetrant weight loss		50	28				293
	DuPont Selar RB/ Polyolefin; 8% Selar RB; barrier prop., laminar technology; extrusion blow molded; 1 L bottle			2.84% penetrant weight loss	with 25% propyl alcohol	50	28				293
	DuPont Selar RB/ Polyolefin; 8% Selar RB; barrier prop., laminar technology; extrusion blow molded; 1 L bottle			2.61% penetrant weight loss	with 50% propyl alcohol	50	28				293
	DuPont Selar RB/ Polyolefin; 8% Selar RB; barrier prop.; extrusion blow molded; bottles			0.12% penetrant weight loss		50	28				293
Laminar, Nylon/PP	DuPont Selar RB 421/ PP; 10% Selar RB 240; barrier prop.; extrusion blow molded		7.1			60	14				293
LDPE/ EVAL Film	0.06/ 0.015 mm thick; EVAL EF-F inside; LDPE outside; film		<0.0035			20		65			266
	0.06/ 0.015 mm thick; EVAL EF-XL inside; LDPE outside; film		<0.0035			20		65			266
	0.06/ 0.025 mm thick; EVAL EF-E inside; LDPE outside; film		<0.004			20		65			266
Nylon	0.0254 mm thick; oriented; film		0.02			20		65			266
	oriented; 0.015 mm thick; PVDC coated; film		0.01			20		65			266
Polyester, PET	0.0254 mm thick; film		0.04			20		65			266
Polyethylene, Fluorinated	barrier prop.; bottles			0.21% penetrant weight loss		50	28				293
Polyethylene, HDPE	DuPont; 1 L bottle			20.30% penetrant weight loss	with 25% methyl alcohol	23	180				293
	DuPont; 1 L bottle			14.99% penetrant weight loss	with 50% methyl alcohol	23	180				293
	DuPont; 1 L bottle			28% penetrant weight loss		50	28				293
	DuPont; 1 L bottle			23.45% penetrant weight loss	with 25% propyl alcohol	50	28				293
	DuPont; 1 L bottle			16.27% penetrant weight loss	with 50% propyl alcohol	50	28				293
	bottles			38.1% penetrant weight loss		50	28				293
	bottles		283			60	14				293
Polyethylene, LDPE	0.051 mm thick; film		16.6			20					266
	film		16.5			20		65			265
			1496			60	14				293
Polypropylene	0.02 mm thick; oriented; film		7			20	14				266
	biaxially oriented; film		7.0			20		65			265
			984			60	14				293

XYLENE (continued)

Material Family	Material Note	Permeability Data			Penetrant Note	Test Conditions					Source
		Perm. Coefficient (cm³·mm/m²·day·atm)	VTR (g·mm/m²·day)	Non-normalized Data		Temp. (°C)	Time (days)	RH (%)	Pressure (kPa)	Test Note	
Polyvinyl Alcohol	Ansell Edmont PVA; supported (lined) glove film; specified by glove film weight			low permeation rate: 0 to 1/2 eyedropper size drops per hour; PVA coating is water soluble; lower detection limit: 0 ppm; permeation rate: <0.9 µg/cm²/min	xylol	23	0.25			ASTM F739	313
Polyvinyl Chloride	bottles			failed		23	180				293
Rubber, Neoprene (CR)	Pioneer Industrial Products Stanzoil N-44; glove film			breakthrough time: 23 minutes; permeation rate: 798 µg/cm²/min	xylol					ASTM F739.85	312
Rubber, Nitrile (NBR)	Ansell Edmont Sol-Vex 37-165; unsupported glove film; 0.54 mm thick			fair permeation rate; 51 to 100 eyedropper size drops per hour; breakthrough time: 75 minutes; permeation rate: <900 µg/cm²/min	xylol	23				ASTM F739	313
	Pioneer Industrial Products Stansolv A-14; glove film			breakthrough time: 92 minutes; lower detection limit: 0.002 ppm; permeation rate: 24 µg/cm²/min	xylol					ASTM F739.85	312
TPE, Vinyl	Pioneer Industrial Products Pylox V-20; glove film			breakthrough time: 4 minutes; permeation rate: 138 µg/cm²/min	xylol					ASTM F739.85	312
TPE, Vinyl coated fabric	Ansell Edmont Wet-Wear 700; clothing; PVC coating with Nylon/Polyester lining			good permeation rate: 6 to 50 eyedropper size drops per hour; breakthrough time: >3 minutes; permeation rate: <90 µg/cm²/min	xylol	23				ASTM F739	313

Permeation Rates

Appendix Two is useful for comparing permeation rates of penetrants through different materials at various temperature ranges. The data is sorted by penetrant with a secondary sort on temperature and a final sort on either permeability coefficient (P) or vapor transmission rate (V). Appendix Two presents data in the most concise form. Only normalized values for permeability coefficient or vapor transmission rate appear. Supporting test information, excepting temperature is not included. For more detailed information, please refer to the chapter containing the appropriate material generic family.

P units are cm^3·mm/m^2·day·atm.

V units are g·mm/m^2·day.

acetic acid

20 - 25°C

Polyphenylene Sulfide (23°C)	0.79 (V)

acetone

20 - 25°C

Polyethylene, LDPE (24°C)	3.9 - 15.8 (V)

>25 - 50°C

Polyester, PET (40°C)	0.87 (V)

air

20 - 25°C

Polyvinylidene Chloride (23°C)	0.03 (P)
HDPE/ EVA/ PVDC/ EVA Film (23°C)	0.1 (P)
LDPE/ EVA/ PVDC/ EVA/ LDPE Film (23°C)	0.1 (P)
Polyvinylidene Chloride (23°C)	0.14 (P)
LDPE/ EVA/ PVDC/ EVA/ LDPE Film (23°C)	0.15 (P)
LDPE/ EVA/ PVDC/ EVA/ LDPE Film (23°C)	0.18 (P)
Polyvinylidene Chloride (23°C)	0.19 (P)
Polyvinylidene Chloride (23°C)	0.2 (P)
LDPE/ EVA/ PVDC/ EVA/ LDPE Film (23°C)	0.21 (P)
LDPE/ EVA/ PVDC/ EVA Film (23°C)	0.24 (P)
Acetal Copolymer	0.87 - 1.3 (P)
Fluoroelastomer, FKM (24°C)	8.6 (P)
Polyphenylene Sulfide	7.9 - 11.8 (P)
Polyethylene, HDPE (20°C)	29.4 (P)
Polyethylene, HDPE (25°C)	30.4 (P)
Rubber, Chlorobutyl (CIIR)	30.7 (P)
Rubber, Butyl (IIR) (23.9°C)	36.4 (P)
Rubber, Chlorobutyl (CIIR) (23.9°C)	38.7 (P)
Rubber, Styrene Butadiene (23.9°C)	138 (P)
TPE, Olefinic (23°C)	140 (P)
TPE, Polyester (21.5°C)	156 (P)
TPE, Olefinic	159 (P)
TPE, Polyester (21.5°C)	207 (P)
TPE, Olefinic	216 (P)
TPE, Olefinic (23°C)	240 (P)
TPE, Olefinic (23°C)	302 (P)
Rubber, Styrene Butadiene (23.9°C)	348 (P)
Rubber, EPDM (23.9°C)	461 (P)
Rubber, Latex (NR) (23.9°C)	496 (P)
Rubber, EPDM (23.9°C)	668 (P)
Rubber, EPDM (23°C)	734 - 907 (P)

>25 - 50°C

Polyethylene, HDPE (30°C)	38.5 (P)
Rubber, Butyl (IIR) (40°C)	51.8 (P)
Polyethylene, HDPE (40°C)	68.9 (P)
Rubber, Nitrile (NBR) (40°C)	95.0 (P)
Polyethylene, HDPE (50°C)	111 (P)
Rubber, Styrene Butadiene (40°C)	397 (P)
Rubber, EPDM (40°C)	683 (P)
Rubber, Latex (NR) (40°C)	1020 (P)
Rubber, Polybutadiene (40°C)	2393 (P)

>50 - 75°C

Rubber, Butyl (IIR) (60°C)	156 (P)
Rubber, Nitrile (NBR) (60°C)	354 (P)
Rubber, Chlorobutyl (CIIR) (65.6°C)	364 (P)
Rubber, Butyl (IIR) (65.6°C)	375 (P)
Rubber, Styrene Butadiene (60°C)	1080 (P)
Rubber, Styrene Butadiene (65.6°C)	1092 (P)
Rubber, EPDM (60°C)	1477 (P)
Rubber, Styrene Butadiene (65.6°C)	2048 (P)
Rubber, Latex (NR) (60°C)	2315 (P)
Rubber, EPDM (65.6°C)	2560 (P)
Rubber, Latex (NR) (65.6°C)	2696 (P)
Rubber, EPDM (65.6°C)	3299 (P)
Rubber, Polybutadiene (60°C)	3810 (P)

>75 - 100°C

Rubber, Nitrile (NBR) (80°C)	855 (P)
Rubber, Chlorobutyl (CIIR) (93.3°C)	1183 (P)
Rubber, Butyl (IIR) (93.3°C)	1195 (P)
Rubber, Styrene Butadiene (80°C)	2091 (P)
Rubber, Styrene Butadiene (93.3°C)	2719 (P)
Rubber, EPDM (80°C)	2851 (P)
Rubber, Latex (NR) (80°C)	3793 (P)
Rubber, Butyl (IIR) (80°C)	3974 (P)
Rubber, Styrene Butadiene (93.3°C)	4346 (P)
Rubber, Latex (NR) (93.3°C)	4574 (P)
Rubber, Polybutadiene (80°C)	5659 (P)
Rubber, EPDM (93.3°C)	7043 (P)
Rubber, EPDM (93.3°C)	7247 (P)

ammonia

≤0°C

Polyethylene, HDPE (-3°C)	32.5 (P)
Fluoropolymer, ECTFE (-1°C)	32.6 (P)
Fluoropolymer, TFE (-3°C)	41.2 (P)
Fluoropolymer, TFE (-2°C)	68.0 (P)

0 - <20°C

Fluoropolymer, FEP (0°C)	29.0 (P)

20 - 25°C

Fluoropolymer, CTFE (25°C)	1.05 (P)
Polyphenylene Sulfide	5.9 (P)
Fluoropolymer, PVDF (23°C)	6.6 (P)
Fluoropolymer, FEP (25°C)	101 (P)
Fluoropolymer, ECTFE (25°C)	113 (P)
Polyethylene, HDPE (25°C)	123 (P)
Fluoropolymer, TFE (25°C)	151 (P)

Fluoropolymer, TFE (25°C)	241 (P)
Polysulfone (23°C)	421 (P)

<u>>50 - 75°C</u>

Fluoropolymer, CTFE (59°C)	24.2 (P)
Fluoropolymer, FEP (66°C)	551 (P)
Fluoropolymer, ECTFE (65°C)	617 (P)
Polyethylene, HDPE (61°C)	623 (P)
Fluoropolymer, TFE (63°C)	755 (P)
Fluoropolymer, TFE (62°C)	1059 (P)

argon

<u>20 - 25°C</u>

TPE, Urethane (TPAU) (20°C)	26.3 (P)
TPE, Urethane (TPAU) (20°C)	43.8 (P)
TPE, Urethane (TPEU) (20°C)	52.5 (P)
TPE, Urethane (TPEU) (20°C)	61.3 (P)
Polyethylene, HDPE (20°C)	66.9 (P)
TPE, Urethane (TPAU) (20°C)	78.8 (P)
TPE, Urethane (TPEU) (20°C)	78.8 (P)
TPE, Urethane (TPAU) (20°C)	105 (P)
TPE, Urethane (TPEU) (20°C)	123 (P)
TPE, Olefinic (23°C)	395 (P)
TPE, Olefinic (23°C)	519 (P)
TPE, Olefinic (23°C)	597 (P)

<u>>25 - 50°C</u>

Polyethylene, HDPE (30°C)	90.2 (P)
Polyethylene, HDPE (50°C)	233 (P)

ASTM Fuel Oil B

<u>20 - 25°C</u>

Nylon 66	0.2 (V)

benzene

<u>20 - 25°C</u>

Polyester, PET (25°C)	0.14 (V)
Polyphenylene Sulfide (23°C)	2.5 (V)

<u>>25 - 50°C</u>

Polyethylene, LDPE (35°C)	236 (V)
Polystyrene (35°C)	472 (V)

carbon dioxide

<u>0 - <20°C</u>

EVOH (5°C)	0.0039 (P)
EVOH (5°C)	0.01 (P)
EVOH (5°C)	0.02 (P)
Nylon 6 (0°C)	0.24 (P)
Fluoropolymer, CTFE (0°C)	2.3 (P)

<u>20 - 25°C</u>

EVOH (23°C)	0.01 (P)
EVOH (23°C)	0.03 (P)
Polyvinyl Alcohol (24°C)	0.04 (P)
Acrylonitrile (24°C)	0.04 - 0.08 (P)
EVOH (23°C)	0.08 (P)
Acrylonitrile Co. (AMA) (24°C)	0.2 (P)
Polyvinylidene Chloride (23°C)	0.47 (P)
Nylon 6 (23°C)	0.55 (P)
Acrylonitrile Copolymer, AMA (22.8°C)	0.64 (P)
HDPE/ EVA/ PVDC/ EVA Film (23°C)	1.1 (P)
LDPE/ EVA/ PVDC/ EVA Film (23°C)	1.1 (P)
LDPE/ EVA/ PVDC/ EVA/ LDPE Film (23°C)	1.1 (P)
Nylon, Amorphous (22.8°C)	1.1 (P)
LDPE/ EVA/ PVDC/ EVA/ LDPE Film (23°C)	1.64 (P)

Nylon, Amorphous (22.8°C)	1.8 (P)
PE/ PVC-PVDC Copolymer Multilayer Film (24°C)	0.39 - 3.2 (P)
Nylon 6 (23°C)	1.8 (P)
Nylon 6 (22.8°C)	1.8 (P)
Polyvinylidene Chloride (24°C)	1.6 - 2.4 (P)
Polyvinylidene Chloride (23°C)	2.13 (P)
Fluoropolymer, PVDF (25°C)	2.2 (P)
LDPE/ EVA/ PVDC/ EVA/ LDPE Film (23°C)	2.2 (P)
Polyvinylidene Chloride (23°C)	2.83 (P)
Nylon 6/66 (23°C)	2.9 (P)
Parylene (25°C)	3.0 (P)
Nylon 66 (23°C)	3.1 (P)
Epoxy (25°C)	3.2 (P)
Nylon 6 (22.8°C)	3.2 (P)
Polyvinylidene Chloride (23°C)	3.15 (P)
LDPE/ EVA/ PVDC/ EVA/ LDPE Film (23°C)	3.24 (P)
Nylon 66 (23°C)	3.5 (P)
Nylon, Amorphous (23°C)	3.8 (P)
Nylon 6 (23°C)	4.1 - 4.6 (P)
Nylon 6/66 (23°C)	4.1 - 4.6 (P)
Fluoropolymer, PVF (25°C)	4.3 (P)
Nylon 66 (23°C)	4.6 (P)
Polyester, PET (25°C)	4.7 (P)
Parylene (25°C)	5.1 (P)
Fluoropolymer, CTFE (25°C)	6.3 (P)
Nylon 66 (23°C)	6.3 (P)
Polyester, PET (25°C)	6.3 (P)
Fluoropolymer, PVDF (23°C)	7.1 (P)
Polyester, PET (25°C)	7.9 (P)
Polyester, PET (25°C)	5.9 - 9.8 (P)
Fluoropolymer, CTFE (25°C)	9.2 (P)
Nylon 66/610 (23°C)	9.4 (P)
Fluoropolymer, CTFE (25°C)	11.8 (P)
Polyvinyl Chloride (24°C)	7.9 - 19.7 (P)
Fluoropolymer, CTFE (25°C)	15.8 (P)
PVC-PVDC Copolymer (25°C)	15.0 - 17.3 (P)
Acetal (23°C)	14.6 - 19.7 (P)
Polypyrrole	22.3 (P)
Polyphenylene Sulfide	29.5 (P)
Polyester, PETG (23°C)	31.5 (P)
Polyethylene, LLDPE	35.6 (P)
Polyester, PCTG (23°C)	50 (P)
Acetal Copolymer	56.7 - 68.5 (P)
Parylene (25°C)	84.2 (P)
Fluoropolymer, ETFE (25°C)	98.4 (P)
ASA (23°C)	101 (P)
Polyethylene, HDPE (23°C)	136 (P)
Polyester, PBT (23°C)	139 (P)
ASA (23°C)	142 (P)
SAN (24°C)	157 (P)
TPE, Urethane (TPAU) (20°C)	175 (P)
ABS (24°C)	157 - 236 (P)
ABS (23°C)	202.6 (P)
ASA (23°C)	203 (P)
Polypropylene (25°C)	208 (P)
Polyethylene, HDPE (25°C)	228 (P)
Fluoropolymer, ETFE (23°C)	232 (P)
ASA (23°C)	233 (P)
Polyethylene, HDPE (24°C)	236 - 276 (P)
Polyethylene, HDPE (20°C)	284 (P)
Polyethylene, HDPE (25°C)	294 (P)
TPE, Vinyl	300 (P)
ABS (23°C)	304 (P)
Polycarbonate (22.8°C)	307 (P)
TPE, Urethane (TPAU) (20°C)	350 (P)
Polystyrene (24°C)	276 - 433 (P)
Polystyrene, GP (23°C)	276 - 433 (P)

Polysulfone (23°C)	374 (P)
Rubber, Styrene Butadiene (24°C)	394 (P)
ABS (24°C)	354 - 472 (P)
Polycarbonate	436 (P)
Polybutylene (22.8°C)	468 (P)
Polystyrene (24°C)	394 - 590 (P)
Polystyrene, GP (23°C)	394 - 590 (P)
Polystyrene, IPS (23°C)	394 - 590 (P)
Polyethylene, MDPE (25°C)	39.4 - 984 (P)
Polystyrene, GP (23°C)	527 (P)
Polybutylene (22.8°C)	561 (P)
Polyethylene, LDPE (24°C)	394 - 787 (P)
TPE, Vinyl (24°C)	39.4 - 1181 (P)
Fluoropolymer, FEP (25°C)	657 (P)
Polycarbonate	677 (P)
Polycarbonate	768 (P)
TPE, Urethane (TPEU) (20°C)	788 (P)
Polyethylene, LDPE	790 (P)
Styrene-Butadieneiene Block Copolymer (23°C)	811 (P)
Polycarbonate	827 (P)
Fluoropolymer, PFA (25°C)	890 (P)
Polyvinyl Chloride (25°C)	959 (P)
Polystyrene, IPS (23°C)	1013 (P)
Styrene-Butadieneiene Block Copolymer (23°C)	1013 (P)
Polyethylene, LDPE (25°C)	1063 (P)
Polyethylene, PE/EVA Copolymer	1100 (P)
TPE, Urethane (TPEU) (20°C)	1138 (P)
Polyurethane (25°C)	1181 (P)
TPE, Polyester (23°C)	1267 (P)
TPE, Urethane (TPAU) (20°C)	1313 (P)
TPE, Olefinic (23°C)	1318 (P)
Styrene-Butadieneiene Block Copolymer (23°C)	1520 (P)
TPE, Polyester (21.5°C)	1555 (P)
TPE, Urethane (TPEU) (20°C)	1576 (P)
TPE, Urethane (TPAU) (20°C)	1751 (P)
TPE, Urethane (TPEU) (20°C)	2014 (P)
TPE, Olefinic (23°C)	2015 (P)
TPE, Vinyl	1400 - 2700 (P)
TPE, Styrenic (23°C)	2303 (P)
TPE, Styrenic (23°C)	2402 (P)
TPE, Styrenic (23°C)	2500 (P)
TPE, Styrenic (23°C)	2539 (P)
TPE, Polyamide (23°C)	2758 (P)
TPE, Polybutadiene	2800 (P)
TPE, Polybutadiene	2900 (P)
TPE, Olefinic (23°C)	3022 (P)
TPE, Polyester (21.5°C)	3024 (P)
TPE, Polyamide (23°C)	3283 (P)
TPE, Styrenic (23°C)	3331 (P)
TPE, Polyamide (23°C)	5122 (P)
TPE, Styrenic (23°C)	5280 (P)
Polymethylpentene (23°C)	5500 (P)
Polymethylpentene (23°C)	6000 (P)
TPE, Styrenic (23°C)	6709 (P)
TPE, Styrenic (23°C)	7020 (P)
TPE, Styrenic (23°C)	7252 (P)
TPE, Styrenic (23°C)	7598 (P)
Rubber, EPDM (23°C)	7516 - 8122 (P)
Rubber, Polybutadiene (24°C)	7874 (P)
TPE, Polyamide (23°C)	11,753 (P)
TPE, Polyamide (23°C)	17,073 (P)
TPE, Styrenic (23°C)	24,657 (P)
Silicone (25°C)	118,110 (P)

<u>>25 - 50°C</u>

EVOH (35°C)	0.03 (P)
Polyimide (30°C)	0.03 (P)
EVOH (35°C)	0.08 (P)

EVOH (35°C)	0.2 (P)
Polyvinylidene Chloride (35°C)	0.44 (P)
Nylon 6 (35°C)	2.61 (P)
Polyimide (30°C)	2.9 (P)
Polyester, PET (35°C)	7.7 (P)
Fluoropolymer, CTFE (50°C)	15.8 (P)
Nylon 6 (50°C)	17.3 (P)
Fluoropolymer, PVDF (30°C)	30.3 (P)
Polyethylene, HDPE (30°C)	344 (P)
Fluoroelastomer, FKM (30°C)	510 (P)
Polyethylene, HDPE (40°C)	527 (P)
Polyethylene, HDPE (50°C)	811 (P)

<u>>50 - 75°C</u>

Fluoropolymer, CTFE (75°C)	98.5 (P)

carbon monoxide

<u>20 - 25°C</u>

Polyethylene, HDPE (20°C)	36.5 (P)

carbon tetrachloride

<u>20 - 25°C</u>

Nylon 66	2 (V)

<u>>25 - 50°C</u>

Polyester, PET (40°C)	0.03 (V)

chlorine

<u>20 - 25°C</u>

Fluoropolymer, PVDF (23°C)	1.2 (P)

chloroform

<u>20 - 25°C</u>

EVOH (20°C)	0.0023 (V)
EVOH (20°C)	0.0024 (V)
LDPE/ EVAL Film (20°C)	<0.0035 (V)
LDPE/ EVAL Film (20°C)	0.01 (V)
LDPE/ EVAL Film (20°C)	0.02 (V)
EVOH (20°C)	0.03 (V)
EVOH (20°C)	0.04 (V)
EVOH (20°C)	0.06 (V)
Nylon (20°C)	0.13 (V)
EVOH (20°C)	0.15 (V)
Nylon (20°C)	0.34 (V)
Polyester, PET (20°C)	7.9 (V)
Polypropylene (20°C)	74.8 (V)
Polyethylene, LDPE (20°C)	138 (V)
Polyethylene, LDPE (20°C)	140.79 (V)

cologne

<u>20 - 25°C</u>

Acetal (23°C)	0.24 (V)

<u>>25 - 50°C</u>

Acetal (38°C)	1.8 (V)

d-limonene

<u>20 - 25°C</u>

Polyvinylidene Chloride (25°C)	0.0088 (V)
Polyvinylidene Chloride (25°C)	0.016 (V)
EVOH (25°C)	0.41 (V)
Polypropylene (25°C)	8.9 (V)
Polyethylene, HDPE (25°C)	149 (V)

dichlorodifluoromethane

20 - 25°C

Polysulfone (23°C)	0.23 (P)

dichlorotetrafluoroethane

20 - 25°C

Polysulfone (23°C)	0.1 (P)

ethane

20 - 25°C

Polyethylene, HDPE (20°C)	90.2 (P)
Polyethylene, HDPE (23°C)	92.9 (P)

ethyl acetate

20 - 25°C

Polyethylene, LDPE (24°C)	11.8 - 118 (V)

>25 - 50°C

Polyester, PET (40°C)	0.03 (V)

ethyl alcohol

20 - 25°C

Acetal (23°C)	0.1 (V)
Acetal (23°C)	0.59 (V)
Polystyrene (24°C)	0.39 - sample failed (V)
Polyethylene, LDPE (24°C)	0.79 - 1.6 (V)

>25 - 50°C

Acetal (38°C)	3.1 (V)

ethylene

20 - 25°C

Polyethylene, HDPE (20°C)	111 (P)

ethylene oxide

20 - 25°C

Polyethylene, LDPE	2100 (P)
TPE, Polybutadiene	25,000 (P)
TPE, Polybutadiene	32,000 (P)

formaldehyde

20 - 25°C

Polyethylene, LDPE (24°C)	0.79 - 2.0 (V)
Polystyrene (24°C)	1.6 - 2.0 (V)
SAN (24°C)	2.0 - 3.9 (V)

Freon 114

20 - 25°C

Fluoropolymer, PVDF (23°C)	0.25 (P)
TPE, Polyester (21.5°C)	233 (P)
TPE, Polyester (21.5°C)	397 (P)
TPE, Polyester (21.5°C)	2419 (P)
TPE, Polyester (21.5°C)	3542 (P)

Freon 115

20 - 25°C

Fluoropolymer, PVDF (23°C)	0.1 (P)

Freon 12

20 - 25°C

Fluoropolymer, PVDF (23°C)	0.16 (P)
Polyethylene, HDPE (23°C)	37.4 (P)
TPE, Polyester (21.5°C)	71 (P)
TPE, Polyester (21.5°C)	104 (P)
TPE, Polyester (21.5°C)	121 (P)
Acetal (23°C)	0.08 (V)

>25 - 50°C

Acetal (38°C)	0.17 (V)
Acetal (38°C)	0.21 (V)

Freon 22

20 - 25°C

TPE, Polyester (21.5°C)	<17.3 (P)
TPE, Polyester (21.5°C)	41 (P)
TPE, Polyester (21.5°C)	51 (P)

Freon 318

20 - 25°C

Fluoropolymer, PVDF (23°C)	0.18 (P)

gasoline

20 - 25°C

Acetal (23°C)	0.04 (V)
HDPE/ EVAL Film	1.4 (V)
HDPE/ EVAL Film	6.4 (V)
Polyethylene, HDPE	25.4 (V)

hair spray

20 - 25°C

Acetal (23°C)	0.31 (V)

>25 - 50°C

Acetal (38°C)	2.4 (V)

helium

0 - <20°C

EVOH (5°C)	1.06 (P)
EVOH (5°C)	1.8 (P)
EVOH (5°C)	2.6 (P)

20 - 25°C

EVOH (23°C)	3.7 (P)
EVOH (23°C)	6.5 (P)
EVOH (23°C)	9.37 (P)
Nylon 66 (23°C)	59.1 (P)
Fluoropolymer, PVDF (23°C)	86.1 (P)
Polyethylene, HDPE (23°C)	97.2 (P)
Fluoropolymer, CTFE (25°C)	142 (P)
Polyethylene, HDPE (20°C)	152 (P)
TPE, Urethane (TPAU) (20°C)	175 (P)
TPE, Urethane (TPEU) (20°C)	175 (P)
TPE, Urethane (TPAU) (20°C)	219 (P)
TPE, Urethane (TPAU) (20°C)	263 (P)
TPE, Urethane (TPEU) (20°C)	263 (P)
TPE, Polyester (21.5°C)	276 (P)
TPE, Polyamide (23°C)	302 (P)
TPE, Urethane (TPAU) (20°C)	306 (P)
TPE, Urethane (TPEU) (20°C)	350 (P)
Fluoropolymer, ETFE (25°C)	354 (P)
TPE, Urethane (TPEU) (20°C)	438 (P)

TPE, Polyamide (23°C)	460 (P)
Fluoropolymer, ETFE (23°C)	591 (P)
Polyvinyl Chloride (25°C)	639 (P)
Fluoroelastomer, FKM (24°C)	771 (P)
Polysulfone (23°C)	772 (P)
TPE, Polyester (21.5°C)	855 (P)
TPE, Polyamide (23°C)	965 (P)
TPE, Polyamide (23°C)	1142 (P)
TPE, Polyester (21.5°C)	1356 (P)
TPE, Polyamide (23°C)	1543 (P)

>25 - 50°C

EVOH (35°C)	5.4 (P)
EVOH (35°C)	9.4 (P)
Polyvinylidene Chloride (35°C)	10.8 (P)
EVOH (35°C)	14.0 (P)
Nylon 6 (35°C)	45.7 (P)
Polyimide (30°C)	55.9 (P)
Polyester, PET (35°C)	70.9 (P)
Polyethylene, HDPE (30°C)	213 (P)
Polyethylene, HDPE (50°C)	466 (P)

>100°C

Fluoroelastomer, FKM (121°C)	15,034 (P)
Fluoroelastomer, FKM (204°C)	57,888 (P)

hexene

>25 - 50°C

Polyester, PET (40°C)	0.05 (V)

hydrochloric acid

20 - 25°C

Polyphenylene Sulfide (23°C)	0.03 (V)

hydrogen

<0°C

Fluoropolymer, CTFE (-16°C)	5.1 (P)
Fluoropolymer, CTFE (-15°C)	5.6 (P)
Fluoropolymer, CTFE (-12°C)	5.9 (P)
Fluoropolymer, ECTFE (-20°C)	10.3 (P)
Fluoropolymer, ECTFE (-21°C)	10.3 (P)
Fluoropolymer, ECTFE (-22°C)	10.4 (P)
Polyethylene, HDPE (-18°C)	27.9 (P)
Polyethylene, HDPE (-16°C)	30.6 (P)
Polyethylene, HDPE (-15°C)	31.9 (P)
Fluoropolymer, FEP (-16°C)	76.8 (P)
Fluoropolymer, FEP (-15°C)	79.3 (P)
Fluoropolymer, FEP (-13°C)	84.4 (P)
Fluoropolymer, TFE (-18°C)	139 (P)
Fluoropolymer, TFE (-17°C)	143 (P)
Fluoropolymer, TFE (-16°C)	149 (P)
Fluoropolymer, TFE (-15°C)	346 (P)
Fluoropolymer, TFE (-14°C)	365 (P)
Fluoropolymer, TFE (-11°C)	395 (P)

0 - <20°C

Fluoropolymer, CTFE (0°C)	21.0 (P)

20 - 25°C

Fluoropolymer, PVDF (23°C)	21.3 (P)
Fluoropolymer, CTFE (25°C)	35.6 (P)
Fluoropolymer, CTFE (25°C)	36.2 (P)
Polyester, PET (25°C)	39.4 (P)
Epoxy (25°C)	43.3 (P)
Parylene (25°C)	43.3 (P)
Fluoropolymer, CTFE (25°C)	64.4 (P)
Polypyrrole	69.9 (P)

Parylene (25°C)	94.5 (P)
Fluoropolymer, ECTFE (25°C)	106 (P)
Fluoropolymer, ECTFE (25°C)	108 (P)
Fluoropolymer, ECTFE (25°C)	109 (P)
Polyethylene, HDPE (23°C)	126 (P)
Polyethylene, HDPE (25°C)	154 (P)
Polyethylene, HDPE (25°C)	156 (P)
Polyethylene, HDPE (25°C)	161 (P)
Polyphenylene Sulfide	165 (P)
TPE, Urethane (TPAU) (20°C)	175 (P)
Parylene (25°C)	213 (P)
Polyethylene, HDPE (20°C)	223 (P)
Polyethylene, HDPE (25°C)	243 (P)
TPE, Urethane (TPAU) (20°C)	263 (P)
TPE, Urethane (TPAU) (20°C)	350 (P)
TPE, Urethane (TPEU) (20°C)	350 (P)
Fluoropolymer, FEP (25°C)	381 (P)
Fluoropolymer, FEP (25°C)	385 (P)
Fluoropolymer, FEP (25°C)	386 (P)
TPE, Urethane (TPAU) (20°C)	394 (P)
TPE, Urethane (TPEU) (20°C)	438 (P)
ASA (23°C)	507 (P)
Fluoropolymer, TFE (25°C)	516 (P)
Fluoropolymer, TFE (25°C)	520 (P)
TPE, Urethane (TPEU) (20°C)	525 (P)
Fluoropolymer, TFE (25°C)	555 (P)
TPE, Urethane (TPEU) (20°C)	613 (P)
Polysulfone (23°C)	709 (P)
Fluoropolymer, TFE (25°C)	1077 (P)
Fluoropolymer, TFE (25°C)	1112 (P)
Fluoropolymer, TFE (25°C)	1173 (P)
Silicone (25°C)	17,716 (P)

>25 - 50°C

Fluoropolymer, CTFE (50°C)	158 (P)
Polyethylene, HDPE (30°C)	294 (P)
Polyethylene, HDPE (40°C)	446 (P)
Polyethylene, HDPE (50°C)	679 (P)

>50 - 75°C

Fluoropolymer, CTFE (67°C)	197 (P)
Fluoropolymer, CTFE (68°C)	204 (P)
Fluoropolymer, CTFE (70°C)	218 (P)
Fluoropolymer, ECTFE (66°C)	576 (P)
Fluoropolymer, ECTFE (67°C)	582 (P)
Fluoropolymer, ECTFE (68°C)	590 (P)
Polyethylene, HDPE (67°C)	740 (P)
Polyethylene, HDPE (67°C)	748 (P)
Polyethylene, HDPE (68°C)	761 (P)
Fluoropolymer, TFE (63°C)	1436 (P)
Fluoropolymer, FEP (67°C)	1550 (P)
Fluoropolymer, FEP (67°C)	1576 (P)
Fluoropolymer, TFE (67°C)	1628 (P)
Fluoropolymer, FEP (68°C)	1637 (P)
Fluoropolymer, TFE (68°C)	1646 (P)
Fluoropolymer, TFE (65°C)	2906 (P)
Fluoropolymer, TFE (67°C)	2994 (P)
Fluoropolymer, TFE (68°C)	3090 (P)

hydrogen sulfide

20 - 25°C

Polyphenylene Sulfide	1.2 (P)
Fluoropolymer, PVDF (23°C)	1.5 (P)

>25 - 50°C

Fluoropolymer, CTFE (50°C)	2.3 (P)

<u>>50 - 75°C</u>

Fluoropolymer, CTFE (75°C)   13.1 (P)

kerosine

<u>20 - 25°C</u>

EVOH (20°C)	>0.0004 (V)
EVOH (20°C)	0.0004 (V)
EVOH (20°C)	<0.0007 (V)
Nylon (20°C)	<0.0007 (V)
EVOH (20°C)	<0.0009 (V)
EVOH (20°C)	0.00098 (V)
EVOH (20°C)	<0.0015 (V)
LDPE/ EVAL Film (20°C)	<0.0035 (V)
LDPE/ EVAL Film (20°C)	<0.004 (V)
Nylon (20°C)	0.01 (V)
Polyester, PET (20°C)	0.01 (V)
Nylon 66	0.08 (V)
Polypropylene (20°C)	1.1 (V)
Polyethylene, LDPE (20°C)	3.8 (V)
Polyethylene, LDPE (20°C)	3.9 (V)

methane

<u>20 - 25°C</u>

Polypyrrole	0.41 (P)
Fluoropolymer, ETFE (23°C)	7.9 (P)
ASA (23°C)	10.1 (P)
ASA (23°C)	11.1 (P)
Polysulfone (23°C)	14.8 (P)
TPE, Urethane (TPAU) (20°C)	17.5 (P)
TPE, Urethane (TPAU) (20°C)	35.0 (P)
TPE, Urethane (TPEU) (20°C)	43.8 (P)
TPE, Urethane (TPAU) (20°C)	52.5 (P)
Polyethylene, HDPE (20°C)	56.7 (P)
TPE, Urethane (TPEU) (20°C)	78.8 (P)
TPE, Urethane (TPAU) (20°C)	96.3 (P)
TPE, Urethane (TPEU) (20°C)	123 (P)
TPE, Urethane (TPEU) (20°C)	158 (P)

methyl alcohol

<u>20 - 25°C</u>

Polyphenylene Sulfide (23°C)	0.12 (V)
Polystyrene (24°C)	0.39 - 2.4 (V)
Polyethylene, LDPE (24°C)	2.4 - 3.2 (V)

methyl ethyl ketone

<u>20 - 25°C</u>

LDPE/ EVAL Film (20°C)	0.0035 (V)
EVOH (20°C)	0.0047 (V)
EVOH (20°C)	0.01 (V)
LDPE/ EVAL Film (20°C)	0.01 (V)
Nylon (20°C)	0.02 (V)
Polyester, PET (20°C)	0.04 (V)
Nylon (20°C)	0.07 (V)
EVOH (20°C)	0.08 (V)
EVOH (20°C)	0.1 (V)
Polypropylene (20°C)	0.24 (V)
Polyethylene, LDPE (20°C)	3.8 (V)

methyl salicylate

<u>20 - 25°C</u>

Nylon 66	0.08 (V)
Acetal (23°C)	0.12 (V)

mineral oils

<u>20 - 25°C</u>

Acetal (23°C)	0 (V)

<u>>25 - 50°C</u>

Acetal (38°C)	0 (V)

motor oils

<u>20 - 25°C</u>

Acetal (23°C)	0 (V)
Nylon 66	0.08 (V)

<u>>25 - 50°C</u>

Acetal (38°C)	0 (V)

n-heptane

<u>20 - 25°C</u>

SAN (24°C)	0.79 - 7.9 (V)
Polyethylene, LDPE (24°C)	118 - 197 (V)

naphtha

<u>20 - 25°C</u>

Nylon 66	2.4 (V)

natural gas

<u>20 - 25°C</u>

Polyethylene, HDPE (23°C)	44.5 (P)

nitogen

<u>>25 - 50°C</u>

nitrogen

<u><0°C</u>

Polyethylene, HDPE (-19°C)	0.95 (P)
Polyethylene, HDPE (-17°C)	0.99 (P)
Polyethylene, HDPE (-10°C)	1.6 (P)
Fluoropolymer, FEP (-9°C)	4.4 (P)
Fluoropolymer, FEP (-7°C)	4.9 (P)
Fluoropolymer, FEP (-5°C)	5.6 (P)
Fluoropolymer, TFE (-25°C)	7.8 (P)
Fluoropolymer, TFE (-23°C)	8.3 (P)
Fluoropolymer, TFE (-17°C)	20.5 (P)
Fluoropolymer, TFE (-14°C)	21.9 (P)

<u>0 - <20°C</u>

Nylon 6 (0°C)	0.08 (P)
Fluoropolymer, ECTFE (11°C)	0.48 (P)
Fluoropolymer, ECTFE (10°C)	0.48 (P)
Fluoropolymer, ECTFE (10°C)	0.53 (P)

<u>20 - 25°C</u>

EVOH (23°C)	0.00039 (P)
EVOH (23°C)	0.0016 (P)
EVOH (23°C)	0.0031 (P)
Polyvinylidene Chloride (23°C)	0.0047 (P)
Fluoropolymer, CTFE (25°C)	0.02 (P)
Polyvinylidene Chloride (23°C)	0.02 (P)
Nylon 6 (23°C)	0.04 (P)
HDPE/ EVA/ PVDC/ EVA Film (23°C)	0.06 (P)
LDPE/ EVA/ PVDC/ EVA/ LDPE Film (23°C)	0.06 (P)
Polyvinylidene Chloride (24°C)	0.04 - 0.08 (P)
Polyvinylidene Chloride (23°C)	0.07 (P)
Acrylonitrile Copolymer, AMA (22.8°C)	0.08 (P)

LDPE/ EVA/ PVDC/ EVA/ LDPE Film (23°C)	0.08 (P)
LDPE/ EVA/ PVDC/ EVA/ LDPE Film (23°C)	0.09 (P)
Polyvinylidene Chloride (23°C)	0.09 (P)
Fluoropolymer, PVF (25°C)	0.1 (P)
LDPE/ EVA/ PVDC/ EVA/ LDPE Film (23°C)	0.11 (P)
LDPE/ EVA/ PVDC/ EVA Film (23°C)	0.12 (P)
PE/ PVC-PVDC Copolymer Multilayer Film (24°C)	0.04 - 0.2 (P)
Polyvinylidene Chloride (23°C)	0.12 (P)
Acrylonitrile Copolymer, AMA (22.8°C)	0.16 (P)
Polyester, PET (23°C)	0.18 (P)
Nylon 6/66 (23°C)	0.2 (P)
Nylon 6 (23°C)	0.28 (P)
Nylon 66 (23°C)	0.28 (P)
PVC-PVDC Copolymer (25°C)	0.05 - 0.59 (P)
Fluoropolymer, CTFE (25°C)	0.33 (P)
Polyester, PET (25°C)	0.28 - 0.39 (P)
Nylon 6 (23°C)	0.35 (P)
Parylene (25°C)	0.39 (P)
Polyester, PET (25°C)	0.39 (P)
Polypyrrole	0.41 (P)
Nylon, Amorphous (23°C)	0.51 (P)
Nylon 66/610 (23°C)	0.61 (P)
Fluoropolymer, CTFE (25°C)	0.98 (P)
Acetal Copolymer	0.87 - 1.3 (P)
Fluoropolymer, ECTFE (25°C)	1.1 (P)
Fluoropolymer, ECTFE (25°C)	1.2 (P)
Fluoropolymer, ECTFE (25°C)	1.3 (P)
Epoxy (25°C)	1.6 (P)
Parylene (25°C)	1.8 (P)
Polyester, PCTG (23°C)	3 (P)
Parylene (25°C)	3.03 (P)
Fluoropolymer, PVDF (23°C)	3.0 (P)
Polyester, PBT (23°C)	3.0 (P)
Fluoropolymer, PVDF (25°C)	3.5 (P)
Polyethylene, LLDPE	3.8 (P)
Polyester, PETG (23°C)	3.9 (P)
SAN (24°C)	3.9 (P)
Fluoroelastomer, FKM (24°C)	4.7 (P)
ABS (24°C)	3.9 - 5.9 (P)
ASA (23°C)	6.1 (P)
ASA (23°C)	7.1 (P)
ASA (23°C)	7.6 (P)
TPE, Urethane (TPAU) (20°C)	8.8 (P)
ASA (23°C)	10.1 (P)
ABS (23°C)	10.1 (P)
Polycarbonate	10.6 (P)
Polycarbonate	11.2 (P)
ABS (24°C)	9.8 - 13.8 (P)
Fluoropolymer, ETFE (25°C)	11.8 (P)
Polycarbonate	12.2 (P)
Polyethylene, HDPE (25°C)	14.0 (P)
Rubber, Chlorohydrin (CO) (21.1°C)	14.7 (P)
Polyethylene, HDPE (25°C)	14.7 (P)
Polyethylene, HDPE (25°C)	15.5 (P)
Rubber, Nitrile (NBR) (21.1°C)	15.6 (P)
Polysulfone (23°C)	15.8 (P)
Polyethylene, HDPE (25°C)	16.5 (P)
Polypropylene (25°C)	16.5 (P)
Fluoroelastomer, FKM (21.1°C)	17.3 (P)
TPE, Urethane (TPAU) (20°C)	17.5 (P)
Polystyrene (24°C)	15.8 - 19.7 (P)
Polystyrene, GP (23°C)	15.8 - 19.7 (P)
Polystyrene, IPS (23°C)	15.7 - 19.7 (P)
Polyethylene, HDPE (20°C)	18.2 (P)
Polyethylene, HDPE (24°C)	15.8 - 23.6 (P)
ABS (23°C)	20.3 (P)
Polyethylene, HDPE (23°C)	20.9 (P)
Polyethylene, HDPE (25°C)	21.3 (P)
Rubber, Butyl (IIR) (21.1°C)	21.6 (P)
Polystyrene (24°C)	19.7 - 23.6 (P)
Polystyrene, GP (23°C)	19.7 - 23.6 (P)
Fluoropolymer, ETFE (23°C)	21.7 (P)
Polycarbonate	22.4 (P)
Polystyrene, GP (23°C)	25.3 (P)
TPE, Polyester (23°C)	25.6 (P)
TPE, Urethane (TPAU) (20°C)	26.3 (P)
TPE, Urethane (TPEU) (20°C)	26.3 (P)
Polyurethane (25°C)	31.5 (P)
TPE, Polyamide (23°C)	33 (P)
Fluoropolymer, FEP (25°C)	33.3 (P)
Fluoropolymer, FEP (25°C)	33.7 (P)
Fluoropolymer, FEP (25°C)	33.8 (P)
TPE, Urethane (TPAU) (20°C)	35.0 (P)
TPE, Urethane (TPEU) (20°C)	35.0 (P)
Styrene-Butadiene Block Copolymer (23°C)	35 (P)
Rubber, Nitrile (NBR) (21.1°C)	39.7 (P)
Polystyrene, IPS (23°C)	40.5 (P)
Polybutylene (22.8°C)	43.3 (P)
TPE, Urethane (TPEU) (20°C)	43.8 (P)
Styrene-Butadiene Block Copolymer (23°C)	45.6 (P)
TPE, Urethane (TPEU) (20°C)	52.5 (P)
Rubber, Chlorohydrin (ECO) (21.1°C)	57 (P)
Rubber, Polysulfide (T) (21.1°C)	57 (P)
Polyethylene, LDPE (24°C)	39.4 - 78.7 (P)
Fluoropolymer, TFE (25°C)	68.6 (P)
Fluoropolymer, TFE (25°C)	68.9 (P)
Fluoropolymer, TFE (25°C)	69.0 (P)
Rubber, Chlorosulf. PE (CSM) (21.1°C)	60.5 - 77.8 (P)
Polyethylene, LDPE (25°C)	70.9 (P)
Polyvinyl Chloride (25°C)	70.9 (P)
Styrene-Butadiene Block Copolymer (23°C)	70.9 (P)
TPE, Polyamide (23°C)	72 (P)
Rubber, EACM (21.1°C)	76 (P)
Rubber, Neoprene (CR) (21.1°C)	77 (P)
Rubber, Propylene Oxide (PO) (21.1°C)	77.8 (P)
Polyethylene, MDPE (25°C)	33.5 - 124 (P)
EU (21.1°C)	82.1 (P)
TPE, Olefinic (23°C)	93 (P)
Fluoropolymer, PFA (25°C)	115 (P)
TPE, Polyester (21.5°C)	121 (P)
Fluoropolymer, TFE (25°C)	124 (P)
Fluoropolymer, FEP (25°C)	126 (P)
Fluoropolymer, TFE (25°C)	128 (P)
Fluoropolymer, TFE (25°C)	133 (P)
TPE, Polyester (21.5°C)	147 (P)
TPE, Olefinic (23°C)	194 (P)
TPE, Polyamide (23°C)	256 (P)
TPE, Olefinic (23°C)	264 (P)
Polymethylpentene (23°C)	400 (P)
Rubber, Styrene Butadiene (21.1°C)	415 (P)
Polymethylpentene (23°C)	470 (P)
Polymethylpentene (23°C)	475 (P)
Polyisoprene Rubber (21.1°C)	529 (P)
Rubber, Latex (NR) (21.1°C)	529 (P)
Rubber, EPDM (21.1°C)	553 (P)
Rubber, EPM (21.1°C)	553 (P)
Rubber, EPDM (23°C)	533 - 648 (P)
TPE, Polyamide (23°C)	657 (P)
Rubber, Polybutadiene (24°C)	787 (P)
TPE, Polyamide (23°C)	1116 (P)
AU (21.1°C)	1382 (P)
Rubber, Polybutadiene (21.1°C)	1728 (P)
Silicone, FVMQ (21.1°C)	14,256 (P)
Silicone (21.1°C)	17,280 (P)

>25 - 50°C

EVOH (35°C)	0.00079 (P)
EVOH (35°C)	0.0031 (P)
EVOH (35°C)	0.01 (P)
Polyimide (30°C)	0.76 (P)
Fluoropolymer, CTFE (50°C)	2.0 (P)
Nylon 6 (50°C)	4.7 (P)
Polyethylene, HDPE (30°C)	29.4 (P)
Polyethylene, HDPE (40°C)	48.6 (P)
Polyethylene, HDPE (50°C)	85.1 (P)

>50 - 75°C

Fluoropolymer, CTFE (68°C)	3.6 (P)
Fluoropolymer, CTFE (69°C)	3.8 (P)
Fluoropolymer, CTFE (70°C)	3.9 (P)
Fluoropolymer, CTFE (75°C)	6.0 (P)
Fluoropolymer, ECTFE (71°C)	21.3 (P)
Fluoropolymer, ECTFE (68°C)	21.7 (P)
Fluoropolymer, ECTFE (72°C)	37.4 (P)
Polyethylene, HDPE (69°C)	128 (P)
Polyethylene, HDPE (68°C)	150 (P)
Polyethylene, HDPE (72°C)	173 (P)
Fluoropolymer, TFE (68°C)	251 (P)
Fluoropolymer, TFE (70°C)	253 (P)
Fluoropolymer, TFE (71°C)	254 (P)
Fluoropolymer, FEP (71°C)	332 (P)
Fluoropolymer, FEP (68°C)	333 (P)
Fluoropolymer, FEP (66°C)	337 (P)
Fluoropolymer, TFE (67°C)	418 (P)
Fluoropolymer, TFE (68°C)	462 (P)
Fluoropolymer, TFE (71°C)	466 (P)

nitrous oxide

20 - 25°C

Fluoropolymer, PVDF (23°C)	22.8 (P)

oxygen

≤0°C

Fluoropolymer, ECTFE (-18°C)	0.48 (P)
Fluoropolymer, ECTFE (-15°C)	0.5 (P)
Polyethylene, HDPE (-16°C)	5.0 (P)
Polyethylene, HDPE (-15°C)	5.2 (P)
Fluoropolymer, FEP (-16°C)	9.0 (P)
Fluoropolymer, FEP (-16°C)	9.1 (P)
Fluoropolymer, TFE (-17°C)	39.8 (P)
Fluoropolymer, TFE (-17°C)	46.1 (P)
Fluoropolymer, TFE (-16°C)	81.2 (P)
Fluoropolymer, TFE (-15°C)	83.7 (P)

0 - <20°C

EVOH (5°C)	0.00039 (P)
EVOH (5°C)	0.0012 (P)
EVOH (5°C)	0.0024 (P)
Polyvinylidene Chloride (5°C)	0.0024 (P)
Polyvinylidene Chloride (5°C)	0.0047 (P)
EVOH (5°C)	0.01 (P)
Nylon MXD6 (5°C)	0.02 (P)
EVOH (5°C)	0.03 (P)
Acrylonitrile Copolymer, AMA (5°C)	0.06 (P)
Nylon 6 (5°C)	0.19 (P)
Nylon 6 (0°C)	0.2 (P)
Polyester, PET (5°C)	0.26 (P)
Fluoropolymer, CTFE (0°C)	0.46 (P)
Nylon 6 (5°C)	0.57 (P)

20 - 25°C

Multilayer, LIM Structure	non-detectable (P)

EVOH (20°C)	0.0002 (P)
EVOH (23°C)	0.0024 (P)
EVOH (20°C)	0.0024 (P)
EVOH (20°C)	0.0026 (P)
EVOH (20°C)	0.003 (P)
EVOH (20°C)	0.0031 (P)
EVOH (20°C)	0.004 (P)
EVOH (23°C)	0.01 (P)
EVOH (20°C)	0.01 (P)
Polyvinylidene Chloride (23°C)	0.01 (P)
PP/ EVAL/ LDPE Film (20°C)	0.01 (P)
Acrylonitrile (24°C)	0.02 (P)
EVAL/ LDPE Film (20°C)	0.02 (P)
EVOH (23°C)	0.02 (P)
EVOH (20°C)	0.02 (P)
Liquid Crystal Polymer (23°C)	0.02 (P)
PET/ EVAL/ LDPE Film (20°C)	0.02 (P)
Polyvinyl Alcohol (24°C)	0.02 (P)
Polyvinyl Alcohol	0.02 (P)
EVOH (20°C)	0.024 (P)
EVOH (20°C)	0.026 (P)
EVOH (20°C)	0.028 (P)
EVOH (20°C)	0.03 (P)
Liquid Crystal Polymer (23°C)	0.03 (P)
Polyvinylidene Chloride (22.8°C)	0.03 (P)
Polyvinylidene Chloride (23°C)	0.03 (P)
Polyvinylidene Chloride (20°C)	0.03 (P)
Polyvinylidene Chloride (23°C)	0.03 (P)
PP/ EVAL/ LDPE Film (20°C)	0.03 (P)
PP/ EVAL/ PP Film (20°C)	0.03 (P)
EVAL/ LDPE Film (20°C)	0.04 (P)
EVOH (20°C)	0.04 (P)
Nylon/ EVAL/ LDPE Film (20°C)	0.04 (P)
Polyvinylidene Chloride (23°C)	0.04 (P)
EVOH (23°C)	0.05 (P)
EVOH (20°C)	0.05 (P)
Nylon/ EVAL/ LDPE Film (20°C)	0.05 (P)
PET/ EVAL/ LDPE Film (20°C)	0.05 (P)
EVOH (20°C)	0.06 (P)
Nylon MXD6 (23°C)	0.06 (P)
Polyvinyl Alcohol (20°C)	0.06 (P)
Polyvinylidene Chloride (23°C)	0.06 (P)
EVOH (20°C)	0.07 (P)
Nylon MXD6 (22.8°C)	0.07 (P)
EVOH (20°C)	0.079 (P)
Acrylonitrile Co. (AMA) (24°C)	0.08 (P)
EVOH (20°C)	0.08 (P)
Nylon/ EVAL/ LDPE Film (20°C)	0.08 (P)
EVOH (20°C)	0.09 (P)
Polyvinyl Alcohol (24°C)	0.09 (P)
Cellulosic Plastic (20°C)	0.1 (P)
EVOH (20°C)	0.1 (P)
Nylon/ EVAL/ LDPE Film (20°C)	0.1 (P)
Polyvinylidene Chloride (23°C)	0.1 (P)
PP/ EVAL/ PP Film (20°C)	0.11 (P)
PET/ EVAL/ LDPE Film (20°C)	0.12 (P)
Nylon (20°C)	0.14 (P)
Nylon/ EVAL/ LDPE Film (20°C)	0.14 (P)
EVOH (20°C)	0.15 (P)
PP/ EVAL/ LDPE Film (20°C)	0.15 (P)
PP/ EVAL/ PET Film (20°C)	0.15 (P)
PP/ EVAL/ PS Film (20°C)	0.15 (P)
EVOH (20°C)	0.16 (P)
Multilayer, LIM Structure	0.16 (P)
PP/ EVAL/ HDPE Film (20°C)	0.16 (P)
PP/ EVAL/ LDPE Film (20°C)	0.16 (P)
PP/ EVAL/ PP Film (20°C)	0.16 (P)

LDPE/ EVA/ PVDC/ EVA/ LDPE Film (20°C)	0.18 (P)
PET/ EVAL/ PP Film (20°C)	0.18 (P)
EVOH (20°C)	0.2 (P)
Nylon (20°C)	0.2 (P)
Nylon 6 (23°C)	0.2 (P)
Nylon 66 (23°C)	0.2 (P)
Nylon/ EVAL/ LDPE Film (20°C)	0.2 (P)
Polyester, PET (20°C)	0.2 (P)
Nylon/ EVAL/ LDPE Film (20°C)	0.21 (P)
Polypropylene (20°C)	0.22 (P)
EVOH (20°C)	0.24 (P)
Multilayer, LIM Structure	0.24 (P)
Fluoropolymer, CTFE (25°C)	0.25 (P)
PP/ EVAL/ PP Film (20°C)	0.25 (P)
EVOH (20°C)	0.26 (P)
Fluoropolymer, CTFE (25°C)	0.26 (P)
PC/ EVAL/ PP Film (20°C)	0.26 (P)
Polypropylene (20°C)	0.26 (P)
PS/ EVAL/ PP Film (20°C)	0.26 (P)
Nylon 6 (20°C)	0.24 - 0.3 (P)
Cellulosic Plastic (20°C)	0.28 (P)
EVOH (20°C)	0.28 (P)
Nylon (20°C)	0.28 (P)
Nylon 66 (23°C)	0.29 (P)
EVOH (20°C)	0.30 (P)
Polyimide	0.3 (P)
Acrylonitrile Copolymer, AMA (23°C)	0.31 (P)
Acrylonitrile Copolymer, AMA (22.8°C)	0.32 (P)
Nylon MXD6 (22.8°C)	0.32 (P)
PP/ EVAL/ LDPE Film (20°C)	0.32 (P)
PP/ EVAL/ PP Film (20°C)	0.34 (P)
Polyvinyl Alcohol (20°C)	0.35 (P)
Nylon 66 (23°C)	0.3 - 0.41 (P)
Polyvinylidene Chloride (24°C)	0.31 - 0.43 (P)
EVOH (20°C)	0.39 (P)
EVOH (22.8°C)	0.39 (P)
LDPE/ EVA/ PVDC/ EVA/ LDPE Film (20°C)	0.39 (P)
Multilayer, LIM Structure	0.39 (P)
HDPE/ EVA/ PVDC/ EVA Film (23°C)	0.4 (P)
LDPE/ EVA/ PVDC/ EVA Film (23°C)	0.4 (P)
LDPE/ EVA/ PVDC/ EVA/ LDPE Film (23°C)	0.4 (P)
Nylon, Amorphous (23°C)	0.41 (P)
HDPE/ EVAL/ PP Film (20°C)	0.42 (P)
Laminar, EVOH/HDPE (23°C)	0.42 (P)
PC/ EVAL/ PP Film (20°C)	0.42 (P)
PS/ EVAL/ PP Film (20°C)	0.42 (P)
Polypropylene (20°C)	0.43 (P)
Polyvinylidene Chloride (23°C)	0.43 (P)
Nylon, Amorphous (22.8°C)	0.47 (P)
Polyvinylidene Chloride (23°C)	0.47 (P)
PE/ PVC-PVDC Copolymer Multilayer Film (24°C)	0.2 - 0.79 (P)
PET/ EVAL/ PP Film (20°C)	0.52 (P)
Fluoropolymer, PVDF (25°C)	0.55 (P)
Polyvinyl Alcohol	0.56 (P)
PP/ EVAL/ PP Film (20°C)	0.58 (P)
LDPE/ EVA/ PVDC/ EVA/ LDPE Film (23°C)	0.59 (P)
HDPE/ EVAL/ LDPE Film (20°C)	0.6 (P)
PP/ EVAL/ PC Film (20°C)	0.6 (P)
PP/ EVAL/ PC Film (20°C)	0.62 (P)
EVOH (20°C)	0.63 (P)
Polyvinylidene Chloride (23°C)	0.63 (P)
Acrylonitrile Copolymer, AMA (22.8°C)	0.64 (P)
Nylon 6 (23°C)	0.61 - 0.71 (P)
Nylon 66 (23°C)	0.61 - 0.71 (P)
Nylon 6 (20°C)	0.57 - 0.8 (P)
Nylon 6 (23°C)	0.7 (P)
HDPE/ EVAL/ PP Film (20°C)	0.71 (P)
Polyvinylidene Chloride (23°C)	0.71 (P)
Polyvinyl Alcohol	0.74 (P)
Nylon (20°C)	0.76 (P)
Nylon 66 (20°C)	0.76 (P)
EVOH (20°C)	0.79 (P)
LDPE/ EVA/ PVDC/ EVA/ LDPE Film (23°C)	0.79 (P)
Nylon 66 (23°C)	0.79 (P)
Cellulosic Plastic (20°C)	0.81 (P)
EVOH (20°C)	0.83 (P)
Nylon 6/66 (23°C)	0.81 - 0.91 (P)
EVOH (20°C)	0.87 (P)
LDPE/ EVA/ PVDC/ EVA/ LDPE Film (23°C)	0.88 (P)
Polyvinyl Alcohol	0.88 (P)
Multilayer, LIM Structure	0.91 (P)
Polyester, PET (23°C)	0.91 (P)
Nylon 6 (23°C)	0.94 (P)
Nylon 6/66	0.94 (P)
PC/ EVAL/ PP Film (20°C)	0.97 (P)
Nylon (20°C)	0.98 (P)
Nylon, Amorphous (22.8°C)	0.98 (P)
Nylon/ EVAL/ PP Film (20°C)	1 (P)
EVOH (20°C)	1.02 (P)
Nylon 6 (23°C)	1.02 (P)
LDPE/ EVAL/ PP Film (20°C)	1.1 (P)
Polyester, PET (20°C)	1.1 (P)
PP/ EVAL/ PP Film (20°C)	1.2 (P)
Fluoropolymer, PVF (25°C)	1.2 (P)
Nylon 6 (22.8°C)	1.2 (P)
Nylon (20°C)	1.3 (P)
Nylon/ EVAL/ PP Film (20°C)	1.3 (P)
PC/ EVAL/ PP Film (20°C)	1.3 (P)
Nylon 66 (23°C)	1.4 (P)
Nylon 6 (22.8°C)	1.4 (P)
HDPE/ PVDC/ PP Film (20°C)	1.4 (P)
LDPE/ EVAL/ PP Film (20°C)	1.4 (P)
LDPE/ PVDC/ PP Film (20°C)	1.4 (P)
Nylon/ PVDC/ PP Film (20°C)	1.4 (P)
PC/ PVDC/ PP Film (20°C)	1.4 (P)
PET/ PVDC/ PP Film (20°C)	1.4 (P)
PP/ PVDC/ PP Film (20°C)	1.4 (P)
PS/ PVDC/ PP Film (20°C)	1.4 (P)
Nylon, Amorphous (23°C)	1.5 (P)
PVC-PVDC Copolymer (25°C)	0.31 - 2.7 (P)
PP/ EVAL/ PP Film (20°C)	1.6 (P)
HDPE/ PVDC/ LDPE Film (20°C)	1.8 (P)
Polyester, PET (25°C)	1.2 - 2.4 (P)
PP/ PVDC/ HDPE Film (20°C)	1.8 (P)
PP/ PVDC/ LDPE Film (20°C)	1.8 (P)
PP/ PVDC/ PC Film (20°C)	1.8 (P)
PP/ PVDC/ PET Film (20°C)	1.8 (P)
PP/ PVDC/ PP Film (20°C)	1.8 (P)
PP/ PVDC/ PS Film (20°C)	1.8 (P)
Nylon 6/ LDPE Film (20°C)	1.6 - 2.0 (P)
Multilayer, LIM Structure	1.8 (P)
Nylon, Amorphous (25°C)	1.97 (P)
Polyester, PET (25°C)	2.0 (P)
Nylon 6 (23°C)	2 (P)
Fluoropolymer, PVDF (23°C)	2.1 (P)
Nylon (20°C)	2.1 (P)
Polyester, PET (20°C)	2.3 (P)
Polypyrrole	2.3 (P)
Polyester, PET (25°C)	2.4 (P)
Acetal Copolymer	2.0 - 2.9 (P)
Polyester, PET (20°C)	2.5 (P)
Fluoropolymer, CTFE (25°C)	2.6 (P)
Fluoropolymer, CTFE (25°C)	2.8 (P)
Nylon 6 (22.8°C)	2.8 (P)

Appendix II - Permeation Rates

Nylon 66/610 (23°C)	2.8 (P)
Polyester, PET (22.8°C)	2.8 (P)
Parylene (25°C)	2.8 (P)
Epoxy (25°C)	2.0 - 3.9 (P)
Polyvinyl Chloride (22.8°C)	3.2 (P)
Nylon (20°C)	3.6 (P)
Nylon 66/610	3.8 (P)
Nylon 66/610 (23°C)	3.8 (P)
Polyester, PET (25°C)	3.9 (P)
Fluoropolymer, CTFE (25°C)	4.7 (P)
Polyvinyl Chloride (20°C)	4.8 (P)
Polyvinyl Chloride (24°C)	2.0 - 7.9 (P)
Polyvinyl Chloride (23°C)	5.2 (P)
Acetal (23°C)	4.7 - 6.7 (P)
Fluoropolymer, CTFE (25°C)	5.9 (P)
Nylon 6/66 (23°C)	5.9 (P)
Nylon 66 (23°C)	6.3 (P)
Polyphenylene Sulfide	5.9 - 7.9 (P)
Nylon (20°C)	7.5 (P)
Polyimide	8.0 (P)
Polyimide	8.7 (P)
HDPE/ EAA/ Nylon/ EAA Film (23°C)	9.5 (P)
Fluoropolymer, ECTFE (25°C)	9.6 (P)
Polyester, PETG (23°C)	9.8 (P)
Polyester, PETG (22.8°C)	10.0 (P)
Polyester, PCTG (23°C)	10 (P)
Fluoropolymer, ECTFE (25°C)	10.2 (P)
Polyphenylene Sulfide	11.8 (P)
Nylon (20°C)	12.4 (P)
Parylene (25°C)	12.6 (P)
Laminar, Nylon/HDPE (23°C)	13.4 (P)
Laminar, Nylon/PP (23°C)	14.2 (P)
Laminar, Nylon/HDPE (23°C)	15.0 (P)
ASA (23°C)	15.2 (P)
Polyester, PBT (23°C)	15.2 (P)
Polyvinyl Alcohol (20°C)	15.2 (P)
Parylene (25°C)	15.4 (P)
ASA (23°C)	18.2 (P)
SAN (24°C)	15.8 - 27.6 (P)
SAN (23°C)	20.3 - 30.4 (P)
TPE, Urethane (TPAU) (20°C)	35.0 (P)
Laminar, Nylon/LDPE (23°C)	35.4 (P)
SAN (24°C)	31.5 - 39.4 (P)
SAN (23°C)	20.3 - 50.7 (P)
Fluoropolymer, ETFE (25°C)	39.4 (P)
Polyethylene, HDPE (23°C)	40.6 (P)
Polyethylene, HDPE (23°C)	43.7 (P)
ABS (23°C)	45.6 (P)
Polyethylene, HDPE (25°C)	49.4 (P)
Polyethylene, HDPE (23°C)	49.6 (P)
Polyethylene, HDPE (25°C)	50.3 (P)
ABS (23°C)	50.7 (P)
ASA (23°C)	50.7 (P)
ABS (24°C)	47.2 - 55.1 (P)
Polypropylene (20°C)	53.2 (P)
ASA (23°C)	55.7 (P)
Polyethylene, HDPE (23°C)	55.9 (P)
Polyethylene, HDPE (24°C)	39.4 - 78.7 (P)
Polyethylene, HDPE (23°C)	59.1 (P)
TPE, Urethane (TPAU) (20°C)	61.3 (P)
Fluoropolymer, ETFE (23°C)	62.6 (P)
Polypropylene (23°C)	64.2 (P)
Polyethylene, HDPE (23°C)	66.0 (P)
Polycarbonate	67.9 (P)
Polyethylene, LDPE (20°C)	68.5 (P)
Polypropylene (25°C)	68.5 (P)
TPE, Urethane (TPEU) (20°C)	70.0 (P)

Polyethylene, HDPE (25°C)	72.8 (P)
Polyethylene, HDPE (20°C)	73.0 (P)
Polyethylene, HDPE (22.8°C)	>75.8 (P)
Polyethylene, HDPE (25°C)	77.0 (P)
Polyethylene, MDPE (23°C)	78.7 (P)
Polypropylene (23°C)	78.7 (P)
Polyurethane (25°C)	78.7 (P)
ABS (23°C)	81.1 (P)
TPE, Urethane (TPAU) (20°C)	87.5 (P)
ABS (24°C)	78.7 - 102 (P)
Polycarbonate	90.6 (P)
Polysulfone (23°C)	90.6 (P)
TPE, Vinyl	93 (P)
Polypropylene (22.8°C)	>99.7 (P)
Fluoropolymer, FEP (25°C)	101 (P)
Polystyrene, GP (23°C)	101 (P)
Polycarbonate (22.8°C)	102 (P)
Polycarbonate	102 (P)
Polystyrene (23°C)	102 (P)
TPE, Urethane (TPEU) (20°C)	105 (P)
Polypropylene (23°C)	107.1 (P)
Fluoropolymer, FEP (25°C)	116 (P)
Polyethylene, LDPE (24°C)	98.4 - 138 (P)
Polystyrene (24°C)	98.4 - 138 (P)
Polystyrene, GP (23°C)	98.4 - 138 (P)
TPE, Urethane (TPAU) (20°C)	123 (P)
Polycarbonate	124 (P)
TPE, Polyester (23°C)	127 (P)
Ionomer	130 (P)
Polyethylene, LLDPE (23°C)	132 (P)
Ionomer	134 (P)
Polystyrene (24°C)	118 - 157 (P)
Polystyrene, GP (23°C)	118 - 157 (P)
Polystyrene, IPS (23°C)	118 - 157 (P)
Ionomer	138 (P)
Polystyrene (22.8°C)	>140 (P)
TPE, Urethane (TPEU) (20°C)	140 (P)
Ionomer	142 (P)
Ionomer	146 (P)
Polyethylene, LDPE	150 (P)
Ionomer	150 (P)
Polybutylene (22.8°C)	152 (P)
Polyethylene, MDPE (25°C)	98.4 - 211 (P)
Polybutylene (22.8°C)	157 (P)
Ionomer	158 (P)
PE Ionomer Copolymer (23°C)	159 (P)
Polystyrene, IPS (23°C)	162 (P)
Styrene-Butadiene Block Copolymer (23°C)	162 (P)
Ionomer	170 (P)
Ionomer	174 (P)
Polyethylene, LDPE (23°C)	177 (P)
PE-Acrylic Acid Copolymer (23°C)	178 (P)
Polyethylene, PE/EVA Copolymer	180 (P)
TPE, Urethane (TPEU) (20°C)	184 (P)
Styrene-Butadiene Block Copolymer (23°C)	192 (P)
Polyethylene, LDPE (25°C)	197 (P)
Polyethylene, LLDPE	199 (P)
TPE, Polyamide (23°C)	204 (P)
Polyethylene, LLDPE	207 (P)
Ionomer	209 (P)
Polyethylene, LDPE (23°C)	218 (P)
Fluoropolymer, TFE (25°C)	222 (P)
Fluoropolymer, TFE (25°C)	223 (P)
Ionomer	229 (P)
TPE, Polyamide (23°C)	230 (P)
Ionomer	233 (P)
Polyethylene, LLDPE	236 (P)

Appendix II - Permeation Rates

Polyethylene, PE/EVA Copolymer	237 (P)
PE-Acrylic Acid Copolymer (23°C)	243 (P)
Polyethylene, ULDPE	256 (P)
Polyethylene, PE/EVA Copolymer	257 (P)
Styrene-Butadiene Block Copolymer (23°C)	263 (P)
Polyvinyl Chloride (25°C)	274 (P)
TPE, Vinyl	190 - 360 (P)
TPE, Olefinic (23°C)	279 (P)
Fluoropolymer, FEP (25°C)	295 (P)
Polyethylene, PE/EVA Copolymer	300 (P)
Fluoropolymer, PFA (25°C)	347 (P)
Polyethylene, PE/EVA Copolymer	356 (P)
Polyethylene, ULDPE	378 (P)
TPE, Polyamide (23°C)	387 (P)
TPE, Vinyl (24°C)	11.8 - 787 (P)
Polyethylene, PE/EVA Copolymer	419 (P)
Polyethylene, PE/EVA Copolymer	435 (P)
Fluoropolymer, TFE (25°C)	442 (P)
Fluoropolymer, TFE (25°C)	451 (P)
Polyethylene, PE/EVA Copolymer	470 (P)
TPE, Olefinic (23°C)	504 (P)
TPE, Polybutadiene	550 (P)
PE-Acrylic Acid Copolymer (23°C)	550 (P)
TPE, Olefinic (23°C)	589 (P)
TPE, Styrenic (23°C)	850 (P)
TPE, Polyamide (23°C)	860 (P)
TPE, Styrenic (23°C)	909 (P)
TPE, Polyamide (23°C)	985 (P)
TPE, Styrenic (23°C)	1059 (P)
TPE, Styrenic (23°C)	1102 (P)
Rubber, Styrene Butadiene (24°C)	1181 (P)
TPE, Styrenic (23°C)	1248 (P)
Polymethylpentene (23°C)	1600 (P)
TPE, Styrenic (23°C)	1646 (P)
TPE, Styrenic (23°C)	1717 (P)
Rubber, EPDM (23°C)	1641 - 1901 (P)
Polymethylpentene (23°C)	1900 (P)
TPE, Styrenic (23°C)	1902 (P)
Polymethylpentene (23°C)	2000 (P)
TPE, Styrenic (23°C)	2177 (P)
TPE, Styrenic (23°C)	3543 (P)
TPE, Styrenic (23°C)	6831 (P)
Silicone (25°C)	19,685 (P)

>25 - 50°C

EVOH (30°C)	0.01 (P)
EVOH (35°C)	0.01 (P)
Polyvinyl Alcohol (35°C)	0.01 (P)
EVOH (50°C)	0.02 (P)
EVOH (35°C)	0.02 (P)
Polyimide (30°C)	0.02 (P)
Cellulosic Plastic (35°C)	0.03 (P)
EVOH (50°C)	0.03 (P)
EVOH (35°C)	0.05 (P)
EVOH (30°C)	0.05 (P)
EVOH (30°C)	0.06 (P)
Liquid Crystal Polymer (38°C)	0.06 (P)
EVOH (50°C)	0.07 (P)
EVOH (35°C)	0.07 (P)
EVOH (35°C)	0.08 (P)
Nylon MXD6 (35°C)	0.11 (P)
Polyvinylidene Chloride (35°C)	0.11 (P)
EVOH (30°C)	0.13 (P)
EVOH (50°C)	0.14 (P)
Liquid Crystal Polymer (38°C)	0.14 (P)
EVOH (50°C)	0.16 (P)
Polyvinylidene Chloride (35°C)	0.17 (P)
Nylon MXD6 (50°C)	0.36 (P)

Nylon (35°C)	0.41 (P)
Polyester, PET (35°C)	0.43 (P)
Polypropylene (35°C)	0.48 (P)
Polyvinylidene Chloride (50°C)	0.5 (P)
Polyvinylidene Chloride (50°C)	0.75 (P)
Acrylonitrile Copolymer, AMA (35°C)	0.79 (P)
Nylon 6 (35°C)	1.3 (P)
Nylon (35°C)	1.6 (P)
Polyester, PET (35°C)	2.01 (P)
Acrylonitrile Copolymer, AMA (50°C)	2.41 (P)
Polyimide (30°C)	2.5 (P)
Nylon 6 (35°C)	3.94 (P)
Polyester, PET (35°C)	5.1 (P)
Fluoropolymer, PVDF (30°C)	5.2 (P)
Nylon 6 (50°C)	5.5 (P)
Polyester, PET (50°C)	6.61 (P)
Polyvinyl Chloride (35°C)	7.4 (P)
Fluoropolymer, CTFE (50°C)	9.2 (P)
Polypropylene (35°C)	79.9 (P)
Polypropylene (35°C)	89.0 (P)
Polyethylene, HDPE (30°C)	93.2 (P)
Fluoroelastomer, FKM (30°C)	95.0 (P)
Polyethylene, HDPE (35°C)	113 (P)
Polyethylene, HDPE (40°C)	142 (P)
Polyethylene, LDPE (35°C)	152 (P)
Polyethylene, HDPE (50°C)	233 (P)
Polyethylene, LDPE (35°C)	293 (P)

>50 - 75°C

Fluoropolymer, CTFE (52°C)	8.2 (P)
Fluoropolymer, CTFE (75°C)	37.4 (P)
Fluoropolymer, ECTFE (55°C)	45.2 (P)
Fluoropolymer, ECTFE (56°C)	46.1 (P)
Polyethylene, HDPE (52°C)	178 (P)
Polyethylene, HDPE (51°C)	218 (P)
Fluoropolymer, FEP (52°C)	452 (P)
Fluoropolymer, FEP (53°C)	465 (P)
Fluoropolymer, TFE (51°C)	471 (P)
Fluoropolymer, TFE (51°C)	478 (P)
Fluoropolymer, TFE (53°C)	884 (P)
Fluoropolymer, TFE (55°C)	1016 (P)

perchloroethylene

20 - 25°C

Acetal (23°C)	0.08 (V)

propane

20 - 25°C

TPE, Polyester (21.5°C)	<17.3 (P)
Polyethylene, HDPE (20°C)	35.5 (P)
TPE, Polyamide (23°C)	236 (P)
TPE, Polyamide (23°C)	789 (P)
TPE, Olefinic (23°C)	1162 (P)
TPE, Olefinic (23°C)	1938 (P)
TPE, Olefinic (23°C)	3332 (P)

propylene

20 - 25°C

Polyethylene, HDPE (20°C)	77.0 (P)

road oil remover

20 - 25°C

Acetal (23°C)	0.01 (V)

>25 - 50°C

Acetal (38°C)	0.07 (V)

shampoo

20 - 25°C

Acetal (23°C)	0.94 (V)

>25 - 50°C

Acetal (38°C)	3.4 (V)

sulfur dioxide

20 - 25°C

Fluoropolymer, PVDF (23°C)	1.5 (P)
Polyethylene, HDPE (20°C)	436 (P)

sulfur hexafluoride

20 - 25°C

Polysulfone (23°C)	0.71 (P)

tar remover

20 - 25°C

Acetal (23°C)	0.01 (V)

>25 - 50°C

Acetal (38°C)	0.07 (V)

tetrachloroethylene

20 - 25°C

Polyethylene, LDPE (24°C)	197 - 295 (V)

toluene

20 - 25°C

Nylon 66	0.08 (V)
Acetal (23°C)	0.24 (V)

trichloroethylene

20 - 25°C

Acetal (23°C)	9.8 (V)

>25 - 50°C

Acetal (38°C)	22.0 (V)

vegetable oils

20 - 25°C

Acetal (23°C)	0 (V)

>25 - 50°C

Acetal (38°C)	0 (V)

water

20 - 25°C

TPE, Polyester (25°C)	207,360 (P)
TPE, Polyester (25°C)	267,840 (P)
Polyphenylene Sulfide (23°C)	0.32 (V)
Nylon 66	1.2 - 2.4 (V)

water vapor

20 - 25°C

Fluoropolymer, CTFE (25°C)	0.005 (V)
Liquid Crystal Polymer (23°C)	0.03 (V)
Polyethylene, HDPE (20°C)	0.03 (V)
Polyethylene, HDPE (25°C)	0.04 (V)
Rubber, EPDM (23°C)	0.06 (V)
PE Ionomer Copolymer (23°C)	0.08 (V)
PE-Acrylic Acid Copolymer (23°C)	0.08 (V)
PE/ PVC-PVDC Copolymer Multilayer Film (24°C)	0.06 - 0.16 (V)
Polyvinylidene Chloride (24°C)	0.1 - 0.12 (V)
Polyethylene, HDPE (24°C)	0.16 (V)
TPE, Olefinic (25°C)	0.16 (V)
Nylon 6/ LDPE Film (20°C)	0.21 (V)
PE-Acrylic Acid Copolymer (23°C)	0.23 (V)
TPE, Olefinic (25°C)	0.23 (V)
Polyethylene, LDPE	0.23 (V)
Acrylonitrile (24°C)	0.24 (V)
Nylon 6 (23°C)	0.24 (V)
Polyethylene, LDPE	0.26 (V)
Polyethylene, LLDPE	0.28 (V)
Polyimide	0.3 (V)
Polyethylene, LDPE	0.31 (V)
Polyethylene, ULDPE	0.31 (V)
Acrylonitrile Co. (AMA) (24°C)	0.35 (V)
Nylon, Amorphous	0.35 (V)
Polyethylene, LLDPE	0.36 (V)
Nylon 66 (23°C)	0.39 (V)
Polyethylene, LDPE	0.46 (V)
Ionomer	0.47 (V)
Polyethylene, LLDPE	0.47 (V)
Polyethylene, ULDPE	0.47 (V)
Polyethylene, LDPE (24°C)	0.39 - 0.59 (V)
TPE, Olefinic (25°C)	0.49 (V)
Ionomer	0.51 (V)
Nylon, Amorphous (25°C)	0.51 (V)
Ionomer	0.55 (V)
Ionomer	0.59 (V)
Ionomer	0.63 (V)
Fluoropolymer, ETFE (25°C)	0.65 (V)
Polyphenylene Sulfide	0.65 (V)
PE-Acrylic Acid Copolymer (23°C)	0.68 (V)
Nylon 66/610 (23°C)	0.7 (V)
Ionomer	0.79 (V)
Nylon 66 (23°C)	0.8 (V)
TPE, Olefinic (25°C)	0.81 (V)
Ionomer	0.95 (V)
Nylon 6 (20°C)	0.8 - 1.2 (V)
Fluoropolymer, PVDF (23°C)	1.02 (V)
Styrene-Butadiene Block Copolymer (23°C)	1.1 (V)
Nylon 66 (23°C)	1.1 - 1.2 (V)
Polystyrene, GP (23°C)	1.2 (V)
Styrene-Butadiene Block Copolymer (23°C)	1.3 (V)
Polystyrene, IPS (23°C)	1.3 (V)
Styrene-Butadiene Block Copolymer (23°C)	1.4 (V)
Polyethylene, PE/EVA Copolymer	1.4 (V)
Nylon 6 (20°C)	1.1 - 1.8 (V)
Nylon 66 (20°C)	1.5 (V)
Polyester, PETG	1.5 (V)
Polycarbonate (23°C)	1.5 (V)
Nylon 6 (23°C)	1.5 - 1.6 (V)
Polyethylene, PE/EVA Copolymer	1.6 (V)
Nylon 6/66 (23°C)	1.5 - 1.8 (V)
Polyimide	1.67 (V)
Polyethylene, PE/EVA Copolymer	1.7 (V)
Polyester, PCTG	1.8 (V)
Acrylonitrile Copolymer, AMA (22.8°C)	2.0 (V)
Polyimide	2.13 (V)
SAN (23°C)	2 - 2.5 (V)
Polystyrene	0.79 - 3.9 (V)
Polystyrene, GP	0.79 - 3.9 (V)
Polystyrene, IPS	0.79 - 3.9 (V)

Polyester, PBT (23°C)	2.5 (V)
Polyethylene, LDPE	2.5 (V)
Polymethylpentene (23°C)	2.5 (V)
ABS (23°C)	2.7 (V)
Acrylonitrile Copolymer, AMA (22.8°C)	3.0 (V)
ASA (23°C)	3 (V)
Polymethylpentene (23°C)	3 (V)
ABS (23°C)	3.1 (V)
Polymethylpentene (23°C)	3.25 (V)
ASA (23°C)	3.5 (V)
Polystyrene (24°C)	3.5 (V)
Polystyrene, GP	3.5 (V)
SAN	1.97 - 5.51 (V)
Polyethylene, PE/EVA Copolymer	3.8 (V)
ABS	2.0 - 6.3 (V)
Polyethylene, PE/EVA Copolymer	4.5 (V)
Polyethylene, PE/EVA Copolymer	4.7 (V)
Polypyrrole	5.4 (V)
TPE, Polybutadiene	7 (V)
Nylon 66 (23°C)	7.5 (V)
Nylon 66 (23°C)	7.9 (V)
TPE, Polybutadiene	9.8 (V)
TPE, Vinyl	10 (V)
TPE, Vinyl	7.9 - 12.9 (V)
TPE, Styrenic (23°C)	47 (V)
TPE, Styrenic (23°C)	54.8 (V)
TPE, Styrenic (23°C)	62.4 (V)
TPE, Styrenic (23°C)	66.1 (V)
TPE, Styrenic (23°C)	83.2 (V)
TPE, Styrenic (23°C)	86.0 (V)
TPE, Styrenic (23°C)	148 (V)
TPE, Styrenic (23°C)	209 (V)
TPE, Styrenic (23°C)	257 (V)
TPE, Styrenic (23°C)	279 (V)
TPE, Styrenic (23°C)	579 (V)

>25 - 50°C

Fluoropolymer, CTFE (37.8°C)	0.0082 - 0.0112 (V)
Polyvinylidene Chloride (30°C)	0.01 (V)
Fluoropolymer, CTFE (37.8°C)	0.0077 - 0.0158 (V)
Fluoropolymer, CTFE (37.8°C)	0.0133 - 0.0163 (V)
Fluoropolymer, CTFE (37.8°C)	0.0119 - 0.0236 (V)
Fluoropolymer, CTFE (37.8°C)	0.0122 - 0.0236 (V)
Polyvinylidene Chloride (38°C)	0.02 (V)
Polyvinylidene Chloride (40°C)	0.02 (V)
Fluoropolymer, CTFE (37.8°C)	0.0171 - 0.0247 (V)
Fluoropolymer, CTFE (37.8°C)	0.0122 - 0.0316 (V)
LDPE/ EVA/ PVDC/ EVA/ LDPE Film (38°C)	0.04 (V)
Polyimide (38°C)	0.04 (V)
Polyvinylidene Chloride (30°C)	0.04 (V)
Fluoropolymer, CTFE (50°C)	0.043 (V)
Liquid Crystal Polymer (38°C)	0.05 (V)
HDPE/ EVA/ PVDC/ EVA Film (38°C)	0.06 (V)
LDPE/ EVA/ PVDC/ EVA/ LDPE Film (38°C)	0.06 (V)
Polyethylene, HDPE (30°C)	0.07 (V)
CTFE/ PVC Film (37.8°C)	0.08 (V)
LDPE/ EVA/ PVDC/ EVA/ LDPE Film (38°C)	0.08 (V)
Nylon 6 (37.8°C)	0.08 (V)
Parylene (37°C)	0.08 (V)
CTFE/ PE/ PVC Film (37.8°C)	0.09 (V)
Polyvinylidene Chloride (40°C)	0.09 (V)
LDPE/ EVA/ PVDC/ EVA Film (38°C)	0.1 (V)
Parylene (37°C)	0.1 (V)
Polyethylene, HDPE (37.8°C)	0.1 (V)
Polyvinylidene Chloride (38°C)	0.1 (V)
Polyvinylidene Chloride (38°C)	0.11 (V)
CTFE/ PE/ PVC Film (37.8°C)	0.12 (V)
Polyethylene, HDPE (37.8°C)	0.12 (V)

Polyvinylidene Chloride (38°C)	0.12 (V)
Polyethylene, HDPE (38°C)	0.13 (V)
Polyethylene, HDPE (40°C)	0.14 (V)
Polyethylene, HDPE (40°C)	0.15 (V)
Polypropylene (40°C)	0.15 (V)
Fluoropolymer, FEP (37.8°C)	0.16 (V)
LDPE/ EVA/ PVDC/ EVA/ LDPE Film (38°C)	0.16 (V)
Polyvinylidene Chloride (38°C)	0.16 (V)
PVC-PVDC Copolymer (37.8°C)	0.08 - 0.24 (V)
Polyethylene, HDPE (38°C)	0.17 (V)
Polyethylene, HDPE (38°C)	0.19 (V)
Polyethylene, MDPE (38°C)	0.22 (V)
Nylon 66 (38°C)	0.23 (V)
HDPE/ EAA/ Nylon/ EAA Film (38°C)	0.24 (V)
Polypropylene (40°C)	0.27 (V)
Polyethylene, MDPE (37.8°C)	0.28 (V)
Polyethylene, HDPE (50°C)	0.32 (V)
Nylon 66 (38°C)	0.34 (V)
Polyethylene, LLDPE (38°C)	0.38 (V)
Cellulosic Plastic (40°C)	0.39 (V)
Nylon (40°C)	0.39 (V)
Polyester, PET (40°C)	0.39 (V)
Polyethylene, LDPE (40°C)	0.39 (V)
Polypropylene (40°C)	<0.39 (V)
Polypropylene (37.8°C)	0.4 (V)
Polyester, PET (37.8°C)	0.39 - 0.51 (V)
Polyethylene, LDPE (40°C)	0.45 (V)
Polyethylene, LLDPE (38°C)	0.46 (V)
Nylon, Amorphous (37.8°C)	0.47 (V)
Polybutylene (37.8°C)	0.47 (V)
Polyester, PET (40°C)	0.47 (V)
Polyethylene, LDPE (37.8°C)	0.39 - 0.59 (V)
Polyester, PET (40°C)	0.51 (V)
Polyethylene, LLDPE (38°C)	0.51 (V)
EVOH (40°C)	0.55 (V)
Nylon, Amorphous (40°C)	0.55 (V)
Polyimide (38°C)	0.56 (V)
Parylene (37°C)	0.59 (V)
Polypropylene (37.8°C)	0.59 (V)
Fluoropolymer, PVDF (38°C)	0.62 (V)
Fluoropolymer, PVDF (38°C)	0.64 (V)
Fluoropolymer, PVDF (38°C)	0.68 (V)
Polyester, PET (37.8°C)	0.71 (V)
Polybutylene (37.8°C)	0.74 (V)
Fluoropolymer, PVDF (38°C)	0.75 (V)
EVOH (40°C)	0.79 (V)
Polyester, PET (37.8°C)	0.8 (V)
Epoxy (37°C)	0.7 - 0.94 (V)
EVOH (40°C)	0.83 (V)
EVOH (40°C)	1.2 (V)
Polyester, PET (40°C)	1.2 (V)
Polyvinyl Chloride (40°C)	1.18 (V)
Polyvinyl Chloride (38°C)	1.18 (V)
EVOH (40°C)	1.2 (V)
Nylon MXD6 (40°C)	1.3 (V)
Fluoropolymer, PVF (37.8°C)	1.3 (V)
EVOH (40°C)	1.5 (V)
Polyester, PET (37.8°C)	1.6 (V)
Polyester, PETG (37.8°C)	1.6 (V)
Polyester, PET (37.8°C)	1.7 (V)
Polyvinyl Chloride (37.8°C)	1.7 (V)
Acrylonitrile Copolymer, AMA (37.8°C)	2.2 (V)
Polyurethane (37°C)	0.94 - 3.4 (V)
EVOH (40°C)	2.4 (V)
Acrylonitrile Copolymer, AMA (40°C)	2.4 (V)
Silicone (37°C)	1.7 - 3.1 (V)
Acrylonitrile Copolymer, AMA (37.8°C)	3.0 (V)

EVOH (40°C)	3.15 (V)
Polystyrene (40°C)	3.35 (V)
Polycarbonate (37.8°C)	3.8 (V)
Polystyrene (37.8°C)	4.0 (V)
Nylon 6 (40°C)	4.02 (V)
Polycarbonate (40°C)	4.33 (V)
Nylon 6 (37.8°C)	5.6 - 5.9 (V)
Nylon (40°C)	6.7 (V)
Polysulfone (38°C)	7.1 (V)
Nylon 6 (37.8°C)	7.1 - 7.7 (V)
Nylon 6 (37.8°C)	7.5 - 7.9 (V)
Nylon 6/66 (37.8°C)	8.7 (V)
Nylon 6 (37.8°C)	9.2 (V)
Nylon 6 (37.8°C)	9.8 (V)
Rubber, Polybutadiene (39°C)	17.7 (V)
TPE, Polyester (38°C)	19.4 (V)
Polyvinyl Alcohol (40°C)	27.9 (V)
TPE, Polyamide (38°C)	31 (V)
TPE, Polyamide (38°C)	34 (V)
TPE, Polyamide (38°C)	38 (V)
TPE, Polyamide (38°C)	67 (V)
TPE, Polyamide (38°C)	89 (V)

>50 - 75°C

Fluoropolymer, CTFE (75°C)	0.12 (V)
Polysulfone (71°C)	27.2 (V)

>75 - 100°C

Fluoropolymer, CTFE (100°C)	0.39 (V)

xylene

20 - 25°C

EVOH (20°C)	<0.0015 (V)
LDPE/ EVAL Film (20°C)	<0.0035 (V)
LDPE/ EVAL Film (20°C)	<0.004 (V)
EVOH (20°C)	0.007 (V)
EVOH (20°C)	0.01 (V)
Nylon (20°C)	0.01 (V)
EVOH (20°C)	0.02 (V)
Nylon (20°C)	0.02 (V)
EVOH (20°C)	0.022 (V)
EVOH (20°C)	0.028 (V)
EVOH (20°C)	0.03 (V)
Polyester, PET (20°C)	0.04 (V)
Polypropylene (20°C)	7 (V)
Polypropylene (20°C)	7.0 (V)
Polyethylene, LDPE (20°C)	16.5 (V)
Polyethylene, LDPE (20°C)	16.6 (V)

>50 - 75°C

Laminar, Nylon/HDPE (60°C)	0.39 (V)
Laminar, EVOH/HDPE (60°C)	0.79 (V)
Laminar, Nylon/HDPE (60°C)	3.5 (V)
Laminar, Nylon/LDPE (60°C)	4.7 (V)
Laminar, Nylon/PP (60°C)	7.1 (V)
Polyethylene, HDPE (60°C)	283 (V)
Polypropylene (60°C)	984 (V)
Polyethylene, LDPE (60°C)	1496 (V)

Permeability of Rubber Glove Films

NATURAL RUBBER GLOVE FILM

Penetrant	Penetrant Note	Temp. (°C)	Time (days)	BTT (min.)	LDL (phr)	Permeation Rate (µg/cm²/min)	Comment	Material Note
Acetaldehyde				2	0.4	78		Pioneer Industrial Products L-118; glove film
		23		7		<900	fair permeation rate; 51 to 100 eyedropper size drops per hour	Ansell Edmont Canners and Handlers 392; unsupported glove film; 0.48 mm thick
Acetic Acid	glacial			21	0.1	12		Pioneer Industrial Products L-118; glove film
	glacial	23		110				Ansell Edmont Canners and Handlers 392; unsupported glove film; 0.48 mm thick
	50% conc.			31	0.02	18		Pioneer Industrial Products L-118; glove film
Acetone				7	0.05	30		"
		23		10		<900	fair permeation rate; 51 to 100 eyedropper size drops per hour	Ansell Edmont Canners and Handlers 392; unsupported glove film; 0.48 mm thick
Acetonitrile		23		4		<9	very good permeation rate; 1 to 5 eyedropper size drops per hour	"
Acrylic Acid		23		80				"
Ammonium Fluoride	40% conc.	23	0.25		0		no permeation detected during a 6 hour test	"
Ammonium Hydroxide	concentrated	23		90				"
	29% conc.			58	1	108		Pioneer Industrial Products L-118; glove film
Amyl Alcohol		23		25		<9	very good permeation rate; 1 to 5 eyedropper size drops per hour	Ansell Edmont Canners and Handlers 392; unsupported glove film; 0.48 mm thick
Aniline				>480	0.008	0	no permeation rate at steady state detected	Pioneer Industrial Products L-118; glove film
		23		25		<9	very good permeation rate; 1 to 5 eyedropper size drops per hour	Ansell Edmont Canners and Handlers 392; unsupported glove film; 0.48 mm thick
Benzaldehyde		23		10		<9	very good permeation rate; 1 to 5 eyedropper size drops per hour	"
Bromopropionic Acid		23		190				"
Butanone (2-)				6		522		Pioneer Industrial Products L-118; glove film
Butoxyethanol (2-)				12	1	162		"
Butyl Alcohol		23		200		<9	very good permeation rate; 1 to 5 eyedropper size drops per hour	Ansell Edmont Canners and Handlers 392; unsupported glove film; 0.48 mm thick
Butyl Cellosolve				12	1	162		Pioneer Industrial Products L-118; glove film
		23		45		<90	good permeation rate; 6 to 50 eyedropper size drops per hour	Ansell Edmont Canners and Handlers 392; unsupported glove film; 0.48 mm thick
Butyrolactone (gamma-)		23		60		<90	good permeation rate; 6 to 50 eyedropper size drops per hour	"
Carbolic Acid				>480	0.06	0	no permeation rate at steady state detected	Pioneer Industrial Products L-118; glove film
Cellosolve				10	0.03	12		"
	solvent	23		25		<9	very good permeation rate; 1 to 5 eyedropper size drops per hour	Ansell Edmont Canners and Handlers 392; unsupported glove film; 0.48 mm thick
Cellosolve Acetate				13				Pioneer Industrial Products L-118; glove film
		23		10		<90	good permeation rate; 6 to 50 eyedropper size drops per hour	Ansell Edmont Canners and Handlers 392; unsupported glove film; 0.48 mm thick
Citric Acid	10% conc.	23	0.25		0		no permeation detected during a 6 hour test	"
Cresol (3-)	m-Cresol			150	5	12		Pioneer Industrial Products L-118; glove film

NATURAL RUBBER GLOVE FILM (cont'd)

Penetrant	Penetrant Note	Temp. (°C)	Time (days)	BTT (min.)	LDL (phr)	Permeation Rate (µg/cm²/min)	Comment	Material Note
Cresol (m-)	3-Cresol			150	5	12		"
Cyclohexyl Alcohol		23		10		<90	good permeation rate; 6 to 50 eyedropper size drops per hour	Ansell Edmont Canners and Handlers 392; unsupported glove film; 0.48 mm thick
Diacetone Alcohol		23		15		<9	very good permeation rate; 1 to 5 eyedropper size drops per hour	"
Diamine				218	0.7	12		Pioneer Industrial Products L-118; glove film
Dibutyl Phthalate		23		200				Ansell Edmont Canners and Handlers 392; unsupported glove film; 0.48 mm thick
Diethanolamine				>480	1.1	0	no permeation rate at steady state detected	Pioneer Industrial Products L-118; glove film
Dimethyl Sulfoxide	DMSO			240	0.004	0	no permeation rate at steady state detected	"
	DMSO	23		180		<0.9	low permeation rate; 0 to 1/2 eyedropper size drops per hour	Ansell Edmont Canners and Handlers 392; unsupported glove film; 0.48 mm thick
Dimethylacetamide	DMAC	23		15		<90	good permeation rate; 6 to 50 eyedropper size drops per hour	"
Dimethylformamide	DMF			67	0.1	246		Pioneer Industrial Products L-118; glove film
	DMF	23		25		<9	very good permeation rate; 1 to 5 eyedropper size drops per hour	Ansell Edmont Canners and Handlers 392; unsupported glove film; 0.48 mm thick
Dioxane		23		5		<900	fair permeation rate; 51 to 100 eyedropper size drops per hour	"
Electroless Copper	MacDermid 9048	23	0.25			0	no permeation detected during a 6 hour test	"
Electroless Nickel	MacDermid J60/61	23	0.25			0	no permeation detected during a 6 hour test	"
Epichlorohydrin		23		5		<900	fair permeation rate; 51 to 100 eyedropper size drops per hour	"
Ethoxyethanol (2-)				10	0.03	12		Pioneer Industrial Products L-118; glove film
Ethoxyethyl Acetate (2-)				13				"
Ethyl Acetate		23		5		<900	fair permeation rate; 51 to 100 eyedropper size drops per hour	Ansell Edmont Canners and Handlers 392; unsupported glove film; 0.48 mm thick
Ethyl Alcohol	ethanol			>480	0.02	0	no permeation rate at steady state detected	Pioneer Industrial Products L-118; glove film
		23		15		<9	very good permeation rate; 1 to 5 eyedropper size drops per hour	Ansell Edmont Canners and Handlers 392; unsupported glove film; 0.48 mm thick
Ethyl Alcohol Amine	monoethanolamine	23		50		<0.9	low permeation rate; 0 to 1/2 eyedropper size drops per hour	"
Ethyl Glycol Ether		23		25		<9	very good permeation rate; 1 to 5 eyedropper size drops per hour	"
Ethylene Glycol				>480		0	no permeation rate at steady state detected	Pioneer Industrial Products L-118; glove film
		23	0.25		0	<0.9	low permeation rate; 0 to 1/2 eyedropper size drops per hour	Ansell Edmont Canners and Handlers 392; unsupported glove film; 0.48 mm thick
Formaldehyde		23		10		<90	good permeation rate; 6 to 50 eyedropper size drops per hour	"
	37% conc.			>480	0.01	0	no permeation rate at steady state detected	Pioneer Industrial Products L-118; glove film
Formalin	solution			>480	0.01	0	no permeation rate at steady state detected	"
Formic Acid	90% conc.	23		150				Ansell Edmont Canners and Handlers 392; unsupported glove film; 0.48 mm thick
Furfural		23		15		<9	very good permeation rate; 1 to 5 eyedropper size drops per hour	"
Hexamethyldisilizane		23		15		<900	fair permeation rate; 51 to 100 eyedropper size drops per hour	"
Hydrazine				218	0.7	12		Pioneer Industrial Products L-118; glove film
	65% conc.	23		150		<9	very good permeation rate; 1 to 5 eyedropper size drops per hour	Ansell Edmont Canners and Handlers 392; unsupported glove film; 0.48 mm thick
Hydrochloric Acid	concentrated	23		290				"
	10% conc.	23	0.25		0		no permeation detected during a 6 hour test	"
	37.5% conc.			211	4	1308		Pioneer Industrial Products L-118; glove film
Hydrofluoric Acid	48% conc.			>480	1	0	no permeation rate at steady state detected	"
	48% conc.	23		190				Ansell Edmont Canners and Handlers 392; unsupported glove film; 0.48 mm thick

NATURAL RUBBER GLOVE FILM (cont'd)

Penetrant	Penetrant Note	Temp. (°C)	Time (days)	BTT (min.)	LDL (phr)	Permeation Rate (µg/cm²/min)	Comment	Material Note
Hydrogen Peroxide	30% conc.	23	0.25		0		no permeation detected during a 6 hour test	"
Hydroquinone	saturated	23	0.25		0	<0.9	low permeation rate; 0 to 1/2 eyedropper size drops per hour	"
Hydroxyethyl Amine (bis(2-))				>480	1.1	0	no permeation rate at steady state detected	Pioneer Industrial Products L-118; glove film
Isobutyl Alcohol		23		15		<9	very good permeation rate; 1 to 5 eyedropper size drops per hour	Ansell Edmont Canners and Handlers 392; unsupported glove film; 0.48 mm thick
Isopropyl Alcohol	isopropanol; IPA			>60		0	no permeation rate at steady state detected	Pioneer Industrial Products L-118; glove film
		23		200		<9	very good permeation rate; 1 to 5 eyedropper size drops per hour	Ansell Edmont Canners and Handlers 392; unsupported glove film; 0.48 mm thick
Lactic Acid	85% conc.	23	0.25		0		no permeation detected during a 6 hour test	"
Lauric Acid	with ethylene oxide; 36% conc.	23	0.25		0		no permeation detected during a 6 hour test	"
Maleic Acid	saturated	23	0.25		0		no permeation detected during a 6 hour test	"
Methyl Alcohol	methanol			>60		0	no permeation rate at steady state detected	Pioneer Industrial Products L-118; glove film
		23		200		<9	very good permeation rate; 1 to 5 eyedropper size drops per hour	Ansell Edmont Canners and Handlers 392; unsupported glove film; 0.48 mm thick
Methyl Cellosolve		23		200		<9	very good permeation rate; 1 to 5 eyedropper size drops per hour	"
Methyl Ethyl Ketone	MEK			6		522		Pioneer Industrial Products L-118; glove film
	MEK	23		5		<900	fair permeation rate; 51 to 100 eyedropper size drops per hour	Ansell Edmont Canners and Handlers 392; unsupported glove film; 0.48 mm thick
Methyl Glycol Ether		23		200		<9	very good permeation rate; 1 to 5 eyedropper size drops per hour	"
Methyl-2-Pyrrolidone (N-)	NMP	23		75		<9	very good permeation rate; 1 to 5 eyedropper size drops per hour	"
Methylamine		23		55		<9	very good permeation rate; 1 to 5 eyedropper size drops per hour	"
Methylphenol (3-)	m-methylphenol			150	5	12		Pioneer Industrial Products L-118; glove film
Methylphenol (m-)	3-methylphenol			150	5	12		"
Morpholine		23		200		<90	good permeation rate; 6 to 50 eyedropper size drops per hour	Ansell Edmont Canners and Handlers 392; unsupported glove film; 0.48 mm thick
Muriatic Acid				211	4	1308		Pioneer Industrial Products L-118; glove film
		23		290				Ansell Edmont Canners and Handlers 392; unsupported glove film; 0.48 mm thick
Nitric Acid	10% conc.	23	0.25		0		no permeation detected during a 6 hour test	"
	50% conc.			233	0.1		permeation rate too large to measure	Pioneer Industrial Products L-118; glove film
Nitrobenzene		23		15		<90	good permeation rate; 6 to 50 eyedropper size drops per hour	Ansell Edmont Canners and Handlers 392; unsupported glove film; 0.48 mm thick
Nitromethane	95.5% conc.	23		10		<90	good permeation rate; 6 to 50 eyedropper size drops per hour	"
Nitropropane	95.5% conc.	23		5		<90	good permeation rate; 6 to 50 eyedropper size drops per hour	"
Octyl Alcohol		23		30		<9	very good permeation rate; 1 to 5 eyedropper size drops per hour	"
Oleic Acid		23	0.25		0		no permeation detected during a 6 hour test	"
Oxalic Acid	saturated	23	0.25		0		no permeation detected during a 6 hour test	"
Palmitic Acid	saturated	23		5				"
Perchloric Acid	60% conc.	23	0.25		0		no permeation detected during a 6 hour test	"
Phenol	saturated			>480	0.06	0	no permeation rate at steady state detected	Pioneer Industrial Products L-118; glove film
		23		90				Ansell Edmont Canners and Handlers 392; unsupported glove film; 0.48 mm thick
Phosphoric Acid	concentrated	23	0.25		0		no permeation detected during a 6 hour test	"
	85% conc.			>480	0.04	0	no permeation rate at steady state detected	Pioneer Industrial Products L-118; glove film

NATURAL RUBBER GLOVE FILM (cont'd)

Penetrant	Penetrant Note	Temp. (°C)	Time (days)	BTT (min.)	LDL (phr)	Permeation Rate (µg/cm²/min)	Comment	Material Note
Potassium Hydroxide	KOH; 50% conc.			>480	0.3	0	no permeation rate at steady state detected	"
	KOH; 50% conc.	23	0.25		0		no permeation detected during a 6 hour test	Ansell Edmont Canners and Handlers 392; unsupported glove film; 0.48 mm thick
Propyl Alcohol		23		200		<9	very good permeation rate; 1 to 5 eyedropper size drops per hour	"
Pyridine		23		10		<900	fair permeation rate; 51 to 100 eyedropper size drops per hour	"
Rule 66	solvent			9	0.002	126		Pioneer Industrial Products L-118; glove film
Sodium Hydroxide	NAOH; 50% conc.			>480	0.1	0	no permeation rate at steady state detected	"
	NaOH; 50% conc.	23	0.25		0		no permeation detected during a 6 hour test	Ansell Edmont Canners and Handlers 392; unsupported glove film; 0.48 mm thick
Sulfuric Acid	battery acid; 47% conc.	23	0.25		0		no permeation detected during a 6 hour test	"
	50% conc.			>480	0.1	0	no permeation rate at steady state detected	Pioneer Industrial Products L-118; glove film
Tannic Acid	65% conc.	23	0.25		0		no permeation detected during a 6 hour test	Ansell Edmont Canners and Handlers 392; unsupported glove film; 0.48 mm thick
Toluene Diisocyanate	TDI			>480	0.1	0	no permeation rate at steady state detected	Pioneer Industrial Products L-118; glove film
	TDI	23		7		<90	good permeation rate; 6 to 50 eyedropper size drops per hour	Ansell Edmont Canners and Handlers 392; unsupported glove film; 0.48 mm thick
Tricresyl Phosphate	TCP	23		45		<0.9	low permeation rate; 0 to 1/2 eyedropper size drops per hour	"
Triethanolamine	TEA			>480	5	0	no permeation rate at steady state detected	Pioneer Industrial Products L-118; glove film
	TEA; 85% conc.	23	0.25		0	<0.9	low permeation rate; 0 to 1/2 eyedropper size drops per hour	Ansell Edmont Canners and Handlers 392; unsupported glove film; 0.48 mm thick
Trihydroxytriethylamine				>480	5	0	no permeation rate at steady state detected	Pioneer Industrial Products L-118; glove film

POLYCHLOROPRENE RUBBER

Penetrant	Penetrant Note	Temp. (°C)	Time (days)	BTT (min.)	LDL (phr)	Permeation Rate (µg/cm²/min)	Comment	Material Note
Acetaldehyde				21	0.1	108		Pioneer Industrial Products Stanzoil N-44; glove film
		23		10		<9000	poor permeation rate; 501 to 5000 eyedropper size drops per hour	Ansell Edmont Neoprene 29-840; unsupported glove film; 0.38 mm thick
		23		17		<9000	poor permeation rate; 501 to 5000 eyedropper size drops per hour	Ansell Edmont Neox; supported (lined) glove film; specified by glove film weight
Acetic Acid	glacial	23		420				Ansell Edmont Neoprene 29-840; unsupported glove film; 0.38 mm thick
	glacial	23		>360				Ansell Edmont Neox; supported (lined) glove film; specified by glove film weight
	50% conc.			>480	0.1	0	no permeation rate at steady state detected	Pioneer Industrial Products Stanzoil N-44; glove film
Acetone				12		210		"
		23		5		<900	fair permeation rate; 51 to 100 eyedropper size drops per hour	Ansell Edmont Neoprene 29-840; unsupported glove film; 0.38 mm thick
		23		10		<900	fair permeation rate; 51 to 100 eyedropper size drops per hour	Ansell Edmont Neox; supported (lined) glove film; specified by glove film weight
Acetonitrile				40		42		Pioneer Industrial Products Stanzoil N-44; glove film
		23		30		<9	very good permeation rate; 1 to 5 eyedropper size drops per hour	Ansell Edmont Neoprene 29-840; unsupported glove film; 0.38 mm thick
		23		90		<0.9	low permeation rate; 0 to 1/2 eyedropper size drops per hour	Ansell Edmont Neox; supported (lined) glove film; specified by glove film weight
Acrylic Acid		23		70				Ansell Edmont Neoprene 29-840; unsupported glove film; 0.38 mm thick
		23	0.25		0	<0.9	low permeation rate; 0 to 1/2 eyedropper size drops per hour	Ansell Edmont Neox; supported (lined) glove film; specified by glove film weight

© *Plastics Design Library*

POLYCHLOROPRENE RUBBER (cont'd)

Penetrant	Penetrant Note	Temp. (°C)	Time (days)	BTT (min.)	LDL (phr)	Permeation Rate (μg/cm²/min)	Comment	Material Note
Ammonium Fluoride	40% conc.	23	0.25		0		no permeation detected during a 6 hour test	Ansell Edmont Neoprene 29-840; unsupported glove film; 0.38 mm thick
	40% conc.	23	0.25		0		no permeation detected during a 6 hour test	Ansell Edmont Neox; supported (lined) glove film; specified by glove film weight
Ammonium Hydroxide	concentrated	23		>360				Ansell Edmont Neoprene 29-840; unsupported glove film; 0.38 mm thick
Ammonium Hydroxide	concentrated	23		>360				Ansell Edmont Neox; supported (lined) glove film; specified by glove film weight
	29% conc.	23		>480	1	0	no permeation rate at steady state detected	Pioneer Industrial Products Stanzoil N-44; glove film
Amyl Alcohol		23		>360		<0.9	low permeation rate; 0 to 1/2 eyedropper size drops per hour	Ansell Edmont Neoprene 29-840; unsupported glove film; 0.38 mm thick
		23	0.25		0	<0.9	low permeation rate; 0 to 1/2 eyedropper size drops per hour	Ansell Edmont Neox; supported (lined) glove film; specified by glove film weight
Aniline				>480	0.005	0	no permeation rate at steady state detected	Pioneer Industrial Products Stanzoil N-44; glove film
		23		35		<9	very good permeation rate; 1 to 5 eyedropper size drops per hour	Ansell Edmont Neoprene 29-840; unsupported glove film; 0.38 mm thick
		23		180		<9	very good permeation rate; 1 to 5 eyedropper size drops per hour	Ansell Edmont Neox; supported (lined) glove film; specified by glove film weight
Aqua Regia		23		45				Ansell Edmont Neoprene 29-840; unsupported glove film; 0.38 mm thick
		23	0.25		0		no permeation detected during a 6 hour test	Ansell Edmont Neox; supported (lined) glove film; specified by glove film weight
Aroclor 1254	50% TCB			>480	1	0	no permeation rate at steady state detected	Pioneer Industrial Products Stanzoil N-44; glove film
Benzene				16		798		"
Bromopropionic Acid		23		180				Ansell Edmont Neoprene 29-840; unsupported glove film; 0.38 mm thick
		23		240				Ansell Edmont Neox; supported (lined) glove film; specified by glove film weight
Butanone (2-)				22		930		Pioneer Industrial Products Stanzoil N-44; glove film
Butoxyethanol (2-)				147	1	30		"
Butyl Acetate				52		318		"
Butyl Alcohol		23		240		<9	very good permeation rate; 1 to 5 eyedropper size drops per hour	Ansell Edmont Neoprene 29-840; unsupported glove film; 0.38 mm thick
		23		>480		<0.9	low permeation rate; 0 to 1/2 eyedropper size drops per hour	Ansell Edmont Neox; supported (lined) glove film; specified by glove film weight
Butyl Cellosolve				147	1	30		Pioneer Industrial Products Stanzoil N-44; glove film
		23		90		<9	very good permeation rate; 1 to 5 eyedropper size drops per hour	Ansell Edmont Neoprene 29-840; unsupported glove film; 0.38 mm thick
		23	0.25		0	<0.9	low permeation rate; 0 to 1/2 eyedropper size drops per hour	Ansell Edmont Neox; supported (lined) glove film; specified by glove film weight
Butyrolactone (gamma-)		23		10		<9	very good permeation rate; 1 to 5 eyedropper size drops per hour	Ansell Edmont Neoprene 29-840; unsupported glove film; 0.38 mm thick
Carbolic Acid				>480	0.2	0	no permeation rate at steady state detected	Pioneer Industrial Products Stanzoil N-44; glove film
Carbon Bichloride	carbon dichloride			28	0.0002	453		"
Carbon Dichloride	carbon bichloride			28	0.0002	453		"
Carbon Tetrachloride				31		1512		"
Cellosolve				352	0.06	18		"
	solvent	23		45		<0.9	low permeation rate; 0 to 1/2 eyedropper size drops per hour	Ansell Edmont Neoprene 29-840; unsupported glove film; 0.38 mm thick
	solvent	23		240		<0.9	low permeation rate; 0 to 1/2 eyedropper size drops per hour	Ansell Edmont Neox; supported (lined) glove film; specified by glove film weight

POLYCHLOROPRENE RUBBER (cont'd)

Penetrant	Penetrant Note	Temp. (°C)	Time (days)	BTT (min.)	LDL (phr)	Permeation Rate (µg/cm²/min)	Comment	Material Note
Cellosolve Acetate				76	0.07	252		Pioneer Industrial Products Stanzoil N-44; glove film
		23		25		<90	good permeation rate; 6 to 50 eyedropper size drops per hour	Ansell Edmont Neoprene 29-840; unsupported glove film; 0.38 mm thick
		23		75		<9	very good permeation rate; 1 to 5 eyedropper size drops per hour	Ansell Edmont Neox; supported (lined) glove film; specified by glove film weight
Chloroform				12	0.2	1368		Pioneer Industrial Products Stanzoil N-44; glove film
Chlorothene				27		1182		"
Citric Acid	10% conc.	23	0.25			0	no permeation detected during a 6 hour test	Ansell Edmont Neoprene 29-840; unsupported glove film; 0.38 mm thick
	10% conc.	23	0.25			0	no permeation detected during a 6 hour test	Ansell Edmont Neox; supported (lined) glove film; specified by glove film weight
Cresol (3-)	m-Cresol			>480	5	0	no permeation rate at steady state detected	Pioneer Industrial Products Stanzoil N-44; glove film
Cresol (m-)	3-Cresol			>480	5	0	no permeation rate at steady state detected	"
Cumene	.			41	0.4	216		"
Cyclohexane				159	0.03	42		"
Cyclohexyl Alcohol		23		150		<9	very good permeation rate; 1 to 5 eyedropper size drops per hour	Ansell Edmont Neoprene 29-840; unsupported glove film; 0.38 mm thick
		23		180		<0.9	low permeation rate; 0 to 1/2 eyedropper size drops per hour	Ansell Edmont Neox; supported (lined) glove film; specified by glove film weight
Diacetone Alcohol		23		300		<0.9	low permeation rate; 0 to 1/2 eyedropper size drops per hour	Ansell Edmont Neoprene 29-840; unsupported glove film; 0.38 mm thick
		23	0.25			<0.9	low permeation rate; 0 to 1/2 eyedropper size drops per hour	Ansell Edmont Neox; supported (lined) glove film; specified by glove film weight
Diamine				>480	0.7	0	no permeation rate at steady state detected	Pioneer Industrial Products Stanzoil N-44; glove film
Dibutyl Phthalate		23		120		<0.9	low permeation rate; 0 to 1/2 eyedropper size drops per hour	Ansell Edmont Neoprene 29-840; unsupported glove film; 0.38 mm thick
		23		300		<9	very good permeation rate; 1 to 5 eyedropper size drops per hour	Ansell Edmont Neox; supported (lined) glove film; specified by glove film weight
Dichloroethane (1,2-)				33	0.09	1482		Pioneer Industrial Products Stanzoil N-44; glove film
Dichloromethane				6		1434		"
Diethanolamine				>480	1.1	0	no permeation rate at steady state detected	"
Diethyl Ether				18	0.1	486		"
Diethylene Dioxide (1,4-)				28		372		"
Dimethyl Sulfoxide	DMSO			0	0.004	0	no permeation rate at steady state detected	"
	DMSO	23		>180		<90	good permeation rate; 6 to 50 eyedropper size drops per hour	Ansell Edmont Neox; supported (lined) glove film; specified by glove film weight
Dimethyl Sulfoxide	DMSO	23	0.25			<0.9	low permeation rate; 0 to 1/2 eyedropper size drops per hour	Ansell Edmont Neoprene 29-840; unsupported glove film; 0.38 mm thick
Dimethylformamide	DMF			110	0.1	246		Pioneer Industrial Products Stanzoil N-44; glove film
	DMF	23		10		<90	good permeation rate; 6 to 50 eyedropper size drops per hour	Ansell Edmont Neoprene 29-840; unsupported glove film; 0.38 mm thick
	DMF	23		60		<90	good permeation rate; 6 to 50 eyedropper size drops per hour	Ansell Edmont Neox; supported (lined) glove film; specified by glove film weight
Dioctyl Phthalate	DOP	23		>360		<0.9	low permeation rate; 0 to 1/2 eyedropper size drops per hour	Ansell Edmont Neoprene 29-840; unsupported glove film; 0.38 mm thick
	DOP	23		120		<0.9	low permeation rate; 0 to 1/2 eyedropper size drops per hour	Ansell Edmont Neox; supported (lined) glove film; specified by glove film weight
Dioxane (1,4-)				28		372		Pioneer Industrial Products Stanzoil N-44; glove film
Electroless Copper	MacDermid 9048	23	0.25			0	no permeation detected during a 6 hour test	Ansell Edmont Neoprene 29-840; unsupported glove film; 0.38 mm thick
	MacDermid 9048	23	0.25			0	no permeation detected during a 6 hour test	Ansell Edmont Neox; supported (lined) glove film; specified by glove film weight

POLYCHLOROPRENE RUBBER (cont'd)

Penetrant	Penetrant Note	Temp. (°C)	Time (days)	BTT (min.)	LDL (phr)	Permeation Rate (µg/cm²/min)	Comment	Material Note
Electroless Nickel	MacDermid J60/61	23	0.25		0		no permeation detected during a 6 hour test	"
	MacDermid J60/61; 90% conc.	23	0.25		0		no permeation detected during a 6 hour test	Ansell Edmont Neoprene 29-840; unsupported glove film; 0.38 mm thick
Epichlorohydrin		23		10		<900	fair permeation rate; 51 to 100 eyedropper size drops per hour	Ansell Edmont Neox; supported (lined) glove film; specified by glove film weight
Ethoxyethanol (2-)				352	0.06	18		Pioneer Industrial Products Stanzoil N-44; glove film
Ethoxyethyl Acetate (2-)				76	0.07	252		"
Ethyl Acetate				34	0.08	1068		"
		23		15		<90	good permeation rate; 6 to 50 eyedropper size drops per hour	Ansell Edmont Neoprene 29-840; unsupported glove film; 0.38 mm thick
		23		200		<90	good permeation rate; 6 to 50 eyedropper size drops per hour	Ansell Edmont Neox; supported (lined) glove film; specified by glove film weight
Ethyl Alcohol	ethanol			>480	0.002	0	no permeation rate at steady state detected	Pioneer Industrial Products Stanzoil N-44; glove film
		23		90		<9	very good permeation rate; 1 to 5 eyedropper size drops per hour	Ansell Edmont Neoprene 29-840; unsupported glove film; 0.38 mm thick
		23		180		<9	very good permeation rate; 1 to 5 eyedropper size drops per hour	Ansell Edmont Neox; supported (lined) glove film; specified by glove film weight
Ethyl Alcohol Amine	monoethanolamine	23	0.25		0	<0.9	low permeation rate; 0 to 1/2 eyedropper size drops per hour	Ansell Edmont Neoprene 29-840; unsupported glove film; 0.38 mm thick
	monoethanolamine	23	0.25		0	<0.9	low permeation rate; 0 to 1/2 eyedropper size drops per hour	Ansell Edmont Neox; supported (lined) glove film; specified by glove film weight
Ethyl Ether				18	0.1	486		Pioneer Industrial Products Stanzoil N-44; glove film
		23		10		<90	good permeation rate; 6 to 50 eyedropper size drops per hour	Ansell Edmont Neoprene 29-840; unsupported glove film; 0.38 mm thick
		23		10		<90	good permeation rate; 6 to 50 eyedropper size drops per hour	Ansell Edmont Neox; supported (lined) glove film; specified by glove film weight
Ethyl Glycol Ether		23		45		<0.9	low permeation rate; 0 to 1/2 eyedropper size drops per hour	Ansell Edmont Neoprene 29-840; unsupported glove film; 0.38 mm thick
		23		240		<0.9	low permeation rate; 0 to 1/2 eyedropper size drops per hour	Ansell Edmont Neox; supported (lined) glove film; specified by glove film weight
Ethylene Dichloride				33	0.09	1482		Pioneer Industrial Products Stanzoil N-44; glove film
Ethylene Glycol				>480		0	no permeation rate at steady state detected	"
		23	0.25		0	<0.9	low permeation rate; 0 to 1/2 eyedropper size drops per hour	Ansell Edmont Neoprene 29-840; unsupported glove film; 0.38 mm thick
		23	0.25		0	<0.9	low permeation rate; 0 to 1/2 eyedropper size drops per hour	Ansell Edmont Neox; supported (lined) glove film; specified by glove film weight
Ethylene Oxide				31	3	60		Pioneer Industrial Products Stanzoil N-44; glove film
Formaldehyde		23		120		<0.9	low permeation rate; 0 to 1/2 eyedropper size drops per hour	Ansell Edmont Neoprene 29-840; unsupported glove film; 0.38 mm thick
		23		120		<9	very good permeation rate; 1 to 5 eyedropper size drops per hour	Ansell Edmont Neox; supported (lined) glove film; specified by glove film weight
	37% conc.			>480	3	0	no permeation rate at steady state detected	Pioneer Industrial Products Stanzoil N-44; glove film
Formalin	solution			>480	3	0	no permeation rate at steady state detected	"
Formic Acid	90% conc.	23	0.25		0		no permeation detected during a 6 hour test	Ansell Edmont Neoprene 29-840; unsupported glove film; 0.38 mm thick
	90% conc.	23	0.25		0		no permeation detected during a 6 hour test	Ansell Edmont Neox; supported (lined) glove film; specified by glove film weight
Freon 12				>480	0.03	0	no permeation rate at steady state detected	Pioneer Industrial Products Stanzoil N-44; glove film
Freon TF				>480	0.03	0	no permeation rate at steady state detected	"
		23		240		<0.9	low permeation rate; 0 to 1/2 eyedropper size drops per hour	Ansell Edmont Neoprene 29-840; unsupported glove film; 0.38 mm thick
		23		120		<9	very good permeation rate; 1 to 5 eyedropper size drops per hour	Ansell Edmont Neox; supported (lined) glove film; specified by glove film weight

POLYCHLOROPRENE RUBBER (cont'd)

Penetrant	Penetrant Note	Temp. (°C)	Time (days)	BTT (min.)	LDL (phr)	Permeation Rate (μg/cm²/min)	Comment	Material Note
Furfural		23		200		<90	good permeation rate; 6 to 50 eyedropper size drops per hour	Ansell Edmont Neoprene 29-840; unsupported glove film; 0.38 mm thick
		23		120		<90	good permeation rate; 6 to 50 eyedropper size drops per hour	Ansell Edmont Neox; supported (lined) glove film; specified by glove film weight
Gasoline				96	0.2	96		Pioneer Industrial Products Stanzoil N-44; glove film
Heptane				124	0.02	12		"
Hexamethyldisilizane		23		50				Ansell Edmont Neoprene 29-840; unsupported glove film; 0.38 mm thick
		23		60				Ansell Edmont Neox; supported (lined) glove film; specified by glove film weight
Hexane				39	0.08	36		Pioneer Industrial Products Stanzoil N-44; glove film
		23		45		<900	fair permeation rate; 51 to 100 eyedropper size drops per hour	Ansell Edmont Neoprene 29-840; unsupported glove film; 0.38 mm thick
Hexane		23		90		<90	good permeation rate; 6 to 50 eyedropper size drops per hour	Ansell Edmont Neox; supported (lined) glove film; specified by glove film weight
Hydrazine				>480	0.7	0	no permeation rate at steady state detected	Pioneer Industrial Products Stanzoil N-44; glove film
	65% conc.	23	0.25			0	no permeation detected during a 6 hour test	Ansell Edmont Neoprene 29-840; unsupported glove film; 0.38 mm thick
	65% conc.	23	0.25			0	no permeation detected during a 6 hour test	Ansell Edmont Neox; supported (lined) glove film; specified by glove film weight
Hydrochloric Acid	concentrated	23	0.25			0	no permeation detected during a 6 hour test	Ansell Edmont Neoprene 29-840; unsupported glove film; 0.38 mm thick
	concentrated	23	0.25			0	no permeation detected during a 6 hour test	Ansell Edmont Neox; supported (lined) glove film; specified by glove film weight
	10% conc.	23	0.25			0	no permeation detected during a 6 hour test	Ansell Edmont Neoprene 29-840; unsupported glove film; 0.38 mm thick
	10% conc.	23	0.25			0	no permeation detected during a 6 hour test	Ansell Edmont Neox; supported (lined) glove film; specified by glove film weight
	37.5% conc.			>480	4	0	no permeation rate at steady state detected	Pioneer Industrial Products Stanzoil N-44; glove film
Hydrofluoric Acid	48% conc.			>480	1	0	no permeation rate at steady state detected	"
	48% conc.	23		60				Ansell Edmont Neoprene 29-840; unsupported glove film; 0.38 mm thick
	48% conc.	23		75				Ansell Edmont Neox; supported (lined) glove film; specified by glove film weight
Hydrogen Peroxide	30% conc.	23		5				Ansell Edmont Neoprene 29-840; unsupported glove film; 0.38 mm thick
	30% conc.	23		7				Ansell Edmont Neox; supported (lined) glove film; specified by glove film weight
Hydroquinone	saturated	23	0.25			0	<0.9	low permeation rate; 0 to 1/2 eyedropper size drops per hour	Ansell Edmont Neoprene 29-840; unsupported glove film; 0.38 mm thick
	saturated	23	0.25			0	<0.9	low permeation rate; 0 to 1/2 eyedropper size drops per hour	Ansell Edmont Neox; supported (lined) glove film; specified by glove film weight
Hydroxyethyl Amine (bis(2-))				>480	1.1	0	no permeation rate at steady state detected	Pioneer Industrial Products Stanzoil N-44; glove film
Isoamyl Acetate				30		312		"
Isobutyl Alcohol		23		10		<0.9	low permeation rate; 0 to 1/2 eyedropper size drops per hour	Ansell Edmont Neoprene 29-840; unsupported glove film; 0.38 mm thick
		23	0.25		0	<0.9	low permeation rate; 0 to 1/2 eyedropper size drops per hour	Ansell Edmont Neox; supported (lined) glove film; specified by glove film weight
Isooctane		23		60		<90	good permeation rate; 6 to 50 eyedropper size drops per hour	Ansell Edmont Neoprene 29-840; unsupported glove film; 0.38 mm thick
		23		360		<0.9	low permeation rate; 0 to 1/2 eyedropper size drops per hour	Ansell Edmont Neox; supported (lined) glove film; specified by glove film weight

POLYCHLOROPRENE RUBBER (cont'd)

Penetrant	Penetrant Note	Temp. (°C)	Time (days)	BTT (min.)	LDL (phr)	Permeation Rate (µg/cm²/min)	Comment	Material Note
Isopropyl Alcohol	isopropanol; IPA			>60	0.0006	0	no permeation rate at steady state detected	Pioneer Industrial Products Stanzoil N-44; glove film
		23	0.25		0	<0.9	low permeation rate; 0 to 1/2 eyedropper size drops per hour	Ansell Edmont Neoprene 29-840; unsupported glove film; 0.38 mm thick
		23	0.25		0	<0.9	low permeation rate; 0 to 1/2 eyedropper size drops per hour	Ansell Edmont Neox; supported (lined) glove film; specified by glove film weight
Isopropyl Benzene				41	0.4	216		Pioneer Industrial Products Stanzoil N-44; glove film
Kerosine				>480	5.0e-05	0	no permeation rate at steady state detected	"
		23		>360		<0.9	low permeation rate; 0 to 1/2 eyedropper size drops per hour	Ansell Edmont Neoprene 29-840; unsupported glove film; 0.38 mm thick
		23	0.25		0	<0.9	low permeation rate; 0 to 1/2 eyedropper size drops per hour	Ansell Edmont Neox; supported (lined) glove film; specified by glove film weight
Lactic Acid	85% conc.	23	0.25		0	<0.9	low permeation rate; 0 to 1/2 eyedropper size drops per hour	Ansell Edmont Neoprene 29-840; unsupported glove film; 0.38 mm thick
	85% conc.	23	0.25		0	<0.9	low permeation rate; 0 to 1/2 eyedropper size drops per hour	Ansell Edmont Neox; supported (lined) glove film; specified by glove film weight
Lauric Acid	with ethylene oxide; 36% conc.	23	0.25		0		no permeation detected during a 6 hour test	Ansell Edmont Neoprene 29-840; unsupported glove film; 0.38 mm thick
	with ethylene oxide; 36% conc.	23	0.25		0		no permeation detected during a 6 hour test	Ansell Edmont Neox; supported (lined) glove film; specified by glove film weight
Maleic Acid	saturated	23	0.25		0		no permeation detected during a 6 hour test	Ansell Edmont Neoprene 29-840; unsupported glove film; 0.38 mm thick
	saturated	23	0.25		0		no permeation detected during a 6 hour test	Ansell Edmont Neox; supported (lined) glove film; specified by glove film weight
Methane Dichloride				6		1434		Pioneer Industrial Products Stanzoil N-44; glove film
Methyl Alcohol	methanol			>60		0	no permeation rate at steady state detected	"
		23		60		<0.9	low permeation rate; 0 to 1/2 eyedropper size drops per hour	Ansell Edmont Neoprene 29-840; unsupported glove film; 0.38 mm thick
		23		15		<0.9	low permeation rate; 0 to 1/2 eyedropper size drops per hour	Ansell Edmont Neox; supported (lined) glove film; specified by glove film weight
Methyl Cellosolve		23		25		<90	good permeation rate; 6 to 50 eyedropper size drops per hour	Ansell Edmont Neoprene 29-840; unsupported glove film; 0.38 mm thick
		23		70		<9	very good permeation rate; 1 to 5 eyedropper size drops per hour	Ansell Edmont Neox; supported (lined) glove film; specified by glove film weight
Methyl Chloroform				27		1182		Pioneer Industrial Products Stanzoil N-44; glove film
Methyl Cyanide				40		42		"
Methyl Ethyl Ketone	MEK			22		930		"
Methyl Glycol Ether		23		25		<90	good permeation rate; 6 to 50 eyedropper size drops per hour	Ansell Edmont Neoprene 29-840; unsupported glove film; 0.38 mm thick
		23		70		<9	very good permeation rate; 1 to 5 eyedropper size drops per hour	Ansell Edmont Neox; supported (lined) glove film; specified by glove film weight
Methyl Iodide				12		3702		Pioneer Industrial Products Stanzoil N-44; glove film
Methylamine		23		270		<90	good permeation rate; 6 to 50 eyedropper size drops per hour	Ansell Edmont Neoprene 29-840; unsupported glove film; 0.38 mm thick
		23		360		<0.9	low permeation rate; 0 to 1/2 eyedropper size drops per hour	Ansell Edmont Neox; supported (lined) glove film; specified by glove film weight
Methylene Chloride				6		1434		Pioneer Industrial Products Stanzoil N-44; glove film
Methylphenol (3-)	m-methylphenol			>480	5	0	no permeation rate at steady state detected	"
Methylphenol (m-)	3-methylphenol			>480	5	0	no permeation rate at steady state detected	"
Mineral Spirits				126	0.002	12		"
	Rule 66	23		90		<9	very good permeation rate; 1 to 5 eyedropper size drops per hour	Ansell Edmont Neoprene 29-840; unsupported glove film; 0.38 mm thick
	Rule 66	23	0.25		0	<0.9	low permeation rate; 0 to 1/2 eyedropper size drops per hour	Ansell Edmont Neox; supported (lined) glove film; specified by glove film weight

POLYCHLOROPRENE RUBBER (cont'd)

Penetrant	Penetrant Note	Temp. (°C)	Time (days)	BTT (min.)	LDL (phr)	Permeation Rate (µg/cm²/min)	Comment	Material Note
Muriatic Acid				>480	4	0	no permeation rate at steady state detected	Pioneer Industrial Products Stanzoil N-44; glove film
		23	0.25		0		no permeation detected during a 6 hour test	Ansell Edmont Neox; supported (lined) glove film; specified by glove film weight
	10% conc.	23	0.25		0		no permeation detected during a 6 hour test	Ansell Edmont Neoprene 29-840; unsupported glove film; 0.38 mm thick
Naphtha				126	0.002	12		Pioneer Industrial Products Stanzoil N-44; glove film
	VM&P	23		15		<900	fair permeation rate; 51 to 100 eyedropper size drops per hour	Ansell Edmont Neoprene 29-840; unsupported glove film; 0.38 mm thick
	VM&P	23	0.25		0	<0.9	low permeation rate; 0 to 1/2 eyedropper size drops per hour	Ansell Edmont Neox; supported (lined) glove film; specified by glove film weight
Nitric Acid	10% conc.	23	0.25		0		no permeation detected during a 6 hour test	Ansell Edmont Neoprene 29-840; unsupported glove film; 0.38 mm thick
	10% conc.	23	0.25		0		no permeation detected during a 6 hour test	Ansell Edmont Neox; supported (lined) glove film; specified by glove film weight
	50% conc.	·		>480	0.0003	0	no permeation rate at steady state detected	Pioneer Industrial Products Stanzoil N-44; glove film
	70% conc.	23		140				Ansell Edmont Neoprene 29-840; unsupported glove film; 0.38 mm thick
	70% conc.	23	0.25		0		no permeation detected during a 6 hour test	Ansell Edmont Neox; supported (lined) glove film; specified by glove film weight
Nitrobenzene				60	1	120		Pioneer Industrial Products Stanzoil N-44; glove film
Nitromethane	95.5% conc.	23		60		<9	very good permeation rate; 1 to 5 eyedropper size drops per hour	Ansell Edmont Neoprene 29-840; unsupported glove film; 0.38 mm thick
	95.5% conc.	23		90		<0.9	low permeation rate; 0 to 1/2 eyedropper size drops per hour	Ansell Edmont Neox; supported (lined) glove film; specified by glove film weight
Nitropropane	95.5% conc.	23		5		<900	fair permeation rate; 51 to 100 eyedropper size drops per hour	Ansell Edmont Neoprene 29-840; unsupported glove film; 0.38 mm thick
	95.5% conc.	23		60		<90	good permeation rate; 6 to 50 eyedropper size drops per hour	Ansell Edmont Neox; supported (lined) glove film; specified by glove film weight
Octyl Alcohol		23		420		<0.9	low permeation rate; 0 to 1/2 eyedropper size drops per hour	Ansell Edmont Neoprene 29-840; unsupported glove film; 0.38 mm thick
		23		>420		<0.9	low permeation rate; 0 to 1/2 eyedropper size drops per hour	Ansell Edmont Neox; supported (lined) glove film; specified by glove film weight
Oleic Acid		23		60		<9	very good permeation rate; 1 to 5 eyedropper size drops per hour	Ansell Edmont Neoprene 29-840; unsupported glove film; 0.38 mm thick
		23		150		<0.9	low permeation rate; 0 to 1/2 eyedropper size drops per hour	Ansell Edmont Neox; supported (lined) glove film; specified by glove film weight
Oxalic Acid	saturated	23	0.25		0		no permeation detected during a 6 hour test	Ansell Edmont Neoprene 29-840; unsupported glove film; 0.38 mm thick
	saturated	23	0.25		0		no permeation detected during a 6 hour test	Ansell Edmont Neox; supported (lined) glove film; specified by glove film weight
Palmitic Acid	saturated	23	0.25		0		no permeation detected during a 6 hour test	Ansell Edmont Neoprene 29-840; unsupported glove film; 0.38 mm thick
	saturated	23	0.25		0		no permeation detected during a 6 hour test	Ansell Edmont Neox; supported (lined) glove film; specified by glove film weight
PCBs	polychlorinated biphenyls; 50% TCB			>480	1	0	no permeation rate at steady state detected	Pioneer Industrial Products Stanzoil N-44; glove film
Pentachlorophenol		23		6		<0.9	low permeation rate; 0 to 1/2 eyedropper size drops per hour	Ansell Edmont Neoprene 29-840; unsupported glove film; 0.38 mm thick
		23		6		<0.9	low permeation rate; 0 to 1/2 eyedropper size drops per hour	Ansell Edmont Neox; supported (lined) glove film; specified by glove film weight
Pentane				38		18		Pioneer Industrial Products Stanzoil N-44; glove film
		23		30		<900	fair permeation rate; 51 to 100 eyedropper size drops per hour	Ansell Edmont Neoprene 29-840; unsupported glove film; 0.38 mm thick
		23		45		<9	very good permeation rate; 1 to 5 eyedropper size drops per hour	Ansell Edmont Neox; supported (lined) glove film; specified by glove film weight

POLYCHLOROPRENE RUBBER (cont'd)

Penetrant	Penetrant Note	Temp. (°C)	Time (days)	BTT (min.)	LDL (phr)	Permeation Rate (μg/cm²/min)	Comment	Material Note
Perchloric Acid	60% conc.	23	0.25		0		no permeation detected during a 6 hour test	Ansell Edmont Neoprene 29-840; unsupported glove film; 0.38 mm thick
	60% conc.	23	0.25		0		no permeation detected during a 6 hour test	Ansell Edmont Neox; supported (lined) glove film; specified by glove film weight
Perchloroethylene				28	0.0002	453		Pioneer Industrial Products Stanzoil N-44; glove film
Petroleum Ether				126	0.002	12		"
Phenol	saturated			>480	0.2	0	no permeation rate at steady state detected	"
		23		180		<90	good permeation rate; 6 to 50 eyedropper size drops per hour	Ansell Edmont Neoprene 29-840; unsupported glove film; 0.38 mm thick
		23		390		<0.9	low permeation rate; 0 to 1/2 eyedropper size drops per hour	Ansell Edmont Neox; supported (lined) glove film; specified by glove film weight
Phosphoric Acid	concentrated	23	0.25		0		no permeation detected during a 6 hour test	Ansell Edmont Neoprene 29-840; unsupported glove film; 0.38 mm thick
	concentrated	23	0.25		0		no permeation detected during a 6 hour test	Ansell Edmont Neox; supported (lined) glove film; specified by glove film weight
	85% conc.			>480	0.04	0	no permeation rate at steady state detected	Pioneer Industrial Products Stanzoil N-44; glove film
Picric Acid	saturated; with ethylene oxide	23		150		<9	very good permeation rate; 1 to 5 eyedropper size drops per hour	Ansell Edmont Neoprene 29-840; unsupported glove film; 0.38 mm thick
	saturated; with ethylene oxide	23		180		<9	very good permeation rate; 1 to 5 eyedropper size drops per hour	Ansell Edmont Neox; supported (lined) glove film; specified by glove film weight
Polychlorinated Biphenyls	PCBs; 50% TCB			>480	1	0	no permeation rate at steady state detected	Pioneer Industrial Products Stanzoil N-44; glove film
Potassium Hydroxide	KOH; 50% conc.			>480	0.3	0	no permeation rate at steady state detected	"
	KOH; 50% conc.	23	0.25		0		no permeation detected during a 6 hour test	Ansell Edmont Neoprene 29-840; unsupported glove film; 0.38 mm thick
	KOH; 50% conc.	23	0.25		0		no permeation detected during a 6 hour test	Ansell Edmont Neox; supported (lined) glove film; specified by glove film weight
Propyl Alcohol		23		150		<0.9	low permeation rate; 0 to 1/2 eyedropper size drops per hour	Ansell Edmont Neoprene 29-840; unsupported glove film; 0.38 mm thick
		23	0.25		0		low permeation rate; 0 to 1/2 eyedropper size drops per hour	Ansell Edmont Neox; supported (lined) glove film; specified by glove film weight
Pyridine				26		702		Pioneer Industrial Products Stanzoil N-44; glove film
Rubber Solvent		23		30		<90	good permeation rate; 6 to 50 eyedropper size drops per hour	Ansell Edmont Neoprene 29-840; unsupported glove film; 0.38 mm thick
		23		60		<90	good permeation rate; 6 to 50 eyedropper size drops per hour	Ansell Edmont Neox; supported (lined) glove film; specified by glove film weight
Rule 66	solvent			332	0.03	0	no permeation rate at steady state detected	Pioneer Industrial Products Stanzoil N-44; glove film
Silicon Etch		23	0.25		0		no permeation detected during a 6 hour test	Ansell Edmont Neoprene 29-840; unsupported glove film; 0.38 mm thick
		23	0.25		0		no permeation detected during a 6 hour test	Ansell Edmont Neox; supported (lined) glove film; specified by glove film weight
Sodium Hydroxide	NAOH; 50% conc.			>480	0.1	0	no permeation rate at steady state detected	Pioneer Industrial Products Stanzoil N-44; glove film
	NaOH; 50% conc.	23	0.25		0		no permeation detected during a 6 hour test	Ansell Edmont Neoprene 29-840; unsupported glove film; 0.38 mm thick
	NaOH; 50% conc.	23	0.25		0		no permeation detected during a 6 hour test	Ansell Edmont Neox; supported (lined) glove film; specified by glove film weight
Stoddard Solvents		23		180		<9	very good permeation rate; 1 to 5 eyedropper size drops per hour	Ansell Edmont Neoprene 29-840; unsupported glove film; 0.38 mm thick
		23	0.25		0		low permeation rate; 0 to 1/2 eyedropper size drops per hour	Ansell Edmont Neox; supported (lined) glove film; specified by glove film weight
Sulfuric Acid	battery acid; 47% conc.	23	0.25		0		no permeation detected during a 6 hour test	Ansell Edmont Neoprene 29-840; unsupported glove film; 0.38 mm thick
	battery acid; 47% conc.	23	0.25		0		no permeation detected during a 6 hour test	Ansell Edmont Neox; supported (lined) glove film; specified by glove film weight
	50% conc.			>480	0.04	0	no permeation rate at steady state detected	Pioneer Industrial Products Stanzoil N-44; glove film

POLYCHLOROPRENE RUBBER (cont'd)

Penetrant	Penetrant Note	Temp. (°C)	Time (days)	BTT (min.)	LDL (phr)	Permeation Rate (µg/cm²/min)	Comment	Material Note
Sulfuric Acid	95% conc.	23		180				Ansell Edmont Neoprene 29-840; unsupported glove film; 0.38 mm thick
	95% conc.	23		>360				Ansell Edmont Neox; supported (lined) glove film; specified by glove film weight
Tannic Acid	65% conc.	23	0.25		0	<0.9	low permeation rate; 0 to 1/2 eyedropper size drops per hour	Ansell Edmont Neoprene 29-840; unsupported glove film; 0.38 mm thick
	65% conc.	23	0.25		0	<0.9	low permeation rate; 0 to 1/2 eyedropper size drops per hour	Ansell Edmont Neox; supported (lined) glove film; specified by glove film weight
Tetrachloroethane (1,1,2,2-)				18		1398		Pioneer Industrial Products Stanzoil N-44; glove film
Tetrachloroethylene (1,1,2,2-)				28	0.0002	453		"
Tetrahydrofuran	THF			11	0.08	4026		"
Toluene	toluol			14		2256		"
Trichloroethane	1,1,1-trichloroethane			27		1182		"
Trichloroethylene	TCE			11		3966		"
Trichlorotrifluoroethane		·		>480	0.2	0	no permeation rate at steady state detected	"
Tricresyl Phosphate	TCP	23	0.25		0	<0.9	low permeation rate; 0 to 1/2 eyedropper size drops per hour	Ansell Edmont Neoprene 29-840; unsupported glove film; 0.38 mm thick
	TCP	23	0.25		0	<0.9	low permeation rate; 0 to 1/2 eyedropper size drops per hour	Ansell Edmont Neox; supported (lined) glove film; specified by glove film weight
Triethanolamine	TEA			>480	5	0	no permeation rate at steady state detected	Pioneer Industrial Products Stanzoil N-44; glove film
	TEA; 85% conc.	23	0.25		0	<0.9	low permeation rate; 0 to 1/2 eyedropper size drops per hour	Ansell Edmont Neoprene 29-840; unsupported glove film; 0.38 mm thick
	TEA; 85% conc.	23	0.25		0	<0.9	low permeation rate; 0 to 1/2 eyedropper size drops per hour	Ansell Edmont Neox; supported (lined) glove film; specified by glove film weight
Trifluoroethanol				>60		0	no permeation rate at steady state detected	Pioneer Industrial Products Stanzoil N-44; glove film
Trihydroxytriethylamine				>480	5	0	no permeation rate at steady state detected	"
Turpentine				>480	0.02	0	no permeation rate at steady state detected	"
Vinyl Acetate				30	0.03	198		"
Xylene	xylol			23		798		"

ACRYLONITRILE-BUTADIENE COPOLYMER

Penetrant	Penetrant Note	Temp. (°C)	Time (days)	BTT (min.)	LDL (phr)	Permeation Rate (µg/cm²/min)	Comment	Material Note
Acetic Acid	glacial			118	0.1	1326		Pioneer Industrial Products Stansolv A-14; glove film
	glacial	23		270				Ansell Edmont Sol-Vex 37-165; unsupported glove film; 0.54 mm thick
	50% conc.	·		>480	0.02	0	no permeation rate at steady state detected	Pioneer Industrial Products Stansolv A-14; glove film
Acetonitrile	·	23		30		<900	fair permeation rate; 51 to 100 eyedropper size drops per hour	Ansell Edmont Sol-Vex 37-165; unsupported glove film; 0.54 mm thick
Acrylic Acid		23		120				"
Ammonium Fluoride	40% conc.	23	0.25		0	·	no permeation detected during a 6 hour test	"
Ammonium Hydroxide	concentrated	23	0.25		0		no permeation detected during a 6 hour test	"
	29% conc.			>480	1	0	no permeation rate at steady state detected	Pioneer Industrial Products Stansolv A-14; glove film
Amyl Acetate		23		60		<90	good permeation rate; 6 to 50 eyedropper size drops per hour	Ansell Edmont Sol-Vex 37-165; unsupported glove film; 0.54 mm thick
Amyl Alcohol		23		30		<0.9	low permeation rate; 0 to 1/2 eyedropper size drops per hour	"
Aniline				72	0.001	18		Pioneer Industrial Products Stansolv A-14; glove film
Aqua Regia		23	0.25		0		no permeation detected during a 6 hour test	Ansell Edmont Sol-Vex 37-165; unsupported glove film; 0.54 mm thick

ACRYLONITRILE-BUTADIENE COPOLYMER (cont'd)

Penetrant	Penetrant Note	Temp. (°C)	Time (days)	BTT (min.)	LDL (phr)	Permeation Rate (µg/cm²/min)	Comment	Material Note
Aroclor 1254	50% TCB			343	1	216		Pioneer Industrial Products Stansolv A-14; glove film
Benzene				27	0.03	582		"
Benzene Chloride				15		960		"
Bromopropionic Acid		23		120				Ansell Edmont Sol-Vex 37-165; unsupported glove film; 0.54 mm thick
Butanone (2-)				6		522		Pioneer Industrial Products Stansolv A-14; glove film
Butoxyethanol (2-)				>480	0.5	0	no permeation rate at steady state detected	"
Butyl Acetate				101	0.1	144		"
		23		75		<900	fair permeation rate; 51 to 100 eyedropper size drops per hour	Ansell Edmont Sol-Vex 37-165; unsupported glove film; 0.54 mm thick
Butyl Alcohol		23	0.25	0		<0.9	low permeation rate; 0 to 1/2 eyedropper size drops per hour	"
Butyl Cellosolve				>480	0.5	0	no permeation rate at steady state detected	Pioneer Industrial Products Stansolv A-14; glove film
		23		90		<9	very good permeation rate; 1 to 5 eyedropper size drops per hour	Ansell Edmont Sol-Vex 37-165; unsupported glove film; 0.54 mm thick
Carbolic Acid				>480		0	no permeation rate at steady state detected	Pioneer Industrial Products Stansolv A-14; glove film
Carbon Bichloride	carbon dichloride			373	0.0002	27		"
Carbon Bisulfide	carbon disulfide			20	0.2	516		"
Carbon Dichloride	carbon bichloride			373	0.0002	27		"
Carbon Disulfide	carbon bisulfide			20	0.2	516		"
		23		30		<900	fair permeation rate; 51 to 100 eyedropper size drops per hour	Ansell Edmont Sol-Vex 37-165; unsupported glove film; 0.54 mm thick
Carbon Tetrachloride				341	1	48		Pioneer Industrial Products Stansolv A-14; glove film
		23		150		<90	good permeation rate; 6 to 50 eyedropper size drops per hour	Ansell Edmont Sol-Vex 37-165; unsupported glove film; 0.54 mm thick
Cellosolve				416	0.03	24		Pioneer Industrial Products Stansolv A-14; glove film
	solvent	23		210		<90	good permeation rate; 6 to 50 eyedropper size drops per hour	Ansell Edmont Sol-Vex 37-165; unsupported glove film; 0.54 mm thick
Cellosolve Acetate				162	0.1	72		Pioneer Industrial Products Stansolv A-14; glove film
		23		90		<90	good permeation rate; 6 to 50 eyedropper size drops per hour	Ansell Edmont Sol-Vex 37-165; unsupported glove film; 0.54 mm thick
Chlorobenzene				15		960		Pioneer Industrial Products Stansolv A-14; glove film
Chlorothene				131	0.05	264		"
Chlorothene VG		23		90		<9000	poor permeation rate; 501 to 5000 eyedropper size drops per hour	Ansell Edmont Sol-Vex 37-165; unsupported glove film; 0.54 mm thick
Chlorotoluene (o-)				52		984		Pioneer Industrial Products Stansolv A-14; glove film
Chlorotoluene (p-)				25		888		"
Chromic Acid	50% conc.			>175	0.1		permeation rate too large to measure	"
	50% conc.	23		240				Ansell Edmont Sol-Vex 37-165; unsupported glove film; 0.54 mm thick
Citric Acid	10% conc.	23	0.25	0			no permeation detected during a 6 hour test	"
Cresol (3-)	m-Cresol			210	5	126		Pioneer Industrial Products Stansolv A-14; glove film
Cresol (m-)	3-Cresol			210	5	126		"
Cumene				271	0.03	48		"
Cyclohexane				>480	0.02	0	no permeation rate at steady state detected	"
Cyclohexyl Alcohol		23	0.25	0		<0.9	low permeation rate; 0 to 1/2 eyedropper size drops per hour	Ansell Edmont Sol-Vex 37-165; unsupported glove film; 0.54 mm thick
Diacetone Alcohol		23		240		<0.9	low permeation rate; 0 to 1/2 eyedropper size drops per hour	"
Diamine				>480	0.7	0	no permeation rate at steady state detected	Pioneer Industrial Products Stansolv A-14; glove film

ACRYLONITRILE-BUTADIENE COPOLYMER (cont'd)

Penetrant	Penetrant Note	Temp. (°C)	Time (days)	BTT (min.)	LDL (phr)	Permeation Rate (μg/cm²/min)	Comment	Material Note
Dibutyl Phthalate		23	0.25		0	<0.9	low permeation rate; 0 to 1/2 eyedropper size drops per hour	Ansell Edmont Sol-Vex 37-165; unsupported glove film; 0.54 mm thick
Dichlorobenzene (1,2-)	o-dichlorobenzene			37		1140		Pioneer Industrial Products Stansolv A-14; glove film
Dichlorobenzene (1,3-)				73	0.3	174		"
Dichlorobenzene (o-)	1,2-dichlorobenzene			37		1140		"
Dichloroethane (1,2-)				16	0.06	1752		"
Diethanolamine				>480	1.1	0	no permeation rate at steady state detected	"
Diethyl Ether				64	0.1	78		"
Diethylamine		23		45		<900	fair permeation rate; 51 to 100 eyedropper size drops per hour	Ansell Edmont Sol-Vex 37-165; unsupported glove film; 0.54 mm thick
Diisobutyl Ketone	DIBK	23		120		<900	fair permeation rate; 51 to 100 eyedropper size drops per hour	"
Dimethyl Sulfoxide	DMSO			0	0.004	6		Pioneer Industrial Products Stansolv A-14; glove film
	DMSO	23		>240		<9	very good permeation rate; 1 to 5 eyedropper size drops per hour	Ansell Edmont Sol-Vex 37-165; unsupported glove film; 0.54 mm thick
Dimethylacetamide	DMAC			28	0.001		permeation rate too large to measure	Pioneer Industrial Products Stansolv A-14; glove film
Dimethylformamide	DMF			35	0.2	246		"
Dioctyl Phthalate	DOP	23		>360		<0.9	low permeation rate; 0 to 1/2 eyedropper size drops per hour	Ansell Edmont Sol-Vex 37-165; unsupported glove film; 0.54 mm thick
Electroless Copper	MacDermid 9048	23	0.25		0		no permeation detected during a 6 hour test	"
Electroless Nickel	MacDermid J60/61	23	0.25		0		no permeation detected during a 6 hour test	"
Ethoxyethanol (2-)				416	0.03	24		Pioneer Industrial Products Stansolv A-14; glove film
Ethoxyethyl Acetate (2-)				162	0.1	72		"
Ethyl Alcohol	ethanol			>480	0.002	0	no permeation rate at steady state detected	"
		23		240		<9	very good permeation rate; 1 to 5 eyedropper size drops per hour	Ansell Edmont Sol-Vex 37-165; unsupported glove film; 0.54 mm thick
Ethyl Alcohol Amine	monoethanolamine	23	0.25		0	<0.9	low permeation rate; 0 to 1/2 eyedropper size drops per hour	"
Ethyl Ether				64	0.1	78		Pioneer Industrial Products Stansolv A-14; glove film
		23		120		<90	good permeation rate; 6 to 50 eyedropper size drops per hour	Ansell Edmont Sol-Vex 37-165; unsupported glove film; 0.54 mm thick
Ethyl Glycol Ether		23		210		<90	good permeation rate; 6 to 50 eyedropper size drops per hour	"
Ethylene Dichloride				16	0.06	1752		Pioneer Industrial Products Stansolv A-14; glove film
Ethylene Glycol				>480		0	no permeation rate at steady state detected	
		23	0.25		0	<0.9	low permeation rate; 0 to 1/2 eyedropper size drops per hour	Ansell Edmont Sol-Vex 37-165; unsupported glove film; 0.54 mm thick
Ethylene Oxide				32	0.3	126		Pioneer Industrial Products Stansolv A-14; glove film
Formaldehyde		23	0.25		0	<0.9	low permeation rate; 0 to 1/2 eyedropper size drops per hour	Ansell Edmont Sol-Vex 37-165; unsupported glove film; 0.54 mm thick
	37% conc.			>480	8	0	no permeation rate at steady state detected	Pioneer Industrial Products Stansolv A-14; glove film
Formalin	solution			>480	8	0	no permeation rate at steady state detected	"
Formic Acid	90% conc.	23		240				Ansell Edmont Sol-Vex 37-165; unsupported glove film; 0.54 mm thick
Freon 12				>480	8	0	no permeation rate at steady state detected	Pioneer Industrial Products Stansolv A-14; glove film
Freon TF				>480	0.01	0	no permeation rate at steady state detected	"
		23	0.25		0	<0.9	low permeation rate; 0 to 1/2 eyedropper size drops per hour	Ansell Edmont Sol-Vex 37-165; unsupported glove film; 0.54 mm thick
Gasoline				>480	0.1	0	no permeation rate at steady state detected	Pioneer Industrial Products Stansolv A-14; glove film
	white	23	0.25		0	<0.9	low permeation rate; 0 to 1/2 eyedropper size drops per hour	Ansell Edmont Sol-Vex 37-165; unsupported glove film; 0.54 mm thick

ACRYLONITRILE-BUTADIENE COPOLYMER (cont'd)

Penetrant	Penetrant Note	Temp. (°C)	Time (days)	BTT (min.)	LDL (phr)	Permeation Rate (µg/cm²/min)	Comment	Material Note
Heptane				2	0.01	0.018		Pioneer Industrial Products Stansolv A-14; glove film
Hexamethyldisilizane		23	0.25		0		no permeation detected during a 6 hour test	Ansell Edmont Sol-Vex 37-165; unsupported glove film; 0.54 mm thick
Hexane				>480	0.08	0	no permeation rate at steady state detected	Pioneer Industrial Products Stansolv A-14; glove film
		23	0.25		0	<0.9	low permeation rate; 0 to 1/2 eyedropper size drops per hour	Ansell Edmont Sol-Vex 37-165; unsupported glove film; 0.54 mm thick
Hydrazine				>480	0.7	0	no permeation rate at steady state detected	Pioneer Industrial Products Stansolv A-14; glove film
	65% conc.	23	0.25		0		no permeation detected during a 6 hour test	Ansell Edmont Sol-Vex 37-165; unsupported glove film; 0.54 mm thick
Hydrochloric Acid	concentrated	23	0.25		0		no permeation detected during a 6 hour test	"
	10% conc.	23	0.25		0		no permeation detected during a 6 hour test	"
	37.5% conc.			>480	0.4	0	no permeation rate at steady state detected	Pioneer Industrial Products Stansolv A-14; glove film
Hydrofluoric Acid	48% conc.		.	134	0.001	30		
	48% conc.	23		120				Ansell Edmont Sol-Vex 37-165; unsupported glove film; 0.54 mm thick
Hydrogen Peroxide	30% conc.	23	0.25		0		no permeation detected during a 6 hour test	"
Hydroquinone	saturated	23	0.25		0	<0.9	low permeation rate; 0 to 1/2 eyedropper size drops per hour	"
Hydroxyethyl Amine (bis(2-))				>480	1.1	0	no permeation rate at steady state detected	Pioneer Industrial Products Stansolv A-14; glove film
Isobutyl Alcohol		23	0.25		0	<0.9	low permeation rate; 0 to 1/2 eyedropper size drops per hour	Ansell Edmont Sol-Vex 37-165; unsupported glove film; 0.54 mm thick
Isooctane		23		360		<0.9	low permeation rate; 0 to 1/2 eyedropper size drops per hour	"
Isopropyl Alcohol	isopropanol; IPA			>480	0.05	0	no permeation rate at steady state detected	Pioneer Industrial Products Stansolv A-14; glove film
		23	0.25		0	<0.9	low permeation rate; 0 to 1/2 eyedropper size drops per hour	Ansell Edmont Sol-Vex 37-165; unsupported glove film; 0.54 mm thick
Isopropyl Benzene				271	0.03	48		Pioneer Industrial Products Stansolv A-14; glove film
Kerosine				>480	0.007	0	no permeation rate at steady state detected	"
		23	0.25		0	<0.9	low permeation rate; 0 to 1/2 eyedropper size drops per hour	Ansell Edmont Sol-Vex 37-165; unsupported glove film; 0.54 mm thick
Lactic Acid	85% conc.	23	0.25		0	<0.9	low permeation rate; 0 to 1/2 eyedropper size drops per hour	"
Lauric Acid	with ethylene oxide; 36% conc.	23	0.25		0		no permeation detected during a 6 hour test	"
Maleic Acid	saturated	23	0.25		0		no permeation detected during a 6 hour test	"
Methyl Alcohol	methanol			118	0.08	18		Pioneer Industrial Products Stansolv A-14; glove film
		23		11		<900	fair permeation rate; 51 to 100 eyedropper size drops per hour	Ansell Edmont Sol-Vex 37-165; unsupported glove film; 0.54 mm thick
Methyl Cellosolve		23	.	11		<90	good permeation rate; 6 to 50 eyedropper size drops per hour	"
Methyl Chloroform				131	0.05	264		Pioneer Industrial Products Stansolv A-14; glove film
Methyl Ethyl Ketone	MEK			6		522		"
Methyl Glycol Ether		23		11		<90	good permeation rate; 6 to 50 eyedropper size drops per hour	Ansell Edmont Sol-Vex 37-165; unsupported glove film; 0.54 mm thick
Methyl Tertiary Butyl Ether	MTBE	23	0.25		0	<0.9	low permeation rate; 0 to 1/2 eyedropper size drops per hour	"
Methylamine		23	0.25		0	<0.9	low permeation rate; 0 to 1/2 eyedropper size drops per hour	"
Methylphenol (3-)	m-methylphenol			210	5	126		Pioneer Industrial Products Stansolv A-14; glove film
Methylphenol (m-)	3-methylphenol			210	5	126		"
Mineral Spirits				>480	0.02	0	no permeation rate at steady state detected	"
	Rule 66	23	0.25		0	<0.9	low permeation rate; 0 to 1/2 eyedropper size drops per hour	Ansell Edmont Sol-Vex 37-165; unsupported glove film; 0.54 mm thick

ACRYLONITRILE-BUTADIENE COPOLYMER (cont'd)

Penetrant	Penetrant Note	Temp. (°C)	Time (days)	BTT (min.)	LDL (phr)	Permeation Rate (µg/cm²/min)	Comment	Material Note
Muriatic Acid				>480	0.4	0	no permeation rate at steady state detected	Pioneer Industrial Products Stansolv A-14; glove film
	10% conc.	23	0.25		0		no permeation detected during a 6 hour test	Ansell Edmont Sol-Vex 37-165; unsupported glove film; 0.54 mm thick
Naphtha				>480	0.02	0	no permeation rate at steady state detected	Pioneer Industrial Products Stansolv A-14; glove film
	VM&P	23	0.25		0	<0.9	low permeation rate; 0 to 1/2 eyedropper size drops per hour	Ansell Edmont Sol-Vex 37-165; unsupported glove film; 0.54 mm thick
Nitric Acid	10% conc.	23	0.25		0		no permeation detected during a 6 hour test	"
	50% conc.			72	0.07	1206		Pioneer Industrial Products Stansolv A-14; glove film
Nitrobenzene				60	1	90		"
Nitromethane	95.5% conc.	23		30		<900	fair permeation rate; 51 to 100 eyedropper size drops per hour	Ansell Edmont Sol-Vex 37-165; unsupported glove film; 0.54 mm thick
Octyl Alcohol		23	0.25		0	<0.9	low permeation rate; 0 to 1/2 eyedropper size drops per hour	"
Oleic Acid		23	0.25		0	<0.9	low permeation rate; 0 to 1/2 eyedropper size drops per hour	"
Oxalic Acid	saturated	23	0.25		0		no permeation detected during a 6 hour test	"
Palmitic Acid	saturated	23		30				"
PCBs	polychlorinated biphenyls; 50% TCB			343	1	216		Pioneer Industrial Products Stansolv A-14; glove film
Pentachlorophenol		23	0.25		0	<0.9	low permeation rate; 0 to 1/2 eyedropper size drops per hour	Ansell Edmont Sol-Vex 37-165; unsupported glove film; 0.54 mm thick
Pentane		23	0.25		0	<0.9	low permeation rate; 0 to 1/2 eyedropper size drops per hour	"
Perchloric Acid	60% conc.	23	0.25		0		no permeation detected during a 6 hour test	"
Perchloroethylene				373	0.0002	27		Pioneer Industrial Products Stansolv A-14; glove film
		23		300		<9	very good permeation rate; 1 to 5 eyedropper size drops per hour	Ansell Edmont Sol-Vex 37-165; unsupported glove film; 0.54 mm thick
Petroleum Ether				>480	0.02	0	no permeation rate at steady state detected	Pioneer Industrial Products Stansolv A-14; glove film
Phenol	saturated			>480		0	no permeation rate at steady state detected	"
Phosphoric Acid	concentrated	23	0.25		0		no permeation detected during a 6 hour test	Ansell Edmont Sol-Vex 37-165; unsupported glove film; 0.54 mm thick
	85% conc.			>480	0.04	0	no permeation rate at steady state detected	Pioneer Industrial Products Stansolv A-14; glove film
Picric Acid	saturated; with ethylene oxide	23		160		<9	very good permeation rate; 1 to 5 eyedropper size drops per hour	Ansell Edmont Sol-Vex 37-165; unsupported glove film; 0.54 mm thick
Polychlorinated Biphenyls	PCBs; 50% TCB			343	1	216		Pioneer Industrial Products Stansolv A-14; glove film
Potassium Hydroxide	KOH; 50% conc.			>480	0.4	0	no permeation rate at steady state detected	"
	KOH; 50% conc.	23	0.25		0		no permeation detected during a 6 hour test	Ansell Edmont Sol-Vex 37-165; unsupported glove film; 0.54 mm thick
Propyl Acetate		23		200		<90	good permeation rate; 6 to 50 eyedropper size drops per hour	"
Propyl Alcohol		23	0.25		0	<0.9	low permeation rate; 0 to 1/2 eyedropper size drops per hour	"
Rubber Solvent	-	23	0.25		0	<0.9	low permeation rate; 0 to 1/2 eyedropper size drops per hour	"
Rule 66	solvent			>480	0.001	0	no permeation rate at steady state detected	Pioneer Industrial Products Stansolv A-14; glove film
Sodium Hydroxide	NAOH; 50% conc.			>480	0.1	0	no permeation rate at steady state detected	"
	NaOH; 50% conc.	23	0.25		0		no permeation detected during a 6 hour test	Ansell Edmont Sol-Vex 37-165; unsupported glove film; 0.54 mm thick
Stoddard Solvents		23	0.25		0	<0.9	low permeation rate; 0 to 1/2 eyedropper size drops per hour	"
Sulfuric Acid	battery acid; 47% conc.	23	0.25		0		no permeation detected during a 6 hour test	"
	50% conc.			>480	0.04	0	no permeation rate at steady state detected	Pioneer Industrial Products Stansolv A-14; glove film
Tannic Acid	65% conc.	23	0.25		0	<0.9	low permeation rate; 0 to 1/2 eyedropper size drops per hour	Ansell Edmont Sol-Vex 37-165; unsupported glove film; 0.54 mm thick
Tetrachloroethene		23		300		<9	very good permeation rate; 1 to 5 eyedropper size drops per hour	"

ACRYLONITRILE-BUTADIENE COPOLYMER (cont'd)

Penetrant	Penetrant Note	Temp. (°C)	Time (days)	BTT (min.)	LDL (phr)	Permeation Rate (μg/cm²/min)	Comment	Material Note
Tetrachloroethylene (1,1,2,2-)				373	0.0002	27		Pioneer Industrial Products Stansolv A-14; glove film
Tetrahydrofuran	THF			17	0.08	4026		"
Toluene	toluol			28	0.002	150		
	toluol	23		10		<900	fair permeation rate; 51 to 100 eyedropper size drops per hour	Ansell Edmont Sol-Vex 37-165; unsupported glove film; 0.54 mm thick
Toluene Diisocyanate	TDI			>480	0.3	0	no permeation rate at steady state detected	Pioneer Industrial Products Stansolv A-14; glove film
Trichloroethane	1,1,1-trichloroethane			131	0.05	264		"
	1,1,1-trichloroethane	23		90		<9000	poor permeation rate; 501 to 5000 eyedropper size drops per hour	Ansell Edmont Sol-Vex 37-165; unsupported glove film; 0.54 mm thick
Trichloroethylene	TCE			9	0.002	372		Pioneer Industrial Products Stansolv A-14; glove film
Trichlorotrifluoroethane				>480	0.01	0	no permeation rate at steady state detected	"
Tricresyl Phosphate	TCP	23	0.25		0	<0.9	low permeation rate; 0 to 1/2 eyedropper size drops per hour	Ansell Edmont Sol-Vex 37-165; unsupported glove film; 0.54 mm thick
Triethanolamine	TEA			>480	5	0	no permeation rate at steady state detected	Pioneer Industrial Products Stansolv A-14; glove film
	TEA; 85% conc.	23	0.25		0	<0.9	low permeation rate; 0 to 1/2 eyedropper size drops per hour	Ansell Edmont Sol-Vex 37-165; unsupported glove film; 0.54 mm thick
Trihydroxytriethylamine				>480	5	0	no permeation rate at steady state detected	Pioneer Industrial Products Stansolv A-14; glove film
Turpentine				>480	0.0009	0	no permeation rate at steady state detected	"
		23		30		<0.9	low permeation rate; 0 to 1/2 eyedropper size drops per hour	Ansell Edmont Sol-Vex 37-165; unsupported glove film; 0.54 mm thick
Vinyl Acetate				30	0.08	402		Pioneer Industrial Products Stansolv A-14; glove film
Xylene	xylol			92	0.002	24		
	xylol	23		75		<900	fair permeation rate; 51 to 100 eyedropper size drops per hour	Ansell Edmont Sol-Vex 37-165; unsupported glove film; 0.54 mm thick

POLYVINYL ALCOHOL

Penetrant	Penetrant Note	Temp. (°C)	Time (days)	BTT (min.)	LDL (phr)	Permeation Rate (μg/cm²/min)	Comment	Material Note
Acetonitrile		23		150		<90	good permeation rate; 6 to 50 eyedropper size drops per hour	Ansell Edmont PVA; supported (lined) glove film; specified by glove film weight
Amyl Acetate		23	0.25		0	<0.9	low permeation rate; 0 to 1/2 eyedropper size drops per hour	"
Amyl Alcohol		23		180		<90	good permeation rate; 6 to 50 eyedropper size drops per hour	"
Aniline		23	0.25		0	<0.9	low permeation rate; 0 to 1/2 eyedropper size drops per hour	"
Benzaldehyde		23	0.25		0	<0.9	low permeation rate; 0 to 1/2 eyedropper size drops per hour	"
Benzene	benzol	23	0.25		0	<0.9	low permeation rate; 0 to 1/2 eyedropper size drops per hour	"
Butyl Acetate		23	0.25		0	<0.9	low permeation rate; 0 to 1/2 eyedropper size drops per hour	"
Butyl Alcohol		23		75		<90	good permeation rate; 6 to 50 eyedropper size drops per hour	"
Butyl Cellosolve		23		120		<90	good permeation rate; 6 to 50 eyedropper size drops per hour	"
Butyrolactone (gamma-)		23		120		<9	very good permeation rate; 1 to 5 eyedropper size drops per hour	"
Carbon Disulfide		23	0.25		0	<0.9	low permeation rate; 0 to 1/2 eyedropper size drops per hour	"
Carbon Tetrachloride		23	0.25		0	<0.9	low permeation rate; 0 to 1/2 eyedropper size drops per hour	"
Cellosolve	solvent	23		75		<90	good permeation rate; 6 to 50 eyedropper size drops per hour	"
Cellosolve Acetate		23	0.25		0	<0.9	low permeation rate; 0 to 1/2 eyedropper size drops per hour	"

POLYVINYL ALCOHOL (cont'd)

Penetrant	Penetrant Note	Temp. (°C)	Time (days)	BTT (min.)	LDL (phr)	Permeation Rate (µg/cm²/min)	Comment	Material Note
Chlorobenzene		23	0.25		0	<0.9	low permeation rate; 0 to 1/2 eyedropper size drops per hour	"
Chloroform		23	0.25		0	<0.9	low permeation rate; 0 to 1/2 eyedropper size drops per hour	"
Chloronaphthalene		23	0.25		0	<0.9	low permeation rate; 0 to 1/2 eyedropper size drops per hour	"
Chlorothene VG		23	0.25		0	<0.9	low permeation rate; 0 to 1/2 eyedropper size drops per hour	"
Citric Acid	10% conc.	23		50				"
Cyclohexyl Alcohol		23	0.25		0	<0.9	low permeation rate; 0 to 1/2 eyedropper size drops per hour	"
Diacetone Alcohol		23		150		<90	good permeation rate; 6 to 50 eyedropper size drops per hour	"
Dibutyl Phthalate		23	0.25		0	<0.9	low permeation rate; 0 to 1/2 eyedropper size drops per hour	"
Diisobutyl Ketone	DIBK	23	0.25		0	<0.9	low permeation rate; 0 to 1/2 eyedropper size drops per hour	"
Dioctyl Phthalate	DOP	23		30		<900	fair permeation rate; 51 to 100 eyedropper size drops per hour	"
Epichlorohydrin		23		300			low permeation rate; 0 to 1/2 eyedropper size drops per hour	"
Ethyl Acetate		23	0.25		0	<0.9	low permeation rate; 0 to 1/2 eyedropper size drops per hour	"
Ethyl Alcohol Amine	monoethanolamine	23	0.25		0	<0.9	low permeation rate; 0 to 1/2 eyedropper size drops per hour	"
Ethyl Ether		23	0.25		0	<0.9	low permeation rate; 0 to 1/2 eyedropper size drops per hour	"
Ethyl Glycol Ether		23		75		<90	good permeation rate; 6 to 50 eyedropper size drops per hour	"
Ethylene Dichloride		23	0.25		0	<0.9	low permeation rate; 0 to 1/2 eyedropper size drops per hour	"
Ethylene Glycol		23		120		<9	very good permeation rate; 1 to 5 eyedropper size drops per hour	"
Freon TF		23	0.25		0	<0.9	low permeation rate; 0 to 1/2 eyedropper size drops per hour	"
Freon TMC		23	0.25		0	<0.9	low permeation rate; 0 to 1/2 eyedropper size drops per hour	"
Furfural		23	0.25		0	<0.9	low permeation rate; 0 to 1/2 eyedropper size drops per hour	"
Gasoline	white	23	0.25		0	<0.9	low permeation rate; 0 to 1/2 eyedropper size drops per hour	"
Hexamethyldisilizane		23	0.25		0		no permeation detected during a 6 hour test	"
Hexane		23	0.25		0	<0.9	low permeation rate; 0 to 1/2 eyedropper size drops per hour	"
Isooctane		23	0.25		0	<0.9	low permeation rate; 0 to 1/2 eyedropper size drops per hour	"
Kerosine		23	0.25		0	<0.9	low permeation rate; 0 to 1/2 eyedropper size drops per hour	"
Lactic Acid	85% conc.	23	0.25		0	<0.9	low permeation rate; 0 to 1/2 eyedropper size drops per hour	"
Methyl Cellosolve		23		30		<90	good permeation rate; 6 to 50 eyedropper size drops per hour	"
Methyl Ethyl Ketone	MEK	23		90		<9	very good permeation rate; 1 to 5 eyedropper size drops per hour	"
Methyl Glycol Ether		23		30		<90	good permeation rate; 6 to 50 eyedropper size drops per hour	"
Methyl Iodide		23	0.25		0	<0.9	low permeation rate; 0 to 1/2 eyedropper size drops per hour	"
Methyl Isobutyl Ketone	MIBK	23	0.25		0	<0.9	low permeation rate; 0 to 1/2 eyedropper size drops per hour	"
Methyl Methacrylate		23	0.25		0	<0.9	low permeation rate; 0 to 1/2 eyedropper size drops per hour	"
Methyl Tertiary Butyl Ether	MTBE	23	0.25		0	<0.9	low permeation rate; 0 to 1/2 eyedropper size drops per hour	"
Methylene Bromide		23	0.25		0	<0.9	low permeation rate; 0 to 1/2 eyedropper size drops per hour	"
Methylene Chloride		23	0.25		0	<0.9	low permeation rate; 0 to 1/2 eyedropper size drops per hour	"
Mineral Spirits	Rule 66	23	0.25		0	<0.9	low permeation rate; 0 to 1/2 eyedropper size drops per hour	"
Morpholine		23		90		<90	good permeation rate; 6 to 50 eyedropper size drops per hour	"
Naphtha	VM&P	23		>420		<0.9	low permeation rate; 0 to 1/2 eyedropper size drops per hour	"
Nitrobenzene		23	0.25		0	<0.9	low permeation rate; 0 to 1/2 eyedropper size drops per hour	"
Nitromethane	95.5% conc.	23	0.25		0	<0.9	low permeation rate; 0 to 1/2 eyedropper size drops per hour	"

POLYVINYL ALCOHOL (cont'd)

Penetrant	Penetrant Note	Temp. (°C)	Time (days)	BTT (min.)	LDL (phr)	Permeation Rate (µg/cm²/min)	Comment	Material Note
Nitropropane	95.5% conc.	23		>360		<0.9	low permeation rate; 0 to 1/2 eyedropper size drops per hour	"
Octyl Alcohol		23	0.25		0	<0.9	low permeation rate; 0 to 1/2 eyedropper size drops per hour	"
Oleic Acid		23		60		<0.9	low permeation rate; 0 to 1/2 eyedropper size drops per hour	"
Pentachlorophenol		23		5		<900	fair permeation rate; 51 to 100 eyedropper size drops per hour	"
Pentane		23	0.25		0	<0.9	low permeation rate; 0 to 1/2 eyedropper size drops per hour	"
Perchloroethylene		23	0.25		0	<0.9	low permeation rate; 0 to 1/2 eyedropper size drops per hour	"
Phenol		23	0.25		0	<0.9	low permeation rate; 0 to 1/2 eyedropper size drops per hour	"
Propyl Acetate		23		120		<9	very good permeation rate; 1 to 5 eyedropper size drops per hour	"
Propylene Oxide		23		35		<90	good permeation rate; 6 to 50 eyedropper size drops per hour	"
Pyridine		23		10		<900	fair permeation rate; 51 to 100 eyedropper size drops per hour	"
Rubber Solvent		23	0.25		0	<0.9	low permeation rate; 0 to 1/2 eyedropper size drops per hour	"
Stoddard Solvents		23	0.25		0	<0.9	low permeation rate; 0 to 1/2 eyedropper size drops per hour	"
Styrene		23	0.25		0	<0.9	low permeation rate; 0 to 1/2 eyedropper size drops per hour	"
Tetrachloroethene		23	0.25		0	<0.9	low permeation rate; 0 to 1/2 eyedropper size drops per hour	"
Tetrahydrofuran	THF	23		90		<90	good permeation rate; 6 to 50 eyedropper size drops per hour	"
Toluene	toluol	23	0.25		0	<0.9	low permeation rate; 0 to 1/2 eyedropper size drops per hour	"
Toluene Diisocyanate	TDI	23	0.25		0	<0.9	low permeation rate; 0 to 1/2 eyedropper size drops per hour	"
Trichloroethane	1,1,1-trichloroethane	23	0.25		0	<0.9	low permeation rate; 0 to 1/2 eyedropper size drops per hour	"
Trichloroethylene	TCE	23	0.25		0	<0.9	low permeation rate; 0 to 1/2 eyedropper size drops per hour	"
Tricresyl Phosphate	TCP	23	0.25		0	<0.9	low permeation rate; 0 to 1/2 eyedropper size drops per hour	"
Triethanolamine	TEA; 85% conc.	23	0.25		0	<0.9	low permeation rate; 0 to 1/2 eyedropper size drops per hour	"
Turpentine		23	0.25		0	<0.9	low permeation rate; 0 to 1/2 eyedropper size drops per hour	"
Xylene	xylol	23	0.25		0	<0.9	low permeation rate; 0 to 1/2 eyedropper size drops per hour	"

POLYVINYL CHLORIDE POLYOL

Penetrant	Penetrant Note	Temp. (°C)	Time (days)	BTT (min.)	LDL (phr)	Permeation Rate (µg/cm²/min)	Comment	Material Note
Acetaldehyde		23		5		<9000	poor permeation rate; 501 to 5000 eyedropper size drops per hour	Ansell Edmont Wet-Wear 600; clothing; nylon netting bonded between two layers of PVC
Acetic Acid	glacial			85	0.1	1.8		Pioneer Industrial Products Pylox V-20; glove film
	glacial	23		6				Ansell Edmont Wet-Wear 600; clothing; nylon netting bonded between two layers of PVC
	glacial	23		30				Ansell Edmont Wet-Wear 550; clothing; PVC stretch outer layer bonded to a lightweight liner
	glacial	23		180				Ansell Edmont Monkey Grip; supported (lined) glove film; specified by glove film weight
	50% conc.			47	0.02	0.36		Pioneer Industrial Products Pylox V-20; glove film
Acetone				<1			permeation rate too large to measure	"
Ammonium Fluoride	40% conc.	23	0.25		0		no permeation detected during a 6 hour test	Ansell Edmont Monkey Grip; supported (lined) glove film; specified by glove film weight

Appendix III - Permeability of Rubber Glove Films

POLYVINYL CHLORIDE POLYOL (cont'd)

Penetrant	Penetrant Note	Temp. (°C)	Time (days)	BTT (min.)	LDL (phr)	Permeation Rate (µg/cm²/min)	Comment	Material Note
Ammonium Hydroxide	concentrated	23		18				Ansell Edmont Wet-Wear 600; clothing; nylon netting bonded between two layers of PVC
	concentrated	23		240				Ansell Edmont Monkey Grip; supported (lined) glove film; specified by glove film weight
	concentrated	23	0.25		0		no permeation detected during a 6 hour test	Ansell Edmont Wet-Wear 550; clothing; PVC stretch outer layer bonded to a lightweight liner
	29% conc.			>480	1	0	no permeation rate at steady state detected	Pioneer Industrial Products Pylox V-20; glove film
Amyl Alcohol		23		12		<0.9	low permeation rate; 0 to 1/2 eyedropper size drops per hour	Ansell Edmont Monkey Grip; supported (lined) glove film; specified by glove film weight
Aniline				>480	0.009	0	no permeation rate at steady state detected	Pioneer Industrial Products Pylox V-20; glove film
		23		180		<9	very good permeation rate; 1 to 5 eyedropper size drops per hour	Ansell Edmont Monkey Grip; supported (lined) glove film; specified by glove film weight
		23		12		<9	very good permeation rate; 1 to 5 eyedropper size drops per hour	Ansell Edmont Wet-Wear 550; clothing; PVC stretch outer layer bonded to a lightweight liner
		23		30		<9	very good permeation rate; 1 to 5 eyedropper size drops per hour	Ansell Edmont Wet-Wear 600; clothing; nylon netting bonded between two layers of PVC
Aqua Regia		23		120				Ansell Edmont Monkey Grip; supported (lined) glove film; specified by glove film weight
Benzene				2		1500		Pioneer Industrial Products Pylox V-20; glove film
Bromopropionic Acid		23		180				Ansell Edmont Monkey Grip; supported (lined) glove film; specified by glove film weight
Butanone (2-)				1			permeation rate too large to measure	Pioneer Industrial Products Pylox V-20; glove film
Butyl Alcohol		23		180		<9	very good permeation rate; 1 to 5 eyedropper size drops per hour	Ansell Edmont Monkey Grip; supported (lined) glove film; specified by glove film weight
		23		40		<90	good permeation rate; 6 to 50 eyedropper size drops per hour	Ansell Edmont Wet-Wear 600; clothing; nylon netting bonded between two layers of PVC
		23		8		<90	good permeation rate; 6 to 50 eyedropper size drops per hour	Ansell Edmont Wet-Wear 550; clothing; PVC stretch outer layer bonded to a lightweight liner
Carbolic Acid				32		78		Pioneer Industrial Products Pylox V-20; glove film
Carbon Tetrachloride		23		25		<900	fair permeation rate; 51 to 100 eyedropper size drops per hour	Ansell Edmont Monkey Grip; supported (lined) glove film; specified by glove film weight
Chromic Acid	50% conc.			>480	0.1	0	no permeation rate at steady state detected	Pioneer Industrial Products Pylox V-20; glove film
	50% conc.	23	0.25		0		no permeation detected during a 6 hour test	Ansell Edmont Monkey Grip; supported (lined) glove film; specified by glove film weight
	50% conc.	23	0.25		0		no permeation detected during a 6 hour test	Ansell Edmont Wet-Wear 550; clothing; PVC stretch outer layer bonded to a lightweight liner
	50% conc.	23	0.25		0		no permeation detected during a 6 hour test	Ansell Edmont Wet-Wear 600; clothing; nylon netting bonded between two layers of PVC
Citric Acid	10% conc.	23	0.25		0		no permeation detected during a 6 hour test	Ansell Edmont Monkey Grip; supported (lined) glove film; specified by glove film weight
Cresol (3-)	m-Cresol			150	5	36		Pioneer Industrial Products Pylox V-20; glove film
Cresol (m-)	3-Cresol			150	5	36		"
Cyclohexane				16		102		"
Cyclohexyl Alcohol		23		360		<0.9	low permeation rate; 0 to 1/2 eyedropper size drops per hour	Ansell Edmont Monkey Grip; supported (lined) glove film; specified by glove film weight
		23		200		<0.9	low permeation rate; 0 to 1/2 eyedropper size drops per hour	Ansell Edmont Wet-Wear 550; clothing; PVC stretch outer layer bonded to a lightweight liner
		23	0.25		0	<0.9	low permeation rate; 0 to 1/2 eyedropper size drops per hour	Ansell Edmont Wet-Wear 600; clothing; nylon netting bonded between two layers of PVC
Diamine				>480	0.7	0	no permeation rate at steady state detected	Pioneer Industrial Products Pylox V-20; glove film
Dibutyl Phthalate		23	0.25		0		no permeation detected during a 6 hour test	Ansell Edmont Wet-Wear 600; clothing; nylon netting bonded between two layers of PVC
Diethanolamine				>480	1.1	0	no permeation rate at steady state detected	Pioneer Industrial Products Pylox V-20; glove film
Diethylene Dioxide (1,4-)				6		1500		"

POLYVINYL CHLORIDE POLYOL (cont'd)

Penetrant	Penetrant Note	Temp. (°C)	Time (days)	BTT (min.)	LDL (phr)	Permeation Rate (µg/cm²/min)	Comment	Material Note
Dimethyl Sulfoxide	DMSO			60	0.004	0	no permeation rate at steady state detected	"
Dimethylacetamide	DMAC			20	0.005		permeation rate too large to measure	"
Dioctyl Phthalate	DOP	23		11				Ansell Edmont Wet-Wear 550; clothing; PVC stretch outer layer bonded to a lightweight liner
	DOP	23	0.25		0	<0.9	low permeation rate; 0 to 1/2 eyedropper size drops per hour	Ansell Edmont Wet-Wear 600; clothing; nylon netting bonded between two layers of PVC
Dioxane (1,4-)				6		1500		Pioneer Industrial Products Pylox V-20; glove film
Electroless Copper	MacDermid 9048	23	0.25		0		no permeation detected during a 6 hour test	Ansell Edmont Monkey Grip; supported (lined) glove film; specified by glove film weight
Electroless Nickel	MacDermid J60/61	23	0.25		0		no permeation detected during a 6 hour test	"
Ethyl Alcohol	ethanol			20		30		Pioneer Industrial Products Pylox V-20; glove film
		23		60		<9	very good permeation rate; 1 to 5 eyedropper size drops per hour	Ansell Edmont Monkey Grip; supported (lined) glove film; specified by glove film weight
Ethyl Alcohol Amine	monoethanolamine	23	0.25		0	<0.9	low permeation rate; 0 to 1/2 eyedropper size drops per hour	
	monoethanolamine	23	0.25		0		no permeation detected during a 6 hour test	Ansell Edmont Wet-Wear 600; clothing; nylon netting bonded between two layers of PVC
	monoethanolamine	23	0.25		0		no permeation detected during a 6 hour test	Ansell Edmont Wet-Wear 550; clothing; PVC stretch outer layer bonded to a lightweight liner
Ethylene Glycol				>480		0	no permeation rate at steady state detected	Pioneer Industrial Products Pylox V-20; glove film
		23		200				Ansell Edmont Wet-Wear 550; clothing; PVC stretch outer layer bonded to a lightweight liner
		23	0.25		0	<0.9	low permeation rate; 0 to 1/2 eyedropper size drops per hour	Ansell Edmont Monkey Grip; supported (lined) glove film; specified by glove film weight
		23	0.25		0	<0.9	low permeation rate; 0 to 1/2 eyedropper size drops per hour	Ansell Edmont Wet-Wear 600; clothing; nylon netting bonded between two layers of PVC
Formaldehyde		23		360		<90	good permeation rate; 6 to 50 eyedropper size drops per hour	"
		23		5		<9	very good permeation rate; 1 to 5 eyedropper size drops per hour	Ansell Edmont Wet-Wear 550; clothing; PVC stretch outer layer bonded to a lightweight liner
		23		80		<9	very good permeation rate; 1 to 5 eyedropper size drops per hour	Ansell Edmont Monkey Grip; supported (lined) glove film; specified by glove film weight
	37% conc.			>480	0.01	0	no permeation rate at steady state detected	Pioneer Industrial Products Pylox V-20; glove film
Formalin	solution			>480	0.01	0	no permeation rate at steady state detected	"
Formic Acid	90% conc.	23		25				Ansell Edmont Wet-Wear 550; clothing; PVC stretch outer layer bonded to a lightweight liner
	90% conc.	23		>360				Ansell Edmont Monkey Grip; supported (lined) glove film; specified by glove film weight
	90% conc.	23		75				Ansell Edmont Wet-Wear 600; clothing; nylon netting bonded between two layers of PVC
Freon TF				11		192		Pioneer Industrial Products Pylox V-20; glove film
		23		6		<900	fair permeation rate; 51 to 100 eyedropper size drops per hour	Ansell Edmont Wet-Wear 550; clothing; PVC stretch outer layer bonded to a lightweight liner
		23		60		<90	good permeation rate; 6 to 50 eyedropper size drops per hour	Ansell Edmont Wet-Wear 600; clothing; nylon netting bonded between two layers of PVC
Gasoline	white	23		5		<900	fair permeation rate; 51 to 100 eyedropper size drops per hour	"
	white	23		6		<90	good permeation rate; 6 to 50 eyedropper size drops per hour	Ansell Edmont Wet-Wear 550; clothing; PVC stretch outer layer bonded to a lightweight liner
Hydrazine				>480	0.7	0	no permeation rate at steady state detected	Pioneer Industrial Products Pylox V-20; glove film
	65% conc.	23	0.25		0		no permeation detected during a 6 hour test	Ansell Edmont Monkey Grip; supported (lined) glove film; specified by glove film weight
	65% conc.	23	0.25		0		no permeation detected during a 6 hour test	Ansell Edmont Wet-Wear 600; clothing; nylon netting bonded between two layers of PVC

POLYVINYL CHLORIDE POLYOL (cont'd)

Penetrant	Penetrant Note	Temp. (°C)	Time (days)	BTT (min.)	LDL (phr)	Permeation Rate (µg/cm²/min)	Comment	Material Note
Hydrochloric Acid	concentrated	23		>300				Ansell Edmont Monkey Grip; supported (lined) glove film; specified by glove film weight
	10% conc.	23		300				Ansell Edmont Wet-Wear 600; clothing; nylon netting bonded between two layers of PVC
	10% conc.	23	0.25		0		no permeation detected during a 6 hour test	Ansell Edmont Wet-Wear 550; clothing; PVC stretch outer layer bonded to a lightweight liner
	10% conc.	23	0.25		0		no permeation detected during a 6 hour test	Ansell Edmont Monkey Grip; supported (lined) glove film; specified by glove film weight
	37.5% conc.			>480	4	0	no permeation rate at steady state detected	Pioneer Industrial Products Pylox V-20; glove film
Hydrofluoric Acid	48% conc.			110	1	0	no permeation rate at steady state detected	"
	48% conc.	23		40				Ansell Edmont Monkey Grip; supported (lined) glove film; specified by glove film weight
	48% conc.	23		75				Ansell Edmont Wet-Wear 550; clothing; PVC stretch outer layer bonded to a lightweight liner
	48% conc.	23		90				Ansell Edmont Wet-Wear 600; clothing; nylon netting bonded between two layers of PVC
Hydrogen Peroxide	30% conc.	23	0.25		0		no permeation detected during a 6 hour test	Ansell Edmont Monkey Grip; supported (lined) glove film; specified by glove film weight
Hydroquinone	saturated	23	0.25		0	<0.9	low permeation rate; 0 to 1/2 eyedropper size drops per hour	"
Hydroxyethyl Amine (bis(2-))				>480	1.1	0	no permeation rate at steady state detected	Pioneer Industrial Products Pylox V-20; glove film
Isoamyl Acetate				5		1602		
Isobutyl Alcohol		23		200		<9	very good permeation rate; 1 to 5 eyedropper size drops per hour	Ansell Edmont Wet-Wear 550; clothing; PVC stretch outer layer bonded to a lightweight liner
		23		120		<9	very good permeation rate; 1 to 5 eyedropper size drops per hour	Ansell Edmont Wet-Wear 600; clothing; nylon netting bonded between two layers of PVC
		23		10		<9	very good permeation rate; 1 to 5 eyedropper size drops per hour	Ansell Edmont Monkey Grip; supported (lined) glove film; specified by glove film weight
Isopropyl Alcohol	isopropanol; IPA			208		0	no permeation rate at steady state detected	Pioneer Industrial Products Pylox V-20; glove film
		23		150		<0.9	low permeation rate; 0 to 1/2 eyedropper size drops per hour	Ansell Edmont Monkey Grip; supported (lined) glove film; specified by glove film weight
Kerosine		23		>360		<0.9	low permeation rate; 0 to 1/2 eyedropper size drops per hour	"
		23		180		<90	good permeation rate; 6 to 50 eyedropper size drops per hour	Ansell Edmont Wet-Wear 600; clothing; nylon netting bonded between two layers of PVC
		23		10		<9	very good permeation rate; 1 to 5 eyedropper size drops per hour	Ansell Edmont Wet-Wear 550; clothing; PVC stretch outer layer bonded to a lightweight liner
Lactic Acid	85% conc.	23	0.25		0	<0.9	low permeation rate; 0 to 1/2 eyedropper size drops per hour	Ansell Edmont Monkey Grip; supported (lined) glove film; specified by glove film weight
Lauric Acid	with ethylene oxide; 36% conc.	23		15				"
Maleic Acid	saturated	23	0.25		0		no permeation detected during a 6 hour test	Ansell Edmont Wet-Wear 600; clothing; nylon netting bonded between two layers of PVC
	saturated	23	0.25		0		no permeation detected during a 6 hour test	Ansell Edmont Wet-Wear 550; clothing; PVC stretch outer layer bonded to a lightweight liner
	saturated	23	0.25		0		no permeation detected during a 6 hour test	Ansell Edmont Monkey Grip; supported (lined) glove film; specified by glove film weight
Methyl Alcohol	methanol			3		18		Pioneer Industrial Products Pylox V-20; glove film
		23		30		<90	good permeation rate; 6 to 50 eyedropper size drops per hour	Ansell Edmont Wet-Wear 600; clothing; nylon netting bonded between two layers of PVC
Methyl Alcohol		23		45		<90	good permeation rate; 6 to 50 eyedropper size drops per hour	Ansell Edmont Monkey Grip; supported (lined) glove film; specified by glove film weight
		23		5		<90	good permeation rate; 6 to 50 eyedropper size drops per hour	Ansell Edmont Wet-Wear 550; clothing; PVC stretch outer layer bonded to a lightweight liner
Methyl Ethyl Ketone	MEK			1			permeation rate too large to measure	Pioneer Industrial Products Pylox V-20; glove film
Methyl Iodide				1			permeation rate too large to measure	"

POLYVINYL CHLORIDE POLYOL (cont'd)

Penetrant	Penetrant Note	Temp. (°C)	Time (days)	BTT (min.)	LDL (phr)	Permeation Rate (µg/cm²/min)	Comment	Material Note
Methylamine		23		60		<9	very good permeation rate; 1 to 5 eyedropper size drops per hour	Ansell Edmont Wet-Wear 600; clothing; nylon netting bonded between two layers of PVC
		23		5		<0.9	low permeation rate; 0 to 1/2 eyedropper size drops per hour	Ansell Edmont Wet-Wear 550; clothing; PVC stretch outer layer bonded to a lightweight liner
		23		135		<9	very good permeation rate; 1 to 5 eyedropper size drops per hour	Ansell Edmont Monkey Grip; supported (lined) glove film; specified by glove film weight
Methylphenol (3-)	m-methylphenol			150	5	36		Pioneer Industrial Products Pylox V-20; glove film
Methylphenol (m-)	3-methylphenol			150	5	36		
Mineral Spirits	Rule 66	23		150		<9	very good permeation rate; 1 to 5 eyedropper size drops per hour	Ansell Edmont Monkey Grip; supported (lined) glove film; specified by glove film weight
Muriatic Acid				>480	4	0	no permeation rate at steady state detected	Pioneer Industrial Products Pylox V-20; glove film
		23		>300				Ansell Edmont Monkey Grip; supported (lined) glove film; specified by glove film weight
Naphtha	VM&P	23		120		<9	very good permeation rate; 1 to 5 eyedropper size drops per hour	"
	VM&P	23		6		<9	very good permeation rate; 1 to 5 eyedropper size drops per hour	Ansell Edmont Wet-Wear 550; clothing; PVC stretch outer layer bonded to a lightweight liner
	VM&P	23		200		<90	good permeation rate; 6 to 50 eyedropper size drops per hour	Ansell Edmont Wet-Wear 600; clothing; nylon netting bonded between two layers of PVC
Nitric Acid	10% conc.	23		285				"
	10% conc.	23	0.25		0		no permeation detected during a 6 hour test	Ansell Edmont Monkey Grip; supported (lined) glove film; specified by glove film weight
	10% conc.	23	0.25		0		no permeation detected during a 6 hour test	Ansell Edmont Wet-Wear 550; clothing; PVC stretch outer layer bonded to a lightweight liner
	50% conc.			114	0.08	0	no permeation rate at steady state detected	Pioneer Industrial Products Pylox V-20; glove film
	70% conc.	23		160				Ansell Edmont Wet-Wear 600; clothing; nylon netting bonded between two layers of PVC
	70% conc.	23		345				Ansell Edmont Monkey Grip; supported (lined) glove film; specified by glove film weight
	70% conc.	23		45				Ansell Edmont Wet-Wear 550; clothing; PVC stretch outer layer bonded to a lightweight liner
Octyl Alcohol		23		>300		<0.9	low permeation rate; 0 to 1/2 eyedropper size drops per hour	Ansell Edmont Monkey Grip; supported (lined) glove film; specified by glove film weight
		23		9		<0.9	low permeation rate; 0 to 1/2 eyedropper size drops per hour	Ansell Edmont Wet-Wear 600; clothing; nylon netting bonded between two layers of PVC
		23		45		<0.9	low permeation rate; 0 to 1/2 eyedropper size drops per hour	Ansell Edmont Wet-Wear 550; clothing; PVC stretch outer layer bonded to a lightweight liner
Oleic Acid		23		90		<9	very good permeation rate; 1 to 5 eyedropper size drops per hour	Ansell Edmont Monkey Grip; supported (lined) glove film; specified by glove film weight
Oxalic Acid	saturated	23	0.25		0		no permeation detected during a 6 hour test	"
Palmitic Acid	saturated	23		75				"
Pentachlorophenol		23		180		<0.9	low permeation rate; 0 to 1/2 eyedropper size drops per hour	"
Pentane				9		102		Pioneer Industrial Products Pylox V-20; glove film
		23		4		<90	good permeation rate; 6 to 50 eyedropper size drops per hour	Ansell Edmont Wet-Wear 550; clothing; PVC stretch outer layer bonded to a lightweight liner
Perchloric Acid	60% conc.	23	0.25		0	.	no permeation detected during a 6 hour test	Ansell Edmont Monkey Grip; supported (lined) glove film; specified by glove film weight
Perchloroethylene		23		4		<90	good permeation rate; 6 to 50 eyedropper size drops per hour	Ansell Edmont Wet-Wear 550; clothing; PVC stretch outer layer bonded to a lightweight liner
Phenol	saturated			32		78		Pioneer Industrial Products Pylox V-20; glove film
		23		35		<9	very good permeation rate; 1 to 5 eyedropper size drops per hour	Ansell Edmont Wet-Wear 550; clothing; PVC stretch outer layer bonded to a lightweight liner
		23		90		<9	very good permeation rate; 1 to 5 eyedropper size drops per hour	Ansell Edmont Wet-Wear 600; clothing; nylon netting bonded between two layers of PVC
		23		75		<9	very good permeation rate; 1 to 5 eyedropper size drops per hour	Ansell Edmont Monkey Grip; supported (lined) glove film; specified by glove film weight

POLYVINYL CHLORIDE POLYOL (cont'd)

Penetrant	Penetrant Note	Temp. (°C)	Time (days)	BTT (min.)	LDL (phr)	Permeation Rate (µg/cm²/min)	Comment	Material Note
Phosphoric Acid	concentrated	23	0.25		0		no permeation detected during a 6 hour test	Ansell Edmont Wet-Wear 550; clothing; PVC stretch outer layer bonded to a lightweight liner
	concentrated	23	0.25		0		no permeation detected during a 6 hour test	Ansell Edmont Monkey Grip; supported (lined) glove film; specified by glove film weight
	concentrated	23	0.25		0		no permeation detected during a 6 hour test	Ansell Edmont Wet-Wear 600; clothing; nylon netting bonded between two layers of PVC
	85% conc.			>480	0.04	0	no permeation rate at steady state detected	Pioneer Industrial Products Pylox V-20; glove film
Picric Acid	saturated; with ethylene oxide	23		40		<9	very good permeation rate; 1 to 5 eyedropper size drops per hour	Ansell Edmont Monkey Grip; supported (lined) glove film; specified by glove film weight
Potassium Hydroxide	KOH; 50% conc.			>480	0.4	0	no permeation rate at steady state detected	Pioneer Industrial Products Pylox V-20; glove film
	KOH; 50% conc.	23	0.25		0		no permeation detected during a 6 hour test	Ansell Edmont Monkey Grip; supported (lined) glove film; specified by glove film weight
	KOH; 50% conc.	23	0.25		0		no permeation detected during a 6 hour test	Ansell Edmont Wet-Wear 600; clothing; nylon netting bonded between two layers of PVC
	KOH; 50% conc.	28	0.25		0		no permeation detected during a 6 hour test	Ansell Edmont Wet-Wear 550; clothing; PVC stretch outer layer bonded to a lightweight liner
Propyl Alcohol		23		90		<9	very good permeation rate; 1 to 5 eyedropper size drops per hour	Ansell Edmont Monkey Grip; supported (lined) glove film; specified by glove film weight
Propylene Glycol		23	0.25		0		no permeation detected during a 6 hour test	Ansell Edmont Wet-Wear 600; clothing; nylon netting bonded between two layers of PVC
		23	0.25		0	<0.9	low permeation rate; 0 to 1/2 eyedropper size drops per hour	Ansell Edmont Wet-Wear 550; clothing; PVC stretch outer layer bonded to a lightweight liner
Pyridine				1			permeation rate too large to measure	Pioneer Industrial Products Pylox V-20; glove film
Rule 66	solvent			42	0.02	60		"
Silicon Etch		23		150				Ansell Edmont Monkey Grip; supported (lined) glove film; specified by glove film weight
Sodium Hydroxide	NAOH; 50% conc.			>480	0.04	0	no permeation rate at steady state detected	Pioneer Industrial Products Pylox V-20; glove film
	NaOH; 50% conc.	23	0.25		0		no permeation detected during a 6 hour test	Ansell Edmont Wet-Wear 550; clothing; PVC stretch outer layer bonded to a lightweight liner
	NaOH; 50% conc.	23	0.25		0		no permeation detected during a 6 hour test	Ansell Edmont Wet-Wear 600; clothing; nylon netting bonded between two layers of PVC
	NaOH; 50% conc.	23	0.25		0		no permeation detected during a 6 hour test	Ansell Edmont Monkey Grip; supported (lined) glove film; specified by glove film weight
Stoddard Solvents		23		360		<0.9	low permeation rate; 0 to 1/2 eyedropper size drops per hour	"
Sulfuric Acid	10% conc.	23		140				Ansell Edmont Wet-Wear 600; clothing; nylon netting bonded between two layers of PVC
	10% conc.	23		60				Ansell Edmont Wet-Wear 550; clothing; PVC stretch outer layer bonded to a lightweight liner
	battery acid; 47% conc.	23	0.25		0		no permeation detected during a 6 hour test	Ansell Edmont Monkey Grip; supported (lined) glove film; specified by glove film weight
	50% conc.			>480	0.04	0	no permeation rate at steady state detected	Pioneer Industrial Products Pylox V-20; glove film
	95% conc.	23		60				Ansell Edmont Wet-Wear 600; clothing; nylon netting bonded between two layers of PVC
	95% conc.	23		5				Ansell Edmont Wet-Wear 550; clothing; PVC stretch outer layer bonded to a lightweight liner
	95% conc.	23		220				Ansell Edmont Monkey Grip; supported (lined) glove film; specified by glove film weight
Tannic Acid	65% conc.	23	0.25		0	<0.9	low permeation rate; 0 to 1/2 eyedropper size drops per hour	"
Tetrachloroethane (1,1,2,2-)				6		2502		Pioneer Industrial Products Pylox V-20; glove film
Tetrahydrofuran	THF			1			permeation rate too large to measure	"
Toluene	toluol			3		2100		"
Toluene Diisocyanate	TDI			>480	0.06	0	no permeation rate at steady state detected	"

POLYVINYL CHLORIDE POLYOL (cont'd)

Penetrant	Penetrant Note	Temp. (°C)	Time (days)	BTT (min.)	LDL (phr)	Permeation Rate (µg/cm²/min)	Comment	Material Note
Trichloroethane	1,1,1-trichloroethane	23		2		<900	fair permeation rate; 51 to 100 eyedropper size drops per hour	Ansell Edmont Wet-Wear 600; clothing; nylon netting bonded between two layers of PVC
	1,1,1-trichloroethane	23		4		<90	good permeation rate; 6 to 50 eyedropper size drops per hour	Ansell Edmont Wet-Wear 550; clothing; PVC stretch outer layer bonded to a lightweight liner
Trichlorotrifluoroethane				11		192		Pioneer Industrial Products Pylox V-20; glove film
Tricresyl Phosphate	TCP	23		200				Ansell Edmont Wet-Wear 550; clothing; PVC stretch outer layer bonded to a lightweight liner
	TCP	23	0.25		0		no permeation detected during a 6 hour test	Ansell Edmont Wet-Wear 600; clothing; nylon netting bonded between two layers of PVC
	TCP	23	0.25		0	<0.9	low permeation rate; 0 to 1/2 eyedropper size drops per hour	Ansell Edmont Monkey Grip; supported (lined) glove film; specified by glove film weight
Triethanolamine	TEA			>480	6	0	no permeation rate at steady state detected	Pioneer Industrial Products Pylox V-20; glove film
	TEA; 85% conc.	23	0.25		0	<0.9	low permeation rate; 0 to 1/2 eyedropper size drops per hour	Ansell Edmont Monkey Grip; supported (lined) glove film; specified by glove film weight
Trifluoroethanol		·		15		1302		Pioneer Industrial Products Pylox V-20; glove film
Trihydroxytriethylamine				>480	6	0	no permeation rate at steady state detected	
Turpentine		23		6		<90	good permeation rate; 6 to 50 eyedropper size drops per hour	Ansell Edmont Wet-Wear 600; clothing; nylon netting bonded between two layers of PVC
		23		10		<9	very good permeation rate; 1 to 5 eyedropper size drops per hour	Ansell Edmont Wet-Wear 550; clothing; PVC stretch outer layer bonded to a lightweight liner
Xylene	xylol			4		138		Pioneer Industrial Products Pylox V-20; glove film

POLYVINYL CHLORIDE POLYOL COATED NYLON POLYESTER BLEND FABRIC

Penetrant	Penetrant Note	Temp. (°C)	Time (days)	BTT (min.)	LDL (phr)	Permeation Rate (µg/cm²/min)	Comment	Material Note
Acetaldehyde		23		3		<9000	poor permeation rate; 501 to 5000 eyedropper size drops per hour	Ansell Edmont Wet-Wear 700; clothing; PVC coating with Nylon/Polyester lining
Acetic Acid	glacial	23		16				"
Ammonium Hydroxide	concentrated	23		11				"
Aniline		23		10		<9	very good permeation rate; 1 to 5 eyedropper size drops per hour	"
Butyl Alcohol		23		35		<9	very good permeation rate; 1 to 5 eyedropper size drops per hour	"
Chromic Acid	50% conc.	23	0.25		0		no permeation detected during a 6 hour test	"
Cyclohexyl Alcohol		23		15		<9	very good permeation rate; 1 to 5 eyedropper size drops per hour	"
Dibutyl Phthalate		23	0.25		0		no permeation detected during a 6 hour test	"
Dioctyl Phthalate	DOP	23		25				"
Ethyl Alcohol Amine	monoethanolamine	23	0.25		0		no permeation detected during a 6 hour test	"
Ethylene Glycol	`	23	0.25		0	<0.9	low permeation rate; 0 to 1/2 eyedropper size drops per hour	"
Formaldehyde		23		30		<9	very good permeation rate; 1 to 5 eyedropper size drops per hour	"
Formic Acid	90% conc.	23		30		·		"
Freon TF		23		15		<900	fair permeation rate; 51 to 100 eyedropper size drops per hour	"
Gasoline	white	23		5		<900	fair permeation rate; 51 to 100 eyedropper size drops per hour	"
Hydrazine	65% conc.	23	0.25		0		no permeation detected during a 6 hour test	"
Hydrochloric Acid	10% conc.	23		55				"
Hydrofluoric Acid	48% conc.	23		5				"
Isobutyl Alcohol		23		7		<9	very good permeation rate; 1 to 5 eyedropper size drops per hour	"
Kerosine		23		75		<90	good permeation rate; 6 to 50 eyedropper size drops per hour	"

POLYVINYL CHLORIDE POLYOL COATED NYLON POLYESTER BLEND FABRIC (cont'd)

Penetrant	Penetrant Note	Temp. (°C)	Time (days)	BTT (min.)	LDL (phr)	Permeation Rate (µg/cm²/min)	Comment	Material Note
Maleic Acid	saturated	23	0.25		0		no permeation detected during a 6 hour test	"
Methyl Alcohol		23		200		<9	very good permeation rate; 1 to 5 eyedropper size drops per hour	"
Methylamine		23		30		<90	good permeation rate; 6 to 50 eyedropper size drops per hour	"
Naphtha	VM&P	23		9		<90	good permeation rate; 6 to 50 eyedropper size drops per hour	"
Nitric Acid	10% conc.	23	0.25		0		no permeation detected during a 6 hour test	"
	70% conc.	23		6				"
Nitropropane	95.5% conc.	23		<5		<90	good permeation rate; 6 to 50 eyedropper size drops per hour	"
Octyl Alcohol		23	0.25		0	<0.9	low permeation rate; 0 to 1/2 eyedropper size drops per hour	"
Pentane		23		11		<90	good permeation rate; 6 to 50 eyedropper size drops per hour	"
Perchloroethylene		23		5		<900	fair permeation rate; 51 to 100 eyedropper size drops per hour	"
Phenol		23		50		<9	very good permeation rate; 1 to 5 eyedropper size drops per hour	"
Phosphoric Acid	concentrated	23	0.25		0		no permeation detected during a 6 hour test	"
Potassium Hydroxide	KOH; 50% conc.	23	0.25		0		no permeation detected during a 6 hour test	"
Propylene Glycol		23	0.25		0		no permeation detected during a 6 hour test	"
Sodium Hydroxide	NaOH; 50% conc.	23	0.25		0		no permeation detected during a 6 hour test	"
Sulfuric Acid	10% conc.	23		120				"
	95% conc.	23		37				"
Trichloroethane	1,1,1-trichloroethane	23		6		<9000	poor permeation rate; 501 to 5000 eyedropper size drops per hour	"
Tricresyl Phosphate	TCP	23	0.25		0		no permeation detected during a 6 hour test	"
Turpentine		23		16		<90	good permeation rate; 6 to 50 eyedropper size drops per hour	"
Xylene		23		>3		<90	good permeation rate; 6 to 50 eyedropper size drops per hour	"

Permeability Units Conversion

There are two related properties derived from Fick's first law and measured to assess the barrier properties of plastic films and similar materials. These properties are the permeability coefficient and the vapor transmission rate.

Fick's first law states that the volume (V) of a substance that penetrates a barrier wall is directly proportional to the area (A) of the wall, partial pressure differential (p) of the penetrant, and time (t); and inversely proportional to the wall thickness (s), if the wall is homogeneous in the direction of penetration. The coefficient P in the equation representing Fick's first law, $V = P \cdot (A \cdot p \cdot t)/s$, is the permeability coefficient.

Fick's first law applies only to permanent gases that obey Henry's law on proportionality of penetrant solubility in the barrier to the partial pressure of the penetrant. Therefore, the permeability coefficient can be measured only for permanent gases, i.e., gases that become liquid at pressures and temperatures far from normal (1 atm and 0°C, respectively). These gases include air, oxygen, argon, and carbon dioxide.

The permeability coefficient can be measured according to ASTM standard D1434. A convenient unit of measurement for the permeability coefficient in the metric system is $(cm^3 \cdot mm)/(m^2 \cdot day \cdot atm)$. Since the permeability coefficient is dependent on temperature, a test temperature must be reported.

The vapor transmission rate (VTR) is measured for the vapors of substances, such as water and acetone, that are liquid at pressures and temperatures close to normal. Vapors do not obey Henry's law, and the vapor transmission rate is not proportional to the pressure differential in Fick's first law. To account for this fact, Fick's first law for vapors is expressed as $W = VTR \cdot (A \cdot t)/s$, where W is the weight of the penetrant.

The vapor transmission rate can be measured according to ASTM standard D3985. A convenient unit of measurement for the vapor transmission rate in the metric system is $(g \cdot mm/ m^2 \cdot day)$. Since the vapor transmission rate is not proportional to the pressure differential, the latter must be stated to make test values of the vapor transmission rate meaningful. (It is customary to substitute a pressure differential with a relative humidity differential in reporting the water vapor transmission rate.) A test temperature must also be reported for the vapor transmission rate because of its dependence on temperature. The vapor transmission rate is influenced by the affinity between the vapor and the barrier material, and by processes that may occur during permeation, such as swelling of the barrier material. Vapor transmission rate values cannot be converted into permeability coefficient values.

The following table gives conversion factors for the common units of measurement of the permeability coefficient and the vapor transmission rate. To convert a value from a common unit to the convenient metric unit for the permeability coefficient, $(cm^3 \cdot mm)/(m^2 \cdot day \cdot atm)$, or for the vapor transmission rate, $(g \cdot mm)/(m^2 \cdot day)$, multiply this value by a factor provided in the corresponding column, taking into account instructions in the Note column.

PERMEABILITY UNITS CONVERSION TABLE

Source Document Unit	Permeability Coefficient Unit (cm³ · mm/m² · day · atm)	Vapor Permeation Rate Unit (g · mm/ m² · day)	Notes
1x10⁻¹⁰ cm³ (STP) · cm/ cm² · sec · cmHg	6.566397e+01		9
1x10⁻¹⁰ cm³ · mm/ cm² · sec · cmHg	6.566397e+00		2, 9
1x10⁻¹⁸ m²/ sec · Pa	8.754480e+18		
1x10⁻²⁰ kg · m/ m² · sec · Pa		8.640000e-10	3
1x10⁻⁸ cm²/ sec · atm	8.640000e+01		
1x10⁻⁸ cm³· cm/ cm² · sec · atm	8.640000e+01		
cm³ (STP) · cm/ cm²· sec · atm	8.640000e+09		
cm³ (STP) · mil/ 100 in²/ day	3.937008e-01		
cm³ (STP) · mm/ m²/ day	1.000000e+00		
cm³ · 0.1 mm/ m² · atm · day	1.000000e-01		
cm³ · 0.5 mm/ 100 in² · day	7.750015e+00		4
cm³ · 100 mm/ m² · day · bar	1.013250e+00		
cm³ · 15m/ m² · atm · day	1.500000e-02		
cm³ · 20m/ m² · day · atm	2.000000e-02		
cm³ · 25m/ m² · atm · day	2.500000e-02		
cm³ · mil/ 100 in² · atm · day	3.397008e-01		
cm³ · mil/ cm² · sec atm	2.194560e+07		
cm³ · mil/ m² · atm · day	2.540000e-02		
cm³ · mm/ cm² · atm · day	1.000000e+01		
cm³ · mm/ cm² · kPa · sec	8.754480e+10		
cm³ · mm/ m² · atm · day	1.000000e+00		
cm³ · mm/ m² · bar · day	1.013250e+00		
cm³ · mm/ m² · day · bar	1.013250e-03		
cm³ · mm/ m² · Pa · day	1.013250e+05		
cm³ · mm/ m² · sec · atm	8.640000e+04		
cm³ · mm/ m² · sec · cmHg	6.566397e+06		
cm³ · N/ m² · bar · day	1.013250e+00		1
cm³/ 100 in² · atm · day	1.550003e+01		1
cm³/ m² · atm · day	1.000000e+01		1, 4
cm³/ m² · day	1.000000e+00		1, 4
cm³/ m² · day · bar	1.013250e+00		1
cm³/ m² · day · bar · 100	1.013250e-02		1
cm³· mil/ 100 in² · bar · day	3.989173e-01		
ft³ · mil/ ft² · psi · day	1.137749e+05		
g · 0.5 mm/ m² · day		5.000000e-01	
g · 100 mm/ m² · day		1.000000e-01	
g · 25m/ m² · day		2.500000e-02	
g · 30m/ m² · day		3.000000e-02	
g · mil/ 100 in² · atm · day		3.937008e-01	3
g · mil/ 100 in² · bar · day		3.989173e-01	3
g · mil/ 100 in² · day		3.397008e-01	
g · mil/ 100 in² · hr		9.448820e+00	
g · mil/ 100 in² · mmHg · day		2.992125e+02	3
g · mil/ day/ 100 in²		3.937008e-01	7
g · mil/ m² · atm · day		2.540000e-02	3
g · mm/ cm² · day		1.000000e+01	
g · mm/ day/ m²		1.000000e+00	6
g · mm/ m² · day		1.000000e+00	
g · mm/ m² · day		1.000000e-03	
g/ 100 in² · day		1.550003e+01	1
g/ m² · day		1.000000e+00	1
grains/ ft² · hr		1.673975e+01	1

Source Document Unit	Permeability Coefficient Unit $(cm^3 \cdot mm/m^2 \cdot day \cdot atm)$	Vapor Permeation Rate Unit $(g \cdot mm/ m^2 \cdot day)$	Notes
$in^3 \cdot mil/ 100\ in^2 \cdot atm \cdot day$	6.451600e+00		
$mg \cdot mil/ in^2 \cdot day$		3.937008e-02	
$mg \cdot mm/ m^2 \cdot Pa \cdot day$		1.013250e+02	3
$ml \cdot mil/ m^2 \cdot atm \cdot day$	2.540000e-02		
$mm^3 \cdot mm/ m^2 \cdot Pa \cdot day$	1.013250e+02		
$mm^3 \cdot mm/ m^2 \cdot s \cdot Pa$	8.754480e-03		
$mm^3/ m \cdot MPa \cdot day$	1.013250e-01		
$N \cdot cm^3/ m^2 \cdot bar \cdot day$		2.540000e-02	8

NOTES TO TABLE

<u>General Note</u> - Values for the permeability coefficient and the vapor transmission rate in most of the above units may be in the range of several powers of magnitude. However, these values are usually given in an easy-to-read decimal format (practical units), with the magnitude factor stated in a table or graph title or in the notes. Care should be taken, when converting, to account for this factor.

1. The conversion factor is applicable only if the film thickness is known; multiply the value-factor product by the film thickness (N) in mm.

2. The original unit, 1 x 10 (to the power of 10) $cm^3/ cm^2/ mm/ sec/ cmHg$, is incorrect, it should be 1 x (10 to the power of 10) $cm^3 \cdot mm/ cm^2/ sec/ cmHg$ or 1 x $10^{10}\ cm^3 \cdot mm/ cm^2 \cdot sec \cdot cmHg$.

3. Unit of pressure (e.g., atm) in the original unit can be ignored for the measurements conducted at normal pressure (1 atm); otherwise the conversion factor is not valid and the value cannot be converted.

4. The conversion factor is applicable only if the pressure differential is known; divide the value-factor product by the pressure in atm.

5. The original unit, $g/ mil/ 100\ in^2/ day$, is incorrect, it should be $g \cdot mil/ 100\ in^2/ day$ or $g \cdot mil/ 100\ in^2 \cdot day$.

6. The original unit, $g/ day/ m^2/ mm$, is incorrect, it should be $g \cdot mm/ day/ m^2$.

7. The original unit, $g/ day/ 100\ in^2/ mil$, is incorrect, it should be $g \cdot mil/ day/ 100\ in^2$.

8. The original unit, $g/ m^2/ mil$, is incorrect, it should be $g \cdot mil/ m^2$; the conversion factor is applicable only if the time is known; divide the value-factor product by the time in days.

9. The original unit has a factor comprising a real number with a positive power of magnitude, which is incorrect. The power of magnitude should be negative. The conversion factor has been revised to account for this error.

ABS See *acrylonitrile butadiene styrene polymer.*

ABS nylon alloy See *acrylonitrile butadiene styrene polymer nylon alloy.*

ABS PC alloy See *acrylonitrile butadiene styrene polymer polycarbonate alloy.*

ABS resin See *acrylonitrile butadiene styrene polymer.*

accelerant See *accelerator.*

accelerator A chemical substance that accelerates chemical, photochemical, biochemical, etc. reaction or process, such as crosslinking or degradation of polymers, that is triggered and/or sustained by another substance, such as a curing agent or catalyst, or environmental factor, such as heat, radiation or a microorganism. Also called accelerant, promoter, cocatalyst.

acetal resins Thermoplastics prepared by polymerization of formaldehyde or its trioxane trimer. Acetals have high impact strength and stiffness, low friction coefficient and permeability, good dimensional stability and dielectric properties, and high fatigue strength and thermal stability. Acetals have poor acid and UV resistance and are flammable. Processed by injection and blow molding and extrusion. Used in mechanical parts such as gears and bearings, automotive components, appliances, and plumbing and electronic applications. Also called acetals.

acetals See *acetal resins.*

acetone A volatile, colorless, highly flammable liquid with molecular formula CH_3COCH_3. Acetone has autoignition temperature 537 °C, mixes readily with water and some other solvents and is moderately toxic. Acetone dissolves most thermoplastics and some thermosets. Used as organic synthesis intermediate, e.g., in the manufacture of bisphenol A and antioxidants, as solvent in paints and acetate fiber spinning and for cleaning of electronic parts. Also called dimethyl ketone, 2-propanone.

acrylate styrene acrylonitrile polymer Acrylic rubber-modified thermoplastic with high weatherability. ASA has good heat and chemical resistance, toughness, rigidity, and antistatic properties. Processed by extrusion, thermoforming, and molding. Used in construction, leisure, and automotive applications such as siding, exterior auto trim, and outdoor furniture. Also called ASA.

acrylic resins Thermoplastic polymers of alkyl acrylates such as methyl methacrylates. Acrylic resins have good optical clarity, weatherability, surface hardness, chemical resistance, rigidity, impact strength, and dimensional stability. They have poor solvent resistance, resistance to stress cracking, flexibility, and thermal stability. Processed by casting, extrusion, injection molding, and thermoforming. Used in transparent parts, auto trim, household items, light fixtures, and medical devices. Also called polyacrylates.

acrylonitrile butadiene styrene polymer ABS resins are thermoplastics comprised of a mixture of styrene-acrylonitrile copolymer (SAN) and SAN-grafted butadiene rubber. They have high impact resistance, toughness, rigidity and processability, but low dielectric strength, continuous service temperature, and elongation. Outdoor use requires protective coatings in some cases. Plating grades provide excellent adhesion to metals. Processed by extrusion, blow molding, thermoforming, calendaring and injection molding. Used in household appliances, tools, nonfood packaging, business machinery, interior automotive parts, extruded sheet, pipe and pipe fittings. Also called ABS, ABS resin, acrylonitrile-butadiene-styrene polymer.

acrylonitrile butadiene styrene polymer nylon alloy A thermoplastic processed by injection molding, with properties similar to ABS but higher elongation at yield. Also called ABS nylon alloy.

acrylonitrile butadiene styrene polymer polycarbonate alloy A thermoplastic processed by injection molding and extrusion, with properties similar to ABS. Used in automotive applications. Also called ABS PC alloy.

acrylonitrile copolymer A thermoplastic prepared by copolymerization of acrylonitrile with small amounts of other unsaturated monomers. Has good gas barrier properties and chemical resistance. Processed by extrusion, injection molding, and thermoforming. Used in food packaging.

acrylonitrile-butadiene-styrene polymer See *acrylonitrile butadiene styrene polymer.*

activation energy An excess energy that must be added to an atomic or molecular system to allow a process, such as diffusion or chemical reaction, to proceed.

adsorption Retention of a substance molecule on the surface of a solid or liquid.

alcohols A class of hydroxy compounds in which a hydroxy group(s) is attached to a carbon chain or ring. Alcohols are produced synthetically from petroleum stock, e.g., by hydration of ethylene, or derived from natural products, e.g., by fermentation of grain. The alcohols are divided in the following groups: monohydric, dihydric, trihydric and polyhydric. Used in organic synthesis, as solvents, plasticizers, fuels, beverages, detergents, etc.

amorphous nylon Transparent aromatic polyamide thermoplastics. Produced by condensation of hexamethylene diamine, isophthalic and terephthalic acid.

annulus test An ozone resistance test for rubbers that involves a flat-ring specimen mounted as a band over a rack, stretched 0 to 100% and subjected to ozone attack in the test chamber. The specimen is evaluated by comparing to a calibrated template to determine the minimum elongation at which cracking occurred.

anthraquinone An aromatic compound comprising two benzene rings linked by two carbonyl (C=O) groups, $C_6H_4(CO)_2C_6H_4$. Combustible. Used as an intermediate in organic synthesis, mainly in the manufacture of anthraquinone dyes and pigments. One method of preparation is by condensation of 1,4-naphthaquinone with butadiene.

antioxidant A chemical substance capable of inhibiting oxidation or oxidative degradation of another substance such as plastic in which it is incorporated. Antioxidants act by terminating chain-propagating free radicals or by decomposing peroxides, formed during oxidation, into stable products. The first group of antioxidants include hindered phenols and amines; the second - sulfur compounds such as thiols.

Ar See *argon.*

area factor The ratio between the total area of pore openings on the surface of a membrane that is in contact with the incoming flow of a penetrant, to the area of this surface.

argon A chemically inert, tasteless, colorless, noncombustible monoatomic gas. Argon is often used to characterize permeability of polymeric films, as carrier gas in gas chromatography, as inert gas shield in welding, in electric bulbs such as neon, lasers and as a process environment. Also called Ar.

aroma barrier A plastic film or its component preventing the escape of aromatic volatiles from foodstuffs or cosmetics seal-packaged in the film.

aromatic polyester estercarbonate A thermoplastic block copolymer of an aromatic polyester with polycarbonate. Has higher heat distortion temperature than regular polycarbonate.

aromatic polyesters Engineering thermoplastics prepared by polymerization of aromatic polyol with aromatic dicarboxylic anhydride. They are tough with somewhat low chemical resistance. Processed by injection and blow molding, extrusion, and thermoforming. Drying is required. Used in automotive housings and trim, electrical wire jacketing, printed circuit boards, and appliance enclosures.

ASA See *acrylate styrene acrylonitrile polymer.*

ASTM C177 An American Society for Testing of Materials (ASTM) standard test method for the measurement of steady-state heat flux in a flat-slab specimen by a guarded-hot-plate apparatus. The method provides for the calculation of thermal transmission properties based on the flux measurements. The measurements are carried out without heat flux reference standards on a variety of solids under different environmental conditions. The apparatus consists of the top and bottom isothermal cold plates with an isothermal heater placed between them, and two specimens placed between the cold plates and the heater, all enclosed in an insulated chamber. The heater has a metered heating core area surrendered by a primary guard.

ASTM D96 An American Society for Testing of Materials (ASTM) standard test method for determining water vapor transmission of materials such as paper, plastic film and sheeting, fiberboards, wood products, etc., that are less than 31 mm in thickness. Two basic methods, the Desiccant Method and the Water Method are used. The specimens have either one side wetted or one side exposed to high humidity and another to low humidity. In the Desiccant Method, the specimen is placed air-tight on a test dish with a desiccant that is weighed to determine the gain of weight due to water vapor transmission. In the Water Method, the water is placed in the dish that is weighed to determine the loss of water due to evaporation through the specimen.

ASTM D256 An American Society for Testing of Materials (ASTM) standard method for determination of the resistance to breakage by flexural shock of plastics and electrical insulating materials, as indicated by the energy extracted from standard pendulum-type hammers in breaking standard specimens with one pendulum swing. The hammers are mounted on standard machines of either Izod or Charpy type. **Note:** Impact properties determined include Izod or Charpy impact energy normalized per width of the specimen. Also called ASTM method D256-84. See also *impact energy.*

ASTM method D256-84 See *ASTM D256.*

ASTM D412 An American Society for Testing of Materials (ASTM) standard methods for determining tensile strength, tensile stress, ultimate elongation, tensile set and set after break of rubber at low, ambient and elevated temperatures using straight, dumbbell and cut-ring specimens.

ASTM D471 An American Society for Testing of Materials (ASTM) standard method for determining the resistance of nonporous rubber to hydrocarbon oils, fuels, service fluids and water. The specimens are immersed in fluids for 22-670 hours at -75 to 250°C, followed by measuring of the changes in mass, volume, tensile strength, elongation and hardness for solid specimens and the changes in breaking strength, burst strength, tear strength and adhesion for rubber-coated fabrics.

ASTM D570 An American Society for Testing of Materials (ASTM) standard method for determining relative rate of water absorption of immersed plastics. The test applies to all kinds of plastics: molded, cast, laminated, etc. The specimens are immersed for 2 to 24 hours or until saturation at ambient temperature, or for 1/2 to 2 hours in boiling water. The absorption is calculated as a percentage of weight gain.

ASTM D638 An American Society for Testing of Materials (ASTM) standard method for determining tensile strength, elongation and modulus of elasticity of reinforced or unreinforced plastics in the form of sheet, plate, moldings, rigid tubes and rods. Five (I-V) types, depending on dimensions, of dumbbell-shaped specimens with thickness not exceeding 14 mm are specified. Specified speed of testing varies depending on the specimen type and plastic rigidity. **Note:** Tensile properties determined include tensile stress (strength) at yield and at break, percentage elongation at yield or at break and modulus of elasticity. Also called ASTM method D638-84. See also *tensile strength.*

ASTM D638, type IV See *ASTM D638.*

ASTM method D638-84 See *ASTM D638.*

ASTM D696 An American Society for Testing of Materials (ASTM) standard test method for the measurement of the coefficient of linear thermal expansion of plastics by using a vitreous silica dilatometer. The test is carried out under conditions excluding any significant creep or elastic strain rate and effects of moisture, curing, loss of plasticizer, etc. The specimen is placed at the bottom of the outer dilatometer tube and the tube is immersed in a liquid bath at a desired temperature.

ASTM D774 An American Society for Testing of Materials (ASTM) standard test method for determining the bursting strength of paper having a bursting strength less than 200 points and a thickness less than 0.6 mm. The bursting strength is determined as the hydrostatic pressure in pascals required to produce rupture when applied at a controlled rate through a rubber diaphragm to a circular area 30.48 mm in diameter of the material held rigidly and initially flat but free to bulge under the increasing pressure.

ASTM D882 An American Society for Testing of Materials (ASTM) standard test method for determining the tensile strength and percentage elongation at break and at yield, elastic modulus, tensile energy to break and other tensile properties of thin plastic film and sheeting of less than 1.0 mm in thickness, according to two different methods. Method A employs a constant rate of specimen grip separation, whereas Method B employs a constant rate of motion of one grip and a variable rate of motion of another, which is attached to a pendulum weighing head.

ASTM D1004 An American Society for Testing of Materials (ASTM) standard test method for determination of the initial tear resistance of flexible plastic films and sheeting. The test is preformed at very low rates of loading, e.g., 51 mm/min, to measure the force required to initiate tearing. The specimen geometry in this test produces a stress concentration in a small area of the specimen. The maximum stress, usually found near the onset of tearing, is recorded in newtons or pounds-force.

ASTM D1006 An American Society for Testing of Materials (ASTM) standard practice for conducting exterior exposure tests of

Glossary of Terms

house and trim paints on new wood. Painted testing panels (boards or plywood) are exposed for several years on vertical fences facing both north and south and visually examined for failures at prescribed intervals (1-6 months).

ASTM D1434 An American Society for Testing of Materials (ASTM) standard test method for determining gas transmission rate, permeance and permeability (for homogeneous materials) of plastic film, sheeting, laminates and plastic-coated papers or fabrics under steady-state conditions. The sample is mounted in a gas transmission cell to form a barrier between 2 chambers. One chamber contains the test gas at a high pressure, and the other chamber receives gas at a lower pressure. The transmission rate is monitored either by the increase in pressure in the receiving chamber (Method M) or by a change in volume of gas (Method V).

ASTM D1708 An American Society for Testing of Materials (ASTM) standard method for determining tensile properties of plastics using microtensile specimens with maximum thickness 3.2 mm and minimum length 38.1 mm, including thin films. Tensile properties include yield strength, tensile strength, tensile strength at break, elongation at break, etc. determined per ASTM D638.

ASTM D1709 An American Society for Testing of Materials (ASTM) standard test method for determining resistance of polyethylene film to impact by the free-falling dart. The impact resistance is measured as the energy that causes 50% failure rate of the film. The energy is calculated as the product of dart weight and dropping height. There are 2 test methods (A and B) using darts with different diameters of their hemispherical head and different dropping heights.

ASTM D1894 An American Society for Testing of Materials (ASTM) standard test method for determining coefficients of starting and sliding friction (static and kinetic coefficients, respectively) of plastic film and sheeting when sliding over itself or other substances under specified conditions.

ASTM D1922 An American Society for Testing of Materials (ASTM) standard test method for determining the resistance of flexible plastic film or sheeting to tear propagation. The resistance is measured as the average force, in grams, required to propagate tearing from a precut slit through a specified length, using an Elmendorf-type pendulum tester and 2 specimens, a rectangular type and one with a constant radius testing length.

ASTM D2167 An American Society for Testing of Materials (ASTM) standard test method for determining in-place density and unit weight of compacted or firmly bonded soil using a rubber balloon apparatus. The calibrated apparatus is used to determine the volume of an excavated soil by measuring the drop in the level of the liquid that fills the apparatus and the void (through expansion of the rubber balloon) when the external pressure is applied.

ASTM D2176 An American Society for Testing of Materials (ASTM) standard test method for determining the folding endurance of paper by the MIT tester. The tester has an oscillating folding head with a clamping jaw and a clamping jaw that can move in a direction perpendicular to the axis of rotation of the folding head. The endurance is reported as the number of double, 135 ° folds made at a given tension before fracture. Also called TAPPI T511.

ASTM D2240 An American Society for Testing of Materials (ASTM) standard method for determining the hardness of materials ranging from soft rubbers to some rigid plastics by measuring the penetration of a blunt (type A) or sharp (type D) indenter of a durometer at a specified force. The blunt indenter is used for softer materials and the sharp indenter - for more rigid materials.

ASTM D2457 An American Society for Testing Materials (ASTM) standard test method for the measurement of specular transparency of clear and colorless thin plastic sheeting. The transmittance is measured as the ratio of the radian flux transmitted by a specimen to that incident on it, in essentially the same direction. The test results are greatly influenced by the design characteristics of the instrument, e.g., the angular width of the receptor aperture that should be <0.1 degree.

ASTM D2752 An American Society for Testing of Materials (ASTM) standard test method for the measurement of the relative degree of openness or degree of fibrilization of milled asbestos fiber by air permeability instruments, the Rapid Surface Area (Method A) and the Dyckerhoff (Method B) apparatus. The resistance to air flow of a compressed specimen of fixed weight and volume is determined. In Method A, a pressure drop due to resistance is measured on a manometer calibrated in specific surface area units. In the Method B, the time required to draw a given volume of air through the specimen under specified conditions of varying hydraulic head is determined.

ASTM D3679 An American Society for Testing of Materials (ASTM) standard specification for extruded single-wall rigid poly(vinyl chloride) siding that establishes its physical requirements (dimensions, weight, weatherability, impact resistance, expansion, shrinkage, and appearance), test methods for physical requirements and marking.

ASTM D3763 An American Society for Testing of Materials (ASTM) standard method for determination of the resistance of plastics, including films, to high-speed puncture over a broad range of test velocities using load and displacement sensors. **Note:** Puncture properties determined include maximum load, deflection to maximum load point, energy to maximum load point and total energy. Also called ASTM method D3763-86. See also *impact energy*.

ASTM method D3763-86 See *ASTM D3763*.

ASTM D3841 An American Society for Testing of Materials (ASTM) standard specification for glass fiber-reinforced polyester construction panels. The specification covers classification, inspection, certification, dimensions, weight, appearance, light transmission, weatherability, expansion, impact resistance, flammability and load-deflection properties of panels and their methods of testing.

ASTM D3985 An American Society for Testing of Materials (ASTM) standard test method for determining the steady-state transmission rate of oxygen gas through a plastic film, sheeting, laminates, coextrusions, or plastic-coated paper or fabric.

ASTM D4275 An American Society for Testing of Materials (ASTM) standard method for determination of butylated hydroxytoluene in ethylene polymers and ethylene-vinyl acetate copolymers by solvent extraction followed by gas chromatographic analysis. Detection of butylated hydroxytoluene is achieved by flame ionization. Butylated hydroxytoluene is a stabilizer used in the manufacture of the ethylene polymers.

ASTM D4434 An American Society for Testing of Materials (ASTM) standard specification for poly(vinyl chloride) sheet roofing, used as a single-ply roof membrane. The material may be unreinforced or reinforced and may contain fibers or fabrics. The specification specifies types, dimensions, mechanical properties, weatherability, resistance to heat aging, appearance, and test methods. The mechanical properties tested include tensile strength and elongation at break, seam strength, tear resistance and tearing strength. The exposure tests include accelerated weathering, water exposure, xenon arc light exposure and fluorescent UV/condensation exposure. Also called ASTM DS D4434.

ASTM D4637 An American Society for Testing of Materials (ASTM) standard specification for unreinforced or fabric-reinforced vulcanized rubber sheet made from EPDM or chloroprene rubber and used as single-ply roof membranes. The specification specifies grades, dimensions, mechanical properties, weatherability, resistance to ozone and heat aging, appearance, and test methods. The mechanical properties tested include tensile strength, set and elongation, seam strength, tear resistance and tearing strength. The exposure tests include water absorption. Also called ASTM DS D4637.

ASTM DS D4434 See *ASTM D4434.*

ASTM DS D4637 See *ASTM D4637.*

ASTM E96 An American Society for Testing of Materials (ASTM) standard test method for determining water vapor transmission of materials such as paper, plastic film and sheeting, fiberboards, wood products, etc., that are less than 31 mm in thickness.Two basic methods, the Desiccant Method and the Water Method are used.The specimens have either one side wetted or one side exposed to high humidity and another to low humidity.In the Desiccant Method, the specimen is placed air-tight on a test dish with a desiccant that is weighed to determine the gain of weight due to water vapor transmission.In the Water Method, the water is placed in the dish that is weighed to determine the loss of water due to evaporation through the specimen.

ASTM E398 An American Society for Testing of Materials (ASTM) standard test method for the determination of water vapor transmission rate of sheet materials with at least one side being hydrophobic, such as plastic film, by a rapid dynamic method. The specimen is mounted between two chambers, one of known relative humidity and another of dry air. The response of an electrical sensor capable of detecting water vapor accumulation in the dry chamber is recorded and used, with the help of a calibrating curve, to determine the water vapor transmission rate. Also called ASTM E398-70.

ASTM E398-70 See *ASTM E398.*

ASTM F1249 An American Society for Testing of Materials (ASTM) standard test method for determining water vapor transmission rate through plastic film and sheeting up to 3 mm in thickness using a pressure-modulated infrared sensor. In addition it provides for determination of the permeance of the film to water vapor and the water vapor permeability coefficient. The specimen is placed as a sealed semi-barrier between two chambers at ambient atmospheric pressure. One chamber is wet and another is dry. As water vapor penetrates through the film from wet chamber into dry one, it is carried by air into the sensor. It measures the fraction of infrared energy absorbed by the vapor and produces an electric signal that is proportional to water vapor concentration

ASTM F372 An American Society for Testing of Materials (ASTM) standard test method for the rapid determination of water vapor transmission rate of flexible barrier films and thin sheeting consisting of single or multilayer synthetic or natural polymers and metal foils including coated materials. The specimen is mounted between two chambers, one of known relative humidity and another of dry air. The time for a given increase in water vapor concentration of the dry chamber is measured by monitoring the differential between two bands in the infrared spectral region, one in which water molecules absorb and the other where they do not. The values obtained are used to calculate the water vapor transmission rate.

atm See *atmosphere.*

atmosphere A metric unit of measurement of pressure equal to $1.013250 \times 1.0E+06$ dynes/cm^2 or $1.013250 \times 1.0E+05$ pascals, which is the air pressure measured at mean sea level. It has a dimension of unit of force per unit of area.Used to denote the pressure of gases, vapors and liquids. Also called atm, standard atmosphere, std atm.

azo A prefix indicating an organic group of two nitrogen atoms linked by a double bond, -N=N-, or a class of chemical compounds containing this group, like azo dyes.

B

bar A metric unit of measurement of pressure equal to $1.0E+06$ dynes/cm^2 or $1.0E+05$ pascals. It has a dimension of unit of force per unit of area. Used to denote the pressure of gases, vapors and liquids.

barrier material Materials such as plastic films, sheeting, wood laminates, particle board, paper, fabrics, etc., with low permeability to gases and vapors. Used in construction as water vapor insulation, food packaging, protective clothing, etc.

bending properties See *flexural properties.*

bending strength See *flexural strength.*

bending stress See *flexural stress.*

benzene An aromatic hydrocarbon with six-atom carbon ring, C_6H_6. Highly toxic and flammable (autoignition point 562 °C). A colorless or yellowish liquid under normal conditions (b.p.80.1°C), soluble in many organic solvents such as ethanol, acetone, tetrachlorocarbon, etc. Used for synthesis of organic compounds.

biodegradation Microorganism-induced degradation of the material that may involve a negative effect such as loss of performance and cracking of an underground pipe or a positive effect such as decomposition of material waste to simple chemical compounds. Usually, the microorganisms such as fungi induce biodegradation by generating the enzymes, proteins that catalyze degradation reactions. Also called microbiological attack.

bisphenol A polyester A thermoset unsaturated polyester based on bisphenol A and fumaric acid.

bleaching Complete loss of color of the material as a result of degradation or removal of colored substances present on its surface. Bleaching can be caused by chemical reactions, radiation, etc.

blistering The formation of bubbles on the surface of a nonmetallic coating or a plastic specimen or article as a result of air or other gases or evaporation of moisture or other volatiles trapped beneath. Blistering is often caused by improper application or excessive mixing of paints, heat and polymer degradation.

blow-up ratio In extrusion blowing of film, the ratio of the extrusion die diameter and the diameter of the tubular film. In blow molding, the ratio between the diameter of a parison and the maximum diameter of the mold cavity.

blown film A plastic film produced by extrusion blowing, wherein an extruded plastic tube is continuously inflated by internal air pressure, cooled, collapsed by rolls and wound up. The thickness of the film is controlled by air pressure and rate of extrusion.

breaking elongation See *elongation.*

bubbling The presence of bubbles of trapped air and/or volatile vapors in nonmetallic coating or plastic specimen or article. Bubbling is often caused by improper application or excessive mixing of paints or degassing.

bursting strength Bursting strength of a material, such as plastic film, is the minimum force per unit area or pressure required to produce rupture. The pressure is applied with a ram or a diaphragm at a controlled rate to a specified area of the material held rigidly and initially flat but free to bulge under the increasing pressure.

C

CA See *cellulose acetate.*

CAB See *cellulose acetate butyrate.*

carbon black A black colloidal carbon filler made by the partial combustion or thermal cracking of natural gas, oil, or another hydrocarbon. There are several types of carbon black depending on the starting material and the method of manufacture. Each type of carbon black comes in several grades. Carbon black is widely used as a filler and pigment in rubbers and plastics. It reinforces, increases the resistance to UV light and reduces static charging.

carbon dioxide A colorless, tasteless gas, CO_2, found in the atmosphere. It is produced as a result of metabolism (e.g., oxidation of carbohydrates) and is used by plants in photosynthesis. Carbon dioxide has low toxicity and is noncombustible. Derived industrially from synthesis gas in ammonia production and from cracking of hydrocarbons. Used widely in refrigeration, carbonated beverages, chemical synthesis, water treatment, medicine, fire extinguishing, and as inert atmosphere.

carbon monoxide A colorless, tasteless gas, CO. Highly flammable (liquid autoignition point 609°C) and toxic. Found in automobile exhaust gases and is a major air pollutant. Manufactured from coke by action of oxygen and carbon dioxide or steam. Used in organic synthesis, synthetic fuels and metallurgy.

cast film Film produced by pouring or spreading resin solution or melt over a suitable temporary substrate, followed by curing via solvent evaporation or melt cooling and removing the cured film from the substrate.

cellulose acetate Thermoplastic esters of cellulose with acetic acid. Have good toughness, gloss, clarity, processability, stiffness, hardness, and dielectric properties, but poor chemical, fire and water resistance and compressive strength. Processed by injection and blow molding and extrusion. Used for appliance cases, steering wheels, pens, handles, containers, eyeglass frames, brushes, and sheeting. Also called CA.

cellulose acetate butyrate Thermoplastic mixed esters of cellulose with acetic and butyric acids. Have good toughness, gloss, clarity, processability, dimensional stability, weatherability, and dielectric properties, but poor chemical, fire and water resistance and compressive strength. Processed by injection and blow molding and extrusion. Used for appliance cases, steering wheels, pens, handles, containers, eyeglass frames, brushes, and sheeting. Also called CAB.

cellulose propionate Thermoplastic esters of cellulose with propionic acid. Have good toughness, gloss, clarity, processability, dimensional stability, weatherability, and dielectric properties, but poor chemical, fire and water resistance and compressive strength. Processed by injection and blow molding and extrusion. Used for appliance cases, steering wheels, pens, handles, containers, eyeglass frames, brushes, and sheeting. Also called CP.

cellulosic plastics Thermoplastic cellulose esters and ethers. Have good toughness, gloss, clarity, processability, and dielectric properties, but poor chemical, fire and water resistance and compressive strength. Processed by injection and blow molding and

extrusion. Used for appliance cases, steering wheels, pens, handles, containers, eyeglass frames, brushes, and sheeting.

centimeter of mercury See *cmHg.*

chain scission Breaking of the chainlike molecule of a polymer as a result of chemical, photochemical, etc. reaction such as thermal degradation or photolysis.

chalking Formation of a dry, chalk-like, loose powder on or just beneath the surface of paint film or plastic caused by the exudation of a compounding ingredient such as pigment, often as a result of ingredient migration to the surface and surface degradation.

channel black Carbon black made by impingement of a natural gas flame against a metal plate or channel iron, from which a deposit is scraped. Used as a reinforcing filler in rubbers. Also called gas black.

chemical saturation Absence of double or triple bonds in a chain organic molecule such as that of most polymers, usually between carbon atoms. Saturation makes the molecule less reactive and polymers less susceptible to degradation and crosslinking. Also called chemically saturated structure.

chemical unsaturation Presence of double or triple bonds in a chain organic molecule such as that of some polymers, usually between carbon atoms. Unsaturation makes the molecule more reactive, especially in free-radical addition reactions such as addition polymerization, and polymers more susceptible to degradation, crosslinking and chemical modification. Also called polymer chain unsaturation.

chemically saturated structure See *chemical saturation.*

chlorendic polyester A chlorendic anhydride-based unsaturated polyester.

chlorinated polyvinyl chloride Thermoplastic produced by chlorination of polyvinyl chloride. Has increased glass transition temperature, chemical and fire resistance, rigidity, tensile strength, and weatherability as compared to PVC. Processed by extrusion, injection molding, casting, and calendering. Used for pipes, auto parts, waste disposal devices, and outdoor applications. Also called CPVC.

chloroethyl alcohol(2-) See *ethylene chlorohydrin.*

chloroform Trichloromethane, $CHCl_3$. Chloroform is a clear, colorless, volatile, nonflammable liquid with characteristic pungent smell. It is toxic and carcinogenic. Derived by chlorination of methane. Formerly used as an anesthetic, it is now used mainly as a solvent and in organic synthesis to manufacture fluorocarbon plastics and insecticides.

chlorohydrins Halohydrins with chlorine as a halogen atom. One of the most reactive of halohydrins. Dichlorohydrins are used in the preparation of epichlorohydrins, important monomers in the manufacture of epoxy resins. Most chlorohydrins are reactive colorless liquids, soluble in polar solvents such as alcohols. **Note:** Chlorohydrins are a class of organic compounds, not to be mixed with a specific member of this class, 1-chloropropane-2,3-diol sometimes called chlorohydrin.

chlorosulfonated polyethylene rubber Thermosetting elastomers containing 20- 40% chlorine. Have good weatherability and heat and chemical resistance. Used for hoses, tubes, sheets, footwear soles, and inflatable boats.

cmHg A metric unit of measurement of pressure equal to 13332.2 dynes/cm^2 or 1333.22 pascals at O°C. One centimeter of mercury is the pressure that would support a column of mercury of length one centimeter and density 12,595 kg/m^3 under the standard acceleration of free fall. Used to denote the pressure of gases, vapors and liquids. Also called centimeter of mercury.

cocatalyst See *accelerator*.

coefficient of friction See *kinetic coefficient of friction*.

coefficient of friction, kinetic See *kinetic coefficient of friction*.

coefficient of friction, static See *static coefficient of friction*.

coextruded film A film made by coextrusion of 2 or more different or similar plastics through a single die with two or more orifices arranged so that the extrudates merge and weld together into a laminar film before cooling. Each ply of coextruded film imparts a desired property, such as impermeability or resistance to some environment and heat-sealability, usually unattainable with a single material.

color The wavelength composition of light, specifically of the light reflected or emitted by the material and its visual appearance (red, blue, ect.). Also called hue, tint, coloration.

color change See *discoloration*.

coloration See *color*.

compatibilizer A chemical compound used to increase the compatibility or miscibility and to prevent the separation of the components in a plastic composition, such as the compatibility of a resin and a plasticizer or of two polymers in a blend. Block copolymers bearing blocks similar to the polymers in the blend are often used as compatibilizers in the latter case.

concentration units The units for measuring the content of a distinct material or substance in a medium other than this material or substance, such as solvent. **Note:** The concentration units are usually expressed in the units of mass or volume of substance per one unit of mass or volume of medium. When the units of substance and medium are the same, the percentage is often used.

conditioning Process of bringing the material or apparatus to a certain condition, e.g., moisture content or temperature, prior to further processing, treatment, etc. Also called conditioning cycle.

conditioning cycle See *conditioning*.

corona discharge treatment Treating the surface of an inert plastic such as polyolefin with corona discharge to increase its affinity to inks, adhesives or coatings. Plastic films are passed over a grounded metal cylinder with a pointed high-voltage electrode above it to produce the discharge. The discharge oxidizes the surface, making it more receptive to finishing. Also called corona treatment.

corona treatment See *corona discharge treatment*.

covulcanization Simultaneous vulcanization of a blend of two or more different rubbers to enhance their individual properties such as ozone resistance. Rubbers are often modified to improve covulcanization.

CP See *cellulose propionate*.

CPVC See *chlorinated polyvinyl chloride*.

cracking Appearance of external and/or internal cracks in the material as a result of stress that exceeds the strength of the material. The stress can be external and/or internal and can be caused by a variety of adverse conditions: structural defects, impact, aging, corrosion, etc. or a combination of thereof. Also called cracks. See also *processing defects*.

cracks See *cracking*.

crazes See *crazing*.

crazing Appearance of thin cracks on the surface of the material or, sometimes, minute frost-like internal cracks, as a result of stress that exceeds the strength of the material, impact, terperature changes, degradation, ect. Also called crazes.

crosslinked polyethylene Polyethylene thermoplastics partially photochemically or chemically crosslinked. Have improved tensile strength, dielectric properties, and impact strength at low and elevated temperatures.

crosslinking Reaction of formation of covalent bonds between chain-like polymer molecules or between polymer molecules and low-molecular compounds such as carbon black fillers. As a result of crosslinking polymers, such as thermosetting resins, may become hard and infusible. Crosslinking is induced by heat, UV or electron-beam radiation, oxidation, etc. Crosslinking can be achieved ether between polymer molecules alone as in unsaturated polyesters or with the help of multifunctional crosslinking agents such as diamines that react with functional side groups of the polymers. Crosslinking can be catalysed by the presence of transition metal complexes, thiols and other compounds.

crystal polystyrene See *general purpose polystyrene*.

crystalline melting point The temperature of melting of the crystallite phase of a crystalline polymer. It is higher than the temperature of melting of the surrounding amorphous phase.

CTFE See *polychlorotrifluoroethylene*.
cycle time See *processing time*.

cyclic compounds A broad class of organic compounds consisting of carbon rings that are saturated, partially unsaturated or aromatic, in which some carbon atoms may be replaced by other atoms such as oxygen, sulfur and nitrogen.

D

d-limonene One of two optical isomers of limonene, a naturally occurring terpene closely related to isoprene. Limonene is a colorless liquid that oxidizes to film in air. Derived from lemon, orange, and other essential oils. Used as flavoring, fragrance, solvent, and wetting agent.

DAP See *diallyl phthalate resins*.

dart impact energy The mean energy of a free-falling dart that will cause 50% failures after 50 tests to a specimen directly stricken by the dart. The energy is calculated by multiplying dart mass, gravitational acceleration and drop height. Also called falling dart impact energy, dart impact strength, falling dart impact strength.

dart impact strength See *dart impact energy*.

decoloration Complete or partial loss of color of the material as a result of degradation or removal of colored substances present in it. Also called decoloring.

decoloring See *decoloration*.

defects See *processing defects*.

deflection temperature under load See *heat deflection temperature.*

degradation Loss or undesirable change in the properties, such as color, of a material as a result of aging, chemical reaction, wear, exposure, etc. See also *stability.*

diallyl phthalate resins Thermosets supplied as diallyl phthalate prepolymer or monomer. Have high chemical, heat and water resistance, dimensional stability, and strength. Shrink during peroxide curing. Processed by injection, compression and transfer molding. Used in glass-reinforced tubing, auto parts, and electrical components. Also called DAP.

diffusion Spontaneous slow mixing of different substances in contact without influence of external forces.

diffusion coefficient Weight of a substance diffusing through a unit area in a unit time per a unit concentration gradient. Also called diffusivity.

diffusivity See *diffusion coefficient.*

dihydric alcohols See *glycols.*

dihydroxy alcohols See *glycols.*

dimethyl ketone See *acetone.*

DIN 53122 A German Standards Institute (Deutsches Institut fuer Normen, DIN) standard test method for determining water vapor transmission of flat materials such as plastic film and sheeting.

DIN 53380 A German Standards Institute (Deutsches Institut fuer Normen, DIN) standard test method for determining gas permeability of flat materials such as plastic film and sheeting.

DIN 53453 A German Standards Institute (DIN) standard specifying conditions for the flexural impact testing of molded or laminated plastics. The bar specimens are either unnotched or notched on one side, mounted on two-point support and struck in the middle (on the unnotched side for notched specimens) by a hammer of the pendulum impact machine. Impact strength of the specimen is calculated relative to the cross-sectional area of the specimen as the energy required to break the specimen equal to the difference between the energy in the pendulum at the instant of impact and the energy remaining after complete fracture of the specimen. Also called DIN 53453 impact test.

DIN 53453 impact test See *DIN 53453.*

discoloration A change in color due to chemical or physical changes in the material. Also called color change.

disperse dyes Nonionic dyes insoluble in water and used mainly as fine aqueous dispersions in dying acetate, polyester and polyamide fibers. A large subclass of disperse dyes comprises low-molecular-weight aromatic azo compounds with amino, hydroxy and alkoxy groups that fix on fibers by forming Van der Waals and hydrogen bonds.

displacement Process of removing one object, e.g., a medium in an apparatus, or its part, and replacing it with another. Also called displacement cycle.

displacement cycle See *displacement.*

drop dart impact See *falling weight impact energy.*

drop dart impact energy See *falling weight impact energy.*
drop dart impact strength See *falling weight impact energy.*

drop weight impact See *falling weight impact energy.*

drop weight impact energy See *falling weight impact energy.*

drop weight impact strength See *falling weight impact energy.*

E

ECTFE See *ethylene chlorotrifluoroethylene copolymer.*

Elmendorf tear strength The resistance of flexible plastic film or sheeting to tear propagation. It is measured, according to ASTM D1922, as the average force, in grams, required to propagate tearing from a precut slit through a specified length, using an Elmendorf-type pendulum tester and 2 specimens, a rectangular type and one with a constant radius testing length.

elongation The increase in gauge length of a specimen in tension, measured at or after the fracture, depending on the viscoelastic properties of the material. **Note:** Elongation is usually expressed as a percentage of the original gauge length. Also called tensile elongation, elongation at break, ultimate elongation, breaking elongation, elongation at rupture. See also *tensile strain.*

elongation at break See *elongation.*

elongation at rupture See *elongation.*
EMAC See *ethylene methyl acrylate copolymer.*

embrittlement A reduction or loss of ductility or toughness in materials such as plastics resulting from chemical or physical damage.

EPDM See *EPDM rubber.*

EPDM rubber Sulfur-vulcanizable thermosetting elastomers produced from ethylene, propylene, and a small amount of nonconjugated diene such as hexadiene. Have good weatherability and chemical and heat resistance. Used as impact modifiers and for weather stripping, auto parts, cable insulation, conveyor belts, hoses, and tubing. Also called EPDM.

epoxides Organic compounds containing three-membered cyclic group(s) in which two carbon atoms are linked with an oxygen atom as in an ether. This group is called an epoxy group and is quite reactive, allowing the use of epoxides as intermediates in preparation of certain fluorocarbons and cellulose derivatives and as monomers in preparation of epoxy resins. Also called epoxy compounds.

epoxies See *epoxy resins.*

epoxy compounds See *epoxides.*

epoxy resins Thermosetting polyethers containing crosslinkable glycidyl groups. Usually prepared by polymerization of bisphenol A and epichlorohydrin or reacting phenolic novolaks with epichlorohydrin. Can be made unsaturated by acrylation. Unmodified varieties are cured at room or elevated temperatures with polyamines or anhydrides. Bisphenol A epoxy resins have excellent adhesion and very low shrinkage during curing. Cured novolak epoxies have good UV stability and dielectric properties. Cured acrylated epoxies have high strength and chemical resistance. Processed by molding, casting, coating, and lamination. Used as protective coatings, adhesives, potting compounds, and binders in laminates and composites. Also called epoxies.

EPR See *ethylene propene rubber.*

ETFE See *ethylene tetrafluoroethylene copolymer*.

ethane An alkane (saturated aliphatic hydrocarbon) with 2 carbon atoms, CH_3CH_3. A colorless, odorless, flammable gas. Relatively inactive chemically. Obtained from natural gas. Used in petrochemical synthesis and as fuel.

ethanediol(1,2-) See *ethylene glycol*.

ethanol See *ethyl alcohol*.

ethene See *ethylene*.

ethers A class of organic compounds in which an oxygen atom is interposed between two carbon atoms in a chain or a ring. Ethers are derived mainly by catalytic hydration of olefins. The lower molecular weight ethers are dangerous fire and explosion hazards. **Note:** Major types of ethers include aliphatic, cyclic and polymeric ethers.

ethyl acetate An ethyl ester of acetic acid, $CH_3CO_2CH_2CH_3$. A colorless, fragrant, flammable liquid. Autoignition temperature 426 deg.C. Toxic by inhalation and skin absorption. Derived by catalytic esterification of acetic acid with ethanol. Used as solvent in coatings and plastics, organic synthesis, artificial flavors, pharmaceuticals.

ethyl alcohol An aliphatic alcohol, CH_3CH_2OH. A colorless, volatile, flammable liquid. Autoignition point is 422°C. Toxic by ingestion. Derived by catalytic hydration of ethylene, fermentation of biomass such as grain, or enzymatic hydrolysis of cellulose. Used as automotive fuel additive, in alcoholic beverages, as solvent for resins and oils, in organic synthesis, cleaning compositions, cosmetics, antifreeze, and antiseptic. Also called ethanol.

ethylene An alkene (unsaturated aliphatic hydrocarbon) with two carbon atoms, $CH_2=CH_2$. A colorless, highly flammable gas with sweet odor. Autoignition point 543°C. Derived by thermal cracking of hydrocarbon gases or from synthesis gas. Used as monomer in polymer synthesis, refrigerant, and anesthetic. Also called ethene.

ethylene acrylic rubber Copolymers of ethylene and acrylic esters. Have good toughness, low temperature properties, and resistance to heat, oil, and water. Used in auto and heavy equipment parts.

ethylene alcohol See *ethylene glycol*.

ethylene copolymers See *ethylene polymers*.

ethylene methyl acrylate copolymer Thermoplastic copolymers of ethylene with <40% methyl acrylate. Have good dielectric properties, toughness, thermal stability, stress crack resistance, and compatibility with other polyolefins. Transparency decreases with increasing content of acrylate. Processed by blown film extrusion and blow and injection molding. Used in heat-sealable films, disposable gloves, and packaging. Some grades are FDA-approved for food packaging. Also called EMAC.

ethylene polymers Ethylene polymers include ethylene homopolymers and copolymers with other unsaturated monomers, most importantly olefins such as propylene and polar substances such as vinyl acetate. The properties and uses of ethylene polymers depend on the molecular structure and weight. Also called ethylene copolymers.

ethylene propene rubber Stereospecific copolymers of ethylene with propylene. Used as impact modifiers for plastics. Also called EPR.

ethylene tetrafluoroethylene copolymer Thermoplastic alternating copolymer of ethylene and tetrafluoroethylene. Has good impact strength, abrasion and chemical resistance, weatherability, and dielectric properties. Processed by molding, extrusion, and powder coating. Used in tubing, cables, pump parts, and tower packing in a wide temperature range. Also called ETFE.

ethylene vinyl alcohol copolymer Thermoplastics prepared by hydrolysis of ethylene-vinyl acetate polymers. Have good barrier properties, mechanical strength, gloss, elasticity, weatherability, clarity, and abrasion resistance. Barrier properties and processibility improve with increasing content of ethylene due to lower absorption of moisture. Ethylene content of high barrier grades range from 32 to 44 mole %. Processed by extrusion, coating, blow and blow film molding, and thermoforming. Used as packaging films and container liners. Also called EVOH.

ethylene-acrylic acid copolymer A flexible thermoplastic with water and chemical resistance and barrier properties similar to those of low-density polyethylene and enhanced adhesion, optics, toughness, and hot tack properties, compared to the latter. Contains 3-20% acrylic acid, with density and adhesion to polar substrates increasing with increasing acrylic acid content. FDA-approved for direct contact with food. Processed by extrusion, blow and film methods and extrusion molding, and extrusion coating. Used in rubberlike small parts like pipe caps, hoses, gaskets, gloves, hospital sheeting, diaper liners, and packaging film.

EVOH See *ethylene vinyl alcohol copolymer*.

extenders Relatively inexpensive resin, plasticizer or filler such as carbonate used to reduce cost and/or to improve processing of plastics, rubbers or nonmetallic coatings.

extrusion coating Coating by extruding a layer of molten resin onto a substrate with sufficient pressure to bond. Used in coating paper and fabrics with polyolefins by extruding a web directly into the roller nip through which the substrate is passing.

extrusion temperature Temperature of the molten thermoplastic maintained in the extruder barrel during the extrusion by means of barrel heating and internal friction of the melt pushed along by a screw or a ram. The temperature may vary along the length of the barrel.

F

falling dart impact See *falling weight impact energy*.

falling dart impact energy See *dart impact energy*.

falling dart impact strength See *falling weight impact energy*.

falling weight impact See *falling weight impact energy*.

falling weight impact energy The mean energy of a free-falling dart or weight (tup) that will cause 50% failures after 50 tests to a directly or indirectly stricken specimen. The energy is calculated by multiplying dart mass, gravitational acceleration and drop height. Also called falling weight impact strength, falling weight impact, falling dart impact energy, falling dart impact strength, falling dart impact, drop dart impact energy, drop dart impact strength.

falling weight impact strength See *falling weight impact energy*.

FEP See *fluorinated ethylene propylene copolymer*.

Fick's first law A physics law that states that the volume (V) of a penetrant, such as gas, that penetrates a barrier wall is directly proportional to the area (A) of the wall, partial pressure differential (p) of the penetrant, and time (t); and inversely proportional to the wall thickness (s), if the wall is homogeneous in the direction of penetration. The coefficient P in the equation representing Fick's first law, V = P · (A · p · t)/s, is the permeability coefficient.

fireproofing agent See *flame retardant*.

five-membered heterocyclic compounds A class of heterocyclic compounds containing rings that consist of five atoms.

five-membered heterocyclic nitrogen compounds A class of heterocyclic compounds containing rings that consist of five atoms, some of which is a nitrogen.

five-membered heterocyclic oxygen compounds A class of heterocyclic compounds containing rings that consist of five atoms, some of which is an oxygen.

flame retardant A substance that reduce the flammability of materials such as plastics or textiles in which it is incorporated. There are inorganic flame retardants such as antimony trioxide (Sb_2O_3) and organic flame retardants such as brominated polyols. The mechanisms of flame retardation vary depending on the nature of material and flame retardant. For example, some flame retardants yield a substantial volume of coke on burning, which prevents oxygen from reaching inside the material and blocks further combustion. Also called fireproofing agent, flame retardant chemical additives, ignition resistant chemical additives.

flame retardant chemical additives See *flame retardant*.

flaw See *processing defects*.

flexural properties Properties describing the reaction of physical systems to flexural stress and strain. Also called bending properties.

flexural strength The maximum stress in the extreme fiber of a specimen loaded to failure in bending. **Note:** Flexural strength is calculated as a function of load, support span and specimen geometry. Also called modulus of rupture in bending, modulus of rupture, bending strength.

flexural stress The maximum stress in the extreme fiber of a specimen in bending. **Note:** Flexural stress is calculated as a function of load at a given strain or at failure, support span and specimen geometry. Also called bending stress.

fluorinated ethylene propylene copolymer Thermoplastic copolymer of tetrafluoroethylene and hexafluoropropylene. Has decreased tensile strength and wear and creep resistance, but good weatherability, dielectric properties, fire and chemical resistance, and friction. Decomposes above 204°C (400°F), releasing toxic products. Processed by molding, extrusion, and powder coating. Used in chemical apparatus liners, pipes, containers, bearings, films, coatings, and cables. Also called FEP.

fluoro rubber See *fluoroelastomers*.

fluoroelastomers Fluorine-containing synthetic rubber with good chemical and heat resistance. Used in underhood applications such as fuel lines, oil and coolant seals, and fuel pumps, and as a flow additive for polyolefins. Also called fluoro rubber.

fluoroplastics See *fluoropolymers*.

fluoropolymers Polymers prepared from unsaturated fluorine-containing hydrocarbons. Have good chemical resistance, weatherability, thermal stability, antiadhesive properties and low friction and flammability, but low creep resistance and strength and poor processibility. The properties vary with the fluorine content. Processed by extrusion and molding. Used as liners in chemical apparatus, in bearings, films, coatings, and containers. Also called fluoroplastics.

fluorosilicones Polymers with chains of alternating silicon and oxygen atoms and trifluoropropyl pendant groups. Most are rubbers.

FMQ See *methylfluorosilicones*.

folding endurance (MIT) See *MIT folding endurance*.

formaldehyde The simplest aldehyde, H_2CO. A readily polymerizable, toxic, skin irritating, carcinogenic gas with strong, pungent odor and autoignition temperature 430°C. Derived by oxidation of methanol or low-boiling olefins. Used as monomer in manufacture of phenolic, acetal, and amino resins; as fertilizer, disinfectant, reducing agent, biocide, sterilant, corrosion inhibitor; in wood products such as plywood, foam insulation, and organic synthesis as an intermediate.

fractional melt index resin Thermoplastics having a low melt index of <1. These resins have higher molecular weights and are harder to extrude because of lower rate and greater force requirements compared to the lower molecular weight resins. They are mainly used for heavy duty applications such as pipe.

FTIR analysis See *Fourier-transform infrared spectrometry*.

furnace black The most common type of carbon black made by burning vaporized heavy oil fractions in a furnace with 50% of the air required for complete combustion. It comes in high abrasion, fast extrusion, high modulus, general purpose, semireinforcing, conducting, high elongation, reinforcing and fast-extruding grades among others. Furnace black is widely used as a filler and pigment in rubbers and plastics. It reinforces, increases the resistance to UV light and reduces static charging.

G

gas black See *channel black*.

gas permeability coefficient A measure of gas permeability of a barrier wall such as plastic film. Gas permeability coefficient, P, is a coefficient in Fick's first law that states that the volume (V) of a substance that penetrates a barrier wall is directly proportional to the area (A) of the wall, partial pressure differential (p) of the penetrant, and time (t); and inversely proportional to the wall thickness (s), if the wall is homogeneous in the direction of penetration. Gas permeability coefficient depends on the test temperature.

general purpose polystyrene General purpose polystyrene is an amorphous thermoplastic prepared by homopolymerization of styrene. It has good tensile and flexural strengths, high light transmission and adequate resistance to water, detergents and inorganic chemicals. It is attached by hydrocarbons and has a relatively low impact resistance. Processed by injection molding and foam extrusion. Used to manufacture containers, health care items such as pipettes, kitchen and bathroom housewares, stereo and camera parts and foam sheets for food packaging. Also called crystal polystyrene.

glycols Aliphatic alcohols with two hydroxy groups attached to a carbon chain. Can be produced by oxidation of alkenes followed by hydration. Also called dihydric alcohols, dihydroxy alcohols.

Graves tear strength A force required to tear completely across a specially designed nicked rubber test specimen, or right-angled test specimen, by elongating it at a specified rate using a power-driven tensile testing machine (Graves machine) as described in the ASTM D624. Expressed in units of force per thickness of specimen.

H

H See *hydrogen.*

halogen compounds A class of organic compounds containing halogen atoms such as chlorine. A simple example is halocarbons but many other subclasses with various functional groups and of different molecular structure exist as well.

halohydrins Halogen compounds that contain a halogen atom(s) and a hydroxy (OH) group(s) attached to a carbon chain or ring. Can be prepared by reaction of halogens with alkenes in the presence of water or by reaction of halogens with triols. Halohydrins can be easily dehydrochlorinated in the presence of a base to give an epoxy compound.

hard clays Sedimentary rocks composed mainly of fine clay mineral material without natural plasticity, or any compacted or indurated clay.

haze The percentage of transmitted light which, in passing through a plastic specimen, deviates from the incident beam via forward scattering more that 2.5°on average (ASTM D883).

HDPE See *high density polyethylene.*

HDT See *heat deflection temperature.*

He See *helium.*

heat deflection point See *heat deflection temperature.*

heat deflection temperature The temperature at which a material specimen (standard bar) is deflected by a certain degree under specified load. Also called heat distortion temperature, heat distortion point, heat deflection point, deflection temperature under load, tensile heat distortion temperature, HDT.

heat distortion point See *heat deflection temperature.*

heat distortion temperature See *heat deflection temperature.*

heat seal temperature Temperature of a thermoplastic film or sheet required to join two or more films or sheets in contact by fusion.

helium A chemically inert, tasteless, colorless, noncombustible monoatomic gas. Helium is often used to characterize permeability of polymeric films, as carrier gas in gas chromatography, as inert gas shield in welding, in electric bulbs such as neon, as heat-transfer medium, in lasers and as a process environment. Also called He.

Henry's law A law that states that the weight of the gas that dissolves in a given quantity of liquid is proportional to the pressure of the gas above the liquid. The law holds true only for equilibrium conditions.

heptane An alkane (saturated aliphatic hydrocarbon) with 6 carbon atoms, $CH_3(CH_2)_4CH_3$.A volatile, colorless, flammable liquid. Autoignition temperature 222°C.Toxic by inhalation. Obtained by fractionation of petroleum. Used as solvent and in organic synthesis. Also called n-heptane.

heterocyclic compounds A class of cyclic compounds containing rings with some carbon atoms replaced by other atoms such as oxygen, sulfur and nitrogen.

high density polyethylene A linear polyethylene with density 0.94-0.97 g/cm^3. Has good toughness at low temperatures, chemical resistance, and dielectric properties and high softening temperature, but poor weatherability. Processed by extrusion, blow and injection molding, and powder coating. Used in houseware, containers, food packaging, liners, cable insulation, pipes, bottles, and toys. Also called HDPE.

high impact polystyrene See *impact polystyrene.*

high molecular weight low density polyethylene Thermoplastic with improved abrasion and stress crack resistance and impact strength, but poor processibility and reduced tensile strength. Also called HMWLDPE.

HIPS See *impact polystyrene.*

HMWLDPE See *high molecular weight low density polyethylene.*

hot fill A process in which containers are filled with a hot liquid. Containers suitable for hot filling should be heat resistant. If they are made of plastic, it should be of a hot-fill grade.

hot tack strength The force required to separate a molten seal in heat-sealable thermoplastic films. It determines the rate at which the film can be sealed. Also called ultimate hot tack strength.

hydrogen A highly flammable diatomic gas, H_2. Occurs on earth mainly in combined form, e.g., with oxygen in water.Autoignition temperature 580°C. Derived by steam reforming, gasification of coal and other methods. Used as hydrogenating and reducing agent in chemical processes and as rocket fuel. Also called H.

hydrophilic starch surface See *hydrophilic surface.*

hydrophilic surface Surface of a hydrophilic substance that has a strong ability to bind, adsorb or absorb water; a surface that is readily wettable with water. Hydrophilic substances include carbohydrates such as starch. Also called hydrophilic starch surface.

hydroxy compounds A broad class of organic compounds that contain a hydroxy (OH) group(s) that is not part of another functional group such as carboxylic group. Also called hydroxyl-containing compounds.

hydroxy group See *hydroxyl group.*

hydroxyl group A combination of one atom of hydrogen and one atom of oxygen, -OH, attached by a single covalent bond to another atom, such as carbon, in a molecule of an organic or inorganic substance. It is a characteristic group of alcohols and hydroxides. Hydroxyl groups on the surface of a material usually make it hydrophilic. Hydroxyl groups are quite reactive, e.g., they readily undergo etherification or esterification. Also called hydroxy group.

hydroxyl-containing compounds See *hydroxy compounds.*

I

ignition resistant chemical additives See *flame retardant.*

impact energy The energy required to break a specimen, equal to the difference between the energy in the striking member of the impact apparatus at the instant of impact and the energy remaining after complete fracture of the specimen. Also called impact strength. See also *ASTM D256, ASTM D3763.*

impact polystyrene Impact polystyrene is a thermoplastic produced by polymerizing styrene dissolved in butadiene rubber. Impact polystyrene has good dimensional stability, high rigidity and good low temperature impact strength, but poor barrier properties, grease resistance and heat resistance. Processed by extrusion, injection molding, thermoforming and structural foam molding. Used in food packaging, kitchen housewares, toys, small appliances, personal care items and audio products. Also called IPS, high impact polystyrene, HIPS, impact PS.

impact property tests Names and designations of the methods for impact testing of materials. Also called impact tests. See also *impact toughness*.

impact PS See *impact polystyrene*.

impact strength See *impact energy*.

impact tests See *impact property tests*.

impact toughness Property of a material indicating its ability to absorb energy of a high-speed impact by plastic deformation rather than crack or fracture. See also *impact property tests*.

inch of mercury See *inHg*.

inHg An English unit of measurement of pressure equal to 3.3864 x 1e+04 dynes/cm^2 or 249.089 pascals at 0°C (32°F). One inch of mercury is the pressure that would support a column of mercury of length one inch and density 12,595 kg/m^3 under the standard acceleration of free fall. Used to denote the pressure of gases, vapors and liquids. Also called inch of mercury.

initial tear resistance The force required to initiate tearing of a flexible plastic film or thin sheeting at very low rates of loading, measured as maximum stress usually found at the onset of tearing. Also called tear resistance, initial.

ionomers Thermoplastics containing a relatively small amount of pendant ionized acid groups. Have good flexibility and impact strength in a wide temperature range, puncture and chemical resistance, adhesion, and dielectric properties, but poor weatherability, fire resistance, and thermal stability. Processed by injection, blow and rotational molding, blown film extrusion, and extrusion coating. Used in food packaging, auto bumpers, sporting goods, and foam sheets.

IPS See *impact polystyrene*.

isophthalate polyester An unsaturated polyester based on isophthalic acid.

Izod See *Izod impact energy*.

Izod impact See *Izod impact energy*.

Izod impact energy The energy required to break a specimen equal to the difference between the energy in the striking member of the Izod-type impact apparatus at the instant of impact and the energy remaining after complete fracture of the specimen. Also called Izod impact, Izod impact strength, Izod.

Izod impact strength See *Izod impact energy*.

J

JIS K6732 A Japanese Standards Association (Nippon Kikaku Kyokai) standard test method for determining tensile properties of plastic materials such as film and sheeting.

JIS P8112 A Japanese Standards Association (Nippon Kikaku Kyokai) standard test method for determining bursting strength of flat materials such as plastic film and paper products.

JIS P8116 A Japanese Standards Association (Nippon Kikaku Kyokai) standard test for determining the resistance of flexible plastic film or sheeting to tear propagation. The resistance is measured as the average force, in grams, required to propagate tearing from a precut slit through a specified length, using an Elmendorf-type pendulum tester.

JIS Z0208 A Japanese Standards Association (Nippon Kikaku Kyokai) standard test method for determining water vapor transmission of flat materials such as plastic film and sheeting.

K

kinetic coefficient of friction The ratio of tangential force, which is required to sustain motion without acceleration of one surface with respect to another, to the normal force, which presses the two surfaces together. Also called coefficient of friction, coefficient of friction, kinetic.

kinetic strip test An ozone resistance test for rubbers that involves a strip-shaped specimen stretched to 23% and relax to 0 at a rate of 30 cycles per minute, while subjected to ozone attack in the test chamber. The results of the test are reported with 2 digits separated with a virgule. The number before the virgule indicates the number of quarters of the test strip which showed the cracks. The number after the virgule indicates the size of the cracks in length perpendicular to the length of the strip.

L

lamellar injection molding Injection molding of individual thermoplastics or their blends, e.g., with liquid-crystal polymers, that produces a lamellar (platelike crystallite) skin texture of the molding for decorative purposes or enhanced surface properties.

LCP See *liquid crystal polymers*.

LDPE See *low density polyethylene*.

linear low density polyethylene Linear polyethylenes with density 0.91-0.94 g/cm^3. Has better tensile, tear, and impact strength and crack resistance properties, but poorer haze and gloss than branched low-density polyethylene. Processed by extrusion at increased pressure and higher melt temperatures compared to branched low-density polyethylene, and by molding. Used to manufacture film, sheet, pipe, electrical insulation, liners, bags and food wraps. Also called LLDPE, LLDPE resin.

linear polyethylenes Linear polyethylenes are polyolefins with linear carbon chains. They are prepared by copolymerization of ethylene with small amounts of higher alfa-olefins such as 1-butene. Linear polyethylenes are stiff, tough and have good resistance to environmental cracking and low temperatures. Processed by extrusion and molding. Used to manufacture film, bags, containers, liners, profiles and pipe.

liquid crystal polymers Thermoplastic aromatic copolyesters with highly ordered structure. Have good tensile and flexural properties at high temperatures, chemical, radiation and fire resistance, and weatherability. Processed by sintering and injection molding. Used to substitute ceramics and metals in electrical components, electronics, chemical apparatus, and aerospace and auto parts. Also called LCP.

LLDPE See *linear low density polyethylene.*

LLDPE resin See *linear low density polyethylene.*

low density polyethylene A branched-chain thermoplastic with density 0.91-0.94 g/cm^3. Has good impact strength, flexibility, transparency, chemical resistance, dielectric properties, and low water permeability and brittleness temperature, but poor heat, stress cracking and fire resistance and weatherability. Processed by extrusion coating, injection and blow molding, and film extrusion. Can be crosslinked. Used in packaging and shrink films, toys, bottle caps, cable insulation, and coatings. Also called LDPE.

M

macroscopic properties See *thermodynamic properties.*

mass spectrometry A method of substance structure analysis based on sending an ionized beam of substance molecules or molecular fragments through a magnetic field to achieve a separation depending on the mass-electric charge ratio of the particles.

MBT See *2-mercaptobenzothiazole.*

mechanical properties Properties describing the reaction of physical systems to stress and strain.

melamine resins Thermosetting resins prepared by condensation of formaldehyde with melamine. Have good hardness, scratch and fire resistance, clarity, colorability, rigidity, dielectric properties, and tensile strength, but poor impact strength. Molding grades are filled. Processed by compression, transfer, and injection molding, impregnation, and coating. Used in cosmetic containers, appliances, tableware, electrical insulators, furniture laminates, adhesives, and coatings.

mercaptobenzothiazole (2-) A nitrogen- and sulfur-containing polyheterocyclic organic thiol used as vulcanization accelerator for rubber. Requires zinc oxide as an activator. Its vulcanizates have a good aging resistance. A yellowish powder with distinctive odor. Combustible. Also called MBT.

methane An alkane (saturated aliphatic hydrocarbon) with one carbon atom, CH$_4$. A colorless, odorless, highly flammable gas. Autoignition temperature 537°C. Reacts with chlorine in light. Occurs as natural and coal gas. Can be obtained synthetically from a mixture of carbon monoxide and hydrogen from steam treatment of hot coal. Used in petrochemical synthesis, for manufacture of carbon black and chlorinated solvents, and as fuel.

methanol See *methyl alcohol.*

methyl alcohol An aliphatic alcohol, CH$_3$OH. A colorless, volatile, flammable liquid. Autoignition point is 464°C. Toxic by ingestion. Derived by catalytic hydrogenation of carbon monoxide, oxidation of natural gas, or gasification of wood. Used as fuel, as solvent for cellulosic and other resins, and in organic synthesis for manufacture of formaldehyde and proteins. Also called methanol.

methylfluorosilicones Silicone rubbers containing pendant fluorine and methyl groups. Have good chemical and heat resistance. Used in gasoline lines, gaskets, and seals. Also called FMQ.

methylphenylsilicones Silicone rubbers containing pendant phenyl and methyl groups. Have good resistance to heat, oxidation, and radiation, and compatibility with plastics.

methylsilicone Silicone rubbers containing pendant methyl groups. Have good heat and oxidation resistance. Used in electrical insulation and coatings. Also called MQ.

methylvinylfluorosilicone Silicone rubbers containing pendant vinyl, methyl, and fluorine groups. Can be additionally crosslinked via vinyl groups. Have good resistance to petroleum products at elevated temperatures.

methylvinylsilicone Silicone rubbers containing pendant methyl and vinyl groups. Can be additionally crosslinked via vinyl groups. vulcanized to high degrees of crosslinking. Used in sealants, adhesives, coatings, cables, gaskets, tubing, and electrical tape.

micron A unit of length equal to 1E-06 meter. Its symbol is Greek small letter mu(μ) or mum.

microtensile specimen A small specimen as specified in ASTM D1708 for determining tensile properties of plastics. It has maximum thickness 3.2 mm and minimum length 38.1 mm. Tensile properties determined with this specimen include yield strength, tensile strength, tensile strength at break and elongation at break.

migration A mass-transfer process in which the matter moves from one place to another usually in a slow and spontaneous fashion. In plastics and coatings, migration of pigments, fillers, plasticizers and other ingredients via diffusion or floating to the surface or through interface to other materials results in various defects called blooming, chalking, bronzing, flooding, bleeding, etc.

mineral acid An inorganic, usually strong, acid such as sulfuric acid (H$_2$SO$_4$).

mineral salt medium A corrosive medium such as aqueous solution, containing mineral or inorganic salt such as sodium chloride (NaCl). Used in material testing, especially of anticorrosive properties.

MIT folding endurance The folding endurance of paper determined by the MIT tester (ASTM D2176) as the number of double, 135° folds made at a given tension before fracture of the specimen. Also called folding endurance (MIT).

modified polyphenylene ether Thermoplastic polyphenylene ether alloys with impact polystyrene. Have good impact strength, resistance to heat and fire, but poor resistance to solvents. Processed by injection and structural foam molding and extrusion. Used in auto parts, appliances, and telecommunication devices. Also called MPE, MPO, modified polyphenylene oxide.

modified polyphenylene oxide See *modified polyphenylene ether.*

modulus of rupture See *flexural strength.*

modulus of rupture in bending See *flexural strength.*

molding defects Structural and other defects in material caused inadvertently during molding by using wrong tooling, process parameters or ingredients. Also called molding flaw. See also *design, etc. Usually preventable.*

molding flaw See *molding defects.*

molecular weight The sum of the atomic weights of all atoms in a molecule. Also called MW.

molecular weight distribution The relative amounts of polymeric molecules of different weights in a specimen. **Note:** The molecular weight distribution can be expressed in terms of the ratio between weight- and number-average molecular weights. Also called polydispersity, MWD, molecular weight ratio.

molecular weight ratio See *molecular weight distribution.*

MPE See *modified polyphenylene ether.*

MPO See *modified polyphenylene ether.*

MQ See *methylsilicone.*

mulch film A film, usually dark colored PVC film, used instead of mulch in agriculture, e.g., to prevent fruit rotting and runners and weed growth in cultivation of strawberrys.

multilayer film A thermoplastic film consisting of two or more different or similar films jointed together, e.g., by coextrusion or lamination, to attain special properties uncharacteristic for a conventional film.

MW See *molecular weight.*

MWD See *molecular weight distribution.*

N

N See *nitrogen.*

n-heptane See *heptane.*

neoprene rubber Polychloroprene rubbers with good resistance to petroleum products, heat, and ozone, weatherability, and toughness.

nitrile rubber Rubbers prepared by free-radical polymerization of acrylonitrile with butadiene. Have good resistance to petroleum products, heat, and abrasion. Used in fuel hoses, shoe soles, gaskets, oil seals, and adhesives.

nitroarylamine A class of aromatic amines containing benzene ring(s) with nitro (NO^2 group substituent(s), such as nitroaniline ($O_2NC_6H_4NH_2$). Used as organic intermediates (e.g., in dye synthesis) and antioxidants in propellants and plastics.

nitrogen A colorless, odorless, combustible diatomic gas, N_2. The major component (about 78 vol%) of earth's atmosphere. Derived from air by fractionation. Used in organic and inorganic synthesis, as inert medium, for food freezing and freeze drying, as food antioxidant, in fertilizers and as a pressurizing gas. Also called N.

nonelastomeric thermoplastic polyurethanes See *rigid thermoplastic polyurethanes.*

nonelastomeric thermosetting polyurethane Curable mixtures of isocyanate prepolymers or monomers. Have good abrasion resistance and low-temperature stability, but poor heat, fire, and solvent resistance and weatherability. Processed by reaction injection and structural foam molding, casting, potting, encapsulation, and coating. Used in heat insulation, auto panels and trim, and housings for electronic devices.

notch effect The effect of the presence of specimen notch or its geometry on the outcome of a test such as an impact strength test of plastics. Notching results in local stresses and accelerates failure in both static and cycling testing (mechanical, ozone cracking, etc.).

notched Izod See *notched Izod impact energy.*

notched Izod impact See *notched Izod impact energy.*

notched Izod impact energy The energy required to break a notched specimen equal to the difference between the energy in the striking member of the Izod-type impact apparatus at the instant of impact and the energy remaining after complete fracture of the specimen. **Note:** Energy depends on geometry (e.g., width, depth, shape) of the notch, on the cross-sectional area of the specimen and on the place of impact (on the side of the notch or on the opposite side). In some tests notch is made on both sides of the specimen Also called notched Izod impact strength, notched Izod impact, notched Izod.

notched Izod impact strength See *notched Izod impact energy.*

nylon Thermoplastic polyamides often prepared by ring-opening polymerization of lactam. Have good resistance to most chemicals, abrasion, and creep, good impact and tensile strengths, barrier properties, and low friction, but poor resistance to moisture and light. Have high mold shrinkage. Processed by injection, blow, and rotational molding, extrusion, and powder coating. Used in fibers, auto parts, electrical devices, gears, pumps, appliance housings, cable jacketing, pipes, and films.

nylon 11 Thermoplastic polymer of 11-aminoundecanoic acid having good impact strength, hardness, abrasion resistance, processability, and dimensional stability. Processed by powder coating, rotational molding, extrusion, and injection molding. Used in electric insulation, tubing, profiles, bearings, and coatings.

nylon 12 Thermoplastic polymer of lauric lactam having good impact strength, hardness, abrasion resistance, and dimensional stability. Processed by powder coating, rotational molding, extrusion, and injection molding. Used in sporting goods and auto parts.

nylon 46 Thermoplastic copolymer of 2-pyrrolidone and caprolactam.

nylon 6 Thermoplastic polymer of caprolactam. Has good weldability and mechanical properties but rapidly picks up moisture which results in strength losses. Processed by injection, blow, and rotational molding and extrusion. Used in fibers, tire cord, and machine parts.

nylon 610 Thermoplastic polymer of hexamethylenediamine and sebacic acid having decreased melting point and water absorption and good retention of mechanical properties. Processed by injection molding and extrusion. Used in fibers and machine parts.

nylon 612 Thermoplastic polymer of 1,12-dodecanedioic acid and hexamethylenediamine having good dimensional stability, low moisture absorption, and good retention of mechanical properties. Processed by injection molding and extrusion. Used in wire jackets, cable sheath, packaging film, fibers, bushings, and housings.

nylon 66 Thermoplastic polymer of adipic acid and hexamethylenediamine having good tensile strength, elasticity, toughness, heat resistance, abrasion resistance, and solvent resistance but low weatherability and color resistance. Processed by injection molding and extrusion. Used in fibers, bearings, gears, rollers, and wire jackets.

nylon 6/66 Thermoplastic polymer of adipic acid, caprolactam, and hexamethylenediamine having good strength, toughness, abrasion and fatigue resistance, and low friction but high moisture absorption and low dimensional stability. Processed by injection molding and extrusion. Used in electrical devices and auto and mechanical parts.

nylon MXD6   Thermoplastic polymer of m-xylyleneadipamide having good flexural strength and chemical resistance but decreased tensile strength.

O

O   See *oxygen*.

olefin resins   See *polyolefins*.

olefinic resins   See *polyolefins*.

olefinic thermoplastic elastomers   Blends of EPDM or EP rubbers with polypropylene or polyethylene, optionally crosslinked.   Have low density, good dielectric and mechanical properties, and processibility but low oil resistance and high flammability.   Processed by extrusion, injection and blow molding, thermoforming, and calendering.   Used in auto parts, construction, wire jackets, and sporting goods. Also called TPO.

OPP   See *oriented polypropylene*.

organic compounds   See *halogen compounds*. Also called organic substances.

organic compounds   See *halogen compounds*.

organic substances   See *organic compounds*.

orientation   A process of drawing or stretching of as-spun synthetic fibers or hot thermoplastic films to orient polymer molecules in the direction of stretching.   The fibers are drawn uniaxially and the films are stretched either uniaxially or biaxially (usually longitudinally or longitudinally and transversely, respectively).   Oriented fibers and films have enhanced mechanical properties.   The films will shrink in the direction of stretching, when reheated to the temperature of stretching.

oriented polypropylene   A grade of polypropylene film hot stretched uniaxially or biaxially (usually longitudinally or longitudinally and transversely, respectively) to orient polymer molecules in the direction of stretching.   Oriented films have enhanced mechanical properties.   They will shrink in the direction of stretching when reheated, e.g., during heat sealing. Also called OPP.

oxazolines   Heterocyclic compounds containing five-membered rings in which one carbon is replaced with an oxygen atom and another with a nitrogen atom.   Oxazolines are colorless liquids soluble in organic solvents and water.   Used as intermediates, e.g., in synthesis of surfactants.

oxygen   A colorless, odorless diatomic gas, O_2. A major component (about 20 vol%) of earth's atmosphere, it is required for cellular respiration and actively supports combustion.   Derived mainly by passing air through a molecular sieve to remove nitrogen.   Used in metallurgy, synthetic gas manufacture, as oxidizing agent for rocket fuel, in wastewater treatment, in medicine and coal gasification.   Also called O.

ozone   An allotropic form of oxygen, O_3.   Unstable gas formed naturally, in air by lightening or in stratosphere by the UV portion of solar radiation, or formed as a result of combustion of fossil fuels, i.e., in exhaust gases from automobiles. O_3 is an active oxidizing agent that accelerates deterioration of rubber.

P

Pa   See *pascal*.

PABM   See *polyaminobismaleimide resins*.

paraffinic plasticizer   Plasticizers for plastics comprising liquid or solid long-chain alkanes or paraffins· (saturated linear or branched hydrocarbons).

partial pressure   The pressure that would be exerted by a gas in a gas mixture if it were present alone.

parts per hundred   A relative unit of concentration, parts of one substance per 100 parts of another.   Parts can be measured by weight, volume, count or any other suitable unit of measure.   Used often to denote composition of a blend or mixture, such as plastic, in terms of the parts of a minor ingredient, such as plasticizer, per 100 parts of a major, such as resin. Also called phr.

parts per hundred million   A relative unit of concentration, parts of one substance per 100 million parts of another.   Parts can be measured by weight, volume, count or any other suitable unit of measure.   Used often to denote very small concentration of a substance, such as impurity or toxin, in a medium, such as air. Also called pphm.

parylene   Thermoplastics made by vapor-phase polymerization of p-xylene.   Hot p-xylene vapors are cooled to condense the monomer and deposit it as a polymer in the form of a thin, uniform coating on a substrate such as paper or fabric.

pascal   An SI unit of measurement of pressure equal to the pressure resulting from a force of one newton acting uniformly over an area of one square meter.   Used to denote the pressure of gases, vapors or liquids and the strength of solids. Also called Pa.

PBI   See *polybenzimidazoles*.

PBT   See *polybutylene terephthalate*.

PC   See *polycarbonates*.

PCT   See *polycyclohexylenedimethylene terephthalate*.

PCTG   See *glycol modified polycyclohexylenedimethylene terephthalate*.

PE copolymer   See *polyethylene copolymer*.

PEEK   See *polyetheretherketone*.

PEI   See *polyetherimides*.

PEK   See *polyetherketone*.

pendant aromatic rings   Aromatic (conjugated unsaturated rings such as those of benzene, C_6H_6) rings attached to the main chain of a polymer molecule.

penetrant   A substance such as gas that penetrates or is capable of penetrating through another substance, usually a solid barrier wall such as plastic film. Also called permeant.

pentaerythritol   A polyol, $C(CH_2OH)_4$, prepared by reaction of acetaldehyde with an excess formaldehyde in alkaline medium.   Used as plasticizer and as monomer in alkyd resins.

perchloroethylene   See *tetrachloroethylene*.

perfluoroalkoxy resins   Thermoplastic polymers of perfluoroalkoxyethylenes having good creep, heat, and chemical resistance and processibility but low compressive and tensile strengths.   Processed by molding, extrusion, rotational molding, and powder coating. Used in films, coatings, pipes, containers, and chemical apparatus linings. Also called PFA.

perm An English unit of measurement of permeability of material in terms of the permeability coefficient. It is equal to the volume of penetrant in cubic feet that penetrates an area of one square foot of a barrier wall one foot thick per day at a pressure differential of one pound-force per square inch.

permanent gas Gases that become liquid at pressures and temperatures far from normal (1 atm and 0°C, respectively). These gases include air, oxygen, argon, and carbon dioxide.

permeant See *penetrant.*

PES See *polyethersulfone.*

PET See *polyethylene terephthalate.*

PETG See *polycyclohexylenedimethylene ethylene terephthalate.*

PFA See *perfluoroalkoxy resins.*

phase transition See *phase transition properties.*

phase transition point The temperature at which a phase transition occurs in a physical system such as material. **Note:** An example of phase transition is glass transition. Also called phase transition temperature, transition point, transition temperature.

phase transition properties Properties of physical systems such as materials associated with their transition from one phase to another, e.g., from liquid to solid phase. Also called phase transition.

phase transition temperature See *phase transition point.*

phenolic resins Thermoset polymers of phenols with excess or deficiency of aldehydes, mainly formaldehyde, to give resole or novolak resins, respectively. Heat-cured resins have good dielectric properties, hardness, thermal stability, rigidity, and compressive strength but poor chemical resistance and dark color. Processed by coating, potting, compression, transfer, or injection molding and extrusion. Used in coatings, adhesives, potting compounds, handles, electrical devices, and auto parts.

phr See *parts per hundred.*

phthalocyanine A nitrogen-containing heterocyclic organic compound, $(C_6H_4C_2N)_2(C_6H_4C_2NH)_2N_4$, belonging to the group of benzoporphyrins and comprising 4 isoindole groups jointed by 4 nitrogen atoms. Readily forms salt complexes with copper, chromium, iron, etc., that are important green and blue dyes and pigments. These pigments have high light and chemical stability. Used in coatings, plastics and textiles.

PI See *polyimides.*

plasticizer A substance incorporated into a material such as plastic or rubber to increase its softness, processability and flexibility via solvent or lubricating action or by lowering its molecular weight. Plasticizers can lower melt viscosity, improve flow and increase low-temperature resilience of material. Most plasticizers are nonvolatile organic liquids or low-melting-point solids, such as dioctyl phthalate or stearic acid. They have to be non-bleeding, nontoxic and compatible with material. Sometimes plasticizers play a dual role as stabilizers or crosslinkers.

plastics See *polymers.*

PMMA See *polymethyl methacrylate.*

PMP See *polymethylpentene.*

polyacrylates See *acrylic resins.*

polyallomer Crystalline thermoplastic block copolymers of ethylene, propylene, and other olefins. Have good impact strength and flex life and low density.

polyamide thermoplastic elastomers Copolymers containing soft polyether and hard polyamide blocks having good chemical, abrasion, and heat resistance, impact strength, and tensile properties. Processed by extrusion and injection and blow molding. Used in sporting goods, auto parts, and electrical devices. Also called polyamide TPE.

polyamide TPE See *polyamide thermoplastic elastomers.*

polyamides Thermoplastic aromatic or aliphatic polymers of dicarboxylic acids and diamines, of amino acids, or of lactams. Have good mechanical properties, chemical resistance, and antifriction properties. Processed by extrusion and molding. Used in fibers and molded parts. Also called PA.

polyaminobismaleimide resins Thermoset polymers of aromatic diamines and bismaleimides having good flow and thermochemical properties and flame and radiation resistance. Processed by casting and compression molding. Used in aircraft parts and electrical devices. Also called PABM.

polyarylamides Thermoplastic crystalline polymers of aromatic diamines and aromatic dicarboxylic anhydrides having good heat, fire, and chemical resistance, property retention at high temperatures, dielectric and mechanical properties, and stiffness but poor light resistance and processibility. Processed by solution casting, molding, and extrusion. Used in films, fibers, and molded parts.

polyarylsulfone Thermoplastic aromatic polyether-polysulfone having good heat, fire, and chemical resistance, impact strength, resistance to environmental stress cracking, dielectric properties, and rigidity. Processed by injection and compression molding and extrusion. Used in circuit boards, lamp housings, piping, and auto parts.

polybenzimidazoles Mainly polymers of 3,3',4,4'-tetraminonbiphenyl (diaminobenzidine) and diphenyl isophthalate. Have good heat, fire, and chemical resistance. Used as coatings and fibers in aerospace and other high-temperature applications. Also called PBI.

polybutylene terephthalate Thermoplastic polymer of dimethyl terephthalate and butanediol having good tensile strength, dielectric properties, and chemical and water resistance, but poor impact strength and heat resistance. Processed by injection and blow molding, extrusion, and thermoforming. Used in auto body parts, electrical devices, appliances, and housings. Also called PBT.

polycarbodiimide Polymers containing -N=C=N- linkages in the main chain, typically formed by catalyzed polycondensation of polyisocyanates. They are used to prepare open-celled foams with superior thermal stability. Sterically hindered polycarbodiimides are used as hydrolytic stabilizers for polyester-based urethane elastomers.

polycarbonate See *polycarbonates.*

polycarbonate polyester alloys High-performance thermoplastics processed by injection and blow molding. Used in auto parts.

polycarbonate resins See *polycarbonates.*

polycarbonates Polycarbonates are thermoplastics prepared by either phosgenation of dihydric aromatic alcohols such as bisphenol A or by transesterification of these alcohols with carbonates, e.g., diphenyl carbonate. Polycarbonates consist of chains with repeating carbonyldioxy groups and can be aliphatic or aromatic. They have

694

very good mechanical properties, especially impact strength, low moisture absorption and good thermal and oxidative stability. They are self-extinguishing and some grades are transparent. Polycarbonates have relatively low chemical resistance and resistance to stress cracking. Processed by injection and blow molding, extrusion, thermoforming at relatively high processing temperatures. Used in telephone parts, dentures, business machine housings, safety equipment, nonstaining dinnerware, food packaging, etc. Also called polycarbonate, PC, polycarbonate resins.

polychlorotrifluoroethylene Thermoplastic polymer of chlorotrifluoroethylene having good transparency, barrier properties, tensile strength, and creep resistance, modest dielectric properties and solvent resistance, and poor processibility. Processed by extrusion, injection and compression molding, and coating. Used in chemical apparatus, low-temperature seals, films, and internal lubricants. Also called CTFE.

polycyclohexylenedimethylene ethylene terephthalate Thermoplastic polymer of cyclohexylenedimethylenediol, ethylene glycol, and terephthalic acid. Has good clarity, stiffness, hardness, and low-temperature toughness. Processed by injection and blow molding and extrusion. Used in containers for cosmetics and foods, packaging film, medical devices, machine guards, and toys. Also called PETG.

polycyclohexylenedimethylene terephthalate Thermoplastic polymer of cyclohexylenedimethylenediol and terephthalic acid having good heat resistance. Processed by molding and extrusion. Also called PCT.

polydispersity See *molecular weight distribution*.

polyester resins See *polyesters*.

polyester thermoplastic elastomers Copolymers containing soft polyether and hard polyester blocks having good dielectric strength, chemical and creep resistance, dynamic performance, appearance, and retention of properties in a wide temperature range but poor light resistance. Processed by injection, blow, and rotational molding, extrusion casting, and film blowing. Used in electrical insulation, medical products, auto parts, and business equipment. Also called polyester TPE.

polyester TPE See *polyester thermoplastic elastomers*.

polyesters A broad class of polymers usually made by condensation of a diol with dicarboxylic acid or anhydride. Polyesters consist of chains with repeating carbonyloxy group and can be aliphatic or aromatic. There are thermosetting polyesters, such as alkyd resins and unsaturated polyesters, and thermoplastic polyesters such as PET. The properties, processing methods and applications of polyesters vary widely. Also called polyester resins.

polyetheretherketone Semi-crystalline thermoplastic aromatic polymer having good chemical, heat, fire, and radiation resistance, toughness, rigidity, bearing strength, and processibility. Processed by injection molding, spinning, cold forming, and extrusion. Used in fibers, films, auto engine parts, aerospace composites, and electrical insulation. Also called PEEK.

polyetherimides Thermoplastic cyclized polymers of aromatic diether dianhydrides and aromatic diamine. Have good chemical, creep, and heat resistance and dielectric properties. Processed by extrusion, thermoforming, and compression, injection, and blow molding. Used in auto parts, jet engines, surgical instruments, industrial apparatus, food packaging, cookware, and computer disks. Also called PEI.

polyetherketone Thermoplastic having good heat and chemical resistance. thermal stability. Used in advanced composites, wire

coating, filters, integrated circuit boards, and bearings. Also called PEK.

polyethersulfone Thermoplastic aromatic polymer having good heat and fire resistance, transparency, dielectric properties, dimensional stability, rigidity, and toughness, but poor solvent and stress cracking resistance, processibility, and weatherability. Processed by injection, blow, and compression molding and extrusion. Used in high temperature applications electrical devices, medical devices, housings, and aircraft and auto parts. Also called PES.

polyethylene copolymer Thermoplastics polymers of ethylene with other olefins such as propylene. Processed by molding and extrusion. Also called PE copolymer.

polyethylene terephthalate Thermoplastic polymer of ethylene glycol with terephthalic acid. Has good hardness, wear and chemical resistance, dimensional stability, and dielectric properties. High-crystallinity grades have good tensile strength and heat resistance. Processed by extrusion and injection and blow molding. Used in fibers, food packaging (films, bottles, trays), magnetic tapes, and photo films. Also called PET.

polyimides Thermoplastic aromatic cyclized polymers of trimellitic anhydride and aromatic diamine. Have good tensile strength, dimensional stability, dielectric and barrier properties, and creep, impact, heat, and fire resistance, but poor processibility. Processed by compression and injection molding, powder sintering, film casting, and solution coating. Thermoset uncyclized polymers are heat curable and have good processability. Processed by transfer and injection molding, lamination, and coating. Used in jet engines, compressors, sealing coatings, auto parts, and business machines. Also called PI.

polymer chain unsaturation See *chemical unsaturation*.

polymers Polymers are high-molecular-weight organic or inorganic compounds the molecules of which comprise linear, branched, crosslinked or otherwise shaped chains of repeating molecular groups. Synthetic polymers are prepared by polymerization of one or more monomers. The monomers are low-molecular-weight substances with one or more reactive bonds or functional groups. Also called resins, plastics.

polymethyl methacrylate Thermoplastic polymer of methyl methacrylate having good transparency, weatherability, impact strength, and dielectric properties. Processed by compression and injection molding, casting, and extrusion. Used in lenses, sheets, airplane canopies, signs, and lighting fixtures. Also called PMMA.

polymethylpentene Thermoplastic stereoregular polyolefin obtained by polymerizing 4-methyl-1-pentene based on dimerization of propylene; having low density, good transparency, rigidity, dielectric and tensile properties, and heat and chemical resistance. Processed by injection and blow molding and extrusion. Used in laboratory ware, coated paper, light fixtures, auto parts, and electrical insulation. Also called PMP.

polyolefin resins See *polyolefins*.

polyolefins Polyolefins are a broad class of hydrocarbon-chain elastomers or thermoplastics usually prepared by addition (co)polymerization of alkenes such as ethylene. There are branched and linear polyolefins and some are chemically or physically modified. Unmodified polyolefins have relatively low thermal stability and a nonporous, nonpolar surface with poor adhesive properties. Processed by extrusion, injection molding, blow molding and rotational molding. Polyolefins are used more and have more applications than any other polymers. Also called olefinic resins, olefin resins, polyolefin resins.

Glossary of Terms

© Plastics Design Library

polyphenylene ether nylon alloys Thermoplastics having improved heat and chemical resistance and toughness. Processed by molding and extrusion. Used in auto body parts.

polyphenylene sulfide High-performance engineering thermoplastic having good chemical, water, fire, and radiation resistance, dimensional stability, and dielectric properties, but decreased impact strength and poor processibility. Processed by injection, compression, and transfer molding and extrusion. Used in hydraulic components, bearings, electronic parts, appliances, and auto parts. Also called PPS.

polyphenylene sulfide sulfone Thermoplastic having good heat, fire, creep, and chemical resistance and dielectric properties. Processed by injection molding. Used in electrical devices. Also called PPSS.

polyphthalamide Thermoplastic polymer of aromatic diamine and phthalic anhydride. Has good heat, chemical, and fire resistance, impact strength, retention of properties at high temperatures, dielectric properties, and stiffness, but decreased light resistance and poor processibility. Processed by solution casting, molding, and extrusion. Used in films, fibers, and molded parts. Also called PPA.

polypropylene Thermoplastic polymer of propylene having low density and good flexibility and resistance to chemicals, abrasion, moisture, and stress cracking, but decreased dimensional stability, mechanical strength, and light, fire, and heat resistance. Processed by injection molding, spinning, and extrusion. Used in fibers and films for adhesive tapes and packaging. Also called PP.

polypyrrole A polymer of pyrrole, a five-membered heterocyclic substance with one nitrogen and four carbon atoms and with two double bonds. The polymer can be prepared via electrochemical polymerization. Polymers thus prepared are doped by electrolyte anion and are electrically conductive. Polypyrrole is used in lightweight secondary batteries, as electromagnetic interference shielding, anodic coatings, photoconductors, solar cells, and transistors.

polystyrene Polystyrenes are thermoplastics produced by polymerization of styrene with or without modification (e.g., by copolymerization or blending) to make impact resistant or expandable grades. They have good rigidity, high dimensional stability, low moisture absorption, optical clarity, high gloss and good dielectric properties. Unmodified polystyrenes have poor impact strength and resistance to solvents, heat and UV radiation. Processed by injection molding, extrusion, compression molding, and foam molding. Used widely in medical devices, housewares, food packaging, electronics and foam insulation. Also called polystyrenes, PS, polystyrol.

polystyrenes See *polystyrene*.

polystyrol See *polystyrene*.

polysulfones Thermoplastics, often aromatic and with ether linkages, having good heat, fire, and creep resistance, dielectric properties, transparency, but poor weatherability, processibility, and stress cracking resistance. Processed by injection, compression, and blow molding and extrusion. Used in appliances, electronic devices, auto parts, and electric insulators. Also called PSO.

polytetrafluoroethylene Thermoplastic polymer of tetrafluoroethylene having good dielectric properties, chemical, heat, abrasion, and fire resistance, antiadhesive properties, impact strength, and weatherability, but decreased strength, processibility, barrier properties, and creep resistance. Processed by sinter molding and powder coating. Used in nonstick coatings, chemical apparatus, electrical devices, bearings, and containers. Also called PTFE.

polyurethane resins See *polyurethanes*.

polyurethanes Polyurethanes (PUs) are a broad class of polymers consisting of chains with a repeating urethane group, prepared by condensation of polyisocyanates with polyols, e.g., polyester or polyether diols. PUs may be thermoplastic or thermosetting, elastomeric or rigid, cellular or solid, and offer a wide range of properties depending on composition and molecular structure. Many PUs have high abrasion resistance, good retention of properties at low temperatures and good foamability. Some have poor heat resistance, weatherability and resistance to solvents. PUs are flammable and can release toxic substances. Thermoplastic PUs are not crosslinked and are processed by injection molding and extrusion. Thermosetting PUs can be cured at relatively low temperatures and give foams with good heat insulating properties. They are processed by reaction injection molding, rigid and flexible foam methods, casting and coating. PUs are used in load bearing rollers and wheels, acoustic clamping materials, sporting goods, seals and gaskets, heat insulation, potting and encapsulation. Also called PUR, PU, urethane polymers, urethane resins, urethanes, polyurethane resins.

polyvinyl chloride Thermoplastic polymer of vinyl chloride, available in rigid and flexible forms. Has good dimensional stability, fire resistance, and weatherability, but decreased heat and solvent resistance and high density. Processed by injection and blow molding, calendering, extrusion, and powder coating. Used in films, fabric coatings, wire insulation, toys, bottles, and pipes. Also called PVC.

polyvinyl fluoride Crystalline thermoplastic polymer of vinyl fluoride having good toughness, flexibility, weatherability, and low-temperature and abrasion resistance. Processed by film techniques. Used in packaging, glazing, and electrical devices. Also called PVF.

polyvinylidene chloride Stereoregular thermoplastic polymer of vinylidene chloride having good abrasion and chemical resistance and barrier properties. Vinylidene chloride (VDC) content always exceeds 50%. Processed by molding and extrusion. Used in food packaging films, bag liners, pipes, upholstery, fibers, and coatings. Also called PVDC.

polyvinylidene fluoride Thermoplastic polymer of vinylidene fluoride having good strength, processibility, wear, fire, solvent, and creep resistance, and weatherability, but decreased dielectric properties and heat resistance. Processed by extrusion, injection and transfer molding, and powder coating. Used in electrical insulation, pipes, chemical apparatus, coatings, films, containers, and fibers. Also called PVDF.

PP See *polypropylene*.

PPA See *polyphthalamide*.

pphm See *parts per hundred million*.

ppm A unit for measuring small concentrations of material or substance as the number of its parts (arbitrary quantity) per million parts of medium consisting of another material or substance.

PPS See *polyphenylene sulfide*.

PPSS See *polyphenylene sulfide sulfone*.

pressure Stress exerted equally in all directions., *processing pressure*

pressure differential See *pressure gradient*.

pressure gradient The rate of decrease of pressure in space at a fixed time, or the magnitude of this decrease. The permeation coefficient of gases through a barrier wall such as plastic film increases with increasing pressure gradient, which is a driving force of the process, and, therefore, should be stated for the coefficient values to be meaningful. Also called pressure differential.

prevulcanization See *scorching.*

process characteristics See *processing parameters.*

process conditions See *processing parameters.*

process media See *processing agents.*

process parameters See *processing parameters.*

process pressure See *processing pressure.*

process rate See *processing rate.*

process speed See *processing rate.*

process time See *processing time.*

process velocity See *processing rate.*

processing additives See *processing agents.*

processing agents Agents or media used in the manufacture, preparation and treatment of a material or article to improve its processing or properties. The agents often become a part of the material. Also called process media, processing aids, processing additives.

processing aids See *processing agents.*

processing defects Structural and other defects in material or article caused inadvertently during manufacturing, preparation and treatment processes by using wrong tooling, process parameters, ingredients, part design, etc. Usually preventable. Also called processing flaw, defects, flaw. See also *cracking.*

processing flaw See *processing defects.*

processing methods Method names and designations for material or article manufacturing, preparation and treatment processes. **Note:** Both common and standardized names are used. Also called processing procedures.

processing parameters Measurable parameters such as temperature prescribed or maintained during material or article manufacture, preparation and treatment processes. Also called process characteristics, process conditions, process parameters.

processing pressure Pressure maintained in an apparatus during material or article manufacture, preparation and treatment processes. Also called process pressure. See also *pressure.*

processing procedures See *processing methods.*

processing rate Speed of the process in manufacture, preparation and treatment of a material or article. It usually denotes the change in a process parameter per unit of time or the throughput speed of material in a unit of weight, volume, etc. per unit of time. Also called process speed, process velocity, process rate.

processing time Time required for the completion of a process in the manufacture, preparation and treatment of a material or article. Also called process time, cycle time. See also *time.*

promoter See *accelerator.*

propane An alkane (saturated aliphatic hydrocarbon) with 3 carbon atoms, $CH_3CH_2CH_3$. A colorless, flammable gas. Autoignition temperature 467°C. Relatively inactive chemically. Obtained from petroleum or natural gas. Used in petrochemical synthesis, as fuel, aerosol propellant, and refrigerant.

propanone (2-) See *acetone.*

propene See *propylene.*

propylene An alkene (unsaturated aliphatic hydrocarbon) with three carbon atoms, $CH_2=CHCH_3$. A colorless, highly flammable gas. Autoignition temperature 497°C. Derived by thermal cracking of ethylene or from naphtha. Used as monomer in polymer and organic synthesis. Also called propene.

PS See *polystyrene.*

PSO See *polysulfones.*

PTFE See *polytetrafluoroethylene.*

PU See *polyurethanes.*

puncture force The minimum force required to puncture a flat plastic material, such as film, or textile with a pointed member, such as pyramid, at a slow rate of loading.
PUR See *polyurethanes.*

PVC See *polyvinyl chloride.*

PVDC See *polyvinylidene chloride.*

PVDF See *polyvinylidene fluoride.*

PVF See *polyvinyl fluoride.*

PVT relationship Pressure-(P) volume-(V) temperature-(T) relationship of Boyle's law stating that the product of the volume of a gas times its pressure is a constant at a given temperature, PV/T=R, where R is Boltzmann constant.

R

Ra See *roughness average.*

reaction injection molding system Liquid compositions, mostly polyurethane-based, of thermosetting resins, prepolymers, monomers, or their mixtures. Have good processibility, dimensional stability, and flexibility. Processed by foam molding with in-mold curing at high temperatures. Used in auto parts and office furniture. Also called RIM.

relative humidity The ratio of the actual vapor pressure of the air to the saturation vapor pressure. Also called RH.

relative humidity gradient The rate of decrease of relative humidity in space at a fixed time, or the magnitude of this decrease. The transmission rate of water vapor through a barrier wall such as plastic film increases with increasing relative humidity gradient, which is a driving force of the process, and, therefore, should be stated for the rate values to be meaningful.

relative viscosity The ratio of solution viscosity to the viscosity of the solvent.
resins See *polymers.*

resorcinol modified phenolic resins Thermosetting polymers of phenol, formaldehyde, and resorcinol having good heat and creep resistance and dimensional stability.

retort Laboratory glassware comprising a spherical container with a long tube in which substances are distilled, an apparatus for extraction or gasification by heating, or an apparatus for sterilization by heating.

RH See *relative humidity.*

rigid thermoplastic polyurethanes Rigid thermoplastic polyurethanes are not chemically crosslinked. They have high abrasion resistance, good retention of properties at low temperatures, but poor heat resistance, weatherability and resistance to solvents. Rigid thermoplastic polyurethanes are flammable and can release toxic substances. Processed by injection molding and extrusion. Also called rigid thermoplastic urethanes, nonelastomeric thermoplastic polyurethanes.

rigid thermoplastic urethanes See *rigid thermoplastic polyurethanes.*

RIM See *reaction injection molding system.*

roughness average A height parameter of surface roughness equal to the average absolute deviation of surface profile from the mean line, calculated as the integrated area of peaks and valleys above and below the mean line, respectively, divided by the length of this line. Also called Ra.

S.

SAN See *styrene acrylonitrile copolymer.*

SAN copolymer See *styrene acrylonitrile copolymer.*

SAN resin See *styrene acrylonitrile copolymer.*

seal initiation temperature The lower limit of a heat-seal temperature range at which a thermoplastic material such as film is beginning to fuse and adhere to itself or other thermoplastic materials.

service life The period of time required for the specified properties of the material to deteriorate under normal use conditions to the minimum allowable level with material retaining its overall usability.

shelf life Time during which a physical system, such as a material, retains its storage stability under specified conditions. Also called storage life.

silicone There are rigid thermoplastic and liquid silicones and silicone rubbers consisting of alternating silicone and oxygen atom chains with organic pendant groups, prepared by hydrolytic polymcondensation of chlorosilanes, followed by crosslinking. Silicone rubbers have good adhesion, flexibility, dielectric properties, weatherability, barrier properties, and heat and fire resistance, but decreased strength. Rigid silicones have good flexibility, weatherability, soil repelling properties, dimensional stability, but poor solvent resistance. Processed by coating, casting, and injection compression, and transfer molding. Used in coatings, electronic devises, diaphragms, medical products, adhesives, and sealants. Also called siloxane.

siloxane See *silicone.*

slip factor A property that characterizes the lubricity of a material such as plastic sliding in contact with another material that is reciprocal of the friction coefficient.

SMA See *styrene maleic anhydride copolymer.*

SMA PTB alloy See *styrene maleic anhydride copolymer PBT alloy.*

softening point Temperature at which the material changes from rigid to soft or exhibits a sudden and substantial decrease in hardness. Also called softening temperature, softening range.

softening range See *softening point.*

softening temperature See *softening point.*

solubility A capacity of one substance to be fully dissolved in another without any phase separation, e.g., precipitation. Usually expressed as a percentage of dissolved substance.

solubility coefficient The volume of a gas that can be dissolved by a unit volume of solvent at a fixed pressure and temperature.

stability The ability of a physical system, such as a material, to resist a change or degradation under exposure to outside forces, including mechanical force, heat and weather. See also *degradation.*

standard atmosphere See *atmosphere.*

starch A polysaccharide, consisting of amylose and amylopectin, found in plants such as potatoes. Gels in water. Used in adhesives, textile sizes, thickeners and in manufacture of biodegradable polymers such as polyesters. The grades include technical and edible.

starch modified low density polyethylene Biodegradable thermoplastic starch-grafted low-density polyethylene.

starch modified polypropylene Biodegradable thermoplastic starch-grafted polypropylene.

starch modified polyurethane Biodegradable thermoplastic starch-grafted polyurethane.

static coefficient of friction The ratio of the force that is required to start the friction motion of one surface against another to the force, usually gravitational, acting perpendicular to the two surfaces in contact. Also called coefficient of friction, static.

std atm See *atmosphere.*

storage life See *shelf life.*

storage stability The resistance of a physical system, such as a material, to decomposition, deterioration of properties or any type of degradation in storage under specified conditions.

STP Standard temperature and pressure equal to 1 atmosphere and 0°C, respectively. Used in measurement of permeability coefficient and other properties dependent on temperature and pressure.

strain The per unit change, due to force, in the size or shape of a body referred to its original size or shape. **Note:** Strain is nondimensional but is often expressed in unit of length per unit of length or percent.

stress cracking Appearance of external and/or internal cracks in the material as a result of stress that is lower than its short-term strength.

stress pattern Distribution of applied or residual stress in a specimen, usually throughout its bulk. Applied stress is a stress induced by an outside force, e.g., by loading. Residual stress or stress memory may be a result of processing or exposure. The stress pattern can be made visible in transparent materials by polarized light.

styrene acrylonitrile copolymer SAN resins are thermoplastic copolymers of about 70% styrene and 30% acrylonitrile with higher strength, rigidity and chemical resistance than polystyrene.

Characterized by transparency, high heat deflection properties, excellent gloss, hardness and dimensional stability. Have low continuous service temperature and impact strength. Processed by injection molding, extrusion, injection-blow molding and compression molding. Used in appliances, housewares, instrument lenses for automobiles, medical devices, and electronics. Also called styrene-acrylonitrile copolymer, SAN, SAN resin, SAN copolymer.

styrene butadiene block copolymer Thermoplastic amorphous block polymer of butadiene and styrene having good impact strength, rigidity, gloss, compatibility with other styrenic resins, water resistance, and processibility. Used in food and display containers, toys, and shrink wrap.

styrene butadiene copolymer Thermoplastic polymers of butadiene and >50% styrene having good transparency, toughness, and processibility. Processed by extrusion, injection and blow molding, and thermoforming. Used in film wraps, disposable packaging, medical devices, toys, display racks, and office supplies.

styrene maleic anhydride copolymer Thermoplastic copolymer of styrene with maleic anhydride having good thermal stability and adhesion, but decreased chemical and light resistance. Processed by injection and foam molding and extrusion. Used in auto parts, appliances, door panels, pumps, and business machines. Also called SMA.

styrene maleic anhydride copolymer PBT alloy Thermoplastic alloy of styrene maleic anhydride copolymer and polybutylene terephthalate having improved dimensional stability and tensile strength. Processed by injection molding. Also called SMA PTB alloy.

styrene plastics See *styrenic resins.*

styrene polymers See *styrenic resins.*

styrene resins See *styrenic resins.*

styrene-acrylonitrile copolymer See *styrene acrylonitrile copolymer.*

styrenic resins Styrenic resins are thermoplastics prepared by free-radical polymerization of styrene alone or with other unsaturated monomers. The properties of styrenic resins vary widely with molecular structure, attaining the high performance level of engineering plastics. Processed by blow and injection molding, extrusion, thermoforming, film techniques and structural foam molding. Used heavily for the manufacture of automotive parts, household goods, packaging, films, tools, containers and pipes. Also called styrene resins, styrene polymers, styrene plastics.

styrenic thermoplastic elastomers Linear or branched copolymers containing polystyrene end blocks and elastomer (e.g., isoprene rubber) middle blocks. Have a wide range of hardnesses, tensile strength, and elongation, and good low-temperature flexibility, dielectric properties, and hydrolytic stability. Processed by injection and blow molding and extrusion. Used in coatings, sealants, impact modifiers, shoe soles, medical devices, tubing, electrical insulation, and auto parts. Also called TES.

sulfur dioxide A colorless, noncombustible gas or liquid with pungent odor, SO_2. Toxic by inhalation, strong irritant. Derived from pyrites or burning sulfur. used in paper pulping, inorganic synthesis, as bleaching agent for oils, for fumigation, as antioxidant, bactericide, and metal refining.

surface roughness Relatively fine spaced surface irregularities, the heights, widths and directions of which establish the predominant surface pattern.

surface tack Stickiness of a surface of a material such as wet paint when touched.

syndiotactic A polymer molecule in which pendant groups and atoms attached to the main chain are arranged in a symmetrical and recurring fashion relative to it in a single plane. .

T

TAPPI T511 See *ASTM D2176.*

tautomeric Pertaining to tautomerism, i.e., isomerism in which migration of a hydrogen atom results in two or more structures, called tautomers that are in equilibrium. For example, enol and keto tautomers of acetoacetate.

tear propagation resistance The force required to propagate a slit in a flexible plastic film or thin sheeting at a constant rate of loading, calculated as an average between the initial and the maximum tear-propagation forces. Also called tear resistance, propagated.

tear resistance, initial See *initial tear resistance.*

tear resistance, propagated See *tear propagation resistance.*

temperature Property which determines the direction of heat flow between objects. **Note:** The heat flows from the object with higher temperature to that with lower.

tensile elongation See *elongation.*

tensile heat distortion temperature See *heat deflection temperature.*

tensile properties Properties describing the reaction of physical systems to tensile stress and strain. See also *tensile property tests.*

tensile property tests Names and designations of the methods for tensile testing of materials. Also called tensile tests. See also *tensile properties.*

tensile strain The relative length deformation exhibited by a specimen in tension. See also *elongation.*

tensile strength The maximum tensile stress that a specimen can sustain in a test carried to failure. **Note:** The maximum stress can be measured at or after the failure or reached before the fracture, depending on the viscoelastic behavior of the material. Also called tensile ultimate strength, ultimate tensile strength, UTS, tensile strength at break, ultimate tensile stress. See also *ASTM D638.*

tensile strength at break See *tensile strength.*

tensile stress The stress is perpendicular and directed to the opposite plane on which the forces act.

tensile tests See *tensile property tests.*

tensile ultimate strength See *tensile strength.*

terephthalate polyester Thermoset unsaturated polymer of terephthalic anhydride.

TES See *styrenic thermoplastic elastomers.*

test methods Names and designations of material test methods. Also called testing methods.

test variables Terms related to the testing of materials such as test method names.

testing methods See *test methods.*

tetrachloroethylene Also called perchloroethylene.

tetrafluoroethylene propylene copolymer Thermosetting elastomeric polymer of tetrafluoroethylene and propylene having good chemical and heat resistance and flexibility. Used in auto parts.

thermal properties Properties related to the effects of heat on physical systems such as materials and heat transport. The effects of heat include the effects on structure, geometry, performance, aging, stress-strain behavior, etc.

thermal stability The resistance of a physical system, such as a material, to decomposition, deterioration of properties or any type of degradation in storage under specified conditions.

thermodynamic properties A quantity that is either an attribute of the entire system or is a function of position, which is continuous and does not vary rapidly over microscopic distances, except possibility for abrupt changes at boundaries between phases of the system. Also called macroscopic properties.

thermoplastic polyesters A class of polyesters that can be repeatedly made soft and pliable on heating and hard (flexible or rigid) on subsequent cooling.

thermoplastic polyurethanes A class of polyurethanes including rigid and elastomeric polymers that can be repeatedly made soft and pliable on heating and hard (flexible or rigid) on subsequent cooling. Also called thermoplastic urethanes, TPUR, TPU.

thermoplastic urethanes See *thermoplastic polyurethanes.*

three-membered heterocyclic compounds A class of heterocyclic compounds containing rings that consist of three atoms.

three-membered heterocyclic oxygen compounds A class of heterocyclic compounds containing rings that consist of three atoms, one or two of which is an oxygen.

time One of basic dimensions of the universe designating the duration and order of events at a given place. See also *processing time.*

toughness Property of a material indicating its ability to absorb energy by plastic deformation rather than crack or fracture.

TPO See *olefinic thermoplastic elastomers.*

TPU See *thermoplastic polyurethanes.*

TPUR See *thermoplastic polyurethanes.*

transition point See *phase transition point.*

transition temperature See *phase transition point.*

tribasic lead maleate A salt of maleic acid, highly effective as heat stabilizer for polymeric materials. Limited to use in applications where toxicity and lack of clarity can be tolerated.

turbidity The cloudiness in a liquid caused by a suspension of colloidal liquid droplets or fine solids.

U

UHMWPE See *ultrahigh molecular weight polyethylene.*

ultimate elongation See *elongation.*

ultimate hot tack strength See *hot tack strength.*

ultimate seal strength Maximum force that a heat-sealed thermoplastic film can sustain in a tensile test without seal failure per unit length of the seal.

ultimate tensile strength See *tensile strength.*

ultimate tensile stress See *tensile strength.*

ultrahigh molecular weight polyethylene Thermoplastic linear polymer of ethylene with molecular weight in the millions. Has good wear and chemical resistance, toughness, and antifriction properties, but poor processibility. Processed by compression molding and ram extrusion. Used in bearings, gears, and sliding surfaces. Also called UHMWPE.

uniaxially oriented A state of material such as polymeric film or composite characterized by the permanent orientation of its components such as polymer molecules or reinforcing fibers in one direction. The orientation is achieved by a number of different processes, e.g., stretching, and is intended to improve the mechanical properties of the material.

units See *units of measurement.*

units of measurement Systematic and non-systematic units for measuring physical quantities, including metric and US pound-inch systems. Also called units.

urea resins Thermosetting polymers of formaldehyde and urea having good clarity, colorability, scratch, fire, and solvent resistance, rigidity, dielectric properties, and tensile strength, but decreased impact strength and chemical, heat, and moisture resistance. Must be filled for molding. Processed by compression and injection molding, impregnation, and coating. Used in cosmetic containers, housings, tableware, electrical insulators, countertop laminates, adhesives, and coatings.

urethane polymers See *polyurethanes.*

urethane resins See *polyurethanes.*

urethane thermoplastic elastomers Block polyether or polyester polyurethanes containing soft and hard segments. Have good tensile strength, elongation, adhesion, and a broad hardness and service temperature ranges, but decreased moisture resistance and processibility. Processed by extrusion, injection molding, film blowing, and coating. Used in tubing, packaging film, adhesives, medical devices, conveyor belts, auto parts, and cable jackets. Also called TPU.

urethanes See *polyurethanes.*

UTS See *tensile strength.*

V

veneer In rubber industry, a thin film applied on a rubber article to protect it against oxygen and ozone attack, act as a migration barrier or for decorative purposes.

Vicat softening point The temperature at which a flat-ended needle of prescribed geometry will penetrate a thermoplastic specimen to a certain depth under a specified load using a uniform rate of temperature rise. **Note:** Vicat softening point is determined according to ASTM D1525 test for thermoplastics such as polyethylene which have no definite melting point. Also called Vicat softening temperature.

Vicat softening temperature See *Vicat softening point.*

vinyl ester resins Thermosetting acrylated epoxy resins containing styrene reactive diluent. Cured by catalyzed polymerization of vinyl groups and crosslinking of hydroxy groups at room or elevated temperatures. Have good chemical, solvent, and heat resistance, toughness, and flexibility, but shrink during cure. Processed by filament winding, transfer molding, pultrusion, coating, and lamination. Used in structural composites, coatings, sheet molding compounds, and chemical apparatus.

vinyl resins Thermoplastics polymers of vinyl compounds such as vinyl chloride or vinyl acetate. Have good weatherability, barrier properties, and flexibility, but decreased solvent and heat resistance. Processed by molding, extrusion, and coating. Used in films and packaging.

vinyl thermoplastic elastomers Vinyl resin alloys having good fire and aging resistance, flexibility, dielectric properties, and toughness. Processed by extrusion. Used in cable jackets and wire insulation.

vinylidene fluoride hexafluoropropylene copolymer Thermoplastic polymer of vinylidene fluoride and hexafluoropropylene having good antistick, dielectric, and antifriction properties and chemical and heat resistance, but decreased mechanical strength and creep resistance and poor processibility. Processed by molding, extrusion, and coating. Used in chemical apparatus, containers, films, and coatings.

vinylidene fluoride hexafluoropropylene tetrafluoroethylene terpolymer Thermosetting elastomeric polymer of vinylidene fluoride, hexafluoropropylene, and tetrafluoroethylene having good chemical and heat resistance and flexibility. Used in auto parts.

vulcanizate Rubber that had been irreversibly transformed from predominantly plastic to predominantly elastic material by vulcanization (chemical curing or crosslinking) using heat, vulcanization agents, accelerants, etc.

vulcanizate crosslinks Chemical bonds formed between polymeric chains in rubber as a result of vulcanization.

W

warpage See *warping.*

warping Dimensional distortion or deviation from the intended shape of a plastic or rubber article as a result of nonuniform internal stress, e.g., caused by uneven heat shrinkage. Also called warpage.

water swell Expansion of material volume as a result of water absorption.

water vapor transmission rate A measure of water vapor (moisture) permeability of a barrier wall such as plastic film. Vapor transmission rate, VTR, is a coefficient in modified Fick's first law that states that the weight (W) of a vapor that penetrates a barrier wall is directly proportional to the area (A) of the wall and time (t), and is inversely proportional to the wall thickness (s); or $W = VTR \cdot (A \cdot t)/s$. The water vapor transmission rate is a characteristic constant for the wall material that is homogeneous in the direction of penetration. It depends on the temperature and relative humidity gradient. Also called WVTR.

weight The gravitational force with which the earth attracts a body.

wetability The degree or extent to which somthing absorbs or can be made to absorb moisture.

whiting A finely divided form of calcium carbonate ($CaCO_3$) obtained by milling high-calcium limestone, marble, shell or chemically precipitated calcium carbonate. Used as an extender filler in plastics and rubbers.

WVTR See *water vapor transmission rate.*

X

xylene An aromatic hydrocarbon comprising benzene ring containing two methyl substituent groups, $C_6H_4Me_2$. It is a colorless, flammable, toxic liquid usually consisting of a mixture of three isomers: ortho-, meta-, and para-xylene. Derived from coal tar and petroleum. Used in aviation fuel, solvent for alkyd resins and coatings, and in the synthesis of phthalic acids.

Table and Graph Index

Acetal Resin - Chapter 1

TABLE 01   Cologne, Shampoo and Hair Spray Permeability Through DuPont Delrin Acetal Resin.

TABLE 02   Gasoline, Freon Propellant, Motor Oil, and Ethyl Alcohol Permeability Through DuPont Delrin Acetal Resin.

TABLE 03   Methyl Salicylate, Nitrogen, Perchloroethylene, Trichloroethylene, Toluene, Carbon Dioxide and Oxygen Permeability Through DuPont Delrin Acetal Resin.

TABLE 04   Mineral Oils, Vegetable Oils, Tar Remover and Road Oil Remover Permeability Through DuPont Delrin Acetal Resin.

TABLE 05   Air and Oxygen Permeability Through Hoechst Celanese Celcon Acetal Copolymer Film.

TABLE 06   Nitrogen and Carbon Dioxide Permeability Through Hoechst Celanese Celcon Acetal Copolymer Film.

Acrylonitrile - Chapter 2

TABLE 07   Gas Permeability and Water Vapor Transmission Through Polyacrylonitrile.

Acrylonitrile-Methyl Acrylate Copolymer - Chapter 3

TABLE 08   Gas Permeability and Water Vapor Transmission Through Acrylonitrile Copolymer.

TABLE 09   Cyclohexanone, Chlorobenzene, Hexane, Butyl Alcohol, Trichloroethene, Methyl Salicylate and Tetrahydrofuran Permeability Through Acrylonitrile Copolymer Bottles.

TABLE 10   Ethyl Acetate, Isopropyl Acetate, Acetone, Butyl Acetate, Toluene, Xylene, Methyl Isobutyl Ketone and Methyl Ethyl Ketone Permeability Through Acrylonitrile Copolymer Bottles.

TABLE 11   Oxygen Permeability at Different Temperatures and Water Vapor Transmission Through BP Chemicals Barex 210 Acrylonitrile Copolymer.

TABLE 12   Water Vapor Transmission and Oxygen Permeability Through BP Chemicals Barex Acrylonitrile-Methyl Acrylate Copolymer.

TABLE 13   Water Vapor Transmission and Oxygen Permeability vs. Humidity Through BP Chemicals Barex Acrylonitrile-Methyl Acrylate Copolymer.

GRAPH 01   Carbon Dioxide Permeability vs. Acrylonitrile Content through Acrylonitrile-Methyl Acrylate Copolymer.

GRAPH 02   Carbon Dioxide and Oxygen Permeability vs. Relative Humidity through Acrylonitrile-Methyl Acrylate Copolymer.

GRAPH 03   Oxygen Permeability vs. Temperature through Acrylonitrile-Methyl Acrylate Copolymer.

Cellulosic Plastic - Chapter 4

TABLE 14   Water Vapor Transmission and Oxygen Permeability Through Coated Cellophane Film.

TABLE 15   Film Properties of Coated Cellophane Film.

Ethylene-Chlorotrifluoroethylene Copolymer - Chapter 6

TABLE 16   Hydrogen Permeability vs. Temperature and Pressure Through Ausimont Halar Ethylene Chlorotrifluoroethylene Copolymer.

TABLE 17   Nitrogen Permeability vs. Temperature and Pressure Through Ausimont Halar Ethylene Chlorotrifluoroethylene Copolymer.

TABLE 18   Oxygen and Ammonia Permeability vs. Temperature and Pressure Through Ausimont Halar Ethylene-Chlorotrifluoroethylene Copolymer.

GRAPH 04   Moisture Vapor Permeability Rate vs. Thickness through Ethylene-Chlorotrifluoroethylene Copolymer.

GRAPH 05   Moisture Vapor Permeability Rate vs. Temperature through Ethylene-Chlorotrifluoroethylene Copolymer.

GRAPH 06   Carbon Dioxide and Oxygen Permeability vs. Temperature through Ethylene-Chlorotrifluoroethylene Copolymer.

GRAPH 07   Nitrogen and Helium Permeability vs. Temperature through Ethylene-Chlorotrifluoroethylene Copolymer.

GRAPH 08   Gas Permeability vs. Temperature through Ethylene-Chlorotrifluoroethylene Copolymer.

Ethylene-Tetrafluoroethylene Copolymer - Chapter 7

TABLE 19   Carbon Dioxide, Nitrogen, Oxygen, Helium and Water Vapor Permeability Through DuPont

TABLE 20   Oxygen, Nitrogen, Carbon Dioxide, Methane and Helium Permeability Through Ausimont Hyflon Ethylene Tetrafluoroehtylene Copolymer.

682

Fluorinated Ethylene-Propylene Copolymer - Chapter 8

TABLE 21	Hydrogen Permeability vs. Temperature and Pressure Through DuPont Teflon Fluorinated Ethylene-Propylene Copolymer.
TABLE 22	Nitrogen Permeability vs. Temperature and Pressure Through DuPont Teflon Fluorinated Ethylene-Propylene Copolymer.
TABLE 23	Oxygen and Ammonia Permeability vs. Temperature and Pressure Through DuPont Teflon Fluorinated Ethylene-Propylene Copolymer.
TABLE 24	Water Vapor, Oxygen, Nitrogen and Carbon Dioxide Permeability Through Fluorinated Ethylene Propylene.
GRAPH 09	Moisture Vapor Permeability Rate vs. Thickness through Fluorinated Ethylene-Propylene Copolymer.
GRAPH 10	Moisture Vapor Permeability Rate vs. Temperature through Fluorinated Ethylene-Propylene Copolymer.
GRAPH 11	Carbon Dioxide and Oxygen Permeability vs. Temperature through Fluorinated Ethylene-Propylene Copolymer.
GRAPH 12	Nitrogen and Helium Permeability vs. Time After Retort through Fluorinated Ethylene-Propylene Copolymer.
GRAPH 13	Gas Permeability vs. Temperature through Fluorinated Ethylene-Propylene Copolymer.

Fluorinated Polyethylene - Chapter 9

TABLE 25	Cyclohexanone, Chlorobenzene, Hexane, Butyl Alcohol, Trichloroethene, Methyl Salicylate and Tetrahydrofuran Permeability Through Fluorinated Polyethylene Bottles.
TABLE 26	Ethyl Acetate, Isopropyl Acetate, Acetone, Butyl Acetate, Toluene, Xylene, Methyl Isobutyl Ketone and Methyl Ethyl Ketone Permeability Through Fluorinated Polyethylene Bottles.
TABLE 27	Kerosine, d-Limonene, 2-Cycle Motor Oil, Pine Oil Cleaner, Diesel Fuel Conditioner and Brakleen Gas Additive Permeability Through Fluorinated Polyethylene Bottles.
TABLE 28	Mineral Spirits, Turpentine, STP Gas Treatment, Paint Thinner, Charcoal Starter and Naphtha Permeability Through Fluorinated Polyethylene Bottles.

Perfluoroalkoxy Resin - Chapter 10

TABLE 29	Gas Permeability of Oxygen, Carbon Dioxide and Nitrogen Through DuPont Company Teflon PFA Perfluoroalkoxy Film.

Polychlorotrifluoroethylene - Chapter 11

TABLE 30	Carbon Dioxide, Hydrogen and Hydrogen Sulfide Permeability Through 3M Kel-F 81 Polychlorotrifluoroethylene Film.
TABLE 31	Hydrogen Permeability vs. Temperature and Pressure Through 3M Kel-F Polychlorotrifluoroethylene.
TABLE 32	Nitrogen Permeability vs. Temperature and Pressure Through 3M Kel-F Polychlorotrifluoroethylene.
TABLE 33	Oxygen and Ammonia Permeability vs. Temperature and Pressure Through 3M Kel-F Polychlorotrifluoroethylene.
TABLE 34	Oxygen, Nitrogen and Carbon Dioxide Permeability Through Allied Signal Aclar Polychlorotrifluoroethylene Film.
TABLE 35	Oxygen, Nitrogen and Helium Permeability Through 3M Kel-F 81 PCTFE Film.
TABLE 36	Water Vapor Permeability Through Allied Signal Aclar Polychlorotrifluoroethylene Film.
TABLE 37	Water Vapor Transmission Through 3M Kel-F 81 Polychlorotrifluoroethylene Film.
TABLE 38	Film Properties of Allied Signal Aclar Polychlorotrifluoroethylene Film.
GRAPH 14	Gas Permeability vs. Temperature through Polychlorotrifluoroethylene.

Polytetrafluoroethylene - Chapter 12

TABLE 39	Hydrogen Permeability vs. Temperature and Pressure Through Carbon Filled DuPont Teflon Polytetrafluoroethylene.
TABLE 40	Hydrogen Permeability vs. Temperature and Pressure Through DuPont Teflon Polytetrafluoroethylene.
TABLE 41	Nitrogen Permeability vs. Temperature and Pressure Through DuPont Teflon Polytetrafluoroethylene.
TABLE 42	Nitrogen Permeability vs. Temperature and Pressure Through Carbon Filled DuPont Teflon Polytetrafluoroethylene.
TABLE 43	Oxygen and Ammonia Permeability vs. Temperature and Pressure Through DuPont Teflon Polytetrafluoroethylene.
TABLE 44	Oxygen and Ammonia Permeability vs. Temperature and Pressure Through Carbon Filled DuPont Teflon Polytetrafluoroethylene.
GRAPH 15	Gas Permeability vs. Temperature through Polytetrafluoroethylene.
GRAPH 16	Gas Permeability vs. Temperature through Carbon Filled Polytetrafluoroethylene.

Polyvinyl Fluoride - Chapter 13

TABLE 45	Water Vapor, Oxygen, Nitrogen and Carbon Dioxide Permeability Through Polyvinyl Fluoride.

Polyvinylidene Fluoride - Chapter 14

TABLE 46	Ammonia, Helium, Chlorine and Hydrogen Permeability Through Solvay Solef Polyvinylidene Fluoride Film.
TABLE 47	Carbon Dioxide, Nitrogen, Oxygen and Water Vapor Permeability Through Solvay Solef 1008 Polyvinylidene Fluoride Film.

TABLE 48   Freon, Nitrous Oxide, Hydrogen Sulfide, and Sulfur Dioxide Permeability Through Solvay Solef Polyvinylidene Fluoride Film.

TABLE 49   Water Vapor, Oxygen, and Carbon Dioxide Permeability Through Atochem Foraflon Polyvinylidene Fluoride Film.

TABLE 50   Water Vapor, Oxygen, Nitrogen and Carbon Dioxide Permeability Through Polyvinylidene Fluoride.

GRAPH 17   Moisture Vapor Permeability Rate vs. Thickness through Polyvinylidene Fluoride.

GRAPH 18   Moisture Vapor Permeability Rate vs. Temperature through Polyvinylidene Fluoride.

GRAPH 19   Carbon Dioxide Permeability vs. Thickness through Polyvinylidene Fluoride.

GRAPH 20   Water Vapor Permeability vs. Thickness through Polyvinylidene Fluoride.

GRAPH 21   Water Vapor Permeability vs. Temperature through Polyvinylidene Fluoride.

GRAPH 22   Nitrogen and Oxygen Permeability vs. Thickness through Polyvinylidene Fluoride.

GRAPH 23   Gas Permeability vs. Thickness through Polyvinylidene Fluoride.

GRAPH 24   Helium and Hydrogen Permeability vs. Thickness through Polyvinylidene Fluoride.

Ionomer - Chapter 15

TABLE 51   Oxygen Gas Permeability Through DuPont Surlyn Zinc Ion Type Ionomoer Film.

TABLE 52   Water Vapor and Oxygen Gas Permeability Through DuPont Surlyn Sodium Ion Type Ionomoer Film.

TABLE 53   Water Vapor Permeability Through DuPont Surlyn Zinc Ion Type Ionomoer Film.

TABLE 54   Film Properties of DuPont Surlyn Sodium Ion Type Ionomoer Film.

TABLE 55   Film Properties of DuPont Surlyn Zinc Ion Type Ionomoer Film.

Parylene - Chapter 16

TABLE 56   Water Vapor, Oxygen, Nitrogen, Carbon Dioxide and Hydrogen Permeability Through Union Carbide Parylene N and Parylene C Parylene Film.

TABLE 57   Water Vapor, Oxygen, Nitrogen, Carbon Dioxide and Hydrogen Permeability Through Union Carbide Parylene D Parylene Film.

TABLE 58   Film Properties of Union Carbide Parylene Films.

Nylon - Chapter 17

TABLE 59   Organic Solvents Permeability Through Oriented and PVDC Coated Nylon Film.

TABLE 60   Oxygen Permeability vs. Relative Humidity Through Oriented Nylon Film.

TABLE 61   Water Vapor Transmission and Oxygen Permeability Through Coated and Uncoated Oriented Nylon Film.

TABLE 62   Film Properties of Coated and Uncoated Oriented Nylon Film.

GRAPH 25   Oxygen Permeability vs. Relative Humidity through Nylon.

Amorphous Nylon - Chapter 18

TABLE 63   Oxygen, Carbon Dioxide, Nitrogen and Water Vapor Permeability Through EMS-American Grilon Grivory G21 Amorphous Nylon Copolymer Film.

TABLE 64   Water Vapor Transmission Through Du Pont Selar PA Amorphous Nylon Barrier Resin.

TABLE 65   Water Vapor, Carbon Dioxide and Oxygen Permeability Through DuPont Selar PA Amorphous Nylon Film.

GRAPH 26   Carbon Dioxide Permeability vs. Relative Humidity through Amorphous Nylon.

GRAPH 27   Oxygen Permeability vs. Relative Humidity through Amorphous Nylon.

GRAPH 28   Oxygen Permeability vs. Temperature through Amorphous Nylon.

Nylon 6 - Chapter 19

TABLE 66   Gas Permeability and Water Vapor Transmission Through BASF Ultramid B Nylon 6 Film.

TABLE 67   Gas Permeability of Carbon Dioxide, Nitrogen, Helium and Water Vapor Transmission Through Oriented Nylon 6.

TABLE 68   Oxygen Permeability vs. Temperature Through Oriented and Non-Oriented Nylon 6.

TABLE 69   Oxygen Gas and Water Vapor Permeability Through BASF Ultramid B4 Nylon 6 Film.

TABLE 70   Oxygen, Nitrogen and Carbon Dioxide Permeability Through Allied Signal Capran 77C Nylon 6 Film.

TABLE 71   Water Vapor and Oxygen Permeability Through Nylon 6 Film.

TABLE 72   Water Vapor Permeability Through Allied Signal Capran 77C Nylon 6 Film.

TABLE 73   Water Vapor, Carbon Dioxide and Oxygen Permeability Through Nylon 6 Film.

TABLE 74   Water Vapor, Oxygen, Nitrogen and Carbon Dioxide Permeability Through Allied Signal Capran 77K PVDC Coated Nylon 6 Film.

TABLE 75   Mechanical Properties of BASF Ultramid B Nylon 6 Film.

TABLE 76   Film Properties of Allied Signal Capran Nylon 6 Film.

Nylon 66 - Chapter 20

TABLE 77	Oxygen, Carbon Dioxide and Water Vapor Permeability Through BASF Ultramid A5 Nylon.
TABLE 78	Oxygen, Carbon Dioxide and Nitrogen Permeability Through DuPont Canada Dartek Nylon.
TABLE 79	Oxygen, Carbon Dioxide, Nitrogen, Helium and Water Vapor Permeability Through DuPont Company Zytel 42 Nylon 66 Film.
TABLE 80	Oxygen Gas and Water Vapor Permeability Through BASF Ultramid A5 Nylon 66 Film.
TABLE 81	Water Vapor Permeability Through DuPont Canada Dartek Nylon 66 Film.
TABLE 82	Liquid Permeability Through DuPont Company Zytel 42 Nylon 66 Bottles.
TABLE 83	Mechanical Properties of BASF Ultramid B Nylon 66 Film.
TABLE 84	Film Properties of DuPont Canada Dartek Nylon 66 Film.

Nylon 6/66 - Chapter 21

TABLE 85	Gas Permeability and Water Vapor Transmission Through BASF Ultramid C Nylon 6/66 Film.
TABLE 86	Water Vapor, Oxygen, Carbon Dioxide and Nitrogen Permeability Through Allied Signal Capran Nylon 6/66 Film.
TABLE 87	Air Conditioning Refrigerants Permeation Loss Through Nylon 6/66 Copolymer.
TABLE 88	Film Properties of Nylon 66/6 Copolymer Film.

Nylon 66/610 - Chapter 22

TABLE 89	Oxygen, Carbon Dioxide, Nitrogen and Water Vapor Permeability Through EMS-American Grilon Grilon XE3303 Nylon 66/610 Copolymer Film.

Nylon MXD6 - Chapter 23

TABLE 90	Oxygen Permeability at Different Temperatures and Water Vapor Transmission Through Nylon MXD6.
TABLE 91	Oxygen Permeability vs. Relative Humidity Through Nylon MXD6.
GRAPH 29	Oxygen Permeability vs. Relative Humidity through Nylon MXD6.

Polycarbonate - Chapter 24

TABLE 92	Oxygen, Carbon Dioxide And Nitrogen Permeability Through Dow Chemical Calibre Polycarbonate.
TABLE 93	Water Vapor, Carbon Dioxide and Oxygen Permeability Through Polycarbonate Film.
TABLE 94	Water Vapor, Oxygen, Nitrogen and Carbon Dioxide Permeability Through Bayer AG Makrolon Polycarbonate Film.

Polybutylene Terephthalate - Chapter 25

TABLE 95	Gas Permeability and Water Vapor Transmission Through BASF AG Ultradur Polybutylene Terephthalate.
TABLE 96	Mechanical and Optical Properties of BASF AG Ultradur Polybutylene Terephthalate Film.

Polyethylene Naphthalate - Chapter 26

TABLE 97	Water Vapor and Oxygen Permeability Through Eastman Chemical Polyethylene Naphthalate (PEN) Homopolymer 14991.
TABLE 98	Film Properties of Eastman Chemical PolyethyleneNaphthalate (PEN) Homopolymer 14991.

Polyethylene Terephthalate - Chapter 27

TABLE 99	Carbon Dioxide, Oxygen and Water Vapor Permeability Through Shell Chemical Cleartuf Polyethylene Terephthalate Polyester (PET).
TABLE 100	Oxygen, Carbon Dioxide, Nitrogen and Hydrogen Permeability Through DuPont Company Mylar Polyester PET Film.
TABLE 101	Organic Solvents Permeability Through Polyester PET Film.
TABLE 102	Oxygen Permeability at Different Temperatures and Carbon Dioxide, Nitrogen and Helium Permeability Through Oriented Polyethylene Terephthalate Polyester.
TABLE 103	Oxygen Permeability vs. Relative Humidity Through Polyester PET Film.
TABLE 104	Water Vapor Transmission and Oxygen Permeability Through Polyethylene Terephthalate Polyester (PET).
TABLE 105	Water Vapor Transmission and Oxygen Permeability Through Coated and Uncoated Polyester PET Film.
TABLE 106	Water Vapor, Acetone, Benzene, Carbon Tetrachloride, Ethyl Acetate and Hexane Permeability Through DuPont Company Mylar Polyester PET Film.
TABLE 107	Water Vapor, Oxygen, Nitrogen and Carbon Dioxide Permeability Through Oriented Polyethylene Terephthalate Polyester.
TABLE 108	Water Vapor, Oxygen and Xylene Permeability Through DuPont Selar PT Polyethylene Terephthalate Polyester Containers.
TABLE 109	Film Properties of Coated and Uncoated Oriented Polyester Film.

TABLE 110   Film Properties of DuPont Company Mylar Polyester PET Film.

Glycol Modified Polycyclohexylenedimethylene - Chapter 28

TABLE 111   Water Vapor, Carbon Dioxide, Oxygen and Nitrogen Permeability Through Eastman Chemical Kodar PCTG 5445 Glycol Modified Polycyclohexylenedimethylene Terephthalate (PCTG) Film.

TABLE 112   Film Properties of Eastman Chemical Kodar PCTG 5445 Glycol Modified Polycyclohexylenedimethylene Terephthalate (PCTG) Film.

Polycyclohexylenedimethylene Ethylene - Chapter 29

TABLE 113   Water Vapor Transmission and Oxygen Permeability Through Polycyclohexylenedimethylene Ethyelene Terephthalate Polyester (PETG).

TABLE 114   Water Vapor, Carbon Dioxide, Oxygen and Nitrogen Permeability Through Eastman Chemical Kodar PETG 6763 Polycyclohexylenedimethylene Ethylene Terephthalate (PETG) Film.

TABLE 115   Film Properties of Eastman Chemical Kodar PETG 6763 Polycyclohexylenedimethylene Ethylene Terephthalate (PETG) Film.

Liquid Crystal Polymer - Chapter 30

TABLE 116   Water Vapor Transmission and Oxygen Permeability Through Hoechst AG Liquid Crystal Polyester.

Polyimide - Chapter 31

TABLE 117   Water Vapor and Oxygen Permeability Through DuPont Company Kapton Polyimide Film.

TABLE 118   Water Vapor, Oxygen, Nitrogen, Carbon Dioxide and Helium Permeability Through Ube Upilex Polyimide Film.

TABLE 119   Film Properties of DuPont Company Kapton Polyimide Film.

TABLE 120   Film Properties of Ube Upilex Polyimide Film.

Polyethylene - Chapter 32

TABLE 121   Film Properties of Dow Chemical Retain Polyethylene Postconsumer Recycle Content Resins.

GRAPH 30   Gas Permeability vs. Density through Polyethylene.

GRAPH 31   Oxygen Permeability vs. Density through Polyethylene.

GRAPH 32   Water Vapor Permeability vs. Density through Polyethylene.

GRAPH 33   Moisture vapor transmission (MVT) vs. Density through Polyethylene.

GRAPH 34   Toluene and FAM Test Fluid Permeability vs. Density through Polyethylene.

GRAPH 35   Methyl Alcohol, Ethyl Acetate, Fuel Oil and Chloroform Permeability vs. Density through Polyethylene.

GRAPH 36   Oxygen Permeability vs. Temperature through Polyethylene.

Low Density Polyethylene - Chapter 33

TABLE 122   Carbon Dioxide, Oxygen, Ethylene Oxide and Water Vapor Permeability Through Low Density Polyethylene Film.

TABLE 123   Oxygen, Nitrogen, Carbon Dioxide and Water Vapor Permeability Through Dow Chemical Low Density Polyethylene.

TABLE 124   Oxygen Permeability at Different Temperatures and Water Vapor Transmission Through Low Density Polyethylene.

TABLE 125   Gas Permeability of Oxygen, Carbon Dioxide, Nitrogen and Helium Through Low Density Polyethylene Film.

TABLE 126   Vapor Transmission of Reagents Through Dow Chemical Low Density Polyethylene.

TABLE 127   Water Vapor Transmission Through Dow Chemical Low Density Polyethylene Extrusion Coating Resins.

TABLE 128   Water Vapor Transmission and Oxygen Permeability Through Low Density Polyethylene Film.

TABLE 129   Water Vapor, Oxygen, Nitrogen and Carbon Dioxide Permeability Through Low Density Polyethylene.

TABLE 130   Xylene and Oxygen Permeability Through Low Density Polyethylene.

TABLE 131   Organic Solvents Permeability Through Low Density Polyethylene Film.

TABLE 132   Mechanical and Optical Properties of BASF AG Lupolen Low Density Polyethylene Film.

TABLE 133   Film Properties of Dow Chemical Low Density Polyethylene Clarity And Liner Film Resins.

TABLE 134   Film Properties of Dow Chemical Low Density Polyethylene Clarity And Liner Film Resins.

TABLE 135   Film Properties of Dow Chemical Low Density Polyethylene Clarity And Liner Film Resins.

TABLE 136   Properties of Dow Chemical Low Density Polyethylene Extrusion Coating Resins.

TABLE 137   Film Properties of Dow Chemical Low Density Polyethylene Fractional Melt Index Film Resins.

TABLE 138   Film Properties of Low Density Polyethylene Film.

TABLE 138   Film Properties of Low Density Polyethylene Film (cont'd).

GRAPH 37   Carbon Dioxide Permeability vs. Thickness through Low Density Polyethylene.

GRAPH 38   Oxygen Permeability vs. Thickness through Low Density Polyethylene.

Linear Low Density Polyethylene - Chapter 34

TABLE 139   Oxygen Gas and Water Vapor Permeability Through Dow Chemical Dowlex Linear Low Density Polyethylene.
TABLE 140   Oxygen Permeability Through DuPont Canada Sclair SL1 and Sclair SL3 Linear Low Density Polyethylene.
TABLE 141   Water Vapor Transmission and Oxygen Permeability Through DuPont Canada Sclair Linear Low Density Polyethylene.
TABLE 142   Water Vapor Transmission Through DuPont Canada Sclair SL1 and Sclair SL3 Linear Low Density Polyethylene.
TABLE 143   Carbon Dioxide and Nitrogen Permeability Through DuPont Canada Sclair SL1 and Sclair SL3 Linear Low Density Polyethylene.
TABLE 144   Film Properties of Dow Chemical Dowlex Linear Low Density Polyethylene.
TABLE 145   Film Properties of Dow Chemical Dowlex Linear Low Density Polyethylene.
TABLE 146   Film Properties of Dow Chemical Dowlex And Next Generation Dowlex Linear Low Density Polyethylene.
TABLE 147   Film Properties of DuPont Canada Sclair Linear Low Density Polyethylene.
TABLE 148   Film Properties of DuPont Canada Sclair SL1 and Sclair SL3 Linear Low Density Polyethylene

Ultra Low Density Ethylene-Octene Copolymer - Chapter 35

TABLE 149   Oxygen Gas and Water Vapor Permeability Through Dow Chemical Attane Ultra Low Density Ethylene Octene Copolymers.
TABLE 150   Film Properties of Dow Chemical Attane Ultra Low Density Ethylene Octene Copolymers.

Medium Density Polyethylene - Chapter 36

TABLE 151   Water Vapor Transmission and Oxygen Permeability Through DuPont Canada Sclair 14D Medium Density Polyethylene.
TABLE 152   Water Vapor, Oxygen, Nitrogen and Carbon Dioxide Permeability Through Medium Density Polyethylene.
TABLE 153   Film Properties of DuPont Canada Sclair 14D Medium Density Polyethylene.

High Density Polyethylene - Chapter 37

TABLE 154   Water Vapor Transmission and Oxygen, Nitrogen and Carbon Dioxide Permeability Through Dow Chemical High Density Polyethylene.
TABLE 155   Oxygen Permeability at Different Temperatures and Water Vapor Transmission Through High Density Polyethylene.
TABLE 156   Water Vapor Transmission Through Hoechst AG Hostalen Polyethylene.
TABLE 157   Water Vapor Transmission and Oxygen Permeability Through DuPont Canada Sclair High Density Polyethylene.
TABLE 158   Water Vapor Transmission and Oxygen Permeability Through High Density Polyethylene.
TABLE 159   Water Vapor, Oxygen, Nitrogen and Carbon Dioxide Permeability Through High Density Polyethylene.
TABLE 160   Oxygen and Carbon Dioxide Permeability at Various Temperatures Through Hoechst AG Hostalen Polyethylene.
TABLE 161   Air and Nitrogen Permeability at Various Temperatures Through Hoechst AG Hostalen Polyethylene.
TABLE 162   Carbon Monoxide, Hydrogen and Helium Permeability at Various Temperatures Through Hoechst AG Hostalen Polyethylene.
TABLE 163   Hydrogen Permeability vs. Temperature and Pressure Through High Density Polyethylene.
TABLE 164   Nitrogen Permeability vs. Temperature and Pressure Through High Density Polyethylene.
TABLE 165   Oxygen and Ammonia Permeability vs. Temperature and Pressure Through High Density Polyethylene.
TABLE 166   Xylene and Oxygen Permeability Through High Density Polyethylene.
TABLE 167   Water Vapor, Carbon Dioxide, Hydrogen, Oxygen, Helium, Ethane, Natural Gas, Freon 12 and Nitrogen Permeability Through Phillips Marlex High Density Polyethylene.
TABLE 168   Argon, Methane, Ethane, Propane, Ethylene, Propylene and Sulfur Dioxide Permeability Through Hoechst AG Hostalen Polyethylene.
TABLE 169   Cyclohexanone, Chlorobenzene, Hexane, Butyl Alcohol, Trichloroethene, Methyl Salicylate and Tetrahydrofuran Permeability Through High Density Polyethylene Bottles.
TABLE 170   Ethyl Acetate, Isopropyl Acetate, Acetone, Butyl Acetate, Toluene, Xylene, Methyl Isobutyl Ketone and Methyl Ethyl Ketone Permeability Through High Density Polyethylene Bottles.
TABLE 171   Kerosine, d-Limonene, 2-Cycle Motor Oil, Pine Oil Cleaner, Diesel Fuel Conditioner and Brakleen Gas Additive Permeability Through High Density Polyethylene Bottles.
TABLE 172   Mineral Spirits, Turpentine, STP Gas Treatment, Paint Thinner, Charcoal Starter and Naphtha Permeability Through High Density Polyethylene Bottles.
TABLE 173   Xylene, Propyl Alcohol, Methyl Alcohol, Xylene/ Propyl Alcohol and Xylene/ Methyl Alcohol Permeability Through High Density Polyethylene Bottles.
TABLE 174   Gasoline Permeability Through High Density Polyethylene.
TABLE 175   d-Limonene (flavor component) Permeability Through High Density Polyethylene.

TABLE 176	Film Properties of Dow Chemical High Density Polyethylene.
GRAPH 39	Gas Permeability vs. Temperature through High Density Polyethylene.

Ethylene-Alpha Olefin Copolymer - Chapter 38

TABLE 177	Film Properties of Dow Chemical Affinity Polyolefin Plastomer.

Ethylene-Vinyl Acetate Copolymer - Chapter 39

TABLE 178	Carbon Dioxide, Oxygen and Water Vapor Permeability Through Ethylene-Vinyl Acetate Copolymer Film.
TABLE 179	Oxygen Gas Permeability Through DuPont Elvax Ethylene Vinyl Acetate Copolymer Film.
TABLE 180	Oxygen Gas Permeability Through DuPont Elvax Ethylene Vinyl Acetate Copolymer Film.
TABLE 181	Water Vapor Permeability Through DuPont Elvax Ethylene Vinyl Acetate Copolymer Film.
TABLE 182	Water Vapor Permeability Through DuPont Elvax Ethylene Vinyl Acetate Copolymer Film.
TABLE 183	Film Properties of Ethylene-Vinyl Acetate Copolymer Film.
TABLE 184	Film Properties of DuPont Elvax Ethylene Vinyl Acetate Copolymer Film.
TABLE 185	Film Properties of DuPont Elvax Ethylene Vinyl Acetate Copolymer Film.
GRAPH 40	Oxygen Permeability vs. Vinyl Acetate Content through Ethylene-Vinyl Acetate Copolymer.
GRAPH 41	Water Vapor Permeability vs. Vinyl Acetate Content through Ethylene-Vinyl Acetate Copolymer.

Ethylene-Vinyl Alcohol Copolymer - Chapter 40

TABLE 186	Gas Permeability of Carbon Dioxide, Nitrogen and Helium at Different Temperatures Through Kuraray Eval F Series Ethylene Vinyl Alcohol Copolymer.
TABLE 187	Gas Permeability of Carbon Dioxide, Nitrogen and Helium at Different Temperatures Through Kuraray Eval H Series Ethylene Vinyl Alcohol Copolymer.
TABLE 188	Gas Permeability of Carbon Dioxide, Nitrogen and Helium at Different Temperatures Through Kuraray Eval E Series Ethylene Vinyl Alcohol Copolymer.
TABLE 189	Oxygen Permeability Through DuPont Selar OH Ethylene Vinyl Alcohol Copolymer.
TABLE 190	Oxygen Permeability at Different Temperatures Through Kuraray Eval Ethylene Vinyl Alcohol Copolymer.
TABLE 191	Oxygen Permeability at Different Temperatures Through Kuraray Eval Ethylene Vinyl Alcohol Copolymer.
TABLE 192	Oxygen Permeability at Different Temperatures Through Kuraray Eval Ethylene Vinyl Alcohol Copolymer.
TABLE 193	Oxygen Permeability vs. Relative Humidity Through Kuraray Eval EF-XL Ethylene Vinyl Alcohol Copolymer Biaxially Oriented Film.
TABLE 194	Oxygen Permeability vs. Relative Humidity Through Kuraray Eval Ethylene Vinyl Alcohol Copolymer Film.
TABLE 195	Oxygen Permeability vs. Relative Humidity Through Kuraray Eval Ethylene Vinyl Alcohol Copolymer Film.
TABLE 196	Oxygen Permeability at 0% Relative Humidity vs. Orientation and Heat Treatment Through Kuraray Eval F Ethylene Vinyl Alcohol Copolymer.
TABLE 197	Oxygen Permeability at 0% Relative Humidity vs. Orientation and Heat Treatment Through Kuraray Eval E Ethylene Vinyl Alcohol Copolymer.
TABLE 198	Oxygen Permeability at 100% Relative Humidity vs. Orientation and Heat Treatment Through Kuraray Eval F Ethylene Vinyl Alcohol Copolymer.
TABLE 199	Oxygen Permeability at 100% Relative Humidity vs. Orientation and Heat Treatment Through Kuraray Eval E Ethylene Vinyl Alcohol Copolymer.
TABLE 200	Organic Solvents Permeability Through Kuraray Eval Ethylene Vinyl Alcohol Copolymer Film.
TABLE 201	Organic Solvents Permeability Through Kuraray Eval EF-XL Biaxially Oriented Ethylene Vinyl Alcohol Copolymer Film.
TABLE 202	Organic Solvents Permeability Through Kuraray Eval F Ethylene Vinyl Alcohol Copolymer Film.
TABLE 203	Organic Solvents Permeability Through Kuraray Eval E Ethylene Vinyl Alcohol Copolymer Film.
TABLE 204	Water Vapor Transmission Through Kuraray Eval Ethylene Vinyl Alcohol Copolymer Film.
TABLE 205	Water Vapor Transmission Through Kuraray Eval Ethylene Vinyl Alcohol Copolymer.
TABLE 206	Water Vapor Transmission Through DuPont Selar OH Ethylene Vinyl Alcohol Copolymer.
TABLE 207	d-Limonene (flavor component) Permeability Through Eval Company Ethylene Vinyl Alcohol Barrier Resin.
TABLE 208	Film Properties of Kuraray Eval Ethylene Vinyl Alcohol Copolymer Film.
GRAPH 42	Oxygen Permeability vs. Relative Humidity through Ethylene-Vinyl Alcohol Copolymer.
GRAPH 43	Oxygen Permeability vs. Relative Humidity through Ethylene-Vinyl Alcohol Copolymer.
GRAPH 44	Oxygen Permeability vs. Relative Humidity through Ethylene-Vinyl Alcohol Copolymer.
GRAPH 45	Oxygen Permeability vs. Relative Humidity through Ethylene-Vinyl Alcohol Copolymer.
GRAPH 46	Oxygen Permeability vs. Ethylene Content through Ethylene-Vinyl Alcohol Copolymer.

GRAPH 47	Oxygen Permeability vs. Moisture Absorption through Ethylene-Vinyl Alcohol Copolymer.
GRAPH 48	Oxygen Permeability vs. Temperature through Ethylene-Vinyl Alcohol Copolymer.
GRAPH 49	Oxygen Permeability vs. Temperature and Moisture Content through Ethylene-Vinyl Alcohol Copolymer.
GRAPH 50	Equilibrium Moisture Abosrption vs. Relative Humidity of Ethylene-Vinyl Alcohol Copolymer.
GRAPH 51	Oxygen Transmission Rate vs. Time After Retort through Ethylene-Vinyl Alcohol Copolymer.
GRAPH 52	Oxygen Uptake vs. Time After Retort through Ethylene-Vinyl Alcohol Copolymer.
GRAPH 53	Carbon Dioxide Permeability vs. Relative Humidity through Ethylene-Vinyl Alcohol Copolymer.

Polyethylene-Acrylic Acid Copolymer - Chapter 41

| TABLE 209 | Oxygen Permeability and Water Vapor Transmission Through BASF AG Lucalen Polyethylene Acrylic Acid Copolymer. |
| TABLE 210 | Film Properties of Dow Chemical Primacor Ethylene Acrylic Acid Copolymer Extrusion Resins. |

Polyethylene-Ionomer Copolymer - Chapter 42

| TABLE 211 | Oxygen Permeability and Water Vapor Transmission Through BASF AG Lucalen Polyethylene Ionomer Copolymer. |

Polypropylene - Chapter 43

TABLE 212	Gas Permeability of Oxygen, Carbon Dioxide, Nitrogen and Helium Through Oriented Polypropylene Film.
TABLE 213	Oxygen Permeability at Different Temperatures and Water Vapor Transmission Through Oriented and Non-Oriented Polypropylene.
TABLE 214	Oxygen Permeability vs. Relative Humidity Through Biaxially Oriented Polypropylene Film.
TABLE 215	Water Vapor Transmission and Oxygen Permeability Through Polypropylene.
TABLE 216	Xylene and Oxygen Permeability Through Polypropylene.
TABLE 217	Water Vapor Transmission and Oxygen Permeability Through Coated and Uncoated Oriented Polypropylene Film.
TABLE 218	Organic Solvents Permeability Through Oriented Polypropylene Film.
TABLE 219	d-Limonene (flavor component) Permeability Through Polypropylene.
TABLE 220	Film Properties of Coated and Uncoated Oriented Polypropylene Film.
GRAPH 54	Oxygen Permeability vs. Relative Humidity through Polypropylene.

Polybutylene - Chapter 45

| TABLE 221 | Water Vapor, Oxygen, Nitrogen and Carbon Dioxide Permeability Through Shell Chemical Duraflex Polybutylene Film. |
| TABLE 222 | Film Properties of Shell Chemical Duraflex Polybutylene Film. |

Polymethylpentene - Chapter 46

TABLE 223	Water Vapor Transmission and Oxygen Permeability Through Mitsui TPX Polymethylpentene Film.
TABLE 224	Nitrogen and Carbon Dioxide Permeability Through Mitsui TPX Polymethylpentene Film.
TABLE 225	Film Properties of Mitsui TPX Polymethylpentene Film.
TABLE 225	Film Properties of Mitsui TPX Polymethylpentene Film (cont'd).

Polyphenylene Sulfide - Chapter 47

| TABLE 226 | Oxygen, Carbon Dioxide, Hydrogen, Ammonia, Hydrogen Sulfide, Oxygen and Air Permeability Through Phillips Ryton Polyphenylene Sulfide Film. |
| TABLE 227 | Water Vapor Transmission and Water, Hydrochloric Acid, Acetic Acid, Benzene and Methyl Alcohol Liquid Permeability Through Phillips Ryton Polyphenylene Sulfide Film. |

Polysulfone - Chapter 48

TABLE 228	Ammonia, Carbon Dioxide, Helium, Hydrogen and Methane Permeability Through Amoco Performance Products Udel Polysulfone.
TABLE 229	Nitrogen, Oxygen, Sulfur Hexafluoride, Dichlorodifluoromethane and Dichlorotetra- fluoroethane Permeability Through Amoco Performance Products Udel Polysulfone.
TABLE 230	Water Vapor Permeability Through Amoco Performance Products Udel Polysulfone.
TABLE 231	Film Properties of Unoriented Amoco Performance Products Udel Polysulfone.

Polyvinyl Alcohol - Chapter 49

TABLE 232	Oxygen and Carbon Dioxide Permeability and Water Vapor Transmission Through Polyvinyl Alcohol.
TABLE 233	Oxygen Permeability Through Air Products and Chemicals Vinex Thermoplastic Polyvinyl Alcohol Copolymer.
TABLE 234	Water Vapor Transmission and Oxygen Permeability Through Polyvinyl Alcohol Film.
TABLE 235	Film Properties of Polyvinly Alcohol Film.

Acrylonitrile-Butadiene-Styrene Copolymer - Chapter 50

TABLE 236	Oxygen, Nitrogen and Carbon Dioxide Permeability and Water Vapor Transmission Through Dow Chemical ABS.
TABLE 237	Oxygen, Nitrogen and Carbon Dioxide Permeability and Water Vapor Transmission Through BASF AG Terluran ABS Film.
TABLE 238	Oxygen, Nitrogen and Carbon Dioxide Permeability and Water Vapor Transmission Through BASF AG Terluran ABS Film.

Acrylonitrile-Styrene-Acrylate Copolymer - Chapter 51

TABLE 239	Nitrogen, Hydrogen and Methane Permeability Through BASF Luran S Acrylate Styrene Acrylonitrile Film.
TABLE 240	Oxygen and Carbon Dioxide Permeability Through BASF Luran S Acrylate Styrene Acrylonitrile Film.
TABLE 241	Water Vapor Transmission Through BASF Luran S Acrylate Styrene Acrylonitrile Film.

Polystyrene - Chapter 52

TABLE 242	Gas Permeability and Water Vapor Transmission Through Dow Chemical Trycite Polystyrene Film.
TABLE 243	Oxygen, Nitrogen, and Carbon Dioxide Permeability and Water Vapor Transmission Through Dow Chemical Styron Polystyrene.
TABLE 244	Oxygen Permeability and Water Vapor Transmission Through Polystyrene.
TABLE 245	Vapor Transmission of Reagents Through Dow Chemical Styron Polystyrene Resin.
TABLE 246	Film Properties of Dow Chemical Trycite Polystyrene Film.
GRAPH 55	Oxygen Permeability vs. Temperature through Polystyrene.
GRAPH 56	Water Vapor Permeability vs. Thickness through Polystyrene.

General Purpose Polystyrene - Chapter 53

| TABLE 247 | Oxygen, Nitrogen and Carbon Dioxide Permeability and Water Vapor Transmission Through BASF AG Polystyrol General Purpose Polystyrene Film. |
| TABLE 248 | Oxygen, Nitrogen and Carbon Dioxide Permeability and Water Vapor Transmission Through Dow Chemical Styron General Purpose Polystyrene. |

Impact Resistant Polystyrene - Chapter 54

| TABLE 249 | Oxygen, Nitrogen and Carbon Dioxide Permeability and Water Vapor Transmission Through Dow Chemical Styron Impact Polystyrene. |
| TABLE 250 | Oxygen, Nitrogen and Carbon Dioxide Permeability and Water Vapor Transmission Through BASF AG Polystyrol Impact Polystyrene Film. |

Styrene-Acrylonitrile Copolymer - Chapter 55

TABLE 251	Gas Permeability and Water Vapor Transmission Through Dow Chemical Tyril Styrene Acrylonitrile Copolymer.
TABLE 252	Oxygen Gas and Water Vapor Permeability Through BASF Luran Styrene Acrylonitrile Copolymer Film.
TABLE 253	Vapor Transmission of Reagents Through Dow Chemical Tyril Styrene Acrylonitrile.
GRAPH 57	Oxygen Permeability vs. Temperature through Styrene-Acrylonitrile Copolymer.
GRAPH 58	Water Vapor Permeability vs. Thickness through Styrene-Acrylonitrile Copolymer.

Styrene-Butadiene Block Copolymer - Chapter 56

TABLE 254	Gas Permeability Through BASF AG Styrolux Styrene Butadiene Block Copolymer.
TABLE 255	Water Vapor Transmission Through BASF AG Styrolux Styrene Butadiene Block Copolymer.
TABLE 256	Mechanical Properties of BASF AG Styrolux Styrene Butadiene Block Copolymer Film.

Polyvinyl Chloride - Chapter 57

| TABLE 257 | Gas Permeability and Water Vapor Transmission Through Polyvinyl Chloride. |
| TABLE 258 | Oxygen Permeability vs. Temperature Through Rigid Polyvinyl Chloride Film. |

TABLE 259	Water Vapor Transmission Through Rigid Polyvinyl Chloride.
TABLE 260	Cyclohexanone, Chlorobenzene, Hexane, Butyl Alcohol, Trichloroethene, Methyl Salicylate and Tetrahydrofuran Permeability Through Rigid Polyvinyl Chloride Bottles.
TABLE 261	Ethyl Acetate, Isopropyl Acetate, Acetone, Butyl Acetate, Toluene, Xylene, Methyl Isobutyl Ketone and Methyl Ethyl Ketone Permeability Through Rigid Polyvinyl Chloride Bottles.
GRAPH 59	Oxygen Permeability vs. Thickness through Polyvinyl Chloride.
GRAPH 60	Water Vapor Permeability vs. Thickness through Polyvinyl Chloride.
GRAPH 61	Water Vapor Permeability vs. Thickness through Polyvinyl Chloride.

Polyvinyl Chloride-Polyvinylidene Chloride - Chapter 58

TABLE 262	Water Vapor, Oxygen, Nitrogen and Carbon Dioxide Permeability Through Polyvinyl Chloride Polyvinylidene Chloride Copolymer.

Polyvinylidene Chloride - Chapter 59

TABLE 263	Air Permeability Through Dow Chemical Saran Wrap Polyvinylidene Chloride Barrier Film.
TABLE 264	Oxygen Permeability Through Dow Chemical Saran Polyvinylidene Chloride Barrier Resin.
TABLE 265	Oxygen Permeability Through Dow Chemical Saran Wrap Polyvinylidene Chloride Barrier Film.
TABLE 266	Oxygen Permeability at Different Temperatures Through Dow Chemical Saran Polyvinylidene Chloride Copolymer.
TABLE 267	Oxygen Permeability vs. Relative Humidity Through Polyvinylidene Chloride Film.
TABLE 268	Oxygen Permeability vs. Relative Humidity Through Dow Chemical Saran Polyvinylidene Chloride Barrier Resin.
TABLE 269	Gas Permeability and Water Vapor Transmission Through Dow Chemical Saran Polyvinylidene Chloride.
TABLE 270	Gas Permeability of Carbon Dioxide, Nitrogen and Helium and Water Vapor Transmission Through Dow Chemical Saran Polyvinylidene Chloride Copolymer.
TABLE 271	Carbon Dioxide Permeability Through Dow Chemical Saran Wrap Polyvinylidene Chloride Barrier Film.
TABLE 272	Nitrogen Permeability Through Dow Chemical Saran Wrap Polyvinylidene Chloride Barrier Film.
TABLE 273	Water Vapor Transmission Through Dow Chemical Saran Wrap Polyvinylidene Chloride Barrier Film.
TABLE 274	Water Vapor Transmission Through Dow Chemical Saran Polyvinylidene Chloride Barrier Resin.
TABLE 275	d-Limonene (flavor component) Permeability Through Dow Chemical Saran Polyvinylidene Chloride Barrier Resin.
TABLE 276	Film Properties of Dow Chemical Saran Polyvinylidene Chloride Barrier Resin.
TABLE 277	Film Properties of Dow Chemical Saran Wrap Polyvinylidene Chloride Barrier Film.
GRAPH 62	Oxygen Permeability vs. Temperature through Polyvinylidene Chloride.
GRAPH 63	Oxygen Permeability vs. Relative Humidity through Polyvinylidene Chloride.
GRAPH 64	Carbon Dioxide and Oxygen Permeability vs. Relative Humidity through Polyvinylidene Chloride.
GRAPH 65	Oxygen Transmission Rate vs. Time After Retort through Polyvinylidene Chloride.
GRAPH 66	Oxygen Uptake vs. Time After Retort through Polyvinylidene Chloride.

Polyethylene/Polystyrene Alloy - Chapter 60

TABLE 278	Water Vapor Transmission Through BASF AG Styroblend WS Polyethylene Polystyrene Blend.

Co-Continuous Lamellae Multilayer Structure - Chapter 61

TABLE 279	Oxygen Permeability Through Multilayer Constructions Made With Lamellar Injection Molding.
TABLE 280	Oxygen Permeability Through Aromatic Nylon Multilayer Constructions Made With Lamellar Injection Molding.

Laminar Multilayer Structure - Chapter 62

TABLE 281	Xylene and Oxygen Permeability Through Nylon/ HDPE Laminar Structure.
TABLE 282	Xylene and Oxygen Permeability Through EVOH/ HDPE, Nylon/ LDPE and Nylon/ PP Laminar Structures.
TABLE 283	Ethyl Acetate, Isopropyl Acetate, Acetone, Butyl Acetate, Toluene, Xylene, Methyl Isobutyl Ketone and Methyl Ethyl Ketone Permeability Through Nylon/ Polyolefin Laminar Structure.
TABLE 284	Cyclohexanone, Chlorobenzene, Hexane, Butyl Alcohol, Trichloroethane, Methyl Salicylate and Tetrahydrofuran Permeability Through Nylon/ Polyolefin Laminar Structure.
TABLE 285	Mineral Spirits, Turpentine, STP Gas Treatment, Paint Thinner, Charcoal Starter and Naphtha Permeability Through Nylon/ Polyolefin Laminar Structure.
TABLE 286	Kerosine, d-Limonene, Motor Oil, Pine Oil, Diesel Fuel Conditioner and Gas Additive Permeability Through Nylon/ Polyolefin Laminar Structure.
TABLE 287	Xylene, Propyl Alcohol and Methyl Alcohol Permeability Through Nylon/ Polyolefin Laminar Structure.

TABLE 288	Xylene, o-Dichlorobenzene, Toluene, Methyl Alcohol and Water Permeability Through Nylon/ HDPE Laminar Structure.
TABLE 289	Naphtha, Trichloroethane, Heptane, Ethyl Acetate and Methyl Ethyl Ketone Permeability Through Nylon/ HDPE Laminar Structure.
TABLE 290	Xylene, Propyl Alcohol and Methyl Alcohol Permeability Through EVOH/ Polyolefin Laminar Structures.
GRAPH 67	Xylene Permeability through Various Bottle Sizes vs. Percentage (%) Weight Concentration of Generic Laminar Multilayer Structure.

Multilayer Films with Ethylene-Vinyl Alcohol Copolymer Barrier - Chapter 63

TABLE 291	Oxygen Permeability vs. Relative Humidity Through PP/ EVOH/ PP Multilayer Film.
TABLE 292	Oxygen Permeability vs. Relative Humidity Through PET/ EVOH/ PP and PC/ EVOH/ PP Multilayer Films.
TABLE 293	Oxygen Permeability vs. Relative Humidity Through PS/ EVOH/ PP, HDPE/ EVOH/ PP and Nylon/ EVOH/ PP Multilayer Films.
TABLE 294	Oxygen Permeability vs. Relative Humidity Through LDPE/ EVOH/ PP, PP/ EVOH/ PET and PP/ EVOH/ PS Multilayer Films.
TABLE 295	Oxygen Permeability vs. Relative Humidity Through PP/ EVOH/ HDPE and PP/ EVOH/ LDPE Multilayer Films.
TABLE 296	Oxygen Permeability vs. Relative Humidity Through PP/ EVOH/ PC and HDPE/ EVOH/ LDPE Multilayer Films.
TABLE 297	Oxygen Permeability vs. Relative Humidity Through Nylon/ EVOH/ LDPE Multilayer Film.
TABLE 298	Oxygen Permeability vs. Relative Humidity Through EVOH/ LDPE Multilayer Film.
TABLE 299	Oxygen Permeability vs. Relative Humidity, Vanilla, Peppermint, Piperonol and Camphor Permeability Through PET/ EVOH/ LDPE Multilayer Film.
TABLE 300	Vanilla, Peppermint, Piperonol and Camphor Permeability Through EVOH/ PE Multilayer Film.
TABLE 301	Vanilla, Peppermint, Piperonol and Camphor Permeability Through PET/ EVOH and Nylon/ EVOH Multilayer Films.
TABLE 302	Gasoline Permeability Through HDPE/ EVOH Multilayer Film.
TABLE 303	Chloroform and Xylene Permeability Through LDPE/ EVOH Multilayer Film.
TABLE 304	Methyl Ethyl Ketone and Kerosine Permeability Through LDPE/ EVOH Multilayer Film.
GRAPH 68	Oxygen Permeability vs. Relative Humidity through Polyvinylidene Chloride coated Nylon/ Ethylene-Vinyl Alcohol Copolymer/ Polyethylene Multilayer Film.
GRAPH 69	Oxygen Permeability vs. Relative Humidity through Oriented Polypropylene/ Ethylene-Vinyl Alcohol Copolymer/ Polyethylene Multilayer Film.
GRAPH 70	Oxygen Permeability vs. Relative Humidity through Ethylene-Vinyl Alcohol Copolymer/ Polyethylene Multilayer Film.

Multilayer Films with Polyvinylidene Chloride Barrier - Chapter 64

TABLE 305	Water Vapor and Oxygen Permeability Through Dow Chemical Saranex LDPE/ EVA/ PVDC/ EVA/ LDPE Multilayer Film.
TABLE 306	Carbon Dioxide and Nitrogen Permeability Through Dow Chemical Saranex LDPE/ EVA/ PVDC/ EVA/ LDPE Multilayer Film.
TABLE 307	Air and Oxygen Permeability Through Dow Chemical Saranex LDPE/ EVA/ PVDC/ EVA/ LDPE Multilayer Film.
TABLE 308	Water Vapor, Carbon Dioxide, Nitrogen, Air and Oxygen Permeability Through Dow Chemical Saranex 23 LDPE/ EVA/ PVDC/ EVA Multilayer Film.
TABLE 309	Water Vapor, Carbon Dioxide, Nitrogen, Air and Oxygen Permeability Through Dow Chemical Saranex 25 HDPE/ EVA/ PVDC/ EVA Multilayer Film.
TABLE 310	Oxygen Permeability vs. Relative Humidity Through PP/ PVDC/ PP Multilayer Film.
TABLE 311	Oxygen Permeability vs. Relative Humidity Through PET/ PVDC/ PP and PC/ PVDC/ PP Multilayer Films.
TABLE 312	Oxygen Permeability vs. Relative Humidity Through PS/ PVDC/ PP, HDPE/ PVDC/ PP and Nylon/ PVDC/ PP Multilayer Films.
TABLE 313	Oxygen Permeability vs. Relative Humidity Through LDPE/ PVDC/ PP, PP/ PVDC/ PET and PP/ PVDC/ PS Multilayer Films.
TABLE 314	Oxygen Permeability vs. Relative Humidity Through PP/ PVDC/ HDPE and PP/ PVDC/ LDPE Multilayer Films.
TABLE 315	Oxygen Permeability vs. Relative Humidity Through PP/ PVDC/ PC and HDPE/ PVDC/ LDPE Multilayer Films.
TABLE 316	Film Properties of Dow Chemical Saranex Multilayer Film.
GRAPH 71	Oxygen Permeability vs. Relative Humidity through Low Density Polyethylene/ Ethylene-Vinyl Acetate Copolymer/ Polyvinylidene Chloride/ Ethylene-Vinyl Acetate Copolymer/ Low Density Polyethylene Multilayer Film.

Multilayer Films - General - Chapter 65

TABLE 317	Oxygen, Nitrogen, Carbon Dioxide and Water Vapor Permeability Through Dow Chemical Saranex PE/ PVC-PVDC Copolymer Multilayer Film.
TABLE 318	Water Vapor and Oxygen Permeability Through Nylon 6/ LDPE and HDPE/ EAA/ Nylon/ EAA Multilayer Films.
TABLE 319	Water Vapor Permeability Through CTFE/ PE/ PVC and CTFE/ PVC Multilayer Films.
TABLE 320	Vanilla, Peppermint, Piperonol and Camphor Permeability Through PET/ PE and Nylon/ PE Multilayer Films.
TABLE 321	Vanilla, Peppermint, Piperonol and Camphor Permeability Through PP/ PE Multilayer Film.
TABLE 322	Film Properties of Dow Chemical Nylopak Multilayer Film.

692

Epoxy Resin - Chapter 66

TABLE 323   Water Vapor, Oxygen, Nitrogen, Carbon Dioxide and Hydrogen Permeability Through Epoxy Resin.

Polypyrrole - Chapter 67

TABLE 324   Water Vapor, Nitrogen, Oxygen, Carbon Dioxide, Methane, and Hydrogen Permeability Through BASF AG Lutamer Polypyrrole Film.

TABLE 325   Physical, Mechanical, Electrical, and Ignition Resistance Properties of BASF AG Lutamer Polypyrrole Film.

Olefinic Thermoplastic Elastomer - Chapter 68

TABLE 326   Air Permeability Through Advanced Elastomer Systems Trefsin 3281 Olefinic Thermoplastic Elastomer.

TABLE 327   Air, Nitrogen, and Oxygen Permeability Through Advanced Elastomer Systems Santoprene Olefinic Thermoplastic Elastomer.

TABLE 328   Gas Permeability of Carbon Dioxide, Argon and Propane Through Advanced Elastomer Systems Santoprene Olefinic Thermoplastic Elastomer.

TABLE 329   Water Vapor Permeability Through Advanced Elastomer Systems Santoprene Olefinic Thermoplastic Elastomer.

Polyamide Thermoplastic Elastomer - Chapter 69

TABLE 330   Oxygen, Carbon Dioxide, Nitrogen and Helium Permeability Through Shore D Hardness 25 and 35 Atochem Pebax Polyether Block Amide Thermoplastic Elastomer Film.

TABLE 331   Oxygen, Carbon Dioxide, Nitrogen, Helium and Propane Permeability Through Shore D Hardness 40 and 55 Atochem Pebax Polyether Block Amide Thermoplastic Elastomer Film.

TABLE 332   Oxygen, Carbon Dioxide, Nitrogen, Helium and Propane Permeability Through Shore D Hardness 63 Atochem Pebax Polyether Block Amide Thermoplastic Elastomer Film.

TABLE 333   Water Vapor Permeability Through Atochem Pebax Polyether Block Amide Thermoplastic Elastomer Film.

TABLE 334   Air Conditioning Refrigerants Permeation Loss Through DuPont Company Zytel FN 726 Polyamide Thermoplastic Elastomer.

Polybutadiene Thermoplastic Elastomer - Chapter 70

TABLE 335   Carbon Dioxide, Oxygen and Water Vapor Permeability Through Japan Synthetic Rubber JSR RB Polybutadiene Thermoplastic Elastomer Stretch Film.

TABLE 336   Carbon Dioxide, Oxygen, Ethylene Oxide and Water Vapor Permeability Through Japan Synthetic Rubber JSR RB Polybutadiene Thermoplastic Elastomer Film.

TABLE 337   Film Properties of Japan Synthetic Rubber JSR RB Polybutadiene Thermoplastic Elastomer Film and Stretch Film.

GRAPH 72   Carbon Dioxide Permeability vs. Thickness through Polybutadiene Thermoplastic Elastomer.

GRAPH 73   Oxygen Permeability vs. Thickness through Polybutadiene Thermoplastic Elastomer.

Polyester Thermoplastic Elastomer - Chapter 71

TABLE 338   Gas Permeability of Various Gases Through DuPont Company Hytrel 4056 Polyester Thermoplastic Elastomer.

TABLE 339   Gas Permeability of Various Gases Through DuPont Company Hytrel 5556 Polyester Thermoplastic Elastomer.

TABLE 340   Gas Permeability of Various Gases Through DuPont Company Hytrel 4056 and Hytrel 7246 Polyester Thermoplastic Elastomer.

TABLE 341   Water Vapor, Carbon Dioxide, Oxygen and Nitrogen Permeability Through Eastman Chemical Kodar Ecdel Copolyester Ether Thermoplastic Elastomer Film.

TABLE 342   Film Properties of Eastman Chemical Kodar Ecdel Copolyester Ether Thermoplastic Elastomer Film.

Thermoplastic Polyester-Polyurethane Elastomer - Chapter 72

TABLE 343   Oxygen Gas and Water Vapor Permeability Through BASF Elastollan C Polyester Urethane Thermoplastic Elastomer.

TABLE 344   Carbon Dioxide and Hydrogen Gas Permeability Through BASF Elastollan C Polyester Urethane Thermoplastic Elastomer.

TABLE 345   Argon and Methane Gas Permeability Through BASF Elastollan C Polyester Urethane Thermoplastic Elastomer.

TABLE 346   Helium and Nitrogen Gas Permeability Through BASF Elastollan C Polyester Urethane Thermoplastic Elastomer.

Thermoplastic Polyether-Polyurethane Elastomer - Chapter 73

TABLE 347   Oxygen Gas and Water Vapor Permeability Through BASF Elastollan 1100 Polyether Urethane Thermoplastic Elastomer.

TABLE 348   Carbon Dioxide and Hydrogen Gas Permeability Through BASF Elastollan 1100 Polyether Urethane Thermoplastic Elastomer.

TABLE 349   Argon and Methane Gas Permeability Through BASF Elastollan 1100 Polyether Urethane Thermoplastic Elastomer.

TABLE 350   Helium and Nitrogen Gas Permeability Through BASF Elastollan 1100 Polyether Urethane Thermoplastic Elastomer.

GRAPH 74   Nitrogen Permeability vs. Temperature through Thermoplastic Polyether-Polyurethane Elastomer.

Styrenic Thermoplastic Elastomer - Chapter 74

TABLE 351	Oxygen Permeability Through Shell Chemical Kraton Neat Styrenic Thermoplastic Elastomer.
TABLE 352	Oxygen Permeability Through Shell Chemical Kraton Styrenic Thermoplastic Elastomer Compounds.
TABLE 353	Carbon Dioxide Permeability Through Shell Chemical Kraton Styrenic Thermoplastic Elastomer Compounds.
TABLE 354	Carbon Dioxide Permeability Through Shell Chemical Kraton Neat Styrenic Thermoplastic Elastomer.
TABLE 355	Water Vapor Transmission Through Shell Chemical Kraton Styrenic Thermoplastic Elastomer Compounds.
TABLE 356	Water Vapor Transmission Through Shell Chemical Kraton Neat Styrenic Thermoplastic Elastomer.

Vinyl Thermoplastic Elastomer - Chapter 75

TABLE 357	Carbon Dioxide, Oxygen and Water Vapor Permeability Through Polyvinyl Chloride Film and Stretch Film.
TABLE 358	Oxygen and Carbon Dioxide Permeability Through Polyvinyl Chloride Polyol.
TABLE 359	Film Properties of Polyvinyl Chloride Film and Stretch Film.

Ethylene-Acrylate Copolymer - Chapter 76

TABLE 360	Nitrogen Permeability Through Ethylene Acrylate Rubber.

Polybutadiene - Chapter 77

TABLE 361	Air Permeability vs. Temperature Through Bayer Taktene 1220 Polybutadiene Rubber.
TABLE 362	Nitrogen, Carbon Dioxide and Water Vapor Permeability Through Polybutadiene Rubber.
TABLE 363	Water Vapor Transmission and Hydrogen, Oxygen, Nitrogen, Carbon Dioxide and Air Permeabiltiy Relative To Butyl Rubber Through Polybutadiene Rubber.
GRAPH 75	Air Permeability vs. Temperature through Polybutadiene.

Butyl Rubber - Chapter 78

TABLE 364	Air Permeability vs. Temperature Through Butyl Rubber.
TABLE 365	Nitrogen Permeability Through Butyl Rubber.
GRAPH 76	Air Permeability vs. Plasticizer (Polaris 45 Oil) Loading through Butyl Rubber.
GRAPH 77	Air Permeability vs. SRF Black Loading through Butyl Rubber.
GRAPH 78	Air Permeability vs. Temperature through Butyl Rubber.

Bromobutyl Rubber - Chapter 79

GRAPH 79	Air Permeability vs. Bromobutyl Concentration through Bromoisobutylene-Isoprene Copolymer.
GRAPH 80	Moisture Vapor Transmission vs. Bromobutyl Concentration through Bromoisobutylene-Isoprene Copolymer.

Chlorobutyl Rubber - Chapter 80

TABLE 366	Air Permeability vs. TemperatureThrough Chlorobutyl Rubber.

Isobutylene Rubber - Chapter 81

TABLE 367	Helium, Hydrogen, Oxygen, Nitrogen, Carbon Dioxide and Air Permeabiltiy Relative To Butyl Rubber Through Polyisobutylene Rubber.

Chlorosulfonated Polyethylene Rubber - Chapter 82

TABLE 368	Nitrogen Permeability Through Chlorosulfonated Polyethylene Rubber.

Polyepichlorohydrin Rubber - Chapter 83

TABLE 369	Nitrogen Permeability Through Epichlorohydrin Rubber.

Epichlorohydrin Copolymer Rubber - Chapter 84

TABLE 370	Nitrogen Permeability Through Epichlorohydrin Copolymer Rubber.

Ethylene-Propylene Copolymer - Chapter 85

TABLE 371	Nitrogen Permeability Through Ethylene-Propylene Rubber.

694

Ethylene-Propylene-Diene Copolymer - Chapter 86

TABLE 372   Air Permeability vs. Temperature Through Ethylene-Propylene-Diene Copolymer Rubber.

TABLE 373   Oxygen, Nitrogen, Carbon Dioxide, Air and Water Vapor Permeability Through DuPont Nordel Ethylene-Propylene-Diene Rubber.

TABLE 374   Nitrogen Permeability Through Ethylene-Propylene-Diene Rubber.

TABLE 375   Oxygen, Nitrogen, Carbon Dioxide and Air Permeabiltiy Relative To Butyl Rubber Through Ethylene-Propylene-Diene Copolymer Rubber.

GRAPH 81   Air Permeability vs. Temperature through Ethylene-Propylene-Diene Copolymer.

Vinylidene Fluoride-Hexafluoropropylene - Chapter 87

TABLE 376   Air, Carbon Dioxide, Helium, Nitrogen and Oxygen Permeability Through Compounded DuPont Viton Fluoroelastomer.

TABLE 377   Nitrogen Permeability Through Fluoroelastomer.

Natural Rubber - Chapter 88

TABLE 378   Air Permeability vs. Temperature Through Natural Rubber.

TABLE 379   Nitrogen Permeability Through Natural Rubber.

TABLE 380   Water Vapor Transmission and Helium, Hydrogen, Oxygen, Nitrogen, Carbon Dioxide and Air Permeabiltiy Relative To Butyl Rubber Through Natural Rubber.

GRAPH 82   Air Permeability vs. Temperature through Natural Rubber.

GRAPH 83   Gas Permeability vs. Mineral Filler Content through Natural Rubber.

Polychloroprene Rubber - Chapter 89

TABLE 381   Hydrogen, Oxygen, Nitrogen, Carbon Dioxide and Air Permeabiltiy Relative To Butyl Rubber Through Neoprene Rubber.

TABLE 382   Nitrogen Permeability Through Neoprene Rubber.

Acrylonitrile-Butadiene Copolymer - Chapter 90

TABLE 383   Air Permeability vs. Temperature Through Bayer Krynac 800 Nitrile Rubber.

TABLE 384   Nitrogen Permeability Through Nitrile Rubber.

TABLE 385   Air Conditioning Refrigerants Permeation Loss Through Nitrile Rubber.

TABLE 386   Carbon Dioxide and Air Permeabiltiy Relative To Butyl Rubber Through Acrylonitrile Butadiene Copolymer Rubber.

TABLE 387   Helium and Hydrogen Permeabiltiy Relative To Butyl Rubber Through Acrylonitrile Butadiene Copolymer Rubber.

TABLE 388   Oxygen and Nitrogen Permeabiltiy Relative To Butyl Rubber Through Acrylonitrile Butadiene Copolymer Rubber.

GRAPH 84   Air Permeability vs. Temperature through Acrylonitrile-Butadiene Copolymer.

Polyisoprene Rubber - Chapter 91

TABLE 389   Nitrogen Permeability Through Polyisoprene Rubber.

Polysulfide Rubber - Chapter 92

TABLE 390   Nitrogen Permeability Through Polysulfide Rubber.

Polyurethane - Chapter 93

TABLE 391   Water Vapor, Oxygen, Nitrogen and Carbon Dioxide Permeability Through Polyurethanes.

TABLE 392   Nitrogen Permeability Through Polyester and Polyether Urethane Rubber.

Propylene Oxide Rubber - Chapter 94

TABLE 393   Nitrogen Permeability Through Polypropylene Oxide Rubber.

Silicone - Chapter 95

TABLE 394   Hydrogen, Oxygen, Nitrogen, Carbon Dioxide and Air Permeabiltiy Relative To Butyl Rubber Through Silicone Rubber.

TABLE 395   Nitrogen Permeability Through Silicone Rubber.

TABLE 396   Water Vapor, Oxygen, Carbon Dioxide and Hydrogen Permeability Through Silicone Rubber.

Methylvinylfluorosilicone - Chapter 96

TABLE 397   Nitrogen Permeability Through Fluorosilicone.

Styrene-Butadiene Copolymer - Chapter 97

TABLE 398   Air Permeability vs. Temperature Through Styrene-Butadiene Rubber.

TABLE 399   Water Vapor Transmission and Helium, Hydrogen, Oxygen, Nitrogen, Carbon Dioxide and Air Permeabiltiy Relative To Butyl Rubber Through Styrene-Butadiene Rubber.

TABLE 400   Nitrogen, Oxygen and Carbon Dioxide Permeability Through Styrene-Butadiene Rubber.

GRAPH 85   Air Permeability vs. Temperature through Styrene-Butadiene Copolymer.

Endnotes to Tables

a) test method: DIN 53455, ISO 527; test temperature: 23°C; relative humidity: 50%; test condition note: standard laboratory atmosphere

b) test method: ASTM D1003; test temperature: 23°C; relative humidity: 50%; test condition note: standard laboratory atmosphere

c) test method: ASTM D1003

d) test method: DIN 67530; test temperature 23°C

e) test method: DIN 53375; test temperature 23°C

f) test method: ASTM D1709; test temperature: 23°C

g) test method: DIN 53373; test temperature: 23°C

h) test method: ASTM D790; strain rate: 5 mm/min.

I) test method: ASTM D882, Type 1; strain rate: 51 mm/min.

j) test method:: ASTM D1709, Method A; test temperature: 23°C

k) test method: ASTM D1922

l) test method: ASTM D1894

m) test method: ASTM D2457; test units: photocell microamps; test name: Gardner

n) test method: ASTM D1525

o) test method: ASTM D882; strain rate: 508 mm/min.

p) test method: ASTM D882; strain rate: 51 mm/min.

q) test method: ASTM D96

r) test method: ASTM D882

s) test method: ASTM D1746

t) test method: ASTM D1709; test note @ 660 mm (26 in.)

u) test method: ASTM F88; test note: 0.21 MPa, 1 second dwell time

v) test method: ASTM F88; test note: @ 163°C, 0.21 MPa, 1 second preheat, 1 second dwell time

w) test method: ASTM D1709, Method A

x) test method: ASTM D1709, Method B

y) test method: Dow Chemical test method

z) test method: ASTM D1922, Method B

aa) test method: ASTM D1922, Method A

ab) test method: Dow Chemical test method; test note: temperature at which 2 lb/in (8.9 N/25.4 microns) heat seal strength is achieved

ac) test method: ASTM D1003-61

ad) test method: JISP8116

ae) property name: quadratone tear strength; test method: JIS K6732

af) test method: JIS P8112

ag) test method: JAS

ah) test temperature: 30°C; test duration: 24 hours

ai) test temperature: 140°C; test duration: 1 hour

aj) test temperature: 20°C; relative humidity: 65%

ak) test note: du Pont test

al) test method: ASTM D1505-66

am) test apparatus: differentiating scanning calorimeter

an) test note: du Pont test; test temperature 105°C

ao) test method: du Pont test; test temperature: 150°C

ap) test method: ASTM D696; test temperature: 30-50°C

aq) test temperature: 25-75°C; test note: sample tested was Mylar 1000A

ar) test method: ASTM D882-80

as) tert load: 1 kilogram

at) test method: ASTM D1004-66

au) test method: ASTM D1922-67

av) test method: ASTM D570-63; test temperature: 23°C; test duration: 24 hours

aw) test method: ASTM D792

ax) test method: ASTM D2117-62T, Method A

ay) test method: ASTM D882-64T, Method A

az) test method: ASTM D1004

ba) test method: ASTM D1922-67

bb) test method: ASTM D744-63T (Muller)

bc) test method: ASTM D1709-62T

bd) test method: ASTM D1709-62T; test name: MIT Fold; test load: 0.5 kg.

be) test method: ASTM D1894-63, Method B

bf)   test method: incline plane

bg)   test method: ASTM D696-44

bh)   test method: ASTM C177

bi)   test temperature: 148.9°C; test duration: 30 minutes

bj)   test method: ASTM D746

bk)   test method: ASTM D882; sample type: ASTM D638-IV sample cutter

bl)   test method: ASTM D1709; test note: 3.8 cm dart @ 66 cm

bm)   test method: du Pont test

bn)   test note: 50% dart drop

bo)   test note: Sentinel sealer

bp)   test method: ASTM D3420

bq)   test method: dynamic thermal analysis; test note: heat and cool rates of 20°C/min.

br)   test method: ASTM D5236

bs)   test method: ASTM D1505

bt)   test name: Mullen; test method: ASTM D774

bu)   test apparatus: T.M. long machine; test note: 1/2" steel ball, velocity 270 ft/sec.

bv)   test method: ASTM D1204; exposure time: 30 minutes; exposure temperature: 149°C

bw)   test method: ASTM D1204; exposure time: 10 minutes; exposure temperature: 149°C

bx)   test name MIT Fold Endurance Test; test method: ASTM D2176

by)   test name MIT Fold Endurance Test; test method: ASTM D2176

bz)   test method: ASTM D1044; test note: CD 10F wheel, 1000 gram weight, 1000 cycles

ca)   test name: crystalline melting point; test apparatus: hot stage microscope

cb)   test method: ASTM D1790

cc)   test method: ASTM D1709, Method A; test note: 66 cm height, 3,8 cm diameter dart

cd)   test apparatus: Taber abraser "A"; test note: CS 10 wheel, 500 gram weight, 100 cycles

ce)   test apparatus: Taber abraser "B"; test note: CS 10F wheel, 500 gram weight, 500 cycles

cf)   test method: ASTM D689

cg)   test method: ASTM D523

ch)   test apparatus: Perkins Bond Tester; sample thickness: 0.05 mm film

ci)   test method: JIS Z1702

cj)   test note: correspond to JIS Z1702

           Endnotes to Tables

ck)   test method: JIZ Z6714

cl)   test method: JIS Z8741

cm)   test method: JIS K6781; test note: at right angle

cn)   test method: Mitsui Petrochemical

co)   test method: ASTM D696

cp)   test method BS874A

cq)   test method: ASTM C351

cr)   test method: ASTM D257

cs)   test method: ASTM D149

ct)   test method: ASTM D150

cu)   test temperature: 160°C; test note: film is taken out of oven, left to stand for 30 min. @ 23°C and then measured

cv)   test temperature: 220°C; test note: film is taken out of oven and left to stand for 30 min. @ 23°C and then measured

cw)   test method: Method A, F50

cx)   test method: ASTM 790

cy)   test method: ASTM D542

cz)   test method: ASTM D1709, modified; test note: 13 mm dart @ 660 mm

da)   test method: ASTM D1505; test note: gradient tube method; specimen note: clear, quenched

db)   test method: ASTM D882 crosshead speed: 25 mm/min.

dc)   source note: data originally from - Licary, J.J and Brads, E.R., Machine Design, May 25, 1967, p. 192.

de)   test method: ASTM D882-56T

df)   test method: ASTM D882-56T; strain rate: 10%/min.

dg)   test method: ASTM D1505-57T

dh)   test apparatus: Abbe Refractometer

di)   test method: ASTM D1894-63

dj)   test method: ASTM D570-57T

dk)   test method: ASTM D1709, method B; test note: 152 cm height, 5.1 cm diameter dart

(11) *Attane Ultra Low Density Ethylene-Octene Copolymers: Performance Plus Compared To LLDPE And EVA Resins In Flexible Packaging*, supplier marketing literature (305-1596-790) - Dow Chemical Company, 1989.

(12) *Trefsin 3281 Food/Medical Grade Thermoplastic Elastomer*, supplier technical report - Advanced Elastomer Systems, 1991.

(15) *Udel Polysulfone Design Engineering Handbook*, supplier design guide (F-47178) - Amoco Performance Products, Inc., 1988.

(25) *Lupolen, Lucalen Product Line, Properties, Processing*, supplier design guide (B 581 e/(8127) 10.91) - BASF Aktiengesellschaft, 1991.

(26) *Polystyrol Product Line, Properties, Processing*, supplier design guide (B 564 e/2.93) - BASF Aktiengesellschaft, 1993.

(29) *Styrolux Product Line, Properties, Processing*, supplier design guide (B 583 e/(950) 12.91) - BASF Aktiengesellschaft, 1992.

(30) *Luran Product Line, Properties, Processing*, supplier design guide (B 565 e/10.83) - BASF Aktiengesellschaft, 1983.

(39) *Handbook Of Properties For Teflon PFA*, supplier design guide (E-96679) - Du Pont Company, 1987.

(60) *Ecdel Elastomers*, supplier design guide (MB-100A) - Eastman Plastics, 1990.

(63) *Kuraray Eval Resin*, supplier design guide (5-2,000-507) - Kuraray Co., Ltd..

(68) *Design Handbook For DuPont Engineering Plastics - Module II*, supplier design guide (F-42267) - DuPont Engineering Polymers.

(70) *Vectra Polymer Materials*, supplier design guide (B 121 BR E 9102/014) - Hoechst AG, 1991.

(78) *Calibre Engineering Thermoplastics Basic Design Manual*, supplier design guide (301-1040-1288) - Dow Chemical Company, 1988.

(89) *Foraflon PVDF*, supplier design guide (694.E/07.87/20) - Atochem S. A., 1987.

(93) *Ultramid Nylon Resins Product Line, Properties, Processing*, supplier design guide (B 568/1e/4.91) - BASF Corporation, 1991.

(94) *Hostalen Polymer Materials*, supplier design guide (HDKR 101 E 9050/022) - Hoechst AG.

(96) *Kel-F 81 PCTFE Engineering Manual*, supplier design guide (98-0211-5944-1 (120.5) DPI) - 3M Industrial Chemical Products Division, 1990.

(97) *Ube Ultra-High Heat-Resistant Polimide Film Upilex*, supplier marketing literature - Ube Industries, Ltd..

(101) *Engineering Properties Of Marlex Resins*, supplier design guide (TSM-243) - Phillips 66 Company, 1983.

(102) *Ryton Polyphenylene Sulfide Resins Engineering Properties Guide*, supplier design guide (1065(a)-89 A 02) - Phillips 66 Company, 1989.

(114) *Hyflon ETFE 700/800 Properties and Application Guide*, supplier design guide - Ausimont USA, Inc..

(121) *Parylene Conformal Coatings Specifications and Properties*, supplier technical report - Union Carbide Specialty Coating Systems, 1992.

(125) *Solvay Polyvinylidene Fluoride*, supplier design guide (B-1292c-B-2.5-0390) - Solvay, 1992.

(130) *Elastollan Design And Processing Guide*, supplier design guide - BASF Corporation, 1993.

(137) *Terluran Product Line, Properties, Processing*, supplier design guide (B 567e/ (8109) 9.90) - BASF Aktiengesellschaft, 1990.

(138) *Aclar Performance Films*, supplier technical report (SFI-14 Rev. 9-89) - Allied-Signal Enineered Plastics, 1989.

(142) *Luran S Acrylonitrile Styrene Acrylate Product Line, Properties, Processing*, supplier design guide (B 566 e / 11.90) - BASF Aktiengesellschaft, 1990.

(143) *Luran S Acrylonitrile Styrene Acrylate Product Line, Properties, Processing*, supplier design guide (B 566 e / 10.83) - BASF Aktiengesellschaft, 1983.

(164) *Bromobutyl Rubber Optimizing Key Properties*, supplier marketing literature - Exxon Chemicals.

(165) *Kodar PETG Copolyester 6763*, supplier technical report (MB-80F/June 1988) - Eastman Plastics, 1988.

(166) *Kodar PCTG Copolyester 5445*, supplier technical report (MB-94/August 1985) - Eastman Plastics, 1988.

(180) *Ultradur Polybutylene Terephthalate (PBT) Product Line, Properties, Processing*, supplier design guide (B 575/1e - (819) 4.91) - BASF Aktiengesellschaft, 1991.

(182) *Topics In Chemistry - BASF Plastics Research And Development*, supplier technical report - BASF Aktiengesellschaft, 1992.

(186) *Lupolen Polyethylene And Novolen Polypropylene Product Line, Properties, Processing*, supplier design guide (B 579 e / 4.92) - BASF Aktiengesellschaft, 1992.

(201) *Delrin Design Handbook For Du Pont Engineering Plastics*, supplier design guide (E-62619) - Du Pont Company, 1987.

(205) *Tefzel Fluoropolymer Design Handbook*, supplier design guide (E-31301-1) - Du Pont Company, 1973.

(210) *Celcon Acetal Copolymer*, supplier design guide (90-350 7.5M/490) - Hoechst Celanese Corporation, 1990.

(216) *Japan Synthetic Rubber JSR RB*, supplier design guide - Japan Synthetic Rubber Company.

Reference numbers correspond to our assigned source document number, if you wish additional information, please contact Plastics Design Library.

(240) *Kraton Thermoplastic Rubber*, supplier design guide (SC:198-89) - Shell Chemical Company, 1989.

(250) *Permeability Of Polymers To Gases And Vapors*, supplier technical report (P302-335-79, D306-115-79) - Dow Chemical Company, 1979.

(251) *Lupolen Features, Applications, Typical Values*, supplier marketing literature (F581 d/e/f) - BASF Aktiengesellschaft, 1992.

(252) *Ultramid Nylon Resins Product Line, Properties, Processing*, supplier design guide (B 568/1e/12.87) - BASF Corporation, 1987.

(253) *Introducing Lamellar Injection Molding Technology - "The LIM Advantage" (Licensing Bulletin)*, supplier marketing literature (304-00383-493 SMG) - Dow Chemical Company, 1993.

(254) *621 Ways To Succeed - 1993-1994 Materials Selection Guide*, supplier technical report (304-00286-1292X SMG) - Dow Chemical Company, 1992.

(255) *Saran Barrier Polymers 1987 Update: Saran Barrier Polymer Dynamics*, supplier technical report (190-383-587) - Dow Chemical Company, 1987.

(256) *Saran Wrap Plastic Film Data Sheets*, supplier technical report (500-(1241, 1242, 1179, 1180, 1243)- (1289, 1289X)) - Dow Chemical Company, 1989.

(257) *Saranex Plastic Film Data Sheets*, supplier technical report (500- (1170, 1184, 1185, 1186, 1161, 1187)- (588, 1289X, 1289)) - Dow Chemical Company, 190.

(258) *Nylopak Plastic Film Data Sheets*, supplier technical report (500-1182-1289X) - Dow Chemical Company, 1989.

(259) *Affinity Polyolefin Plastomers*, supplier marketing literature (305-01953-893 SMG) - Dow Chemical Company, 1993.

(260) *Affinity Polyolefin Plastomer Data Sheets*, supplier technical report (305- (01963, 01965, 01966, 01967, 01968)- 893) - Dow Chemical Company, 1993.

(261) *Trycite Film Data Sheets*, supplier technical report (500- (1349, 1350, 1351)) - Dow Chemical Company, 1992.

(262) *Styron Polystyrene Resins For Applications Requiring Impact Resistance*, supplier design guide (301-471-1281) - Dow Chemical Company, 1981.

(263) *General Purpose Styron*, supplier design guide (301-678-1085) - Dow Chemical Company, 1985.

(264) *Gas Barrier Properties Of Eval Resins - Technical Bulletin No. 110*, supplier technical report - Eval Company of America.

(265) *Eval Films The Ultimate Laminating Film For Barrier Packaging Applications - Technical Bulletin No. 160*, supplier technical report - Eval Company of America.

(266) *Chemical And Solvent Barrier Properties Of Eval Resins - Technical Bulletin No. 180*, supplier technical report - Eval Company of America.

(267) *Eval Films For Flexible Barrier Packaging*, supplier marketing literature (94-2-1,000) - Kuraray Co., Ltd., 1994.

(268) *Eval Film Properties Comparison*, supplier technical report - Kuraray Co., Ltd..

(269) *Mylar Technical Information- Physical, Thermal Properties*, supplier technical report (H-32181) - DuPont Company.

(270) *Mylar Polyester Film*, supplier technical report (E-99499) - DuPont Company, 1988.

(271) *Guide to Excellence- Research And Development*, supplier marketing literature - DuPont Company.

(272) *Kapton Polyimide Film- Safe Handling*, supplier technical report (E-72084) - DuPont Company, 1988.

(273) Kreuz, J. A., Milligan, S. N., Sutton, R. F., *Kapton Polyimide Film- Advanced Flexible Dielectric Substrates For FPC/TAB Applications*, supplier technical report (H-24917) - DuPont Company, 1990.

(274) *Hytrel Polyester Elastomer - Gas Permeability (HYT-506B)*, supplier technical report (E-37763) - DuPont Company, 1984.

(275) *Zytel FN Flexible Nylon Alloy Products and Properties Guide*, supplier technical report (H-14079-1) - DuPont Company, 1990.

(276) *Dartek Film Data Sheets*, supplier technical report (H-27768) - DuPont Canada, 1990.

(277) *Sclair Linear Polyethylene Resins For Film Packaging*, supplier marketing literature - DuPont Canada.

(278) *Sclairfilm Polyolefin Film - SL-1 and SL-3 Laminating Film*, supplier technical report (H-27763) - DuPont Canada, 1990.

(279) *Surlyn Ionomer Resin Increases Packaging Efficiency and Package Performance*, supplier marketing literature (E-54995) - DuPont Company.

(280) *Surlyn Ionomer Resin Selector Guide*, supplier technical report (E-48623 (1/90)) - DuPont Company, 1986.

(281) *Elvax Ethylene Vinyl Acetate Copolymer Resins*, supplier technical report (E-45625) - DuPont Company, 1983.

(282) *Santoprene Thermoplastic Elastomer Physical Properties*, supplier technical report (AES-1015) - Advanced Elastomer Systems, 1990.

(283) *Vinex Thermoplastic Polyvinyl Alcohol Copolymer Resins Data Sheets*, supplier technical report (152-(9107, 9108, 9109)) - Air Products and Chemicals, 1991.

(284) *Capron Nylon Resins For Films - Operating Manual*, supplier technical report (SFF-08) - Allied Signal Inc., 1992.

(285) *Capran Nylon Films*, supplier technical report - Allied Signal Inc..

(286) *American Mirex Corporation Rigid PVC Films*, supplier technical report - American Mirrex Corporation, 1990.

(287) *Pebax*, supplier design guide - Atochem, 1987.

(288) *Chemical Resistance of Halar Fluoropolymer*, supplier technical report (AHH) - Ausimont.

(289) *Guide Data Makrolon (Mechanical, Thermal, Electrical and Other Properties)*, supplier technical report (KU 46.100a/e) - Bayer AG, 1992.

(290) *A New Clear Polyester Container For Hot fill And Retort Applications*, supplier technical report (H-44606) - DuPont Company, 1992.

(291) Fetell, Arthur I., *Polyolefin + Nylon Yields Barrier Container*, Food & Drug Packaging, trade journal - Edgell Communications, 1986.

(292) *Selar PA 3426 Barrier Resin*, supplier technical report (E-73974) - DuPont Company, 1985.

(293) *Selar RB Barrier Resins - Resin Blend Technical Information*, supplier technical report (H-42016) - DuPont Company, 1992.

(294) *High Barrier Amorphous Nylon Resins and Extensions of the Laminar Technology*, supplier technical report (E-73971) - DuPont Company, 1985.

(295) *Selar Barrier Resin Selector Guide*, supplier marketing literature (H-38769-1) - DuPont Company, 1992.

(296) *Barex Barrier Resins - Barrier Properties*, supplier technical report (Bx-555) - BP Chemicals Inc., 1992.

(297) *Cleartuf PET Packaging Resins*, supplier technical report (SC: 1820-94) - Shell Chemical Company, 1994.

(298) *Polysar Butyl Rubbers Handbook*, reference book - Miles Polysar.

(299) *Elastomers Technical Information - Elastomer Permeability*, supplier technical report (TI-20) - Exxon Chemical Company, 1974.

(300) *Elastomers Technical Information - Factors in the Gas Permeability of Elastomers*, supplier technical report (TI-28) - Exxon Chemical Company, 1974.

(301) Fusco, James V., Hous, Pierre, *Butyl and Halobutyl Rubbers*, Rubber Technology, *Third Edition*, reference book - Van Nostrand Reinhold Company, Inc., 1987.

(302) *Designed For The Future*, supplier technical report - Mitsui.

(303) *Gas Permeability Of Kraton Rubbers*, supplier technical report (SC:941-87) - Shell Chemical, 1987.

(304) *Duraflex Polybutylene Specialty Resins Properties Guide*, supplier technical report (SC:867-87) - Shell Chemical, 1988.

(305) *The Engineering Properties Of Viton Fluoroelastomer*, supplier design guide (E-46315-1) - DuPont Company, 1987.

(306) Adam, S.J., David, C.E., *Permeation Measurement Of Fluoropolymers Using Mass Spectrometry And Calibrated Standard Gas Leaks*, 23rd International SAMPE Technical Conference, conference proceedings - SAMPE, 1991.

(307) *Product Data Bulletin - Grivory G21*, supplier marketing literature (GV8-104) - EMS - American Grilon Inc..

(309) *Elastomer Selection and Service Guide*, supplier design guide - Seals Eastern Inc., 1977.

(310) *Teflon - A Performance Guide For The Chemical Processing Industry*, supplier technical report (E-21623-2) - DuPont Company.

(311) *Phone Conversation With DuPont Technical Service*, (800-441-7111) - DuPont Company, 1994.

(312) *Pioneer Industrial Gloves Chemical Resistance Guide - Degradation + Permeation Test Data*, supplier technical report (0788) - Pioneer Industrial Products, 1988.

(313) *Ansell Edmont Chemical Resistance Guide - 5th Edition*, supplier technical report (CRG-GC-REV.960) - Ansell Edmont Industrial, 1990.

(314) *Eastman PEN Homopolymer Data and Fact Sheets*, supplier technical report (df/M: 215 052.230) - Eastman Chemical Company, 1994.

(315) Sidwell, J.A., *Food Contact Polymeric Materials*, review report (ISSN: 0889-3144), RAPRA Technology Ltd., 1992.

Trade Name Index

Aclar (Allied Sig.)

    CTFE 37

Aclar 22C/ PE/ PVC (Allied Sig.)

    CTFE/ PE/ PVC Film 381

Aclar 33C/ PVC (Allied Sig.)

    CTFE/ PVC Film 381

Aclar 88A/ PE/ PVC (Allied Sig.)

    CTFE/ PE/ PVC Film 381

Affinity (Dow)

    POP 211

Attane (Dow)

    ULDPE 187

Barex (BP Chem.)

    Acrylonitrile Copol. 7

Calibre (Dow)

    Polycarbonate 113

Capran (Allied Sig.)

    Nylon 6 85

    Nylon 6/66 105

Celcon (Hoechst Cel.)

    Acetal Copol. 1

Cellophane (Dow)

    Cellulosic 15

Cleartuf (Shell)

    PET 125

Dartek (DuPont Can.)

    Nylon 66 95

Delrin (DuPont)

    Acetal 1

Dowlex (Dow)

    LLDPE 175

Duraflex (Shell)

    Polybutylene 275

Ecdel (Eastman)

    Polyester TPE 407

Elastollan (BASF)

    TPAU 411

    TPEU 415

Elvax (DuPont)

    EVA 215

Eval (Eval Co.)

    EVOH 225

EVOH (Eval Co.)

    EVOH 225

Exxon Butyl (Exxon)

    Butyl Rubber 437

Exxon Chlorobutyl (Exxon)

    CIIR 443

Foraflon (Atochem)

    PVDF 53

Grilon (Emser)

    Nylon 66/610 109

Grivory (Emser)

    Amorphous Nylon 79

Halar (Ausimont)

    ECTFE 19

HDPE (Dow)

    HDPE 195

Hostalen (Hoechst AG)

    HDPE 195

Hyflon (Ausimont)

    ETFE 25

Hypalon (DuPont)

    CSM 447

Hytrel (DuPont)

    Polyester TPE 407

JSR (Jap. Synth.)

    Polybutadiene TPE 401

Kapton (DuPont)

    Polyimide 145

Kel-F (3M)

    CTFE 37

Kodar PCTG (Eastman)

    Polyester PCTG 137

Kodar PETG (Eastman)

    PETG 139

Kraton (Shell)

    Styrenic TPE 419

Krylene (Bayer)

    SBR 485

Krynac (Bayer)

    Nitrile Rubber 467

LDPE (Dow)

    LDPE 157

Lucalen (BASF AG)

    EAA 259

    PE Ionomer Copolymer 263

Lupolen (BASF AG)

    EVA 215

    LDPE 157

Luran (BASF AG)

    SAN 307

Luran S (BASF AG)

    ASA 295

Lutamer (BASF AG)

    Polypyrrole 389

Makrolon (Bayer)

    Polycarbonate 113

Marlex (Phillips)

    HDPE 195

Mylar (DuPont)

    PET 125

Nordel (DuPont)

    EPDM 455

Nylopak (Dow)

    HDPE/ EAA/ Nylon/ EAA Film 381

Parylene (Union Carbide)

    Parylene 69

Pebax (Atochem)

    Polyamide TPE 395

Polysar Butyl (Bayer)

    Butyl Rubber..................437

Polystyrol (BASF AG)

    GPPS 303

    IPS 305

Primacor (Dow)

    EAA 259

Retain (Dow)

    PE 149

Ryton (Phillips)

    PPS 281

Santoprene (Adv. Elast.)

    TPO 391

Saran (Dow)

    PVDC 321

Saran MA (Dow)

    PVDC 321

Saran Wrap (Dow)

    PVDC 321

Saranex (Dow)

    HDPE/ EVA/ PVDC/ EVA Film.....367

    LDPE/ EVA/ PVDC/ EVA Film 367

    LDPE/ EVA/ PVDC/ EVA/ LDPE Film 367

    PE/ PVC-PVDC Copolymer Multilayer Film 367

SBR (Exxon)

    SBR..............................485

Sclair (DuPont Can.)

    HDPE 195

    LLDPE..........................175

    MDPE...........................191

Sclairfilm (DuPont Can.)

    LLDPE..........................175

SclairTak (DuPont Can.)

    LLDPE..........................175

Selar OH (DuPont)

    EVOH...........................225

Selar PA (DuPont)

    Amorphous Nylon 79

Selar PT (DuPont)

    PET 125

Selar RB/ HDPE (DuPont)

    Nylon/HDPE Laminar Structure... 343

Selar RB/ Polyolefin (DuPont)

    EVAL/Polyolefin Laminar Structure 343

    Nylon/Polyolefin Laminar Structure 343

Selar RB 215/ HDPE (DuPont)

    Nylon/HDPE Laminar Structure... 343

Selar RB 215/ LDPE (DuPont)

    Nylon/LDPE Laminar Structure ... 343

Selar RB 300/ HDPE (DuPont)

    Nylon/HDPE Laminar Structure... 343

Selar RB 421/ HDPE (DuPont)

    EVAL/HDPE Laminar Structure .. 343

Selar RB 421/ PP (DuPont)

    Nylon/PP Laminar Structure........ 343

Solef (Solvay)

    PVDF 53

Styroblend WS (BASF AG)

    PE/PS 339

Styrolux (BASF AG)

    Styr. Butad. Block Copol. 311

Styron (Dow)

    GPPS 303

    IPS............................... 305

    PS 299

Surlyn (DuPont)

    Ionomer 63

Taktene (Bayer)

    Polybutadiene 433

Teflon (DuPont)

    FEP 27

    PFA 35

    TFE.............................. 45

Tefzel (DuPont)

    ETFE 25

Terluran (BASF AG)

    ABS 291

TPX (Mitsui)

    PMP.............................. 277

Trefsin (Adv. Elast.)

    TPO 391

Trycite (Dow)

    PS 299

Tyril (Dow)

    SAN ...:........................ 307

Udel (Amoco)

    Polysulfone 288

Ultradur (BASF AG)

    PBT................................ 117

Ultramid (BASF)

    Nylon 6.......................... 85

    Nylon 66......................... 95

    Nylon 6/66...................... 105

Ultramid B4/ Lupolen 3020 D (BASF)

    Nylon 6/LDPE Multilayer Film...... 381

Upilex (Ube)

    Polyimide 145

Vectra (Hoechst AG)

    LCP............................... 143

Vinex (Air Products)

    PVA............................... 287

Vistalon (Exxon)

    EPDM 455

Viton (DuPont)

    Fluoroelastomer (FKM) 459

Zytel (DuPont)

    Nylon 66......................... 95

Zytel FN (DuPont)

    Polyamide TPE 395